Science in the Ancient World

Science in the Ancient World

From Antiquity through the Middle Ages

Russell M. Lawson

An Imprint of ABC-CLIO, LLC
Santa Barbara, California • Denver, Colorado

Library of Congress Cataloging-in-Publication Data

Names: Lawson, Russell M., 1957– author.
Title: Science in the ancient world : From Antiquity through the Middle Ages/ Russell M. Lawson.
Description: Santa Barbara, California : ABC-CLIO, [2021] | Includes bibliographical references and index.
Identifiers: LCCN 2021023080 (print) | LCCN 2021023081 (ebook) | ISBN 9781440873522 (hardcover) | ISBN 9781440873539 (ebook)
Subjects: LCSH: Science, Ancient—Encyclopedias. | Science, Medieval—Encyclopedias.
Classification: LCC Q124.95 .L394 2021 (print) | LCC Q124.95 (ebook) | DDC 509.3—dc23
LC record available at https://lccn.loc.gov/2021023080
LC ebook record available at https://lccn.loc.gov/2021023081

ISBN: 978-1-4408-7352-2 (print)
978-1-4408-7353-9 (ebook)

25 24 23 22 21 1 2 3 4 5

This book is also available as an eBook.

ABC-CLIO
An Imprint of ABC-CLIO, LLC

ABC-CLIO, LLC
147 Castilian Drive
Santa Barbara, California 93117
www.abc-clio.com

This book is printed on acid-free paper ∞

Manufactured in the United States of America

For Ben, Dave, and Ri

Contents

Preface xi

Introduction xv

PART 1 The Beginnings of Science

1 The Scientific Mentality before the Onset of Civilization (Prehistory–4000 BCE) 3

2 Civilization and Science in Ancient Mesopotamia (4000–300 BCE) 11

3 Civilization and Science in the Nile River Valley (3100–300 BCE) 20

PART 2 Magic, Religion, and Science

4 Magic, Superstition, and Science in the Ancient World (4000 BCE–500 CE) 31

5 Ancient Religion and Science (4000 BCE–500 CE) 41

PART 3 Science in Ancient Central, Southern, and Eastern Asia

6 Science in Ancient Central and Southern Asia (3000 BCE–500 CE) 57

7 Science in Ancient Eastern Asia (2100 BCE–400 CE) 76

PART 4 The Ancient Greeks

8 Ionian Greek Science (750–300 BCE) 89

9 Science in Magna Graecia (750–300 BCE) 103

10 Classical Greek Science (500–300 BCE) 113

11 Athenian Science (500–300 BCE) 131

PART 5 Geographical and Astronomical Knowledge

12 Exploration and Geographical Knowledge in the Ancient World (4000 BCE–500 CE) 159

13 Phoenician and Carthaginian Science (1000–200 BCE) 183

14 Astronomy in the Ancient and Medieval Worlds (4000 BCE–1500 CE) 192

PART 6 The Expansion of Greek Science

15 Hellenistic Greece (331–30 BCE) 219

16 Greek Philosophy and the Philosophy of Science (750–100 BCE) 232

17 Greek Science in the Roman Empire (750 BCE–500 CE) 247

PART 7 Greco-Roman Science

18 Medical Science in Ancient Greece (750–100 BCE) 265

19 Medical Science in the Roman Empire (300 BCE–500 CE) 278

20 Mathematical Science in Ancient Greece and Rome (750 BCE–200 CE) 290

21 Science during the Pax Romana (31 BCE–180 CE) 297

PART 8 Applied Science in the Ancient and Medieval Worlds

22 Engineering and Technology in the Ancient and Medieval Worlds (4000 BCE–1500 CE) 315

23 Social and Behavioral Sciences in the Ancient and Medieval Worlds (500 BCE–1500 CE) 341

PART 9 Science in the Later Roman and Byzantine Empires

24 Late Roman Science (180–500 CE) 369

25 Byzantine Science (300–1450 CE) 395

PART 10 The Impact of Ancient Science upon Medieval Asia, Africa, and Europe

26 Science in Medieval Mesopotamia and the Levant (600–1500 CE) 419

27 Science in Moorish Spain (700–1450 CE) 444

28 Science in Medieval Africa (500–1500 CE) 452

29 Science in Medieval Central and Southern Asia (500–1500 CE) 461

30 Science in Medieval East and Southeast Asia (400–1644 CE) 474

31 Science and Philosophy during the European Middle Ages (500–1300 CE) 481

PART 11 Expansion of Science into New Worlds

32 Exploration and Geographic Knowledge in the Medieval World (500–1500 CE) 505

33 Science in the Americas (500–1500 CE) 516

PART 12 The Impact of Ancient and Medieval Science upon Modern Science

34 Conclusion: The Scientific Revolution 531

Chronology 539

Bibliography 557

Index 579

Preface

Science in the Ancient World: From Antiquity through the Middle Ages is an expansion of an earlier work published by ABC-CLIO in 2004, *Science in the Ancient World: An Encyclopedia.* The present work spans prehistory to 1500 CE, examining thousands of years of history in four world regions: Asia, Africa, Europe, and the Americas. Highlights of this period include the onset of civilization and science in Mesopotamia and Egypt, the accomplishments of the ancient Greeks between the years 700 BCE and 100 CE, the adaptation of Greek science by the Romans, the spread of Greek science during the Hellenistic Age, the expansion of Islamic power and commensurate scientific knowledge, and the development of science and philosophy in ancient China and India.

The work combines geographic region with a chronological approach. Each region and period of science will focus on the history of science, including agriculture, alchemy, architecture, astrology, astronomy, biology, botany, chemistry, engineering, exploration, geography, hydraulics, institutions of science, marine science, mathematics, medicine, meteorology, military science, myth and religion, philosophy, philosophy of science, physics, psychology, and social sciences. In all of these fields, theory and application are explored, and the leading individuals and schools of thought, the centers of intellectual activity, and the greatest accomplishments and inventions are highlighted.

The book includes a consideration of the impact of ancient science on medieval science throughout the world, as well as a consideration of the impact of ancient science on the coming of the European Renaissance, which resulted in the European Scientific Revolution. The basic premise of *Science in the Ancient World* is that the most creative time and place in human history was the first millennium BCE in the region of the Mediterranean Sea, encompassing southern Europe, western Asia, and northern Africa. The development of science by the ancient Greeks, building on the work of the ancient Mesopotamians and Egyptians and then adopted by the ancient Romans, set the stage for a fundamental understanding of the cosmos, world, humans, and natural environment that would influence subsequent periods of thought—the Middle Ages and modern age.

Science in the Ancient World: From Antiquity through the Middle Ages differs from other such books in that it has one voice: one author, rather than a collection of authors; it is aimed toward a more general audience (assuming very little scientific knowledge, hence secondary school and general university students, general readers) as opposed to university science majors and graduate students. It encompasses in one volume the whole of ancient into medieval science (up to 1500 CE), rather than providing an exclusive focus only on ancient science (prior to 500 CE); it covers the world, not just ancient Greece and Rome but medieval Europe, ancient and medieval Africa, ancient and medieval Asia, and the Americas. It includes a host of additional topics not included in typical scholarly works on science but based on the author's definition of science, which echoes that of ancient and medieval thinkers, who believed that science was not only empirical but was also sometimes connected to religion and magic, was often the same as philosophy, and was the general pursuit of knowledge in all aspects of life and thought. In addition, as it is the author's belief that the best way to learn about the past is through the voice of the participants themselves, *Science in the Ancient World: From Antiquity through the Middle Ages* includes copious transcriptions from primary sources in the public domain.

NOTE ON CHRONOLOGY

Scholars and skeptics during the past several centuries have questioned the meaning and purpose behind traditional chronologies. Not only do we find absolute time questioned by scholars, who argue that relative time is more akin to human experience, and not only do we have fascinating discoveries showing that time is shaped by motion and gravity but also our age is one that has discarded traditional chronologies in favor of modern methods to trace human experience. Historians have long known that the traditional chronology of the Christian church, which divides human experience into BC, before Christ, and AD, *anno Domini*, "in the year of our Lord," is based on a misapprehension as to when Jesus of Nazareth was born. We know that 1 AD does not represent the year of Christ's birth. He was perhaps a young lad of four, five, or six years old in the year 1 AD. At the beginning of the third millennium AD, many intellectuals were up in arms about putting so much emphasis on a chronological system based on a religious perspective not adhered to by a growing number of people worldwide. It has therefore become widely accepted to adopt a new dating system—to alter slightly the old—in favor of not blatantly basing the current year and past chronology according to the birth of a man two thousand years ago. In the search to avoid controversy, historians and other intellectuals have adopted a new format to trace the passage of human time. Unwilling to rock the boat too much and discard the dating system that has been used for well over a thousand years and is centered on the birth of Jesus of Nazareth, yet unwilling as well to offend anyone, intellectuals have decided upon the notion of the *Common Era*, CE, and the notion of BCE, *before the Common Era*. What exactly Common Era means is not altogether clear; if by it we mean the common human experience, it makes no sense to have a Common Era and "before the Common Era," since during that point between BCE and CE, the typical commonalities of human existence did not change. Nothing particularly notable changed in human existence from 1 BCE to 1 CE. Nevertheless, to adhere to current scholarly standards, *Science in the Ancient World* adopts the chronological framework of BCE/CE.

The scope of this book begins with the study of science in *antiquity*, which is a broad epoch in the history of humankind generally accepted by modern scholars and teachers. In terms of chronological limiting, the ancient world begins before the onset of civilization in the fourth millennium (about 3500) BCE and includes a prehistoric, precivilized period of hundreds of thousands of years when human thought was rudimentary yet sometimes revolutionary. During the fourth millennium BCE, civilization emerged in ancient Iraq, which the Greeks called Mesopotamia, centered on the lower Tigris and Euphrates Rivers, and in ancient Egypt, centered on the lower Nile River in North Africa. The ancient world comes to an end with the decline of the Roman Empire during the fifth and sixth centuries CE, roughly about 500 CE.

Thinkers during the first six centuries of the first millennium CE recognized that the transition that occurred in the Mediterranean world from the third to the sixth centuries CE was the end of one epoch and the beginning of another. *Science in the Ancient World* adopts this chronological limit as well. Renaissance thinkers conceived of a middle period between the end of the greatness of the ancient world and their own period of achievement and discovery, which began in the fourteenth century. This period of about eight hundred years has been variously called the Dark Ages, the Middle Ages, and the medieval period. Since the tripartite ancient/medieval/modern categorization is generally adopted in the academic world, *Science in the Ancient World* embraces this mode of defining historical epochs.

Introduction

Ancient science was the intellectual pursuit to understand the origins and workings of nature and humanity. *Science* is a term that encompasses many methods and varied disciplines over time. Science has engaged human thought for millennia. The questions that scientists ask tend to remain largely the same even as the answers differ according to time and culture. The strange and sometimes simple explanations that the ancient Greeks and Romans gave for natural phenomena appear less absurd to the modern mind when one considers that the answers of today may appear ridiculous to observers a thousand years from now. Among ancient scientists—the Mesopotamians, Persians, Indians, Chinese, Egyptians, Greeks, Romans, Africans, and Americans—the Greeks were far and away the leaders in scientific inquiry because they asked the most penetrating questions, many of which still elude a complete answer.

There is a temptation to view the past according to the standards and precepts of the present. The historian encounters countless similarities when comparing modern and ancient science; clearly the building blocks of today's science were formed two thousand to three thousand years ago in the ancient Mediterranean. Nevertheless, the development of ancient science occurred in a preindustrial age before the dawn of Islam or Christianity, during a polytheistic, superstitious time. Magic and astrology were considered as legitimate as medicine and astronomy. The earth was the center of a finite universe; the planets twinkled like gods watching from above; the moon governed the fertility of nature and woman. Fertility symbols and statuettes of priestesses and mother goddesses dot the archaeological finds from the dozens of millennia BCE, reminding us of the power women once had in ancient societies before the coming of male gods reflecting male dominance. Rhea, Cybele, Artemis, Hera, Isis, and Ishtar were early fertility goddesses who represented the universal mother image, who brought life, love, and death to her children the humans.

Ancient humans were animists who believed in a spiritual component to natural phenomena and pantheists who saw in nature something warm, maternal, and universal. They lived in an environment wholly alive: the surrounding woods, mountains, and streams filled with life and spirit. Nature was an unpredictable extension of self. Humans sought to charm the spirits of nature that were mysterious yet very familiar. Nature embraced early humans; it was all they knew. Humans formed communities to seek the best means to yield life and happiness from the environment, which they were dependent upon yet in competition with for food and shelter.

The most rudimentary form of scientific thought occurred at some vague point in the distant past when the ancient human began to develop a sense of self in the surrounding whole, to see that self and other humans as similar, and to be aware of life—and of death. This awareness of self, of mortality, of birth, of the future, of the past, allowed ancient humans to detach themselves not only from nature but from the moment as well, to forge a weak notion of the past-present-future continuum of time, to gain a nascent historical perspective. A human who may not have known *time* still knew the passing of days and the changing of seasons, the growth of youth and the decline of the aged. But existence was sufficiently precarious to accept the here and now and not be overwhelmed with anticipation of the future.

The transition from awareness to conceptualization of life, self, and nature depended in part on advancements in the human community. A secure existence with plentiful food guaranteed a growing anticipation of the future and reflection upon the past, a sense of belonging, of love and being loved. Food—that is, economic security—brought the freedom to speculate on self and others, on the community and those outside of it, on nature and survival, on control and being controlled. To explain existence, questions were asked and answers attempted and contemplated. The once vague sense of self became a clear sense of being. The intuitive recognition of the maternal spirit world matured into a desire to understand it. Rudolph Otto called this object of awareness the *numinous*—the awesome, majestic, sublime *Other* of which humans feel a part and are called upon to respond to and to know.

Human science began in pursuit of understanding the numinous and its manifestations. Science was initially not very different from religion. Ancient scientists were religious leaders, priests who doubled as scientists in searching for signs of the divine in nature. If the motions of the planets determined the future, the secrets of life and death, then the ancient thinker had to turn to the study of the heavens for religious purposes. The first ornithologists were prognosticators who sought messages from the gods in the flight habits of birds. Soothsayers gained familiarity with animal organs in the search for abnormal lobes and other intestinal aberrations. Early humans also turned to the study of flora for the best building materials and palatable food, such as grains for bread and edible roots and flowers. Flower petals, stalks, and roots, as well as tree bark, leaves, twigs, and roots, largely comprised the ancient *materia medica*, the potions and teas used to relieve pain, stop bleeding, reduce symptoms, and calm the hysterical.

In some ways, ancient scientists would be scarcely recognizable to contemporary, twenty-first-century scientists. The scientists described and portrayed in this book were priests, government officials, kings, emperors, slaves, merchants, farmers, and aristocrats. They wrote history, biography, and essays. They were artists, explorers, poets, musicians, abstract thinkers, and sensualists. The demands upon scientific study then were different from those of today. The study of astrology was necessary to know one's fate, the future. Astronomy and mathematics were essential to forming calendars to fit the cycles of nature and seasons of the year.

The ancient scientist was often seeking a practical result rather than pursuing scientific thought for its own sake. At the same time, the ancient scientist was something of a wise person, a community savant who was expected to know—or at least to have thought about, or investigated—all things natural, spiritual, and human. The Greeks called such a thinker *polumathes*, which is the origin for our word *polymath*, someone who is learned in many fields of knowledge.

Ancient scientists pursued some of the same goals as their modern counterparts do. Modern physicists and chemists seek to know the basic particles that constitute matter in the universe; ancient Stoics and Epicureans hypothesized the same particles and sought the same knowledge of the movement and patterns of atoms. Albert Einstein, the theoretical physicist, wanted to know the mind of God, the ultimate secrets of the universe, a continuing search that was inaugurated three thousand years before, in the ancient Mediterranean. Einstein would have liked Plato; Niels Bohr, the twentieth-century Danish physicist, would have found a friend in Aristotle. What are the abstract patterns present in the universe? Mathematicians today and millennia ago have been united in the quest to find out, to impose the rational human mind upon the most complex and least concrete inquiries. Psychologists today still work in the shadow of the great psychologists of the past, though the present concern to know the human mind and the nature of personality is a more secular pursuit than it once was. Political scientists today still rely on the initial systematic inquiries into human government that Plato and Aristotle made in the fourth century BCE. Students at modern medical schools take the Hippocratic Oath, recognizing that even though the techniques of medicine have changed from the days of Hippocrates and Galen, the ultimate goals and humanitarian concerns have not. In short, the college arts and science curricula and professional scientific careers of today are not a recent development. Rather, the moderns who are in pursuit of knowledge of humans and the universe continually ascend the intellectual and methodological building blocks constructed during antiquity.

The ancient Greeks did not have a word with the precise meaning of "science," which today means a methodical, concrete, objective, efficient, puzzle-solving approach to understanding natural phenomena. The Greek word with the closest equivalency was *episteme*, "to know." The Greek scientist was someone who *knew*. To the ancient mind, science involved much that we would identify as artistic, abstract, subjective, mythical, and emotional. Especially in the last two centuries, modern science has elevated science to the unique plateau of the search for objective knowledge. Scientists detach themselves from the environment, seeking in the here and now the means, intellectually and pragmatically, to reduce, dismantle, control, and reconstruct life. The ancient scientist associated with and attempted to recapture nature, which was an extension of self, something from the collective past experienced in the present moment. Science at present is secular and materialistic, seeking the transcendent—the origins and meaning of life—by the reconstruction of natural history. Ancient science reveled in the spiritual, the oneness of life and being. Science and religion, reason and faith, were rarely discordant in the history of science until recent times. Today's phrase *natural theology* implies that there is something religious in nature and there is something natural in religion, both of which describe the ancient scientific mindset precisely.

The modern scientific mindset is utilitarian, coercive, and technical; it embraces changes and focuses on things rather than ideas. The ancient scientific mindset was rather primitive, focusing on values, sentiments, morality, unity, the static and changeless, and the organic and alive. Science today is progressive and historicist, focusing on what is *becoming* through the movement of time. Science in antiquity focused on *being*, what *is*, regardless of the passage of time. Finally, modern science is generally a professional discipline practiced by scientists with terminal college degrees. It is well organized under prescribed methods, esoteric forms of communication, and agreed-upon theories. Modern scientists join together under the umbrella of a precise system of thought and methodology that explains clearly what the role of the scientist is in the accumulation and utilization of scientific knowledge over time. Ancient scientists were amateurs, polymaths, and generalists who were rarely well organized and who adhered to general philosophical schools of thought that were inclusive of any well-educated, thoughtful individual.

Science in the Ancient World: From Antiquity through the Middle Ages pays respect to the modern definition and practice of science while meeting the ancients on their own terms. Ancient philosophy and science were usually indistinguishable because of the worldview of the ancient thinker. Aristotle was a leading scientist of antiquity, yet he was a leading philosopher as well. Hippocrates was the great student of medicine, yet much of his work was theoretical and speculative, not empirical, focused on understanding rather than cures. Many ancient scientists were first and foremost soldiers and explorers who engaged in science on the side or out of utter necessity. Lucretius, the Roman Epicurean, was a scientist who recorded his ideas in verse. Other ancient scientists were devoted to the study of magic and astrology. Science and superstition often complemented each other in the ancient world. Moreover, ancient science was inclusive of all intellectual pursuits, not only the hard sciences. Historical inquiry, for example, was as valid an object of scientific inquiry as physics was.

The first civilizations in the history of humankind emerged during the last four millennia BCE. The *Oxford English Dictionary* defines civilization as a "civilized condition or state; a developed or advanced state of human society." This is sufficiently vague that a variety of more precise definitions have branched out from the original. All of them have some reference to the Latin root of the word, *civis*: "citizen." A citizen implies one who is part of a body politic; hence "civilization" generally refers to a level of society wherein the citizen has certain rights and responsibilities incumbent upon a particular role in the community. Citizenship requires a settled existence, which relies on the domestication of agriculture and livestock, the accumulation of surplus wealth, domestic and international trade, a social structure based on the distribution of wealth, a political structure that administers and protects wealth, and a system of writing to record the production, consumption, and distribution of wealth. Citizens might be farmers, tradespeople, artisans, and scribes. This organized division of labor requires a glue to bond it together into a working whole. The glue, in ancient as well as modern societies, has been professionals in political, social, cultural, and religious institutions who are themselves not producers but who administer production, distribution, and storage of wealth; who are engaged in the educational, social, and cultural systems built upon such wealth; and who express the collective thoughts and feelings of the citizenry through literature, art, music, drama, and thought. Philosophic and scientific inquiry was a necessary condition for the development of civilization.

Science, the systematic acquisition of knowledge to enable humans to understand natural and human phenomena, supported the thoughts, structures, and institutions of society in the ancient Near East, where civilization first began. Mesopotamian and Egyptian engineers built ziggurats and pyramids fitted with huge blocks of stone arrayed with incredible precision. Engineers in Mesopotamia designed and implemented a complicated network of canals and dikes for flood control and irrigation. Agriculture appeared as early as 10,000 BCE along the banks of the Tigris and Euphrates Rivers. The people who eventually immigrated to and took control of Mesopotamia, the Sumerians, were agriculturalists who learned when to plant and how to develop techniques to increase yields. Likewise, the early inhabitants of the Nile River valley discovered agriculture—perhaps independently, perhaps learning it from the Sumerians—but without the need for dikes and canals. The Nile rose and fell in such a fashion as to guide farmers when to plant and harvest. Surplus food required a means of record keeping. Scribes invented a form of writing—cuneiform in Sumeria, hieroglyphics in Egypt—to track daily economic, social, and political activities. Sumerian astronomers developed calendars based on the phases of the moon to help in the preparation of almanacs to provide meteorological information for farmers. Egyptian astronomers developed solar calendars. Metallurgists of the ancient Near East figured out how to heat copper and tin to extreme temperatures to produce bronze, an effective metal for tools and weapons. Inventors built wheeled vehicles for transport and wooden and papyrus reed ships for river and ocean navigation and trade. Sumerian sailors followed the coast and the stars as they sailed the Persian Gulf and Arabian Sea to the Indus River, where they traded with, and helped to stimulate, the emerging civilization of the Indus River valley (around 2500 BCE).

The Indus River valley, or Harappan, civilization lasted for about a millennium, during which it featured many of the accomplishments found in Mesopotamia and Egypt. The Harappan people lived in fine, well-designed, and well-constructed cities (such as Mohenjo Daro). They invented writing, discovered (or learned from others) how to make bronze, and had a sophisticated agricultural system that included the production of cotton. It is possible that the Indus River civilization spread east through land or ocean contacts to influence the origins of the Yellow River civilization of China. This civilization emerged during the mid-second millennium BCE: it featured writing, bronze tools and weapons, dynastic leaders, and sophisticated agriculture. Meanwhile, halfway around the world, the Olmec civilization developed in Central America toward the end of the second millennium. The Olmecs were a warlike people who lived on the eastern Mexican coast of the Gulf of Mexico. One of their many achievements was the (apparently) independent development of a system of hieroglyphic writing.

There was very little difference, at the onset of scientific thought in the ancient Near East and the Mediterranean, between the soothsaying and prognosticating of the shaman, prophet, magician, and astrologer, on the one hand, and the early attempts of the naturalist and philosopher, on the other, to examine the natural environment to assess the course of things from past to future. Science began when, during the third, second, and first millennia BCE, the astrologer became astronomer, the soothsayer became biologist, and the magician became chemist. Mesopotamian scientists, such as at Babylon, studied the phases of the moon, the movement of the planets, and the course of the stars to arrive at accurate star charts, a vision of the zodiac, and lunar calendars. Mesopotamian mathematicians developed algebra, geometry, quadratic equations, the Pythagorean theorem, and the sexagesimal system of numbers. Egyptian astronomers (according to the Greek writer Diogenes Laertius) developed the first solar calendar, studied the motions of the planets and stars, and predicted lunar and solar eclipses. Egyptian mathematicians excelled at geometry, approximating the value of pi (π)—the ratio of the circle's circumference to its radius.

The ancient Near East was the site of a number of other flourishing civilizations that emerged in the wake of the Sumerians and Egyptians. The Hittites of Asia Minor were a warlike people who nevertheless developed a system of writing, had an organized government, and were the first people to learn to make iron tools and weapons. Babylon, along the Euphrates River, was the center of a dynamic civilization that developed from Sumerian origins in Mesopotamia. The Babylonians made significant discoveries in astronomy, mathematics, and social organization. The law code of the Babylonian king Hammurabi (r. 1792–1750 BCE) shows progress toward the development of a more civil society. Toward the end of the second millennium, along the shores of the eastern Mediterranean at what is now Lebanon, several seagoing city-states emerged, their wealth and sophisticated culture based on trade. These Phoenicians developed a system of writing, later adopted by the Greeks. They had the best naval technology of the time and, being explorers, had the most up-to-date knowledge of geography before the Greeks. The Phoenicians explored the entire extent of the Mediterranean and beyond, into the European and African coasts of the Atlantic Ocean. They also came to know the Red Sea and the East African coast of the Indian Ocean. Meanwhile, during the second millennium BCE, the Hebrews were developing a dynamic civilization in Palestine. Influenced by the Mesopotamian and Egyptian cultures, the Hebrews developed an astonishing culture centered upon their interpretation of the cosmos encompassed by the one god: omniscient, omnipotent, and omnipresent Yahweh.

Other civilizations of the Near East that developed during the first millennium BCE included the Lydians, Assyrians, Chaldeans, and Persians, stretching from the Aegean Sea east to the Indus River. The Lydians dominated Asia Minor for a brief time from their capital at Sardis. They are noteworthy for developing the first system of coinage. The Assyrians and Chaldeans centered their respective power around Mesopotamia particularly at the cities of Assur and Babylon. Chaldean astrologers became famous (and infamous) during subsequent centuries. The Persian Empire existed for about two hundred years, from the sixth to the fourth centuries. The Persian gift was for organization and logistics. They developed the largest and most efficient empire before the Romans.

Scholars have long debated whether early civilizations, such as the Indus River Valley, Yellow River Valley, and the Olmec, of Mexico, developed in isolation or through cultural contacts. Science today is an international work in progress, one of the great forces of unification crossing cultural, political, geographic, and linguistic boundaries. Has it always been this way? Clearly Renaissance European scientists had the mentality of cross-cultural scientific exchange as did the scientists of Constantinople, the Arab world, and western Europe before them; all shared a deep interest in especially Aristotelian science. Greek was the lingua franca of the ancient Mediterranean world from 500 BCE to 500 CE; the Greek language was heavily dominated by philosophic and scientific terms. The scientists of the Roman Empire were typically Greek, as were the teachers and physicians. The greatest period of scientific achievement before the European Renaissance occurred during a thousand-year period mostly in the eastern Mediterranean, at scientific capitals in Europe, Asia, and Africa. Scientists and philosophers at Magna Graecia, Athens, Miletus, Byzantium, Antioch, and Alexandria conversed in Greek about a variety of scientific topics, ranging from mathematics to physics to chemistry to biology. Many scientists were great travelers, spreading information from one place to another. But the Greeks were probably not the first people to engage in the sharing and communication that is the hallmark of the pursuit of knowledge. The Phoenicians, no doubt, spread their knowledge of world geography to a variety of peoples. Some scholars believe that before the Phoenicians, other civilizations of the Near East developed naval technology that allowed the exploration of the Indian, Atlantic, perhaps even the Pacific Oceans. If so, cultural exchanges included the sharing of scientific knowledge. Perhaps the similarities of culture and science throughout the ancient world in Asia, Europe, Africa, and America were due not to coincidence or isolated parallel development but rather to intrepid explorers who brought knowledge of one people to another, hence beginning the process of world scientific achievement that we know so well today.

Science in the Ancient World: From Antiquity through the Middle Ages is comprehensive yet realistic about the fact that the Mediterranean was the focus of the most scientific activity during antiquity. Scientific accomplishments prior to the first millennium BCE are fully discussed, yet the greatest period of activity was between the years 600 BCE and 200 CE. During this eight hundred-year period, the Greeks adopted the discoveries of their Egyptian, Phoenician, and Mesopotamian predecessors while advancing their own highly original theories and observations of the natural and human world. A host of thinkers—Thales, Anaximander, Anaximenes, Xenophanes, Pythagoras, Hecataeus, Heraclitus, Hippo, Empedocles, Hippocrates, Aristotle, Theophrastus, Erasistratus, Eratosthenes, Euclid, and Archimedes—brought physical and life sciences, mathematics, social sciences, and philosophy to a level hitherto not, and subsequently rarely, achieved. The Romans, who came to control North Africa, the Middle East, and most of Europe by the end of the first millennium BCE, were themselves not scientifically inclined. They recognized, however, the Greek achievement in thought, so they adopted Greek philosophy and learning, which became truly Graeco-Roman. Most scientists henceforth in imperial Rome, the Later Roman Empire, the Byzantine Empire, and medieval Europe, commented on, edited, condensed, and explicated the works of the thinkers of the Greek Hellenic and Hellenistic eras. The questioning of the knowledge of the ancients during the European Renaissance resulted in the Scientific Revolution, when such thinkers as Copernicus, Kepler, Galileo, Boyle, and Newton reinterpreted physics, astronomy, mathematics, and chemistry. The Enlightenment in Europe and America of the eighteenth century set the stage for further scientific advances during the nineteenth and twentieth centuries.

People, rather than cities or schools of thought, engage in science. Ancient science, much more than modern science, was an individual endeavor. It is somewhat anachronistic to categorize and conceptualize ancient scientists according to modern expectations of scientists, who are professional puzzle solvers working often in teams to generate solutions to intricate and esoteric problems of interest to a very few. Hence *Science in the Ancient World: From Antiquity through the Middle Ages* focuses heavily on individual scientists and their lives, discoveries, methods, tools, and writings. The primary sources for ancient science are the writings of individual scientists, the works of ancient historians, and the observations of philosophical commentators. The best way to study ancient science is to study the ancient literary sources.

The earliest Greek scientists left behind few writings or kept their ideas to themselves and a few select disciples. Fortunately, there was enough interest among later scientists and philosophers to record the work of their predecessors by way of indicating either difference or debt. Without Plato's dialogues we would scarcely know of Socrates's theories and arguments. Aristotle was such a universal thinker that he not only wrote about most scientific topics but painstakingly recorded the views of earlier thinkers as well. From Plato and Aristotle, we learn about the first Ionian scientists from western Turkey and the Aegean Isles—Thales, Anaxagoras, Anaximenes, Anaximander—who initiated the scientific quest to seek rational explanations of natural phenomena.

Greek philosophers and scientists of western Anatolia—Ionian Greece, according to the ancients—used the Mesopotamian and Egyptian accomplishments in science and mathematics to attempt to understand natural phenomena. Thales of Miletus (625–545 BCE), the first Ionian Greek scientist, was, according to later commentators, influenced by Babylonian and Egyptian scientists. He was an astronomer who reputedly wrote books on solstices and equinoxes, using the sundial as an instrument of science; wrote a book on navigating at sea according to the stars, especially the North Star; could predict solar and lunar eclipses; knew enough about engineering to divert a river from its original course; believed that the fundamental substance of the universe was water; and generally broke from superstition to try to explain according to reason the nature of physical phenomena. Thales was the teacher of Anaximander of Miletus (610–540 BCE), who conceived of the universe as an infinite fire and the earth in the shape of a cylinder, and believed that water generated life, in the form of a fishlike creature that eventually crawled on land. Anaximenes of Miletus (585–525 BCE), the student of Anaximander, believed that the universal substance was air; he was able to distinguish planets from stars; and he was a meteorologist who explained rainbows as caused by an interaction of light and moisture. Anaxagoras of Clazomenae (500–428 BCE), another student of Anaximander, believed that the causative principle of the universe was mind (*nous*). Two other Ionian Greek scientists were noteworthy. Leucippus lived in the fifth century; he expanded the ideas of Empedocles (495–435 BCE) that there were four primary elements, earth, air, fire, and water; and conceived of the universe as being made up of invisible particles in constant motion: atoms. Hippocrates of Cos (460–377 BCE) was the first physician to focus exclusively on physical and natural causes of disease. Hippocrates advocated the theory that disease is a product of an imbalance in the four humors of the body: blood, phlegm, black bile, and yellow bile.

Aristotle and his contemporaries also continued to rely heavily on the teachings of Greeks living in Magna Graecia, in southern Italy. Some Ionian Greek scientists emigrated to Magna Graecia, where a dynamic community of inquiry developed, led by Xenophanes of Colophon (570–478 BCE) and Pythagoras (570–490 BCE). Xenophanes, like other Ionian scientists, argued for cause-and-effect explanations instead of relying on the gods (which did not exist) or other supernatural causes to understand the universe. Pythagoras, whose life is shrouded in myth, possibly conceived of the earth as a sphere and believed that reality is best approximated through mathematics. These philosophers of what Diogenes Laertius called the Italian school tended less toward idealism and more toward materialism in their philosophic and scientific explanations. These early Greek thinkers set the standard for later scientists by asking questions that required deep speculative thought and concrete analysis to answer. Doubt was an important scientific tool to the Greek intellectual. Doubt of earlier theories drove Athenian scientists. Doubt of pat answers, of a *received tradition*, drove Epicureans, Skeptics, and Cynics. Doubt that spurred the continuing search set the standard for all subsequent scientists.

Greek science during the Hellenic Age (750–330 BCE) explored metaphysical and physical sciences, life sciences, social sciences, and behavioral sciences. The Ionian Greek scientists of the seventh, sixth, and fifth centuries spawned scientific inquiry throughout the Mediterranean world, especially at Athens in the fifth and fourth centuries. The student of Socrates, Plato, and his student, Aristotle, developed sophisticated theories of the nature of reality, cause and effect, and explanations of natural and human phenomena. Plato (427–347 BCE) eschewed empirical observation, advocating the use of intuition, reason, and mathematics to conceive of the fundamental truths of the universe. Like his teacher, Socrates (469–399 BCE), Plato used deductive logic and the method of dialectical inquiry to arrive at logical and rational explanations of all things. Plato's student, Aristotle (384–322 BCE), however, broke from his teacher to develop an inductive approach to science. Whereas Plato believed in deep contemplation to arrive at an understanding of reality, Aristotle believed in a more practical approach of observing particular phenomena to arrive at an understanding of reality; once universal truths are known through induction, then the thinker can engage in the process of deduction to arrive at other, particular truths—a method called the *syllogism*. Aristotle was one of the great polymaths of all time, mastering the subjects of physics, metaphysics, meteorology, astronomy, zoology, biology, ethics, government, history, geography, and psychology. He spawned a school of thought, the Peripatetic, that would, along with Plato's followers, the Academics, dominate philosophy and science for two thousand years.

The Ionian school of medicine at Cos, initiated by Hippocrates, investigated a wide variety of diseases and speculated on their causes. The Greeks were fascinated by flora and its potential for healing agents; Theophrastus (370–286 BCE), a student of Aristotle, was an early botanist. The origins of modern social and behavioral sciences can be found in the writings of Greek scientists of over two thousand years ago. Historians such as Herodotus (490–430 BCE) and Polybius (208–126 BCE) were geographers, ethnographers, and explorers. History itself was considered a science by the Greeks. The Greek *polis* (city-state) inspired commentators and analysts to debate its origin and significance. Athenians took the lead. Aristotle's *Politics* and *Athenian Constitution* brought the study of politics and society to the realm of science. Plato's *Republic* analyzed the concept of justice, and his dialogues, particularly *Timaeus* and *Phaedo,* initiated the examination of the self, the *psyche*, the study of which excited many subsequent Greek thinkers, such as the Academics, Neoplatonists, and Christians.

Another of Aristotle's pupils was the Macedonian conqueror Alexander the Great (356–323 BCE), who, in conquering the Persian Empire—including the modern countries of Turkey, Palestine, Egypt, Iraq, Iran, Afghanistan, and Pakistan—brought Aristotelian philosophy and science to northern Africa and western and southern Asia. This Hellenistic Age (331–31 BCE) included some of the greatest scientific accomplishments in human history. The center of Hellenistic science was at Alexandria in northern Egypt, founded by Alexander in 331 BCE. Alexandria featured a 500,000-volume library, which encouraged the work of such thinkers as the mathematician Euclid, who flourished at the beginning of the third century and wrote *Elements*; and Eratosthenes (276–195 BCE), who wrote *Geographica,* in which he developed the measurements of latitude and longitude and mapped the known world, and *Chronographia*, in which he developed a system of timekeeping. Eratosthenes provided an extremely accurate measure of the circumference of the earth (23,300 miles, off only by 2,000 miles). Alexandria was also a center of medical knowledge, where the anatomist Herophilus (325–255 BCE), who flourished at the beginning of the third century BCE, made the connection between the pulse and the heart, while his student Erasistratus (275–194 BCE) studied the brain and human nervous system. Other Hellenistic scientists included Archimedes of Syracuse (287–212 BCE), who used mathematics to devise various forms of technology, especially military technology.

Hellenistic astronomers made important contributions to the understanding of the earth, planets, and meteorological phenomena. Aristarchus of Samos (310–230 BCE) was the first astronomer to hypothesize the heliocentric (sun-centered) universe. He understood the concept of parallax, that the distance of the stars from the earth interferes with accurate observations of stellar phenomena. Aristarchus also tried to estimate the distance of the earth to the sun. Hipparchus (190–120 BCE), who had an astronomical research center at Rhodes, estimated the distance of the earth to the moon. He used Eratosthenes's measurements of the earth's circumference to establish a set of coordinates based on a 360-degree spherical earth. He established the system of deferents and epicycles that was used by Claudius Ptolemy (100–170 CE).

During the half dozen centuries before and after the birth of Christ, science became truly an international inquiry, with research centers and scholars working in southern Europe, western Asia, and northern Africa. Greeks working at Alexandria, Antioch, Constantinople, and Athens dominated scientific achievements. Few

Romans became involved in science; those who did, such as Pliny the Elder (23–79 CE), were interested more in applied than in theoretical science. Others, such as the Epicurean Lucretius (98–55 BCE), expounded on Greek science and philosophy.

Claudius Ptolemy of Alexandria was arguably the leading scientist of the Roman period. The Romans themselves were great soldiers rather than thinkers. Having conquered the lands of the Mediterranean by the end of the first century BCE, the Romans adopted Greek philosophy, such as Stoicism and Epicureanism, and Greek science. Greek scientists such as Ptolemy, Strabo, Plutarch, Galen, and Asclepiades were the leading minds of the Roman Principate (30 BCE–180 CE). Ptolemy was a geographer, mathematician, and astronomer who, next to Aristotle, had a profound impact on medieval science and philosophy. Ptolemy's *Almagest* provides a detailed theoretical exposition on the geocentric universe. Astronomers had long been perplexed by the visible phenomena of the universe, such as the sun rising and setting and the wanderers (planets) of the night sky. Ptolemy's theory of epicycles provided a logical mathematical explanation. Strabo (63 BCE–21 CE) was a synthesizer of geographical knowledge. The *Essays* by Plutarch (46–120 CE) provide a variety of scientific disquisitions. Galen of Pergamon (130–200 CE) was the finest physician of imperial Rome. Relying heavily on the theories of Hippocrates, Galen also studied the works of earlier Roman physicians, such as Asclepiades (120–40 BCE). Galen believed that the health of the body relied on the "vital spirit" of the person.

Other areas of ancient science besides the Greco-Roman Mediterranean world occurred in South Asia, East Asia, and America. Along the Indus River, a Bronze-Age culture focused on sophisticated town planning, monumental architecture, writing, and irrigation. Vedic philosophy, based on texts such as the Upanishads and Bhagavad Gita, emphasized an all-pervasive being, infinite and absolute. Philosophers and scientists of the Vedic Age (1500–500 BCE) were mathematicians, physicians, astronomers, metallurgists, and chemists. Individual thinkers included the linguist Panini (fl. fourth century BCE); the physician Suśruta, author of *The Compendium of Suśruta*; the physicist Kanada, a contemporary and counterpart to Greek materialists who advocated an atomic theory of reality; and gymnosophists, such as Calanus (c. 398–323 BCE), ascetic philosophers who reputedly influenced many of their Greek counterparts. In East Asia, political stability under the Xia (2100–1600 BCE), Shang (1600–1046 BCE), Zhou (1046–256 BCE), and Han (206 BCE–220 CE) dynasties allowed for thinkers such as K'ung fu-tzu (Confucius, 551–479 BCE) to introduce a system of thought based on rational, ritual behavior and conformance to the inherent order of the universe. The Confucian mentality focused on the state, humans, and heaven working together in harmony. One of the first great astronomers in world history, Gan De (400–340 BCE), compiled a star chart. The materialist philosopher Wang Chong (25–100 CE) was likewise an astronomer and meteorologist. Zhang Heng (78–139 CE) was, like his counterpart in Greece, Aristotle, a polymath, mastering a variety of fields of inquiry. Cai Yong (132–192 CE) combined the study of music and mathematics. Chinese historians included Sima Qian (145–90 BCE), who authored *Records of the Grand Historian*. Civilization in America dates back from the fourth to the first millennia BCE. Norte Chico, Olmec, and Maya civilizations engaged in monumental architecture, mathematics, astronomy, and sophisticated hydraulics and engineering practices.

Peoples in diverse locations throughout the world engaged in a variety of sciences during the first millennium CE. The Pax Romana of the Roman Empire came to an end during the third century CE; so, too, did the framework of peace and political order that maintained a fertile environment in which science could grow and flower. Science at the end of the ancient world during the decay of the Roman Empire involved generally unoriginal thinkers who looked back to the glory days of ancient Greece, from which they continued to draw inspiration and theories.

Science during the Later Roman Empire (200–500 CE) and Middle Ages (500–1300 CE) was limited to commentaries, syntheses, explications, and disseminations of the works of ancient scientists. Clement of Alexandria, for example, in the early third century wrote *Miscellanies*, in which he discussed his Greco-Roman scientific predecessors. Oribasius of Pergamon's fourth-century *Synopsis* synthesized and commented on the work of Galen; medieval Arab scholars used the *Synopsis* to understand Galen's medical thought. The fourth-century philosopher Themistius (317–388 CE) was a notable commentator on Aristotle's work. Themistius's *Paraphrase* had a profound impact on the medieval understanding of Aristotle. The sixth-century commentator Simplicius wrote *Commentary on the Physics* and *Commentary on the Heavens*, in which he discussed scientists such as Hippo, Hippasus, Diogenes of Apollonia, Zeno of Elea, Parmenides, Melissus, Xenophanes, and Thales.

Ancient thought, culture, and institutions had a profound impact on the subsequent centuries of the European Middle Ages, the European Renaissance (1300–1600), the Scientific Revolution (1500–1700), the Enlightenment (1650–1800), and the modern world. The decline and transition of the Roman Empire during the fourth and fifth centuries CE served to bring forward ancient thought to the scattered kingdoms of western Europe and the Byzantine Empire of the eastern Mediterranean. The Byzantine Empire was a Greek civilization still beholden to Greek language, ideas, and culture. Kingdoms of western Europe, such as that of the Franks, adopted the Latin language, and Roman customs, institutions, and thought. The king of the Franks, Charlemagne, for example, had himself declared emperor of the Romans in 800 CE. Several centuries later, Otto the Great founded the Holy Roman Empire. Meanwhile Byzantines called Constantinople the "Second Rome," and emperors such as Justinian (527–565) considered themselves heirs to the traditions and power of Augustus Caesar. Medieval thinkers such as Boethius and Thomas Aquinas adopted the intellectual structures of Greek philosophy. Late medieval thought, following upon developments in Islamic science, was dominated by Aristotelian science. Renaissance thinkers continued the emphasis upon Platonic and Aristotelian thought and also embraced ancient Stoicism, Skepticism, mysticism, and astrology. Catalysts of the Scientific Revolution, such as Copernicus and Galileo, were heavily influenced by Aristotle and Claudius Ptolemy, among others. Indeed, the intellectual and scientific paradigms of the ancient world have only recently been replaced by new assumptions and theories and hitherto unimagined experiments and research technologies.

The problem of when and how—and even if—the Roman Empire declined and fell is complicated by the varied dimension of cultural change in the fourth and fifth centuries CE. In both the Western European Empire—and subsequent European kingdoms—and the Eastern Roman Empire—and subsequent Byzantine Empire—the polytheistic, superstitious, pantheistic pagans, who watched constantly for divine signs to indicate the course of the future, became monotheistic and similarly superstitious Christians who conceived of a variety of supernatural forces of both good and evil that waged war over the Christian soul. There were more similarities than differences between paganism and Christianity, so it was common to find Christians who, like the philosopher Boethius, could not quite rid themselves of their pagan proclivities, and pagans who, like the emperor Constantine, were sufficiently attracted to Christianity to approach full conversion.

Augustine of Hippo likewise studied the Greco-Roman classics before converting to Christianity. He developed a fascinating combination of assumption and syllogism, logic and piety, reason and faith, which describes the religious thought not only of Augustine but of the Middle Ages in general.

Meanwhile, Byzantine scholars at Constantinople, Alexandria, and Gaza continued to read, teach, and transmit to posterity the great writings in philosophy and science of the ancient past, as well as a host of more recent commentaries on, especially, Aristotle. Byzantine scholars were typically Christian (though studying and teaching pagan authors) and were usually lesser thinkers, hence rarely subsequently known, compared to the Greek masters. Muslim students of Greek philosophy and science in western Asia, North Africa, and Spain studied and commented upon ancient literature and retained numerous writings from the ancient world unknown to the Latin West. The beginnings of Muslim interest in Greek science occurred during the Abbasid dynasty, of the eighth century.

The Muslims Avicenna and Averroes had a central role in bringing an Aristotelian Renaissance to western Europe. Arab scholars made contributions to scientific knowledge in their own right. The Iraqi polymath Ibn al-Haytham (965–1039), for example, was an empirical scientist who wrote *Book of Optics* in 1021. The English scientific empiricist Roger Bacon (1214–1294) was one of many European scientists influenced by his contemporary Arab counterparts.

The greatest Aristotelian of the European Middles Ages, the Christian philosopher Thomas Aquinas (1225–1274), attempted to reconcile Greek philosophy and science with Christian theology. The writings of Muslim commentators on Aristotle had become known through Latin translations by the early to mid-thirteenth century. Aquinas therefore had at his disposal a vast corpus of Aristotle's works. He made great use of them in his own writings, in particular the *Summa Theologica.* Aquinas relied heavily on Aristotelian methods to arrive at logical deductions about the existence and nature of God and God's works. Repeatedly Aquinas referred to Aristotle as simply "the Philosopher." Like the Philosopher, Aquinas used logical syllogisms of common everyday things, such as wood and fire, to arrive at correct answers to the questions he posed throughout the *Summa.* Aquinas's use of science was, of course, limited by his methodology—it was not empirical—and by his focus on Christian theology. His successors in the European Renaissance were quick to point out his shortcomings, as they attempted to use ancient scientific literature as the basis for a full study of all natural phenomena.

The Aristotelian conception of nature had such a hold over the medieval mind that a revolution was required to bring about change. The revolutionary was a Polish churchman, Nicolaus Copernicus (1473–1543), who published a book in 1543, *On the Revolution of the Heavenly Spheres.* Copernicus was largely beholden to Aristotle and his disciples, such as Claudius Ptolemy, in that Copernicus accepted the finite universe of sun, moon, earth, five planets, and outside starry vault of the fixed stars. But Copernicus believed that this conception of the universe worked better if one assumed it was heliocentric rather than geocentric. Copernicus's theory of a sun-centered universe caused such a stir that his book was condemned by the church but vigorously supported by courageous scientists such as Galileo Galilei (1564–1642). Galileo, in *Dialogue on the Two Chief World Systems* (1632), provided empirical arguments, based on his observations using the telescope, to support the heliocentric universe. But more, Galileo showed that the entire ancient conception of the universe was incorrect. His studies of the sun revealed sunspots when it had long been assumed that the sun was a perfect, unchanging heavenly body. His observations of the moon revealed craters rather than the ideal, godlike moon of the ancient astronomers. Galileo discovered the moons of Jupiter, contradicting the ancient notion that the universe had a set number of heavenly bodies moving in set spherical paths. He turned his telescope to the stars and discovered the Milky Way was a vast, seemingly infinite realm not limited to a starry vault. Galileo and other astronomers, such as Johannes Kepler (1571–1630), showed that ancient science had many limitations as well as many errors to correct.

The European Scientific Revolution that began in the 1500s and 1600s featured other discoveries that ushered in a new conception of science. The voyages of Christopher Columbus (1451–1506) across the Atlantic revealed that the ancient conception of the earth as having only three continents (Europe, Asia, Africa) and two oceans (Atlantic, Indian) was in error. Peter Ramus (1515–1572) challenged Aristotelian logic. Pierre Gassendi (1592–1665) and Robert Boyle (1627–1691) resurrected the ancient atomic theory, arguing that the universe is comprised of invisible corpuscles in constant motion. About the same time, Englishman Francis Bacon (1561–1626), in *Novum Organum,* proposed a "new science" based on purely empirical methods. Frenchman René Descartes (1596–1650) provided the mathematical underpinnings for Bacon's empirical approach. The greatest seventeenth-century thinker was Isaac Newton (1643–1727), whose *Principia Mathematica* (1687) provided an empirical, mathematical approach to explaining the universe. Newton proposed three laws of motion that explained the fundamental principle of universal gravitation. Newton's laws showed why the planets orbited the sun in elliptical motion, how the competing push and pull of gravitation kept the planets and other space phenomena in the position moving according to the same rate and direction, and why time and space were universal constants.

Scientists such as Newton built upon the existing foundation of ancient and medieval science. The discoveries of Copernicus, Galileo, Kepler, and Harvey would have perhaps scarcely occurred without the initial work of Ptolemy of Alexandria, Aristotle of Athens, and Galen of Pergamon. Modern scientists continue to work in the shadows of the columns and stoa of ancient schools of thought.

PART I

The Beginnings of Science

1

The Scientific Mentality before the Onset of Civilization (Prehistory–4000 BCE)

Ancient humans were religious thinkers who believed in spirits and deities, who buried loved ones and comrades in expectation of seeing them again, and who believed there was a spiritual component of life: a soul. Ancient humans journeyed in an environment wholly alive to them. The surrounding woods, mountains, and streams were filled with life and spirit. Nature was an unpredictable extension of self. Ancient humans tried to charm or to manipulate a particular spirit of nature that remained otherwise mysterious yet very familiar. Nature embraced early humans; it was all they knew. Nature was a parent, filling their stomachs with game and providing shelter from ice storms. Ancient humans were at once completely dependent upon this grand parent yet at the same time in competition with it. Some days game was scarce, the wind icy off the slopes, the way treacherous. The humans joined together into a community to seek the best means to yield life and happiness from the embrace of nature.

Ancient humans were hunters, aware that a successful hunt was largely in their hands; it was their responsibility. They had names and families, a place in the community. They were aware of other surrounding peoples with whom they could trade for flint or copper. They filled their days with work, making strong axes with copper heads, daggers with flint blades, grass sheaths and quivers, arrows quick in flight. They lived actively, skillfully, defending themselves and their people; they were proud, strong, and intelligent. But they were also humbled by nature. Such was ancient human life: living akin to nature, dependent if in slight conflict with it; yet in the end, for all their work, skill, and intentions, all humans died in nature's grasp.

One discovers an unexpected sophistication in the lives of ancient humans during the ages before the discovery of agriculture. Humans formed sickles from bone to harvest wild grain for dinner, which might include a feast of bison or salmon. Women prettied themselves with coiffures attractive to gods (and men). Men spent the evening around the hearth fashioning the tools of their trade, particularly bows, quivers, arrows, awls, spear-throwers, spears, knives, and axes. Dancers swayed in religious procession to the music of flutists, who used bone to fashion their instruments. Cave dwellers cooked meat over open flames and used the radiating light to sketch and to paint dens deep in the earth. After a good day's hunt, the community of hunters divided the spoils of mammoth fur, skin, meat, and tusks for food, clothing, tools and weapons, and artistic expression. Ivory was a favorite medium for amulets, decorative tools, and religious implements.

Geologists and paleontologists of the nineteenth century discovered the antiquity of the earth and humankind, that humans began to develop much earlier than the short chronology imposed by a literal reading of the Bible had previously suggested, that the antiquity of humankind—and its earlier, ancestral, forms—go back hundreds of thousands, even millions, of years. One of the ancestors of *Homo sapiens*, *Homo erectus*, began using tools, a necessity of survival that all subsequent humans employed. In the Middle Pleistocene period, perhaps 500,000 years ago, human ancestors used bone and stone to make hand axes and scrapers. And at some point in the Middle Pleistocene, ancestral humans developed fire.

Scholars believe that the fire drill (bow drill) was the discovery by which humans were able to artificially induce fire for cooking, light, warmth, and protection. Fire allowed for more sophistication in the development of primitive tools and weapons; wooden spear tips were hardened by fire. The fire drill, a means of creating fire artificially by friction, was developed perhaps initially in ancient India, from which it spread by travel and communication to other places, such as Mesopotamia. The fire drill as initially invented was little more than two dry sticks being rubbed together until friction caused a spark that could be nestled in a bed of wood shavings to create fire. There were various additions to this basic technique found in all of the early human societies throughout the world.

By the time of the beginning of the Late Pleistocene, around 100,000 BCE, when ancestral humans were represented by Neanderthals, humans began to recognize themselves as humans, began to sense their own being in the entirety of their surroundings, began to be aware of life, to be aware of death, to be aware of pleasure and pain and the means to increase the former and reduce the latter. This is awareness of self as a separate entity.

Paleontological evidence of an expanded brain capacity in Neanderthal humans suggests more sophisticated cognitive abilities. Whether or not the Neanderthal communicated verbally is not known, but *Homo sapiens* had the jaw structure to allow clear verbal communication. At some point in the distant past, perhaps as early as 100,000 BCE, early humans developed the means by which they communicated concepts, ideas, facts, and simple observations by spoken words. Language was an invention, a form of technology, a technique to survive, to direct one's life. It allowed for comparison, analysis, planning, and expression. It helped to engender further thought processes, perhaps further inventions. Significantly, too, Neanderthals were the first to bury their dead, which means that they thought about death and about the possibilities of what might come after death. Their having buried implements with the dead suggests the frail notion they had of the afterlife. Also, during the many millennia after 100,000 BCE of the Late Pleistocene, there were new tools fashioned that expressed a concrete desire and enabled techniques to somehow or other manipulate the environment. This manipulation of the environment first occurred with fire; then other techniques were added: more sophisticated tools, hunting strategies, and magic. Modern *Homo sapiens* by 40,000 BCE became artists, as cave drawings illustrate. They understood depth and perspective, were perhaps engaged in a form of self-expression, and apparently felt that said drawings had important ceremonial uses—for example, to cause a successful hunt, thus to manipulate the environment.

Expanding brain size, burial of the dead, tool use, verbal communication, painting, magic and ceremony: all occurred during the thousands of years of the Late Pleistocene. These developments indicate a new sense of human identity, an awareness of human life and death. During these years, humans, unlike other lifeforms, broke away from complete dependence on the natural environment. Once this emotional detachment occurred, then humans tried to fashion a physical and structural reality by means of tools, shelter, and magic. Humans of the Late Pleistocene no longer had a sense of unquestioning oneness with nature, unlike other forms of animal life. The mental-emotional break from the natural environment was the first step in the direction of dramatic changes in human society and technology that are usually designated the Neolithic Age (about 10,000 BCE) and the onset of Civilization (about 4000 BCE).

A developing scientific mentality accompanied the increasing distance of humans from their dependence upon and identity with the natural environment. Humans' examining life, nature, and time and its movement; analyzing and attempting to control life and death; and naming and identifying things, including themselves, are all mental activities associated with science. Survival required examining the whole to yield understanding of the parts at the same time as breaking aspects of experience into units from which to analyze the whole.

PALEOLITHIC

Scientists categorize human development according to material culture, social and political institutions, sophistication of thought and culture, and level of technology. The latter is the traditional means by which scholars have designated the origins of civilization in Europe, Asia, Africa, and elsewhere. Precivilized societies did not practice metallurgy, hence tools and weapons were limited and very primitive. Precivilized peoples relied on stone, wood, bone, and ivory for tools; metals, if used at all, were for decorative purposes. Some anthropologists have identified humans as *tool users*, which indeed reflects the skill and success at adapting to the natural environment of ancient, precivilized, peoples.

Ancient myths describe a sociological and technological level of ancient peoples that scholars term *Paleolithic*, referring to the Old Stone Age. The first great advances in human technology after the development of fire and of human communication were in the use of stone, bone, ivory, and wood that occurred in the Paleolithic Age. Paleolithic tools were often astonishingly beautiful and very effective. They required forethought, analysis, trial and error, and the ability to repeat successful experiments that yielded the right kind of tool.

Some aspects of precivilized culture were anything but primitive. Art and sculpture from paleolithic Europe, for example, show that these people had a keen eye for nature, conceived and portrayed gods and goddesses, and even began to execute portraits of other humans. Social organization allowed for successful hunts, a rudimentary social hierarchy, and basic rituals and taboos by which to set rules of proper behavior.

NEOLITHIC

The *Neolithic* (New Stone Age) differed from previous times in that humans made revolutionary advances in thought, social organization, and adaptation to the environment. In particular, an agricultural revolution took place, based on the occurrence sometime during this period of one of the greatest scientific discoveries in human history. Some person or group of people living at or near the Tigris and Euphrates River valleys in Asia (a place subsequently known to the Greeks as *Mesopotamia*) around 10,000 BCE used observation and hypothesis to perform an experiment. Perhaps they had noticed that suddenly, after the winter, land that had been barren of food produced wild barley or wild oats, and they wondered how such a miracle occurred. It would have been obvious to such people that animals gave birth to young in the spring, that trees rejuvenated and flowered in the spring, that berries became plentiful on vines and bushes when the days grew longer and the sun became hotter. How did the plant emerge from the soil, the egg appear in the nest, the woman become pregnant with new life? These appeared to be questions linked by a common miracle of newness, of birth, of growth. Ancient fertility cults devoted to plentiful foodstuffs, animal procreation, and female fertility indicate that long before civilization, humans had discovered the idea of fertility, of providing an environment suitable for growth, of the relationship of male and female to conception, pregnancy, and birth. Could there be a connection between male semen being implanted in a healthy, fertile female and a seed from a plant being planted in rich soil?

The domestication of plants—the realization of the process of planting the seed; waiting for germination; cultivating the plant; harvesting the mature grain, vegetable, or fruit; collecting the seed; and then planting as before—ushered in the so-called Neolithic Age. How purposeful was the process of initiation is unclear. Perhaps accident played a large role in the initial discovery; yet much of science has occurred serendipitously. Some anonymous gatherer of wild grain used powers of observation and reason to figure out the process, which resulted in a complete change in all aspects of life.

Agriculture, although rudimentary and haphazard at first, nevertheless involved a scientific process of planning, implementing, controlling, and producing results. As harvests became plentiful, and surplus food was produced and stored, Neolithic humans had a basic understanding of and experienced a general control over their environment, which is the essence of science. Surplus allowed for a break from day-to-day survival; it allowed for a conception of the future based on planning the necessary foodstuffs for a coming winter or period of drought. With more food, there was no longer the need for yearly migrations in search of food. Unlike their forebears, Neolithic peoples were not nomads. Scholars refer to the region where this first occurred as the *Fertile Crescent.* Agriculture allowed for the first stable communities to be made permanent where the soil was rich, moisture was plentiful, and enemies were at a distance. These first villages were at places such as Çatal Hüyük, in eastern Turkey, and Jericho, near the Dead Sea. These small towns supported a population of well over one thousand people, who lived in sun-dried mud brick homes that looked out upon narrow avenues that crisscrossed at right angles. Mud brick walls surrounded the town. Townspeople developed a sense of commonality, a sense of community, which also implied an awareness of the foreign. Restrictions, exclusiveness, control over property, struggle for more territory, and the beginnings of trade all characterized Neolithic society.

Surplus food gave these people a sense of well-being and wealth, opportunities to rest and plan the future, and a sense of time's passing in the seasons, the planting, and the harvesting. Writing was a consequence of the need to keep track of surplus goods from year to year—the first pictographs simply recorded agricultural data. This sense of time, the continuum of past, present, and future, was a prerequisite for scientific thought. Science is fundamentally an intellectual activity involving a historical perspective, wherein the data of experience (natural and human) are accumulated and examined in the present in order to speculate upon, perhaps predict, the future. The sense of time incumbent upon agriculture was therefore fundamental to the emergence of science.

The beginnings of life sciences in the ancient world were associated with interest in magic and the belief that the divine often used natural phenomena such as animals to prophesy and fulfill what has been fated to occur. Ancient humans sought to feel the spiritual forces inherent in nature and to encourage their presence and their benevolence. The shaman in primitive societies is the religious leader who practices magical arts to bring about the good harvest, the successful hunt, the safety of the tribe, and the healing of the sick. The shaman's work is protoscientific, as is magic itself. The first botanists were scientists interested in creating a *materia medica* of plants and herbs that could be used in healing, especially in association with incantations and spells. The first biologists studied life in relation to the divine, seeing in the natural the supernatural, assuming that humans have a direct link with the gods. The first zoologists were soothsayers who examined sacrificial animals to determine the future and priests who conceived of deities with animal characteristics. In time the supernatural and magic became dissociated from the natural and real, and a true life science was born.

The change from the Paleolithic to the Neolithic occurred at various times throughout the world, just as the onset of civilization varied according to place and time. In Europe, the Neolithic occurred about five thousand years later than in the Fertile Crescent. Neolithic Europe lasted for thousands of years—in some places, such as northern Europe, much longer than in southern Europe along the Mediterranean Sea. Humans of the Neolithic, just like those of the Paleolithic, before the advent of record keeping, are anonymous, unknown—almost. A few serendipitous finds in archaeology and anthropology have yielded some fascinating accounts of random individuals in the precivilized past. A great example is a

Neolithic person whose body was discovered in the Ötzal Alps in 1991; he lived thousands of years before, but his corpse had been preserved by ice; he has been nicknamed Ötzi, the Iceman.

The Iceman was middle aged when he died about 3300 BCE in southern Europe. He was killed by an arrow shot from an assailant high up in the mountains. He was armed with a stone-edged dagger, a copper ax, a stone scraper, borer, and arrowheads. He sharpened his tools and weapons with a pressure flaker. He possessed a small satchel with herbs for healing, including a birch fungus useful to treat intestinal worms. Tattoos covered his body. Researchers have recently suggested that the tattoos were purposeful incisions along the lines of acupuncture treatment for ailments.

The worldview of Ötzi would have been contradictory. On the one hand, he lived a simple existence unencumbered by the concerns of our own times. He felt tugged toward nature, upon which he was dependent, to provide him with all things that he knew in life. He acquired food and shelter from nature; there were few alternatives, nothing like the choices of today. His rule of life was to survive and exist in the only ways he knew how, the ways of his ancestors. Such dependence on nature—on the climate, on opportunities to capture food, on the apparent whims of the forces of weather, on the dangers of animals and the dark—leads inevitably to a deep respect for nature. But at the same time, he possessed an identity as a human, living in a growingly sophisticated human community with farming, trade, tools, and medical techniques. He made his tools; he medicated himself; he found his food; he made his shelter—and he was aware of it. He had a sense of personal identity, an "I" or "ego"—he doubtless had a name and lived in or visited small communities in the Alps. He was aware of his individuality and personal identity yet heavily dependent on that nature outside of himself. The contradiction inherent in Ötzi's life was expressed in many ways: as a hunter he would have been successful in part because of his skill and weapons, but not if he could not find game. He might form a garment out of skin to keep out the cold, but then an even colder night might come. He could build a sturdy shelter out of skins or grass, using ingenuity and artisanry, and then the wind would blow it down. He could live actively, skillfully, defending himself and his people; then a random act might strike him down. Early humans like Ötzi had increasing awareness about life and nature, but such awareness brought a host of fears: of death, of attack, of disease, of hunger, of the night, of evil forces. To respond to such fear is to join in community. There is strength in numbers. Thus as a human gains awareness based on detachment from nature, the human seeks to replace the former unity with community and society.

Ötzi and his companions in the European Neolithic had increasing human awareness, but there was still much they did not know. Time was simply the passing of days and of the seasons; it was aging from youth to adulthood to middle age and maybe even old age. They had a limited view on the larger human community outside of their own region. They were simple, localistic, and generally ignorant. They knew few alternatives to their own existence. Their inventiveness was limited to trial and error. The past was very vague, the future even vaguer. Most of their knowledge was negative: how to avoid (*not* experience) pain, hunger, suffering, death. They were aware of the unknown but spent most of their life avoiding it, not penetrating it. Life was a series of events, most of which could potentially kill them. They spent their life avoiding, not pursuing, which left little room for creativity. Their inventiveness was born of the search to survive. Self-expression was for survival. Art was utilitarian, usually wrapped up in magic, to appease good or evil spirits. Yet their universe was largely amoral. Whatever we might take to be good or evil, beautiful or ugly, Ötzi would have doubtless been neutral. Because the future was so unclear, because Ötzi and his companions lived day to day and death was a daily, frequent visitor, the acceptance of this part of life—death—was much easier than for humans today, who have a sense of life expectancy and concrete visions of the future. The ability to see a possible future based on a remembered past, to see nature as outside oneself, to form judgments about nature, about self, and about others takes time over the course of human existence for humans to master. Perhaps in Ötzi's Neolithic world of increasing food, stable villages, communities with trade and rudimentary customs, the sense of control over one's life, one's future, was growing, becoming more apparent. The Neolithic revolution meant that hunger would be less problematic, that survival would be less day to day, and a perception of passing time—past, present, and future—might be slowly acquired. Once the future seemed a bit more secure, Ötzi thought more about it. Possessing greater control over the day to day engenders a desire to control life and nature in all of its forms. Developments in the human community and human technology release the individual and the group for speculation and conceptualization, such as the conceptualization of life, its purpose and end, and the forces of the universe—spirits, deities—that might play a role in existence.

Further north and later in time than Ötzi's time, in the British Isles, the growing sophistication of Neolithic society, characterized by greater protectiveness and a sense of security, in part due to increasing knowledge about the natural environment as expressed through magic, yielded monolithic cultures. The most famous, at Stonehenge, England, dates from 2800 to 1500 BCE. Scholars continue to disagree on how the massive monoliths that make up Stonehenge could have been transported to the site using Neolithic technology. The medieval chronicler Geoffrey of Monmouth in *History of the Kings of Britain* hypothesized that the magician Merlin knew the proper engineering techniques to take the stones down (they were originally located in Ireland, Geoffrey said, and moved to England by ship). Merlin was known, according to Geoffrey, as a man of supreme wit "whether it be in foretelling that which shall be or in devising engines of artifice." Merlin claimed that "in these stones is a mystery, and a healing virtue against many ailments," which suggests the possibility that the builders intended for the monoliths to have magical abilities to heal the sick (trans. Evans). Geoffrey also noted that kings of England were buried at Stonehenge; buried remains from the Neolithic period have been found at Stonehenge.

Diodorus Siculus, a first-century BCE Greek polymath, recorded a tradition of an ancient people who resided

> in the regions beyond the land of the Celts. . . . This island . . . is situated in the north and is inhabited by the Hyperboreans, who are called by that name because their

> home is beyond the point whence the north wind (Boreas) blows; and the island is both fertile and productive of every crop, and since it has an unusually temperate climate it produces two harvests each year. Moreover, the following legend is told concerning it: Leto was born on this island, and for that reason Apollo is honoured among them above all other gods; and the inhabitants are looked upon as priests of Apollo, after a manner, since daily they praise this god continuously in song and honour him exceedingly. And there is also on the island both a magnificent sacred precinct of Apollo and a notable temple which is adorned with many votive offerings and is spherical in shape. (Trans. Oldfather)

Was this temple Stonehenge? Neolithic monolithic sites such as Stonehenge are found throughout the world, such as at Malta (Ħaġar Qim), Ireland (Brownshill Dolmen), France (Carnac Stones), Turkey (Göbekli Tepe), Easter Island (Moai), Spain (Dolmen of Menga), and Bolivia (Tiwanaku).

FURTHER READING

Ernst Cassirer, *An Essay on Man* (New Haven, CT: Yale University Press, 1944).

Nora Chadwick, *The Celts* (Harmondsworth, England: Penguin Books, 1970).

Gordon Childe, *What Happened in History* (Harmondsworth, England: Penguin Books, 1946).

Diodorus Siculus, *The Library of History of Diodorus Siculus*, trans. C. H. Oldfather, vol. 1. (London: Loeb Classical Library, 1933), http://penelope.uchicago.edu/Thayer/E/Roman/Texts/Diodorus_Siculus/home.html.

George Dvorsky, "The Final Days of Ötzi the Iceman Revealed through New Analysis of His Tools," Gizmodo, June 20, 2018, https://gizmodo.com/final-days-of-otzi-the-iceman-revealed-through-new-anal-1826982899.

Sebastian Evans, trans., *Geoffrey of Monmouth* (London: J. M. Dent, 1904).

Geoffrey of Monmouth, *History of the Kings of Britain*, trans. Lewis Thorpe (Harmondsworth, England: Penguin Books, 1966).

Jacquetta Hawkes, *Prehistory. History of Mankind: Cultural and Scientific Development*, vol. 1, part 1 (New York: Mentor Books, 1965).

N. Joly, "The Early History of Fire," *Popular Science Monthly* 10 (1876), https://en.wikisource.org/wiki/Popular_Science_Monthly/Volume_10/November_1876/The_Early_History_of_Fire.

Robert Redfield, *The Primitive World and Its Transformations* (Ithaca, NY: Cornell University Press, 1957).

Meilan Solly, "What Ötzi the Iceman's Tattoos Reveal about Copper Age Medical Practices," *Smithsonian*, September 10, 2018, https://www.smithsonianmag.com/smart-news/what-otzi-icemans-tattoos-reveal-about-copper-age-medical-practices-180970244.

Cristian Violatti, "Neolithic Period," *Ancient History Encyclopedia*, https://www.ancient.eu/Neolithic.

John N. Wilford, "Lessons in Iceman's Prehistoric Medicine Kit," *New York Times*, December 8, 1998, https://www.nytimes.com/1998/12/08/science/lessons-in-iceman-s-prehistoric-medicine-kit.html.

Leonard Woolley, *The Beginnings of Civilization: History of Mankind: Cultural and Scientific Developments*, vol. 1, part 2 (New York: Mentor Books, 1965).

2

Civilization and Science in Ancient Mesopotamia (4000–300 BCE)

THE BEGINNINGS OF CIVILIZATION

Archaeologists tell us that the first towns appeared in the Near East (today's Middle East) perhaps eleven thousand to twelve thousand years ago in what is typically known as the Fertile Crescent. Jericho, near the Dead Sea, and Çatal Hüyük, in eastern Anatolia, were small, walled communities with ordered streets, and their economy depended upon agriculture. Such Neolithic villages ensured much more stability for the inhabitants than the previous hunting and gathering way of life did. Slowly, with hard work and an accommodating climate, agricultural surplus led to population increase and a higher standard of living. Surplus meant wealth, some of which was available for trade. Agricultural surplus and trade yielded a nascent class system: farmers and merchants, laborers and landowners. As the few acquired more rights for less work and suffering, they engaged in intellectual pursuits: a class of priests emerged who associated the success of the community with a patron (male or female) deity. As the mediators of humans to the gods, the priests achieved tremendous power. This, and the need for an organized system to ensure that fields were irrigated, led to government. The Mesopotamians were the first to develop symbols, markings to indicate natural, human, and physical names that earlier humans had devised in oral communication. Mesopotamians invented the idea of using a stylus in wet clay to make various markings indicating symbols to record harvests and other events. The desire to record laws and the reigns of kings as well as more mundane administrative and trade data led to a primitive form of writing, cuneiform, used for record keeping and, in time, to the creation of such heroic tales as the *Epic of Gilgamesh*. At the end of the fourth millennium BCE, the people of the Tigris and Euphrates River valleys had developed the first civilization.

There are a variety of possible definitions of the word "civilization." Typically, civilization refers to a society of structured institutions, an organized and efficient system of living. This is an *objective* way of defining civilization. But civilization can also connote an intellectual, moral, *subjective* way of living: one based on what people think about themselves and the surrounding cosmos. Sophisticated ideas and frameworks of thought to explain human society and the natural environment clearly accompanies the structured institutions of advanced societies. Even more, some thinkers have fashioned other approaches to civilization. Sigmund Freud, for example, argued civilization is actually repressive toward human happiness. Generally, however, civilization such as was found in the first places of sophisticated human societies includes the objective as well as the subjective definitions.

The first sophisticated society that had both objective and subjective expressions of civilization was Sumer in the land of Mesopotamia. The scattered city-states of Sumer during the fourth and third millennia BCE, during the Early Bronze Age, were the *first* in a great many human accomplishments. Cities such as Ur, Eridu, Kish, Lagash, and Nippur had tens of thousands of people at their height. They were busy metropolises surrounded by thick walls and ramparts; the "Seven Sages" laid the foundations for the walls of Uruk, according to the *Epic of Gilgamesh.* There existed extensive trade, the production of crafts, specialization of labor, and a multilayered social structure. The government was as organized as the economy; priests served as conduits for divine instructions from patron deities; scribes recorded the decisions of gods and men. Writing, created to account for surplus wealth and trade, developed into an expression of human hopes, fears, and aspirations. Once humans had reached the capacity to guarantee a surplus of food year after year, to gain greater control over life and ensure survival, they were free to speculate on human existence and the mysteries of nature.

Irrigation of land to produce surplus crops has been a driving force in the history of civilization. The agricultural revolution that began in the Fertile Crescent of western Asia began because early agriculturalists had the basic realization that dry soil did not support plants. Where the climate is such that the land is parched, humans must artificially water or flood the land to ensure moisture for the growing plants. Irrigation probably first took place along the banks of the Tigris and Euphrates Rivers in Mesopotamia. The Tigris and Euphrates flood in the spring, bringing the water from melting snow of the mountains in eastern Turkey. These rivers have low banks, hence rising water often escapes the riverbed and floods the land. Such flooding is unpredictable. The early inhabitants of Mesopotamia decided to exercise human will over the flooding waters, building an intricate series of dikes and canals to channel the water in the direction of fields requiring extra moisture. These dikes and canals also were useful during the dry and hot seasons of the year; farmers would purposely channel water from the rivers into their dry fields, providing the moisture necessary for crops to grow.

The Mesopotamians were superb builders. Pottery was discovered around 6000 BCE in Mesopotamia, followed by the potter's wheel two millennia later. Mesopotamia was the region where two other technological and engineering discoveries were made. Remains of buildings dating back to 2100 BCE show a pitched-brick vault of sun-dried bricks narrowing in to form a conical enclosure. The Mesopotamians were perhaps also responsible for the first monumental architecture: the ziggurat at Ur, dedicated to the moon god Sin, which had four stories or huge platforms built one upon another, leading to a pinnacle and a rectangular enclosure.

THE FIRST SCIENTISTS

The worldview of the astrologist and soothsayer, that by observing nature one can prepare for what is to come, is not far different from the worldview of the scientist, that by observing nature one can understand natural phenomena, which allows one to be better prepared for what is to come. The ancient Sumerians of the Tigris and Euphrates River valleys were the first scientists in world history because they were the first to take this intellectual leap from the superstitious to the scientific worldview. The ancient Mesopotamians developed the first explanations of the origins and makeup of the earth, gods, and humans. These were not expressed scientifically but rather in myth. The *Epic of Gilgamesh* reveals one of the earliest recorded attempts to explain human origins and how sophisticated human society came about. The *Epic* reveals a growing sense of awareness that humanity exists in a natural environment infused with magic and mystery that needs explanation. That humans can understand the mystery of their surroundings reveals a growing confidence in human intellect. That the nature of the universe involves deities with human forms indicates an advanced sense of the human perspective of itself: humans are concerned and affiliated with the origins and mysteries of the universe. Humans interact with the gods, and some humans are descended from the gods.

By the time the *Epic of Gilgamesh* was written, in the third millennium BCE, the detachment of humans from nature and the ability of humans to examine the natural environment from such a detached perspective were such that the universe and its supernatural forces were increasingly seen as humanlike. Not only did the gods act like humans, having the same moral standards or lack thereof, but nature itself was coming to be perceived by human ethical standards of good and evil.

The earliest accounts of science were necessarily attached to mythical stories. One story identifies a gardener named Shukallituda who finds his crops repeatedly destroyed by heat and wind and so prays to heaven for help. Receiving help in the form of a new idea, he determines to plant his garden under shade trees. The idea works: the shade helps the vegetables survive in the hot climate. Shukallituda identified a problem, conceived of a hypothetical solution, tested it, and achieved a desirable result that could be repeated again elsewhere. Other clay cuneiform documents that survive from Iraq indicate that four thousand years ago, Sumerian farmers had an empirical approach to farming and used almanacs to record data and to give advice. Sumerian agriculture involved simple bronze tools, spades, hoes, scythes, and sickles. Ploughs had ploughshares and seed feeders: devices to allow the immediate dropping of seed into the furrow.

Mesopotamians were the first great practitioners of medical science. The Sumerians of the third millennium BCE collected herbs and substances thought to aid in the healing of the body; at one point a physician recorded this data on a clay tablet, which describes what medicine to take for what illness, though it does not record results and is more a manual than a text of clinical observations. The physician used salt and sodium nitrate, the oils of plants, and teas made from boiling herbs such as myrtle and thyme. Some medicines were salves used externally, others were taken internally, washed down with beer. This feat of preparing a materia medica was notable in that no magical spells or incantations reinforced the use of medicine. The physician's scientific approach used various materials to achieve a physical result: the healing of the human body.

The Sumerians were the first to build oceangoing vessels and to sail the seas. Their ideas spread to Egypt, India, the Mediterranean, and the Aegean. Mesopotamian and Egyptian shipbuilders fashioned the hull of the ship to the keel without ribs, instead using mortise-and-tenon joints to join planks to keel. Papyrus rope bound through mortises helped provide stability that would otherwise have been provided by the ribs of the ship. Because of this shell mode of construction, such ships were not seaworthy, and ancient Mesopotamian and Egyptian shipbuilders devised a method of constructing seaworthy ships from papyrus reeds. The ancient geographer Strabo briefly described how these boats were constructed: reed beds "supply reeds from which all kinds of reed-vessels are woven. Some of these vessels, when smeared all over with asphalt, can hold water, whereas the others are used in their bare state. They also make reed-sails, which are similar to rush-mats or wicker-work" (trans. Jones).

The asphalt of Mesopotamia was a type of petroleum distillate, also called naphtha. Strabo recorded how the ancient Mesopotamians used this substance:

> Babylon produces also great quantities of asphalt, concerning which Eratosthenes states that the liquid kind, which is called naphtha, is found in Susis, but the dry kind, which can be solidified, in Babylonia; and that there is a fountain of this latter asphalt near the Euphrates River; and that when this river is at its flood at the time of the melting of the snows, the fountain of asphalt is also filled and overflows into the river; and that there large clods of asphalt are formed which are suitable for buildings constructed of baked bricks. Other writers say that the liquid kind also is found in Babylonia. Now writers state in particular the great usefulness of the dry kind in the construction of buildings, but they say also that boats are woven with reeds and, when plastered with asphalt, are impervious to water. The liquid kind, which they call naphtha, is of a singular nature; for if the naphtha is brought near fire it catches the fire; and if you smear a body with it and bring it near to the fire, the body bursts into flames; and it is impossible to quench these flames with water (for they burn more violently), unless a great amount is used, though they can be smothered and quenched with mud, vinegar, alum, and bird-lime. It is said that Alexander, for an experiment, poured some naphtha on a boy in a bath and brought a lamp near him; and that the boy, enveloped in flames, would have been nearly burned to death if the bystanders had not, by pouring on him a very great quantity of water, prevailed over the fire and saved his life. Poseidonius says of the springs of naphtha in Babylonia, that some send forth white naphtha and others black; and that some of these, I mean those that send forth white naphtha, consist of liquid sulphur (and it is these that attract the flames), whereas the others send forth black naphtha, liquid asphalt, which is burnt in lamps instead of oil. (Trans. Jones)

MIDDLE BRONZE AGE

A dynamic variety of cultures were in competition from the end of the third millennium to the middle of the second millennium in the ancient Near East. The Akkadian Empire controlled much of Mesopotamia and the surrounding regions at the close of the third millennium. Similarly, the Amorites of western Asia took control of much of the Near East at the end of the third millennium; they increasingly came under the sway of Sumerian culture in Mesopotamia. There was a renaissance of culture centered at the city of Ur; the ziggurat of Ur was built around this time. Scribal schools flourished, as did religion and monumental architecture. The first written law code was promulgated during the reign of Ur-Nammu (c. 2112–2195 BCE).

BABYLON

Babylon was for thousands of years the symbol of worldliness, grandeur, magic, and astrology. An old Mesopotamian city, it emerged to political and cultural prominence at the beginning of the second millennium BCE. The city, situated at the lower Euphrates River, fascinated ancient peoples with tales of walls sixty miles in circumference, two hundred feet high, and fifty feet wide. Babylon, which the Elder Pliny called the headquarters of the Chaldeans, the infamous astrologers of the ancient world, was the capital of the First Dynasty, which began in 1894 BCE. The most famous king of early Babylon was Hammurabi (1792–1750 BCE). Indeed, the Code of Hammurabi, one of the first law codes in the history of humankind, records a society where law has overcome brute force and blood vengeance in the settling of disputes. Under Hammurabi, laws were codified to deal with agriculture, water rights, the use of metal as a means of exchange, relations among master and slave and among equals, builders who built weak houses or leaky boats, and thieves of the shadoof and water wheel (used in irrigation).

The Babylonian Code of Hammurabi reveals that Babylon had a dynamic medical community. As early as the third millennium BCE, extensive materia medica existed that indicated what medicines were useful for individual illnesses. *Materia medica* was pharmaceutical knowledge of basic salves and analgesics to take for injuries and pain. Many of the laws of the Code deal with payments to physicians and penalties for malpractice. Physicians performed surgery, drained tumors, and set bones; veterinarians performed surgery on draft animals. Physicians, one gathers, were jack-of-all-trades scientists and astrologers. Of course, magic and omens tempered the scientific efficacy of Babylonian medicine. Early documents indicate that the magician (*ashipu*) was often the same as the physician (*asu*). Nevertheless, medical observations gave the Babylonians a keen awareness of many diseases and suggestions for how to treat them. The Babylonians surpassed the Sumerians in their sophisticated observation and diagnosis of disease.

IRON AGE

The Late Bronze Age of the second millennium BCE in the ancient Near East was a volatile time of competing empires and warfare. Babylon fell to the Kassites, and the Hurrians and Hittites struggled for control of much of the region west of Mesopotamia. The Hittites were the culture that introduced the ancient world to iron metallurgy. Hittites developed writing, based on cuneiform, in part to record agricultural produce, taxes, laws, and judicial decisions. The Hittites developed a sophisticated code of laws in the seventeenth century BCE similar to the Mesopotamian Code of Hammurabi. The law code introduced fines for crimes such as manslaughter and murder. Befitting an agricultural society, laws dealt with property boundaries and prices for livestock. There were laws as well to reinforce custom, such as respect for family relations and to counter rejection of custom, as through sexual abuse, adultery, and misuse of magic; laws also governed market prices and crimes against the state. For example, laws restricted what Hittite physicians could charge for healing a free person or a slave, and what merchants could charge for copper and bronze items. Indeed, the Hittite empire had a hierarchical administrative and justice system dispersed from the ultimate power of the king to provincial governors and magistrates. There were local courts where community elders made judicial decisions.

Hittite medicine was based on magic, salves and oils, and a basic pharmacopeia. For example, a tablet from the late second millennium discovered at the Hittite capital of Hattasu included a magical charm provided by a "Hurrian woman doctor" to use "the fine oil of Azzari" to anoint the commanding general, his horse, and his chariot, apparently to make him impregnable in battle (Casson, p. 6).

During the second millennium BCE, the city of Assur in Upper Mesopotamia began to expand power, which led to a small empire by the end of the second millennium. During the first millennium, this Assyrian Empire grew in strength to conquer most of the ancient Near East. Its greatest period was from 911 to 609 BCE. Records were discovered in the temple of Assur that provide information on the Assyrian penchant for superstition, for omens from the flights of birds, sacrificial animals, and the heavens. Moreover, the Assyrians made lists "of plants, trees, animal, gods, place names, a multiplication table, [and] an astronomical text" (Casson, p. 9). The library of Ashurbanipal (668–631 BCE) had texts dealing with magic, divination, and oneiromancy as well as those relating to scholarly and educational pursuits.

The Assyrian Empire fell to the Chaldeans toward the end of the seventh century; the Chaldean or Neo-Babylonian Empire ruled for the next century. It reached its height under Nebuchadnezzar II (604–562 BCE), covering the area from Anatolia to Egypt east to Iran. It was under Nebuchadnezzar that Chaldean soothsayers, astrologers, and magicians grew in fame throughout the ancient world.

PHYSICAL SCIENCES AND MATHEMATICS

The Mesopotamians made some notable discoveries in the hard sciences and mathematics. Sumerian chemists discovered a technique to smelt copper and tin to produce bronze, which revolutionized tool use and warfare. The initial interest in astrology became a more practical interest in astronomy: the Mesopotamians became adept at tracing the movement of the planets, identifying the constellations of the night sky, and predicting the phases of the moon. Astronomical observations led the Babylonians to remarkably accurate lunar calendars. The Babylonians of the second and first millennia BCE observed the heavens with the assumption that the movements of planets and stars and the phases of the moon had an impact on earthly events. The appearance of a new moon was greeted with wonder and gladness, as a foreshadowing of good things. The gradual appearance of the moon and its horns informed the prognosticator of the course of events. The moon lasting to the thirtieth day foreshadowed evil. Likewise other phenomena indicated good and evil according to the prognosticator's interpretation: eclipses; halos; the simultaneous appearance of moon and sun; the conjunction of the planet Jupiter and the moon; the appearance or disappearance and place on the horizon of Venus, Mercury, and Mars; the appearance of thunderstorms during certain celestial events; and earthquakes.

Beginning in the eighth century BCE, Babylonian astronomers kept ephemerides of lunar, planetary, and stellar phenomena. They studied the course of the stars to arrive at accurate star charts, a vision of the zodiac, and lunar calendars. This included accurate lunar and solar eclipse charts from 450 BCE on. Pliny the Elder claimed that the Babylonian day began at sunrise and that earthquakes resulted when three planets (Mars, Jupiter, Saturn) were aligned with the sun. According to Epigenes, Babylonians recorded astronomical observations on clay tablets. The astronomer Ptolemy clearly knew of Babylonian mathematics and astronomy as well. Indeed, Ptolemy's geocentric system, which relied on the inner planets Mercury and Venus orbiting the sun on epicycles as the sun orbited the earth, was remarkably similar to Babylonian astronomical schemes. The most famous Babylonian astronomer was Berossus, a priest of Baal who lived at Cos in the Aegean Sea in the third century BCE. According to Vitruvius,

> Berosus, who came from the state, or rather nation, of the Chaldees, and was the pioneer of Chaldean learning in Asia, [argued that] the moon is a ball, one half luminous and the rest of a blue colour. When, in the course of her orbit, she has passed below the disc of the sun, she is attracted by his rays and great heat, and turns thither her luminous side, on account of the sympathy between light and light. Being thus summoned by the sun's disc and facing upward, her lower half, as it is not luminous, is invisible on account of its likeness to the air. When she is perpendicular to the sun's rays, all her light is confined to her upper surface, and she is then called the new moon. (Trans. Morgan)

By this theory Berossus explained the phases of the moon. Berossus, according to Vitruvius, was an expert in astrology, "which concerns the influences of the twelve signs, the five stars, the sun, and the moon upon human life." He and other Chaldeans knew "the art of casting nativities, which enables them to declare the past and the future by means of calculations based on the stars" (trans. Morgan).

Babylonian mathematicians developed many of the initial ideas and techniques that would characterize Greek mathematics of the first century BCE. Babylonian mathematics, clearly meant for practical arithmetical problems and solutions rather than theoretical purposes, included algebra, quadratic equations, coefficients, number progressions, linear algebra, and simple geometry. They developed quadratic equations and solutions to geometric problems such as the Pythagorean theorem (long before Pythagoras). They worked with coefficients, squares, cubes, square roots, and the radii of circles. Place-value notation in numbers was developed by the Babylonians, who first used zero as a place notation toward the end of the first millennium BCE during the Hellenistic Age. Babylonian mathematicians divided the circle into 360 degrees. This was based on the sexagesimal system of base 60; digits could be determined by a fraction of 60 or its multiples.

Babylonians conceived of numbers as concrete amounts, not as conceptual symbols. Cuneiform texts survive that indicate solutions to problems of engineering and building. Geometry, more abstract, was not as developed in Babylon. Nevertheless, the Pythagorean theorem was first used in Babylon as early as the first centuries of the second millennium BCE. Texts also indicate that Babylonian mathematicians solved problems involving hexagons, triangles, trapezoids, and the radii of circles. Babylonians understood the principle that a line that extends from a right angle to the hypotenuse of a triangle creates two equal triangles. Geometry was also used in surveying. One of the oldest maps in the world, surviving on a clay tablet, is of the Euphrates flowing through the ancient city of Babylon.

Mesopotamian science had a clear impact on the development of Greek science during the first millennium BCE. The Mesopotamians anticipated the mathematical theorems of Pythagoras and Euclid. Babylonian astronomy and mathematics influenced Thales, the first notable Greek scientist. His notion that the primal element is water was derived from Sumerians of an earlier millennium. Ancient cultures adopted the Mesopotamian system of timekeeping and figuring the yearly calendar. Babylonian scholars developed a lunar calendar of thirteen months. Mesopotamian astronomy and astrology came to be represented by the Chaldeans and magi of the ancient Near East.

The Chaldeans of ancient Mesopotamia had quite the reputation for soothsaying, astrology, and astronomy among ancient writers. Diodorus Siculus, in his encyclopedic *Universal History*, wrote,

> The Chaldaeans, belonging as they do to the most ancient inhabitants of Babylonia, have about the same position among the divisions of the state as that occupied by the priests of Egypt; for being assigned to the service of the gods they spend their entire life in study, their greatest renown being in the field of astrology. But they occupy themselves largely with soothsaying as well, making predictions about future events, and in some cases by purifications, in others by sacrifices, and in others by some other charms they attempt to effect the averting of evil things and the fulfilment of the good. They are also skilled in soothsaying by the flight of birds, and they give out interpretations of both dreams and portents. They also show marked ability in making divinations from the observation of the entrails of animals, deeming that in this branch they are eminently successful. . . . The Chaldaeans have of all men the greatest grasp of astrology, and that they bestowed the greatest diligence upon the study of it. But as to the number of years which, according to their statements, the order of the Chaldaeans has spent on the study of the bodies of the universe, a man can scarcely believe them; for they reckon that, down to Alexander's crossing over into Asia [333 BCE], it has been four hundred and seventy-three thousand years, since they began in early times to make their observations of the stars. (Trans. Oldfather)

Strabo, the ancient geographer, wrote of the Chaldeans:

> In Babylonia a settlement is set apart for the local philosophers, the Chaldaeans, as they are called, who are concerned mostly with astronomy; but some of these, who are not approved of by the others, profess to be genethlialogists [astrologers]. There is also a tribe of the Chaldaeans, and a territory inhabited by them, in the

> neighbourhood of the Arabians and of the Persian Sea, as it is called. There are also several tribes of the Chaldaean astronomers. For example, some are called Orcheni, others Borsippeni, and several others by different names, as though divided into different sects which hold to various different dogmas about the same subjects. And the mathematicians make mention of some of these men; as, for example, Cidenas and Naburianus and Sudinus. Seleucus of Seleuceia is also a Chaldaean, as are also several other noteworthy men. (Trans. Jones)

Even after the fall of the Chaldean empire, Chaldean soothsayers and prognosticators continued their work throughout the ancient Near East and west into Mediterranean lands. The Persians conquered the Chaldeans in 550 BCE and controlled the region of Iran, Mesopotamia, and other lands of the ancient Near East until the conquests of Alexander of Macedon in the 330s. Subsequently Mesopotamia was controlled by Greek kings, the Romans, and then the Parthians.

FURTHER READING

Lionel Casson, *Libraries in the Ancient World* (New Haven, CT: Yale University Press, 2001).

Diodorus Siculus, *The Library of History of Diodorus Siculus*, trans. C. H. Oldfather, vol. 1 (London: Loeb Classical Library, 1933), http://penelope.uchicago.edu/Thayer/E/Roman/Texts/Diodorus_Siculus/home.html.

Epic of Gilgamesh, trans. N. K. Sandars (London: Penguin Books, 1972).

William Hallo and William Simpson, *The Ancient Near East* (New York: Harcourt Brace Jovanovich, 1971).

Harry A. Hoffner Jr., *The Laws of the Hittites: A Critical Edition* (Leiden, Netherlands: Brill, 1997).

Samuel Noah Kramer, *History Begins at Sumer* (Philadelphia: University of Pennsylvania Press, 1980).

O. Neugebauer, *The Exact Sciences in Antiquity* (New York: Dover Books, 1969).

Joan Oates, *Babylon* (London: Thames and Hudson, 1986).

Jack Sasson et al., eds., *Civilizations of the Ancient Near East*, 4 vols. (New York: Charles Scribner's Sons, 1995).

Strabo, *Geography*, trans. H. L. Jones, 8 vols. (Cambridge, MA: Harvard University Press, 1917–1932), http://penelope.uchicago.edu/Thayer/E/Roman/Texts/Strabo/home.html.

R. Campbell Thompson, *Assyrian and Babylonian Literature: Selected Transactions*, with a critical introduction by Robert Francis Harper (New York: D. Appleton and Company, 1901).

Vitruvius, *The Ten Books on Architecture*, trans. Morris H. Morgan (Cambridge, MA: Harvard University Press, 1914).

Leonard Woolley, *The Beginnings of Civilization: History of Mankind: Cultural and Scientific Developments*, vol. 1, part 2 (New York: Mentor Books, 1965).

3

Civilization and Science in the Nile River Valley (3100–300 BCE)

Sun worship and the concern for understanding the environment were natural in a land wherein the people were dependent upon nature and its natural rhythms for their survival. The archaic and classical Greeks hailed Egyptian civilization as a prime source for Greek cultural and scientific achievements. Founders of Greek thought, such as Thales and Pythagoras, were thought to have visited Egypt. Plutarch wrote that Solon learned of Atlantis from Pseuophis of Heliopolis and Sonchis of Sais. Others, such as Herodotus and Plato, later made the same pilgrimage to discover the secrets of Egypt. Ancient writers were generally in agreement that Egyptian culture was the earliest of the Mediterranean civilizations.

Egyptian existence was built upon the Nile River, its annual floods, and the ever-present sun. Herodotus (490–430 BCE), when he journeyed to Egypt in the fifth century BCE, discovered that the Nile River was the basis of Egyptian society and culture. He called Egypt "the gift of the Nile." Land in low-lying areas naturally received sufficient moisture as well as a fertile layer of silt from the flooding Nile. Land rising above the banks of the Nile required a technique to irrigate the soil. The Egyptians used an irrigation tool called a shadoof, which was a simple machine operated by the farmer.

The origins of Egyptian history are shrouded in legend and myth. The historian and traveler Diodorus Siculus, having visited Egypt in the first century BCE, where he made a record of their ideas, religion, and philosophy, provided in his *Universal History* an account, derived from the Egyptians, of the origins of the gods and attributes of civilization. Egyptians knew of the importance of the human mastery of fire in the development of civilization, and they assigned its origins, according to Diodorus, to Hephaestus, "their first king," later deified as the smith god. Isis, the fertility goddess, the same writer declared, "discovered the fruit of both wheat and barley which grew wild over the land along with the other plants but was still unknown to man"; Osiris "also devised the cultivation of these fruits, all men were glad to change their food." Osiris also showed humans how to cultivate grapes and ferment wine. Diodorus learned on his journey to Egypt that "Isis also established laws . . . in accordance with which the people regularly dispense justice to one another and are led to refrain through fear of punishment from illegal violence and insolence; and it is for this reason also that the early Greeks gave Demeter," the comparable Greek fertility goddess, "the name Thesmophorus [law giver], acknowledging in this way that she had first established their laws." Egyptian priests also told Diodorus that

> it was by Hermes . . . that the common language of mankind was first further articulated, and that many objects which were still nameless received an appellation, that the alphabet was invented, and that ordinances regarding the honours and offerings due to the gods were duly established; he was the first also to observe the orderly arrangement of the stars and the harmony of the musical sounds and their nature, to establish a wrestling school, and to give thought to the rhythmical movement of the human body and its proper development. He also made a lyre and gave it three strings, imitating the seasons of the year; for he adopted three tones, a high, a low, and a medium; the high from the summer, the low from the winter, and the medium from the spring. The Greeks also were taught by him how to expound (*hermeneia*) their thoughts, and it was for this reason that he was given the name Hermes.

The imaginative Egyptian priests and their willing listener Diodorus Siculus further claimed that the Nile,

> at the time of the rising of Sirius, which is the season when the river is usually at flood, breaking out of its banks inundated a large section of Egypt and covered especially that part where Prometheus was governor; and since practically everything in this district was destroyed, Prometheus was so grieved that he was on the point of quitting life wilfully. Because its water sweeps down so swiftly and with such violence the river was given the name Aëtus [eagle]; but Heracles, being ever intent upon great enterprises and eager for the reputation of a manly spirit, speedily stopped the flood at its breach and turned the river back into its former course. Consequently certain of the Greek poets worked the incident into a myth, to the effect that Heracles had killed the eagle which was devouring the liver of Prometheus. The river in the earliest period bore the name Oceanê, which in Greek is Oceanus; then because of this flood, they say, it was called Aëtus, and still later it was known as Aegyptus after a former king of the land.

Isis also, according to the Egyptian priests that Diodorus interviewed, had an Ascelpian character:

> The Egyptians say that she was the discoverer of many health-giving drugs and was greatly versed in the science of healing; consequently, now that she has attained immortality, she finds her greatest delight in the healing of mankind and gives aid in their sleep to those who call upon her, plainly manifesting both her very presence and her beneficence towards men who ask her help. In proof of this, as they say, they advance not legends, as the Greeks do, but manifest facts; for practically the entire inhabited world is their witness, in that it eagerly contributes to the honours of Isis because she manifests herself in healings. For standing above the sick in their sleep she gives them aid for their diseases and works remarkable cures upon such as submit themselves to her; and many who have been despaired of by their physicians because of the difficult nature of their malady are restored to health by her, while numbers who have altogether lost the use of their eyes or of some other part of their

body, whenever they turn for help to this goddess, are restored to their previous condition. (Trans. Oldfather)

Clearly the Egyptians had a deep belief in the power of amulets and magical sayings, and religious superstition to help them achieve a blessed afterlife. A priest of the New Kingdom, one Ani (c. 1250 BCE), recorded hymns and prayers and his beliefs about life and death in the *Book of the Dead.* Ani assumed that deities such as Osiris, Isis, and Ra personified natural phenomena. In a paean to Osiris, for example, Ani calls the god "he who is lord of the stars that never change." In the *Book of the Dead,* Osiris is the sun, Isis is the moon, and the world is created and will perish. The universe is spherical, and the stars are on fire. The bright lights of the heavens determine human destiny. Much of Egyptian culture focused on the afterlife, as shown by, for example, the practice of mummification.

ASTRONOMY

The Egyptians, living in a land of the distant horizon, the black night of countless stars, and the hot summer days with the sun high overhead, made a host of astronomical observations, though most of them were done for simplistic purposes. Manetho, the third-century-BCE chronologist of Egyptian history, gave names to the five planets known to the ancients: Cronos or Saturn, "the shining star"; Zeus or Jupiter, "the radiant star"; Ares or Mars, "the fiery star"; Aphrodite or Venus, "the fairest"; Hermes or Mercury, "the glittering star" (trans. Waddell). The Egyptians were particularly concerned with tracing the rising of Sirius, the Dog Star, which indicated the rise of the Nile, in July, when the Egyptians began their new year. Indeed, Egyptian astronomers devised an accurate 360-day solar calendar, which was later adapted by the Greeks. Herodotus heard directly from priests, who were the astronomers, about their calendar and astronomical observations. Herodotus declared that the Egyptian technique of adding five intercalary days every year was the most efficient technique. The motions of other stars and planets indicated other meteorological events. According to the Late Roman compiler Diogenes Laertius, Egyptian astronomers understood lunar and solar eclipses, counting 373 solar and 832 lunar eclipses. He noted that they worshipped Osiris the sun and Isis the moon; that the world was created and would perish; that the universe was spherical; that the stars were on fire. The Egyptians conceived of the transmigration of souls. Rain, they believed, was "caused by change in the atmosphere." (trans. Oldfather) Herodotus claimed that the Egyptians were the first people to develop the pseudoscience of astrology: the study of stars as determiners of human destiny.

Egyptians used basic arithmetic and fractions largely for practical purposes. In geometry, the Egyptians engaged in simple calculations and measurements of shapes: isosceles triangles, the circle, and varied rectangular shapes, such as the trapezoid. Scholars are unsure whether Egyptian mathematicians calculated the value of pi (π), the ratio of the circumference of a circle to its diameter. One wonders how the massive exactness of the pyramids could have been made without the knowledge that comes from a clear understanding of geometry.

MANETHO (EARLY THIRD CENTURY BCE)

Astronomy and chronology are complementary, as seen in the work of the Egyptian priest Manetho, who wrote a detailed and sophisticated chronology of the history of Egypt, a chronology that is much used and cited by early historians. Manetho identified the Greek gods as being originally Egyptian and significantly believed the first Egyptian (god) was Hephaestus, the Greek god of fire and metallurgy, one of the earliest expressions of science and a prerequisite of civilization. Manetho was extremely concerned with the antiquity of Egypt and tried reckoning its age according to lunar and solar years. An Egyptian monk, Panodôrus, who lived at the end of the fourth century CE, relied on Manetho for his chronological interpretation of Egyptian history:

> From the creation of Adam, indeed, down to Enoch, *i.e.*, to the general cosmic year 1282, the number of days was known in neither month nor year; but the Egregori (or "Watchers"), who had descended to earth in the general cosmic year 1000, held converse with men, and taught them that the orbits of the two luminaries, being marked by the twelve signs of the Zodiac, are composed of 360 parts. Observing the moon's orbit which is nearer the earth, smaller, and more conspicuous, as it has a period of thirty days, men decided that it should be reckoned as a year, since the orbit of the sun also was filled by the same twelve signs of the Zodiac with an equal number of parts, 360. So it came to pass that the reigns of the Gods who ruled among them for six generations in six dynasties were reckoned in years each consisting of a lunar cycle of thirty days. The total in lunar years is 11,985, or 969 solar years. By adding these to the 1058 solar years of the period before their reign, they reach the sum total of 2027 years. (Trans. Waddell)

Manetho's *History of Egypt* and other writings only survive in fragments recorded by later commentators such as the chronologists Julius Africanus, Eusebius of Caesarea, and George Syncellus. In his works Manetho provides fascinating if sometimes dubious tidbits on Egyptian history, including the history of science. His work provided the basis for the historical reconstruction of the chronology of dynasties in ancient Egypt. Manetho identified the gods of the Mediterranean with Egyptian rulers, claiming that the first, Hephaestus, ruled for over 700 years; followed by Helios (the sun), who ruled for over eight years; followed by Agathodaemon and then Cronos, who ruled for over 40 years; followed by Osiris and Isis, who ruled jointly for 35; then followed by Typhon, Orus, Ares, Anubis, Heracles, Apollo, Ammon, Tithoes, Sosus, and Zeus. After the rule of such mythical kings, gods, and demigods, Manetho identified dynasties of human rulers, beginning with Menes, succeeded by his son, Athothis, who "built the palace at Memphis; and his anatomic works are extant, for he was a physician." Kenkenes succeeded Athothis, and Uenephes succeeded Kenkenes; Uenephes was a builder of pyramids (perhaps those of Sakara). The kings of the First Dynasty, Manetho reckoned, ruled for 252 years. The Second Dynasty lasted 297 to 302 years; the Third Dynasty, 198 to 214 years. During the reign of the first pharaoh of the Third Dynasty, Necherochis, Manetho recorded a lunar eclipse, using the words, "the moon waxed beyond reckoning," which terrified an army of Libyans into surrender to Egyptian forces. This eclipse occurred in 2654 BCE. Necherochis's successor, Sesorthos, Manetho wrote, "was styled Asclepios in Egypt because of his medical skill." The Fourth Dynasty lasted 277 years and included the reign of Suphis (Cheops), known for erecting the Great Pyramid (trans. Waddell).

PYRAMIDS

Imhotep was the legendary builder of the first Egyptian step pyramid (ziggurat) as well as a physician who was so honored by the Egyptians that he became the deity of learning and giver of science, rather like the Greek Prometheus. The Egyptians credited Imhotep with the development of mathematics, the calendar, architecture, and medicine. In time, Imhotep's shrine became a place where the sick went to be healed. Historically, Imhotep's pyramid, built around 2700 BCE at Saqqara in the Third Dynasty, was a series of mastabas (six in all) built up to form a hierarchical pyramid. Imhotep's ziggurat became a model for the building of more sophisticated pyramids as time passed.

The historian and geographer Herodotus, who visited Egypt in the mid-fifth century BCE, provided in Book 2 of his *Histories* a concise account of the building of the Great Pyramid during the reign of Cheops (also known as Khufu) in the twenty-sixth century BCE. The pharaoh used 100,000 men to drag limestone blocks from distant quarries; for part of the way, the blocks were transported by barge on the Nile. Indeed, Cheops had a canal dug that flooded with the rising Nile, allowing the barges to bring the limestone blocks right to the building site. The Great Pyramid took twenty years to build. It was built in stages by layers, like a step pyramid, the laborers using complicated levers to raise the blocks up each level. The massive limestone blocks were fitted exactly, then polished. The height of the pyramid is equal to each of its four sides at the base. According to Herodotus, the laborers of this astonishing work lived on a simple vegetarian diet. Scores of pyramids were built throughout Egyptian history covering the Old, Middle, and New Kingdoms, as well as the three Intermediate Kingdoms. Pyramids dotted the Nile, some as far south as Sudan, built during the Napatan (900–270) culture of the Second Kingdom of Kush. The pyramids found at Gebel Barkal overlooking the Nile in Sudan featured a new building technique, the corbel vault (seventh century).

MEDICINE AND LIFE SCIENCES

Egyptians developed an extensive materia medica; they were "a race of druggists," according to Homer: the most able physicians were knowledgeable about botany. Herodotus claimed they were a race of doctors, there being a physician for every type of illness. Their knowledge of medicine was based on folklore more than empirical observation. The Egyptian practice of medicine relied on healthy doses of magic. Here, in medicine as well as in other branches of science, the Egyptians believed in a clear connection between the natural and the supernatural. In Egypt, as elsewhere in the ancient Near East, religious concerns and rituals came before the practice of medicine. According to Herodotus, Egyptians believed that good health required a monthly purging using strong emetics. Egyptian dentists and physicians could pull teeth, deliver babies, set bones, and prescribe medicine for healing. Their knowledge of anatomy and physiology was rudimentary, though there are records of physicians examining patients and making diagnoses. One fragment of Manetho declares that solar eclipses have "a baneful influence upon men in their head and stomach" (trans. Waddell).

Herodotus also reported that the Egyptians were knowledgeable about many plants. The Egyptians had discovered the qualities of the lotus plant: they harvested it; ate the root, which was a sweet, like a fruit; dried the head, a poppy; and made bread out of it. They also collected a plant called *kiki*, which produced something like castor oil and used it like olive oil, though Herodotus found the smell annoying. The papyrus plant was harvested to make paper and used in constructing reed boats, but it was also good to eat when baked. Herodotus also observed or heard about crocodiles, ibis, hippopotamuses, and elephants. He correctly reported on the characteristics of the crocodile, pointing out how its lower jaw is stationary while the upper jaw opens wide to gorge itself on food. Always in search of the remarkable, Herodotus claimed that the crocodile allowed the sandpiper to pick leeches out of its open mouth. Herodotus was fascinated by fowl and provided an interesting description of the ibis. He was also credulous enough to report on the phoenix, though confessing that he had never seen one himself.

EMBALMING

Herodotus observed on his visit to Egypt the different practices of mummification. The most extensive, done for the wealthy, involved removing the brain through the nostrils and clearing out the intestines through an incision made in the abdomen. The body was then filled with spices, particularly myrrh, and soaked or "pickled" in natron (an embalming fluid made of sodium carbonate) for seventy days, after which the body was wrapped in white linen with gum used as the adhesive. The less wealthy had "oil of cedar" injected through the anus into the intestinal cavity, after which they were soaked in natron as well (trans. Selincourt). The very poor had the intestines cleaned with a strong emetic, after which they, too, were kept in natron for seventy days. The practice of mummification indicated the Egyptian belief in the afterlife. Herodotus heard from Egyptian priests that they believed in a cycle of reincarnation that lasted for three millennia, during which the soul migrated from one animal to another, up the chain of being, before returning to inhabit the human body again.

Manetho wrote a scientific treatise, *Epitome of Physical Doctrines*, which survives only in one passage, recorded by Eusebius of Caesarea:

> The Egyptians say that Isis and Osiris are the Moon and the Sun; that Zeus is the name which they gave to the all-pervading spirit, Hephaestus to fire, and Demeter to earth. Among the Egyptians the moist element is named Ocean and their own River Nile; and to him they ascribed the origin of the Gods. To Air, again, they give, it is said, the name of Athena. Now these five deities,—I mean Air, Water, Fire, Earth, and Spirit,—traverse the whole world, transforming themselves at different times into different shapes and semblances of men and creatures of all kinds. In Egypt itself there have also been born mortal men of the same names as these deities: they were called Hêlios, Cronos, Rhea, as well as Zeus, Hêra, Hêphaestus, and Hestia. Manetho writes

> on this subject at considerable length, while Diodorus gives a concise account. (Trans. Waddell)

Here, Manetho added to the traditional four elements of earth, air, fire, and water, as per the Greeks, a fifth: spirit. In his lost works, Manetho also recorded the reliance of Egyptians on a medicinal concoction, *kyphi*, which had sixteen ingredients: "Honey, wine, raisins, cyperus, resin, myrrh, aspalathus, seselis, mastic, bitumen, thryon, dock, as well as of both junipers (arceuthids—one called the greater, the other the less), cardamom, and reed" (trans. Waddell).

According to Diogenes Laertius, Egyptian meteorologists understood the cause of rain to be an atmospheric change. The geographer Herodotus's notion of the meteorology of Egypt, which he doubtless got from Egyptian priests, was extremely primitive: he believed that the strong African winds blew the sun off course along the upper Nile Valley, making it very dry and affecting the level of the Nile downstream in Egypt.

THE SEARCH FOR IMMORTALITY

The Egyptians, meanwhile, expanded the human search for personal immortality and individual recognition both in the present and in the future. Although the mass of Egyptians lived anonymous lives, the pharaohs and their families and close supporters became convinced of their own personal immortality, their greatness spanning the epochs, and they demanded monuments to direct future individuals not to forget *this one or that one* commemorated in a monument or buried within a pyramid. The self-centered arrogance of the fourteenth-century BCE monotheist Akhenaten (Amenhotep IV, r. 1379–1362 BCE) revealed a clear identification of himself with the divine, of his own inner light with the great light of the sun.

Amenhotep IV ruled during the Eighteenth Dynasty of Egyptian pharaohs, when Thebes was the religious capital of the New Kingdom dedicated to the worship of the sun, Amen. Upon assuming the throne in 1379 BCE, Amenhotep changed his name to Akhenaten, moved the religious center south to a new capital at Amarna, and devoted himself to the worship of the one god, Aten, the disk of the sun. Akhenaten's *Hymn to Aten* is a pious account of the god written by his own true worshipper, the pharaoh. His philosophy: Aten is the universal source of all light, warmth, and truth; only Akhenaten understands this and therefore knows his true self as a reflection of the divine. Aten is with each human at the moment of conception; he nourishes the child in the womb; he determines the child's length of days and aims of life; he watches over each of his children, each human. Aten is the beginning and end of all things, the sum of time, the universal presence that chooses the sun's disk as his incarnation, visible to humans, especially pharaoh. As the sun's rays reach deep into a person's being, so, too, does the Aten. Akhenaten feels the presence of the god within himself. It strengthens him to act on the behalf of the Aten, spreading his worship notwithstanding opposition and trials.

Even after its repeated conquest by the Nubians, Assyrians, Persians, Greeks, and Romans, Egypt remained an important center of thought and culture. During the ancient and medieval periods, the city of Alexandria was a premier center of science in the Mediterranean world.

FURTHER READING

E. A. Wallis Budge, trans., *Papyrus of Ani: Egyptian Book of the Dead* (New York: Dover Books, 1967).

Diodorus Siculus, *The Library of History of Diodorus Siculus*, trans. C. H. Oldfather, vol. 1 (London: Loeb Classical Library, 1933), http://penelope.uchicago.edu/Thayer/E/Roman/Texts/Diodorus_Siculus/home.html.

Paul Dunbavin, *Under Ancient Skies: Ancient Astronomy and Terrestrial Catastrophism* (Nottingham, UK: Third Millennium Publishing, 2005).

Adolf Erman, *Life in Ancient Egypt*, trans. H. M. Tirard (New York: Dover Books, 1894).

The Fragments of Manetho, trans. W. G. Waddell (Cambridge, MA: Harvard University Press, 1940), http://penelope.uchicago.edu/Thayer/E/Roman/Texts/Manetho/home.html.

Gebel Barkal and the Sites of the Napatan Region, UNESCO, https://whc.unesco.org/en/list/1073.

William Hallo and William Simpson, *The Ancient Near East* (New York: Harcourt Brace Jovanovitch, 1971).

Herodotus, *The Histories*, trans. Aubrey de Selincourt (Harmondsworth, England: Penguin Books, 1972).

Thor Heyerdahl, *Early Man and the Ocean: A Search for the Beginning of Navigation and Seaborne Civilizations* (New York: Vintage Books, 1980).

"Irrigation," *Encyclopædia Britannica*, vol. 12 (1962).

O. Neugebauer, *The Exact Sciences in Antiquity* (New York: Dover Books, 1969).

J. J. O'Connor and E. F. Robertson, *History of Mathematics*, School of Mathematics and Statistics, University of St. Andrew's Scotland, http://www-history.mcs.st-andrews.ac.uk/history/References/Heron.html.

Elizabeth Riefstahl, *Thebes in the Time of Amunhotep III* (Norman: University of Oklahoma Press, 1964).

Robert M. Schoch, *Voyages of the Pyramid Builders: The True Origins of the Pyramids from Lost Egypt to Ancient America* (New York: Penguin Books, 2004).

Piotr O. Scholz, *Ancient Egypt: An Illustrated Historical Overview* (Hauppauge, NY: Barron's Education Press, 1997).

Jon Manchip White, *Everyday Life in Ancient Egypt* (New York: Capricorn Press, 1967).

PART 2

Magic, Religion, and Science

4

Magic, Superstition, and Science in the Ancient World (4000 BCE–500 CE)

Ancient humans used collective experiences, senses, thoughts, dreams, fantasies, and fears by which to understand existence: they conceived of explanations to account for the way things were in human and natural existence. This is the essence of myth. Mythology is an expression of human existence, an ever-changing mirror of the perceptions of the human condition. The *etymology* (origin) of the word *mythology* is based on *mythos,* Greek for "story" or "word," and *logia or logos*: Greek for "word," "idea," "expression." Myths are based in a real past that is imprecise in time, a reality of human behavior that transcends a particular time and place. Science, too, has these attributes. The close association of myth and science is clearly seen when considering ancient science. The ancients struggled to explain the natural environment and their own humanness with respect to the divine. Initial explanations involved gods and heroes, fantastic occurrences and remarkable events. Ancient peoples would explain human behavior, natural phenomena, by personalizing it as representing behavior and phenomena by a god or goddess. In ancient Iraq, ancient Greece, ancient Rome, ancient Egypt, and ancient India—indeed, throughout the world—nature deities were given personalities, physical characteristics, a mythological history. In many respects these different peoples shared the same ideas, the same deities. Beneath the fantastic and remarkable elements of oral tradition, the poems sung by the ancient bard, was a real attempt to understand and portray nature and self. Joseph Campbell, the leading student of myth in modern times, calls myth "a mask of God." His book *The Power of Myth* outlines the functions of myth in ancient societies. Myth, he argues, involves mystical, cosmological, sociological, and pedagogical functions. *Myth is mystical*: the universe is seen as "a holy picture." Myths are allegories for what is truth—for example, stories to explain nature. Myths reveal the most basic repressed human fears and emotions; they are archetypes of human feelings, expressions, and attempts to understand existence. *Myth is cosmological*: seeking factual foundations for life's mystery, something hidden within existence—that is, searching for a causative explanation for things. *Myth is sociological*: "supporting and validating a certain social order"—relationships of parts to a whole, the rituals, traditions, and customs of a people. *Myth is pedagogical*: teaching how to live—the means by which society communicates and finds solutions to different problems (Campbell, pp. xviii, 39).

Ancient societies used myths to explain the origins of, and the relationship of humans to, the universe. The Sumerians engaged in speculation into the relation of humans to the divine, the ultimate purpose of existence, and the role of the individual in the whole. Notwithstanding that human population dramatically rose during the third millennium BCE in Mesopotamia, the Sumerians were able to isolate individual human achievements and personality traits. In the *Epic of Gilgamesh*, for example, we read about an individual human who is complex, unique, and tragic. Gilgamesh's search for eternal life and happiness is a general human search made singular to the life of one man. Gilgamesh, moreover, is part god, and he can compete on the gods' level, which indicates a growing awareness of the significance and value of the individual person.

THE GREEKS

Greece is a landscape of deep valleys, massive ridges, hidden forests, and daunting mountains. The gods of Greece were therefore those of the mountains, forests, rivers, rocky coasts of the seas, and the isles of the Aegean and Adriatic. Scholars believe that before the coming to power of the Mycenaean Greeks, the early Greeks and people of Crete and the islands worshipped nature deities dominated by fertility goddesses. Athena, Aphrodite, Hera, Artemis, Hestia, Demeter, and Rhea were once, in the mythical past, the primary objects of worship among the Greeks. Then Rhea's son, Zeus, took power from his father, Cronos, and brought to power as well his brothers, Poseidon and Hades. Zeus ruled through the terrible thunderbolt, which only he controlled. It cast fear into gods and humans alike. Poseidon personified the sea; Hades ruled (became) the underworld. The fertility goddesses dwindled in power or assumed different guises—for example, Athena became a masculine virgin; Hera became a powerless shrew and nag of her brother and husband, Zeus; Aphrodite was the vixenish goddess of love, always getting gods and humans into uncomfortable affairs of lust and adultery; Demeter retained her role as a fertility goddess of the grain.

The poems of Homer, Homeric hymns, and poems of Hesiod illustrate how the ancient bards personified natural and human phenomena by means of the gods and goddesses. Hesiod was a theogonist who lived in the eighth century BCE. In his attempt to understand the origins of the divine, he explored the beginnings of humankind as well. Hesiod's *Works and Days*, an account of the pastoral life, and *Theogony*, an account of the gods, are poetry rather than science. At times, however, poetry and science mix. Often it takes the poet to imagine the possibilities inherent in life and nature.

Hesiod claimed legitimacy from the inspiration coming to him as he tended sheep at the foot of Mount Helicon in Boeotia, in eastern Greece. The daughters of

Zeus, the Muses, particularly Calliope, the Muse of poetry, paid a visit to the shepherd, inspiring in him divine song of the purposes and ways of humankind. Hesiod sought the origins of human crafts, such as metallurgy and agriculture, and found an answer in the story of Prometheus, the Titan who rebelled against Zeus and gave civilizing fire to humans. Hesiod explained suffering and evil with the story of Pandora, the wife of Prometheus's brother, Epimetheus. Pandora, and women in general, unleashed troubles on man, according to Hesiod.

The Greeks were not alone among ancient peoples in wrestling with the mysteries and complexities of nature and human behavior. Nor were they unique in using what appear to be childish stories to explain things. Hesiod was among the best at providing such explanations. He had the curiosity and propensity of a scientist during a time, the eighth century BCE, of poets, centaurs, nymphs, heroes, and gods.

To Homer, an anonymous Greek poet who lived sometime during the beginning of the first millennium BCE, humans are journeyers. Homer's *Iliad* and *Odyssey*, composed orally around 1000 BCE but penned during the eighth century, present the Greeks as wanderers, explorers, and discoverers. Greek history was built upon the words *in search of.* Because of the Greek quest to know, the heritage of Western civilization during the past two millennia is an unending search for answers to inspired questions.

One of the long-standing mysteries in ancient history involves Homer: who he was, when he lived, and whether or not his poems relate the episodes of real events. The word *Homer* means "one who is led"—literally, "blind." Homer may not refer to a person but rather to the frequent occurrence in the ancient world of bards who were blind. To the ancients, however, Homer was a real person who came from several possible places. The people of Smyrna, Chios, and Colophon all claimed Homer as a favorite son. The Roman emperor Hadrian decided to settle the matter once and for all by submitting the question of Homer's origins to the Oracle of Apollo at Delphi. The oracle responded, "Do you ask me of the obscure race and country of the heavenly siren? Ithaca is his country, Telemachus his father, and Epicasta, Nestor's daughter, the mother that bare him, a man by far the wisest of mortal kind" (trans. Evelyn-White).

Notwithstanding his uncertain origins, Homer's poems have been admired for thousands of years. The anthropomorphic portrait of the *Iliad* and *Odyssey* make Homer's work a pious testament to the wonder and beauty of the universe. Homer presents a moving portrait of universal human characteristics that is at the same time very down to earth and majestically physical. The *numinous*, or transcendent divine, in Homer is the pathos of death; the beauty of the human body; the anguish of war; the wondrous rhetoric of great heroes; the awesome human strength, will, and stubborn refusal to bend or to surrender; the elevation of humans to equality among natural phenomena; the treasure of love and life amid the unceasing roar of armies and cries of the vanquished; the endless human struggle against time and destiny; the elegant dignity of humankind. Plutarch, in his *Life and Poetry of Homer*, stated this well:

> But if any one should say that Homer was a master of painting, he would make no mistake. For some of the wise men said that poetry was speaking painting, and

painting silent poetry. Who before or who more than Homer, by the imagination of his thoughts or by the harmony of his verse, showed and exalted gods, men, places, and different kinds of deeds? For he showed by abundance of language all sorts of creatures and the most notable things—lions, swine, leopards. Describing their forms and characters and comparing them to human deeds, he showed the properties of each. He dared to liken the forms of gods to those of men. Hephaestus prepared Achilles' shield; he sculptured in gold, land, sky, sea, the greatness of the Sun and the beauty of the Moon and the host of the stars crowning all. He placed on it cities in different states and fortunes, and animals moving and speaking. Who has more skill than the artificer of such an art? (Revised, Goodwin)

THE MYTHOLOGICAL PAST

Homer's *Iliad* and *Odyssey* are stories that describe a precivilized, protoscientific culture. Homer's story has no clear antecedent, no set time frame. It is an episodic, vaguely chronological account of a shadowy past time. The poet provides no clear explanation of the events of the past. Causation is directed by anonymous fate, an act of the faceless divine. The river Ocean surrounds the disk of the earth that is filled with mysterious forces, nymphs, satyrs, demons, and spirits of nature, rarely seen but present. Heaven's vault hosts the movement of the sun and stars, which emerge from the river Ocean in the east and submerge below its wine-dark waters in the west. Constellations associated with the mythological past move across the sky: the Pleiades, Hyades, the Great Bear and the Little Bear, the hunter Orion. In contrast, under the disk of earth to an uncertain distance is the unending gulf of Tartarus.

The Homeric poems describe a time of wealthy warriors and impoverished peasants, with few specialized workers. Seers, soothsayers, and healers form the rudiments of a middle class, along with bards; artisans who fashioned items from clay, wood, and stone; and metalworkers. Homer's world was a Bronze Age world; the poet took great care to describe the beauty of bronze shields, weapons, and cauldrons. Also portrayed in the poems are the sleek black ships of the Greeks, powered by a single square sail as well as the strength of the warriors who at sea became rowers. Greek mariners rarely crossed the sea out of sight of land, yet the poems imply that extended sea voyages were not improbable.

Iliad

The *Iliad* opens with an account of the god Apollo, who sent plague to the disobedient Greek host. Apollo was the god not only of plague but of healing and medicine, of music and crafts. Apollo sired Asclepius, who was taught the healing art by Chiron the centaur. Asclepius in turn passed on his knowledge to his sons Podalirios and Machaon, who in the *Iliad* healed Menelaos from an arrow wound by putting balm on it. Later, according to the Homeric Cycle of myth, the sons of Asclepius healed Heracles's son Philoctetes, considered the master of archery among the Greek host. The *Iliad* portrays human consciousness as the presence of the divine. But what is more, Homer described distinct humans, singular characters of a certain place and time. The Homeric heroes were individual actors forging their own destinies within the vague confines of fate. Heroes such as Achilles battle not only humans but gods that personify natural forces. For example, Achilles fights the river Scamander to a draw. The *Iliad* overall presents a wonderful picture of the drama and pathos of humans and nature.

Odyssey

The *Odyssey* provides a narrative account of the adventures of Odysseus, king of Ithaca, as he spends years sailing the Mediterranean in search of a way home to his wife Penelope and son Telemachus. Odysseus reaches home because of fate, which plays a huge role in both of the Homeric poems. Fate is the ultimate dictator of all things human, natural, and divine. Although the Greeks gave names to the trio of fates (Lachesis, Clotho, and Atropos), the operation of fate was an anonymous power not unlike the dominance of natural law in modern times.

Odysseus makes it home to Ithaca because of fate but also because of his intelligence—"wit," according to Homer, the ability to think oneself out of scrapes. Odysseus outthinks his opponents by weighing a given situation through observation and analysis and then giving a measured, commonsense response. He engages in on-the-spot reasoning. He uses his mind to corral his passions and emotions. Though not a scientist, nevertheless Odysseus serves as a patron for all subsequent Greek thinkers who will have the curiosity to go in search of knowledge.

The ancient Greeks of the first millennium BCE penetrated ever deeper into human mysteries. They realized that their own civilization was vastly different from those of the peoples to the north, east, and west. Greek writers and philosophers contrasted their culture, which included sophisticated literary discourse, the reflective art of mathematics, the physical and biological sciences, and technological and artistic achievements, with surrounding "barbarian" cultures. How did the Greek achievement come about? Thoughtful Greeks, in search of answers, believed that in the distant past, in a golden age of gods and heroes, humans were somehow granted rational thought that enabled them to adapt to and seek to control their environment. Hesiod believed that civilization was the result of the gift of fire. Prometheus the Titan, against the will of Zeus, the king of the gods, taught humans the uses of fire, which resulted in the civilizing arts and sciences. The myth of Prometheus had, perhaps, a Sanskrit origin: in some Indian epics, Pramanthu invented the tool called the fire drill.

Myths such as that centering upon Prometheus, recorded by Hesiod and Aeschylus (525–456 BCE), a fifth-century Athenian playwright, provide a written account of the scientific origins of human experience. That Prometheus provided humans with fire refers to a distant collective memory that at some point in the past, humans did not use fire, but then its uses—to warm, light, cook, and protect—were discovered by humans. Myths are rarely chronological, hence they will not enable a determination of *when* a discovery happened.

Aeschylus, in *Prometheus Bound*, was explicit in assigning to Prometheus the role of the paradoxical benefactor of humankind. According to Aeschylus, Prometheus was a primeval Titan with human characteristics who defied the plan of the eternal mind and power, the god Zeus, to maintain humans in a primitive, animalistic state. Prometheus saved humans from their fate of ignorance and innocence. "I found them all helpless at first," Prometheus says of humans, "and made them able to reflect and use their wits." Humans were aimless, blind creatures who "acted in every matter without intelligence, till I revealed to them the risings of the stars and settings hard to judge," astronomy and astrology. "And then I found for them the art of using numbers, that master science," mathematics, "and arrangement of letters" in poetry and prose, "and discursive memory." Prometheus taught humans how to use chariots, how to sail the seas, "soothing medicines," and "the ways of prophecy": divination and magic. Prometheus, in short, "gave all arts and sciences to men" (trans. Warner).

Prometheus, whose name means "foresight," could not give humans what they really needed, his own gift of looking into the future. Aeschylus echoed Hesiod in blaming Prometheus for dooming humans to live in confusion, driven by fate into an uncertain future. Though humans had tasted the power to control material existence, to understand the workings of the universe, their uncertainty about the future guaranteed their impotence. Prometheus unwittingly cursed humans as he cursed himself. His punishment was to endure ceaseless torment, chained to the rocks of the distant Caucasus Mountains and visited daily by a vulture that ate his liver, which regenerated during the night, hence leading to an endless cycle of torture.

The myth of Prometheus was a fascinating attempt on the part of reflective Greeks to try to understand the varied contradictions in their society. Rational thought and science led to an apparent knowledge of humans and the universe that, while it solved some problems of human existence and provided some of the comforts of civilized living, failed to release humans from war, disease, famine, and other forms of suffering. The curse of Prometheus, according to the Greeks, is the temptation to assume that science will provide the answers, will lead to a golden age. But the Greeks discovered, with the Hebrew poet of Ecclesiastes, that in much knowledge is much suffering.

MAGIC

Magic was widely used in the ancient world as a means of manipulating or bringing about forces of nature. Magic involved astrology, soothsaying, the reliance upon omens, sacrifice, spells, and incantations. Magic, of course, developed in Paleolithic and Neolithic times, when ancient humans sought to feel the spiritual forces inherent in nature and to encourage their presence and their benevolence. The shaman in primitive societies is the religious leader who practices magical arts to bring about the good harvest, the successful hunt, the safety of the tribe, and the healing of the sick. The shaman's work is protoscientific, as is magic itself. Hence in ancient civilized societies, such as Mesopotamia and Egypt, soothsayers and diviners had the important role of divining what fate had in store for a person or the community at large. There was much empathy in this science—feeling what nature might bring, hoping to see a sign, observing nature, and practicing sacrifice to discover what the predetermined future held. Magic was often associated with influencing the divine with respect to the afterlife. Egypt, the first society to develop a clear sense of a blessed afterlife, often relied on magical incantations (as prescribed in the Egyptian *Book of the Dead*) to influence the judge Osiris by altering the weight of the heart, making it lighter and not weighed down with sin. Soothsayers and diviners were especially prevalent throughout the ancient world and in ancient literature.

The Chaldeans, the wise men of ancient Babylon, were infamous practitioners of magic who grew legendary among ancient historians and commentators. They were known to be experts at foretelling the future because of their expertise in astrology. They were practitioners of medicine and were often identified as physicians in early law codes. The Chaldeans and magi are good examples of the close association of magic and science in the ancient world.

The earliest people to observe and study nature with an intent to acquire knowledge of natural (and supernatural) phenomena were soothsayers. The first scientists—soothsayers, diviners, astrologers, prognosticators—examined natural phenomena looking for clues as to the way of things. The task of Sumerian and Babylonian astrologers was to scan the sky, studying at night the cycles of the moon and the movements of planets and during the day the constantly changing position of the sun, searching for omens to indicate the future of nature and humans. Thunderstorms, earthquakes, floods, drought, pestilence, war, sudden death, the random occurrence, and the slip of the tongue all held a hidden meaning for prophets and seers. Gudea, a Sumerian king of the third millennium BCE, interpreted dreams and inspected livers of sacrificed animals to gain insights into the will of the gods. The gods spoke through the organs of sacrificial animals, particularly the livers of goats or bulls. The soothsayer was an expert at reading what the liver had to tell a people. Many ancient leaders, warriors, and kings had their minds changed and their actions altered by the shape of a liver.

Mopsus, who journeyed with the Greek Argonauts to Colchis, surpassed "all others in the art of augury from birds" (trans. Rieu). The *Iliad* begins with the advice of the diviner "skilled in the flight of birds," in reading the hidden message of fate inherent in the flight of birds. (trans. Fitzgerald) According to Plutarch, Roman diviners used a staff called a *lituus* to mark the sky conceptually for ease in watching the flight of birds. The fundamental assumption of the soothsayer and seer was that the gods spoke through natural phenomena: animal organs, bird habits, thunderstorms, earthquakes, and so on. The gods spoke through nature under different circumstances of time and place. The common universal law of nature, as it were, was that nature was a means of communication, a way for humans to discover the intent and will of the gods—in particular, *fate*. An individual's fate could not be changed, but knowledge nevertheless gave one a sense of power, a slight control over one's destiny. Apollonius of Rhodes tells the story of Idmon, who joined the voyage of the *Argo* even though "his own bird-lore had told him he would die" on the voyage. Apollo "had taught him the prophetic art, how birds should be observed, and how to find omens in burnt offerings." Later, as Argonauts met in harmony with the Mariandyni, "at this moment Fate intervened and Idmon son of Abas met his predestined end. He was a learned soothsayer, but not all his prophetic lore could save him now" (trans. Rieu).

That such credulity in believing that animal sacrifice foretells future events continued to be believed by even the most sophisticated, scientific people was revealed in the first-century BCE Vitruvius's *On Architecture*, in which he wrote,

> Our ancestors, when about to build a town or an army post, sacrificed some of the cattle that were wont to feed on the site proposed and examined their livers. If the livers of the first victims were dark-coloured or abnormal, they sacrificed others, to see whether the fault was due to disease or their food. They never began to build defensive works in a place until after they had made many such trials and satisfied themselves that good water and food had made the liver sound and firm. If they continued to find it abnormal, they argued from this that the food and water supply found in such a place would be just as unhealthy for man, and so they moved away and changed to another neighbourhood, healthfulness being their chief object. (Trans. Morgan)

ANCIENT CELTS

In northern Europe and ancient Britain, precivilized peoples were similarly polytheistic and anthropomorphic, relying on bards and soothsayers, such as the Druids, to understand and to interpret the natural environment. Epic poetry in ancient Britain was verbal, passed from one bard to another among illiterate peoples. The topics of the poetry were great heroes, the supernatural, and natural phenomena.

An example of a Celtic nature poem sung by an Irish bard, a *filid*, is "The Mystery," composed by the Druid Amergin. "The Mystery" is in a collection of poems, *Lebor Gabála, The Book of Invasions*, which describes the invasions of Ireland by warriors. Ultimately, the poet believes that nature is "The Mystery," which the poet himself, as the "I," or ego, symbolizes:

The Mystery

I am the wind which breathes upon the sea,
I am the wave of the ocean,
I am the murmur of the billows,
I am the ox of the seven combats,
I am the vulture upon the rocks,
I am the beam of the sun,
I am the fairest of plants,
I am the wild boar in valour,
I am a salmon in the water,
I am a lake in the plain,
I am a word of science,
I am the point of the lance of battle,
I am the God who created in the head the fire.
Who is it who throws light into the meeting on the mountain?
Who announces the ages of the moon?
Who teaches the place where couches the sun?
(If not I)

(Trans. Hyde)

Such verse reveals an ability to intuit the various phenomena found in the natural environment, to put them into sublime words so that others will remember and pass down such knowledge. The divine in Celtic verse is mysterious, uncertain, appearing supernaturally often by means of magic.

Cornelius Tacitus, author of the short book, *Germania,* in the early second century CE, wrote of the customs of the peoples of northern Europe, the Germans, describing their illiterate, oral culture based on myths and magic. "In the traditional songs which form their only record of the past," he wrote, "the Germans celebrate an earth-born god called Tuisto. His son Mannus is supposed to be the fountain-head of their race and himself to have begotten three sons who gave their names to three groups of tribes—the Ingaevones, nearest the sea; the Herminones, in the interior; and the Istaevones, who comprise all the rest" (trans. Mattingly). Tacitus wrote that Germans considered eloquence to be on a par with martial strength and worshipped a god, which Tacitus identified as Hercules, who signified the ability to fight and to sing. In time, in the Later Roman Empire, the Gauls of what is today France were known for their exceptional ability at eloquence. Indeed, Romans often hired Gallic tutors to teach panegyric, a popular form of verse in the third and fourth centuries CE. Symmachus, a fourth-century orator, received his training in oratory from a Gallic teacher.

Julius Caesar, who was the first Roman to invade Britain, as well as the conqueror of Gaul, wrote about the customs of these northern Europeans in his book, *Gallica.* Caesar wrote of the importance of the Druids in Celtic and Gallic culture. The Druids were leaders in religion, education, and government. The intellectuals of the Celts, the Druids were professors to many young aspirants. "It is said," Caesar wrote, "that these pupils have to memorize a great number of verses—so many, that some of them spend twenty years at their studies. The druids believe that their religion forbids them to commit their teachings to writing" because they do "not want their doctrine to become public property, and in order to prevent their pupils from relying on the written word and neglecting to train their memories" (trans. Handford). Hippolytus, the Late Roman commentator, recorded a tradition that Greek Pythagoreans influenced the Druids:

> And the Celtic Druids investigated to the very highest point the Pythagorean philosophy, after Zamolxis, by birth a Thracian, a servant of Pythagoras, became to them the originator of this discipline. Now after the death of Pythagoras, Zamolxis, repairing there, became to them the originator of this philosophy. The Celts esteem these as prophets and seers, on account of their foretelling to them certain (events), from calculations and numbers by the Pythagorean art; on the methods of which very art also we shall not keep silence, since also from these some have presumed to introduce heresies; but the Druids resort to magical rites likewise. (Trans. MacMahon)

FURTHER READING

Aeschylus, *Prometheus Bound,* trans. Rex Warner, in *Ten Greek Plays* (Boston: Houghton Mifflin, 1957).

Apollonius of Rhodes, *The Voyage of Argo,* trans. E. V. Rieu (Harmondsworth, England: Penguin Books, 1971).

Walter Burkert, *Structure and History in Greek Mythology and Ritual* (Berkeley: University of California Press, 1979).

A. R. Burn, *The World of Hesiod* (Harmondsworth, England: Penguin Books, 1936).

Caesar, *The Conquest of Gaul*, trans. S. A. Handford (Harmondsworth, England: Penguin Books, 1951).

Joseph Campbell, *The Power of Myth* (New York: Anchor Books, 1991).

The Epic of Gilgamesh, trans. N. K. Sandars (London: Penguin Books, 1972).

M. I. Finley, *The World of Odysseus* (Harmondsworth, England: Penguin Books, 1972).

Robert Graves, *The Greek Myths*, vol. 1 (Harmondsworth, England: Penguin Books, 1960).

Jaspar Griffin, "Greek Myth and Hesiod," in *Oxford History of the Classical World*, ed. John Boardman, Jasper Griffin, and Oswyn Murray (Oxford: Oxford University Press, 1986).

Sir Thomas Heath, *Aristarchus of Samos* (New York: Dover Books, 1981).

Hesiod, *Homeric Hymns, Epic Cycle, Homerica*, trans. Hugh G. Evelyn-White (Cambridge, MA: Harvard University Press, 1936).

Hesiod, *Theogony and Works and Days*, trans. Richard Lattimore (Ann Arbor: University of Michigan Press, 1959).

Hippolytus, *Refutation of the Heresies*, trans. J. H. MacMahon, in *Ante-Nicene Fathers*, vol. 5, ed. Alexander Roberts, James Donaldson, and A. Cleveland Coxe (Buffalo, NY: Christian Literature Publishing Co., 1886); revised and edited for New Advent by Kevin Knight, http://www.newadvent.org/fathers/050101.htm.

Homer, *Iliad*, trans. Robert Fitzgerald (New York: Doubleday, 1989).

Homer, *Odyssey*, trans. Robert Fitzgerald (New York: Random House, 1990).

Douglas Hyde, *The Story of Early Gaelic Literature* (London: Unwin, 1905).

Thorkild Jacobsen, *The Treasures of Darkness: A History of Mesopotamian Religion* (New Haven, CT: Yale University Press, 1976).

Karl Kerenyi, *Prometheus: Archetypal Image of Human Existence* (New York: Pantheon Books, 1963).

Plutarch, *Life and Poetry of Homer*, in *Essays and Miscellanies: The Complete Works*, corrected and revised by William W. Goodwin, vol. 3, https://www.gutenberg.org/files/3052/3052-h/3052-h.htm.

Jack Sasson et al., eds., *Civilizations of the Ancient Near East*, 4 vols. (New York: Charles Scribner's Sons, 1995).

Tacitus, *The Agricola and the Germania*, trans. H. Mattingly and S. A. Handford (Harmondsworth, England: Penguin Books, 1970).

Oliver Taplin, "Homer," in *Oxford History of the Classical World*, ed. John Boardman, Jasper Griffin, and Oswyn Murray (Oxford: Oxford University Press, 1986).

Vitruvius, *The Ten Books on Architecture*, trans. Morris H. Morgan (Cambridge, MA: Harvard University Press, 1914).

J. E. Zimmerman, *Dictionary of Classical Mythology* (New York: Harper and Row, 1971).

5

Ancient Religion and Science (4000 BCE–500 CE)

Ancient peoples were religious peoples. They were typically polytheistic, believing in many deities. Polytheism relies on an assumption that nature is infused with the divine. To recognize the divine in nature is to propitiate the gods, for by gaining their favor one will enjoy more happiness in this world and perhaps avoid disaster. To worship transcendent deities, to assume that supernatural beings cause and control nature, logically requires the study of nature. Hence ancient peoples were natural theologians as a result of their religious assumptions. The practice of religion (Latin, *religio*) meant a reverent response, a sense of awe and piety, toward the divine. Nature, directly controlled by the gods, continually revealed the divine character.

The Mesopotamians and other peoples of the ancient Near East looked at the divine and the universe with astonishment and fear; if piety was a result, so, too, was curiosity and the quest to know. Clues for the inquiring Sumerian were found in the apparent operation of natural phenomena. Fate had its own inherent inner laws that captivated and puzzled humans. There is nothing a scientist likes better than a puzzle. Fate seemed less bewildering once the Sumerian realized that nature provided hints of the course of what will be. The gods, too, knew the future, even if they could not change it. A goal of religious rites was to appeal to the gods for direction as to the course of future events.

For example, Sumerian culture owed its foundation to water and its uses. The Sumerian epic of creation, *Enuma Elish*, describes Apsu, the deified personification of the original waters of creation. Adad was the god of storms; Ea, of waters; Ennugi, of irrigation; Ningirsu, of the fertility of irrigated soil; Ninurta, of freshwater wells. The Sumerians clearly thought a lot about water. Rain and river water led to plentiful crops, watched over by Nisaba, goddess of crops. Ninhursag, the mother goddess, also oversaw all plants and animals, all foodstuffs. Once the crops were harvested, Shulpae oversaw the happy feasting; Siduri, the goddess of wine, joined in. The miracles of water, food, plenty, drink, feasts, and full bellies had to be explained: How else but because of the benevolence of the gods? In Sumerian myths, Enki was a Prometheus-like figure of wisdom and benefactor of humankind, providing knowledge of the arts and crafts.

The Egyptians were like the Mesopotamians: dependent upon the climate, particularly upon the Nile and its annual floods and the ever-present sun. Various gods had the sun as their domain of power: Amun, Re (the sun at noon), Aten (the sun's disk), Khepri (the sun in the east), Atum (the sun in the west). Other deities ruled various regions of Egypt according to the location of the Nile (upper or lower Egypt). Fertility gods and goddesses were frequently associated with the Nile: Hapy was the god of the Nile flooding during summer. Nun and Nefertem were deities of the waters of the deep that were present at the creation. Sobek was the god of the swamp. The Egyptians, devoted to understanding nature upon which they were so dependent, worshipped gods of science and learning. Imhotep, who was reputedly a builder during the age of the pyramids in the third millennium BCE, in time became a god of healing and magic. Selqet was his female counterpart: the goddess of healing. Seshat and Thoth were deities of learning, scribes, and wisdom.

The *Papyrus of Ani*, or *Book of the Dead*, shows the complete intertwining of science and religion:

> Adoration of Rāa when riseth he in horizon eastern of heaven. Behold Osiris, the scribe of the holy offerings of the gods all, Ani! Saith he, Homage to thee. . . Thou risest thou shinest, making bright thy mother, crowned as king of the gods, doeth to thee mother Nut [with] her two hands the act of worship. Receiveth thee Manu with content, embraceth thee Maat at the double season. May he give splendour and power together with triumph, [and] a coming forth as a soul living to see Horus of the double horizon, to the ka of Osiris, the scribe Ani, triumphant before Osiris. Saith he, Hail gods all of the Soul Temple, [ye] weighers of heaven [and] earth in the balance, givers of food [and] abundance of meat! [Hail] Tatunen, One, maker of mankind [and of] the substance of the gods of the south, north, west, [and] east! Ascribe praise to Rā, the lord of heaven, the Prince, Life, Strength, Health Creator of the gods. Adore ye him in his Presence beautiful in his rising in the . . . boat. Shall worship thee the beings of the heights, shall worship thee the beings of the depths. Write for thee Thoth [and] Maat day every. Thine enemy [is] given to the fire, the evil one hath fallen; his arms [are] bound, removed hath Rā his legs; the sons of impotent revolt never [again] shall they rise up! (Trans. Budge)

Thoth, according to Wallis Budge, the translator of the *Papyrus of Ani*, "was the personification of intelligence." He was self-created and self-existent and was the "heart of Ra." He invented writing, letters, the arts and sciences, and he was skilled in astronomy and mathematics. Among his many titles are "Lord of Law," "Maker of Law," and "Begetter of Law." "He justified Osiris against his enemies, and he wrote the story of the fight between Horus, the son of Osiris, and Set. As 'lord of Law' he presides over the trial of the heart of the dead, and, as being the justifier of the god Osiris against his enemies, he is represented in funereal scenes as the justifier also of the dead before Osiris." Thoth is also "the counter of the heavens and the stars, and of all that therein is." Thoth "was self-produced, and was the great god of the earth, air, sea and sky." Thoth recited, it was thought, "words and compositions . . . on behalf of the deceased" that "preserved the latter from the influence of hostile powers and made him invincible in the 'other world.'" Thoth, "as the chronologer of heaven and earth, . . . became the god of the moon; and as the reckoner of time, he obtained the name . . . 'the measurer'; in these capacities he had the power to grant life for millions of years to the deceased." The god Ptah was a creative force that fulfilled the "mandates of Thoth the divine intelligence." Ptah's name "means the 'opener,' and he was identified by the Greeks with [Hephaistos] and by the Latins with Vulcan." Khnemu "worked with Ptah in carrying out the work of creation ordered by Thoth . . . ; his name means 'to mould,' 'to model.' His connexion with the primeval water caused him to be regarded as the chief god of the inundation and lord of the cataract at Elephantine" on the Nile River. Maat was Thoth's wife, and worked with Ptah and Khnemu in acts of creation.

There was a clear connection between the vast numbers and representations and powers of Egyptian deities and those of the Greeks of Homer's and Hesiod's times. Among the Greeks, Hesiod's *Theogony* provided an account of the origins and powers of the divine. Uranos, the sky, and Gaia, the earth, bore the Titans, including Cronos (time), Prometheus (foresight), and Epimetheus (hindsight). Cronos rebelled against his father Uranos to establish his own rule, only to lose it to the wielder of the thunderbolt: Zeus. Zeus thereafter ruled gods and humans from his throne atop Mount Olympus.

Hesiod's works illustrate how the ancient mind anthropomorphized the divine and personified natural phenomena. Human behavior was understandable if the gods themselves engaged in slander, theft, rape, murder, adultery, violence, jealousy, and so on. The gods personified human emotions, such as love and anger; propensities, such as rumor and indolence; behavior, such as war and creativity. If such anthropomorphism lessened humans' responsibility for their own actions, it nevertheless did not avoid or whitewash the issues of human psychology, hence introducing them for debate, as it were, among contemporaries and later thinkers.

The gods also personified natural phenomena. The myth of Persephone's capture by Hades, the god of the underworld, explained the changing seasons. Human and natural fertility were explained by the existence of Artemis, the goddess of childbirth and the fecundity of nature; Demeter, the goddess of the harvest, of grain; and Selene, the goddess of the monthly cycle of the moon. Parents explained thunderstorms to their children by noting Zeus's frequent anger. Plague occurred when sent by the god of rats, Apollo. Lies and thievery were explained by the presence of Hermes, the sleight of hand. Hesiod, believing in the age-old decline in culture from "the good old days," argued that his time was an Iron Age of evil and despair, unlike the previous ages of humankind: the Golden Age of humans that were like gods; the Silver Age of hubris, when Zeus punished human arrogance; the Bronze Age, when humans were giants; and the Heroic Age of heroes descended from the gods but destroyed by the Trojan War.

The *Homeric Hymns* and the *Iliad* and *Odyssey* describe a worldview that is entirely anthropomorphic. Gods and goddesses mingle constantly with humans, so much so that humans never are entirely sure who is a god and who is a mortal. Perhaps the stranger at the door is a god sent to test humankind, as the characters of the *Iliad* and *Odyssey* constantly wonder and worry. The gods symbolize the conscience and the subconscious mind, so when Achilles thought to destroy Agamemnon in book 1 of the *Iliad*, Athena stilled his anger and brought reason and patience to him. The *Odyssey* describes the dichotomy of civilization: thought, personified by Odysseus; savagery, by the Cyclops. Odysseus repeatedly uses his wits, ruses, deceptions, and analytical thought to get himself out of scrapes. Odysseus has wisdom, that is, he exercises reason over passions, he has forethought and hindsight, and he has patience when others are impulsive. Divinities of wisdom, thought, and wit logically aid Odysseus. Athena watches over him, as does Hermes (Odysseus's great-grandfather). Indeed, we find in Greek mythology an emphasis on deities that personify human thought in one way or another. Besides Athena, Apollo is a god of wisdom, of the lyre, of dreams, of prophecy, and of seers. Prometheus teaches humans creativity, analytical thinking, inventiveness, and healing. Asclepius, the son of Apollo, learns from Chiron the centaur; Asclepius becomes the principal god of healing. Some myths have Asclepius joining Jason and the crew of the *Argo* on their voyage to Colchis to steal the golden fleece. Perhaps this is because, besides Asclepius, Chiron was supposed to have taught Jason and Heracles as well. Apollo and Asclepius as healers were often called Paean. Myth also hypothesized another healer by the same name (often called Paeon) who cared for injured gods, such as Ares and Aphrodite, who stupidly fought against the Achaeans at Troy.

Fate in Homer's poems is anonymous, unknowable, and it requires tremendous will and strength of character for one to forge one's own life in time. Homer put a premium on the characteristics of the individual hero. In so doing, he created a personality standard of the heroic individual who strives against fate by means of the peculiar human trait of *arete*, bravery. The hidden presence of gods poetically describes the unconscious mind of individuals who typically heed the advice of this inner voice and act accordingly.

DAWN OF PHILOSOPHY

Greek religion of the second millennium at Crete and the Greek Peloponnesus was a primitive form of anthropomorphism: gods and nature both were organic and sentimental; life, nature, and the supernatural were one. With the Archaic Age, and the dawn of philosophy, some aspects of Greek religion became more sophisticated, as philosophers such as Xenophanes rejected the gods, who acted like so many spoiled humans. Increasingly such philosophers of *being* as Thales, Anaximander, Parmenides, and Plato conceived of the divine in metaphysical terms. Theology (literally, the "study of God") for these Greek scientists and philosophers was an intellectual study to understand nature insofar that nature is an extension of self, something from the collective past of humans and nature experienced in the present moment.

At the same time, Greeks increasingly looked within to find God (*theos*). One might argue that the first great psychology occurred as the result of Greeks being initiated into the mysteries of the many fertility cults of the ancient Mediterranean. The mystical rites of the Gnostics, for example, involved initiates finding the presence of god, hidden in nature, within themselves. The divine and supernatural lived within a person; the mystery cults helped the individual to discover the divine presence. In so discovering god within, the initiate found salvation (*salus*), a release from temporal cares, a guarantee of release upon death to a spiritual and metaphysical realm of being. Some of the more famous mysteries were the rites of Isis, the Egyptian goddess; Cybele, an Anatolian mother goddess; the savior gods Adonis and Serapis; Demeter, one of the ancient Olympians; and Mithras, an Asian import. By the time of the third and fourth centuries CE, during the declining centuries of the Roman Empire, Neoplatonic philosophers were worshipping natural phenomena, as did their ancestors of centuries past, but with a new twist. Julian the philosopher/emperor worshipped Helios the sun because Helios was the most visible manifestation for the unseen, unknowable, anonymous, and distant One. Studying natural phenomena was the means to find the truth.

Early Near East monotheists expressed a more sophisticated view of a similar idea. The Hebrews recognized that Yahweh expresses his will through natural law, yet he is not bound to the set order of the universe. The Egyptian pharaoh Akhenaten (Amenhotep IV, r. 1379–1362 BCE), also conceived of a single god, Aten, the disk of the sun. In his "Hymn to the Aten," Akhenaten sang, "Splendid you rise in heaven's lightland, O living Aten, creator of life! When you have dawned in eastern lightland, You fill every land with your beauty. You are beauteous, great, radiant, high over every land; Your rays embrace the lands, to the limit of all that you made" (trans. Lichtheim).

Akhenaten's "Hymn to the Aten" sounds remarkably similar to the writings in the Hebrew Old Testament. Whereas ancient polytheistic religions typically portray the divine as humanlike and hence superficial, as powerful but not all-powerful, as being bound by fate just as humans are, Akhenaten's Aten and the Hebrew Yahweh are blinding, omniscient, omnipotent, all-encompassing. When Moses asks God, in the book of Exodus, who He is, God replies, "I am that I am." This is the most powerful proclamation of the divine in human history, yet it is a proclamation made to an individual man who is terrified out of his wits at a particular time and place. "I am" encompasses all things, all time, all life, all thought. And yet "I am" can become instantly knowable to anyone, not just to Moses. The Hebrew Yahweh had similar attributes to Aten, yet Yahweh was at the same time more visibly grander and powerful yet more personal. The prophet Elijah discovered that Yahweh was less apt to be made known in thunder, lightning, and earthquakes than in the quiet moment, the gentle breeze. King David, the writer of Psalms, knew a personal God who cares for each individual, whether king or commoner, and who is a gentle shepherd to each of his children. David's interaction with Yahweh in the Psalms reveals anguish, fear, loneliness, hope, courage, and love; the Psalms portray the breadth of human emotions and the search of the individual to find peace and redemption from pain, suffering, sin, and death. The Psalms provide a complete psychological portrait of humankind.

OLD TESTAMENT

During the second millennium BCE, Mesopotamians of Semitic origin migrated west to a land that they called Canaan. The book of Genesis, the initial chapter of the Old Testament, describes the transformation of these Semites into Hebrews by focusing upon the story of a man, Abram, and his wife, Sarai, who, embracing the God Yahweh, became the ancestors of the people of Israel. The Hebrews were one of the most creative and dynamic of the early peoples of the Near East. Their civilization, centered at Jerusalem, was oriented around an all-powerful, all knowing God—the only God, the creator, father, protector, and judge of his children, the chosen people of Israel.

Although the Hebrews, under the leadership of Moses, continued to feel the terror of the unknown, their fear was mitigated by their realization that Yahweh cared for them, that he was Fate himself, that he controlled all things, and that he could circumvent the laws of nature should he so desire. The Hebrews, as a result, developed a sophisticated sense of natural theology, that is, a pious search to understand nature qua God's creation. The apocryphal book Wisdom of Solomon, for example, counters the supposed wisdom of the credulous pagan with the understanding of nature of the pious Hebrew:

> For all men who were ignorant of God were foolish by nature; and they were unable from the good things that are seen to know him who exists, nor did they recognize the craftsman while paying heed to his works; but they supposed that either fire or wind or swift air, or the circle of the stars, or turbulent water, or the luminaries of heaven were the gods that rule the world. If through delight in the beauty of these things men assumed them to be gods, let them know how much better than these is their Lord, for the author of beauty created them. (Wisdom of Solomon 13:1–3 [Revised Standard Version])

The author of Wisdom, an Alexandrian Jew of the first century BCE, was clearly knowledgeable about Greek science—for example, Empedocles's theory of the four elements.

Natural theology accommodated science, the search for knowledge, but only within the confines of faith. The book of Wisdom honors God for giving humans the ability to use science:

> For both we and our words are in his hand, as are all understanding and skill in crafts. For it is he who gave me unerring knowledge of what exists, to know the structure of the world and the activity of the elements; the beginning and end and middle of times, the alternations of the solstices and the changes of the seasons, the cycles of the year and the constellations of the stars, the natures of animals and the tempers of wild beasts, the powers of spirits and the reasonings of men, the varieties of plants and the virtues of roots; I learned both what is secret and what is manifest, for wisdom, the fashioner of all things, taught me. (Wisdom of Solomon 7:16–22 [RSV])

Diviners and Physicians

The focus of the Old Testament on God's chosen people can give the impression that the Hebrews were the only people. The books of Genesis and Exodus do, however, describe the interaction of the Hebrews with other peoples of the ancient Near East, such as the Egyptians and Mesopotamians. The culture of the Near East during the second and first millennia BCE was superstitious, priests and prophets being diviners of fate and of the will of the gods as well as serving as healers and interpreters of dreams. Upon explaining the pharaoh's dreams, the prisoner and outcast Joseph grew to be the most powerful man in Egypt. Daniel was able to interpret the dreams of King Nebuchadnezzar when the host of "magicians," "enchanters," "sorcerers," and "Chaldeans" could not. In the apocryphal book of Tobit, the angel of God tells Tobias, a young man seeking to wed Sarah, that he can expel demons by using the heart and liver of a fish, and make the blind see by anointing eyes with the gall of the fish. The Hebrews also believed in the magical, spiritual significance of numbers, such as 1, unity; 3, heaven; 4, the corners of the earth and directions of the winds; 7, the product of 3 and 4 and the number of planets, sun, and moon.

In the Old Testament, illness is often the result of sin, requiring spiritual cleansing as well as physical healing. The book of Leviticus, originally written in the early first millennium BCE, provides a detailed description of skin diseases (lumped under the vague diagnosis of leprosy). The priest inspected the rash or boils as would a physician and then pronounced his opinion not on how to cure the illness but rather on whether the patient was unclean and therefore unable to enter the sanctuary of the temple in Jerusalem. Centuries later, in the second century BCE in the apocryphal book Ecclesiasticus (Sirach), a Hellenized Jew who was clearly imbued with a healthy respect for physicians—characteristic of the Hellenistic Age—wrote,

> Honor the physician with the honor due him, according to your need of him, for the Lord created him; for healing comes from the Most High, and he will receive a gift from the king. The skill of the physician lifts up his head, and in the presence of great men he is admired. The Lord created medicines from the earth, and a sensible man will not despise them. Was not water made sweet with a tree in order that his power might be known? And he gave skill to men that he might be glorified in his marvelous works. By them he heals and takes away pain; the pharmacist makes of them a compound. His works will never be finished; and from him health is upon the face of the earth. (Sirach 38:1–8 [RSV])

The writer called upon physicians to "pray to the Lord that he should grant them success in diagnosis and in healing, for the sake of preserving life" (Sirach 38:14 [RSV]).

The Hebrews were generally suspicious of healers who advocated cures divorced from the healing power of God (Yahweh). The Old Testament mentions a materia medica that included balsam for wounds and mandrake for infertility. However, healing was entirely up to God and his prophets. For example, in 2 Kings, in the Old Testament, the prophet Elisha heals the Aramaean Naaman through the agency of God in order to show the power of God. The temple in Jerusalem was a place where the sick could go for divine healing, though at the time of Jesus of Nazareth, the sick, who were considered unclean by the Pharisees, were forbidden to enter the temple. The Old Testament and the New Testament identify mental illnesses as caused by demons. The most prevalent physical illness was leprosy, which was loosely used by the Hebrew and Christian writers alike to mean a variety of illnesses, chiefly of the skin. The New Testament, like the Old Testament, rarely discusses physicians and medical practice, the few exceptions being to treat them with suspicion.

Psychology

The Old Testament is a unique book that describes the identification of self with God. Yahweh always has a direct interest in his people in general and each person in particular. This powerful sense of the divine nature of the human psyche grew more sophisticated with the passing centuries. Through intimate interaction with God, humans developed a more clear sense of individual identity and all that life as a unique human entails: suffering, sin, and redemption; the struggle between God's will and individual free will; love and hate; trust and mistrust. The Old Testament, in short, provides a series of detailed individual and group psychological portraits that encompass the whole of human experience. Psalm 139 reads, for example,

> O Lord, thou hast searched me and known me! Thou knowest when I sit down and when I rise up; thou discernest my thoughts from afar. Thou searchest out my path and my lying down, and art acquainted with all my ways. Even before a word is on my tongue, lo, O Lord, thou knowest it altogether. Thou dost beset me behind and before, and layest thy hand upon me. Such knowledge is too wonderful for me; it is high, I cannot attain it. (Psalms 139:1–6 [RSV])

Believing that Yahweh knew the Hebrews individually and collectively, and that they in turn knew Yahweh and his works, the Hebrews were confident in their religious beliefs but also their scientific observations. The opening lines of Genesis betray this certainty that humans could, when inspired by God, know the nature of things, the origin of the universe and human existence. Every ancient civilization had stories of the origins and/or process of the universe. The Hebrew description was poetic, similar to those found in other cultures of the Near East and Mediterranean, and extremely sophisticated in its brevity yet clear in describing what should be indescribable: "In the beginning God created the heavens and the earth. The earth was without form and void, and darkness was upon the face of the deep; and the Spirit of God was moving over the face of the waters" (Genesis 1:1–2 [RSV]).

Philo of Alexandria (20 BCE–40 CE)

Philo, a Jew living in the Diaspora in Egypt, was the Hellenistic philosopher who joined Greek and Hebrew philosophy into one system of thought. Living in Alexandria, Philo was therefore influenced by Hellenistic Greek thought, but at the same time he believed in the teachings of the Jewish Torah. He was a product of the Greek educational system, the trivium and quadrivium. Heavily influenced by the pre-Socratics as well as philosophers such as Plato, but also by the writings of the Torah, especially the Pentateuch (the first five books of the Old Testament), Philo believed that the transcendent creative force of the universe, known to the Greeks as the mind, the infinite, and the Logos, was the same as the god of Abraham, Moses, and David. Philo argued that the writings and teachings of Moses were the inspiration for Greek philosophy, ranging from Homer's anthropomorphism to Plato's ideal forms. Moses was himself influenced by Egyptian philosophers, Philo wrote in *A Treatise on the Account of the Creation of the World*, from whom he learned that there had to be "an active cause, and a passive subject; and that the active cause is the intellect of the universe, thoroughly unadulterated and thoroughly unmixed, superior to virtue and superior to science, superior even to abstract good or abstract beauty; while the passive subject is something inanimate and incapable of motion by an intrinsic power of its own, but having been set in motion, and fashioned, and endowed with life by the intellect, became transformed into that most perfect work, this world." Philo argued that stories in the Torah (and by implication, Egyptian and Greek myth) were allegories for God and his creative activity. God was the uncreated essence, the creator of all things; Philo wrote that God created not over time, but "at once, not merely by uttering a command, but by even thinking of it" (trans. Yonge).

Philo believed philosophers and scientists played an important role in society and culture. Philosophers are healers of souls, he wrote in *On a Contemplative Life*, "which are under the mastery of terrible and almost incurable diseases, which pleasures and appetites, fears and griefs, and covetousness, and follies, and injustice, and all the rest of the innumerable multitude of other passions and vices, have inflicted upon them." Philosophers such as the Greeks saw that the four basic elements as well as the principles of mathematics derives from the mind of God, the Logos: God "is superior to the good, and more simple than the one, and more ancient than the unit," he wrote. "The elements are inanimate matter, and immovable by any power of their own, being subjected to the operator on them to receive from him every kind of shape or distinctive quality which he chooses to give them." The philosopher, by studying God's creation—therefore the mind of God—is someone who understands the basic laws and ethics of the universe. Moses, in his account of creation, Philo wrote in *A Treatise on the Account of the Creation of the World*, believed that "the law corresponds to the world and the world to the law, and that a man who is obedient to the law, being, by so doing, a citizen of the world, arranges his actions with reference to the intention of nature, in harmony with which the whole universal world is regulated" (trans. Yonge).

Philo borrowed ideas from the Greeks but himself influenced the Christian writers of the New Testament and theologians who developed the concept of the Trinity: thinkers and writers such as Paul of Tarsus, author of Epistles, and the Gospel of John, both in the New Testament.

NEW TESTAMENT

The New Testament, the Greek writings that make up the second part of the Christian Bible, does not seem upon initial consideration to have anything to do with science. The New Testament is filled with miracles, prophecy, theology, revelation, and the acts of God. Yet the Gospels, history, epistles, and prophecy of the New Testament mirror the times in which it was written, during the first-century-CE Roman Empire. The writers of the New Testament were Jews who had adopted the teachings of Jesus of Nazareth and become Christians. They came from the eastern Mediterranean, particularly the region of Palestine, and the Roman protectorates of Galilee and Judaea. Palestine at that time was a culturally diverse region, with many languages, customs, and religious beliefs. The land had been Hellenized, that is, imbued with Greek culture, in the wake of the fourth-century-BCE conquests of Alexander the Great. The authors of the New Testament knew Hebrew and Aramaic, but they wrote in Greek, hence some of their ideas inevitably involved Greek ideas, including the ideas of science and philosophy.

Dreams

The Gospel of Matthew sets the tone for how ancient science is portrayed throughout the New Testament. Joseph, the father of Jesus and husband of Mary, learns from a dream that he is to marry his betrothed even though she is already pregnant. The Greeks typically thought that dreams reflected divine messages—the gods sometimes formed what appears in retrospect to have been a rudimentary conception of consciousness and the conscience. The psychoanalyst Carl Jung, a student of classical mythology and of early Christianity, argued that dreams reflected the psychic underpinnings of reality. Hence when Joseph dreamed that he must take his young child and wife to Egypt, it was a real answer to the daunting problem of the Judaean King Herod's vengeance. Likewise, the three wise men, the Magi from the East, were told in a dream to depart for home and avoid a return trip to Herod's court. Again, the dream provided a real solution to a perplexing issue.

The Magi were influenced not only by dreams but by the stars as well. The Gospel of Matthew describes how the Magi were led to Bethlehem by a majestic star that appeared to their west. Ancient magi were Mesopotamian (Chaldean), or perhaps Arabian, astrologers and magicians who believed that the position of stars and planets indicated the course of things. It required tremendous faith in the stars to make a long and hazardous journey merely to follow a bright heavenly body. But then, the ancient world usually considered stars and planets, the sun and the moon, to have a divine, spiritual presence. Hence the names of the planets are the names of Roman gods.

Jesus the Logos

The Greco-Roman world was one of superstition and belief in the supernatural and magical. Jesus of Nazareth was not the first holy man to foretell the future, heal the sick, and command natural phenomena to suit his will. Ancient shamans, soothsayers, prognosticators, and the like claimed to know how to unlock, manipulate, and read the implicit, spiritual powers contained in natural phenomena. That Jesus could calm the winds on the Sea of Galilee by a verbal command or walk on water at will was astonishing to the typical human who did not know the ways of magic. Jesus prophesied the future, such as the destruction of Jerusalem in 70 CE, as so many other seers of his time claimed to do. The Christian faithful believe that Jesus, as the Logos, used such tricks to reveal his complete power over nature. The Logos was a common idea in Greek and Roman philosophical and scientific circles. The Logos was the "word," the transcendent idea of creation that linked the human mind to immortal, eternal truth. The Gospel of John specifically identifies Jesus as the Logos, hence identifying him with the same creative force understood by Philo, Plato, the Stoics, the Pythagoreans, and the Gnostics.

Medicine

Jesus was also a healer, and the New Testament gives wonderful insights into the practical medical techniques of the ancient Mediterranean. Ointments of varying types were used to treat illnesses of the skin, to relieve pain, or to soothe one's body and mind. Myrrh was not only an ointment used to prepare corpses for burial but was also a salve used for skin ailments or, when mixed with wine, a general analgesic. The writer of the third Gospel, traditionally identified as Luke, a physician from Antioch, tells the story of the woman who anointed Jesus with an expensive ointment held in an alabaster container. Jesus reprimands the Pharisee who questions her act of love, pointing out that he had not welcomed Jesus to his home by anointing his hair with such ointment.

Besides the use of salves, oil made from olives or other seeds/fruits, and wine, the sick had few remedies to rely upon, save prayer or the intervention of a holy man or other such healer. Christianity focused on God the healer, hence the ill must approach God through faith, fasting, and prayer. A physician trained in the ways of the Greeks—for example, a follower of the Hippocratic school—would be looked upon with suspicion by the common folk who were healed by Jesus. For example, the Gospel of Mark gives the account of a woman who was healed of a hemorrhage by merely touching the garment of Jesus. According to Mark, she had been ill for years and had spent her money on the advice of physicians to no avail. Others, with diseases for which there was no apparent cure, were similarly attracted to Jesus by the hope of instantaneous healing. Leprosy, a degenerative disease for which there was no cure, was a common complaint of the time. Leprosy was contagious, hence lepers were pushed to the outskirts of society. As in the Old Testament, the writers of the New Testament used the word *lepros* indiscriminately for a variety of different skin diseases. In an age when degenerative illness was a sign of the consequences of sin, lepers were particularly shunned as unclean, both physically and spiritually. But Jesus touched and healed them nevertheless.

Jesus the Healer

Jesus of Nazareth, the son of a carpenter, was not the typical faith healer. The New Testament Gospels introduce a subtle and sophisticated approach to healing that bridges the gap between practical medical techniques, on the one hand, and healing magic, on the other. The latter relied upon the manipulation of nature, in this case the diseased body. At first glance, Jesus appears to be this kind of healer. On the other hand, the practical Hippocratic approach to healing involved a balanced approach to living, adopting the dictum of moderation in all things. Rest, exercise, and moderation in diet and drink had long-term healing qualities—they make up what is called preventive medicine today. Hippocrates taught that the healthy person lives in harmony with nature. The Hippocratic Oath calls for the physician to respect the rights of the patient, to feel compassion and to sympathize with the patient's condition. In the first chapter of the Gospel of Mark, a leper approaches Jesus and begs him for healing. Jesus, "feeling sorry for him," heals the man. This sounds in part like the Hippocratic relationship of physician and patient, yet Hippocrates and his students never expected a miraculous disappearance of illness, especially such a terribly wasting disease as leprosy (Mark 1: 41, RSV).

To the Christian faithful of the past and present, Jesus was a completely unique healer for his time in his method of healing through empathy. He did not use magical arts to heal; rather, he had the ability to feel completely what the other person felt, whether anguish, depression, fatigue, or pain. His was a full expression of love and empathy revealed in the touch of his hand and the soft words of counsel and command. His healing required the patient to have faith. If the patient could not believe in Jesus, hence empathize with him, the healing would not occur. This kind of healing through faith, love, and empathy was new to the ancient world. Indeed, it is new to our world. Yet in recent years, more students of the body and mind are understanding the incredible power of emotion, belief, and love in the eradication of disease.

RELIGION AND SCIENCE IN THE ANCIENT WORLD

Science is hardly religious, and religion is rarely considered science. Yet the aims of the scientist and the religious thinker are quite similar: to seek what is true, to discover what is the source of truth, to find what is timeless and absolute, and to discover the means of contentment. Naturalists, physicists, and psychologists have long had the same goals as the religious believer, whether that believer be an animist, polytheist, or monotheist. Christians, Jews, Zoroastrians, Hindus, Buddhists, and Taoists seek by intellectual means the nature of existence, the nature of reality, the moral absolute, the best way to live, the secret to life. Often religious thinkers, such as followers of the teachings of Siddhartha Gautama, the founder of Buddhism in ancient India, engage in a difficult process of psychic self-analysis to arrive at a sense of what must be true. Often the contemplator of nature and existence, like the Taoist, a follower of the teachings of Lao Tze in the *Tao Te Ching* of ancient China, tries to understand a hidden reality that is scarcely revealed to humans blinded by time. All thinkers realize the quandary in existence of the conflict of the apparent Good and the apparent Evil, which we can witness in animal existence through the ages or conceive of as an eternal, even supernatural, conflict, as do the adherents of the teachings of Zoroaster and the holy book of Zoroastrianism, the *Avesta*. The compulsion of self-desire contrasted with the demands of society, which political scientists, sociologists, and historians have long studied, is succinctly and poetically expressed in one of the great works of ancient Indian Hinduism, the *Bhagavad Gita*. Religion may not be science, per se, but the same qualities that go into scientific contemplation are present in religious thinking.

FURTHER READING

Apollonius of Rhodes, *The Voyage of Argo*, trans. E. V. Rieu (Harmondsworth, England: Penguin Books, 1971).

The Bhagavad Gita, or The Message of the Master, trans. Yogi Ramacharaka (Chicago: Yogi Publication Society, 1907).

E. A. Wallis Budge, trans., *The Book of the Dead: The Papyrus of Ani in the British Museum*(London: British Museum, 1895).

E. R. Dodds, *Pagan and Christian in an Age of Anxiety* (New York: W. W. Norton, 1965).

The Gospel of Buddha: Compiled from Ancient Records, trans. Paul Carus (Chicago: Open Court Publishing Co., 1915).

Michael Grant, *Jesus: An Historian's Review of the Gospels* (New York: Charles Scribner's Sons, 1977).

Marian Hillar, "Philo of Alexandria, 20 BCE–40 CE," *Internet Encyclopedia of Philosophy*, https://www.iep.utm.edu/philo.

Thorkild Jacobsen, *The Treasures of Darkness: A History of Mesopotamian Religion* (New Haven, CT: Yale University Press, 1976).

Morton Kelsey, *Healing and Christianity* (New York: Harper and Row, 1973).

M. Lichtheim, *Ancient Egyptian Literature*, vol 2 (Berkeley: University of California Press, 1976), http://web.archive.org/web/19990221040703/http://puffin.creighton.edu/theo/simkins/tx/Aten.html

The New Oxford Annotated Bible with the Apocrypha (Oxford: Oxford University Press, 1977).

Harry M. Orlinsky, *Ancient Israel* (Ithaca, NY: Cornell University Press, 1960).

Oxford Companion to the Bible (Oxford: Oxford University Press, 1993).

James B. Pritchard, ed., *Ancient Near Eastern Texts Relating to the Old Testament* (Princeton, NJ: Princeton University Press, 1969).

The Sacred Books of China: The Texts of Taoism, trans. James Legge (Oxford: Clarendon Press, 1891).

Rudolph Steiner, *Christianity as a Mystical Fact* (New York: Anthroposophic Press, 1947).

The Works of Philo Judaeus, trans. C. D. Yonge, vols. 1, 4 (London: George Bell and Sons, and Henry G. Bohn, 1854, 1855).

PART 3

Science in Ancient Central, Southern, and Eastern Asia

6

Science in Ancient Central and Southern Asia (3000 BCE–500 CE)

INDO-EUROPEAN INVASIONS (3000–1000 BCE)

Europe and Asia to the southern regions of the Mediterranean, Middle East, and lands southeast and southwest of the Caspian Sea experienced invasions of nomadic warriors from northern Europe and Asia during the third and second millennia BCE. These nomads, who have been variously identified as the Dorians who invaded Greece, the Sea Peoples who tried to invade Egypt, the Philistines who invaded Palestine, and the peoples who conquered the Hittites of Turkey, were also peoples who migrated into southern Asia, through what is today Afghanistan, Iran, Pakistan, and India.

It is possible though far from certain that among these invaders was a perceived holy man, Zoroaster, who taught a dualistic religion that conceived of a good contrasted with an evil force. The good, Ahuramazda, was an omniscient and omnipotent god of light helping humans struggle against the darkness. Each human had to choose between good and evil. The good were blessed in the afterlife, and the evil were damned.

When the Persian Empire began under Cyrus the Great in 559 BCE, the Persians were Zoroastrians who worshipped Ahuramazda. Cyrus conquered the Median Empire (central Asia), the Babylonian Empire (Mesopotamia), and the Lydian (Phrygian) Empire of Anatolia. Under Darius I, Persia was an empire that stretched from the Indus River to Egypt and Asia Minor. The Persian Empire existed for about two hundred years—from 559 to 333 BCE. The Persian gift was for organization and logistics, developing the largest and most efficient empire of its time. Accounts of the origins of the people of Iran, the Persians, and their history and intellectual and scientific development derive from ancient Greek and Roman sources, Arabic sources, and Persian sources. These sources identify the religious intellectuals of the Persian Empire as the magi.

THE MAGI OF PERSIA

Magic is a word derived by the Greeks from the class of holy and wise men of Persia and the East, the magi, and their founder, Zoroaster. Diogenes Laertius, the author of *Lives of the Philosophers,* claimed that Zoroaster was the first magician. The Roman scientist Pliny likewise blamed magic on the magi, specifically Zoroaster. Pliny condemned magic (and the magi) not because it was false but because it was often erroneous, leading people to believe in nonsensical potions and spells. Pliny indicated that a Persian who accompanied Xerxes on his expedition against the Greeks in 480 BCE, Othanes, wrote a book on magic that had a profound impact on the Greeks.

Among the magi were the wise men, identified in the New Testament book of Matthew, who followed an eastern star from Anatolia to Judaea, where they found the Christ child lying in a manger. Mediterranean writers, such as Hermippus, Eudemius, Xanthus, and Theopompus, who wrote extensively on these magi typically identified them as followers of Zoroaster, the Persian mystic who explained the world according to an eternal battle between good (*Ahura Mazda*) and evil (*Ahriman*). For Zoroastrians, as well as for the magi, astrology was extremely important, for by identifying the stars and knowing their movements, even predicting their movements, one might know, predict, the future. According to Diogenes Laertius, the magi believed that they had the ability to pull shapes from the air, hence to see what others could not. The historian Dinon claimed that Zoroaster means literally "star worshipper," which fits the worldview of the astrologer and the magician. Matthew 2:10–11 describes how the magi followed the star until it stopped over a house in Bethlehem, wherein was the newborn Jesus; the magi brought gifts and worshipped the child whom they associated with the star. Some Asian goddesses, such as Astarte and Ishtar, were identified with astral phenomena—one wonders whether the magi stopped to worship the Virgin Mary as well. Other Mediterranean writers associated magi with Chaldeans, who were astrologers and soothsayers of Babylon, and gymnosophists, who were the wise men of India. Clearchus of Soli, according to Diogenes Laertius, claimed that gymnosophists were latter-day magi. Aristotle wrote about the Asian magi, claiming that they were even more ancient than Egyptian astrologers and magicians.

Pliny the Elder, who could be very superstitious, nevertheless condemned the magi for ridiculous magical tricks and fantasies. He believed that the magi, and their leader Zoroaster, originated the use of magic in Asia, from which it spread to Greece. The magi employed their dark arts in reputed healings, poisonings, and fortune-telling. Pliny recalled a magus he had once known, who claimed that a certain herb (*cynocephalia*) was useful to prevent black magic from being employed against oneself, but if one pulled it from the ground, roots and all, it was lethal. Pliny, nevertheless, did not doubt that the magi knew the use of magic. To block their spells, he advised the use of asbestos.

The often credulous medieval European encyclopedist Isidore of Seville wrote,

> The first of the magi was Zoroaster, king of the Bactrians, whom Ninus, king of the Assyrians, slew in battle, and of whom Aristotle writes that on the evidence of his works it is clear that he composed 2,000,000 verses. . . . The magi are they who are

> usually called *malefici* because of the greatness of their guilt. They throw the elements into commotion, disorder men's minds, and without any draught of poison they kill by the mere virulence of a charm. . . . They summon demons, and dare to work such juggleries that each one slays his enemies by evil arts. They use blood also, and victims, and often touch dead bodies. (Trans. Brehaut)

The magi were perhaps the religious leaders under the Achaemenian Dynasty of Persian kings initiated by Cyrus in the sixth century BCE. Zoroaster himself might date from this time. Some scholars believe that Zoroaster was a Mede, and that the magi as a whole were derived from a Median caste of religious leaders. According to Ibn Abi Usaibi'ah, in discussing the possible origins of medicine, whether divine or developed by humans, claimed that "as for the Magi, they report that Zoroaster, who they claim was their prophet, brought books on four sciences in twelve thousand volumes of buffalo hide, one thousand of them relating to medicine" (trans. Kopf).

Other Arab commentators, such as Al-Biruni, in *The Chronology of Ancient Nations*, provided the Persian account of their own origins:

> the Persians and Magians think that the duration of the world is 12,000 years, corresponding to the number of the signs of the zodiac and of the months; and that Zoroaster, the founder of their law, thought that of those there had passed, till the time of his appearance, 3,000 years, intercalated with the day-quarters; for he himself had made their computation, and had taken into account that defect, which had accrued to them on account of the day-quarters, till the time when they were intercalated and were made to agree with real time. From his appearance till the beginning of the AEra Alexandri [conquests of Alexander], they count 258 years; therefore they count from the beginning of the world till Alexander 3,258 years. However, if we compute the years from the creation of Gayomarth, whom they hold to be the first man, and sum up the years of the reign of each of his successors—for the rule (of Iran) remained with his descendants without interruption—this number is, for the time till Alexander, the sum total of 3,354 years. So the specification of the single items of the addition does not agree with the sum total. . . . A section of the Persians is of opinion that those past 3,000 years which we have mentioned are to be counted from the creation of Gayomarth; because, before that, already 6,000 years had elapsed—a time during which the celestial globe stood motionless, the natures (of created beings) did not interchange, the elements did not mix—during which there was no growth, and no decay, and the earth was not cultivated. Thereupon, when the celestial globe was set a-going, the first man came into existence on the equator, so that part of him in longitudinal direction was on the north, and part south of the line. The animals were produced, and mankind commenced to reproduce their own species and to multiply; the atoms of the elements mixed, so as to give rise to growth and decay; the earth was cultivated, and the world was arranged in conformity with fixed norms. (Trans. Sachau)

From this account, one gathers that the Persians were concerned, if confused, about their origins, and the beginnings of creation itself.

The holy scripture of the magi and their adherents, the Zoroastrians, was the Avesta, a vast compilation of mostly liturgical texts comprising centuries. The Avesta, like most religious scriptures, posits a supreme being, Ahuramazda, the creator and sustainer of all things. The Avesta is a dialogue between the worshipper and

Ahuramazda, the worshipper asking questions about laws, rules, and so on and receiving answers from Ahuramazda. There are sections governing the work of physicians. The first physician, according to the Avesta, was Thrita, whom Ahuramazda sent to earth with knowledge of thousands of healing herbs. Other healers, besides herbalists, were those who used the knife (surgeons) and those who used the holy word (priests).

THE ACHAEMENID EMPIRE

Sources are scattered and vague respecting scientific activity during the initial Persian Empire under the Achaemenid Dynasty. Under Cyrus and his successors, the Persians expanded their power from the Indus River west to Egypt and Anatolia. At the end of the sixth century BCE, Ionian Greek city-states in Anatolia (Turkey) resisted Persian control; Greek city-states in mainland Greece assisted the Ionians. Darius I (r. 522–486) put down the revolt in Anatolia and decided to invade Greece (490); the invasion failed, but his successor, Xerxes (r. 486–465), tried again in 480; he also failed. The Ionian Greek city-states were free for about a century, but after the Peloponnesian War (431–404), Persia again reestablished control, which led Alexander the Great, in 336, to decide to free the Greek city-states. This led to the conquest of the entire Persian empire by Alexander.

It was during the time of Persian rule, around 500 BCE, that an explorer, Scylax of Caryanda, wrote the *Periplus*. Scylax was an Anatolian Greek who, according to Herodotus, sailed under the auspices of Darius I:

> A great part of Asia was discovered by Darius, who, wishing to ascertain the place where the river Indus (the only river, after a single exception, which produces crocodiles) falls into the sea, dispatched Scylax of Caryanda. Proceeding from the city of Caspatyrus, and the Pactyian territory, they sailed down the river in an easterly direction to the sea, and then continuing their voyage on the sea towards the west, in the thirtieth month they arrived at the place [the Red Sea] from which the Egyptian king dispatched the Phoenicians. After their voyage, Darius subdued the Indians, and opened the navigation of this sea [Arabian Sea].
>
> *(The Athenaeum)*

One source, Suidas, a Byzantine commentator of the tenth century, wrote that "Scylax of Caryanda, a mathematician and musician, wrote a periplus of the coasts beyond the pillars of Hercules, a book respecting Heraclides, king of the Mylassians, a description of the circuit of the earth, and an answer to the history of Polybius."

Most of the *Periplus* does not survive. The part that does is a description of the Atlantic coast of northwestern Africa, the Mediterranean, the Black Sea, and the Sea of Azov. Scylax provided a description of landforms, peoples, distances between noteworthy places, and facts about localities and local traditions.

In lieu of good literary sources on the history of Persian science, historians must rely on the sources of other commentators; one of the best ancient sources was the *Histories* of Herotodus. He implied, in his discussion of Persian religious beliefs, that the Persians recognized the four basic elements identified by the Greeks: earth, air, fire, and water. The Persians educated young men to be scribes and for the priesthood, which involved scientific practices. Herodotus discussed the Persian educational system during the Achaemenid Empire, which, as transcribed by the geographer Strabo, involved the teaching of boys

> from five years of age to twenty-four [who] are trained to use the bow, to throw the javelin, to ride horseback, and to speak the truth; and they use as teachers of science their wisest men, who also interweave their teachings with the mythical element, thus reducing that element to a useful purpose, and rehearse both with song and without song the deeds both of the gods and of the noblest men. And these teachers wake the boys up before dawn by the sound of brazen instruments, and assemble them in one place, as though for arming themselves or for a hunt; and then they divide the boys into companies of fifty, appoint one of the sons of the king or of a satrap as leader of each company, and order them to follow their leader in a race, having marked off a distance of thirty or forty stadia. They require them also to give an account of each lesson, at the same time training them in loud speaking and in breathing, and in the use of their lungs, and also training them to endure heat and cold and rains, and to cross torrential streams in such a way as to keep both armour and clothing dry. (Trans. Jones)

Herodotus described the magi, claiming that they were experts at oneiromancy.

Magi served as physicians, but their knowledge was limited; for example, Cyrus had to send for Egyptian physicians to heal eye ailments. But the knowledge of Egyptian physicians turned out to be insufficient, according to Herodotus, during the reign of Darius I. When Darius severely sprained his ankle, experienced great pain, and could not sleep, his Egyptian physicians were at a loss as to what to do, and they feared the king's wrath. Then Darius learned of a slave named Demokedes, from Croton, in Italy, who had been brought to Persia as a prisoner. Demokedes was Greek, the son of a priest of Ascelpios; he had worked at Athens and at Samos, where, in the service of the tyrant Polycrates, he had been made prisoner by the Persians when they conquered the island. The slave appeared before Darius and, though fearful and reluctant, was forced to apply his herbs and healing methods to heal the king's ankle. He was rewarded with the post of the court physician; he became wealthy and sat at the king's table. As physician, he was able to heal Darius's wife of a tumor on the breast. Sent with a small contingent of Persians to reconnoiter Greece for a possible invasion, Demokedes escaped and made his way back to Greece and then to Croton, where he lived out his life.

Mesopotamian astronomers under the Persians worked to understand lunar and solar activities in order to create an accurate calendar. They accurately predicted lunar and solar years and understood the movement of the stars and orbits of planets (such as Mars, Venus, and Saturn). Under Darius I, the astronomer Naburimanni, from Babylon, "made a study of lunar eclipses and arrived at calculations more accurate than those of Ptolemy" (Ghirshman, pp. 203–4). He derived complex tables for eclipses. Kidinnu was a fourth-century Chaldean astronomer and mathematician who "discovered the precession of the equinoxes and arrived at an exact calculation of the length of the year, making an error of only seven minutes, sixteen seconds" (Ghirshman, p. 204). Vettius Valens, the second-century CE astrologist, claimed that in his work he "did not compute eclipse tables himself but used Hipparchus for the sun, Sudenas and Kitenas [Kidinnu] for the moon" (quoted in Neugebauer, pp. 175–76).

Al-Biruni, the Persian astronomer, described the perspective of ancient Persians and other ancient people of central and southern Asia respecting the development of timekeeping and calendars according to the moon and sun:

> People distinguish two kinds of years—the Solar year and the Lunar year. They have not used other stars for the purpose of deriving years from them, because their motions are comparatively hidden, and can hardly ever be found out by eyesight; but only by astronomical observations and experiments. Further (they used only sun and moon for this purpose), because the changes of the particles of the elements and their mutual metamorphoses, as far as time and the state of the air, plants and animals, etc., are concerned, depend entirely upon the motions of these two celestial bodies, because they are the greatest of all, and because they excel the other stars by their light and appearance; and because they resemble each other. Afterwards people derived from these two kinds of years other years. (Trans. Sachau)

"The Persians," he continued, "reckoned their year as 365 days, and neglected the following fractions until the day-quarters had summed up in the course of 120 years to the number of days of one complete month, and until the fifth parts of an hour, which, according to their opinion, follow the fourth parts of a day." He wrote as well that the intercalation "was an affair settled under the special patronage of their kings at a meeting of the mathematicians, literary celebrities, historiographers, and chroniclers, priests, and judges,—on the basis of an agreement of all those regarding the correctness of the calculation, after all the persons I have mentioned had been summoned to the royal court from all parts of the empire, and after they had held councils in order to come to an agreement." Such were the calendars of the historical ancient Persians. But the first, legendary dynasty of Persian kings, the Pishdadian Dynasty, Al-Biruni, claimed, "reckoned the year as 360 days, and each month as 30 days, without any addition and subtraction; that they intercalated one month in every sixth year, which they called 'intercalary month,' and two months in every 120th year; the one on account of the five days (the Epagomense), the other on account of the quarter of a day; that they held this year in high honour, and called it the 'blessed year,' and that in it they occupied themselves with the affairs of divine worship and matters of public interest." The "heathen Arabs," Al-Biruni, claimed, using a lunar calendar, observed "the difference between their year and the solar year, which is 10 days 21½ hours, to speak roughly, and adding it to the year as one month as soon as it completed the number of days of a month." The people of India, Al-Biruni learned, "use the appearance of new-moon in their months, that they intercalate one lunar month in every 976 days, and that they fix the beginning of their era to the moment when a conjunction takes place in the first minute of any zodiacal sign." He also reported, however, that Indians had quite a diversity in how they recorded time. Some used "one revolution of the sun, starting from a point of the ecliptic and returning to it. This is the solar year." Others used "360 risings of the sun. This is called the middle-year, because it is longer than the lunar year and shorter than the solar year." Still others used "12 revolutions of the moon, starting from the star *Alsharatan* (*i.e.,* the head of Aries) and returning to it. This is their lunar year, which consists of

327 days and nearly 7⅔ hours." Finally, some used "12 lunations. This is the lunar year" (trans. Sachau).

After the conquests of Alexander and the division of the Persian Empire into various Hellenistic kingdoms, Greek culture came to dominate central Asia. Science was a dynamic mixture of Greek, African, and Asian learning in Asia Minor, Egypt, Mesopotamia, Afghanistan, and western India.

By the end of the first millennium BCE, the Roman Empire dominated throughout the Mediterranean world, conquering the Hellenistic kingdoms. In response to the Greek Hellenistic kingdoms as well as the Roman rise to power, successors of the Persians, Iranians who called themselves Parthians, built a whole new empire in central Asia. This Parthian Empire lasted from 247 BCE to 224 CE, when the Sasanian Empire replaced it. The Sasanian Empire came to an end in 651 due to Muslim conquests.

One of the most significant centers of learning during the Sasanian era was the city of Gundeshapur, founded initially by the third-century ruler Shapur I (r. 240–270) for Roman prisoners; indeed, it was initially designed like a Roman military camp. Here, the Academy of Gundeshapur was founded. This academy was the center of learning during the Sasanian period. It had a massive library, including a medical academy, and was a center for Zoroastrian learning. A medical hospital was created, reputedly the best of its time. Under the sixth-century ruler Khosrow I (r. 531–579), the academy was a center of mathematics, astronomy, philosophy, and medicine. Scientists came from the west, east, and south: the Byzantine Empire, India, and China. Nestorian Christians exiled from the Byzantine Empire found a home at Gundeshapur. Scholars at the Academy of Gundeshapur were influenced by such works as the *Sushruta Samhita*, the ancient Indian text of medicine and surgery, which inspired physicians at the academy to engage in sophisticated research and surgery.

ANCIENT SOUTHERN ASIA

Civilization appeared in the Indus Valley about 2500 BCE, perhaps even centuries before. These early Harappan peoples of the upper Indus Valley had the typical characteristics of civilization. Cultivation of the soil along the Indus River, which required an organization consistent with an established government, yielded agricultural surplus, which was the basis for wealth, trade, loose class divisions, and government institutions. People were divided by occupation, with the dominant numbers involved in agriculture, though there were scattered artisans, merchants, and shopkeepers. Trade extended along the river into the Indian Ocean; there is evidence of trade with the civilizations of Mesopotamia. The island of Bahrain in the Persian Gulf was a crossroads of trade between Mesopotamian and Harappan merchant sailors. The Harappan people developed a system of writing, which has not been deciphered. Most of the writing is found on clay tablets and pots. The writing system appears to have been in daily use by the inhabitants and not restricted to government officials. The writing script complements pictographs of animals, especially cattle and other quadrupeds. Each tablet, among the thousands discovered, appears somewhat unique, perhaps denoting an individual identifier for a person. The people of the Indus Valley used numbers as well to record trade.

The remains of urban sites, the most famous being Harappa and Mohenjo Daro, were excavated in the early twentieth century. There are hundreds of excavated sites along the upper Indus and tributaries. Clearly, this ancient civilization was not just centered on the Indus Valley. Other large cities included Lurewala, Ganweriwala, Rakhigarhi, and Dholavira, in regions ranging from Afghanistan to the Punjab south to the Sabarmati and Narmada Rivers. These cities, which are three thousand to four thousand years old, reveal a level of sophistication consistent with other civilized centers of the world at the time in North Africa, western Asia, and eastern Asia. The cities were well planned and laid out in orderly avenues, with distinct government buildings, fortifications, and neighborhoods. Some houses had two stories and a system of plumbing; city streets had drainage canals for waste and water. There is no evidence of monumental architecture. The people of the ancient Indus River were polytheistic, though there was not a clear pantheon of gods; they believed in a divine creation. A class of priests organized religious theory and devotion, though they were not as sophisticated as those of some other early civilizations. Scholars today date the Harappan civilization according to three distinct epochs ranging from about 3000 to 1300 BCE.

VEDIC AGE

During the middle of the second millennium BCE, as part of the widespread migrations throughout Europe and Asia from north to south, a warlike people, the Indo-Aryans, swept through upper India, destroying the Harappan civilization and spreading south and east through the Indian peninsula. The Indo-Aryans of India were related to those who migrated into Persia and had similar religious beliefs to Zoroastrianism: the notion of the eternal battle between good and evil. The presence of evil in Hinduism and Buddhism (a derivative of Hinduism) is, as in Judaism, Christianity, and Zoroastrianism, located in the temporal experiences of the body. The Indo-Aryans brought a belief system, Hinduism, and the worship of gods such as Vishnu and Shiva. Hindus believed in endless cycles of life (samsara, wanderings, endless suffering), each subsequent life determined by karma resulting from actions in previous/present existence. Hindus believed that a person can escape from the evils of time and the body by purification over multiple lifetimes, that eventually results in the experience of Moksha (discovery of self; connection of individual self with Universal Self, overall universal oneness, universal consciousness) or Nirvana (liberation from self, liberation of consciousness of self).

The holy books or scriptures of the Hindus were the Vedas, collections of hymns and myths written in Sanskrit. Hindu religion was anthropomorphic and moralistic. The people of the Vedic period, covering about a thousand years from the mid-second to the mid-first millennia BCE, lived in kingdoms. A distinct caste system developed based on set divisions in society: the Brahmans (priests), Kshatriyas (warriors), Vaishyas (merchants), and Sudras (workers). The Hindus of the Vedic Age developed regional devotional cults (bhakti), worshipping gods such as Shiva, Vishnu, Brahma, Krishna, and Rama, among hundreds of others.

The Vedas were the initial holy scriptures of Hinduism. *Veda* literally means "complete knowledge," so a person who studies them commensurately possesses knowledge. The *Rig Veda*, *Sama Veda*, *Yajur Veda*, and *Atharva Veda* primarily comprise hymns to Hindu deities. They provide a general cosmology of the nature of the universe. The *Rig Veda* and *Sama Veda* have hundreds of hymns, predominantly to the god of fire Agni and the storm god Indra. The *Yajur Veda* has formulas for rituals and sacrifices to the gods. The *Atharva Veda* has a variety of magic charms, many of which are to prevent or cure disease. The *Atharva Veda* reveals the concern of people during the Vedic Age, as in all ages, for the means to find and maintain health. Other religious texts, especially the Upanishads and the Puranas, emerged at the end of the Vedic Age. The Upanishads are works describing Hindu religious practices: how the individual soul unites with Brahma, the supreme deity, and karma, the individual actions in life that impact future existence. The Puranas are works that describe the gods, universe, and cosmology of the Hindu worldview. Some of this content is very detailed and evocative of scientific knowledge. The Vedas as a whole provide information on medicine and hygiene. The *Rig Veda* describes particulars of the Hindu materia medica, listing hundreds of medicines, and how humans learn from observing animals. In Hymn 2 of the *Atharva Veda*, "Purusha, Primeval Man or humanity personified," the singer of the hymn, in describing the first human, Purusha, provides a detailed anatomy of humans, their aims and aspirations, and the universe in which they live. The Puranas identify the founder of Indian medicine as Dhanvantari.

Toward the end of the Vedic Age, physicians, schools, and medical practices, divorced from hymns and magic, began to appear, though precisely when is unclear. Legendary healers provided teachings and writings about medicine. One of the first was Atreya, a legendary physician whose teachings were transmitted to pupils and posterity by written works, generally called the *Compendium*, which consisted of specific volumes such as the *Bheda* and *Agnivesa*. Susruta was a legendary surgeon. He or his school of thought also wrote a compilation, the *Compendium* or *Samhita*. According to Hindu mythology, Indra provided medical and surgical knowledge to humans by means of an ancient sage, Dhanvantari, who passed the knowledge to Atreya and Susruta. Atreva and Susruta are considered to be the founders of the Ayurveda system of medicine. There is also a tradition that these two teachers represented two schools: Atreya taught at the university in Taxila, in western India, and Susruta taught at the university in Benares, in eastern India. Students of Atreya included Charaka, who described human anatomy in the *Compendium*. The *Compendium* of Susruta/Charaka provides an extensive description of the human skeletal structure, anatomy, and physiology. It is a sophisticated account of surgery and midwifery as well. Dozens of surgical instruments are described, as are a variety of surgical techniques such as incising, excising, suturing, extracting, puncturing, evacuating, and scarifying. There are descriptions of complex ophthalmological operations, such as the couching of cataracts; intestinal surgery; removing stones from the urethra; using forceps when delivering newborns; and rhinoplasty. Medicated wine is presented as anesthesia to dull the sensations of the patient. A variety of therapies are examined. Students learned about human anatomy by watching and participating in the dissection of human bodies.

Students practiced surgical techniques on models that simulated the human body. Medicine was not completely divorced from superstition, as a physician's prognosis for a patient's recovery relied on "messengers, omens and dreams" (trans. Bhishagratna). Charaka was an advocate of humors, as was Hippocrates, though he counted three—phlegm, bile, and air—rather than the Hippocratic four.

According to the *Compendium* of Susruta, the Ayurveda was originally part of the *Atharva Veda* but was separated from it into eight components: the Salya-Tantram, which deals with surgery; the Shalakya-Tantram, which focuses on diseases of the head (mouth, nostrils, ears); the Kaya-Chikitsa, focusing on a variety of general diseases, such as dysentery, leprosy, insanity, and fever; the Bhuta Vidya, dealing with incantations to rid people of "demonical diseases"; the Kaumara-Bhritya, focusing on infant health and disease; the Agada-Tantram, providing information on antidotes to venomous bites and stings; the Rasayana-Tantrum, "the science of rejuvenation," that is, how to live longer and retain better memory and vigor; and the Vajeekarana-Tantram, focusing on aphrodisiacs to ensure fertility. Regarding diseases,

> the Purusha (man) is the receptacle of any particular disease, and that which proves a source of torment or pain to him, is denominated as a disease. There are four different types of disease such as, Traumatic or of extraneous origin (Agantuka), Bodily (Sharira), Mental (Manasa) and Natural (Svabhavika). A disease due to an extraneous blow or hurt is called Agantuka. Diseases due to irregularities in food or drink, or incidental to a deranged state of the blood, or of the bodily humours acting either singly or in concert, are called Sharira. Excessive anger, grief, fear, joy, despondency, envy, misery, pride, greed, lust, desire, malice, etc. are included within the category of mental (Manasa) distempers; whereas hunger, thirst, decrepitude, imbecility, death, sleep, etc. are called the natural (Svabhavika) derangements of the body. The Mind and the Body are the seats of the abovesaid distempers according as they are restricted to either of them, or affect both of them in unison. (Trans. Bhishagratna)

Medical students should have the following qualities, according to the *Compendium* of Susruta:

> Such . . . a student, belonging to one of the three twice-born castes such as, the Brahmana, the Kshatriya, and the Vaishya, and who should be of tender years, born of a good family, possessed of, a desire to learn, strength, energy of action, contentment, character, self-control, a good retentive memory, intellect, courage, purity of mind and body, and a simple and clear comprehension, command a clear insight into the things studied, and should be found to have been further graced with the necessary qualifications of thin lips, thin teeth and thin tongue, and possessed of a straight nose, large, honest, intelligent eyes, with a benign contour of the mouth, and a contented frame of mind, being pleasant in his speech and dealings, and usually painstaking in his efforts. A man possessed of contrary attributes should not be admitted into the sacred precincts of medicine. (Trans. Bhishagratna)

At the same time during the Vedic Age, Indian mathematicians were at work to devise elaborate altars for religious worship. The designs and mathematics for such altar construction are found in the four *Sulba Sutras*, written in Sanskrit, which make up part of the *Vedangas*. The *Sulba Sutras* provide precise measurements for geometric constructions of altars of various designs. They also provide a sense, along with all of Vedic literature, of the pursuit of knowledge of self. Practically, the *Sulba Sutras* provide geometric and arithmetic knowledge. The Pythagorean theorem is revealed, which has led some commentators to believe the accuracy of the stories that Pythagoras learned from the Egyptians and Indians. The *Sutras* provide formulas for converting squares into rectangles and circles, and converting a circle into a square. In arithmetic, the *Sulba Sutras* provides the formula for discovering square roots. Some historians of mathematics believe that the Hindus anticipated the Greeks and Muslims in the development of algebra. These formulas were put into practice in creating a variety of different kinds of altars of varying shapes.

The *Sulba Sutras* were in the tradition of Indian mathematical and astronomical development during the first millennium BCE. Indians used a decimal system for weights and measures, engaged in astronomical observations, and developed simple geometric constructs. By about 1000 BCE, certain cultural groups in India, notably the Jains, worked with numbers and geometry and speculated on infinity. The Indians, like most ancient peoples, developed a lunar calendar. Their astronomical knowledge was influenced by Babylonians in understanding the lunar zodiac. Hindu astronomical writings included computations describing eclipses, charts of astronomical observations, and ephemerides. Contact with Greco-Roman and Mesopotamian civilization at the beginning of the Common Era led to sophisticated lunar and solar calendars that included the nineteen-year Metonic cycle.

Toward the end of the Vedic Age, the Upanishads were written, extending the focus of the Vedas especially toward personal contemplation and the relationship of the self to the Self, Atman, and to the ultimate reality, Brahman: all-pervasive being, infinite, absolute. The *Khandogya-Upanishad* describes the understanding of the basis of reality: "The essence of all beings is the earth, the essence of the earth is water, the essence of water the plants, the essence of plants man, the essence of man speech, the essence of speech the Rig-veda, the essence of the Rig-veda the Sama-veda, the essence of the Sama-veda the udgitha (with is Om). That udgitha (Om) is the best of all essences, the highest, deserving the highest place" (vol. 1, ed. Müller).

About the same time as the Upanishads were written, the *Mahabharata* was written, perhaps around 400 to 300 BCE. The most significant part of the *Mahabharata* was the Bhagavad Gita, which details the conversation between the Avatar Krishna and the warrior Arjuna. The Bhagavad Gita is notable from a behavioral science point of view, in that it describes the conflict the individual goes through between the expectations of the public and the personal morality of the self.

Also around the end of the Vedic Age in northern India, the philosopher Siddhartha Gautama (563–483) introduced a new focus on achieving enlightenment that broke from the traditional ascetism of Hinduism. Though hardly a scientist, nevertheless Gautama's Buddhism provided the same focus found in so many other places in the world around 500 BCE: that of people seeking the nature of truth and turning to a contemplation of the great spiritual ideas to discover it.

Other great thinkers of the mid-first millennium included Kanada, founder of Vaishesika philosophy. Kanada, like the Greeks Leucippus and Democritus, was an atomist. Such thinkers believed that atoms were incredibly minute particles in constant motion. "Atoms cannot exist in an uncombined state in creation," in the words of Benoy Kumar Sarkar. These philosophers examined the properties of matter, such as "elasticity, cohesiveness, impenetrability, viscosity, fluidity, [and] porosity."

During the fourth century, India was impacted by the invasion of Alexander of Macedon along the Indus River valley. Beginning in 326 BCE, Alexander conquered local kings in present-day Afghanistan and Pakistan, and he established several Greek colonies in these regions. Much was made among the ancient historians and geographers of India of the campaigns of Alexander, his battle with the Indian chieftain Porus, his exploration of the Indus River, and his interviews with Indian philosophers. Plutarch, for example, in *Life of Alexander*, recorded this episode:

> He captured ten of the Gymnosophists who had done most to get Sabbas to revolt, and had made the most trouble for the Macedonians. These philosophers were reputed to be clever and concise in answering questions, and Alexander therefore put difficult questions to them, declaring that he would put to death him who first made an incorrect answer, and then the rest, in an order determined in like manner; and he commanded one of them, the oldest, to be the judge in the contest. The first one, accordingly, being asked which, in his opinion, were more numerous, the living or the dead, said that the living were, since the dead no longer existed. The second, being asked whether the earth or the sea produced larger animals, said the earth did, since the sea was but a part of the earth. The third, being asked what animal was the most cunning, said: "That which up to this time man has not discovered." The fourth, when asked why he had induced Sabbas to revolt, replied: "Because I wished him either to live nobly or to die nobly." The fifth, being asked which, in his opinion, was older, day or night, replied: "Day, by one day"; and he added, upon the king expressing amazement, that hard questions must have hard answers. Passing on, then, to the sixth, Alexander asked how a man could be most loved; "If," said the philosopher, "he is most powerful, and yet does not inspire fear." Of the three remaining, he who was asked how one might become a god instead of man, replied: "By doing something which a man cannot do"; the one who was asked which was the stronger, life or death, answered: "Life, since it supports so many ills." And the last, asked how long it were well for a man to live, answered: "Until he does not regard death as better than life." So, then, turning to the judge, Alexander bade him give his opinion. The judge declared that they had answered one worse than another. "Well, then," said Alexander, "thou shalt die first for giving such a verdict." "That cannot be, O King," said the judge, "unless thou falsely saidst that thou wouldst put to death first him who answered worst." (Trans. Perrin)

Whereas some Greek authors, such as Arrian, praised Calanus as an Indian philosopher who emolliated himself, Megasthenes the geographer wrote that Indian philosophers did not practice such things. He cited the philosopher Mandanis the gymnosophist as a man who refused to respond to the demands of Alexander.

The historian of science of ancient India, like all historians of ancient India, confronts a paucity of materials. Ancient Indian philosophers were noteworthy in penning, in Sanskrit, metaphysical accounts of existence and beautiful poetry, but they were rarely historians of their own people. As a result, some of the best sources to undercover ancient Indian existence come from outside observers, especially the

Greeks, who, after the journeys and conquest of Alexander of Macedon in the fourth century BCE, were introduced to the people of the Indus River valley and sought information of India further east.

One of these observers was Megasthenes, who at the end of the fourth century was a diplomat in the service of Alexander's successor in the region, Seleucus I (r. 305–281). The Hellenistic king sent Megasthenes on a diplomatic mission to meet with Chandragupta Maurya (r. 322–297), who in the wake of Alexander's conquests, founded a dynasty that would control upper India from 322 to 185 BCE. His capitol, Pataliputra, in the upper Ganges valley, would be an important city in northeastern India until 1200 CE. Seleucus sent Megasthenes on diplomatic missions to the upper Indus and upper Ganges regions, where Megasthenes met with Chandragupta Maurya, who, like his successors, ruled from Pataliputra for centuries. During his stay (of an uncertain length) at the court of Chandragupta Maurya (whom the Greeks called Sandrakottos) at Pataliputra, Megasthenes wrote an account of his visits to India, the *Indica*, fragments of which survive. The *Indica* provides information on ancient Indian geography, society, beliefs, and science. Megasthenes relied on Indian philosophers, Brahmins, for his sources of information.

Megasthenes's journey, taken probably around 302 BCE, began at Bactria, part of the empire of Seleucus I. From Bactria he journeyed to Kabul, traveled down the Kabul River, and took an overland route to the Hydaspes River, where Alexander had met the Indian king Porus in battle in 326 BCE. He took the Royal Road east through the Punjab, crossing the many tributaries of the Indus River, until he came to the Ganges River, which he paralleled to Pataliputra. He noted in *Indica* that the rivers that feed the Indus and Ganges rise in the north, their sources in mountains that "stretch along the northern frontier." More specifically, he wrote, "To account for the rivers being so numerous, and the supply of water so superabundant, the native philosophers and proficients in natural science advance the following reasons: They say that the countries which surround India—those of the Skythians and Baktrians, and also of the Aryans—are more elevated than India, so that their waters, agreeably to natural law, flow down together from all sides to the plains beneath, where they gradually saturate the soil with moisture, and generate a multitude of rivers." He learned from his philosophic sources that the general shape of India was a rhomboid, with unequal sides, west and east, surrounded by the Indian Ocean, stretching from the Indus to the Ganges Rivers. He learned as well that the Indian subcontinent stretches far to the south, beyond the Tropic of Cancer, so that "the gnomon of the sundial may frequently be observed to cast no shadow, while the constellation of the Bear [the Big Dipper] is by night invisible, and in the remotest parts even Arcturus," one of the brightest stars in the northern hemisphere, "disappears from view." Pataliputra is just north of the Tropic of Cancer, so it is possible that Megasthenes personally made these astronomical observations. Moreover, very unusual to Mediterranean astronomers was the strange phenomenon of shapes casting a southern shadow, south of the Tropic of Cancer. He learned about, and probably experienced, the Indian monsoons, writing that the double rainfall in the course of the year yielded plentiful crops: "The rains of the summer season," he wrote, "are wont to fall every year at a stated period with surprising regularity." The soil contained gold, silver, copper, iron, and tin, useful for "making articles of use and ornament, as well as the implements and accoutrements of war" (trans. McCrindle).

Megasthenes was naturally intrigued by the Indian philosophers who had the same inquisitiveness as the Greek philosophers, so he made a particular study of their origins, habits, and ideas. Scholars in recent centuries have debated whether the civilization of the Indus River valley emerged in isolation from other civilizations or because of contact with the people, for example, of the Tigris and Euphrates and Nile Rivers. Megasthenes argued that India independently developed arts and sciences: "As was the case with the Greeks . . . in like manner as with them, the arts and other appliances which improve human life were gradually invented, Necessity herself teaching them to an animal at once docile and furnished not only with hands ready to second all his efforts, but also with reason and a keen intelligence." Megasthenes also wrote that "all that has been said regarding nature by the ancients [of Greece] is asserted also by philosophers out of Greece, on the one part in India by the Brachmanes [Brahmins], and on the other in Syria by the people called the Jews" (trans. McCrindle).

Whereas the Greeks explained the origins of their philosophy and science according to legendary godlike founders, such as Prometheus and Asclepios, likewise the Indians had their fanciful notions of intellectual origins. "The men of greatest learning among the Indians," Megasthenes wrote, "tell certain legends, of which it may be proper to give a brief summary." They had the legend of a Dionysus-like figure who taught them wine making "as well as other arts conducive to human well-being. He [Dionysus] was, besides, the founder of large cities, which he formed by removing the villages to convenient sites . . . and introduced laws and courts of justice" (trans. McCrindle). For such greatness, this mortal was honored as a god. Megasthenes implied that the Greek myth of the god Dionysus derived, therefore, from India. Two centuries later, the Greek historian Diodorus Siculus learned from Egyptian priests that Dionysus was actually Osiris, that Osiris had traveled to India, had taught the Indians cultivation, had founded cities, and so impressed the people that they claimed that Osiris, that is, Dionysus, was native to their land.

Megasthenes learned as well from their "ancient philosophers" the custom of no slavery allowed, only equality among citizens. Laws treated all equally, he thought, even if property was unevenly divided. He also learned that their ancient cities, founded by Dionysus and Heracles, practiced democracy, as did some Greek cities. The philosophers taught Megasthenes that there were seven castes in Indian society, of which the philosophers were at the pinnacle; their numbers were few "but in point of dignity preeminent over all." Philosophers were religious leaders, and also "when, gathered together at the beginning of the year, they forewarn the assembled multitudes about droughts and wet weather, and also about propitious winds, and diseases, and other topics capable of profiting the hearers" (trans. McCrindle).

Most of Megasthenes's *Indica* has been lost; only fragments remain, and these are preserved by other, later ancient writers who quoted from his book. This includes the third-century writer Claudius Aelian, who, in *On the Characteristics of Animals*, included this comment, based on Megasthenes:

> The Indians cure the wounds of the elephants which they catch, in the manner following:—They treat them in the way in which, as good old Homer tells us, Patroklos treated the wound of Eurypylos,—they foment them with lukewarm water. After this they rub them over with butter, and if they are deep allay the inflammation by applying and inserting pieces of pork, hot but still retaining the blood. They cure ophthalmia with cows' milk, which is first used as a fomentation for the eye, and is then injected into it. The animals open their eyelids, and finding they can see better are delighted, and are sensible of the benefit like human beings. In proportion as their blindness diminishes their delight overflows, and this is a token that the disease has been cured. The remedy for other distempers to which they are liable is black wine; and if this potion fails to work a cure nothing else can save them. (Trans. McCrindle)

Aelian was often given to nonsense, and although Megasthenes was a serious author, often the philosophers, who were his sources, might advertently or inadvertently mislead with spurious information. For example, Megasthenes reported, inaccurately, that Indian elephants were far larger than African elephants.

The Roman geographer Strabo also relied on Megasthenes for information on Indian philosophers and included this excerpt based on the *Indica*:

> Speaking of the philosophers, he (Megasthenes) says that such of them as live on the mountains are worshippers of Dionysos, showing as proofs that he had come among them, the wild vine, which grows in their country only, and the ivy, and the laurel, and the myrtle, and the box-tree, and other evergreens, none of which are found beyond the Euphrates, except a few in parks, which it requires great care to preserve. They observe also certain customs which are Bacchanalian. Thus they dress in muslin, wear the turban, use perfumes, array themselves in garments dyed of bright colours; and their kings, when they appear in public, are preceded by the music of drums and gongs. But the philosophers who live on the plains worship Herakles. . . . Megasthenes makes a different division of the philosophers, saying that they are of two kinds—one of which he calls the Brachmanes [Brahman], and the other the Sarmanes [*Śramaṇa*]. The Brachmanes are best esteemed, for they are more consistent in their opinions. From the time of their conception in the womb they are under the guardian care of learned men, who go to the mother and, under the pretence of using some incantations for the welfare of herself and her unborn babe, in reality give her prudent hints and counsels. The women who listen most willingly are thought to be the most fortunate in their children. After their birth the children are under the care of one person after another, and as they advance in age each succeeding master is more accomplished than his predecessor. The philosophers have their abode in a grove in front of the city within a moderate-sized enclosure. They live in a simple style, and lie on beds of rushes or (deer) skins. They abstain from animal food and sexual pleasures, and spend their time in listening to serious discourse, and in imparting their knowledge to such as will listen to them. The hearer is not allowed to speak, or even to cough, and much less to spit, and if he offends in any of these ways he is cast out from their society that very day, as being a man who is wanting in self-restraint. After living in this manner for seven-and-thirty years, each individual retires to his own property, where he lives for the rest of his days in ease and security. They then array themselves in fine muslin, and wear a few trinkets of gold on their fingers and in their ears. They eat flesh, but not that of animals employed in labour. They abstain from hot and highly seasoned food. They marry as many wives as they please, with a view to have numerous children, for by having many wives greater advantages are enjoyed, and, since they have no slaves, they have more need to have children around them to attend to their

> wants. The Brachmanes do not communicate a knowledge of philosophy to their wives, lest they should divulge any of the forbidden mysteries to the profane if they became depraved, or lest they should desert them if they became good philosophers: for no one who despises pleasure and pain, as well as life and death, wishes to be in subjection to another, but this is characteristic both of a good man and of a good woman. Death is with them a very frequent subject of discourse. They regard this life as, so to speak, the time when the child within the womb becomes mature, and death as a birth into a real and happy life for the votaries of philosophy. On this account they undergo much discipline as a preparation for death. They consider nothing that befalls men to be either good or bad, to suppose otherwise being a dream-like illusion, else how could some be affected with sorrow, and others with pleasure, by the very same things, and how could the same things affect the same individuals at different times with these opposite emotions? Their ideas about physical phenomena, the same author tells us, are very crude, for they are better in their actions than in their reasonings, inasmuch as their belief is in great measure based upon fables; yet on many points their opinions coincide with those of the Greeks, for like them they say that the world had a beginning, and is liable to destruction, and is in shape spherical, and that the Deity who made it, and who governs it, is diffused through all its parts. They hold that various first principles operate in the universe, and that water was the principle employed in the making of the world. In addition to the four elements there is a fifth agency, from which the heaven and the stars were produced. The earth is placed in the centre of the universe. Concerning generation, and the nature of the soul, and many other subjects, they express views like those maintained by the Greeks. They wrap up their doctrines about immortality and future judgment, and kindred topics, in allegories, after the manner of Plato. Such are his [Megasthenes] statements regarding the Brachmanes. Of the Sarmanes those who are held in most honour are called the Hylobioi. They live in the woods, where they subsist on leaves of trees and wild fruits, and wear garments made from the bark of trees. They abstain from sexual intercourse and from wine. They communicate with the kings, who consult them by messengers regarding the causes of things, and who through them worship and supplicate the deity. Next in honour to the Hylobioi are the physicians, since they are engaged in the study of the nature of man. They are simple in their habits, but do not live in the fields. Their food consists of rice and barley-meal, which they can always get for the mere asking, or receive from those who entertain them as guests in their houses. By their knowledge of pharmacy they can make marriages fruitful, and determine the sex of the offspring. They effect cures rather by regulating diet than by the use of medicines. The remedies most esteemed are ointments and plasters. All others they consider to be in a great measure pernicious in their nature. This class and the other class practise fortitude, both by undergoing active toil, and by the endurance of pain, so that they remain for a whole day motionless in one fixed attitude. Besides these there are diviners and sorcerers, and adepts in the rites and customs relating to the dead, who go about begging both in villages and towns. Even such of them as are of superior culture and refinement inculcate such superstitions regarding Hades as they consider favourable to piety and holiness of life. Women pursue philosophy with some of them, but abstain from sexual intercourse. (Trans. McCrindle)

Based on Megasthenes's *Indica*, Hippolytus, the late Roman commentator, learned about the teachings of the Indian Brahmins and wrote:

> There is also with the Indians a sect composed of those philosophizing among the Brachmans. They spend a contented existence, abstain both from living creatures

and all cooked food, being satisfied with fruits; and not gathering these from the trees, but carrying off those that have fallen to the earth. They subsist upon them, drinking the water of the river Tazabena. But they pass their life naked, affirming that the body has been constituted a covering to the soul by the Deity. These affirm that God is light, not such as one sees, nor such as the sun and fire; but to them the Deity is discourse, not that which finds expression in articulate sounds, but that of the knowledge through which the secret mysteries of nature are perceived by the wise. And this light which they say is discourse, their god, they assert that the Brachmans only know on account of their alone rejecting all vanity of opinion which is the soul's ultimate covering. These despise death, and always in their own peculiar language call God by the name which we have mentioned previously, and they send up hymns (to him). But neither are there women among them, nor do they beget children. But they who aim at a life similar to these, after they have crossed over to the country on the opposite side of the river, continue to reside there, returning no more; and these also are called Brachmans. But they do not pass their life similarly, for there are also in the place women, of whom those that dwell there are born, and in turn beget children. And this discourse which they name God they assert to be corporeal, and enveloped in a body outside himself, just as if one were wearing a sheep's skin, but that on divesting himself of body that he would appear clear to the eye. But the Brachmans say that there is a conflict in the body that surrounds them, (and they consider that the body is for them full of conflicts); in opposition to which, as if marshalled for battle against enemies, they contend, as we have already explained. And they say that all men are captive to their own congenital struggles, viz., sensuality and in chastity, gluttony, anger, joy, sorrow, concupiscence, and such like. And he who has reared a trophy over these, alone goes to God; wherefore the Brachmans deify Dandamis [Mandamis], to whom Alexander the Macedonian paid a visit, as one who had proved victorious in the bodily conflict. But they bear down on Calanus as having profanely withdrawn from their philosophy. But the Brachmans, putting off the body, like fishes jumping out of water into the pure air, behold the sun. (Trans. MacMahon)

In the period after the Common Era, especially around the time of the Gupta Empire (320–550), there were developments in mathematics and astronomy in India. Aryabhata of Paliputra, during the sixth century, used and systematized Hindu algebra into a new system. He discovered the area of a triangle, circle, and trapezium; the length of the radius of a circle; and the value of pi (π). As an astronomer, he knew that the earth rotates on its axis. The polymath Varaha-mihira (c. 505–587) admitted that Hindu astronomy was dependent upon the Greeks. The Hindus, like most ancient and medieval thinkers, were intrigued by astrology as well. According to Al-Biruni, in *The Chronology of Ancient Nations*, "The Hindus divided the globe, in conformity with their 27 Lunar Stations, into 27 parts, each Station occupying nearly 13 degrees of the ecliptic. From the stars entering these Stations . . . , they derived their astrological dogmas as required for every subject and circumstance in particular" (trans. Sachau).

FURTHER READING

Alberuni's India: An Account of the Religion, Philosophy, Literature, Geography, Chronology, Astronomy, Customs, Laws and Astrology of India about A. D. 1030, trans. Edward C. Sachau (London: Kegan Paul, Trench, Trubner, & Co., 1910).

Al-Biruni, *The Chronology of Ancient Nations*, trans. C. Edward Sachau (London: William H. Allen and Co., 1879).

The Athenaeum, A Magazine of Literary and Miscellaneous Information (London: Longman, Hurst, Rees, and Orme, 1808).

Kaviraj Kunja Lal Bhishagratna, ed. and trans., *An English Translation of the Sushruta Samhita*, vol. 1: *Sutrasthanam* (Calcutta, India: By the author, 1907).

Ernest Brehaut, *An Encyclopedist of the Dark Ages: Isidore of Seville* (New York: Columbia University, 1912).

Glyn Daniel, *The First Civilizations: The Archaeology of Their Origins* (New York: Thomas Y. Crowell, 1968).

Diodorus Siculus, *The Library of History of Diodorus Siculus*, trans. C. H. Oldfather, vol. 1 (London: Loeb Classical Library, 1933), http://penelope.uchicago.edu/Thayer/E/Roman/Texts/Diodorus_Siculus/home.html.

Diogenes Laertius, *Lives of the Philosophers*, trans. R. D. Hicks, 2 vols. (Cambridge, MA: Harvard University Press, 1931, 1938).

Walter A. Fairservis, *Harrapan Civilization and Its Writing* (Leiden, Netherlands: Brill, 1992).

R. Ghirshman, *Iran: From the Earliest Times to the Islamic Conquest* (Harmondsworth, UK: Penguin Books, 1954).

Herodotus, *The Histories*, trans. Aubrey de Selincourt (Harmondsworth, England: Penguin Books, 1972).

Hippolytus, *Refutation of the Heresies*, translated by J. H. MacMahon, from *Ante-Nicene Fathers*, vol. 5, ed. Alexander Roberts, James Donaldson, and A. Cleveland Coxe (Buffalo, NY: Christian Literature Publishing Co., 1886); revised and edited for New Advent by Kevin Knight, http://www.newadvent.org/fathers/050101.htm.

F. Rudolf Hoernle, *Studies in the Medicine of Ancient India, Part 1: Osteology or the Bones of the Human Body* (Oxford: Clarendon Press, 1907).

Ibn Abi Usaibi'ah, *History of Physicians*, trans. L. Kopf (Bethesda, MD: National Library of Medicine, 1971).

Narain Singh Kalota, *India as Described by Megasthenes* (Delhi, India: Concept Publishing, 1978).

Michael Kublin, and Hyman Kublin, *India* (New York: Houghton Mifflin Co., 1991).

J. W. McCrindle, trans., *Ancient India as Described by Megasthenes and Arrian* (London: Trubner & Co., 1877).

Firouzeh Mirrazavi, "Academy of Gundishapur," *Iran Review*, http://www.iranreview.org/content/Documents/_Academy_of_Gundishapur.htm.

F. Max Müller, ed., *The Sacred Books of the East*, vol. 1: *The Upanishads* (New York: Christian Literature Co., 1897).

F. Max Müller, ed., *The Sacred Books of the East*, vol. 4: *The Zend-Avesta* (Oxford: Clarendon Press, 1895).

O. Neugebauer, *The Exact Sciences in Antiquity* (New York: Dover Books, 1969).

Oxford Companion to the Bible (Oxford: Oxford University Press, 1993).

Pliny the Elder, *Natural History,* trans. John F. Healy (London: Penguin Books, 1991).

Plutarch's Lives, trans. Bernadotte Perrin, vol. 7 (London: William Heinemann, 1919).

John F. Price, *Applied Geometry of the Sulba Sutras*, n.d. http://chaturpata-atharvan-ved.com/spiritual-books-section/spiritual-books/acharya-literature/scientist-acharya-of-ancient-india/SulbaSutras-Applied-Geometry-by-John-Price-EN.pdf.

Vivian Robson, *The Fixed Stars and Constellations in Astrology* (Whitefish, MT: Kessinger Publishing, 2003).

R. J. Ruben, "Otology at the Academy of Gondishapur, 200–600 CE," *Otology and Neurotology* 38 (2017): 1540–45.

Benoy Kumar Sarkar, *Hindu Achievements in Exact Science: A Study in the History of Scientific Development* (London: Longmans, Green and Co., 1918).

Rüdiger Schmitt, "Democedes," *Encyclopaedia Iranica*, http://www.iranicaonline.org/articles/democedes.

Upinder Singh, *A History of Ancient and Early Medieval India: From the Stone Age to the 12th Century* (Delhi, India: Pearson Education, 2008).

Strabo, *Geography*, trans. H. L. Jones, 8 vols. (Cambridge, MA: Harvard University Press, 1917–1932), http://penelope.uchicago.edu/Thayer/E/Roman/Texts/Strabo/home.html.

Christopher Warnock, *The Mansions of the Moon: A Lunar Zodiac for Astrology and Magic* (Morrisville, NC: Lulu Press, 2019).

7

Science in Ancient Eastern Asia (2100 BCE–400 CE)

Science in ancient China, from the beginnings of the Yellow River civilization to the Han Dynasty, emerged in relative isolation from the more dynamic scientific cultures of western Asia, Europe, and northern Africa. The first Chinese civilization developed along the Yellow River in the second millennium BCE. The Xia Dynasty, from 2100 to 1600, and the Shang Dynasty, from 1600 to 1046, featured a culture of cities based on surplus agriculture that resulted in a sophisticated society that included the invention of writing as well as bronze. Since bronze is known to have been developed in Thailand during the fourth millennium BCE, it is possible that the Chinese borrowed this technology from the Thai people. The Yellow River is prone to flooding because it is on a higher plane than the surrounding countryside; dwellers of the Yellow Valley developed the means to control the river with dikes and canals. This required, of course, techniques of irrigation and governmental organization, which the Shang people devised. One technology that was in use involved workers trudging on rotating paddles that were attached to an axle and then to a gear that would bring water up from a stream on multiple pallets. Another irrigation technique was the use of the windlass—a manual crank tied to a basket that was drawn up and emptied into a canal leading to the fields. Chinese civilization expanded to the Yangtze River valley during the Zhou Dynasty, 1046 to 256. Dikes, canals, and irrigation ditches required sophisticated civil engineering, at which the Chinese excelled. Chinese roads traversed the thousands of miles of the varied lands of different dynasties. The Chinese were excellent artisans, mechanics, and, clearly, expert wall builders: the Great Wall of China—the exemplary model—was built during various periods, particularly during the reign of the emperor Qin Shi Huang (r. 221–210).

During the first millennium BCE, sophisticated Chinese thought developed, including theories based on speculation and observation to explain human and natural history. Chinese intellectuals of the first millennium BCE developed an understanding of an uncreated universe with neither a precise beginning nor an end. The universe was a cosmic, organic whole, amoral, without a providential supernatural agency. Time was not computed from a set beginning, which did not exist. Time, rather, was cyclical, and the elements of time and existence were simultaneous. There was for these Chinese intellectuals, at the same time, a sense of an organic universal process in the universe that ties natural and human existence. This is illustrated by the cosmic dialectic of yin-yang. The primal forces of the feminine (yin) were symbolized by the moon, cold, water, earth, nourishment, autumn, and winter; those of the masculine (yang) were symbolized by the sun, fire, heat, spring, and summer. Present also in the yin-yang dichotomy was an elementary understanding of wave physics: one comes and the other goes, as one flourishes, the other diminishes. The scientist and philosopher Wang Chong (27–100 CE) wrote, "The Yang having reached its climax retreats in favour of Yin; the Yin having reached its climax retreats in favour of Yang" (Needham, vol. 4, p. 7). Nature operates according to action-reaction. Yin-yang was initially introduced in the *I Ching,* the *Book of Changes.* The *I Ching* presents a worldview that is an organic, eternal, self-generating whole. The cosmic process is cyclical, timeless. Professors of the point of view of yin-yang were the shamans of Chinese society, called *Ju.* They used the *I Ching* as a means of divination, to predict the cosmic forces at work that influence human time. Over the course of the first millennium BCE, these intellectuals became teachers of what is good, what is right, which anyone can learn. They promoted the idea of the common experience of humans, of the common educational basis of humans, and of human interaction with the cosmic forces of nature.

Such was the philosophy developed in the sixth century BCE, by the Ju/shaman known as Lao Tzu (Lao Tze), the founder of Taoism. Reputedly Lao Tzu wrote the *Tao Te Ching,* the book of the way. The philosophical, mystical concept of the Tao concerned Lao Tze and his followers. To them, words could not express the inexpressible (Tao):

> We look at it, and we do not see it, and we name it "the Equable." We listen to it, and we do not hear it, and we name it "the Inaudible." We try to grasp it, and do not get hold of it, and we name it "the Subtle." With these three qualities, it cannot be made the subject of description; and hence we blend them together and obtain The One. Its upper part is not bright, and its lower part is not obscure. Ceaseless in its action, it yet cannot be named, and then it again returns and becomes nothing. This is called the Form of the Formless, and the Semblance of the Invisible; this is called the Fleeting and Indeterminable. We meet it and do not see its Front; we follow it, and do not see its Back. When we can lay hold of the Tâo of old to direct the things of the present day, and are able to know it as it was of old in the beginning, this is called (unwinding) the clue of Tâo. (Trans. Legge, 1891)

The Way is the unseen, unknown whole; it explains the relationship between humans and nature, where the latter dominates the former. Humans must seek harmony with nature, an organic approach to life; clearly this philosophy fits well with the complementary forces of the yin-yang. In its essence, Taoism is naturalistic, not humanistic, teaching that humans must conform to nature above all else, which leads to contentment. It is intuitive, existential, and individualistic; selfless yet self-absorbed; altruistic for the good of self.

Before the Confucian period of the mid-first millennium, the Ju were involved in divination, in keeping records of the movements of planets and stars. Hence they had a historical as well as astronomical/astrological role in society: to explain how human affairs related to the cosmic order. There were rituals that helped humans engage the cosmic whole, and the Ju were in involved in said rituals. During the Zhou period of the first century millennium, "the source of the authority for ritualized behavior gradually was transferred from the super-rational to the purely rational," writes Frederick Mote. The Ju were at the center of this transformation. "Ritual came to be philosophically conceived as something that contributed to the harmony of the cosmos" (Mote, p. 31).

CONFUCIUS (551–479 BCE)

Confucius (K'ung-Fu-tzu), reformer, teacher, and sage, advocated a universe in which the Ju discovers through knowledge of self and nature the harmony within and without. Confucius taught the ways of the sage, the teacher of what is good, what is right, and what is the correct ritual behavior. Confucius believed in rational, ritual behavior and propriety. Human affairs, he taught, must follow and imitate the moral order and rationality of the cosmos. The order of the cosmos was reflected in the order of society. Confucius supported the Zhou Dynasty and sought through his teaching to influence government leaders to rule with virtue in accordance with the moral order of the universe. Not only must the rulers of society practice virtue but the people must as well. Confucius advocated a common educational basis for all citizens, emphasizing the common experiences of all humans. Confucius taught filial piety, respect for ancestors and tradition. He taught *Li*, ritual propriety and rational behavior: ritual behavior is rational. All people must follow a set way of living: ritual decorum. Passive acceptance rather than active pursuit and acquisition, poverty, and obscurity are sage-like. The greatest virtue, he taught, is *Jen*: love, benevolence, goodness, well-being for all. Jen is an active rather than passive virtue. As Confucius said in *Analects*, "Do not do to others what you would not like yourself" (trans. Waley).

Confucius's disciples included Mencius (372–289) and Hsün Tzu (also known as Xunzi, c. 310–235), who continued the teacher's preoccupation with how humans, both individuals and the group, relate to the state and thereby to the cosmic order. Mencius believed that the ruler's mandate of heaven must derive ultimately from the people; human order, which is a reflection of the cosmic order, therefore derives from the people. Mencius had an optimistic view of humanity. He believed that human civilization was a result of the good qualities in humanity. Hsün Tzu, on the other hand, had a more negative point of view. He believed civilization provided order, imposed it on humans given to disorder: "Submitting to imposed norms, if one has the intelligence to do so, and finding value in so doing is the essence of Hsün Tzu's social and political thought" (Mote, p. 65).

The Chinese of the first millennium, like the Romans and Greeks of the same period, believed that the writing of history is a means to preserve the moral and political order. Historical writing was didactic, a teaching tool for correct behavior. Confucius traditionally was the first great historian of China. The Chinese associated Confucius with the writing of the first history of China, the *Book of*

Documents (*Shû King*), an account of events in China from the mid-third to the mid-first millennia BCE. Modern scholars disagree with the assertions of K'ung An-kuo (c. 156–74) and Sima Qian, two historians of the Han Dynasty, that assigned Confucius the authorship of the *Book of Documents*. In all likelihood, the Ju, the recorders of history, collected the various random historical documents that constitute the *Book of Documents*. Amid the scattered remains of the past recorded in the *Book of Documents* are themes that had a tremendous impact on Chinese thought. The most important theme was the *t'ien ming*, or *mandate of heaven*, which asserted that rulers must rule according to virtue, consistent with the order of the universe, or be replaced.

Confucius was also traditionally accorded authorship of the second great Chinese book of history, the *Spring and Autumn Annals*, *Ch'un ch'iu*. This book is a fragmented chronicle of the period from 722 to 481 BCE, examining political and natural occurrences during those years. The *Spring and Autumn Annals* was prized by Confucius's followers, who wrote various commentaries on the master's work. One of these is the first narrative history in China, the *Tso Chuan*, *The Commentary of Tso* (or, *Zuo Zhuan, The Commentary of Zuo*), which is a didactic narrative of political events focusing on the Confucian ideals of virtue, proper ritual behavior, and conformity to the mandate of heaven. Although in these works there is a sense of the supernatural at work in the universe, they are primarily secular historical and philosophical works. The *Tsu Chuan* is a book describing cause and effect in human history. Other histories of this time include the *Kuo yü, Conversations from the States*, a narrative, didactic story; *Chan-kuo ts'e, Intrigues of the Warring States*; and most significantly, *Shih Chi, Records of the Historian.*

SIMA QIAN (145–90 BCE)

One of the great accomplishments of ancient Chinese thought was in the field of history, which is revealed by an examination of *Shih Chi, Records of the Historian*, by Sima Qian (Ssu-Ma Ch'ien). Born in a town in the Yellow River valley, Sima Qian practiced Confucian filial piety, dutifully following in the footsteps of his father, who was the Grand Historian of the Han Dynasty in the latter second century BCE. Sima Qian believed the *Shih Chi* was a continuation of Confucius's *Spring and Autumn Annals*: it was a didactic, fact-based narrative that focused on political events, especially the actions of Chinese rulers. He divided the book into five sections: annals of dynasties; chronological tables; treatises on human as well as scientific and social scientific topics; the aristocracy; and biographies. Sima Qian was noteworthy in his focus on the individual in his history. He "first saw," writes Burton Watson, "that the life story of the individual could be given independent existence and could reveal in its form and development a meaning larger than that of single deeds" (Watson, 1958, p. 126). These life stories were told in the fashion of a struggle for virtue, which Sina Qian thought was the basic theme of all human existence. His goal was "to discover exactly where one stood in this process of growth and decay and how to determine what ritual steps, what policies, what moral attitudes were most appropriate to such a period in history" (Watson, 1962, p. 98).

Indeed, the Chinese Grand Historian was the Grand Astronomer and Astrologer as well: the keeper of records of time, the reign of an emperor, and the cyclical patterns of the planets and stars in the heavens and how human existence mirrors those patterns.

CONFUCIANISM AND THE HAN DYNASTY

As reflected in Sima Qian's *Shih Chi*, Confucianism was the dominant philosophy of the Han Dynasty, 202 BCE–220 CE. Confucian intellectuals such as Sima Qian believed that the state, humans, and heaven work together in harmony. Indeed, they believed that Han emperors represented this trinity. Han emperors, Confucian thinkers taught, had the mandate of heaven because of their moral leadership. Confucian historians tended to see the mandate operating in a cyclical pattern wherein the beginning of a dynasty was virtuous but the end of a dynasty was corrupt. Politics involved the moral order of the universe and how individual actions reflected this order. Confucian morality, ritual propriety, and education were essential in Han culture. Han emperors ruled by means of a well-trained Confucian bureaucracy trained in Confucian schools and a central university. Confucian sages had the ears of Han emperors. Entrance into the bureaucracy was accomplished by means of extensive tests on Confucian writings and theory. This continued until the end of Confucian influence in China around 1900. Indeed, for over two millennia, Confucianism helped maintain the order and status quo of Han and subsequent dynasties in China.

CHINESE ASTRONOMY AND MATHEMATICS

Chinese astronomy featured the development of lunar and solar calendars, star catalogues and charts, records of astronomical observations, the creation of sundials, and the philosophy and observations of astrology. Chinese astronomers observed comets, sunspots, novae, and meteors—although accurate explanations eluded them. The Chinese, like ancient cosmologists in Europe, western Asia, and northern Africa, lived their lives according to the patterns of the stars and movements of the planets, moon, and sun. Astronomical movements could predict the seasons, as well as when floods would occur, seeds should be planted, and harvest might be done.

Here is one Han Dynasty poet's versified conception of the nightly observations of celestial phenomena:

> Brilliant is the sky o'er-head,
> Splendid there the stars are spread.
> Grand the sun and moon move on,
> All through you, one man alone.
> The sun and moon move in their orbits;
> The stars keep to their paths;
> The four seasons observe their turns,
> And all the people are truly good.

> Oh! such music as I speak of Corresponds to the power of Heaven,
> Leading to worth and excellence; And all listen to it. (Trans. Legge, 1879)

As early as the beginning of the first millennium BCE, the Chinese constructed palaces according to celestial models. An example is the Mingtang, the "Hall of Numinous Brightness," built during the Zhou Dynasty. Here, according to one ancient source, "the Mingtang is the *taimiao* (Grand Ancestral Temple) of the Son of Heaven, wherein the Emperor sacrifices to his ancestors in the company of the Supernal Lord. . . . The Mingtang is that wherein the unification of all things by Heaven and Earth is manifest." Another ancient source described "the yin and yang of the *Mingtang* are the means by which the kingly ruler responds to Heaven" (Pankenier).

Clearly the mandate of heaven by which the ancient Chinese emperor ruled was considered more than just empty words but a reflection of the great celestial patterns of the ages. Indeed, Chinese emperors convinced their subjects—or at least the astronomers of the ancient world—that their rule, their power, was aligned to the rule and power of the heavens. The imperial astronomer of the Han Dynasty, Feng Hsiang Shih, wrote that by the will of the emperor the imperial astronomer "distinguishes," "orders" the planets and stars, "so that he can make a general plan of the state of the heavens. He takes observations of the sun at the winter and summer solstices, and of the moon at the spring and autumn equinoxes, in order to determine the succession of the four seasons" (Needham, vol. 3, p. 190). Imperial astrologers were charged by the emperor with studying the stars, planets, sun, and moon in order to predict events on earth. Other astronomical officials in the Han Dynasty kept track of meteorology and time. Calendars were frequently made to mark the ascension of a ruler, since emperors, having the mandate of heaven, required a calendar marking the specifics of the movement of the heavens to fit the specifics of their rule.

Unnamed Chinese astronomers were at work in the second millennium, creating the earliest Chinese calendar, which dated to 1300 BCE—it was an extremely accurate solar calendar. Important Chinese astronomers included Gan De (Kan Te, 400–340 BCE), who wrote *Treatise on Jupiter.* Regarding Jupiter, "every 12 years Jupiter returns to the same position in the sky; every 370 days it disappears in the fire of the Sun in the evening to the west, 30 days later it reappears in the morning to the east" (O'Connor and Robertson "Gan De"). Gan De also wrote *Thien Wen Hsing Chan, Astronomical Star Prognostication.* Chinese scholar Shih Shen in the fourth century BCE developed a catalogue of 800 stars in his book, *Thien Wen, Astronomy.* Wu Hsien also published a fourth-century star chart. One of the earliest calendars, devised in the fourth century, was the *Hsia Hsiao Cheng, The Lesser Annuary of the Hsia Dynasty.* This was something of a farmer's almanac, focusing on meteorology, stars, animals, and moon phases. A few centuries later appeared the *Yu Ling, Monthly Ordinances,* which examined months of the year and important ceremonies. First-century-BCE astronomer Luoxia Hong (130–70) created a calendar for the emperor Wu-ti that was based on 29- or 30-day months with an intercalary month added every seven years. He constructed an armillary sphere, perhaps the first one every made, which allowed the emperor (and another observer) to visualize the equatorial line; lines of the tropics, arctic, and antarctic; and the ecliptic through which the zodiac moves. About two hundred years later, Han astronomer Zhang Heng (78–139 CE) created an even more sophisticated armillary sphere, one powered by water. He described the geocentric universe thus: "The sky is like a hen's egg, and is as round as a crossbow pellet, the Earth is like the yolk of the egg, lying alone at the centre. The sky is large and the Earth small" (O'Connor and Robertson, "Zheng Heng"). A Confucian scholar, Zhang Heng served as imperial chief astrologer and reformed the calendar in 123 CE. He created a primitive seismograph that registered an earthquake in 138 CE. Another Han astronomer, Liu Hong (129–210), published a highly accurate lunar calendar in 187. Cai Yong (132–192 CE) was a versatile polymath of the Han Dynasty. Chinese astrology was heavily based on the lunar calendar. Calendars were revised in subsequent centuries as emperors came and went. In the fifth century CE, mathematician and astronomer Zu Chongzhi (429–501) created an extremely accurate solar calendar that required an intercalary month every 144 years.

During the first millennium BCE, the use of counting boards and a decimal system developed in China. Chinese mathematicians understood the Pythagorean theorem, the concept of zero, and also quadratic equations. The first great Chinese mathematical works included the *Nine Chapters on the Mathematical Art*, or *Jiuzhang suanshu*, which contains math problems, land surveys, and measures for grain. It was first composed between the tenth and second centuries BCE. Many commentaries, such as by Xu Yue (160–227) and Liu Hui (220–280), were written during the Han Dynasty and afterward. Liu Hui, in the third century CE, gained an approximate knowledge of π. Another early work was the *Chou Pei Suan Ching, The Arithmetical Classic of the Gnomon and the Circular Paths of Heave*n. This book discussed the Pythagorean theorem, the gnomon, circle, square, heights, distances, measuring the sun's shadow, latitudes, the sun's diameter, right-angled triangles, fractions, multiplication, division, common denominators, and square roots. Other Chinese mathematical works included *Book on Numbers and Compilations* (the *Suan Shu Shu*), written about 200 BCE. In the fifth century CE, an obscure Chinese mathematician, Sun Zi, published *Mathematical Manual*, which included an intricate description of how to use a counting board for multiplication.

CHINESE MEDICINE

Ancient Chinese medicine had a lot of similarities to ancient Greek medicine but probably only because of parallel development rather than actual cultural contact. Whereas the Greeks relied on the Hippocratic method and the diagnosis of illness based on the imbalance of the four humors (blood, yellow bile, black bile, and phlegm) in illness, the Chinese relied on the balance of yin-yang, in which there were three circulation tracks of yin and three of yang over which disease coursed over time. Such ideas were promoted by Bian Que (c. 401–310 BCE), who, according to the historian Sina Qian, conceived of four ways of diagnosing a patient: listening to the patient's internal organs, such as the heart; detecting odors given off by the patient; discovering the patient's medical history; and using the sense of touch in examination, such as feeling the patient's pulse. He advocated the use of acupuncture, cauterization, drugs, plasters, massage, and exercise for therapy. Bian Que wrote a standard work, *Nei Ching, the Yellow Emperor's Manual of Corporeal.* It was the basic manual from which most subsequent manuals of Chinese medicine derived. Much of the book was in dialogue form. Later medical treatises, such as *Su Wen, The Plan Questions and Answers,* and *Ling Shu, The Vital Axis,* relied on the work of Bian Que. The Chinese also believed, as did the Hippocratic school, that meteorological events such as wind, rain, heat, cold, and arid conditions affected disease.

Many significant Chinese physicians practiced and wrote during the Han Dynasty. For example, Zhang Zhongjing (Zhang Ji, 150–219), an expert of fever, wrote *Treatise on Febrile Diseases.* Hua Tho (190–265) practiced "medical gymnastics, massage and physiotherapy" (Needham, vol. 6, p. 52). He was reputedly one of the first physicians to use an alcohol-based anesthesia. After the Han Dynasty, in the third-century Western Jin Dynasty, Wang Shuhe (180–270) was a commentator on the Han Dynasty physician Zhang Zhongjing's *On Cold Damage.* Wang Shuhe was an imperial physician who wrote *Mai Jing,* t*he Pulse Classic,* which discussed numerous diseases and how the patient's pulse could gauge prognosis. He analyzed the pulse in internal and external organs, and conceived of the yin-yang of the pulse.

Traditional Chinese medicine involved superstition, philosophy, alchemy, and acupuncture and often was associated with Taoism. Acupuncture began in China in the mid-first millennium BCE. The first catalog of the loci, or places of puncture on the body, appeared in the *Divine Pivot, Ling shu.* There were initially 360 such loci on the body, to correspond with the number of bones known in the human body and number of days in the solar year. There were dozens and dozens of books written on acupuncture in ancient and medieval China. One of the most complete, written by Huangfu Mi (215–282), was *Huang ti chia i ching,* t*he Systematic Classic of Acupuncture & Moxibustion,* in 280 CE.

Toward the beginning of the Common Era, Taoism became less philosophical, more mystical; less spiritual, more material. Many adherents believed that the Tao could be gained by physical, magical processes, which led to an alchemical approach to Taoism. The second-century-BCE Taoist tract *Huainan-tzu* presented a Taoist interpretation of the origins of the universe:

> *Yin-yang* resulted from the concentration of the essence of heaven and earth. The essence of *Yin-yang* by its concentration formed the four seasons, and the essence of the four seasons by its distribution formed the multitude of things. The accumulation of the hot elements in *Yang* originated fire, the essence of which became the sun. The accumulation of the cold elements in *Yin* created water, the essence of which became the moon. By the interaction of the sun and moon, the heavenly bodies were produced. While the heaven received the stars and planets, the earth received water and dust. (Leicester, p. 55)

Alchemists and philosophers of ancient China believed there were five elements: earth, fire, water, metal, and wood. The five-element theory founder or systematizer Tsou Yen (Zou Yan), who lived in the late fourth or early third century BCE, was an alchemist and philosopher. Sina Qian, in *Records of the Historian*, wrote of Tsou Yen:

> He examined deeply into the phenomena of the increase and decrease of the Yin and the Yang. . . . First he had to examine small objects, and from these he drew conclusions . . . about large ones, until he reached what was without limit. . . . He began by classifying China's notable mountains, great rivers and connecting valleys; its birds and beasts; the fruitfulness of its waters and soils, and its rare products; and from this extended . . . his survey to what is beyond the seas, and men are unable to observe. . . . He made citations of the revolutions and transmutations of the Five Powers . . . , arranging them until each found its proper place. (Needham, vol. 2, pp. 232–33)

Yin-yang and the five elements provided the basis for reality, according to Tsou Yen and other so-called naturalists of ancient China.

GE HONG (283–343)

Ge Hong lived in the fourth century CE, during a time of warring states in the wake of the Han Dynasty. An alchemist, he had a macrobiotic approach, in which "with the aid of botany, zoology, mineralogy and alchemy, it is possible to prepare drugs or elixirs which will prolong life, giving longevity . . . or immortality" (Needham, vol. 6, p. 58). He sought to reconcile Taoism and Confucianism. A soldier and military commander in his youth, in time he became a recluse and scholar. He wrote two books, one on Taoism, *Inner Chapters of the Master who Embraces Simplicity*, and a counterpart on Confucianism, *Outer Chapters of the Master who Embraces Simplicity*. In the former, he wrote on how a person can achieve immortality through Taoist contemplation, merging the self with the *Tao*, or essence in the universe. He believed there was a vital energy within a person that could stave off illness. To achieve immortality, however, a person must ingest alchemical medicines, such as those based on cinnabar, the source of mercury, or gold. His philosophy was that gold would never disintegrate, so if it was taken within the human body, neither would the human body disintegrate. To prepare for such immortality, he believed one must study and practice the tenets of Confucianism.

Chinese alchemists such as Ge Hong sought the means of the transmutation of common metals to gold, or cinnabar and gold to medicines, not for material wealth but, rather, to ward off sickness and extend life or to develop supernatural qualities. Gold-infused medicine was the means by which a person's body could be strengthened, be made immortal.

SOUTHEAST ASIA

Ancient society in Southeast Asia was heavily influenced by other cultures, whether Chinese to the north or Indian to the west. Ancient Funan, covering most of Southeast Asia centered on the Mekong River, developed rice agriculture, extensive trading, complex political structures, and writing in the first century BCE. These ancient peoples had a port situated at the Straits of Malacca that served as a depot for trade from the west, such as India and even Rome, and from the east, such as Indonesia and the Philippines. Ancient Cambodians at the beginning of the Christian era had writing, architecture, politics, and astronomical science influenced by Indian culture. Cambodian people had trade links with India and the Roman Empire as well, perhaps by means of the Indians.

The people in the scattered islands of Polynesia in the South Pacific several thousand years ago had developed a ship technology that allowed for extensive movement, trade, and exploration. They sailed in double canoes, two long and narrow canoes lashed together with a temporary platform situated between them. It is possible that they got this idea from American Indians of the Northwest American coast, who had developed identical marine technology and knew how to brave the waters of the North Pacific. The prevailing winds and currents of the waters off North America proceeded southwest toward Polynesia. Some scholars believe that the Polynesian islands were originally settled by North American Indians.

FURTHER READING

The Analects of Confucius, trans. Arthur Waley (Abingdon, UK: Routledge, 2005).

W. G. Beasley and E. G. Pulleyblank, eds., *Historians of China and Japan* (London: Oxford University Press, 1961).

David Chandler, *A History of Cambodia*, 4th ed. (New York: Routledge, 2018).

The Chinese Classics: With a Translation, Critical and Exegetical Notes, Prolegomena, and Copious Indexes, vol. 4, part 1: *The First Part of the She-King, or the Lessons from the States; and the Prolegomena* (Hong Kong: Lane, Crawford & Co., 1871).

Christopher Cullen, "The Suan shu shu 'Writings on Reckoning,'" Neeham Research Institute, http://www.nri.cam.ac.uk/suanshushu.html.

John K. Fairbank, Edwin O. Reischauer, and Albert M. Craig, *East Asia: Tradition and Transformation* (Boston: Houghton Mifflin, 1978).

Thor Heyerdahl, *Early Man and the Ocean: A Search for the Beginning of Navigation and Seaborne Civilizations* (New York: Vintage Books, 1980).

Keith Knapp, "Ge Hong (Ko Hung, 283–343 C. E.)," *Internet Encyclopedia of Philosophy*, https://www.iep.utm.edu/gehong.

Henry Leicester, *The Historical Background of Chemistry* (New York: Dover Books, 1971).

Frederick W. Mote, *Intellectual Foundations of China* (New York: Knopf, 1971).

Joseph Needham, *Science and Civilisation in China, Volume 2: History of Scientific Thought* (Cambridge: Cambridge University Press, 1956).

Joseph Needham, *Science and Civilisation in China, Volume 3: Mathematics and the Sciences of the Heavens and the Earth* (Cambridge: Cambridge University Press, 1959).

Joseph Needham, *Science and Civilisation in China, Volume 4: Physics and Physical Technology, Part 3: Civil Engineering and Nautics* (Cambridge: Cambridge University Press, 1971).

Joseph Needham, *Science and Civilisation in China, Volume 6: Biology and Biological Technology, Part 4: Medicine* (Cambridge: Cambridge University Press, 2000).

O. Neugebauer, *The Exact Sciences in Antiquity* (New York: Dover Books, 1969).

J. J. O'Connor, and E. F. Robertson, "Gan De," http://mathshistory.st-andrews.ac.uk/Biographies/Gan_De.html.

J. J. O'Connor, and E. F. Robertson, "Liu Hong," http://mathshistory.st-andrews.ac.uk/Biographies/Liu_Hong.html.

J. J. O'Connor, and E. F. Robertson, "Luoxia Hong," http://mathshistory.st-andrews.ac.uk/Biographies/Luoxia_Hong.html.

J. J. O'Connor, and E. F. Robertson, "Sun Zi," http://mathshistory.st-andrews.ac.uk/Biographies/Sun_Zi.html.

J. J. O'Connor, and E. F. Robertson, "Zhang Heng," http://mathshistory.st-andrews.ac.uk/Biographies/Zhang_Heng.html.

J. J. O'Connor, and E. F. Robertson, "Zu Chongzhi," http://mathshistory.st-andrews.ac.uk/Biographies/Zu_Chongzhi.html.

David W. Pankenier, "Cosmic Capitals and Numinous Precincts in Early China," *Journal of Cosmology* 9 (2010): 2030–40, http://journalofcosmology.com/AncientAstronomy100.html.

Vivian Robson, *The Fixed Stars and Constellations in Astrology* (Whitefish, MT: Kessinger Publishing, 2003).

The Sacred Books of China: The Texts of Confucianism: Part 1: The Shû King, trans. James Legge (Oxford: Clarendon Press, 1879).

The Sacred Books of China: The Texts of Taoism, trans. James Legge (Oxford: Clarendon Press, 1891).

Qian Sima, *Records of the Grand Historian*, trans. Burton Watson (New York: Columbia University Press, 2011).

The Tso chuan: Selections from China's Oldest Narrative History, trans. Burton Watson (New York: Columbia University Press, 1989).

[Wang Shuhe,] *The Pulse Classic*, World Digital Library, Library of Congress, https://www.wdl.org/en/item/18717.

Christopher Warnock, *The Mansions of the Moon: A Lunar Zodiac for Astrology and Magic* (Morrisville, NC: Lulu Press, 2019).

Burton Watson, *Early Chinese Literature* (New York: Columbia University Press, 1962).

Burton Watson, *Ssu-Ma Ch'ien: Grand Historian of China* (New York: Columbia University Press, 1958).

PART 4

The Ancient Greeks

8

Ionian Greek Science (750–300 BCE)

The Ionian school of thought featured the first sophisticated scientist-philosophers of the ancient world, most of whom hailed from Miletus or other small cities of western Turkey, which the Greeks called Ionia. It was the Homeric influence on the Ionian refugees from Greece that set in motion the pattern of thought central to this region of the Mediterranean. Beginning in the twelfth century BCE, the remnants of Mycenaean towns and cities fled at the Dorian approach, migrating east to the extreme peninsulas of the mainland, the islands of the Aegean, and the western coast of Turkey. These people, subsequently called the Ionians, founded Ionian city-states such as Athens, Chios, Samos, Halicarnassus, Cos, Colophon, Ephesus, and Miletus. Tradition has it that the original inhabitants of Miletus were Carians, the natives of Anatolia or Asia Minor (present-day Turkey). Invaders from Athens, however, themselves responding to invasions of the Dorians further west, crossed the Aegean Sea and attacked the Carians of Miletus. The victorious Athenians established Miletus as one of twelve cities of this region inhabited by the Greeks from eastern Greece and the Aegean. These twelve Ionian cities became collectively called Ionia. Ionian Greek city-states had a reputation throughout the ancient world for thought and culture, as opposed to the Dorian states of western Greece. The Ionians and Dorians in coming centuries always considered themselves different. They had similar traditions and language, they worshipped similar gods and goddesses, but the level and sophistication of their respective cultures differed. The Dorians were more militaristic, and their cities focused on war and defense rather than art, poetry, and science. The Ionians, on the other hand, were devoted to thought, culture, investigating nature, and expressing their ideas. The Ionians in time became the leaders of Greek science during the Archaic, Classical, and Hellenistic Ages of Greece. The cities of Athens, Miletus, Cos, and Chios became centers of philosophy and science. Homer, Thales, Anaxagoras, Anaximander, Hippocrates, Socrates, and Plato were representative of Ionian philosophers, scientists, and physicians. Sixth-century-BCE Miletus was one of the most dynamic centers of speculative and scientific thought in the history of humankind. Diogenes Laertius's *Lives of the Philosophers* identifies the

Ionian school of thought, which in the sixth century was really the Milesian School. The first great scientists and philosophers of ancient Greece came from Miletus. These included Thales, Anaximander, Anaximenes, and Hecataeus. Among Ionian city-states, Miletus was the early leader.

Homer's works, though epic poetry, create a worldview of an inhabitable world known by men who are thinkers, doers, creators, and individualists. Ionian thinkers embraced Homer's subtle call to find explanations for all that *is*. They replaced Homer's explanation of natural and human phenomena according to fate, the will of the gods, and human interaction with the divine with more realistic explanations of physical phenomena exclusive of direct divine action and intervention.

Plutarch, the Greek philosopher and historian of the Pax Romana, believed that the great intellectual accomplishments by Greeks and Romans of the previous millennia had come about by their working in the shadow of Homer. He wrote, "Homer appears to have been at home in the whole sphere and art of logic, and to have supplied many incentives, and as it were seeds of all kinds of thought and action to his posterity, not to poets alone, but to the authors of historical and scientific works." Homer had "sound knowledge on matters of fact." Plutarch's essay on *The Life and Poetry of Homer* provides a full accounting of how Homer's *Iliad* and *Odyssey* influenced Greek and Roman thought. Homer, Plutarch argued, used "poetic art" to inspire in his readers "a certain intellectual pleasure" that resulted in an increased desire to "find the truth," that is, the "nature of reality," as Homer divided it into "natural, ethical, and dialectical" investigations (revised, Goodwin).

Plutarch, who studied Plato and Aristotle in depth, believed (as he wrote in the *Life and Poetry of Homer*) that Homer's poems influenced the Athenians in their understanding of how human actions relate to fate. Homer argued

> that not all things happen by fate, but some things are in the power of men, the choice of whom is free. The same man in a way acts as he desires and falls into what he does not desire. And this point of view he has clearly expounded in many places, as in the beginning of each of his poems: in the "Iliad" saying the wrath of Achilles was the cause of the destruction of the Greeks and that the will of Zeus was fulfilled; in the "Odyssey" that the comrades of Odysseus went to their destruction by their own folly. For they had offended by touching the sacred oxen of the Sun, although they could have abstained from doing so. (Revised, Goodwin)

The Greek philosophers before and during Aristotle's time were the first political scientists, though Plutarch believed that a poet, not a scientist, inspired them. Although there is no Homeric Greek word for "law," Plutarch argued, in *The Life and Poetry of Homer*, that the word *righteousness*, which means "to distribute well," is the basis for law, "because it distributes . . . equal parts to all or to each according to his worth," which was a Homeric ideal. Homer believed that the great lawgiver was Zeus and that humans must necessarily follow Zeus's will. Homer also, Plutarch wrote, conceived of the different parts of a "civil polity." Homer believed that "in every city it is sanctioned by the law that there is to be a meeting of a council to consider before the popular assembly is called together." The shield of Achilles, in which the god Hephaestus conjectured two cities, included one that "is ruled democratically, since they have no leader, yet all by their own will conduct themselves according to the laws" (revised, Goodwin).

If we assume the validity of Plutarch's appraisal of Homer's significance, it can be argued then that the pre-Socratic philosophers, including the Ionians, sought knowledge by means of Homer. Thales, for example, sought the *first cause*; he rejected supernatural cause and arbitrariness in the universe for universal physical causes.

THALES (C. 626–545 BCE)

Ancient Greek writers such as Plutarch identified Thales of Miletus as the first great scientist of the Mediterranean world. Thales was one of the famed Seven Sages of the Hellenic world; of the seven, Thales was the natural philosopher. He hailed from Miletus, a natural place for such questioning and answering to occur. Questioning is often a product of uncertainty. Sixth-century Miletus was a city at war variously against outsiders such as the Lydians. It was a busy port city situated on the Meander River and the Gulf of Latmos. Milesian ships traveled about the Aegean and beyond, particularly north through the Hellespont and Bosporus to the Black (Euxine) Sea. Milesian trading colonies dotted the shores of the Black Sea. Miletus was also a restive place domestically, as economic classes fought for control. Tyrants ruled sixth-century Miletus.

About the time of the birth of Thales, Miletus began to mint coins made of electrum, which yielded efficient and increased trade. The Milesians got the idea from the Lydian city of Sardis, located about one hundred miles to the northeast. The Lydians were ruled by kings, the most notable being Croesus. According to Herodotus, Croesus made Sardis a center of philosophy—exactly how is not clear. It does not seem unreasonable, however, that Croesus would have employed the services of Milesian scientists. Croesus recruited Thales according to Herodotus; the philosopher diverted the Halys River so that the Lydian army could get across during a time of war. The Lydians themselves appear to have produced few scientists—a notable exception being the fifth-century historian Xanthus—hence Croesus's reliance upon Milesian scientists.

Diogenes Laertius, in *Lives of the Philosophers*, related one tradition that Thales was a native Phoenician who immigrated to Miletus; if a native of Phoenicia, this explains his interest and skill at celestial navigation. Other traditions indicate that he learned from the Chaldeans of Mesopotamia as well as from Egyptian scientists; some sources make him an Egyptian immigrant from Phoenician colonies in the western Mediterranean. Ionians such as Thales combined Babylonian and Egyptian knowledge of astronomy and mathematics with the Greek speculative habit of questioning. Legend has it that he predicted a solar eclipse in 585 BCE. Plutarch recorded in his *Life of Solon* that the two men knew each other, and that Thales tried to convince Solon that a philosopher must not have distractions in the way of wife and children, a view of which Plutarch was highly critical.

Thales, like Socrates, never wrote anything, hence, to know his thought one must rely on the reports of other ancient writers. Diogenes Laertius, who had a twofold division of Hellenic philosophy, the Ionian and the Italian, lauded Thales as the founder of the former by means of his influence on Anaximander. The initial element, Thales believed, is water, from which all else springs. Plutarch, who concluded in the *Life of Solon* that "Thales alone had raised philosophy above mere practice into speculation" (trans. Dryden), believed that Homer inspired Thales to this conclusion.

Diogenes Laertius also presented Thales with the title of first ancient astronomer because of two books he reputedly wrote: *On the Solstices* and *On the Equinoxes*. Thales was the first to identify Ursa Minor, Little Bear, the basis of sailing the Mediterranean by the North Star. Some Greeks believed Thales explained this in his book *Nautical Astronomy*. If Diogenes was correct, Thales measured the movement of the sun on the zodiac from "solstice to solstice," that is, its southern progression toward the Tropic of Capricorn from June to December and its northern progressions toward the Tropic of Cancer from December to June. From Thales's solar observations, he arrived at an estimate of the size of the sun to be 1/720 its diameter at any one time on the zodiac. Thales made similar studies of the moon's path of movement and relative size.

Diogenes Laertius ascribed to Thales the ability to measure the height of the pyramids of Egypt, to argue for the immortality of the soul, to identify souls in other natural phenomena, and to be the first to proclaim, "Know thyself." Proclus, a Neoplatonist living in the Later Roman Empire, believed that "Thales was the first to demonstrate that a circle is bisected by its diameter" as well as the first to demonstrate the "theorem . . . that when two straight lines intersect with one another the angles at the vertex are equal" (trans. Barnes). Aristotle claimed that Thales believed that water holds the earth in place, though he ignored the equally significant question of what holds water in place. Thales thought that water served as the primal element because it was a fluid that could congeal into a solid or be dispersed into a gas. The late Roman commentator Hippolytus wrote that Thales "said that some such thing as water is the generative principle of the universe, and its end—for that out of this, solidified and again dissolved, all things consist, and that all things are supported on it; from which also arise both earthquakes and changes of the winds and atmospheric movements, and that all things are both produced and are in a state of flux corresponding with the nature of the primary author of generation—and that the Deity is that which has neither beginning nor end" (trans. MacMahon).

Thales was clearly a Greek thinker of amazing breadth and originality. The ancient sources indicate that his predecessors in scientific inquiry, the Egyptians and Mesopotamians, influenced the scope and technique of his investigations. He was not the first to consider the origins of the universe, but perhaps he was the first to divorce such investigation from superstition. Indeed, he apparently refused to adopt the Mesopotamian (Chaldean) propensity to assign magical explanations to natural phenomena. Modern scholars tend to doubt that Thales had the knowledge and observational skills to predict an eclipse, yet Babylonian astronomers, from whom he apparently learned, were able to predict lunar eclipses. Besides, Herodotus noted that Thales predicted the *year* rather than the day or week of the solar eclipse, which makes the prediction a little less astonishing. Thales believed that the earth was a round arched disk, perhaps hollow underneath.

The ancients believed that Thales either discovered or learned from others (such as the Egyptians) the technique of using the gnomon, or sundial, to determine the equinoxes and solstices. Herodotus claimed that Thales followed the Egyptian solar calendar rather than the Greek lunar calendar.

In short, Thales was a thinker and scientist learned in the theories and techniques of Egyptian, Mesopotamian, and Phoenician philosophers, astronomers, and mathematicians. He is significant in the history of science for his ability to bring the work of his Asian and African predecessors to the Greeks. Thales tempered the superstition of the Chaldean and Egyptian priests with the rational mindset of the Ionian Greeks. Most important, Thales was not content to merely describe natural phenomena; rather, he sought the *cause* of things in the workings of nature. Hippolytus wrote that Thales, "having been occupied with an hypothesis and investigation concerning the stars, . . . looking towards heaven, alleging that he was carefully examining supernal objects, fell into a well; and a certain maid, by name Thratta, remarked of him derisively, that while intent on beholding things in heaven, he did not know, what was at his feet" (trans. MacMahon).

Thales inspired and taught other Milesians, notably Anaximander and Anaximenes. All three were more significant for the questions that they asked than for the answers they attempted. All three speculated on the ultimate cause in a temporal sense and the limitless being in the transcendent sense. Each man assumed that an ultimate reality must exist but they differed on its nature and substance. Thales thought that water forms the essence; for Anaximander it is the infinite; for Anaximenes it is air. Such explanations appear simple today. But the questions that produce the search for answers are far from simple. What is the cause of all things? Whence comes the cause? What is its nature? Is there existence prior to the cause? Is time limited? Why is there change, and how does an effect yield a further cause? Are we humans brought forth by the cause? If so, how? The questions were unending; so, too, were the answers.

Miletus was just north of Didyma, a shrine to the god Apollo, Apollo Didymaeus. Apollo was the god of healing, wisdom, and learning. The Milesians clearly took to heart the influence of their divine neighbor. One anecdote from the life of Thales informs us of the special relationship of the city with Apollo. Once fishermen brought up a golden tripod from the sea. Not sure what to do with the miraculous treasure, they sent to the oracle at Delphi to ask Apollo who deserved such treasure. Apollo said to give it "to the wisest." The fishermen assumed Apollo meant Thales, but Thales modestly disagreed. After going the rounds to all of the Seven Sages, the tripod returned to Thales, who wisely sent it to Delphi, for only a god "is the wisest." This story will sound remarkably familiar to readers of Plato's *Apology*, in which Apollo calls Socrates the wisest, an honor that Socrates declines, saying that he knows, on the contrary, that he does not know. Thales, in questioning and searching for answers, was a precursor to Socrates, just as Miletus, the sixth-century center of science, anticipated Athens, the dominant center of science in the fifth century BCE.

ANAXIMANDER (C. 610–546 BCE)

Anaximander was the student of Thales, the son of Praxiades of Miletus, and a leader in the early development of the Ionian school of thought. He continued his teacher's search to discover the one source of all things, be it material or spiritual. Thales had argued for the origin of all things in water; Anaximander sought the unity of all things in the uncreated, unlimited infinite (*apeiron*). Anaximander engaged in the contradictory pursuit of imposing limits and definitions on the limitless that by its very nature is indefinable. Yet he was the first ancient scientist to try to explain all phenomena according to a single principle. The infinite is unknowable, anonymous, absolute, without beginning or end.

Anaximander reputedly wrote a book, *On Nature*, but nothing survives from it except by paraphrase and a quote or two found in the commentators of later antiquity. Plutarch, Simplicius, Hippolytus, and Diogenes Laertius, among others, recorded the general outline of Anaximander's principle of the infinite and its necessary corollaries, that there has, is, and will always exist an unending cycle of creation and destruction, for what purpose is unclear. This infinite is an anonymous, impersonal force, not a personal deity, rather an amoral, eternal, absolute that is truly unrecognizable to humans. The infinite as a creative principle made the earth, which is a cylinder that rests in air surrounded by the moon, sun, planets, and stars, all made of fire. Parenthetically Anaximander would add that an infinite universe must accommodate an infinite number of worlds. We see the heavenly bodies because the fire protrudes through gaps or holes in the air. The heavenly bodies vary in size, though they are all equally distant from the earth. The sun, the source of warmth, acting upon water produces life. The first form of life, rather like a fish, crawled upon land and eventually ended up looking remarkably like humans. Not surprisingly, Anaximander rarely had fish for dinner.

Hippolytus provided a concise summary of Anaximander in his *Refutation of All Heresies*:

> Anaximander, then, was the hearer of Thales. Anaximander was son of Praxiadas, and a native of Miletus. This man said that the originating principle of existing things is a certain constitution of the Infinite, out of which the heavens are generated, and the worlds therein; and that this principle is eternal and undecaying, and comprising all the worlds. And he speaks of time as something of limited generation, and subsistence, and destruction. This person declared the Infinite to be an originating principle and element of existing things, being the first to employ such a denomination of the originating principle. But, moreover, he asserted that there is an eternal motion, by the agency of which it happens that the heavens are generated; but that the earth is poised aloft, upheld by nothing, continuing (so) on account of its equal distance from all (the heavenly bodies); and that the figure of it is curved, circular, similar to a column of stone. And one of the surfaces we tread upon, but the other is opposite. And that the stars are a circle of fire, separated from the fire which is in the vicinity of the world, and encompassed by air. And that certain atmospheric exhalations arise in places where the stars shine; wherefore, also, when these exhalations are obstructed, that eclipses take place. And that the moon sometimes appears full and sometimes waning, according to the obstruction or opening of its (orbital) paths. But that the circle of the sun is twenty-seven times larger than the moon, and that the sun is situated in the highest (quarter of the firmament); whereas the orbs of the fixed stars in the lowest. And that animals are produced (in moisture) by evaporation from the sun. And that man was, originally, similar to a different animal, that is, a fish. And that winds are caused by the separation of very rarified exhalations of the atmosphere, and by their motion after they have been condensed. And that rain arises from earth's giving back (the vapours which it receives) from the (clouds) under the sun. And that there are flashes of lightning when the wind coming down severs the clouds. (Trans. MacMahon)

The ancients claimed for Anaximander other accomplishments. The geographer Agathemerus wrote that Anaximander was the first to draw a map of the earth, which must have been on a long and narrow parchment. Hecataeus reputedly used Anaximander's map as the basis for his own. Diogenes Laertius recorded that Anaximander was the first to introduce to the Greeks the sundial, or gnomon, with which to gauge time by day and to approximate the summer and winter solstices and the fall and spring equinoxes. Herodotus, however, assigned these scientific accomplishments to the Mesopotamians, from whom, perhaps, Anaximander took his own ideas. Many of Anaximander's ideas were carried forward by his disciple, fellow Milesian Anaximenes, who sought the origins and nature of existence in air.

The climate of the Aegean Sea influenced Greek views of the gods and nature. Boreas was the source of the prevailing north winds—the Etesian winds—that blew from May to October. The Zephyr, the west wind that blew in like a stampeding horse, was husband to the goddess Iris, the rainbow. The Greek western shores of the Aegean were drier than the Turkish eastern shores of the Aegean; the North Aegean was cooler than the South Aegean. The first Greeks to try to explain meteorological phenomena included Anaximander and his student Anaximenes. Anaximander believed that wind blowing against clouds caused lightning; the winds themselves derived from air. Rainbows, notwithstanding that they are caused by vapor in sunlight, were considered a clear sign of a coming storm. Anaximenes had a more advanced view: He explained hail and snow as rainwater cooling and solidifying as it falls to earth, and rainbows as caused by sunlight hitting water vapor.

ANAXIMENES (C. 586–526 BCE)

Anaximenes of Miletus was the student of Anaximander and the teacher of Anaxagoras. He was an important figure in the development of the Ionian school because he continued the focus of inquiry established by his predecessors Thales and Anaximander. Anaximenes, like his teacher, believed that there was a constant indivisible source of all being, all that existed. Whereas Thales believed water was the primal element and Anaximander thought it was the infinite, Anaximenes pointed to air as the source of all things. Air is always changing, always in movement, he argued; its condition of relative heat or cold generates other founding substances such as water, fire, and earth. Thales and Anaximenes were the first Greek students of chemistry.

Like the work of his predecessors, nothing survives from Anaximenes's own pen. Later commentators, notably Hippolytus, Plutarch, and Aristotle, summarized and assessed his beliefs. Anaximenes believed that the earth, planets, sun, and moon were literally "suspended in mid-air"—all heavenly bodies were round, flat disks that ride upon the all-encompassing air. The earth existed first, moisture from which became rarified and produced the bright, fiery heavenly bodies. The planets and sun orbited about the flat, inhabited top of the earth, but they never passed underneath. The sun was sometimes hidden (night) because it was obscured by highlands to the west. Anaximenes understood the planets ("wanderers") to behave differently from the stars; the former were moved about by the air, and the latter were fixed to the vault of heaven. Anaximenes also believed that the stars were of much greater distance from the earth. The sun gave off heat because of its relatively close proximity to the earth.

The late Roman commentator Hippolytus, again, provided a concise summary of Anaximenes' beliefs:

> Anaximenes, who himself was also a native of Miletus, and son of Eurystratus, affirmed that the originating principle is infinite air, out of which are generated things existing, those which have existed, and those that will be, as well as gods and divine (entities), and that the rest arise from the offspring of this. But that there is such a species of air, when it is most even, which is imperceptible to vision, but capable of being manifested by cold and heat, and moisture and motion, and that it is continually in motion; for that whatsoever things undergo alteration, do not change if there is not motion. For that it presents a different appearance according as it is condensed and attenuated, for when it is dissolved into what is more attenuated that fire is produced, and that when it is moderately condensed again into air that a cloud is formed from the air by virtue of the contraction; but when condensed still more, water, (and) that when the condensation is carried still further, earth is formed; and when condensed to the very highest degree, stones. Wherefore, that the dominant principles of generation are contraries—namely, heat and cold. And that the expanded earth is wafted along upon the air, and in like manner both sun and moon and the rest of the stars; for all things being of the nature of fire, are wafted about through the expanse of space, upon the air. And that the stars are produced from earth by reason of the mist which arises from this *earth*; and when this is attenuated, that fire is produced, and that the stars consist of the fire which is being borne aloft. But also that there are terrestrial natures in the region of the stars carried on along with them. And he says that the stars do not move under the earth, as some have supposed, but around the earth, just as a cap is turned round our head; and that the sun is hid, not by being under the earth, but because covered by the higher portions of the earth, and on account of the greater distance that he is from us. But that the stars do not emit heat on account of the length of distance; and that the winds are produced when the condensed air, becoming rarified, is borne on; and that when collected and thickened still further, clouds are generated, and thus a change made into water. And that hail is produced when the water borne down from the clouds becomes congealed; and that snow is generated when these very clouds, being more moist, acquire congelation; and that lightning is caused when the clouds are parted by force of the winds; for when these are sundered there is produced a brilliant and fiery flash. And that a rainbow is produced by reason of the rays of the sun failing on the collected air. And that an earthquake takes place when the earth is altered into a larger (bulk) by heat and cold. These indeed, then, were the opinions of Anaximenes. (Trans. MacMahon)

The impact of heat and cold on matter plays an important role in the physical universe. Plutarch recorded Anaximenes's view that hard substances were cold and fluid ("slack") substances were hot. Anaximenes's argument that the cooling of air is the cause of much that exists laid the groundwork for later generations to seek the chemical properties of the basic substances of existence: earth, air, fire, water. Atomists especially appreciated Anaximenes's work. One can see his influence on the thinking of Lucretius the Epicurean. Anaximenes also had a direct impact on Anaxagoras and, through him, the philosophers and scientists of fifth-century Athens.

ANAXAGORAS (C. 510–428 BCE)

Anaxagoras, a native of the Ionian town of Clazomenae, was one of the last members of the Ionian school. He was influenced by his predecessors Thales, Anaximander, Anaximenes, and the shadowy Linus of Thebes, about whom Diogenes Laertius wrote that he "composed a poem describing the creation of the world, the causes of the sun and moon, and the growth of animals and plants." Anaxagoras called Linus's principle of unity, "mind" (*nous*).

Anaxagoras was a questioner who sought explanations for the ultimate origins of all phenomena. Anaxagoras spent his youth during a period of Persian occupation of western Turkey; after the Persian Wars he migrated to Athens at the invitation of Pericles the great general and leader. Plutarch claimed in his *Life of Pericles* that Anaxagoras, who was known by contemporaries as "mind," helped to form the philosophic mind of Pericles. Some historians believe that Pericles's concubine Aspasia, a native Milesian inclined toward philosophy, was influential in bringing Anaxagoras to Athens. Anaxagoras, the student of the Milesian Anaximenes, had a major impact on the development of scientific speculation in Athens, as his students included Archelaus, who was the teacher of Socrates.

Anaxagoras reputedly wrote a treatise, *Physics*, in which he stated his belief that all existence is encompassed by mind/*nous*, an infinite and transcendent presence of which humans partake and by which humans can know reality. Anaxagoras "was the first of the philosophers," wrote Plutarch, "who did not refer the first ordering of the world to fortune or chance, nor to necessity or compulsion, but to a pure, unadulterated intelligence, which in all other existing mixed and compound things acts as a principle of discrimination, and of combination of like with like" (trans. Dryden). One can clearly see the influence of Anaximander and Anaximenes on Anaxagoras's conception of infinity: "Air and aether," he wrote, forms a "surrounding mass" that "is infinite in quantity." The "seeds of all things" are infinite in number and variety, he claimed, anticipating the atomists; hence there are infinite worlds, phenomena, humans, forms of life, and so forth. The glue, as it were, that holds this infinite multiplicity together is mind, which insures that "all things" share "all things" (trans. Burnet).

Anaxagoras's concept of mind is a highly original idea similar to the Hebrew Yahweh, Akhenaten's Aten, and the Stoic Logos: a universal force, the first cause, the knower of all things yet unique in its singularity, its aloofness from human concerns. "Nous set in order all things," Anaxagoras wrote, "that were to be, and all things that were and are not now and that are." "Nous," he argued, "ever is." Likewise there is no true "coming into being and passing away; for nothing comes into being or passes away, but there is mingling and separation of things that are" (trans. Burnet).

As an astronomer and physicist, Anaxagoras made some startling discoveries. He was the first to understand the true nature of lunar and solar eclipses because he was the first to understand that the moon produced no light of its own, but merely reflected the light of the sun. Anaxagoras also had a remarkable interpretation of the origins of the universe. In the beginning, "all things were together" and nothing was "distinguishable." Mind made the whole to revolve; as it turned, various parts were "separated off by the force and swiftness." The ongoing revolution of all things grew ever larger, though mind knew it all. In time the air and aether were separated and formed two realms: "The rare is separated off from the dense, the warm from the cold, the light from the dark, and the dry from the moist." The earth derived from "the dense and the moist and the cold and the dark," while the aether derived from "the rare and the warm and the dry" and the light. Echoing Thales, Anaxagoras believed that the moist produced the earth, and the heavy "earth stones" went out toward the "further part of the aether," producing the stars, planets, sun and moon. The heavenly bodies, he agreed with his teacher Anaximenes, were stones of fire. Anaxagoras broke from most ancient astronomers in his belief that the order of the heavenly bodies orbiting the earth was the Moon, Sun, Mercury, Venus, Mars, Jupiter, and Saturn. Anaxagoras was a student of the climate of the Aegean, the prevailing north winds—the Etesian winds, especially from May to October, that cooled the north and left the south warmer and the West Aegean drier. Anaxagoras argued that "we call rainbow the reflection of the sun in the clouds. Now it is a sign of storm; for the water that flows round the cloud causes wind or pours down in rain." Anaxagoras also said, "With the rise of the Dogstar men begin the harvest; with its setting they begin to till the fields. It is hidden for forty days and nights" (trans. Burnet).

One can see a direct influence of the thought of Anaxagoras upon Socrates and Plato, hence all of subsequent Western philosophy and science. Plato continually reinforced the belief of Anaxagoras that "what appears is a vision of the unseen," and "from the weakness of our senses we are not able to judge the truth," Socrates, in the *Phaedo,* claimed that Anaxagoras set him on the path of seeking the essence of what is in mind. Plutarch wrote (*Natural Questions)* that Plato and Democritus, following Anaxagoras, believed that plants were animals "fixed in the earth." Anaxagoras claimed that humans were superior to animals "because we use our own experience and memory and wisdom and art" (trans. Burnet).

Anaximenes's student Anaxagoras, his student Archelaus, and Archelaus's student Socrates brought Ionian thought to Athens in the fifth century BCE. Anaxagoras's belief that the first cause and ongoing universal providence was mind (*nous*), which is incorporeal and transcendent, was shared by Socrates, who came to these beliefs by means of Archelaus. The Ionian philosophers at Athens countered Sophists, such as Protagoras, who advocated a completely human-centered view. The more speculative Ionians believed that reality was not necessarily what humans perceived nor how they perceived it. Our individual minds allow us to peer through the window of the soul to mind, though the window is often cloudy and our observations inexact. It takes rigorous years of thought and study to train the individual mind to be able to perceive qualities of Mind, or Good, as Socrates and Plato called the ultimate reality.

HECATAEUS (C. 550–476 BCE)

Another Ionian philosopher-scientist was Hecataeus of Miletus, the son of Hegesander, an Ionian Greek and a contemporary of Xenophanes who, like Xenophanes, wrote one of the first works critical of traditional Greek mythology. This was his *Genealogies*, in which Hecataeus tried to provide a rational account of myth and legend, including an account of his own supposed descent (after sixteen generations) from a supposed god. The gods were not eternal but created, he argued. Hecataeus sought to separate legend and fabulous stories from fact. Hecataeus, perhaps inspired by Miletus's role as a trading city, also wrote a book that he titled *Periegesis* ("journey"). The *Periegesis* was an account of the world, which Hecataeus divided into two continents, Europe and Asia (Africa being an extension of the latter). It was a historical and geographical account of Europe and Asia based loosely on the map of the world drawn by Anaximander around 550 BCE. Hecataeus was also involved in civic affairs, giving advice to the Milesians about their external affairs with the Persian Empire, the dominant force of the region circa 500 BCE.

Hecataeus's works did not survive time. Herodotus, who wrote his *Histories* about fifty years after Hecataeus, paraphrased a few passages from Hecataeus, and clearly relied on him for information on history and geography. Herodotus related that Hecataeus went to Egypt and talked to the priests of Amun at Thebes about their land, its people, and history. Like Herodotus, Hecataeus, lacking documents and the convenience of libraries, had to travel about, interviewing people, to arrive at the bases of knowledge with which he would write an account of the human journey (*Periegesis*) based on his own journeys. Herodotus used Hecataeus's work in his account of the Persian Wars. According to Diogenes Laertius, Hecataeus wrote a treatise *On the Egyptian Philosophy.*

HELLANICUS (491–405 BCE)

Little is known of the life of Hellanicus of Lesbos save that he came from the city of Mytilene, on the island of Lesbos, in the Aegean Sea. An Ionian Greek, his many works have not survived time. Dionysius of Halicarnassus, a historian of the first century BCE, paraphrased a few of his comments from his books on the Etruscans and Italy. Thucydides, in book 1 of the *Peloponnesian War*, commented on Hellanicus's *Attic History*, deriding it as inaccurate and incomplete. Hellanicus was apparently a scholar of note, penning dozens of books, mostly chronologies of earlier times. He was one of the first historians to try to make sense of human events according to a temporal process. He established chronologies accepted by ancient historians for the lives of heroes and the occasion of the war at Troy. Yet he was still bound to a mythological past, assuming that the likes of Heracles founded some of the first cities throughout the Mediterranean region. Hellanicus and other chronologists used the mythical wanderings of Heracles to identify the origins of place names in the Greek world.

HERODOTUS (490–430 BCE)

Herodotus of Halicarnassus is often called the "Father of History," but he could as well have been given the epithets "Father of Geography" and "Father of Anthropology." Herodotus sought to write a complete history of the world in which he intertwined natural history and human history. The result was his *Histories*, one of the first professed factual accounts of human experience.

Herodotus was a boy living in his native Halicarnassus when the Greeks defeated the invading Persians at Salamis and Plataea in 480 and 479 BCE. The war began in part because the Greek cities of western Turkey—Ionian Greek city-states such as Halicarnassus—revolted from Persian control. Darius I, king of Persia, put down the rebellion and decided to punish Greece as a whole. After he failed, his successor Xerxes tried and failed as well. Herodotus set himself the task of explaining the improbable Greek success. The Persians had a massive empire stretching from the Aegean Sea to the Indus River, from the Caspian Sea to Egypt. They vastly outnumbered the Greeks, perhaps ten to one. And yet the Greeks defeated the Persians on several occasions. Why? Herodotus spent years searching for an answer, traveling throughout the Persian Empire, going to Phrygia (in central Turkey), the Black Sea, the Tigris and Euphrates River valleys, the Persian Gulf, Palestine, and Egypt. He studied the Persians, Phrygians, Phoenicians, Egyptians, Arabians, Indians, and Scythians. Lacking many literary sources, he acquired information by personal observation and oral interviews. He called his work histories (Greek *historia*), meaning "researches" and "inquiries." Herodotus's *Histories* involved past kings, politics, and war. Moreover, Herodotus described the culture, society, institutions, beliefs, and legends of the peoples of Europe, Asia, and Africa—the world as he knew it. He portrayed human experience in light of natural history, in particular geography: landscape, climate, flora, fauna, rivers, deserts, and mountains. In the course of his *Histories*, Herodotus had much to say about science.

Herodotus wrote about the origins, extent, and characteristics of the Nile River in Africa. The Nile rose year after year at the summer solstice, when the rays of the sun are directly overhead at the Tropic of Cancer. At this time, the Nile inundated lower Egypt, leaving behind rich soil upon receding, and constantly altering the landscape of the delta, the silty mouth of the river. Herodotus, fascinated by the Nile delta, theorized that Egypt north of Memphis was comparatively new. The Nile was in a sense annually creating an altogether different Egypt.

Herodotus's experience was that most rivers rose in the spring. It was perplexing to discover that the Nile rose at the summer solstice, the beginning of summer. What explained the cause of its unusual summer flood? Herodotus disagreed with the three most prevalent theories. The first was that north winds blew against the northern-flowing waters, backing up the current and causing the river to spill over its banks. The second theory, advocated by Hecataeus, was that the Nile's source was the "river Ocean" encircling the earth south of Africa; the Nile's behavior was therefore dependent on the ocean. The third theory, that the Nile's rise was caused by melting snow, seemed absurd to Herodotus; for the further south one went toward the source of the Nile, the hotter it got. Where, then, was the snow? Herodotus's own view reflected the limitations of his astronomy. He believed that strong winds drove the sun before them, moving it away from the Nile's source, depriving the river of its heat, and therefore limiting evaporation and resulting in a rising river.

Herodotus was ambitious to describe the geography of all three continents: Europe, Asia, and Africa. He argued that those countries at the earth's extremes,

India, Arabia, Ethiopia, and the Tin Islands (Britain), produced the biggest, most fragrant, richest items; animals were larger, people were gigantic, gold was profuse, and spices caused a general fragrance in the air. At the western extreme of the earth, Libya, a land of sun and sand, Mount Atlas rose from the parched earth: "It is so high that according to report the top cannot be seen, because summer and winter it is never free of cloud. The natives . . . call it the Pillar of the Sky." At the opposite extreme of the earth in the east lay the Caucasus, "the longest and loftiest of all mountain ranges" (trans. Selincourt). The Caucasus bounds the western shores of the Caspian Sea. Unlike the Mediterranean, which flows into the Atlantic, which is (along with the Indian) the outer ocean of the world, the Caspian is a self-contained, inland sea. To Herodotus, the world-encircling ocean marks the *circumference* of a disk-shaped earth.

The *Histories* is encyclopedic in its description of the peoples that made up Herodotus's immediate past, the fifth century BCE. Perhaps because he was a traveler, a tourist in faraway countries, his reports of distant peoples and their strange customs have the tone of fascination, not condescension. Many Greeks of Herodotus's time and afterward referred to any non-Greek as a barbarian. But one does not find such bigotry in Herodotus. He was credulous, to be sure, but objective in his way too. An anthropologist (student of *anthropos*, "human"), Herodotus observed without preconception and recorded what he heard, saw, and felt. He did not purposefully mislead, even if his information was fanciful and often inaccurate. He sought to understand the Egyptians, Persians, Scythians, and others on their terms, not on his own.

Herodotus provided fascinating information about Egyptian astronomy and medicine; the strange creatures, such as the crocodile and hippopotamus, of Africa; the remote Scythians of northern Asia; and the flora and fauna of the diverse regions of the world. In the history of science, Herodotus made his mark on the study of the earth (geography), the study of human culture (anthropology), and inquiries into the human experience (history).

LEGACY

Ionian thought reached fruition through the efforts of Plato and Aristotle. Though the former was a speculative philosopher and the latter was an analytical scientist, Plato and Aristotle agreed in the essential spiritual core of the universe, that *being*, acting through the mind and soul, unites humans together into a great whole, that the First Cause is an unknown, unmoved mover that transcends all human thoughts and actions.

FURTHER READING

Jonathan Barnes, trans., *Early Greek Philosophy* (London: Penguin Books, 1987).

John Boardman, Jasper Griffin, and Oswyn Murray, *Oxford History of the Classical World* (Oxford: Oxford University Press, 1986).

John Burnet, *Early Greek Philosophy* (London: Adam and Charles Black, 1892).

Diogenes Laertius, *Lives of Eminent Philosophers*, trans. R. D. Hicks, 2 vols. (Cambridge, MA: Harvard University Press, 1931, 1938).

Michael Grant, *Readings in the Classical Historians* (New York: Scribner's, 1992).

Thomas Heath, *Aristarchus of Samos* (New York: Dover Books, 1981).

Herodotus, *The Histories*, trans. Aubrey de Selincourt (Harmondsworth, England: Penguin Books, 1972).

Hippolytus, *Refutation of the Heresies*, trans. J. H. MacMahon, in *Ante-Nicene Fathers*, vol. 5, ed. Alexander Roberts, James Donaldson, and A. Cleveland Coxe (Buffalo, NY: Christian Literature Publishing, 1886); revised and edited for New Advent by Kevin Knight, http://www.newadvent.org/fathers/050101.htm.

Henry M. Leicester, *The Historical Background of Chemistry* (New York: Dover Books, 1971).

O. Neugebauer, *The Exact Sciences in Antiquity* (New York: Dover Books, 1969).

Plutarch, *The Life and Writings of Homer*, in *Essays and Miscellanies: The Complete Works*, revised by William W. Goodwin, vol. 3, https://www.gutenberg.org/files/3052/3052-h/3052-h.htm.

Plutarch, "Life of Pericles," in *The Lives of the Noble Grecians and Romans*, trans. John Dryden (New York: Random House, 1992).

9

Science in Magna Graecia (750–300 BCE)

Magna Graecia, "Greater Greece," was a region in southern Italy settled by Greek colonists in the eighth and seventh centuries BCE. This region of the foot of Italy, along with neighboring Greek states across the Strait of Messina in Sicily, developed the leading scientific culture of the western Mediterranean in the first millennium BCE. Some Ionian philosophers fled the Aegean region for southern Italy because of the Persian threat in the early sixth century BCE. Acragas and Syracuse, in Sicily, and Thourioi, Croton, Elea, and Tarentum, in Italy, were cities founded originally as colonies of the more established Greek city-states of Corinth, Athens, and Rhodes. Magna Graecia was the home of some of the greatest philosophers and scientists of the ancient world, such as Pythagoras, Xenophanes, and Parmenides.

PYTHAGORAS (570–490 BCE)

The most famous of these philosopher-scientists was Pythagoras. Along with Thales and Anaximander, Pythagoras ranks as one of the founders of Greek philosophy. Legends told of Pythagoras emphasized his great wisdom, his ability to foretell future events, his relationship with the gods (particularly the god of wisdom, Apollo), and his understanding and control of nature. Like other holy men, many miracles and aphorisms were associated with Pythagoras that cannot be confirmed by his own words, as he apparently never recorded them, or they were irrevocably lost. Later writers, such as Plutarch, Aristotle, Porphyry, and Diogenes Laertius, recorded the bases of his philosophy. He believed in the transmigration of souls, which necessitated a vegetarian diet because belief in reincarnation necessitated a refusal to abide by the norm of animal sacrifice to the gods. Likewise, Pythagoras chose to wear clothes of linen rather than animal skin. A fragment of the writings of Xenophanes portrays humorously Pythagoras's belief in reincarnation: "Once, they say, he was passing by when a dog was being ill-treated. 'Stop!' he said, 'don't hit it! It is the soul of a friend! I knew it when I heard its voice'."

(Burnet). Plutarch believed that Pythagoras adopted his view toward animals from Homer:

> To this is related also another doctrine of Pythagoras, namely, that the souls of the dead pass into other forms of bodies. This did not escape Homer's notice, for he made Hector talking with horses, and Antilochus and Achilles himself not only talking with them but listening to them, and a dog recognizing Odysseus before men, even before his intimates. What other thing is he establishing but a community of speech and a relation of soul between men and beasts? Besides, there are those who ate up the oxen of the Sun and after this fell into destruction. Does he not show that not only oxen but all other living creatures, as sharers of the same common nature, are beloved by the gods? (Revised, Goodwin)

Pythagoras was a metaphysician who emphasized the transcendent nature of being and believed the best way to approach an understanding of this invisible, universal truth was through mathematics: number is the essence of being. Plutarch argued, in *The Life and Poetry of Homer,* that Pythagoras followed Homer in his appreciation for number:

> For Pythagoras thought number had the greatest power and reduced everything to numbers—both the motions of the stars and the creation of living beings. And he established two supreme principles,—one finite unity, the other infinite duality. The one the principle of good, the other of evil. For the nature of unity being innate in what surrounds the whole creation gives order to it, to souls virtue, to bodies health, to cities and dwellings peace and harmony, for every good thing is conversant with concord. The nature of duality is just the contrary,—to the air disturbance, to souls evil, to bodies disease, to cities and dwellings factions and hostilities. For every evil comes from discord and disagreement. So he demonstrates of all the successive numbers that the even are imperfect and barren; but the odd are full and complete, because joined to the even they preserve their own character. Nor in this way alone is the odd number superior, but also added to itself it generates an even number. For it is creative, it keeps its original force and does not allow of division, since *per se* the mind is superior. But the even added to itself neither produces the odd nor is indivisible. And Homer seems to place the nature of the one in the sphere of the good, and the nature of the dual in the opposite many times. . . . He always makes use of the uneven number as the better. For making the whole world to have five parts, three of these being the mean, he divides it. (Revised, Goodwin)

A believer in the immortality of the soul, like Plato, Plutarch (also a believer) argued that Pythagoras derived this insight from Homer.

Pythagoras was reputedly of Ionian origin, possibly born in Samos to the engraver Mnesarchus. As an adult, Pythagoras migrated to Croton, in southern Italy. Herodotus, a famous traveler to such places as Egypt, believed that Pythagoras traveled to Egypt to learn the mysteries of the transmigration of souls. Porphyry, in *Life of Pythagoras,* claimed, "As to his knowledge, it is said that he learned the mathematical sciences from the Egyptians, Chaldeans and Phoenicians; for of old the Egyptians excelled, in geometry, the Phoenicians in numbers and proportions, and the Chaldeans of astronomical theorems, divine rites, and worship of the Gods; other secrets concerning the course of life he received and learned from the Magi." He learned the various writing of the Egyptians: the "epistolic, the hieroglyphic, and symbolic" (trans. Guthrie). He was a student of physiognomy.

Diogenes Laertius named his teacher as Pherecydes. He astonished his contemporaries with his extravagant claims, one of which was that he had been blessed with being able to remember his past lives. Hence he recalled being Aethalides, the son of Hermes; then Euphorbus, at the time of the Trojan War; followed by Hermotimus; and then Pyrrhus, a fisherman. After Pyrrhus, he became Pythagoras. In this last life (that we know of!) Pythagoras opened a school and got involved in politics in Croton, but he died in Metapontum.

Number

Changing seasons, unending cycles, allow humans to conceive of number, a quantitative aspect to life. Pythagoras found numerical harmony in music, shapes, and quantities of things. He reputedly discovered that, regarding a right angle, should one take the angle of each of the two sides forming the right angle, square each angle and add them, the sum will equal the third angle of the hypotenuse. Such theorems as the Pythagorean, abstract conceptions based on measurement of lines and spaces, gave rise among his followers to a general approach to life, ergo, that reality is not concrete, rarely seen, yet can be approximated by mathematics. Time, for example, is not seen, not "real," but movement can be traced according to set measurements, numbers, which gives us an idea of some metaphysical reality that we call time. Pythagoras believed that "mathematical disciplines and speculations . . . are intermediate between the physical and the incorporeal realm." He had a mystical view about numbers, such as the tetractys, four rows of ten digits arranged in order, over which Pythagoreans would swear, "I call to witness him who to our souls expressed The *Tetractys*, eternal Nature's fountain-spring" (trans. Guthrie).

*
* *
* * *
* * * *

According to Porphyry,

> As the geometricians cannot express incorporeal forms in words, and have recourse to the descriptions of figures, as that is a triangle, and yet do not mean that the actually seen lines are the triangle, but only what they represent, the knowledge in the mind, so the Pythagoreans used the same objective method in respect to first reasons and forms. As these incorporeal forms and first principles could not be expressed in words, they had recourse to demonstration by numbers. Number one denoted to them the reason of Unity, Identity, Equality, the purpose of friendship, sympathy, and conservation of the Universe, which results from persistence in Sameness. For unity in the details harmonizes all the parts of a whole, as by the participation of the First Cause. Number two, or *Duad*, signifies the two-fold reason of diversity and inequality, of everything that is divisible, or mutable, existing at one time in one way, and at another time in another way. After all these methods were not confined to the Pythagoreans, being used by other philosophers to denote unitive powers, which contain all things in the universe, among which are certain reasons of equality, dissimilitude and diversity. These reasons are what they meant by the terms *Monad* and *Duad*, or by the words uniform, biform, or diversiform.

> The same reasons apply to their use of other numbers, which were ranked according to certain powers. Things that had a beginning, middle and end, they denoted by the number Three, saying that anything that has a middle is triform, which was applied to every perfect thing. They said that if anything was perfect it would make use of this principle and be adorned, according to it; and as they had no other name for it, they invented the form *Triad*; and whenever they tried to bring us to the knowledge of what is perfect they led us to that by the form of this *Triad*. So also with the other numbers, which were ranked according to the same reasons. All other things were comprehended under a single form and power which they called *Decad*, explaining it by a pun as decad, meaning comprehension. That is why they called Ten a perfect number, the most perfect of all as comprehending all difference of numbers, reasons, species and proportions. For if the nature of the universe be defined according to the reasons and proportions of members, and if that which is produced, increased and perfected, proceed according to the reason of numbers; and since the *Decad* comprehends every reason of numbers, every proportion, and every species, why should Nature herself not be denoted by the most perfect number, Ten? Such was the use of numbers among the Pythagoreans. (Trans. Guthrie)

It is entirely possible that Pythagoras was the first to hypothesize the spherical shape of the earth based on geometry. Diogenes Laertius asserted that Pythagoras thought that the earth was perfect, that the sphere was the most perfect shape, hence the earth must be a sphere. But a perfect sphere would naturally be the center of all things; hence Pythagoras had a geocentric conception of the universe. Pythagoras conceived that motions of the planetary spheres caused a noise due to rubbing—the "music of the spheres"—or, as the Late Roman commentator put it, "the world sins, and . . . its system corresponds with harmony" (trans. MacMahon). Porphyry credulously describes Pythagoras's ability "to hear the harmony of the Universe, and [he] understood the universal music of the spheres, and of the stars which move in concert with them, and which we cannot hear because of the limitations of our weak nature." Further, "Pythagoras affirmed that the nine Muses were constituted by the sounds made by the seven planets, the sphere of the fixed stars, and that which is opposed to our earth, called 'anti-earth.' He called *Mnemosyne*, or Memory, the composition, symphony and connexion of them all, which is eternal and unbegotten as being composed of all of them" (trans. Guthrie). His ability with music, according to Porphyry, was such that he used it for his own mental acumen and was able to heal people by music. Porphyry's account of Pythagoras includes absurdities about heaven and earth, such as this:

> Beans were interdicted, it is said, because the particular plants grow and individualize only after (the earth) which is the principle and origin of things, is mixed together, so that many things underground are confused, and coalesce; after which everything rots together. Then living creatures were produced together with plants, so that both men and beans arose out of putrefaction whereof he alleged many manifest arguments. For if anyone should chew a bean, and having ground it to a pulp with his teeth, and should expose that pulp to the warm sun, for a short while, and then return to it, he will perceive the scent of human blood. Moreover, if at the time when beans bloom, one should take a little of the flower, which then is black, and should put it into an earthen vessel, and cover it closely, and bury in the ground for ninety days, and at the end thereof take it up, and uncover it, instead of the bean he will find either the head of an infant, or the pudenda of a woman. (Trans. Guthrie)

Pythagorean School

Much of what is attributed to Pythagoras derived from his followers, which included philosophers and scientists such as Alcmaeon, Philolaus, Hippo, Ecphantus, and Hippasus. Hippolytus claimed that Pythagoras,

> being astonished at the management of the entire fabric [of the universe], required that at first his disciples should keep silence, as if persons coming into the world initiated in (the secrets of) the universe; next, when it seemed that they were sufficiently conversant with his mode of teaching his doctrine, and could forcibly philosophize concerning the stars and nature, then, considering them pure, he enjoins them to speak. This man distributed his pupils in two orders, and called the one esoteric, but the other exoteric. And to the former he confided more advanced doctrines, and to the latter a more moderate amount of instruction. (Trans. MacMahon)

Alcmaeon (fl. 500 BCE)

Alcmaeon was a younger contemporary of Pythagoras, perhaps his student. Like Pythagoras he lived in Croton, in southern Italy. Alcmaeon was a naturalist and physician besides being a philosopher. Diogenes Laertius, in the *Lives of the Philosophers*, claimed that Alcmaeon wrote a natural history, the first of its kind, and that like Pythagoras he believed in the transcendent soul and the spiritual makeup of the stars, planets, and moon. He was known as a teacher true to the basic teachings of the Pythagorean school. He believed that the spiritual is not still but, rather, in constant movement. The soul is eternal, unlike the body. Human bodily existence is transient rather than transcendent, subject to time, hence to death. Alcmaeon was possibly the first Greek to conceive of the indivisible eternity (rather than transmigration) of the human soul. Aristotle quoted Alcmaeon as saying "that men die because they cannot attach the beginning to the end" (trans. Barnes).

Alcmaeon as a physician was more the philosopher, believing that all life exists in pairs, or opposites—wet and dry, good and bad, angry and happy—and that illness in humans derives from an imbalance of these opposites. The well person does not want to be more hot than cold or vice versa. Believing thus, he anticipated the physics of Heraclitus and Empedocles and the medical theories of Hippocrates, the contemporary of Socrates. Aristotle's student, Theophrastus, quoted Alcmaeon as believing that sound is caused by echoes in the ear canal, that the nose brings air to the brain to detect scent, that taste is dictated by the tongue and its saliva, and that the eyes have water that helps reflect an image that is external to them. In short, "all the senses are somehow connected to the brain. That is why they are incapacitated if it is moved or displaced; for it obstructs the passages through which the senses work" (trans. Barnes).

Alcmaeon surprisingly made his mark also as an astronomer. A Pythagorean, he doubtless believed in the sphericity of the earth and its place in the center of the universe surrounded by planets and the fixed stars. The Pythagoreans also believed in the geocentric universe. He advocated (and perhaps conceived of) the theory that "the planets have a motion from west to east, in a direction opposite to that of the fixed stars," which move east to west (trans. Heath). Such a conclusion about the movement of the stars demanded patient observation and a willingness to engage the major problem of ancient planetary astronomy: the retrograde motion of the planets, the "wanderers."

Philolaus of Croton (470–385 BCE)

Philolaus of Croton, in Italy, was one of the leading Pythagorean philosophers of the ancient world. Indeed, many of the ideas attributed to Pythagoras might have originated with Philolaus. His writings, such as the *Bacchae* and *On the World*, exist only in fragments preserved by later commentators. Philolaus hypothesized the "counter-world" that brings to ten the number of heavenly bodies: sun, moon, earth, and six planets. But as nine is a number less perfect than ten (the product of 1, 2, 3, and 4), there must be another heavenly body, the "counter-earth," hidden behind the sun, hence unseen to earthlings. Philolaus also advocated the idea of the central fire around which the heavenly bodies (even the earth) orbit; this was an early version of a heliocentric universe. Pythagoreans such as Philolaus hypothesized other contrasts in nature, such as the infinite and the finite, and even and odd numbers—as well as an even-odd number. Number is the most elevated of all truths; geometry is the form of thought that can conceive of the purity of mathematical forms. Proclus, the mathematician of the Late Roman Empire, wrote that Philolaus assigned geometric forms to the traditional Greek gods. Indeed, the four elements also had their divine counterparts: Cronos, the Titan father of Zeus, is water; Ares, the god of war, is fire; Hades, the god of the underworld, is earth; Dionysus, the god of wine, represents the warm, moist nature of air.

Philolaus, like other Pythagoreans, was concerned with health, seeing good health as the product of a balance and believing that animal flesh was to be avoided. Iamblichus recorded Philolaus's belief that animals were defined by their brain, the source of thought; the heart, the seat of the soul; the navel, the basis of growth; and the genitals, the source of regeneration. Enjoining the principle that like attracts like, Philolaus argued that animals were conceived in and lived in heat, as semen is hot, the womb is hot, and whatever enters into the human body is expelled in a heated form: cool air is exhaled as warm breath; cool water is expelled as hot urine.

Hippo of Croton (fl. Sixth Century BCE)

Hippo of Croton was a Pythagorean philosopher who wrote on human health and the essential elements in nature. He believed moisture was the most important phenomenon, the excess or insufficiency of which caused ill health. A well person or animal has a good balance of moisture in the body. Animals acquire water from ponds, rivers, streams, and wells, all of which derive ultimately from the sea, the depth of which is greater than all else, hence the source of all water. Hippolytus, the Late Roman commentator, wrote that "Hippo, a native of Rhegium, asserted as originating principles, coldness, for instance water, and heat, for instance fire.

And that fire, when produced by water, subdued the power of its generator, and formed the world. And the soul, he said, is sometimes brain, but sometimes water; for that also the seed is that which appears to us to arise out of moisture, from which, he says, the soul is produced" (trans. MacMahon).

Hippasus, who lived in the fifth century but about whom little is known, reportedly discovered the mathematical harmonies of the musical scales, the dodecahedron (a twelve-sided figure), and irrational geometric constructs. Ecphantus of Syracuse, also a shadowy figure, according to Hippolytus,

> affirmed that it is not possible to attain a true knowledge of things. He defines, however, as he thinks, primary bodies to be indivisible, and that there are three variations of these, viz., bulk, figure, capacity, from which are generated the objects of sense. But that there is a determinable multitude of these, and that this is infinite. And that bodies are moved neither by weight nor by impact, but by divine power, which he calls mind and soul; and that of this the world is a representation; wherefore also it has been made in the form of a sphere by divine power. And that the earth in the middle of the cosmical system is moved round its own centre towards the east. (Trans. MacMahon)

There were, perhaps, two members of the Pythagorean school who were women. Pythagoras's wife, Myia, and his daughter (perhaps), Theano, were reputedly Pythagorean philosophers, though little is known about their lives and work.

XENOPHANES (570–478 BCE)

As significant as Pythagoras and his school was Xenophanes, a poet whose elegies and satires portrayed philosophers and their theories as well as provided insights into the nature of the universe. Xenophanes, from the Ionian town of Colophon, was the son of Orthomenes; like his fellow Ionians Thales, Anaximander, Anaximenes, and Anaxagoras, Xenophanes believed in a transcendent cause of all things and was uncomfortable attributing phenomena to the proclivities of anthropomorphic gods. At some point during his life, Xenophanes emigrated to Magna Graecia, where he lived at Elea; a tradition grew over time that he was the teacher of Parmenides—their philosophies are indeed similar. In short, Xenophanes was an original philosopher, perhaps the first skeptic; Hippolytus, the Late Roman commentator, wrote that Xenophanes believed "that there is no possibility of comprehending anything, expressing himself thus:—*For if for the most part of perfection man may speak, / Yet he* knows *it not himself, and in all attains surmise*" (trans. MacMahon).

Xenophanes challenged contemporary beliefs about nature and the divine: "Homer and Hesiod have ascribed to the gods all things that are a shame and a disgrace among mortals, stealings and adulteries and deceivings of one another." "But mortals deem that the gods are begotten as they are, and have clothes like theirs, and voice and form." "Yes, and if oxen and horses or lions had hands, and could paint with their hands, and produce works of art as men do, horses would paint the forms of the gods like horses, and oxen like oxen, and make their bodies in the image of their several kinds" (trans. Burnet). Plutarch appears to have disagreed with Xenophanes's condemnation of Homer, arguing in the *Life and Poetry of Homer* that Homer believed that

> poetry requires gods who are active; that he may bring the notion of them to the intelligence of his readers he gives bodies to the gods. But there is no other form of bodies than man's capable of understanding and reason. Therefore he gives the likeness of each one of the gods the greatest beauty and adornment. He has shown also that images and statues of the gods must be fashioned accurately after the pattern of a man to furnish the suggestion to those less intelligent, that the gods exist. . . . If it is necessary to ask how he knew that God was an object of the intelligence, it was not directly shown, as he was using poetic form combined with myth. (Revised, Goodwin)

Xenophanes himself proposed a singular, all-powerful god of thought who exercises providence over the universe. "There never was nor will be a man who has clear certainty as to what I say about the gods and about all things; for even if he does chance to say what is right, yet he himself does not know that it is so. But all are free to guess" (trans. Burnet).

Xenophanes's verse obscures many of his beliefs about the universe, or rather, his beliefs were obscure and verse was the best way to portray them. According to Hippolytus, Xenophanes believed that the universe is neither finite nor infinite, a contradiction that may be resolved by considering some of his ideas on natural history. Xenophanes wrote that "all things come from the earth, and in earth all things end"; that "all things are earth and water that come into being and grow"; that "this limit of the earth above is seen at our feet in contact with the air; below it reaches down without a limit" (trans. Burnet). These passages seem to mark Xenophanes as a precursor of the materialists, who likewise hypothesized an ongoing creation and destruction of the universe. Xenophanes also was indebted to Thales, perhaps, in assigning to water (that is, mud) a creative force. Xenophanes also held the contrary view that above the earth was a limited region of air, sun, moon, planets, and stars, while below the earth was an infinite void. This appears to reflect a view of a flat earth. More perplexing was his belief that clouds of fire form the heavenly bodies, and that the sun only appears to be one phenomenon but is in reality a multitude, indeed an infinite number. How? Xenophanes, according to the commentator Aetius, believed "that there are many suns and moons according to the regions, divisions, and zones of the earth; and at certain times the disc lights upon some division of the earth not inhabited by us and so, as it were stepping on emptiness, suffers eclipse." Also, Xenophanes "maintains that the sun goes forward *ad infinitum*, and that it only appears to revolve in a circle owing to its distance" (trans. Heath). In short, the flat earth is infinitely long, over which are infinite numbers of suns moving east to west; as one appears, moves by, and disappears from the eyes of the observer, the sun seems to be one body appearing and reappearing, but the senses fool us.

Diogenes Laertius's summary of his thought sheds some light:

> His doctrine was, that there were four elements of existing things; and an infinite number of worlds, which were all unchangeable. He thought that the clouds were produced by the vapour which was borne upwards from the sun, and which lifted them up into the circumambient space. That the essence of God was of a spherical

> form, in no respect resembling man; that the universe could see, and that the universe could hear, but could not breathe; and that it was in all its parts intellect, and wisdom, and eternity. He was the first person who asserted that everything which is produced is perishable, and that the soul is a spirit. He used also to say that the many was inferior to unity. Also, that we ought to associate with tyrants either as little as possible, or else as pleasantly as possible. (Trans. Yonge)

Xenophanes's ideas appear ludicrous today, but if we give him the benefit of the doubt, we see that he relied on observation more than blind faith in the same old Homeric gods. Xenophanes took a fresh approach to the phenomena, and though he was in error, his was the same scientific approach of his contemporaries and forebears, the Ionians.

PARMENIDES (FL. 450 BCE)

Parmenides of Elea, reputedly the student of Xenophanes, wrote in verse; in one poem, *On Nature*, he presented his views on the ungenerated *being*, the cause and essence of all things. Diogenes Laertius, in *Lives of the Philosophers*, claimed that Parmenides was the first to hypothesize the spherical nature of the earth, "and that it was situated in the centre of the universe. He also taught that there were two elements, fire and earth; and that one of them occupies the place of the maker, the other that of the matter. He also used to teach that man was originally made out of clay; and that they were composed of two parts, the hot and the cold; of which, in fact, everything consists. Another of his doctrines was, that the mind and the soul were the same thing." Moreover, "he also used to say that philosophy was of a twofold character; one kind resting on certain truth, the other on opinion" (trans. Yonge). Parmenides possibly got the idea from Pythagoras, who was born in Samos but migrated to southern Italy, where he collected many disciples who influenced the development of philosophy, mathematics, and science for centuries to come. Pythagoras believed in the transmigration of souls and in the idea that *number* forms the basis of the universe. This theory of the mathematical foundations of reality would influence such important Greek philosophers as Plato and the mathematician Euclid.

Hippolytus summed Parmenides's thought thus:

> For Parmenides likewise supposes the universe to be one, both eternal and unbegotten, and of a spherical form. And neither did he escape the opinion of the great body (of speculators), affirming fire and earth to be the originating principles of the universe—the earth as matter, but the fire as cause, even an efficient one. He asserted that the world would be destroyed, but in what way he does not mention. The same (philosopher), however, affirmed the universe to be eternal, and not generated, and of spherical form and homogeneous, but not having a figure in itself, and immoveable and limited. (Trans. MacMahon)

Parmenides and his followers, Zeno of Elea and Melissus of Samos, founded the Eleatic school of philosophy in the fifth century. Zeno was chiefly known for his paradoxes. For example, he tried to show that even emptiness has being and that motion is at rest. This is true only if one realizes that at each singular moment (the many parts of which make up the whole of movement), time, hence motion, are indeed still, as in still photography. Diogenes Laertius described his "chief doctrines" thus: "That there were several worlds, and that there was no vacuum; that the nature of all things consisted of hot and cold, and dry and moist, these elements interchanging their substances with one another; that man was made out of the earth, and that his soul was a mixture of the before-named elements in such a way that no one of them predominated" (trans. Yonge). Melissus of Samos was a man of affairs, a statesman and general, who was also a philosopher. Melissus wrote *On Nature or On What Exists,* which survives only in fragments preserved by the Late Roman commentator Simplicius. Like Parmenides, Melissus concerned himself with arguments to prove that being is without cause, not created, infinite, and eternal. "His doctrine was," according to Diogenes, "that the Universe was infinite, unsusceptible to change, immoveable, and one, being always like to itself, and complete; and that there was no such thing as real motion, but that there only appeared to be such. As respecting the Gods, too, he denied that there was any occasion to give a definition of them, for that there was no certain knowledge of them" (trans. Yonge).

The Eleatic school, as well as other philosophers of Magna Graecia, such as the Pythagoreans, laid the groundwork for Socrates and his followers in the fifth and fourth centuries.

FURTHER READING

Jonathan Barnes, trans., *Early Greek Philosophy* (London: Penguin Books, 1987).

John Burnet, ed. and trans., *Early Greek Philosophy* (London: Adam and C. Black, 1930).

Diogenes Laertius, *The Lives and Opinions of Eminent Philosophers*, trans. C. D. Yonge (London: Henry Bohn, 1853).

Thomas Heath, *Aristarchus of Samos* (New York: Dover Books, 1981).

Hippolytus, *Refutation of the Heresies*, trans. J. H. MacMahon, in *Ante-Nicene Fathers*, vol. 5, ed. Alexander Roberts, James Donaldson, and A. Cleveland Coxe (Buffalo, NY: Christian Literature Publishing, 1886); revised and edited for New Advent by Kevin Knight, http://www.newadvent.org/fathers/050101.htm.

Plutarch, *The Life and Writings of Homer*, in Plutarch, *Essays and Miscellanies: The Complete Works*, corrected and revised by William W. Goodwin, vol. 3, https://www.gutenberg.org/files/3052/3052-h/3052-h.htm.

Porphyry, *Life of Pythagoras*, trans. Kenneth S. Guthrie (1920), http://www.tertullian.org/fathers/porphyry_life_of_pythagoras_02_text.htm.

Bertrand Russell, *A History of Western Philosophy* (New York: Simon and Schuster, 1945).

10

Classical Greek Science (500–300 BCE)

The Greek Classical Age was a period of supreme intellectual achievement that occurred over the space of several centuries. The Classical Age was epitomized by the city-state of Athens and the three most significant philosophers and scientists of the ancient world: Socrates (469–399), Plato (425–346), and Aristotle (384–322). Independent poleis (city-states) characterized Greek civilization. They were competitive and aggressive; hence, war was common among the Greeks. Yet this competitiveness was expressed not only in war but in trade and thought as well. Herodotus, in the *Histories*, described the Greeks as a free people who showed what freedom can accomplish among humans if given the chance. Athens was the most democratic of the Greek city-states, which might explain why the Athenians produced such outstanding work in science and philosophy.

The Greek Classical Age occurred after many centuries of change in the Greek world. The history of ancient Greece is typically divided into three chronological divisions: Archaic (800–500 BCE), Classical (500–331 BCE), and Hellenistic (331–31 BCE). The so-called Archaic Age was a time of new beginnings. Hitherto Greece had a violent and primitive past. At the beginning of the Archaic Age, Greece was emerging from a period of cultural, social, and economic darkness. Its so-called Dark Ages occurred after centuries of Mycenaean domination of the Balkan Peninsula and Peloponnesus. Mycenaean kings such as Agamemnon, Menelaus, Nestor, and Odysseus (as identified in Homer's *Iliad* and *Odyssey*) dominated society by means of the sword. The Mycenaean Age was a time of unending war. Invasions of Greece by an even more warlike people, the Dorians, destroyed Mycenaean dominance and forced the vanquished to strongholds in eastern Greece, the Aegean islands, and the Aegean shores of Asia Minor. These people of the Aegean region, as opposed to the mainland Greeks of the west and south, were the Ionians. The intervening centuries were dark; literate, urban culture disappeared. Few records survive from the Dark Ages, and those that do, such as the *Iliad* and the *Odyssey*, were initially composed orally and not recorded in written script until centuries later.

The Greek Dark Ages ended at the beginning of the eighth century BCE, when the historical record reveals that literacy, cities, trade, and sophisticated institutions returned to Greece. Such civilized attributes included scientific thought. By the eighth century, the Greek polis (the city-state) spread throughout Greece, the Aegean, and elsewhere. Greek poleis were independent regional urban powers that dotted Greece and fiercely promoted and defended their respective interests. The city of Athens controlled the region of Attica, for example, and competed with other city-states both Ionian and Dorian, such as Thebes, Corinth, and Sparta. The Greek polis was dynamic and wealthy, actively engaged in trade and in developing vibrant new institutions, and it sought to extend its economic and political power throughout the Mediterranean.

The mountainous peninsula of Greece would not easily host agriculture, though we find in the *Works and Days* of Hesiod, written about 700 BCE, a portrait of the pastoral existence of the farmer and herder. Homer's *Odyssey* emphasizes the agricultural wealth of landowners at the beginning of the first millennium; the details of production of grain, hogs, wine, and beef could hardly have been based on mere poetic imagination. Greece was fit more for the cultivation of wine and olives, the latter requiring little effort to cultivate properly, the former perfectly adapted to the climate of the Mediterranean. The rocky soil hosted enough successful viticulture, olive production, farming of grains, and raising of cattle and goats for milk and cheese that the population of Greece expanded.

At the dawn of the Archaic Age, the Greek poleis were so populated that they sent groups of citizens on expeditions to discover and settle regions in other lands where a healthy trade could be guaranteed. This phenomenon of Greek colonization resulted in Greek poleis being spread west to Spain, France, Italy, and Sicily; south to North Africa; southeast to Egypt; east to Anatolia and Palestine; and northeast to the Black Sea region. Greek colonization led to the spread of Greek culture and ideas into Asia, Africa, and eastern and western Europe. Most important, the Greeks gained a greater awareness of geography and other peoples—new experiences and new ideas stimulated thought and led to the first scientific revolution in human history.

Historians of science refer to the Archaic Age philosophers and scientists of the seventh and sixth centuries BCE as the *pre-Socratics*, that is, those who preceded the great Athenian philosopher Socrates. The number of pre-Socratic philosophers and scientists is astonishing. Many were Ionians who came from western Asia Minor on the eastern shores of the Aegean Sea. They appear, at least according to the beliefs of the philosopher Plutarch, to have been influenced by the poems of Homer, who believed "that the world [cosmos] is one and finite. For if it had been infinite, it would never have been divided in a number having a limit" (revised, Goodwin). Thales of Miletus, his student Anaximander, Anaximenes of Miletus, and Xenophanes of Colophon were the first Ionian thinkers. Like Homer, they sought to explain the universe by imposing the structure of the human mind upon it, to conceptualize it so as to understand it. They broke from primitive anthropomorphism to conceive of a universe of cause-and-effect relationships, of infinite existence and movement, of a primary universal cause. At the same time, they sought the essence of all things in water, air, fire, or mind. Anaxagoras, who moved to Athens and influenced Pericles, advocated the latter idea, that mind *(nous)* was the infinite and universal, the cause of all that existed. The fifth-century philosopher Melissus of Samos, a follower of Parmenides, who lived on an island off the coast of Turkey in the Aegean Sea near Miletus, argued in *On Nature or On What Exists* that the universe is ungenerated, has always existed, that "nothing" cannot exist. This is a fascinating leap into the metaphysical beyond, to assume that there is something, everywhere, eternal and infinite; that what we might consider *nothing* is absurd; for even the word itself is *something*.

Pre-Socratic philosophers took the Homeric concept of *arete*—individual free will in the face of overwhelming odds or fate—and fashioned it into a more universal presence of mind (*nous*) making itself known in each individual by means of the soul (*psyche*), an incorporeal presence, a being (*ousia*) that transcends time and place. Truth is a mental state scarcely achieved by humans living in time. But humans at least have the ability to recognize what they typically lack, and they have the freedom to pursue truth so as to achieve a sense of well-being that, more than strength of body, is a firm anchor in the storms of time. Plato and the Academy, and Aristotle and the Peripatetics also hypothesized a transcendent truth expressed as knowledge, the Logos, the word spoken in the creation throughout all time, recognized by the individual knower at a given point in time.

Diogenes Laertius, the Late Roman biographer of philosophers and scientists hypothesized the Italian school of Pythagoras, Xenophanes, and Democritus in contrast to the Ionian school of Thales, Anaximander, and Anaximenes. Not everyone in the Italian school lived in Italy. Diogenes used a geographic convenience to separate the philosophers of "becoming" from the philosophers of "being." The founder of the Italian school, Pythagoras, was more a thinker who focusing on "being"; yet it was from his home in Croton that Pythagoras founded a school of thought that would bring forth many of the greatest philosophers of the ancient world, most of whom focused on "becoming," change and movement, rather than "being," the stable and transcendent counter to change.

HERACLITUS OF EPHESUS (FL. EARLY FIFTH CENTURY BCE)

Heraclitus had the reputation among ancients of being one of the most difficult philosophers to comprehend, either because of brilliant or muddled thinking. He was, perhaps more than any other thinker, almost modern in his emphasis on the contradictions and absurdities of existence out of which emerges a sort of truth. Sextus Empiricus, the skeptic, could respect such vagaries and inconsistencies, and approved of some of Heraclitus's comments in *On Nature.* Aristotle, however, was not as approving and condemned Heraclitus's "childish" astronomy, which included the view that the sun is as we see it, only about a foot in diameter (trans. Heath). Hippolytus, the Late Roman commentator, quoted long excerpts from Heraclitus's writings, in which are found ideas of an eternal fire that is the ultimate reality. Fire is the creator and destroyer, and through it all change occurs.

Fire condenses into water and then into earth and vice versa. Fire is the source for the lights of the sky—the sun, moon, planets, and stars being concave repositories of this divine fire. The repository in which lies the sun's fire is extinguished every evening and kindled again in the morning; how the repository moves from west to east unseen over the course of the night to begin a new day is unclear.

Hippolytus wrote that

> Heraclitus, a natural philosopher of Ephesus, surrendered himself to universal grief, condemning the ignorance of the entire of life, and of all men; nay, commiserating the (very) existence of mortals, for he asserted that he himself knew everything, whereas the rest of mankind nothing. But he also advanced statements almost in concert with Empedocles, saying that the originating principle of all things is discord and friendship, and that the Deity is a fire endued with intelligence, and that all things are borne one upon another, and never are at a standstill; and just as Empedocles, he affirmed that the entire locality about us is full of evil things, and that these evil things reach as far as the moon, being extended from the quarter situated around the earth, and that they do not advance further, inasmuch as the entire space above the moon is more pure. So also it seemed to Heraclitus. (Trans. MacMahon)

Heraclitus put an emphasis on change, on things becoming one thing and then another. There was a lot of the Stoic and the skeptic in Heraclitus, who inspired a host of philosophers and scientists to question, to doubt, to search, to seek answers in nature and in self. Diogenes Laertius provided a succinct summary of his philosophy:

> Everything happens according to destiny, and that all existing things are harmonized, and made to agree together by opposite tendencies; and that all things are full of souls and daemones. He also discussed all the passions which exist in the world, and used also to contend that the sun was of that precise magnitude of which he appears to be. One of his sayings too was, that no one, by whatever road he might travel, could ever possibly find out the boundaries of the soul, so deeply hidden are the principles which regulate it. He used also to call opinion the sacred disease; and to say that eye-sight was often deceived. Sometimes, in his writings, he expresses himself with great brilliancy and clearness; so that even the most stupid man may easily understand him, and receive an elevation of soul from him. And his conciseness and the dignity of his style, are incomparable. Moreover, fire is an element, and that it is by the changes of fire that all things exist; being engendered sometimes by rarity, some times by density. But he explains nothing clearly. He also says, that everything is produced by contrariety, and that everything flows on like a river; that the universe is finite, and that there is one world, and that that is produced from fire, and that the whole world is in its turn again consumed by fire at certain periods, and that all this happens according to fate. That of the contraries, that which leads to production is called war and contest, and that which leads to the conflagration is called harmony and peace; that change is the road leading upward, and the road leading downward; and that the whole world exists according to it. For that fire, when densified becomes liquid, and becoming concrete, becomes also water; again, that the water when concrete is turned to earth, and that this is the road down; again, that the earth itself becomes fused, from which water is produced, and from that everything else is produced; and then he refers almost everything to the evaporation which takes place from the sea; and this is the road which leads upwards.

> Also, that there are evaporations, both from earth and sea, some of which are bright and clear, and some are dark; and that the fire is increased by the dark ones, and the moisture by the others. But what the space which surrounds us is, he does not explain. He states, however, that there are vessels in it, turned with their hollow part towards us; in which all the bright evaporations are collected, and form flames, which are the stars; and that the brightest of these flames, and the hottest, is the light of the sun; for that all the other stars are farther off from the earth; and that on this account, they give less light and warmth; and that the moon is nearer the earth, but does not move through a pure space; the sun, on the other hand, is situated in a transparent space, and one free from all admixture, preserving a well proportioned distance from us, on which account it gives us more light and more heat. And that the sun and moon are eclipsed, when the before-mentioned vessels are turned up wards. And that the different phases of the moon take place every month, as its vessel keeps gradually turning round. Moreover, that day and night, and months and years, and rains and winds, and things of that kind, all exist according to, and are caused by, the different evaporations. For that the bright evaporation catching fire in the circle of the sun causes day, and the predominance of the opposite one causes night; and again, from the bright one the heat is increased so as to produce summer, and from the dark one the cold gains strength and produces winter; and he also explains the causes of the other phenomena in a corresponding manner. But with respect to the earth, he does not explain at all of what character it is, nor does he do so in the case of the vessels; and these were his main doctrines. (Trans. Yonge)

The pre-Socratic Greek philosophers were the founders of chemical thinking in that they tried to uncover the basic substances of the universe. Their joint conclusion was that the universe was composed of several foundational substances: water, air, fire, earth, and ether, organized in perhaps infinite quantities (Anaximander) by a creative and sustaining mind (Anaxagoras) that is the essence of all being. The materialists and atomists Empedocles, Heraclitus, Leucippus, and Democritus, on the other hand, argued for the material rather than the metaphysical foundations of the universe—that change and conflict, matter in motion, and invisible particles, atoms, formed a universe that is always in the process of becoming. The fifth-century philosopher Empedocles argued that there were four elements—earth, air, water, and fire. The fourth-century philosopher Leucippus argued that invisible atoms combined and recombined to form all matter.

IDEALISTS

Socrates, like his predecessor Pythagoras, did not write anything; what we know of Socrates comes from his students Plato and Xenophon. Plato wrote in dialogue form featuring the highly sophisticated conversations of Socrates, his students, and sometimes his opponents, particularly the Sophists. The Sophists were paid teachers of wisdom who often focused less on the pursuit of truth and more on ways to contrive a good argument. Many, such as Protagoras, were humanists—that is, they restricted knowledge to human experience—which was the opposite of what Socrates and Plato believed. Neither agreed with the theories of the atomists, such as Leucippus and Democritus, who tried to explain nature according to physical substances rather than ideas.

By today's standards, Socrates and Plato were philosophers more than scientists. They built a philosophy around a conception of the divine attributes of the universe, what Socrates called *ideal forms*. These ideas, such as justice, beauty, and the good, resembled the pre-Socratic notions of the mind, being, and Logos. Truth is hidden from humans, who see only the shadows of what is real. The search for reality is therefore worthwhile, though it requires intense contemplation and intuition, years of study and mastery of the theories of music, logic, mathematics, justice, and epistemology. Socrates and Plato speculated somewhat on the nature of the universe, the shape of the world, the soul and its immortality, human thought, psychology, and politics, but they lacked the empirical focus that usually defines science.

Most importantly, Plato's works reveal a human conception of the universe that reveals just how far civilization had taken human thought from its ancestral Paleolithic and Neolithic focus on anthropomorphism and the magical supernatural. Plato's dialogues show that Plato and his forebears, the Ionian philosophers, replaced archaic anthropomorphism with a new sense of human judgment. Plato declared the cosmos to be good, and in doing so implied that first, the cosmos is a reflection of human morality, and second, that humans have the power to make such a pronouncement. It took tremendous confidence, or arrogance, to make a pronouncement that what had for so long been considered an amoral universe was now a moral, that is, a *good* universe. The universe is good because humans know what good is, and the universe fits the human qualifications of *the good*. Plato and his contemporaries analyzed the universe from their detached perspective and were in awe, in amazement, at its immensity, power, and beauty, qualities that suggested that it was accommodative and necessary for humans, a perfect home. But a home is a shelter; it is not a part of a person. The universe is the home of the human, who can partake of it, enjoy it, gain shelter from it, gain knowledge from and of it, and use it for human ends. Platonic thinking influenced the philosophers and scientists of the ancient world, particularly the Neoplatonists, and scores of others in Europe, Asia, Africa, and America for centuries.

Unlike Plato, his pupil Aristotle was a scientist interested in experimentation, observation, collection and categorization of data, and inductive and deductive reasoning. He used these tools of reason and analysis to master the topics of metaphysics, physics, logic, statecraft, biology, botany, and astronomy. Aristotle was a prolific writer. His *Metaphysics*, *Politics*, *Poetics*, *Physics*, and *Nichomachean Ethics* were and still are classic treatises on those subjects. His school, the Lyceum, and his followers, the Peripatetics, beginning with Theophrastus, continued to influence Greek and European science for years to come. First the Romans, then the Byzantines, the Arabs, and the medieval churchmen found inspiration and knowledge in Aristotle's writings.

MATERIALISTS

Meanwhile the Italian school countered the idealism of the Platonists and Aristotelians with a materialist psychology. Developed by Leucippus and Democritus, this materialism took the form of invisible atoms constituting all existence. Even the mind, thoughts, dreams, soul, and being were composed of atoms constantly in movement. Life held them together into an individual being; death released them to scatter and form other arrangements. Mortality was nothingness for the individual but not for the universe, which is constantly transforming and remaking itself. Epicureans such as Lucretius believed the aim of life was achieving contentment, a pleasant sense of self that is generally immune from fear of the future, unfazed by mental, emotional, or physical pain. Stoics, such as Marcus Aurelius, conceived of a material Logos that created and governed (if distantly) the universe. Humans share in the Logos by means of the mind and soul; but identity is restricted to the here and now, its future limited by death.

In a remarkable and stunning insight that anticipated modern science by two thousand years, Greek philosophers hypothesized the existence of atoms. For some, the countless mysteries of life and nature could not be adequately explained by reference to the gods or to obscure, ephemeral, metaphysical forces such as mind, Logos, number, and the infinite. Speculation on the spiritual world can be unending but unsubstantiated by the senses and experience. The atomists demanded that explanations of existence be reduced to common sense and the evidence of the senses. This radical empiricism required a disbelief in anything that could not be seen, heard, smelled, touched, or tasted. Everything else must be the product of overactive imaginations. The atomists allowed themselves one general supposition: There must be a force that causes the movement in things detected by humans. This force must be the basis of cause, of the nature of things, hence the fundamental force of being and matter in the universe. But this force of being, cause, and matter must result from a thing—something material, composed of matter—not an idea.

Two ancient schools of thought, the Epicurean and the Stoic, embraced the science of atoms. The most famous atomists were Leucippus, Democritus, Epicurus, Zeno, and Lucretius. The forerunners of the atomists, such as Empedocles, had hypothesized the existence of material elements, four in number, which the atomists embraced as the basic atomic phenomena: earth, air, fire, and water. Democritus argued that only two things really existed: atoms and void; but the former combines into various other forms, initially the four elements and eventually everything else in existence. Atoms have no quality, for example, no color or taste, though they form into things of color and taste. Lucretius, in *On the Nature of Things* went further, giving to atoms specific shapes that allowed them to combine and recombine to form all things. Some were more dense than others, some of a distinctive texture, some with hook-like features that were highly resistant to change. The Epicureans believed that atoms swerved at times without warning, being totally random in their movement. The randomness of atomic movement (of electrons) formed the basis of the modern *Uncertainty Principle* advanced by the German physicist Werner Heisenberg. Random movement and collision are the essence of causation and, because unpredictable, free will. Lucretius anticipated Galileo in arguing that atoms constantly moved at a uniform rate through a void of no resistance that did not alter the rate of movement. Like the corpuscular theorists of the Scientific Revolution, such as Boyle and Descartes, the ancient atomists believed that atoms were completely material, without mind or soul, of infinite number, producing infinite and recurrent possibilities.

EMPEDOCLES (495–435 BCE)

Empedocles was one of the greatest pre-Socratic Greek scientists. A native of Acragas, on the isle of Sicily, he was a philosopher, physician, mystic, and materialist. Diogenes Laertius claimed that he was the student of Parmenides. His work survives only in fragments, and his life is not well known. The best source for Empedocles's life, notwithstanding the plentiful anecdotes it contains, is Diogenes Laertius. Plutarch, in his *Life and Poetry of Homer*, argued that Homer anticipated Empedocles in the development of the four elements: earth, air, water, and fire. Plutarch wrote that

> the poet seems to have signified this enigmatically in the conflict of the gods, in which he makes some help the Greeks and some the Trojans, showing allegorically the character of each. And he set over against Poseidon Phoebus, the cold and wet against the hot and dry: Athene to Ares, the rational to the irrational, that is, the good to the bad. Hera to Artemis, that is, the air to the moon, because the one is stable and the other unstable. Hermes to Latona, because speech investigates and remembers, but oblivion is contrary to these. Hephaestus to the River God, for the same reason that the sun is opposed to the sea. (Revised, Goodwin)

According to Hippolytus, "Empedocles believed that from the elements all are things generated and to these things are destroyed through the two affective forces, love and strife: the former attracts, the latter repulses." Empedocles, anticipating Aurelius Augustine's dualism of the City of God and the City of Man, conceived of a transcendent realm dominated by love and a physical realm dominated by strife. Everything, even the truth, is in a process of becoming that never ends. He hypothesized a materialist foundation to the universe that was built upon by Leucippus and Democritus. Hippolytus wrote that Empedocles

> advanced likewise many statements respecting the nature of demons, to the effect that, being very numerous, they pass their time in managing earthly concerns. This person affirmed the originating principle of the universe to be discord and friendship, and that the intelligible fire of the *monad* is the Deity, and that all things consist of fire, and will be resolved into fire; with which opinion the Stoics likewise almost agree, expecting a conflagration. But most of all does he concur with the tenet of transition of souls from body to body, expressing himself thus:—
>
> *For surely both youth and maid I was,*
> *And shrub, and bird, and fish, from ocean stray'd.* (Trans. MacMahon)

His books, in particular *On Nature* and *Purifications*, were written in obscure verse that tended to hide rather than reveal. In these works, he dealt with the divine; the origins of humans; a basic astronomy of the relation of sun, earth, and moon; the nature of the earth; biology and zoology; the means of sensory perception; and medicine. His astronomy was primitive, advocating the notion that the universe is a crystal sphere within which is a dark hemisphere of air and a light hemisphere of fire, which gives us the phenomenon of day and night. Diogenes Laertius wrote that Empedocles "asserts that the sun is a vast assemblage of fire, and that it is larger than the moon. And the moon is disk-shaped; and that the heaven itself is like crystal; and that the soul inhabits every kind of form of animals and plants" (trans. Yonge). Aristotle (according to Heath) attributed to Empedocles the original theory that light travels from one point to the next over a certain period of time. He also studied the attraction and repulsion of magnets and the nature of sound and vision. As a physician, Empedocles established a medical school at Magna Graecia that was the counterpart to the more famous Hippocratic school at Cos, in the Aegean.

LEUCIPPUS (FL. FIFTH CENTURY)

Leucippus, of the fifth century BCE, was the first scientist to develop a complete theory for a universe based on the existence of an infinite number of atoms. He influenced generations of physicists and chemists in their search to understand the basic structure, the infinitesimal particles, of the universe. Leucippus did not have sophisticated equipment, nor did he practice empirical science. He relied on the teaching of his predecessors, made logical assumptions, and arrived at a theory to explain the structure of matter. Over two millennia later, physicists such as Ernst Rutherford and Niels Bohr confirmed the existence of atoms, which vindicated Leucippus's initial theoretical stab in the dark.

Leucippus was an Ionian Greek, born sometime after 500 BCE somewhere in western Turkey or the islands of the Aegean Sea. Nothing is really known about his life, save that he was influenced by the materialist philosopher Empedocles and the idealists Parmenides and Anaxagoras. Only one line survives from his writings. In *On Mind*, he argued that "no thing happens in vain, but everything for a reason and by necessity" (trans. Barnes). To the ongoing debate in classical Greece about the nature of the universe, whether it was finite or infinite, fundamentally material or spiritual, created or uncreated, Leucippus contributed his view that the universe was infinite, uncreated, and in constant motion and change. In this he agreed with Empedocles. Beyond the growth and decay that one witnesses in nature and everyday life, Leucippus posited the existence of invisible material bodies, atoms (Greek *atomos*), that are always moving, changing, combining and recombining and are of varying shapes and sizes. Atoms are small enough that we cannot see them, but we feel their presence. How else, Leucippus wondered, can we explain matter, its changes, its variety, unless we realize the existence of atoms? What separates a tree from a stone, a fish from a human, if not the underlying differences in these invisible particles? To speak of particles moving and changing implies some substance or arena in which they move and change. Leucippus argued for the presence of a void or space devoid of particles, of anything. Nevertheless, the void exists. It does not represent "being," the basic fundamental of existence, but rather "nonbeing," the opposite of existence, the opposite of anything. *Nothing* therefore exists, Leucippus argued. Being cannot exist without nonbeing; atoms cannot exist without the void in which they move and act.

Leucippus naturally used his atomic theory to explain the origins and extent of the universe. According to Diogenes Laertius, Leucippus argued that atoms of like form are attracted to the center of a "vortex," a swirling mass; atoms of unlike form are thrust outside the vortex. The vortex eventually forms a central spherical body, such as the earth; other atoms form the stars and other planets. Leucippus believed that such a process occurs infinitely, leading to a host of other worlds and stars. The planets orbit the earth, which is at the center of a solar system in which the sun orbits at the extreme. The brightness of the sun, stars, and moon derives from their movement, the friction of which, one gathers, induces fire. Leucippus hypothesized that the earth is broad in the middle, like a drum or (as Thomas Heath says) tambourine. The earth swirls around at the center on an axis that tilts toward the south. The northern regions of the earth are damp and frozen, the southern regions hot and on fire.

Hippolytus, the Late Roman commentator, summed Leucippus's thought thus:

> Leucippus . . . affirms things to be infinite, and always in motion, and that generation and change exist continuously. And he affirms plenitude and vacuum to be elements. And he asserts that worlds are produced when many bodies are congregated and flow together from the surrounding space to a common point, so that by mutual contact they made substances of the same figure and similar in form come into connection; and when thus intertwined, there are transmutations into other bodies, and that created things wax and wane through necessity. But what the nature of necessity is, (Parmenides) did not define. (Trans. MacMahon)

Diogenes Laertius wrote,

> And his principal doctrines were, that all things were infinite, and were interchanged with one another; and that the universe was a vacuum, and full of bodies; also that the worlds were produced by bodies falling into the vacuum, and becoming entangled with one another; and that the nature of the stars originated in motion, according to their increase; also, that the sun is borne round in a greater circle around the moon; that the earth is carried on revolving round the centre: and that its figure resembles a drum; he was the first philosopher who spoke of atoms as principles. (Trans. Yonge)

Leucippus initiated a school of thought that would have a major impact on subsequent science and philosophy not only in the ancient world but over all subsequent centuries to the present. His most famous disciples were Democritus, Epicurus, and Zeno. From Leucippus's basic supposition, the philosophies of Epicureanism and Stoicism would spring. His would be a philosophy in constant opposition to the idealism of Plato and Christian philosophers. Leucippus's atomic theory would be resurrected two thousand years later by the English scientist Robert Boyle, who called atoms, "corpuscles." Boyles's corpuscular theories went a long way toward sustaining the scientific revolution of the sixteenth, seventeenth, and eighteenth centuries. Ironically, over these past two millennia, knowledge of Leucippus has dwindled to almost nothing, while knowledge of his theory has come to dominate the world of science.

DEMOCRITUS (C. 460–370)

Ancient writers believed that Democritus was the student of Leucippus. They are associated together as the first philosophers to hypothesize that invisible material objects—atoms—make up the universe. Although, like those of Leucippus, little survives of Democritus's writings, the theory of atoms was widely discussed by ancient authors; hence much was known about Democritus and his theories even if only fragments of his writings survive time. The breadth of his interests is astonishing. Diogenes Laertius, in his brief biography, lists Democritus's writings covering works on science, ethics, mathematics, medicine, poetry, nutrition, and more. Hippolytus argued that Democritus associated with Indian gymnosophists, Egyptian priests, and Chaldean magi. Democritus made his mark especially in the application of his theories of the universe to everyday morality and the simple goal of living a good life.

Democritus, according to tradition, was born in northern Greece and later relocated to Athens. Like most philosophers of the time, he is said to have journeyed to Egypt. He was a contemporary of Socrates, which gives rise to some interesting speculations on what the discussions of these two men, holding such opposite beliefs, might have been. Whereas Socrates represented the school of thought that reality was immaterial, that thought was the essence of all things, Democritus represented the school of thought that being was material, that invisible, irreducible particles constituted existence, even what was normally considered the transcendent and ephemeral, such as thoughts, the soul, and the divine. His lost works include books on Pythagoras, mind, causes, geometry, and numbers, which took issue with the Pythagorean hypotheses that abstractions represented reality, which could only be known by means of mathematical reasoning. If *number* dominates the universe, what is the role and significance of flesh-and-blood humans or of a world comprising rock, dirt, plants, and animals? The Pythagoreans, argued Democritus, sought to reduce the universe to *one*, a singularity from which all else derives. But what does the universe really reveal to us? Multiplicity. There are infinite things, uncountable phenomena, multitudes of causes and effects leading to more causes and effects. How can an invisible, ephemeral number, a single being, be the sum of all things, the *cause* from which all else springs?

Democritus and his associate or teacher Leucippus argued that if being existed, it was material. All ideas, spirits, numbers, divinities, and souls, as well as the phenomena of nature, are formed by irreducible substances of the four elements—earth, air, fire, and water—that are invisible yet universally present, in constant motion, undergoing continual change, combining and recombining into infinite forms. Atoms are distinct particles detached from an emptiness or void in which they move, combine, disintegrate, and recombine. How are atoms attached to each other? Rather like puzzle pieces, atoms have distinct and unique shapes, even "hooks" that allow an attachment, if not a merging, of one to another. Aristotle, in *Metaphysics*, condemned the atomist belief that three distinctions in the "shape, order, and positions" of atoms form the bases of all reality (trans. Ross).

To Democritus, the possibilities of "shape, order, and positions" are infinite, hence infinity is the sine qua non of all things. Creation and destruction, if they exist, occur within the overall enduring pattern of infinite creations and destructions. Time is meaningless if by time we mean the register of events in chronological, linear order from beginning to end. Neither *beginning* nor *end* has meaning in an infinite universe. There must then be infinite worlds, suns, stars, moons, men, cities, and so on. Hippolytus described Democritus's theory thus:

> He maintained worlds to be infinite, and varying in bulk; and that in some there is neither sun nor moon, while in others that they are larger than with us, and with

> others more numerous. And that intervals between worlds are unequal; and that in one quarter of space (worlds) are more numerous, and in another less so; and that some of them increase in bulk, but that others attain their full size, while others dwindle away and that in one quarter they are coming into existence, while in another they are failing; and that they are destroyed by clashing one with another. And that some worlds are destitute of animals and plants, and every species of moisture. And that the earth of our world was created before that of the stars, and that the moon is underneath; next (to it) the sun; then the fixed stars. And that (neither) the planets nor these (fixed stars) possess an equal elevation. And that the world flourishes, until no longer it can receive anything from without. (Trans. MacMahon)

The implications are profound for a point of view that makes our solar system part of a vast whole and not alone and singular. If the heavenly bodies are not unique, they can potentially hold a lesser status, be brought down from the realm of the divine. Hence for Democritus the planets, stars, and sun are material, differing in heat and size. Anticipating Galileo's discoveries, Democritus argued (as did Anaxagoras before him) that the moon had mountains and valleys and was not a shining deity. And the geocentric view of the solar system is absurd in an infinite universe.

Diogenes Laertius wrote that Democritus believed that

> atoms and the vacuum were the beginning of the universe; and that everything else existed only in opinion. That the worlds were infinite, created, and perishable. But that nothing was created out of nothing, and that nothing was destroyed so as to become nothing. That the atoms were infinite both in magnitude and number, and were borne about through the universe in endless revolutions. And that thus they produced all the combinations that exist; fire, water, air, and earth; for that all these things are only combinations of certain atoms; which combinations are incapable of being affected by external circumstances, and are unchangeable by reason of their solidity. Also, that the sun and the moon are formed by such revolutions and round bodies; and in like manner the soul is produced; and that the soul and the mind are identical: that we see by the falling of visions across our sight; and that everything that happens, happens of necessity. Motion, being the cause of the production of everything, which he calls necessity. The chief good he asserts to be cheerfulness; which, however, he does not consider the same as pleasure; as some people, who have misunderstood him, have fancied that he meant; but he understands by cheerfulness, a condition according to which the soul lives calmly and steadily, being disturbed by no fear, or superstition, or other passion. . . . Everything which is made he looks upon as depending for its existence on opinion; but atoms and the vacuum he believes exist by nature. (Trans. Yonge)

One wonders, What does have meaning in a material universe of constant motion and apparently random circumstances? Democritus's response is that one must find one's particular place in the universe: this is the means of individual contentment. He counseled acceptance of the contrariness and suffering of life—and acceptance of the joys and pleasures of life as well. Extremes in behavior, thought, and feelings lead to discontent. One must discover the middle, where resides the greatest chance of peace. The universe is amoral and requires humans to fashion their own moral system, which is personal and limited to their time on earth. Death, and the release of one's atoms into oblivion, will come soon enough.

Democritus was quite a geographer as well. He wrote on the size and shape of the earth, the possibility of circumnavigating continents, the causes of terrestrial phenomena, history, and meteorology. He had a profound impact on subsequent thinkers. The ancient world had enough strife, pain, and bitterness to encourage a withdrawal to self and the circumscription of truth and morality to one's own life, thoughts, and feelings. The philosophy of Epicurus and his disciples, such as Lucretius, as well as that of Zeno and his fellow Stoics owed much to Democritus.

PROTAGORAS (C. 485–415 BCE)

Democritus's students included Protagoras, who had the nickname "Wisdom," according to Diogenes Laertius, who also observed that Protagoras

> was the first person who asserted that in every question there were two sides to the argument exactly opposite to one another. And he used to employ them in his arguments, being the first person who did so. . . . And he used to say that nothing else was soul except the senses, . . . and that everything was true. And another of his treatises he begins in this way: "Concerning the Gods, I am not able to know to a certainty whether they exist or whether they do not. For there are many things which prevent one from knowing, especially the obscurity of the subject, and the shortness of the life of man." (Trans. Yonge)

Plato features Protagoras in some of his dialogues, such as the *Protagoras*. He is famous for the remark, "Man is the measure of all things." A clear humanist, Diogenes Laertius claimed Protagoras was the first Sophist to take money for his teaching, and "he was also the first person who gave a precise definition of the parts of time." He invented the type of argument known as sophistry. Besides his treatise on the gods, he wrote on mathematics, on the "Original Condition of Man," a volume on the republic, another on justice, and "two books of Contradictions" (trans. Yonge).

DIOGENES OF APOLLONIA (LATE FIFTH CENTURY BCE)

Another scientist influenced by Democritus was Diogenes of Apollonia, a shadowy figure in the history of science, in part because nothing is known about his life but also because his philosophy and science are obscure. Diogenes Laertius claimed that he was a student of Anaximenes. We only know of his ideas from later commentators. His most well-known book, now lost, was *On Nature*. Diogenes argued that air was the fundamental element upon which all else depended. He believed that to understand the role of air in nature was to understand the distinct characteristics of animals, plants, and humans.

Diogenes accepted in principle the existence of the four elements—earth, air, fire, and water—but so tied were they one to another that a single element must be the source and essence of the other three. This single element was air, which Diogenes believed was the root of life and thought (for without air one dies and does not think). Air forms the soul (*psyche*) of each animal, particularly humans. But the anatomy of humans, animals, and plants differs, hence their respective use of air differs, as do their intelligence and souls.

Aristotle and his student Theophrastus discussed Diogenes's anatomical and psychological ideas at length. The key to human intelligence, he argued, were the blood vessels that were the conduits for the movement of air throughout the human body. Diogenes had a clear, if superficial, knowledge about the extent and location of vessels (veins and arteries). The origins of the two central vessels are in the spleen and liver. The physician, seeking to correct an imbalance of the body's humors, will typically lance the spleen or liver vessel to bleed the patient. According to Theophrastus, Diogenes argued that the quantity and movement of air determines pain and pleasure, intelligence and dullness, courage and fear. An intelligent person, as opposed to a brute animal, nonthinking plant, stupid louse, or ignorant infant, has great quantities of light, fresh, dry air circulating throughout the body, inhabiting the mind, generating quick thoughts and clear judgments.

Diogenes was clearly a man of his times, echoing the materialist theories of Leucippus and Democritus and the developing notion of the four humors of the Hippocratic school. His influence on subsequent thought was limited, however, because of questionable theories described in confusing prose. As Simplicius complained in *Commentary on the Physics*, "Diogenes of Apollonia . . . wrote for the most part in a muddled fashion" (trans. Barnes). Diogenes Laertius, however, claimed that the opening line of his treatise was, "It appears to me that he who begins any treatise ought to lay down principles about which there can be no dispute, and that his exposition of them ought to be simple and dignified." Diogenes Laertius wrote that "his principle doctrines" were "that the air was an element; that the worlds were infinite, and that the vacuum also was infinite; that the air, as it was condensed, and as it was rarified, was the productive cause of the worlds; that nothing can be produced out of nothing; and that nothing can be destroyed so as to become nothing; that the earth is round, firmly planted in the middle of the universe, having acquired its situation from the circumvolutions of the hot principle around it, and its consistency from the cold" (trans. Yonge).

EPICURUS (341–271 BCE)

Epicurus gave his name to Epicureanism, a Greek philosophy that gained tremendous popularity in Rome. Epicurus was an Ionian Greek who lived part of his life in Athens. His teachers, according to Diogenes Laertius, were Nausiphanes and Naucydes. Epicurus gained his own followers upon developing a system of thought based on the atomism of Leucippus and Democritus. He taught that the universe was materialistic, formed exclusively by indestructible, irreducible particles (atoms) that were infinite and in constant motion. Since all is dominated by an impersonal, anonymous force of matter, humans, according to Epicurus, must respond with the search for individual peace of mind (Greek *ataraxia*).

Diogenes Laertius, in his extensive biography of Epicurus in *Lives of Eminent Philosophers*, claimed that Epicurus pursued his own path of learning when his teachers could not explain to him the concept of chaos. Not that the idea of chaos necessarily bothered him—what did bother him was being unprepared to meet the varied contingencies of life with a calm and disinterested passivity. Epicurus taught that the pursuit of pleasure—meaning the absence of pain—was what mattered in life. Death was inevitable, but since there was no sentient afterlife, only disintegration into atoms, death should cause no fear. The gods existed, but they were impotent in a materialistic universe; humans had no fear of retribution for offending the powers of the universe—an atom simply cannot be offended.

Diogenes Laertius wrote further that Epicurus wrote thirty-seven volumes of natural philosophy, discussing atoms and vacuum, among other topics. In these volumes, Epicurus discussed "the canonical [philosophy], which serves as an introduction to science. The physical [philosophy] embraces the whole range of speculation on subjects of natural philosophy," including "production, and destruction, and nature." Of the senses, he wrote,

> The senses are devoid of reason, nor are they capable of receiving any impressions of memory. For they are not by themselves the cause of any motion, and when they have received any impression from any external cause, then they can add nothing to it, nor can they subtract anything from it. Moreover, they are out of the reach of any control; for one sensation cannot judge of another which resembles itself; for they have all an equal value. Nor can one judge of another which is different from itself; since their objects are not identical. In a word, one sensation cannot control another, since the effects of all of them influence us equally. Again, the reason cannot pronounce on the senses; for we have already said that all reasoning has the senses for its foundation. Reality and the evidence of sensation establish the certainty of the senses; for the impressions of sight and hearing are just as real, just as evident, as pain. . . . In fact, every notion proceeds from the senses, either directly, or in consequence of some analogy, or proportion, or combination. Reasoning having always a share in these last operations. The visions of insanity and sleep have a real object, for they act upon us; and that which has no reality can produce no action. (Trans. Yonge)

Epicurus argued, Diogenes said, that the "operations of the senses" provided humans with preconceptions of reality; "these preconceptions . . . furnish us with certainty." Of his physics, Diogenes quoted a supposed letter from Epicurus to Herodotus:

> First of all, we must admit that nothing can come of that which does not exist; for, were the fact otherwise, then everything would be produced from everything, and there would be no need of any seed. And if that which disappeared were so absolutely destroyed as to become non-existent, then everything would soon perish, as the things with which they would be dissolved would have no existence. But, in truth, the universal whole always was such as it now is, and always will be such. For there is nothing into which it can change; for there is nothing beyond this universal whole which can penetrate into it, and produce any change in it. . . . Now the universal whole is a body; for our senses bear us witness in every case that bodies have a real existence; and the evidence of the senses, as I have said before, ought to be the rule of our reasonings about everything which is not directly perceived. Otherwise, if that which we call the vacuum, or space, or intangible nature, had not a real existence, there would be nothing on which the bodies could be contained, or across which they could move, as we see that they really do move. Let us add to this reflection that one cannot conceive, either in virtue of perception, or of any analogy founded on perception, any general quality peculiar to all beings which is not either an attribute, or an accident of the body, or of the vacuum. . . . Now, of bodies, some

are combinations, and some the elements out of which these combinations are formed. These last are indivisible, and protected from every kind of transformation; otherwise everything would be resolved into non-existence. They exist by their own force, in the midst of the dissolution of the combined bodies, being absolutely full, and as such offering no handle for destruction to take hold of. It follows, therefore, as a matter of absolute necessity, that the principles of things must be corporeal, indivisible elements. The universe is infinite. For that which is finite has an extreme, and that which has an extreme is looked at in relation to something else. Consequently, that which has not an extreme, has no boundary; and if it has no boundary, it must be infinite, and not terminated by any limit. The universe then is infinite, both with reference to the quantity of bodies of which it is made up, and to the magnitude of the vacuum; for if the vacuum were infinite, the bodies being finite, then, the bodies would not be able to rest in any place; they would be transported about, scattered across the infinite vacuum for want of any power to steady themselves, or to keep one another in their places by mutual repulsion. If, on the other hand, the vacuum were finite, the bodies being infinite, then the bodies clearly could never be contained in the vacuum. . . . The atoms which form the bodies, these full elements from which the combined bodies come, and into which they resolve themselves, assume an incalculable variety of forms, for the numerous differences which the bodies present cannot possibly result from an aggregate of the same forms. Each variety of forms contains an infinity of atoms, but there is not for that reason an infinity of atoms; it is only the number of them which is beyond all calculation. . . . The atoms are in continual state of motion. . . . Among the atoms, some are separated by great distances, others come very near to one another in the formation of combined bodies, or at times are enveloped by others which are combining; but in this latter case they, nevertheless, preserve their own peculiar motion, thanks to the nature of the vacuum, which separates the one from the other, and yet offers them no resistance. The solidity which they possess causes them, while knocking against one another, to re-act the one upon the other; till at last the repeated shocks bring on the dissolution of the combined body; and for all this there is no external cause, the atoms and the vacuum being the only causes. . . . But, again, the worlds also are infinite, whether they resemble this one of ours or whether they are different from it. For, as the atoms are, as to their number, infinite, as I have proved above, they necessarily move about at immense distances; for besides, this infinite multitude of atoms, of which the world is formed, or by which it is produced, could not be entirely absorbed by one single world, nor even by any worlds, the number of which was limited, whether we suppose them like this world of ours, or different from it. There is, therefore, no fact inconsistent with an infinity of worlds. (Trans. Yonge)

Hippolytus, the Late Roman commentator, in *Refutation of the Heresies*, summed Epicurus's thought thus:

Epicurus, however, advanced an opinion almost contrary to all. He supposed, as originating principles of all things, atoms and vacuity. He considered vacuity as the place that would contain the things that will exist, and atoms the matter out of which all things could be formed; and that from the concourse of atoms both the Deity derived existence, and all the elements, and all things inherent in them, as well as animals and other (creatures); so that nothing was generated or existed, unless it be from atoms. And he affirmed that these atoms were composed of extremely small particles, in which there could not exist either a point or a sign, or any division; wherefore also he called them atoms. Acknowledging the Deity to be eternal and incorruptible, he says that God has providential care for nothing, and that there is no such thing at all as providence or fate, but that all things are made by chance. For

> that the Deity reposed in the intermundane spaces, (as they) are thus styled by him; for outside the world he determined that there is a certain habitation of God, denominated the intermundane spaces, and that the Deity surrendered Himself to pleasure, and took His ease in the midst of supreme happiness; and that neither has He any concerns of business, nor does He devote His attention to them. As a consequence on these opinions, he also propounded his theory concerning wise men, asserting that the end of wisdom is pleasure. Different persons, however, received the term pleasure in different acceptations; for some (among the Gentiles understood) the passions, but others the satisfaction resulting from virtue. And he concluded that the souls of men are dissolved along with their bodies, just as also they were produced along with them, for that they are blood, and that when this has gone forth or been altered, the entire man perishes; and in keeping with this tenet, (Epicurus maintained) that there are neither trials in Hades, nor tribunals of justice; so that whatsoever any one may commit in this life, that, provided he may escape detection, he is altogether beyond any liability of trial (for it in a future state). In this way, then, Epicurus also formed his opinions. (Trans. MacMahon)

Epicurus was hardly a scientist but rather a philosopher who formed a pattern of life in response to a particular scientific paradigm popular in Greece toward the end of the Classical Age and beginning of the Hellenistic Age. Such were the conflicts and uncertainty of these last few centuries BCE that Epicurus's teachings continued to influence other thoughtful individuals looking for a way out of the conundrum of trying to explain self in what appeared to be an increasingly cold and impersonal universe. His most famous disciple was the Roman poet Lucretius.

EPICUREAN SCHOOL

Epicurus's students, according to Diogenes Laertius, included Metrodorus of Athens, who wrote on topics such as physicians, Sophism, and wisdom; and Hermarchus of Mitylene, who wrote on mathematics and philosophy. The Epicurean philosophy was a largely amoral system of thought that influenced Greece and Rome particularly in the last several centuries BCE. Epicurus gave his name to a philosophy built upon several radical (for the time) assumptions. Epicureans adopted the theories of Leucippus and Democritus that atoms, indestructible, irreducible, infinitely moving material particles, composed all things. Mind, soul, gods, love—anything emotional and spiritual—have an atomic basis. Atoms were, however, insentient, hence impersonal and amoral. Assuming such a universe without apparent purpose, first Democritus and then Epicurus developed a moral philosophy in response. There was no need to fear anything if the phenomenon causing the most fear, death, was irrelevant. One died without knowing the result, as the body, mind, and soul disintegrated into multitudinous atoms that would eventually go to form some other material item. Why fear the retribution of the gods when the gods were impotent (even if they did exist) and could not reward or punish?

Epicurus taught that one must focus exclusively on the self and its ability to find meaning, purpose, and happiness. He argued that it was easier to find happiness without death and retribution hanging over a person. Time became, in a sense, irrelevant, as the duration of life hardly mattered when disintegration and nonexistence was the end of all human life and endeavor. Epicurus sought peace of mind, a stable, passive, almost emotionless state of lack of pain. Pleasure is the key to happiness, Epicurus argued; pleasure is simply the lack of pain. Life is a series of moments devoted to peace, lack of conflict, lack of pain, lack of emotion, lack of intense pleasure—a true middle state.

Knowledge of who first proposed the idea of fundamental material elements making up all existence is lost in time. One tradition has it that the Egyptians conceived of four elements as the basis of all things, which inspired Greek thinkers to enlarge on the theory. The earliest Greek scientists conceived of earth, air, fire, and water as being the fundamental elements of the universe. Thales argued for water being the primal element; Anaximenes thought it was air. Empedocles thought that love and strife defined and moved the four elements. Aristotle conceived of a ranking of the elements in the universe: earth has a downward movement because it is heaviest; water is next in weight; air is third, being much lighter; fire is fourth, having an inherent upward movement. Aristotle argued for a fifth element, ether (aether) to provide the quintessence (fifth essence). Ether did not partake of the material nature of the other four, nor did it experience condensation and rarefaction. Since ether was the epitome of perfection, it had a circular motion and was the stuff through which the heavenly bodies roamed. Later scientists, such as Leucippus, Democritus, and Epicurus, conceived of a more fundamental source of the material four elements: atoms. Each element, they thought, is composed of atoms, invisible, constantly moving, material foundations of all things.

FURTHER READING

Aristotle, *Metaphysics*, trans. W. D. Ross, in *The Works of Aristotle*, 2 vols. (Chicago: Encyclopedia Britannica, 1952).

Jonathan Barnes, trans., *Early Greek Philosophy* (London: Penguin Books, 1987).

Diogenes Laertius, *The Lives and Opinions of Eminent Philosophers*, trans. C. D. Yonge (London: Henry Bohn, 1853).

Sir Thomas Heath, *Aristarchus of Samos* (New York: Dover Books, 1981).

Hippolytus, *Refutation of the Heresies*, trans. J. H. MacMahon, in *Ante-Nicene Fathers*, vol. 5, ed. Alexander Roberts, James Donaldson, and A. Cleveland Coxe (Buffalo, NY: Christian Literature Publishing, 1886); revised and edited for New Advent by Kevin Knight, http://www.newadvent.org/fathers/050101.htm.

Henry M. Leicester, *The Historical Background of Chemistry* (New York: Dover Books, 1971).

Lucretius, *The Nature of the Universe*, trans. R. E. Latham (Harmondsworth, England: Penguin Books, 1951).

Plutarch, *Essays and Miscellanies: The Complete Works*, corrected and revised by William W. Goodwin, vol. 3, https://www.gutenberg.org/files/3052/3052-h/3052-h.htm.

Bertrand Russell, *A History of Western Philosophy* (New York: Simon and Schuster, 1945).

11

Athenian Science (500–300 BCE)

The center of science in the Archaic and Classical periods of ancient Greece was the city of Athens, located in eastern Greece on the Attic Peninsula. Athens harked back to the Mycenaean period of the second millennium BCE. It withstood the Dorian invasions of the twelfth and eleventh centuries sufficiently to emerge as the center of culture on the mainland in the ninth and eighth centuries. The Athenians at the beginning of the Archaic Age, circa 800 BCE, identified with the Greeks of the Aegean islands and western Turkey—the Ionians—and not with the Dorians to the south and west. Athens prided itself on its intellectual and cultural stature, a preeminence rarely matched for over a thousand years.

The foundations of Athenian intellectual greatness came during the Archaic Age, from 800 to 500 BCE. During this time, Athens went through dramatic changes in government, society, economy, and thought. Increasing agricultural production and surplus allowed for the growth of population and trade, so much so that Athens, like other Greek city-states (poleis), sent citizens abroad to found new colonies. Wealth and trade led to the rise of the middle class, which demanded more input into local government, which resulted in the world's first *rule of the people*, democracy, as well as a greater role in military affairs. The development of the heavily armed infantry, the hoplite soldier, signified the social and economic ferment of Archaic Age Athens.

Social, economic, and political change always accompanies intellectual change. At the same time that Athenian hoplites were leading Greeks in the defense of Hellas from Persian invasion, and as the Athenian empire grew and its political and economic dominance carried forth into the Aegean islands and Ionia, there were dramatic changes in philosophy and science. Around the time that Pericles came to power in Athens, an Ionian philosopher arrived from Asia Minor, Anaxagoras of Clazomenae. Anaxagoras brought with him the general beliefs of Ionian philosophers—a confidence in human thought, a questioning attitude, a search for the causes of natural phenomena, and an attempt to explain the nature of humans and the universe—as well as his own specific ideas, the chief of which was his belief that mind (*nous*) governed all things. This notion that a transcendent, spiritual force—an idea—was the ultimate reality of which all being partakes was revolutionary for its time. Many Athenians distrusted Anaxagoras's ideas and thought he was impious, atheistic, and dangerous. Pericles, the great Athenian who became Anaxagoras's pupil, was the exception; he took the philosopher's truths to heart, according to ancient writers such as Plutarch, and developed into a generally good man who could weather the storms of human existence.

DEMOCRACY

At the beginning of the Peloponnesian War, Pericles, a general, gave a funeral oration, as recorded by the historian Thucydides. In the speech, Pericles praised Athens as a democracy, where the people ruled and equality among citizens existed. Athens was an open society, he said, where goods were freely exchanged in the agora, or marketplace, and ideas were freely exchanged in the open air and bright light of free speech and movement. Athens, said Pericles, "is an education to Greece" insofar as the political and intellectual freedom brought about by democracy led to dramatic advances in culture, art, literature, philosophy, and science (trans. Warner).

The fact that scholars debate whether Athens, an imperialistic state that denied rights to women and slaves, should be called a democracy should not prevent us from recognizing the incredible intellectual accomplishments of Periclean Athens. Fifth-century Athenians made a study of the science of government. Ancient political and social scientists such as Solon, Thucydides, Pericles, Socrates, and Plato studied the nature of government and society not according to random acts of the personalities of gods and humans but according to the actions and movements of people, the *demos*, over time. These thinkers and others explained the role of virtue (*arete*) and restraint (*sophrosyne*) in the functioning of the state, that the mind and the will—not brute passions—govern a people. The arrogant pursuit of power in the individual was overwhelmed by the common needs of the whole as acted out in the Council of 500, the assembly of free citizens, the court system and trial by jury, and the institution of ostracism, meant to curtail the power of demagogues. The key to such political and social success was simply freedom; as Aeschylus proclaimed in *The Persians*, Athenians "call themselves no man's slaves or subjects" (trans. Collard).

THE AGE OF PERICLES

During the mid-fifth century, Pericles and his mistress, Aspasia, herself a philosopher, gathered among them the leading minds of Athens and Ionia to create one of the most dynamic expressions of civilization ever seen in human history. It was Pericles who executed the plan to rebuild the Athenian Acropolis, which was the religious center of the city. The Persian invasions of the early fifth century had destroyed many of the buildings. After the Greek defeat of the king of Persia, Xerxes, and the establishment of Greek freedom, Athens took the lead in forming the Delian League, designed to be an alliance of Ionian Greek city-states to protect western Turkey and the Aegean from Persian attack. Under Pericles's leadership, Athens controlled the Delian League and demanded tribute from subject Greek city-states in return for Athenian military and naval protection. These funds, acquired by means of the Athenian Empire, were used to rebuild the Acropolis.

The architectural, engineering, and artistic genius that went into the construction of the temples of the Athenian Acropolis continue to inspire and generate awe. The grandest temple, of course, is the Parthenon, named for Athena, the virgin (*parthenos*) goddess and patroness of Athens. Even after 2,400 years, the Parthenon continues to dominate the Acropolis. The architects were Ictinus and Callicrates; Pheidias oversaw the whole. The Parthenon is a monument to grace and beauty—tons of marble sculpted, molded, and set in just the right way. The design was sophisticated and intricate, the quality of its construction and engineering a marvel. Pericles also worked with the Milesian urban planner Hippodamus (498–408 BCE), who redesigned the Athenian port city of Piraeus. Hippodamus reputedly wrote the *Urban Planning Study for Piraeus*. Aristotle wrote that Hippodamus was also a political scientist who visualized the ideal city. Hippodamus also planned the structure of the Athenian colony of Thourioi (Thurii), in southern Italy, and was thought to have helped redesign the city of Rhodes.

Athens was home to many other thinkers of varying stripes. Herodotus (490–430), originally from Halicarnassus in western Turkey, spent time in Athens promoting his *Histories*; for the first time, Athenians heard of the geography, history, and peoples of the eastern Mediterranean. One thinker was not impressed: Thucydides (460–400) composed his *History of the Peloponnesian War* using an exact, chronological style in contrast to Herodotus's more rambling account. Thucydides used history as a route to uncover general patterns in human behavior. Playwrights such as Aeschylus, Sophocles, and Euripides likewise used the stage to develop themes of human behavior and its consequences, the relation of humans to the divine, and the conflict between good and evil. Xenophon (430–355), a student of Socrates, wrote of other places, such as Persia, to the east. Antisthenes (445–365), also a student of Socrates and a founder of the Cynic school of thought, preached philosophy and poverty at Athens. Diogenes Laertius wrote that "he taught that the wise man was sufficient for himself." He was, Diogenes said, "the original cause of the apathy of Diogenes, and the temperance of Crates, and the patience of Zeno" (trans. Yonge). He wrote on rhetoric, law, zoology, physiognomy, logic, natural philosophy, and Homer.

Aspasia was a well-known thinker and philosopher. Originally from Miletus, a leading center of science during the fifth century BCE, Aspasia was possibly associated with Anaximander and Archelaus, Pericles's and Socrates's teachers, respectively. A hetaera, one of the elegant, promiscuous consorts of the rich and famous of Athens, she was Pericles's companion for many years, holding forth in salon fashion among the intellectual elite of mid-fifth-century Athens. She reputedly opened a school of philosophy, having not only well-born women but some of Athens's most important men as her students.

SOLON (640–560 BCE)

The Greeks considered Solon the Athenian statesman to be one of the Seven Sages because of his wisdom in framing laws for Athens. In his *Lives of the Philosophers*, Diogenes Laertius claimed that of the Seven Sages, all were statesmen save Thales, who was the only naturalist. Plutarch's *Life of Solon* points out, however, that if Solon was not a naturalist, he was a political thinker who devised laws to promote equality and democracy in sixth-century Athens. Solon was a political scientist, the first and perhaps greatest in antiquity, one who applied reason to law and government.

Little is known of Solon's life except through anecdote. The sources indicate that he was born in Salamis, the little island south of Athens. He came to Athens at an early age and gained a reputation for wisdom. He was a merchant who earned great wealth and came to see that Athens's hope for political and economic greatness lay in production, trade, and the sea. As archon, an executive office in early Athens, Solon suggested that Athenians open their city and citizenship to foreign craftspersons and traders, hence to rid the city of its reliance upon land and decreasing the power of the landed aristocracy in the process. A city of merchants, craftspersons, and laborers necessarily moved toward a more open society politically—an important precondition for democracy in Athens. Such was Solon's reputation that he gained the respect of the Athenians to reform their law code to conform to their more middle-class society. The focus of his laws was on fairness, equitability, and reason. His legislation reformed the calendar, based advancement and participation in government on merit and wealth, matched a sense of public responsibility to the accumulation of wealth, relieved the burdens of the poor by canceling debts, and established the institutions upon which democracy in Athens would be based.

Aristotle, in his *On the Athenian Constitution*, praised Solon for his government reforms, in particular three of them: "First and most important, the prohibition of loans on the security of the debtor's person; secondly, the right of every person who so willed to claim redress on behalf of any one to whom wrong was being done; thirdly, the institution of the appeal to the jury courts; and it is to this last, they say, that the masses have owed their strength most of all, since, when the democracy is master of the voting-power, it is master of the constitution" (trans. Kenyon).

According to ancient writers, upon completing his legal reforms, Solon decided to leave the city for ten years to allow the reforms to take shape without the constant presence of their creator. He traveled to Egypt, as all wise men of antiquity apparently did, spoke with the Egyptian priests, and learned from them as well. He traveled to Lydia, where he met King Croesus, who was wealthy, powerful, and vain. Croesus sought compliments from all who visited him, but wise Solon refused to accommodate the king. Instead, he tried to convince Croesus that happiness had nothing to do with wealth but, rather, with a useful life and honorable death. Croesus, according to Herodotus, learned this lesson later, when he was captured by Cyrus, the king of Persia. According to anecdotal accounts, Solon visited just about everywhere on his ten-year sojourn, making a point to visit the other sages like himself and giving advice wherever he went. He became the legendary legal sage of antiquity, whose life showed posterity that wisdom, reason, sound judgment, and a clear understanding of human experience has much to do with the science of government. Aristotle was impressed, and so, too, were the framers of the U.S. Constitution centuries later.

Solon was a poet who wrote his own epitaph, as it were, indicating his accomplishments as a lawgiver:

> I gave to the mass of the people such rank as befitted their need,
> I took not away their honour, and I granted naught to their greed;
> While those who were rich in power, who in wealth were glorious and great,
> I bethought me that naught should befall them unworthy their splendour and state;
> So I stood with my shield outstretch, and both were safe in its sight,
> And I would not that either should triumph, when the triumph was not right.
>
> *(Trans. Kenyon)*

ARCHELAUS (FIFTH CENTURY BCE)

Archelaus was the student of Anaxagoras and the teacher of Socrates. He was a transitional figure in the movement of the Ionian school of thought to Athens, where it would find its greatest exponents in Socrates, Plato, and Aristotle. Archelaus was probably born at Athens, though some accounts claim he was a native of Miletus. At any rate, he had the same scientific interests as the Milesians. This debt is clearly seen in Archelaus's ideas that heat and cold are the two basic principles of matter. "The originating principle of motion," Hippolytus records Archelaus as stating, "is the mutual separation of heat and cold, and that heat is moved, and that the cold remains at rest" (trans. MacMahon). He further argued, as reported by Hippolytus, "that water, being dissolved, flows towards the centre, where the scorched air and earth are produced, of which the one is borne upwards and the other remains beneath" (trans. MacMahon). Archelaus anticipated Aristotle by arguing that hot is movement and cold is rest. Like Anaxagoras, he believed that mind (*nous*) is the initial cause and being. Hippolytus noted that Archelaus believed

> that the earth is at rest, and that on this account it came into existence; and that it lies in the centre, being no part, so to speak, of the universe, delivered from the conflagration; and that from this, first in a state of ignition, is the nature of the stars, of which indeed the largest is the sun, and next to this the moon; and of the rest some less, but some greater. And he says that the heaven was inclined at an angle, and so that the sun diffused light over the earth, and made the atmosphere transparent, and the ground dry; for that at first it was a sea, inasmuch as it is lofty at the horizon and hollow in the middle. And he adduces, as an indication of the hollowness, that the sun does not rise and set to all at the same time, which ought to happen if the earth was even. (Trans. MacMahon)

Archelaus assigned thought to animals as well as humans, though by degree of complexity one excels the other in the development of the attributes of civilization. Hippolytus wrote,

> He affirms that the earth, being originally fire in its lower part, where the heat and cold were intermingled, both the rest of animals made their appearance, numerous and dissimilar, all having the same food, being nourished from mud; and their existence was of short duration, but afterwards also generation from one another arose unto them; and men were separated from the rest (of the animal creation), and they appointed rulers, and laws, and arts, and cities, and the rest. And he asserts that mind is innate in all animals alike; for that each, according to the

difference of their physical constitution, employed (mind), at one time slower, at another faster. (Trans. MacMahon)

How precisely Archelaus influenced Socrates is unclear. Hippolytus argued that Socrates was a pupil of Archelaus. Diogenes Laertius claimed that philosophy took a turn with Socrates because of his development of ethics, a field of thought he learned from Archelaus. More importantly, Socrates believed mind was the essence of all things and reality is immaterial and transcendent, which were ideas clearly developed by the Ionian school of thought, in particular by Anaxagoras.

SOCRATES (469–399 BCE)

Notwithstanding its manifold limitations, Athenian freedom allowed a commoner such as Socrates to become one of the great thinkers of all time; to develop sophisticated ideas on metaphysics, government, society, and philosophy; and to influence scores of Athenians. So influential was he that those jealous of his intellect sought to get rid of him and succeeded in doing so. Socrates inspired the work of countless other philosophers and scientists, such as Plato, Aristotle, and Xenophon. He was a philosopher of *being* perfectly in line with Ionian thinkers such as Anaxagoras and Archelaus. Socrates believed that reality is incorporeal and transcendent, perceived by very few thinkers who have the intellect and discipline to penetrate the generally unknowable ideal forms of the universe. Socrates did not write anything, but we know of his thought by means of his disciples Plato and Xenophon. Socrates, according to his student Plato, elevated the workings of the state and varied roles of the people into an abstract conceptual framework wherein the practical aspects of rights and wrongs, of justice and injustice, were understandable according to the idea of justice itself. How, Socrates asked, can we know whether or not justice is accomplished at the Athenian courts if we are ignorant of what true justice is, regardless of time and place? With Socrates we find the beginnings of a rigorous examination of thought and practice that formed the basis for the social and political theories of today's social sciences.

Plato's dialogues indicate that Socrates believed that constant questioning would lead to the truth. His technique in conversation, teaching, and debate was to pose a question, responding to the answer with another question, and so on. Each question became more penetrating and was posed in such a way to yield a desired response from the respondent. Plato's dialogues show that Socrates had already deduced the truth that he was carefully and patiently trying to elicit from his conversants. This technique of assuming an a priori truth that one will certainly arrive at through a mental process is the hallmark of deductive reasoning, of which Socrates was the greatest practitioner.

Socrates was unafraid to challenge the beliefs and knowledge of others in his pursuit of truth and his goal of revealing this truth and the correct methods by which to achieve a knowledge of the truth to others. His challenging intellectual demeanor eventually got him into trouble after the Athenians had lost the Peloponnesian War to the Spartans and Athenian democracy was crumbling, at the end of the fifth century. Socrates was put on trial for corrupting the youth of

Athens, charges that he vigorously denied and showed to be absurd in his trial, as recorded by Plato in the *Apology.* Nevertheless, Socrates was convicted and sentenced to die, which he could avoid by exile. Athenian to the end, Socrates chose to end his life by drinking the hemlock provided by the executioner, discoursing with his disciples on life and death to the last possible moment.

The shadow of Anaxagoras and the Ionian philosophers and scientists hung over Athens during the fifth century. This was in part because the aggressive militaristic and imperialistic actions of the Athenians contradicted their professed beliefs in democracy and equality. Perhaps more important, Anaxagoras's pupil, Archelaus, was the teacher of Socrates, who was the teacher of Plato. Hence the great thinkers throughout Hellas of the fifth century were centered at Athens. Socrates had a reputation as a Sophist, though his disciple Plato vehemently disagreed with the assertion, pointing out that Socrates did not take money for his teachings, unlike true Sophists such as Protagoras. In the dialogue *Protagoras*, Plato imagined or re-created a debate between Protagoras and Socrates, in which Socrates's dialectic method of interrogatives and logical deduction overwhelmed Protagoras's reliance upon rhetoric. Protagoras signified the use of the spoken word to persuade—to create, as it were, a momentary truth—whereas Socrates advocated the view, which Anaxagoras would have found acceptable, that truth is everlasting for all times and places and only those thinkers who train their mind to seek the truth will be able to find it.

PLATO (427–347 BCE)

Greek thinkers of the sixth and fifth centuries BCE, such as Thales, Pythagoras, Anaxagoras, Xenophanes, and Socrates, sought to explain the universe by imposing the structure of the human mind upon it, to conceptualize it so to understand it. They broke from the primitive anthropomorphism of their forebears to create a universe that resembled the human mind—a universe with a sound ethical structure, one with a logical, orderly, harmonious whole. The Greek philosopher Plato built a philosophy around his conception of the divine attributes of the universe, which he called *ideal forms.*

Plato was a product of the aristocratic Athenian society, which was naturally suspicious of democracy, which often appeared to be mob rule. This was his experience as a young man watching his friend and mentor Socrates put on trial for simply teaching the truth. Many of Plato's most famous dialogues feature Socrates teaching his students, confronting his enemies, defending himself, and preparing for death by drinking hemlock. Socrates died in 399 BCE, after which Plato traveled and learned more about Pythagoras and his teachings. When he returned to Athens after about ten years, he opened a school called the Academy. Students at the Academy studied philosophy, logic, music, astronomy, and mathematics. Plato believed that knowledge derived from intuition and reason. He eschewed empirical, scientific thinking, arguing that mere observation reveals only the shadows of reality. Plato was not a scientist, yet his thought was so broad and penetrating that he had a founding role in many disciplines of study that would in time become empirical sciences.

Political Science

Plato, his mentor Socrates, and his student Aristotle could arguably be considered the founders of what today we call political science. They formed a systematic approach to government, employing a methodology based on reason, logic, and observation. Most important, for Plato was to consider what government *should be* and then to follow with implementation of ideas upon what *is*. Plato argued in his dialogue *Republic* that the "universal principle" of justice could be mirrored in human society by building a community where "everyone ought to perform the one function . . . for which his nature best suited him" (trans. Cornford). Plato extended community justice to the self, arguing that just individuals achieve order and tranquility in their conduct by self-mastery and discipline, bringing themselves into tune with the universe, forming an inward musical harmony.

Plato's ideal society, the republic, has often been criticized for creating a totalitarian state based on the absolute control of the *guardians*: those who, like Socrates, approached knowledge of the ideal forms. Plato, a product of the Athenian democracy of the fifth century, sought to free humans from the problems inherent in the free exchange of ideas, by forcing them to conform to the *truth*, the ideal forms as understood by the guardians. Plato thus erected something akin to an ideal city, one reflecting the truth. What would be the point of freedom in a society where the truth is known? Freedom only matters in a society where the truth is *not known*—hence comes the demand for the freedom to pursue it.

The truth, as Plato conceived it, is transcendent, invisible, generally unknown; material, corporeal existence hints at the truth, as the mind informed by the senses can only approximate an understanding of truth. Humans only experience the shadows of what is real. But the shadows provide a beginning for those exceptional thinkers, such as Socrates, who can bring the mind out of its inherent darkness to approach the light of truth. The technique to discover reality first involves mathematics, understanding the principle of *number*, which is based on infinite individual units that are distinct wholes. Geometry, the shapes that represent transcendent forms, help the mind to grasp abstract ideas, which sets the stage for dialectic. Socrates was the master of dialectic, the technique of question and answer, of drawing out the truth by a series of ever more penetrating interrogatives.

Through geometry and dialectic, for example, Plato discovered (after Pythagoras) that the earth is a sphere. The sphere is the most perfect shape, and as the earth is the most perfect solid, the earth therefore is a sphere. Astronomy is important, but only insofar as mathematics and dialectic inform us of the true nature of the universe that we only dimly perceive with our senses.

Plato believed that the wisest persons in a society should rule it. In the *Republic*, these were the guardians, both male and female, who trained for years to recognize and know the truth and lived in communal fashion, unconcerned with the material conditions of life. Plato also conceived of a society where the single ruler, the king, would be a philosopher as well, therefore ruling according to justice. This idea of the philosopher-king was one of Plato's greatest legacies. Alexander the Great, the student of Aristotle, sought to rule as a philosopher. Marcus Aurelius (121–180 CE), a

Stoic, was emperor of Rome. Many other Roman emperors thought of themselves as philosophers, even if the fact did not match the image. Julian (331–363 CE) was the ideal philosopher-king: he ruled according to philosophy (Neoplatonism), he modeled himself on the philosophers of past and present, and he exercised the wisdom of humility. Plato himself tried to put his ideas into action in the person of Dionysius II, tyrant of Syracuse (r. 367–357, 346–344). As in most such cases, however, power was more persuasive than philosophy, and Plato failed miserably.

Academy

The words *academy*, *academe*, and *academic* derive from the fourth-century Athenian school founded by the Greek philosopher and scientist Plato. Plato founded his academy in the years after he had been a student of Socrates, had experienced his teacher's death in 399, and had traveled to places throughout the Mediterranean, ending up at Syracuse, where he had failed in trying to make Dionysus, the tyrant of the city, into a philosopher-king. Plato decided that if he could not turn kings into philosophers, then at least he could train the sons of Athens in the art of thinking, the understanding of what is real and true, the best way to live, and the art of citizenship.

The Academy was named for a local god and was dedicated to Zeus's daughters, the Muses. Men and women were equally admitted to the Academy; the two requirements were a good understanding of mathematics, particularly geometry, and wealth—the school was free but relied on donations of wealthy alumni. Mathematics formed the core of the curriculum: arithmetic, geometry, and related subjects such as astronomy and music. Plato used Socrates's technique of the dialogue supplemented by lectures and discussions. Two women who attended the academy, according to Diogenes Laertius, were Axiothea and Lastheneia. Plato also claimed that there was a female Athenian philosopher who predated Socrates, called Diotima.

Upon Plato's death in 346, the directorship of the school was assumed by Speusippus (408–339 BCE). Diogenes Laertius related that Speusippus "investigated in his school as was common to the several sciences; and [he] endeavoured, as far as possible, to maintain their connection with each other." He also allowed the "female pupils of Plato, Lasthenea of Mantinea, and Axiothea of Phlius," to become his students as well. Diogenes recorded a variety of his dialogues dealing with justice, philosophy, the soul, *Essays on the Genera and Species of Examples*, and a commentary on mathematicians (trans. Yonge). Xenocrates (396–314 BCE), who succeeded Speusippus, was the epitome of the philosopher: dedicated to wisdom, chaste, poor by choice. He lived the life of a philosopher whose behavior and motives were beyond reproach. Among the anecdotes that Diogenes Laertius wrote of Xenocrates was this: "On one occasion, when a sparrow was pursued by a hawk, and flew into his bosom, he caressed it, and let it go again, saying that we ought not to betray a suppliant"; and "to one who had never learnt music, or geometry, or astronomy, but who wished to become his disciple, he said, 'Be gone, for you have not yet the handles of philosophy.'" He wrote, according to Diogenes, "six books on

Natural Philosophy; six on Wisdom; . . . one volume on the Indefinite; . . . one on Species; . . . one on Ideas, one on Art; two on the Gods; two on the Soul; one on Knowledge; one on the Statesman; one on Science; one on Philosophy; one on the School of Parmenides." The six volumes of natural philosophy included, "the Principal, one; one treatise on Genus and Species; one on the doctrines of the Pythagoreans; two books of Solutions; seven of Divisions; several volumes of Propositions." In addition, he wrote "six books on Mathematics; two more books on subjects connected with the Intellect; five books on Geometry; . . . one of Contraries; one on Arithmetic; one on the Contemplation of Numbers; one on Intervals; six on Astronomy," among many others (trans. Yonge). He was head of the Academy—the scholarch—for twenty-five years. He was succeeded by Polemon, who controlled the Academy from 313 to 269; then Crates, scholarch from 269 to 264; and then the skeptic Arcesilaus (315–240 BCE), who in the third century was head of what was subsequently called the Middle Academy. Diogenes described Arcesilaus's skeptical thought thus: he was "the first man who professed to suspend the declaration of his judgment, because of the contrarieties of the reasons alleged on either side. He was likewise the first who attempted to argue on both sides of a question, and who also made the method of discussion, which had been handed down by Plato, by means of question and answer, more contentious than before" (trans. Yonge). Lacydes of Cyrene succeeded Arcesilaus as scholarch (241–215).

Students of the Academy taught others the Socratic approach to knowledge and life, engendering a school of thought focusing on transcendent realities perceived by human reason and intuition that were subsequently termed "academic." Many of the great mathematical accomplishments of the age were initiated by the academics. During the third to the second centuries, under Arcesilaus, Carneades (214–139 BCE), founder of the New Academy, and his successor, the Carthaginian Clitomachus (187–110 BCE), academic thought became more skeptical, even cynical. During this time, the Academy gained a reputation for sophistry and splitting hairs over minute philosophical issues (as opposed to the reputation of Aristotle's successors [see below], who would focus on practical solutions to the many questions of life and nature). Carneades, for example, could not find anything in the daily happenings of life in Athens—or anywhere else, for that matter—that resembled *reality*, which is unseen and unknown. Carneades continued Plato's emphasis on the hazy understanding of reality in a physical world. Academics of the first century, such as Philo of Larissa (154–84 BCE) and Antiochus of Ascalon (125–68 BCE), were less concerned with finding said reality and more concerned with their reputation compared to other schools of thought. Significantly, Antiochus of Ascalon was a teacher of Marcus Tullius Cicero; hence the Academy began to influence Roman philosophy and culture, as had other schools of thought—namely, the Epicurean and the Stoic—before it.

ARISTOTLE (384–322 BCE)

Plato's most famous student, Aristotle, attended the Academy but never became head of the school. Instead, Aristotle eventually (334 BCE) opened a rival school at Athens, the Lyceum, the curriculum of which was based on Aristotle's inductive scientific approach rather than Plato's deductive, rational, and intuitive approach.

Aristotle was the greatest scientist of the ancient world. A student of Plato, Aristotle, was the teacher of Alexander the Great and the founder of the Peripatetic school of thought. Aristotle mastered all objects of inquiry, including metaphysics, physics, logic, politics, ethics, poetry, zoology, biology, astronomy, geography, natural history, psychology, and magic and astrology. His vast writings included *Metaphysics*, *Physics*, *Nicomachean Ethics*, *Politics*, and *Poetics*. Aristotle was one of the first empirical thinkers, though he generally relied on tried-and-true methods of science: observation, collection and categorization of specimens, analysis of data, induction, and deduction. In some places of his works, such as *History of Animals*, Aristotle describes experiments such as dissection. He collected information in a variety of ways, including interviewing supposed experts on subjects: fishermen on fish, shepherds on goats, apothecaries on drugs, soothsayers on animal behavior. He was willing to consider the work of the poets, such as Homer, as sources of scientific information. Aristotle's mastery of the subjects he studied gained him the reputation in subsequent centuries as an infallible guide to natural phenomena and philosophy. After 1500 CE, in light of new discoveries by Copernicus, Galileo, Newton, and others, many of Aristotle's theories were rejected, though his influence on modern science is undeniable.

Aristotle was born in the small town of Stagira, in Thrace, a primitive outpost of Greek culture east of Macedonia. His father was a wealthy court physician to the kings of Macedonia. Hence Aristotle spent his early years at Pella, the capital city of King Amyntas (r. 393–370) and his successor King Philip II (382–336). Aristotle, seeking to follow in his father's footsteps as a scientist and physician, journeyed south to Athens in 366. He became a student at the Academy, Plato's school in Athens. Aristotle became the philosopher's most famous student. At the Academy, Aristotle fit in as a wealthy aristocrat, but his Thracian/Macedonian background plagued him among condescending Athenians. In the end Aristotle's superior intellect silenced all criticism.

From Plato, Aristotle learned of the idea of universal truth, which Socrates termed "the Good." Plato and his teacher, Socrates, believed that the Good and other transcendent ideals such as Justice and Beauty could not be known or seen but, rather, were beyond human conception: what we call good, justice, and beauty are mere shadows of the truth. Aristotle, however, questioned whether such truths were beyond human comprehension: perhaps they were every bit a part of human experience. Plato taught his students at the Academy that the best means to approach an understanding of truth was through reason, the study of mathematics and music, intuition, and intense and deep contemplation. Aristotle, less the mystical, more the pragmatic thinker, broke from his teacher by adopting the scientific approach to human behavior, natural philosophy, natural science, ethics, and metaphysics, but not before learning from Plato the concept of *being* (Greek *ousia*), the divine essence from which all things derive. Aristotle did not abandon this religious interpretation of the ultimate reality but brought science to bear to discover and to understand it. For Aristotle, then, science is a pious act focused on discovering the nature of goodness, justice, virtue, and being. Human experience is an essential matter for study, since the better human beings echo *being* itself.

Upon Plato's death, Aristotle departed what was no doubt a hostile and competitive situation among Plato's students, each jockeying to take the place of the master. Aristotle, however, journeyed to a small kingdom in Asia Minor (present Turkey) where he became court philosopher to King Hermias, who ruled from 350 to 341. Aristotle married the king's daughter but soon fled (with his wife) upon the tragic assassination of the king. Aristotle ended up back in Macedonia in 343, this time as tutor to the royal prince Alexander. Legend has it that Philip enticed Aristotle to return to Pella, an intellectual and cultural backwater compared to Athens, with a tempting salary and a promise: Stagira having been destroyed and its population enslaved in one of Philip's campaigns, Philip proposed that in return for Aristotle's services the king would rebuild the town and bring the inhabitants out of slavery. Aristotle agreed to the terms.

Alexander eventually became king of Macedonia in 336, upon his father's assassination, and then spent the next thirteen years of his life conquering Greece, Asia Minor, Palestine, Egypt, Iran, Iraq, and Afghanistan—all of which made up the Persian Empire. Alexander was a warrior, conqueror, and megalomaniac who suspected himself to be the heroic son of the king of the gods, Zeus. And yet, strangely, Aristotle, who eschewed the life of a warrior, had been his teacher for three years during the impressionable ages from thirteen to sixteen. Indeed, below the surface of Alexander's actions, one can see hints that Alexander had adopted the life of a philosopher, that he thought of himself as a scientist, even a physician. Alexander, for example, composed letters to Aristotle that included samples of plant and animal life that the king had gathered for his teacher's collection. So at some point during those three years from 343 to 340, Aristotle did have an impact on Alexander the Great. In the meantime, Aristotle had left Macedonia for Athens, where he opened the Lyceum, named for a grove in Athens sacred to Apollo. The philosopher eventually broke with Alexander over the death of Aristotle's grandnephew Callisthenes (c. 360–327), a philosopher and historian who accompanied Alexander's expedition; Callisthenes was implicated in a plot to assassinate the king and was executed. Even so, the Athenians associated Aristotle with Alexander, who was very unpopular in Athens. Upon Alexander's death in 323, the Athenians felt free enough to throw off the shackles imposed on them by Alexander, and one shackle was represented by Alexander's former teacher. Aristotle became very unpopular in Athens and was eventually forced to flee the city and abandon his school. He died soon after, in 322 BCE.

Logic

Aristotle is perhaps best known today as a logician. He created a system of thought based on "first principles," fundamental assumptions that one cannot doubt—the famous *a priori* truths. Whereas Plato believed that one must accomplish knowledge of truth by means of reason and intuition, Aristotle believed the philosopher must observe particular phenomena to arrive at an understanding of reality, a scientific technique known as *induction.* Once truth is known through induction from the particular to the universal, the philosopher can engage in the process of *deduction* from the basis of the universal to arrive at other particular truths. Aristotle's system of logic is known as *syllogism*. For example, assume the following:

(1) Each human can know things; each human is a knower (A)

(2) What humans collectively know (A) we call knowledge (B)

(3) Knowledge (B) is synonymous to reality, the truth (C)

Therefore, each individual human can know what is the truth. If A=B and B=C, then A=C.

Metaphysics

Metaphysics is the study of reality that transcends the physical world. Once again a priori truths are the basis for metaphysical studies. Aristotle assumed that there is a *First Cause,* an "unmoved mover," that he defined as *actuality,* in contrast to *potency,* the potential, which represents movement. Aristotle argued that all reality can be explained according to cause and effect, act and potential. For example, time is an actual phenomenon, and it has existence as a form or essence (*ousia*). Time acts upon human movement, providing a temporal context in which humans are born, live, and die, all the while measuring their lives according to the standard of time. Aristotle further argued in *Metaphysics* that one must distinguish between art and experience. Art as essence is based on abstract thought—what the Greeks termed the *logos*—whereas experience is based on a series of particular events occurring in time.

Poetics

In *Poetics*, Aristotle argued that poetry (art) explores universals, how things *ought to be*, while history (*historia*) explains the particulars of human existence, how things *are*. Wisdom represents the unification of art and experience. The Arab commentator Ibn Abi Usaibi'ah recorded the following anecdote about Aristotle in regard to *Poetics*:

> A group of scholars were thus greatly assisted by the science of those orators, grammarians and stylists, but another group, headed by Epicurus and Pythagoras, opposed them bitterly, claiming that there was no need of all their sciences in any domain of philosophy. They considered grammarians as kindergarten teachers, poets as tellers of nonsense and lies and stylists as authors of intrigues, partiality and contradiction. When Aristotle heard of this controversy, he was fired with zeal and took up the defense of the grammarians, orators and stylists, proclaiming the worthiness of their cause. He maintained that philosophy could not do without them, language being one of the instruments of their science, that man's advantage over animals is his capacity for speech, and the better his speech, the truer his claim to humanity; the same goes for eloquence in meaningful phrases and punctuation of style, as well as for the judicious choice of concise and elegant words. He added that since philosophy is the noblest of all the sciences, its expression should be

appropriately clad in the truest words, most eloquent terms and most concise phrases; it should also be the farthest from doubt, error, uncouthness of speech, ugliness of pronunciation and stammering, for all these shortcomings extinguish the light of wisdom, cloud its message, preclude its fulfillment, embarrass the listener, spoil its meaning and spread doubt. (Trans. L. Kopf)

Physics

Aristotle, less metaphysical than the pre-Socratic philosophers of Greece, wrote *Physics* to explore what the fundamental principles of nature were. Aristotle identified four natural causes of movement in nature: the material substance of an object; the class *(genos)* to which it belongs; the agent that moves the object; and the ultimate goal *(telos)* of said movement. Aristotle believed most movement to be finite since it was linear between different points. Put another way, the four elements—earth, air, fire, and water—move in a linear path and thus are finite. Each element has its set, natural place. Earth moves toward the center, while fire tends upward. The only form of infinity in Aristotle's mind was the endlessness of the circle. He conceived of a fifth element, ether (*aether*), as having a circular path: thus it is in infinite motion. Natural science, he wrote, is concerned with physical movement from the "originating principles" of nature. Aristotle associated nature with the first cause. His unmoved mover was an amorphous divine force of creation, which established the laws through which movement—plant, animal, human existence—occurred.

Aristotle's categorizations of the four causal determinants in nature had a profound impact on the formation of a vocabulary of science. His notion of type or class is the basis for our notion that a species in nature comprises a set genus. Aristotle's idea of goal or purpose forms the philosophical concept of teleology, the study of the end of natural phenomena, the future pole or stopping point of time itself.

Aristotle and his student Theophrastus wrote seminal accounts on meteorology in which they ascribed to astronomical phenomena the causes of meteorological phenomena. The rising of the Pleiades, the phases of the moon, the appearance of the moon upon rising, and the appearance of the horizon upon sunrise and sunset all have an impact on, and give warning about, the weather.

Behavioral Science

Aristotle was one of the first scientists of human behavior, the human *psyche*. His study of *Ethics* is based on correct behavior in the human community: selfless motives lead to virtuous actions, as opposed to selfish motives, which result in vice. Society clearly benefits from actions done for their own sake and not from ulterior motives. The best society promotes virtuous actions by means of the body politic—this was Aristotle's argument in his treatise *Politics*. Aristotle studied human and animal behavior to see how they were similar and different. He believed that by observation, analysis, deduction, and induction, an understanding of patterns in human behavior can be acquired.

Aristotle wrote treatises on dreams, memory, the senses, prophecy, sleep, and the soul. Aristotle believed that the soul was the "moving Principle" of essence or physical being. The soul is the actuality within the potency of the body. It is the unmoved mover within each individual human. The mind (*nous*) is an expression of the soul. Aristotle argued that each human soul was part of a universal whole, a world soul, the ultimate actuality, the first cause. The idea of a soul, of course, seems to be outside the realm of scientific study. But Aristotle believed that philosophy and science were completely united; that one could not understand nature without understanding thought or understand movement without understanding being. For Greek philosophers, the "defining principle," the meaning of any phenomenon, is the *logos*, which literally means "word." The human soul, the world soul, is therefore the logos.

Aristotle's study of dreams provided a rational explanation of what the ancients often considered a supernatural phenomenon. No matter if a person is asleep or awake, he argued, the same quality of sense perception is active in acquiring information and making sense of the world. Aristotle was sufficiently skeptical of dreams to wonder about their cause and significance. In his treatise *On Prophesying by Dreams*, he provided a balanced view of dreams, wondering why God would speak to humans through dreams yet realizing that at times dreams do seem to reflect reality. One possibility he explored was that as dreams often recall some of the details of the previous waking hours, likewise dreams might foretell actions of the next day insofar as one might (unconsciously) perform certain actions one had dreamed about the preceding night—hence the dream comes true. Aristotle argued that the only thing "divine" about a dream is that it is part of nature, which is itself the creation of God, hence divine. That events turn out according to one's dream is either coincidence or the result of the subtle impact of a dream on an individual's actions. Carl Jung, the great modern student of dreams, clearly had the same understanding and lived his life in accordance with what his dreams predicted for the coming day. Aristotle recognized also that amid the host of dreams humans have, a few might by coincidence end up occurring, which has no supernatural or scientific significance. Animals dream, as do slaves and other "inferior" (in Aristotle's mind) humans, hence dreams are more a product of nature and not sent by a god. But nature is itself divine, Aristotle argued; in this sense dreams are divine as well. Although Aristotle did not agree with the interpretation of dreams by materialists such as Democritus, that dreams are caused by the emanation of atoms that cause images upon the brain, he did agree with their point of view that such an atomic rendering of dreams could in no way be prophetic of the future.

Along with dreams, Aristotle was interested in magic as an object of inquiry. He wrote about the Asian magi, claiming that they were even more ancient than Egyptian astrologers and magicians.

Zoology

Aristotle was very interested in plant and animal life and relied on his student Alexander for specimens of the flora and fauna of Asia. Aristotle was a collector and cataloguer of flora and fauna. Aristotle's contributions to zoological study were several treatises, *History of Animals*, *On the Parts of Animals*, *On the Motion of Animals*, *On the Gait of Animals*, and *On the Generation of Animals*. These works compare humans with animals, assuming that the latter, like the former, participated in Being and had similar behavior and physical attributes. Aristotle, like his teacher Plato, conceived of a chain of being that allowed for the categorization of animals according to their physical and spiritual characteristics. In *Parts of Animals,* Aristotle noted that although animals made for a less profound area of study than the metaphysical did, nevertheless an inquiry thereof was accessible to anyone willing to explore natural history. Consistent with his Platonic background, Aristotle studied animals for the sake of understanding the whole of natural history. He assumed that the source of all good and beauty was the same source of animal and biological phenomena. Hence even animals mirror the divine. Aristotle's *History of Animals*, *Parts of Animals*, and *Generation of Animals* assign to animals human characteristics of behavior, such as treachery and courage, the desire to engage in conflict and war, though humans are superior to animals because of their deliberative capacity of thought, especially the ability of humans to recall and analyze past experience. Nevertheless animals, he argued, teach and learn. Animals can learn the healing properties of substances for self-healing. Some animals, such as pigeons, are monogamous. Animals can exhibit empathy toward others. Some animals, such as the crane, are highly intelligent. Aristotle hypothesized that animal life is part of a grand unity of life, a reflection of metaphysical Being. Yet he was unwilling to go as far as some thinkers in their comparison of humans and animals. Any similarity to human behavior and emotions in animals appears to be due to "analogy," similar physical attributes, and not because of any real human characteristics in animals. For example, animals and humans sleep, during which their sense-perception is active just as when they are awake. Animals, like humans, dream. Animals can also display behavioral characteristics such as courage, fear, obsequiousness, and the like, but these qualities differ from similar human behavior because animals lack a rational, deliberative capacity. Animals have a different place on the great chain of being. Aristotle discussed the bone structure of vertebrates as being linked into a whole system, which includes cartilage and muscles; likewise, the movement of blood throughout the body is part of a unified whole. The male animal's role in reproduction, according to Aristotle, is an active, creative role that is not permanently linked to the resulting product, just as God creates material things but is not Himself present in His creation. All animals are drawn to, and enjoy, copulation, he declared. In the *Generation of Animals*, as in other works, Aristotle maintained a sense of humility with regard to some of the mysteries of animal generation, characteristics, and behavior. He confessed in *History of Animals*, "Nature passes from lifeless things up to animal life by such gradual degrees that the continuity obscures boundaries and puzzles us how to classify intermediate forms" (trans. Wheelwright). Such humility was justified, as he made some obvious errors in his analysis of animal life, such as his theory of spontaneous generation. Nevertheless, the breadth of his knowledge on this subject was astonishing. He studied mammals, birds, fish, crustaceans, and insects in detail.

Ethics

How, one might ask, can ethics, the ultimate basis of behavior, the set of rules that establish the good, be understood according to science? Aristotle believed that the tools of science—observation, categorization, logic, induction—can be brought to bear on the study of human behavior. The scientist studies human behavior in its incredible variety of contexts to arrive at general laws of how humans act and how they should act. How humans act is the realm of the scientist; how humans should act is the realm of the philosopher. Once again, Aristotle combined science and philosophy into one organized study. Aristotle believed that the ultimate end of human existence was happiness, which occurred when humans conformed to "the good." The good is accomplished when humans exercise reason in accordance with virtue. But what is virtue? Aristotle studied human behavior to arrive at a definition: virtue is an action performed for its own sake, that is, an action performed for the sake of the good, an action performed out of principle. Vice, the opposite of virtue, derives from actions committed for selfish reasons, for personal motives. Actions that occur not because of the search for power, wealth, fame, and security are virtuous actions.

Politics

Aristotle's contribution to understanding government, *The Politics*, applied his philosophical methods and assumptions to the understanding of statecraft. He argued that the state was, as it were, the actual, while the citizens were the potential. The latter were the parts (the particulars) that made up the whole, the universal body politic. Aristotle conceived of a pluralistic society operating according to natural laws based in part on reason and necessity, a social compact among people to promote security and serve the needs of survival. Within this concept of the state (which represents virtue), people move, act, and struggle for power and wealth. Nature has created a hierarchical state of being; in society this is realized through unequal ranks of people according to intelligence, property, gender, and personal freedom. Aristotle argued, based on his experience at Athens, that slavery was justified because of the inferior intellect of slaves. Likewise, he assumed that women lacked the same cognitive abilities of males and therefore should not participate in democracy. In *The Athenian Constitution*, Aristotle provided a detailed analysis of Athenian democracy, providing details into the life and political science of the great Athenian lawgiver Solon.

In short, the overall goal of the state is the good of the whole, the achievement of general happiness for all people. To Aristotle, the state is the agency through which the citizen exercises virtue, which is how the individual attains the good, that is, happiness.

Astronomy

Aristotle explored his ideas on astronomy in *On the Heavens* and *Meteorology*. Based on observation, Aristotle established the spherical nature of the earth. Viewing a lunar eclipse, Aristotle detected a slight curvature of the shadow of the earth on the moon's surface. He also believed, following Plato, that the sphere was the most perfect surface, hence the logical shape of the earth. Much of Aristotle's thought on astronomy, however, was erroneous. Observation with the naked eye was insufficient for the study of the nature of the stars and planets. His theories, though logical, were not empirically based, hence suspect. Observation tells us that the sun rises and sets, moves across the sky of an earth that appears perfectly still. If the earth were moving, would we not feel the motion? Further support for the theory that the earth is at rest, the stable center of the solar system (the universe), was the direction of falling objects. Would a rock fall to the ground in a straight line if the earth were moving? Aristotle assumed that the earth must be the center of things toward which terrestrial objects moved. Even extraterrestrial, planetary objects, have form yet are nevertheless perfect spheres, perhaps heavenly beings, that move in perfect spherical orbits around the stable earth. The universe is finite, the extreme edge being the starry vault, the fixed stars that move in an unchanging pattern around the planetary solar system.

Aristotle's ideas, such as the model of the geocentric universe, described above, were advocated and defended for centuries after the philosopher's death. Aristotle also gave public lectures of his philosophy and science, which were taken down by students and eventually published in a similar format. Theophrastus, who took over the helm of the Lyceum, organized Aristotle's papers and writings and pursued Aristotle's theories and investigations in the physical and metaphysical worlds.

THEOPHRASTUS (370–286 BCE)

Theophrastus, a student of both Plato and Aristotle, was a native of the Isle of Lesbos and an early Peripatetic philosopher, famous for his work in flora. Theophrastus, like his teacher Aristotle, was at the same time a deep yet practical thinker. A polymath, his creative output was immense. Vitruvius, the Roman architect, quoted Theophrastus's view of the importance of learning: "The man of learning is the only person in the world who is neither a stranger when in a foreign land, nor friendless when he has lost his intimates and relatives; on the contrary, he is a citizen of every country, and can fearlessly look down upon the troublesome accidents of fortune. But he who thinks himself entrenched in defences not of learning but of luck, moves in slippery paths, struggling through life unsteadily and insecurely" (trans. Morgan). Diogenes Laertius listed dozens of titles that came from Theophrastus's pen. These included works on astronomy, logic, physics, meteorology, morality, zoology, psychology, political science, mathematics, history, poetry, music, and commentaries on earlier philosophers, such as Anaxagoras, Anaximenes, Archelaus, Democritus, Diogenes, Empedocles, Plato, and Aristotle. The list is so astonishing in his breadth and extent that it is reproduced here:

> Three books of the First Analytics; seven of the Second Analytics; one book of the Analysis of Syllogisms; one book, an Epitome of Analytics; two books, Topics for referring things to First Principles; one book, an Examination of Speculative Questions about Discussions; one on Sensations; one addressed to Anaxagoras; one on the Doctrines of Anaxagoras; one on the Doctrines of Anaximenes; one on the

Doctrines of Archelaus; one on Salt, Nitre, and Alum; two on Petrifactions; one on Indivisible Lines; two on Hearing; one on Words; one on the Differences between Virtues; one on Kingly Power; one on the Education of a King; three on Lives; one on Old Age; one on the Astronomical System of Democritus; one on Meteorology; one on Images or Phantoms; one on Juices, Complexions, and Flesh; one on the Description of the World; one on Men; one, a Collection of the Sayings of Diogenes; three books of Definitions; one treatise on Love; another treatise on Love; one book on Happiness; two books on Species; on Epilepsy, one; on Enthusiasm, one; on Empedocles, one; eighteen books of Epicheiremes; three books of Objections; one book on the Voluntary; two books, being .an Abridgment of Plato's Polity; one on the Difference of the Voices of Similar Animals; one on Sudden Appearances; one on Animals which Bite or Sting; one on such Animals as are said to be Jealous; one on those which live on Dry Land; one on those which Change their Colour; one on those which live in Holes; seven on Animals in General; one on Pleasure according to the Definition of Aristotle; seventy-four books of Propositions; one treatise on Hot and Cold; one essay on Giddiness and Vertigo and Sudden Dimness of Sight; one on Perspiration; one on Affirmation and Denial; the Callisthenes, or an essay on Mourning; one on Labours; one on Motion; three on Stones; one on Pestilences; one on Fainting Fits; one the Megaric Philosopher; one on Melancholy; one on Mines; two on Honey; one a collection of the Doctrines of Metrodorus; . . . two books on those Philosophers who have treated of Meteorology; on Drunkenness; . . . twenty-four books of Laws, in alphabetical order; ten books, being an Abridgment of Laws; one on Definitions; one on Smells; one on Wine and Oil; eighteen books of Primary Propositions; three books on Lawgivers; six books of Political Disquisitions; a treatise on Politicals, with reference to occasions as they arise, four books; four books of Political Customs; on the best Constitution; one, five books of a Collection of Problems; on Proverbs; one on Concretion and Liquefaction; one on Fire; two on Spirits; one on Paralysis; one on Suffocation; one on Aberration of Intellect; one on the Passions; one on Signs; . . . two books of Sophisms; one on the Solution of Syllogisms; two books of Topics; two on Punishment; one on Hair; one on Tyranny; three on Water; one on Sleep and Dreams; three on Friendship; two on Liberality; three on Nature; eighteen on Questions of Natural Philosophy; two books, being an Abridgment of Natural Philosophy; eight more books on Natural Philosophy; one treatise addressed to Natural Philosophers; two books on the History of Plants; eight books on the Causes of Plants; five on Juices; one on Mistaken Pleasures; one, Investigation of a proposition concerning the Soul; one on Unskilfully Adduced Proofs; one on Simple Doubts; one on Harmonics; one on Virtue; one entitled Occasions or Contradictions; one on Denial; one on Opinion; one on the Ridiculous; two called Soirees; two books of Divisions; one on Differences; one on Acts of Injustice; . . . one on Praise; one on Skill; three books of Epistles; one on Self-produced Animals; one on Selection; one entitled the Praises of the Gods; one on Festivals; one on Good Fortune; one on Enthymemes; one on Inventions; one on Moral Schools; one book of Moral Characters; one treatise on Tumult; one on History; one on the Judgment Concerning Syllogisms; one on Flattery; one on the Sea; one essay, addressed to Cassander, Concerning Kingly Power; one on Comedy; one on Meteors; one on Style; one book called a Collection of Sayings; one book of Solutions; three books on Music; one on Metres; the Megades; one on Laws; one on Violations of Law; one a collection of the Sayings and Doctrines of Xenocrates; one book of Conversations; one on an Oath; one of Oratorical Precepts; one on Riches; one on Poetry; one being a collection of Political, Ethical, Physical, and amatory Problems; one book of Proverbs; one book, being a Collection of General Problems; one on Problems in Natural Philosophy; one on Example; one on Proposition and

> Exposition; a second treatise on Poetry; one on the Wise Men; one on Counsel; one on Solecisms; one on Rhetorical Art, a collection of sixty-one figures of Oratorical Art; one book on Hypocrisy; six books of a Commentary of Aristotle . . . ; sixteen books of Opinions on Natural Philosophy; one book, being an Abridgment of Opinions on Natural Philosophy; one on Gratitude; one called Moral Characters; one on Truth and Falsehood; six on the History of Divine Things; three on the Gods; four on the History of Geometry; six books, being an Abridgment of the work of Aristotle on Animals; two books of Epicheiremes; three books of Propositions; two on Kingly Power; one on Causes; one on Democritus; one on Calumny; one on Generation; one on the Intellect and Moral Character of Animals; two on Motion; four on Sight; two on Definitions; one on being given in Marriage; one on the Greater and the Less; one on Music; one on Divine Happiness; one addressed to the Philosophers of the Academy; one Exhortatory Treatise; one discussing how a City may be best Governed; one called Commentaries; one on the Crater of Mount Etna in Sicily; one on Admitted Facts; one on Problems in Natural History; one, What are the Different Manners of Acquiring Knowledge; three on Telling Lies; one book, which is a preface to the Topics; one addressed to Aeschylus; six books of a History of Astronomy; one book of the History of Arithmetic relating to Increasing Numbers; one called the Acicharus; one on Judicial Discourses; . . . one volume of Letters to Astyceron, Phanias, and Nicanor; one book on Piety; one called the Evias; one on Circumstances; one volume entitled Familiar Conversations; one on the Education of Children; another on the same subject, discussed in a different manner; one on Education, called also, a treatise on Virtue, or on Temperance; one book of Exhortations; one on Numbers; one consisting of Definitions referring to the Enunciation of Syllogisms; one on Heaven; two on Politics; two on Nature, on Fruits, and on Animals. And these works contain in all two hundred and thirty-two thousand nine hundred and eight lines. (Trans. Yonge)

Diogenes Laertius described Theophrastus as "a man of extraordinary acuteness, who could both comprehend and explain everything" (trans. Yonge). Theophrastus's *Book of Signs* examined clouds and other environmental conditions with which to predict the weather. Theophrastus thought that animal behavior indicated weather changes. The quacking of a duck, the calling of a crow, the preening of a hawk forecast the weather. Flies biting reveal that rain is nigh. High winds accompany a dog rolling about on the ground. Theophrastus agreed with Aristotle (in the *Meteorologica*) about the eight principal winds observed at Athens.

Upon the death of Aristotle, Theophrastus took over leadership of the Lyceum, Aristotle's school in Athens. The Lyceum had a wonderful garden that Aristotle enjoyed strolling through as he taught; Theophrastus maintained it and used it as an herbarium. Plutarch called Theophrastus "the most curious inquirer, and the greatest lover of history" (trans. Dryden). Proclus, the fifth-century-CE writer, recorded Theophrastus's beliefs that Chaldean astrologers could predict the future, including the time of one's death. Whether they predicted Theophrastus's own death is not known, though Diogenes Laertius did transcribe his extensive will.

Theophrastus's *Enquiry into Plants* provides a full description of the flora of the ancient Mediterranean; it is an indispensable tool for those researching the history of botany. Theophrastus classified and described trees, shrubs, and flowers, building a materia medica of those plants with medicinal properties. Like Aristotle, Theophrastus was interested in classifying plants; his *Enquiry* provided detailed discussion of plant types and morphology. All aspects of plants—roots, leaves, buds, fruit, propagation, uses, and types according to place and climate—were described in detail. Theophrastus provided encyclopedic accounts of trees, shrubs, flowers, wild and cultivated plants, herbs, edible and inedible plants, cereals, spice-bearing plants, and the odors of plants. The writing was dry but detailed and without parallel in scope for centuries.

THE PERIPATETIC SCHOOL

After Theophrastus's death in 287 BCE, Strato (335–269) assumed leadership of the Lyceum and the Peripatetic philosophers. Strato was more an empiricist than Theophrastus was and focused the studies of the Lyceum down this more experimental path. A Peripatetic is one who moves about or walks while engaged in discussion or disputation. Aristotle, imitating Socrates, enjoyed teaching in such a way at the Lyceum. For centuries Aristotle's disciples were known by the master's teaching style. Peripatetic philosophy as Aristotelian philosophy was focused on a concrete approach to the acquisition of knowledge as opposed to that of the Academics, the followers of Plato, who believed that such knowledge was elusive and involved a spiritual, intuitive approach to discovering what was real. For the Peripatetics, observation, collection of data, analysis, and experimentation would provide the thinker with a good understanding of reality, which is present before us, not hidden in some ethereal realm of being. Strato, according to Diogenes Laertius, was "surnamed the Natural Philosopher, from his surpassing all men in the diligence with which he applied himself, to the investigation of matters of that nature." He wrote treatises on a variety of topics, including "Philosophy, . . . Human Nature, the Generation of Animals, Mixtures, Sleep, Dreams, Sight, Perception, Pleasure, Colours, Diseases, Judgments, Powers, Metallic Works, Hunger, and Dimness of Sight, Lightness and Heaviness, Enthusiasm, Pain, Nourishment and Growth, Animals whose Existence is Doubted, Fabulous Animals, Causes, a Solution of Doubts, . . . [and] the Examination of Inventions" (trans. Yonge).

Other Peripatetic philosophers included Demetrius of Phalerum (350–280), a student of Theophrastus who was not only a philosopher but the ruler of Athens, appointed by the Macedonian king Cassander (c. 355–297). Diogenes wrote that Demetrius was "a man of great learning and experience on every subject" (trans. Yonge). He wrote on politics, history, rhetoric, laws, Homer, and justice.

EUDOXUS (C. 408–355)

Eudoxus was a student of Plato who became a famous philosopher and scientist in his own right during the early decades of the fourth century BCE. Eudoxus wrote *Phainomena*, one of the seminal statements of the ancient world on astronomy. Although the book was filled with erroneous information, arguing for a geocentric universe comprised of twenty-seven spheres, Eudoxus believed in a completely rational approach devoid of superstition and mysticism. He had his own school at Athens, where he taught his students his geometric ideas on proportion, volume, the circle, pyramids, and cones. He reputedly influenced both Euclid and Archimedes. According to Diogenes Laertius, he wrote *Voyage round the World.*

THE LEGACY OF ATHENS

The impact of classical Athens upon subsequent science and philosophy is, of course, vast. Plato and Aristotle, Epicurus and Zeno, the Academy and the Lyceum—these provided the chief philosophical and scientific divisions during the Later Roman Empire, medieval Europe, the Renaissance, the Enlightenment, and even more recent years. One can see over the course of Western thought a dichotomy between two great philosophical systems: the Platonic, with its focus on subjective thought, intuition, and pure reason; and the Aristotelian, with its focus on empiricism, observation and the use of the senses, and finding truth in nature. Solon, Anaxagoras, Socrates, Plato, Aristotle, Zeno, and Epicurus sought answers to the most penetrating questions: What is the nature of being? Is there a soul, and of what is it made? What is the role of the divine in our lives? What is the good, and is there a contrasting presence of evil? What is the duty of humankind? Is there a universal code of ethics upon which to mold behavior? Is the state necessary, and if so, what is its purpose? What is virtue, and is it possible to obtain? Is there an objective truth? What role does the self have in the acquisition of knowledge? These questions and so many more, asked by the Athenians of the ancient world, are perhaps its greatest legacy.

Plato's legacy of thought reached forward two millennia to modern times. Plato influenced subsequent Greek and Roman thought by means of his writings, which were widely circulated, as well as by means of the Academy. The Academy existed for centuries after Plato's death, sometimes exercising very little influence on philosophy but at other times being a center of thought in the Mediterranean world. Neoplatonist philosophers and scientists dominated thought at the end of the Roman Empire. Medieval philosophers such as Augustine, Boethius, and Anselm were heavily influenced by Plato. During the European Renaissance, Platonic thought experienced a revival; the philosopher Ficino is an example. Platonic thought was an important inspiration to the nineteenth-century Romantics. Among scientists, Plato's impact has been more checkered. Science does not easily accommodate a philosophy that limits the reality of physical phenomenon, that proclaims that all a scientist does in studying nature is to understand the mere shadows of what is real. And yet subjective thought is part of scientific thinking. Plato has had a major impact on science, even if it is not altogether clear.

Aristotle's teaching and writings, along with Plato's, dominated philosophic and scientific thought for over two thousand years. During the Hellenistic Age (330–30 BCE) and during the period of Imperial Rome (30 BCE–476 CE), ancient philosophers, orators, botanists, physicians, astronomers, astrologers, and teachers, both pagan and Christian, began their studies with one of the many surviving texts of Aristotle's works. The greatest astronomer of the centuries after the birth of Christ, Claudius Ptolemy, relied heavily on Aristotle's conceptions of the earth and heavens. Aristotle's theories of motion, his reliance upon deductive reasoning, attracted disciples during the European Middle Ages, not only in the primitive feudal environment of Western Europe but in Eastern Europe as well, especially at the great capital of the Byzantine Empire, Constantinople. Muslim scientists came into contact with Aristotle's writings, had the Greek translated into Arabic, and circulated them through the Muslim world, from Iran to Egypt to Morocco to southern Spain. Muslim scholars such as Averroes became Aristotelians. Eventually, after 1100 CE, Western European intellectuals translated Aristotle's works into Latin, which attracted the attention of theologians such as Thomas Aquinas. Aquinas's massive treatises on philosophy, theology, and science, in particular the *Summa Theologica*, were heavily dependent on Aristotle, whom Aquinas referred to simply as *the Philosopher*. During subsequent centuries of the Renaissance, intellectuals divided themselves into competing camps, the Platonists and Aristotelians. Aristotle's theories on astronomy and motion became the starting point for Renaissance scientists such as Galileo, who disproved but still owed much to Aristotle's theories. Even up to 1700 in Europe and America, intellectuals and college curricula relied heavily on Aristotle's teachings. Today's courses in logic still go back to the fourth-century-BCE teachings of Aristotle.

During the centuries after Plato and Aristotle, Athens continued to host some of the great philosophers and scientists in the Mediterranean world: Chrysippus, Carneades, Arcesilaus, Zeno, and Epicurus. During the third century, Epicurus (341–271) used Athens as his philosophical retreat as he developed the theories that became the philosophy of Epicureanism. Only Zeno rivaled Epicurus. Zeno (333–262) taught among the stoa of Athens, hence his followers called themselves Stoics. Like the Epicureans the Stoics rejected Platonic idealism for a materialist philosophy that explained the universe according to atoms.

The Athenians, the "education of Greece" in Pericles's words, developed, by means of the varied schools of thought, an approach to learning that the Greeks called the *paideia*, the education of the body and mind. Today, it is called the liberal arts. Traditionally, the liberal arts were based in the trivium (grammar, logic, and rhetoric) and the quadrivium (arithmetic, geometry, music, and astronomy). Liberal arts involve the study of those subjects that open the mind and help bring about a free people. The seven objects of inquiry—grammar (literature, languages), logic (philosophy, deductive and inductive thinking), rhetoric (history, humanities), arithmetic (numerical reasoning and inductive thinking), geometry (spatial reasoning and deductive thinking), music (arts, studies in culture and human expression), and astronomy (the hard sciences)—form the essence of the liberal arts: to question, to seek, to learn, to know, to accept others and oneself. Reflected in the trivium and quadrivium are spiritual seeking, emotional and natural thinking, focus on the intellect and past tradition, finding solutions to the manifold problems of life—in short, self-discovery. The Greeks believed that such an education system resulted in thoughtful, analytical, and articulate people.

FURTHER READING

Aristotle, *On the Athenian Constitution*, trans. Frederic G. Kenyon (London: G. Bell and Sons, 1891).

Aristotle, *On Sleep and Sleeplessness; On Prophesying by Dreams; On Memory and Reminiscence*, trans. J. I. Beare, in *The Parva Naturalia* (Oxford: Clarendon Press, 1908).

Renford Bambrough, ed. and trans., *The Philosophy of Aristotle* (New York: New American Library, 1963).

Jonathan Barnes, *Aristotle* (Oxford: Oxford University Press, 1982).

Jonathan Barnes, trans., *Early Greek Philosophy* (London: Penguin Books, 1987).

Jacques Brunschwig and Geoffrey Lloyd, *Greek Thought: A Guide to Classical Knowledge* (Cambridge, MA: Harvard University Press, 2000).

Christopher Collard, trans. *Aeschylus: Persians and Other Plays* (Oxford: Oxford University Press, 2008).

Diogenes Laertius, *The Lives and Opinions of Eminent Philosophers*, trans. C. D. Yonge (London: Henry Bohn, 1853).

Diogenes Laertius, *Lives of Eminent Philosophers*, trans. R. D. Hicks, 2 vols. (Cambridge, MA: Harvard University Press, 1931, 1938).

Will Durant, *The Life of Greece* (New York: Simon and Schuster, 1939).

Michael Grant, *Readings in the Classical Historians* (New York: Scribner's, 1992).

R. M. Hare, *Plato* (Oxford: Oxford University Press, 1982).

Thomas Heath, *Aristarchus of Samos* (New York: Dover Books, 1981).

Hippolytus, *Refutation of the Heresies*, trans. J. H. MacMahon, in *Ante-Nicene Fathers*, vol. 5, ed. Alexander Roberts, James Donaldson, and A. Cleveland Coxe (Buffalo, NY: Christian Literature Publishing, 1886); revised and edited for New Advent by Kevin Knight, http://www.newadvent.org/fathers/050101.htm.

Ibn Abi Usaibi'ah, *History of Physicians*, trans. L. Kopf (Bethesda, MD: National Library of Medicine, 1971).

Carl Jung, *Man and His Symbols* (New York: Doubleday, 1964).

O. Neugebauer, *The Exact Sciences in Antiquity* (New York: Dover Books, 1969).

Plato, *The Last Days of Socrates*, trans. Hugh Tredennick (Harmondsworth, England: Penguin Books, 1959).

Plato, *The Republic of Plato*, trans. Francis Cornford (Oxford: Oxford University Press, 1945).

Plutarch, "Dion" and "Alcibiades," in *The Lives of the Noble Grecians and Romans*, trans. John Dryden (New York: Random House, 1992).

Plutarch, *The Life and Writings of Homer*, in Plutarch, *Essays and Miscellanies: The Complete Works*, trans. William W. Goodwin, vol. 3, https://www.gutenberg.org/files/3052/3052-h/3052-h.htm.

Plutarch, *Life of Solon*, trans. Ian Scott-Kilvert (Harmondsworth, England: Penguin Books, 1960).

The Portable Plato, trans. Benjamin Jowett (Harmondsworth, England: Penguin Books, 1976).

Charles A. Robinson Jr., *Athens in the Age of Pericles* (Norman: University of Oklahoma Press, 1959).

Charles B. Schmitt, *Aristotle and the Renaissance* (Cambridge, MA: Harvard University Press, 1983).

Theophrastus, *Enquiry into Plants and Minor Works on Odours and Weather Signs*, trans. Arthur Hort, 2 vols. (Cambridge, MA: Harvard University Press, 1916).

Thucydides, *The Peloponnesian War*, trans. Rex Warner (Harmondsworth, England: Penguin Books, 1972).

William Turner, "Aristotle," *The Catholic Encyclopedia* (New York: Encyclopedia Press, 1913).

Vitruvius, *The Ten Books on Architecture*, trans. Morris H. Morgan (Cambridge, MA: Harvard University Press, 1914).

Philip Wheelwright, ed. and trans., *Aristotle* (New York: Odyssey Press, 1951).

PART 5

Geographical and Astronomical Knowledge

12

Exploration and Geographical Knowledge in the Ancient World (4000 BCE–500 CE)

The first explorer in the Western tradition was Odysseus. According to Homer, Odysseus, driven by the gods, fate, and his own will to survive, wandered the Mediterranean for years seeking the way home to Ithaca, his kingdom, and his faithful wife, Penelope. Odysseus's travels took him, Homer tells us, from the Aegean Sea to the coasts of North Africa, then to Sicily and the Straits of Messina, and as far as the "Ocean stream," the Atlantic Ocean. Odysseus's journey was longer, more daring, and more fantastic than his earlier counterpart, Jason, who with his Argonauts explored the Black Sea to its eastern limit at Colchis in quest of the golden fleece.

Doubters today who consider Homer's epic imaginary, conceding only that if Odysseus lived, his journey would have hardly reached the Atlantic, have few counterparts among the ancients, who never doubted that Odysseus and his heroic comrades, Achilles and Agamemnon, lived; nor did the ancients ever doubt the veracity of their magnificent deeds. Such was the spell that Homer cast upon the ancient Mediterranean world that subsequent explorers, such as Alexander of Macedon, subsequent writers and historians, such as Herodotus, and subsequent mariners, such as Pytheas of Massilia, lived, worked, and thought from the initial foundation of Homer's tale. It is hardly surprising to discover that the Greeks in the millennium after Homer had the aggressive activity of Achilles, the hardy strength and guile of Diomedes, the self-sacrifice of Patrokolos, the patience and perseverance of Penelope, and the inner strength—the will that comes from self-awareness, the daring extension of self into the world at large—of Odysseus.

The history of the Mediterranean world was built around the words *in search of.* Odysseus was in search of stability, order, and home in a world that was a round disk surrounded by the river Ocean, where in the daunting Mediterranean Sea, humans were subject to the whims of nature as personified by the gods and goddesses. Herodotus of Halicarnassus was in search of the past and how it played upon the present. Socrates was in search of truth, inner wisdom, and human awareness. Alexander of Macedon was in search of power and glory but also, even more, the knowledge of the unknown. Pytheas of Massilia was in search of minerals, wealth, and their sources in the mysterious lands to the north. Eudoxus the Cyzican (408–355) was in search of routes to India.

Yet the Greeks were probably not the first to explore the oceans of the earth. There is compelling evidence that ancient peoples had taken to the sea in prehistoric times, before written records preserved the memory of their voyages. The prevailing winds and currents of the world's oceans could make explorers of people who cast adrift from shores in rafts, canoes, reed ships, and ships with wooden hulls. The winds and currents of the oceans move in set directions because of the rotation of the earth; the relative heat of the equator, as opposed to northern and southern latitudes; and the position of land relative to the seas. Equatorial currents between thirty degrees north and south latitude flow from east to west, whereas the ocean currents in northern and southern latitudes between thirty and sixty degrees flow from west to east; from sixty to ninety degrees, ocean currents flow east to west. Prevailing winds and currents would take intentional or unintentional ocean travelers from the coast of northwestern Africa to the Caribbean region of America; from the coast of South America to the islands of the South Pacific; from the coast of northern East Asia to the lands of northwestern America; from the coast of northeastern America to northwestern Europe.

The mysteries of seas and oceans intrigued ancient humans even as they tried to penetrate the vastness of the earth's waters and conquer their intuitive fear of what lay hidden beneath the "wine dark sea." The Hebrews believed that God formed the waters of the deep stocked with sea creatures and monsters during the first three days of creation. Noah became the first shipbuilder and navigator when called upon by God to build an ark—one of fantastic dimensions to withstand the forty days and nights of rain that resulted in the inundation of the earth and the death of all animals and humans except those on the ark. Gilgamesh found the key to ageless youth on the ocean floor after visiting the Sumerian Noah, Upnashitim. Homer placed Hades in the river Ocean west of the Pillars of Heracles, the Strait of Gibraltar. Odysseus was driven by the angry god of the sea, Poseidon, all around the mysterious Mediterranean. The sea came closest to defeating that most resourceful man in Homer's poems. The Greeks were so certain that the ocean was a place of mystery—like the dark forests and the high mountains—that they populated it with countless Nereides, maidens of the sea, as well as varied sea monsters—such as Charybdis, the whirlpool that engulfed many of Odysseus's men, and the clashing rocks that met Jason's *Argo.*

The modern student looks mostly in vain for the beginnings of oceanography in the ancient world. The mysteries of the ocean remained mostly mysteries, the stuff of fantasy and legend. Among the Greeks, there were a few thinkers who tried to move from fantasy to fact regarding the ocean. Thales thought that water was the basic element of all things. His student Anaximander went further, declaring that humans evolved from the sea. The third-century-CE Roman writer Censorinus declared that Anaximander thought that humans derived from fish. Plutarch wrote that some Greeks revered Poseidon because they believed that humans originally came from the sea. Xenophanes argued that the oceans produced meteorological events, such as the wind and rain. He recognized fossils of sea creatures for what they were and speculated on the inundation of the earth that had occurred in the past and would occur again to destroy humankind.

Ancient peoples clearly navigated rivers, bays, gulfs, seas, and oceans. Ancient centers of civilization along the Tigris, Euphrates, Nile, and Indus Rivers engaged in not only river traffic but oceangoing sailing as well. Along the Nile River, Egyptian shipbuilders constructed reed ships from the prolific papyrus reed that grew at the mouth of the Nile. Trade in these reed ships was extensive. At Tassili, in the Sahara Desert, present-day Algeria, near the Libyan border, ancient rock paintings show reed boats. Clearly there was a trading connection between ancient Egypt and North Africa to the west. The Egyptians and Mesopotamians also engaged in extensive trade; some of this would be overland, but much would be by sea, as both cultures used papyrus reed ships. These were not just small boats but also large ships that could hold up to a hundred tons of cargo.

The Egyptians built reed ships from two large rolls and one medium roll of papyrus reeds. Two large rolls, each thirty feet long, were placed parallel with a small reed bundle in-between. Reed ropes pulled the three bundles taut. Each of the large bundles supported a mast, the two masts placed at angles reaching a point above the center of the ship. A trapezoidal single sail, wider up top than below, was attached to the dual mast. The bow and stern of the boat were curved upward, held in place by ropes secured to the ship. Navigators used a twenty-foot flexible steering oar to guide the ship. Thor Heyerdahl, the foremost archaeologist and historian of the ancient reed ship, has described the strong possibility of ancient reed ships being able to navigate the Mediterranean, Red, and Arabian Seas and perhaps the Indian and Atlantic Oceans as well.

The Old Testament records examples of such reed ships. The baby Moses was set adrift in a small reed boat; Noah built the ark of cypress and reeds; Isaiah chapter 18 refers to the papyrus ships of Kush. Although over time wooden cypress ships replaced reed ships, Pliny the Elder recorded reed ships sailing the Indian Ocean as far as Sri Lanka and the mouth of the Ganges River.

The reed ship was "wash-through," in which the boat was naturally buoyant and almost impossible to sink. Ocean water made the reed bundles swell. The sea might crash over the ship but could not but briefly submerge it. Other ancient peoples built similar kinds of vessels from bamboo and balsa wood. Ancient Peruvians built balsa wood rafts for sailing the Pacific Ocean. Balsa wood is light and buoyant; such a raft could also be briefly submerged but not sunk. The principle of wash-through boats was different from the hulled ships built by the Phoenicians, Chinese, Greeks, and Romans, in which a hollow hull was used to displace the water; the problem was that when the hulls leaked or cracked or broke, the ship would sink.

Egyptian river ships, such as royal barges, were made of wood. Herodotus described boats of the Nile as made of acacia wood cut into planks for the hull and decking. No ribs were used, which made the boats useful only for river navigation. The Egyptians used the papyrus plant to caulk the ships and to make sails. Rather than building the ship first by laying the keel and joining ribs to it, Egyptian shipbuilders, like the Mesopotamians before them, joined the hull of the ship to the keel without ribs. Such ships were hardly seaworthy, hence the reliance upon reed ships for oceangoing traffic.

Whether the ships were made of papyrus reeds or wood, ancient peoples knew more about the winds and currents of the seas and oceans than we, accustomed as we are to steel ships and modern radar, might expect. Often the first ships used in navigation were extremely small—the size of a pleasure yacht today—which was not a disadvantage at sea, where smaller ships can ride more effectively the dips and crests of the waves. Although ancient Greek mariners often feared to leave sight of land, it was actually safer for a ship to be in the open sea, away from sandbars, reefs, and other obstructions found along the coast. Ancient navigators hoped for cloudless nights on the open sea so that they could use Polaris (the North Star), the clustered Pleiades next to Orion, Sirius (the Dog Star), and other well-known constellations in the zodiac to keep track of their relative position. During the day, keeping track of the position of the sun on the horizon helped to gauge latitude.

GEOGRAPHY

European geography begins with Homer, whose geographic knowledge derived from the explorations of the Egyptians, the mariners of ancient Crete, and the Phoenicians, who explored the western Mediterranean, sailed beyond the Strait of Gibraltar, and colonized Spain, North Africa, and West Africa. Homer, of course, was a poet rather than a geographer. The Homeric worldview was formed by myth and legend, but these usually contain a kernel of truth. Odysseus's journey was one many Phoenician mariners made during the early centuries of the first millennium BCE. The Phoenicians left behind few records of their voyages to North Africa, Sicily, southern Spain, and the Atlantic Ocean beyond the Strait of Gibraltar. Tales of Phoenician adventures became part of the epic cycle of Mediterranean bards, which explained how Homer borrowed so much from the Phoenician experience to mold the character and exploits of Odysseus. That Greeks were, by 800 BCE, making voyages from the Aegean Sea to the Black Sea by way of the Hellespont and Bosporus made the story of Jason and the Argonauts, which featured mythological renderings of real places and peoples, all the more believable. Writings (*graphia*) about the world (*geo*) were initially based on tales of historical voyages made by anonymous explorers who nevertheless acquired knowledge of the Mediterranean world and passed it on, however fancifully.

Homer's *Iliad* and *Odyssey*, written during the eighth century BCE and based on earlier oral stories, provide plentiful if often erroneous information about the world. The poems describe a round disk, Earth, comprised of three ambiguous continents: Europe, Asia, and Africa (Libya), surrounded by the river Ocean. Homer's knowledge of the extent of the Mediterranean, from Gibraltar to the Dardanelles and beyond, to the Atlantic and Black Seas, respectively, derived from the sailors of Egypt, Phoenicia, Carthage, Crete, and Greece. Homer's contemporary, Hesiod, wrote of similar mythic places, such as the Elysian Fields, an island in the Ocean west of the Pillars of Heracles.

Greeks gained greater knowledge and experienced new cultures when, in the eighth and seven centuries, city-states sent colonizing expeditions to Anatolia, North Africa, Spain, southern France, Italy, and the shores of the Black Sea. By the seventh century BCE, the Greeks had explored and colonized the Atlantic and Mediterranean shores of the Iberian Peninsula. Herodotus of Halicarnassus reported that colonists from Thera colonized Cyrene, in North Africa. Herodotus, along with Hecataeus of Miletus and Hellanicus of Lesbos, made the first attempts to provide factual information—geography and history—divorced from myths of the lands and peoples of the Mediterranean. These historians and geographers based their accounts on their own travels as well as on travelers' tales of distant lands provided by Phoenician, Cyrenian, Egyptian, Byzantine, and Massilian traders and sailors.

Greek shipbuilders advanced merchant and warship construction by basing the construction of the ship on a keel-and-rib design. Third-century poet Apollonius of Rhodes, who described the *Argo,* claimed that the shipbuilders bound a strong rope about the sides to provide additional holding power to the planks to withstand being battered by the swells of the sea. The single mast was secured in a mast box and pulled taut by ropes secured fore and aft. Greek and Phoenician ships relied heavily on rowers—the Phoenicians developed the bireme, a two-row system of oars; the Phoenicians as well as the Greeks added a third row: the trireme.

The first-century geographer Strabo attributed to the Phoenician people of Sidon in the eastern Mediterranean extensive knowledge of marine science, writing,

> The Sidonians, according to tradition, are skilled in many beautiful arts, as the poet also points out; and besides this they are philosophers in the sciences of astronomy and arithmetic, having begun their studies with practical calculations and with night-sailings; for each of these branches of knowledge concerns the merchant and the ship-owner; as, for example, geometry was invented, it is said, from the measurement of lands which is made necessary by the Nile when it confounds the boundaries at the time of its overflows. This science, then, is believed to have come to the Greeks from the Aegyptians; astronomy and arithmetic from the Phoenicians; and at present by far the greatest store of knowledge in every other branch of philosophy is to be had from these cities. (Trans. Jones)

GEODESY

The Greeks invented the science of geodesy. The Ionian Greeks of western Anatolia (Turkey) were the first to hypothesize that the earth was not a flat disk: Anaximander of Miletus, during the sixth century BCE, who reputedly drew the first map of the world, believed the earth was a cylinder. Parmenides, a century later, speculated that earth must have zones of heat and cold: torrid, frigid, and temperate zones. The Athenian Plato, in his dialogue *Phaedo*, hypothesized, after Pythagoras, that since the sphere is the most perfect form, this must be the earth's shape. Plato's student, the greatest scientist of classical Greece, Aristotle, provided empirical proof of earth's sphericity by observing the earth's shadow cast upon the surface of the moon during a lunar eclipse. However, Aristotle still had a limited view of earth, underestimating the extent of Asia, believing that the Caspian Sea flowed into the outer Ocean and that India marked the eastern extreme of the continent. Aristotle had the genius to conceive of a hemispheric world, north and south, perhaps basing his ideas on the reports of Phoenician and Carthaginian sailors.

The first-century-CE Greek biographer and scientist Plutarch, ever on the lookout to substantiate myth with fact, studied the nature and origin of stories about the Isles of the Blessed and Calypso's island of Ogygia. This latter island is featured in Homer's *Odyssey.* Homer placed the island, where Calypso held Odysseus against his will for seven years, somewhere in the West, which made sense since Calypso was the daughter of Atlas. Later legends placed Ogygia in the Atlantic, where Cronos, a god of Phoenician origin, held sway. In his dialogue "Concerning the Face which Appears in the Orb of the Moon," Plutarch discussed in detail the myth, trying to put the best face on what might be true about it. He hypothesized that Ogygia lay "five days off from Britain as you sail westward," that the route was difficult going, in part because the sea is "congealed," and it was as well a land of the midnight sun (trans. Cherniss and Helmbold). Pytheas discovered such conditions on his voyage, and it is possible Plutarch used it as evidence for his discussion of Ogygia. Pytheas claimed to have visited Thule, which the Stoic Seneca also briefly mentioned in his play *Medea.*

Plutarch, in *Life of Sertorius*, also reported the legend of the Isles of the Blessed. Sertorius, a Roman general who was in Spain around 80 BCE, met with mariners who claimed to have just returned from islands 1,200 miles off the coast of Spain. Why they left such a wonderful place of abundant fruits and continuous mild breezes is a mystery, but they convinced Sertorius (and Plutarch) of their reality and made the Roman long exceedingly to sail there to live his life in perfect repose.

The first-century polymath Elder Pliny also provided detailed descriptions of these isles in a failed attempt to make fiction fact. Pliny referred to the Hesperides, inhabited by the daughters of the setting sun, as the "Islands of the Ladies of the West" (trans. Rackham). He also discussed uncritically the island of Atlantis. Plato wrote of Atlantis in the *Timaeus.* The Athenian lawgiver Solon, during his travels, heard from the Egyptians about Atlantis, a large island hosting an advanced civilization that was destroyed by a tidal wave. From Plato's account, it is not clear whether Atlantis, a vast island civilization that mysteriously vanished, was an actual place, the legends of which Plato had read, or merely the product of his (or another's) fertile imagination.

MASSILIAN EXPLORERS

There are also anecdotal reports in the ancient literature that the Greeks of Massilia, in the western Mediterranean, and Caria, in southwestern Turkey, made important voyages of discovery. The Massilian seafarer Euthymenes sailed south along the West African coast circa 500 BCE. He claimed to have reached the Senegal River, about latitude 15° north. He argued that he discovered the source of the Nile in the Atlantic Ocean flowing into the African continent. The second-century-CE travel writer, Pausanias, described the voyage of Euphemus the Carian, who was caught in a storm, or the trade winds, off the coast of North Africa and blown west to an island in the Atlantic with unique inhabitants he had never before seen. Pausanias writes,

> Euphemus the Carian said that on a voyage to Italy he was driven out of his course by winds and was carried into the outer sea, beyond the course of seamen. He affirmed that there were many uninhabited islands, while in others lived wild men. The sailors did not wish to put in at the latter, because, having put in before, they had some experience of the inhabitants, but on this occasion they had no choice in the matter. The islands were called Satyrides by the sailors, and the inhabitants were red haired, and had upon their flanks tails not much smaller than those of horses. As soon as they caught sight of their visitors, they ran down to the ship without uttering a cry and assaulted the women in the ship. At last the sailors in fear cast a foreign woman on to the island. Her the Satyrs outraged not only in the usual way, but also in a most shocking manner. (Trans. Jones)

Another Carian, Scylax, around 500 BCE explored from the Indus River to the Red Sea, the northwest coast of Africa, and throughout the Mediterranean world, from the Strait of Gibraltar to the Sea of Azov.

PYTHEAS OF MASSILIA

The most famous Massilian discoverer was Pytheas. His voyage into the Atlantic, if it took place, occurred about 300 BCE and covered perhaps 7,000 miles. Pytheas reputedly circumnavigated the British Isles. His immediate object was tin, but he appears to have broadened his quest into a full-scale exploring expedition. Subsequent explorers journeyed to the "tin islands" visited by Pytheas.

Little is known about Pytheas's journeys and even less about his life. The account of his journey to the North Atlantic is lost. But ancient authors and geographers, such as Strabo and Polybius, wrote about Pytheas and provided some details of his journey. Pytheas was an ad hoc scientist, an explorer who under the circumstances of penetrating the unknown had to adopt a scientific methodology of observation, forming and testing hypotheses, and recording data. The third-century-BCE geographer Eratosthenes used Pytheas's account, which was condemned by the geographer Strabo as full of lies. The few details that do survive provide compelling evidence that Pytheas of Massilia did indeed journey to the North Atlantic.

Massilia in Pytheas's time was a city situated at the mouth of the Rhone River in what is today southern France. Massilia was a seafaring city engaged in shipbuilding and trade. Carthage, across the Mediterranean in North Africa, dominated the trade of the western Mediterranean. Mariners such as Pytheas frequently looked for the rare opportunity to sail undetected through the Strait of Gibraltar. Somehow or other Pytheas was able to do so around 300 BCE. Knowing of the Carthaginian tin trade with the British Isles, Pytheas and crew made their way up the coast of Spain and hugged the shores of France until they reached England. He claimed to have circumnavigated the Isles, making measurements and taking notes, and to have explored the islands on foot.

The most tantalizing and incredible part of Pytheas's journey was his exploration of Thule. Strabo quoted Pytheas that Thule "is a six days' sail north of Britain, and is near the frozen sea" (trans. Jones). Thule was never clearly defined by the ancients or later authors; even as late as the sixteenth century CE, the son and biographer of Christopher Columbus, Ferdinand, recorded a visit by his father to the North Atlantic, where he heard of Thule. Candidates for Thule are northern Scotland, the Orkney Islands, the Shetland Islands, Iceland, or Scandinavia; some have claimed Thule to be Greenland or Baffin Island in Canada! Pytheas's description of the environs of Thule lead one to speculate it was Norway or Iceland. Pytheas described the icy fog that enveloped the sea in that arctic region; his description could have only been gained by experience. Pytheas wrote regarding the sea around Thule that "there was no longer either land properly so-called, or sea, or air, but a kind of substance concreted from all these elements, . . . a thing in which . . . the earth, the sea, and all the elements are held in suspension; and this is a sort of bond to hold all together, which you can neither walk nor sail upon" (trans. Jones).

As a scientist, Pytheas's knowledge of celestial navigation helped him on the outward and return voyages. The report of his exploits included precise measurements of the British coastline, which he estimated to be about 40,000 *stadia*, or 4,600 miles in extent. Pytheas believed that there was a relationship between the moon and ocean tides. He made calculations of the position of the North Star and also used the sundial to determine the solstices and equinoxes. Like most inquisitive Greeks of his time, Pytheas's questions resulted in reasonable answers based on experience and observations made on the spot.

ALEXANDER OF MACEDON (356–323 BCE)

One might not think of Alexander of Macedon, the conqueror of the Persian Empire, as a scientist, yet such was the influence of his teacher Aristotle, his tutor in Alexander's early adolescence, that Alexander balanced his aggressive pursuit of martial glory with the aggressive pursuit of knowledge, including scientific information. This was completely unlike his father, the relatively barbarian king Philip II. Alexander also often strived to prove to the Greeks and others that he was not a barbarian from the backwater of Macedon, but a true Greek intellectual. The *unknown* intrigued Alexander—the unknown of the world, flora and fauna, the culture of peoples, the nature of the gods, and perhaps even the nature of the self. Alexander's deep-seated sense of *arete*, of heroic virtue, drove him forward into other places, conquering other peoples. Along the way he studied the places he visited the—rivers, plains, mountains, coasts, cities—and sought to conquer them intellectually as well as physically. These places included Anatolia, what is today Turkey; the eastern Mediterranean, what is today Lebanon, Syria, Israel, Jordan; the Nile River valley, what is today Egypt; the Tigris and Euphrates rivers flowing into the Persian Gulf, what is today Iraq, what the Greeks called Mesopotamia; Persia, what is today Iran; Bactria and Sogdiana, what is today Afghanistan; and the Indus River, what is today Pakistan. There is an astonishing scope to these conquests, accomplished in just ten years. And Alexander wanted more. He sought to discover Scythia, what would today be the region of the Caucasus Mountains northward; he sought to explore the Arabian Peninsula, the Arabian Sea, and the Red Sea; he sought to explore the Black Sea and its tributaries; he sought to go further west, across the Adriatic into the Italian peninsula. No doubt his vision of his own power, had he lived, would have contrasted sharply with the Romans of Italy and the Carthaginians of the western Mediterranean.

Alexander's vision formed the Hellenistic Age, a period of restless intellectual inquiry that covered three hundred years, from Alexander's death in 323 to the death of Cleopatra in 31 BCE. Alexander's vision of a united empire under his control included cultural interchange including basic institutions, such as marriage, clothing, language, habits, and a fascinating combination of Greek with Anatolian, Semitic, Egyptian, Mesopotamian, Persian, and Indian thought and culture. He seems to have envisioned cultural and intellectual diversity long before it caught on in the rest of the world. He conceived of the notion of training 30,000 Persian youth in Greek ways to create a fascinating Greco-Persian culture. Intrigued by the peoples he met on his journeys, he often asked for guides and philosophers to accompany him; hence did he learn of the Indian philosophers, the Gymnosophists.

One mark of an explorer is the reputation that they leave behind after their death. Alexander's mark on the imagination of the places that he conquered, and places that he might have conquered, long outlived him. He was remembered for hundreds of years among Muslim peoples in Asia and Africa. Muslim savants and intellectuals often compared their knowledge and learning to that of the great conqueror. He was known—and perhaps feared—in China. His legacy was such that all subsequent conquerors and thinkers sought to emulate this man in his power, glory, and intellect. Julius Caesar was once in despair for failing to equal the achievements of Alexander. The Greek philosopher Plutarch was so inspired by Alexander that it helped him to conceive of his "design," which "is not to write histories, but lives," as he declared in the *Life of Alexander*: "And the most glorious exploits do not always furnish us with the clearest discoveries of virtue or vice in men; sometimes a matter of less moment, an expression or a jest, informs us better of their character and inclinations, than the most famous sieges, the greatest armaments, or the bloodiest battles whatsoever. Therefore as portrait painters are more exact in the lines and features of the face, in which the character is seen, than in the other parts of the body, so I must be allowed to give my more particular attention to the marks and indications of the souls of men, and while I endeavour by these to portray their lives, may be free to leave more weighty matters and great battles to be treated of by others" (trans. Dryden). Michel de Montaigne, the Renaissance philosopher, thought of Alexander as one of the three greatest men of antiquity, along with Epaminondas of Thebes and the Athenian philosopher Socrates. Napoleon thought of himself as not only a general and conqueror but a thinker and philosopher as well, in imitation of Alexander. Hence a young Macedonian conqueror of the fourth century BCE caused such an influential change in cultural and intellectual history during ensuing centuries.

Accompanying Alexander, during the conquest of the Persian Empire from 334 to 323 BCE, were a host of scientists who recorded his journeys, collected specimens, and made observations about the different peoples of Asia.

Aristotle's grandnephew Callisthenes accompanied the king, serving as official historian. Other historians, such as Aristander of Telmessus and Cleitarch of

Colophon, accompanied Alexander, as well as Sophists, such as Anaxarchus of Abdera; Indian sages, such as Calanus; and soothsayers, such as Aristobulus of Cassandreia. His interest in geographic exploration inspired Alexander to send Nearchus to explore the Arabian Sea. Archias of Pella explored the Arabian Peninsula as far as Bahrain, which was then a trade and seafaring center. Androsthenes of Thasos sailed farther west, along the southern coast of the Arabian Peninsula. Another captain sent by Alexander, Hieron, reportedly circumnavigated the Arabian Peninsula into the Red Sea. Alexander also sent unnamed explorers to explore the Caspian Sea. Arrian (88–175), the Greek biographer of Alexander wrote that Alexander had many questions regarding the Caspian: whether it was linked to the Black Sea, or connected to outer ocean, and if not, what rivers fed into it.

Alexander has been the object of much debate among ancient and modern historians, who wonder what exactly his motives were in planning and carrying out his ten-year conquest of the Persian Empire. Some see Alexander driven by an idealist desire to unite all humans under his sole rule. Others argue that Alexander was a megalomaniac conqueror like so many other tyrants before and since. A possible interpretation for the historian of science is that Alexander's conquests were fueled by a deep personal motivation for glory and power as well as a strong desire to confront and conquer the unknown. Arrian's *Anabasis* quotes unnamed sources (but perhaps Nearchus) that Alexander used a specific word, *pothos*, to describe this longing to explore the unknown. Arrian cited several examples of Alexander's use of the word. Upon the death of King Philip Alexander, trying to solidify his hold on power, marched north to the Danube River to put down rebellions. Upon reaching the Danube, Alexander was "seized with a longing to go beyond the river" (trans. Brunt and Robson). In the *Indica*, Arrian quoted Nearchus: "Alexander had a vehement desire [*pothos*] to sail the sea which stretches from India to Persia" (trans. Brunt and Robson). This was the origin of his plan to send Nearchus to explore the Arabian Sea and Persian Gulf. Again, when Alexander had returned from his conquests and the disastrous journey across the Gedrosian Desert of southern Asia, upon reaching Persepolis he "was seized with a desire to sail down by the Euphrates and Tigris into the Persian Sea; and to see the outlets of these rivers into the sea, as he had seen the outlet of the Indus, and the ocean near it" (trans. Brunt and Robson). Arrian cited various writers who claimed that Alexander had a longing to see and conquer much more, including Arabia and Africa, both of which he proposed to circumnavigate, the latter all the way to the Pillars of Heracles at the Strait of Gibraltar. He was also interested in Sicily and Italy, having heard of the Romans' growing strength, and also longed to explore the Euxine (Black) Sea and conquer the savage Scythians to the north.

THE VOYAGE OF NEARCHUS

Nearchus of Crete (c. 360–312 BCE) was Alexander's admiral of an expedition that in 325 BCE explored the coastal waters of the Arabian Sea and Persian Gulf, from the Indus River to the Tigris and Euphrates Rivers. The landscape and some of the peoples Nearchus encountered were hostile. The naval expedition paralleled a land expedition of Macedonian troops led by Alexander through the harsh Gedrosian Desert. Besides providing logistical support for Alexander's journey, Alexander ordered Nearchus to make observations of the landscape and seascape, the peoples along the way, the natural productions, and remarkable phenomena. Nearchus's observations became the basis for Arrian's *Indica*, appended to the account of Alexander's campaigns into the Persian Empire, the *Anabasis*. Nearchus was an ad hoc scientist, an explorer and adventurer who by the circumstances of time and place was confronted with a terra incognita, a land unknown to himself and his contemporaries. Mere survival required that he make accurate observations and quick judgments about the diverse and unexpected experiences he encountered.

Nearchus, the son of Androtimus, had come to the Macedonian court at Pella from Crete when Alexander was still a youth. Nearchus was several years older, yet the two became friends. He was one of Alexander's companions who went into exile in response to the break between Alexander and his father, King Philip (r. 359–336) in 337. During the Persian campaign, Nearchus served as a satrap (governor) of Phrygia; later, having rejoined Alexander, he served as admiral of the Macedonian fleet. The ancient sources are vague on Nearchus and his life, but one assumes that as a Cretan, he followed the tradition of his people in learning early how to sail the Mediterranean. Such experience was useful when Alexander ordered Nearchus to explore the Arabian Sea.

THE INDICA

Arrian's *Indica* is wandering and inexact, the typical work of a geographer who relied heavily on the accounts of others as well as on tradition and hearsay. Nearchus was the prime source for Arrian's account of the Indus River, Arabian Sea, and Persian Gulf. Arrian supplemented his information with other, less useful sources, such as Nearchus's navigator, Onesicritus, and geographers, such as Herodotus and Eratosthenes. The *Anabasis* as well as the *Indica* repeated the errors of geography that were taught by Aristotle and were assumed true, without compelling evidence to the contrary, by Alexander and Nearchus. For example, Aristotle taught and Arrian recorded as fact, based on Nearchus, that the Taurus Mountains (of Asia Minor) extended east across Asia to India and beyond to the "Eastern Ocean," which washed the eastern shores of India (trans. Brunt and Robson). Arrian used Nearchus's accounts of zoology to confirm or deny the legends promulgated by more fanciful authors such as Megasthenes. Nearchus claimed to have seen a tiger skin but not a tiger, so he could not substantiate the story that they were more swift than a horse and could fell a single elephant on their own. Nearchus did witness a parrot, and its chattering like a human, of which Arrian was skeptical. And he reported on the snakes of India and how the Greeks had no antidotes for snakebites but Indian physicians did, and these physicians also were useful healers for other physical complaints.

Nearchus also reported on the strange and unexpected position of the stars and the sun. As they made preparations for the voyage near the mouth of the Indus in the late summer, the men noticed that "at midday . . . everything seemed shadowless" (trans. Brunt and Robson). As they were near the Tropic of Cancer soon after the summer solstice, the rays of the sun at noon would appear almost directly overhead.

After negotiating the shoals, tides, and maze of channels at the mouth of the Indus River and waiting for favorable winds to carry them forth, they began their journey along the coast, in what the Greeks called the Erythraeum Sea. The forbidding coastline challenged the ability of the rowers aboard the barges and triremes of the flotilla. The men were preoccupied with finding food and fresh water and avoiding shipwreck in shoal water. Nearchus kept a precise account of the voyage, noting the islands, the shape of the coast, and distances traveled in "stades" (*stadia*). Local pilots guided the Greeks, pointing out landmarks and providing historical and scientific tidbits. From such information, an accurate map could eventually be made. At the Tomerus River, on today's Pakistani coast, they confronted a native host who were extremely primitive. "Those captured were hairy, not only their heads but the rest of their bodies; their nails were rather like beasts' claws." The pilots told Nearchus that these people "used their nails . . . as if they were iron tools; with these they tore asunder their fishes, and even the less solid kinds of wood; everything else they cleft with sharp stones; for iron they did not possess. For clothing they wore skins of animals, some even the thick skins of the larger fishes" (trans. Brunt and Robson). The pilots called these paleolithic people the "fish-eaters," their homeland, *Ichthyophagi*. Shortly thereafter they met a band of friendly fish eaters, who offered the Greeks mutton, which "had a fishy taste, like the flesh of the sea-birds, since even the sheep feed on fish; for there is no grass in the place." Eventually coming to a harbor where fishermen used primitive boats to harvest the sea, they took on board a pilot who knew the coastline. He was "a Gadrosian called Hydraces" (trans. Brunt and Robson). Near today's border of Pakistan and Iran, they raided a small village for food, which was mostly made from fish flour.

ICHTHYOPHAGI

Using the reports of his pilots as well as his own observation, Nearchus arrived at as assessment the economy of the fish eaters:

> These Fish-eaters live on fish; and hence their name; only a few of them fish, for only a few have proper boats and have any skill in the art of catching fish; but for the most part it is the receding tide which provides their catch. Some have made nets also for this kind of fishing; most of them about two stadia in length. They make the nets from the bark of the date-palm, twisting the bark like twine. And when the sea recedes and the earth is left, where the earth remains dry it has no fish, as a rule; but where there are hollows, some of the water remains, and in this a large number of fish, mostly small, but some large ones too. They throw their nets over these and so catch them. They eat them raw, just as they take them from the water, that is, the more tender kinds; the larger ones, which are tougher, they dry in the sun till they are quite sere and then pound them and make a flour and bread of them; others even make cakes of this flour. Even their flocks are fed on the fish, dried; for the country has no meadows and produces no grass. They collect also in many places crabs and oysters and shell-fish. There are natural salts in the country;

> from these [fish] they make oil. Those of them who inhabit the desert parts of their country, treeless as it is and with no cultivated parts, find all their sustenance in the fish; but a few of them sow part of their district, using the corn as a relish to the fish, for the fish form their bread. The richest among them have built huts; they collect the bones of any large fish which the sea casts up, and use them in place of beams. Doors they make from any flat bones which they can pick up. But the greater part of them, and the poorer sort, have huts made from the fishes' backbones. (Trans. Brunt and Robson)

These were the backbones of whales, of course; Nearchus reported that whales much larger than those of the Mediterranean appeared around the ships, terrifying the sailors. Nearchus inspired them by leading them into naval combat against the whales. The pilots informed Nearchus that "some of these whales go ashore at different parts of the coast; and when the ebb comes, they are caught in the shallows; and some even were cast ashore high and dry; thus they would perish and decay, and their flesh rotting off them would leave the bones convenient to be used by the natives for their huts. Moreover, the bones in their ribs served for the larger beams for their dwellings; and the smaller for rafters; the jawbones were the doorposts, since many of these whales reached a length of five-and-twenty fathoms." This 150-foot creature, if real, was perhaps the blue whale. Locals informed the Greeks about another mystery of the deep—an island off the coast of Iran that was "sacred to the Sun": anyone who chanced to set foot upon it disappeared. Nearchus, scowling at such superstition, himself disembarked on the island, showing it was but a "fairy-tale" (trans. Brunt and Robson).

When the fleet reached Cape Sharita at the Strait of Hormuz, the entrance to the Persian Gulf, Arrian claimed, based on Nearchus's authority, that they briefly rendezvoused with Alexander in the region of Carmania before setting forth, sailing along the eastern shore of the Persian Gulf. Nearchus's descriptions of this region were not as detailed as he desired because there were few good harbors and beaches at which to make port. Eventually they reached the Euphrates River, up which they rowed to Babylon. Discovering that Alexander expected a rendezvous at the city of Susa, east of the Tigris, the fleet set sail once again, descended the Euphrates, ascended the Pasitigris River, and ended their voyage.

Nearchus's voyage opened a completely unknown region to Greek conquest, trade, and culture. His description of the South Asian coast was sufficiently accurate for other mariners to base their itineraries on. Later writers continued to rely on Nearchus. Biographers of Alexander, such as Diodorus and Plutarch, relied on Nearchus. Scientists such as Strabo and Pliny the Elder did as well. Nearchus's fame was richly deserved. But one wonders about those nameless pilots, the local guides, who were the true explorers, upon whose knowledge and experience Nearchus relied and could not do without.

HELLENISTIC GEOGRAPHERS

Alexander opened up a whole new world for Greek scientists. The scientific center of Alexandria in Egypt, founded by Alexander in 331, sponsored a host of geographers during the ensuing Hellenistic Age.

Eratosthenes of Cyrene (276–195 BCE) provided some of the most accurate estimates of land surface measurements of his time. The geographer Strabo praised him for being more realistic than other geographers, such as Polybius (208–126), who allowed fancy to take hold of reason. Eratosthenes divided the earth into climatic zones: the poles, arctic, temperate, tropics (Cancer, Capricorn), equator. He believed that the tropics, where the solstices occurred, have more extreme heat than the equator, which was higher in elevation and had more rain. He doubted Homer's geographic knowledge, even if others, such as Strabo, were more credulous.

Eratosthenes was the first scientist to map the earth according to latitude and longitude, profoundly influencing Hipparchus and Ptolemy. He devised the idea of the meridian, the abstract focal point from which all geographic measurements can proceed. In his *Chronography*, Eratosthenes devised a system of abstract chronology as well, basing the measure of passing years according to the Olympic games occurring every four years beginning in 776 BCE.

Eratosthenes, who believed that the landmass of Europe, Asia, and Africa was exclusive to the Northern Hemisphere, estimated the extent of Europe, Asia, and Africa to be 77,000 stadia. The unknown Southern Hemisphere must be inhabited, though by whom Eratosthenes was uncertain. It is possible that Eratosthenes, who advocated the theory of a greater ratio of water to land, believed that a western voyage across the Atlantic would return the voyager to Asia.

Hipparchus (190–120 BCE), an astronomer of Nicaea, in northwestern Turkey, was a geodesist who used latitude and longitude in a grid pattern at precise intervals to measure the surface of the earth. He also established a set of coordinates based on a 360-degree spherical globe. Hipparchus was influenced by what he could learn from Babylonian astronomers, particularly Babylonian ephemerides. In his use of the deferent-epicycle system of the geocentric universe and cartographical coordinates to measure the earth, Hipparchus had a profound influence on other astronomers and geographers—notably Ptolemy.

Posidonius (135–50 BCE), originally from Syria, was a Stoic philosopher and Greek thinker who was very influential in first-century Rome. A student of Panaetius (185–110), who first introduced Roman intellectuals to Stoic thought, Posidonius continued the Stoic influence upon the Romans particularly through his friendship with Marcus Tullius Cicero. He was known to be the teacher of Gnaeus Pompey. Posidonius particularly made his mark in the science of geodesy, imitating Eratosthenes in his estimate of the circumference of the earth based on the use of the gnomon to measure the angle of the sun's shadow at Alexandria, which he compared to a similar measurement made at Rhodes. He calculated the circumference of earth at 180,000 stades (a little over 20,000 miles, which is remarkably accurate). Much traveled, Posidonius reputedly journeyed all over the Mediterranean in search of scientific data. A polymath, he wrote on meteorology and history and penned a periplus based on his travels. Posidonius was also a student of geology (volcanoes) and meteorology (tides).

STRABO (63 BCE–21 CE)

The geographer Strabo was the Greek writer of *Geography,* an influential treatise on history and geography. Strabo was a Stoic philosopher influenced by

Athenodorus (74–7), a contemporary Stoic philosopher; Xenarchus, a notable first-century-BCE Peripatetic philosopher; and the first-century geographer Tyrannion. Strabo spent time in Alexandria, journeyed throughout the Near East, and ascended the Nile River to Ethiopia. His *Geography* was based in part on his personal experiences and revealed him to have been a polymath interested in a variety of topics of inquiry.

Strabo's concept of the earth was typical for his time. He unquestioningly relied on Homer for the geographic essentials. He envisioned three continents—Europe, Asia, and Africa—which combined to form a great island surrounded by the vast encircling river Ocean. The sun orbited the earth, appearing out of Ocean in the east and setting in Ocean in the west. The Caspian Sea flowed into the outer ocean. The Isles of the Blessed lay to the west, in the Atlantic. Ocean was a river, flowing in and about, which explained the tides.

Strabo went beyond Homer, however, in his hypothesis that the world could be circumnavigated by sailing west from Europe to Asia on the same latitude. The distance, and only the distance, would prevent the fulfillment of the journey. Strabo had the support of the Stoics Posidonius and Athenodorus partly because the greater the water on the surface of the earth, the Stoics thought, the stronger was the bond that held the planets and stars in place. Strabo indicated Hipparchus's disagreement with both ideas, the former idea because of the possibility of an interrupted ocean, the latter idea because of its absurdity. Strabo condemned Eratosthenes for his criticism of Homer, arguing that Homer's myths were meant to bring his hearers and readers to an awareness of the core of truth found in the *Iliad* and *Odyssey.*

The reader of Strabo, in short, finds an author who was skeptical of many of the stories told of other lands and peoples. He used reason to sort through anecdote to arrive at plausible cause and effect. For example, Strabo spent much time in book 1 of his *Geography* examining Posidonius of Rhodes and his stories, particularly one about Eudoxus of Cyzicus. Posidonius gave as fact the account of Eudoxus's voyages in and about Africa, which Strabo, checking for internal consistency in the stories, branded as spurious. Strabo also doubted the accounts of Pytheas that Polybius and other writers had passed along as true.

The structure of Strabo's *Geography* is as follows: Books 1 and 2 provide useful summaries of the lost works of writers such as Eratosthenes while allowing Strabo to engage in long dissertations defending writers such as Homer and attacking others, such as Posidonius of Rhodes. Books 3 and 4 analyze the peoples, places, and sources of information regarding Spain and Gaul. Book 5 provides a description of Italy, including an extensive discussion of the city of Rome. Books 6 through 10 discuss Magna Graecia, Sicily, Germany, and Greece, while Books 11 through 14 discuss Asia Minor and the Near East. Books 15, 16, and 17 describe India and Africa.

ROMAN GEOGRAPHY

The Elder Pliny (23–79 CE) relied extensively on previous writers to pen his vast *Natural History*, which includes much on world geography from a narrow

Roman perspective. He had traveled throughout the empire as a soldier and administrator; during his travels he tirelessly recorded his observations and local accounts of natural phenomena. Added to this was his broad reading and collection of useful facts from past experts in the sciences. His geographical descriptions of western Europe were therefore very accurate. The overall extent of Europe, its size and limits, was unknown to Pliny, but he was not afraid to speculate. He assumed that Europe was wonderful enough to be half the size of the known world. Based on the work of the Greek geographer Hecataeus (550–476), Pliny credulously described the Hyperboreans, who lived in northern Europe near the source of the winds and who were reputedly extremely happy in their imagined utopia. Pliny's knowledge of Africa and Asia was, of course, limited, and Pliny made constant use of fabulous stories, exaggerating places, peoples, and landmarks, making them larger and grander than what was real. His imagination told him of strange peoples at the farthest extent of the world: people with no heads but with faces on their chests, others with one large foot that gave them shade on particularly hot days. When Pliny used sources penned by actual observers, such as the scientific observations based on Alexander's expedition to Asia, Herodotus's stories of the eastern Mediterranean, and Polybius's account of North and West Africa, his descriptions were better, if dated. Pliny, like Strabo, wrote at length about India, relying heavily on the works of Nearchus of Crete, Onesicritus of Astypalaea (c. 360–290), and Aristobulus of Cassandreia (c. 375–301), all of whom accompanied Alexander on his expedition and journey to India.

Pomponius Mela, a first-century-CE Spaniard who authored the Latin *Chorographia*, provided a general overview of the geography of the Roman Empire during the first century. He wrote the book in the 40s, during the reign of Claudius and after that emperor's conquest of Britain, and restricted his observations to the "known world," that is, to the world as the Romans conceived of it. Pomponius Mela knew, of course, that the world was spherical, and he conceived of a Southern Hemisphere to balance the Northern. But since this Southern Hemisphere (*Antichthones*) was unknown, he refrained from describing it. Writing, therefore, of the world as Romans knew it, Pomponius Mela described the world from a Mediterranean perspective, focusing on what the Romans knew of the three recognized continents: Europe, Africa, and Asia. The author used his rhetorical skills to compose an imaginary travelogue of the world, going from place to place.

Other Roman geographers included Theodosius of Tripolis (160–90), who developed mathematical geography, in which he calculated the amount of time in terms of daylight and darkness for various places at different seasons at particular latitudes. He was thought by Vitruvius, the architect, to have invented a forerunner of the sextant to determine latitude. Julius Caesar (100–44), in the *Gallic War,* described the peoples, landscapes, and rivers of Gaul and Britain based in part on firsthand observations. Little is known of the first-century-CE scientist Agathemerus except that he wrote the treatise *Geography*, which has survived in fragments. Perhaps he lived at the beginning of the Common Era, making him a contemporary of Strabo. Agathemerus appears to have been a Roman. He was intrigued by Greek geographers who hypothesized the spherical nature of the earth and believed that Delphi, sacred to Apollo, formed the center. He lauded

Democritus for his conception of an earth more wide than long. He claimed that Anaxagoras was the first to draw a world map. Agathemerus had the benefit of having the works of Hecataeus and Herodotus before him and the increased knowledge of the world that was brought about by Alexander's conquests and the explorations of Pytheas of Massilia and Nearchus of Crete.

CONTINENTS, ZONES, AND SEAS

Greek geographers developed various theories to explain the relation of continents to oceans and different climatic zones. The first Greek scientists, such as Pythagoras, upon determining that the earth was spherical, hypothesized the existence of five climatic zones on the earth's surface: the two poles, two temperate zones, and the equator, which legend and imagination told them was an impassable zone of fire. Time and experience taught scientists like Polybius that such a zone of fire was absurd. Greek geographers had different conceptions about the relation of ocean and land on the earth. Aristotle and others, such as Strabo, Pliny, Crates, and Pomponius Mela, argued that there was a world-encircling ocean surrounding the three joined continents of Europe, Asia, and Africa. Herodotus, Plato, Hipparchus, and Ptolemy believed the opposite, that land dominated seas on the earth's surface. Ptolemy, inspired by Marinus of Tyre (70–130), argued that the Indian Ocean was an inland sea and that Asia extended south and then west to join with Africa. Second-century-BCE geographer Crates of Mallus, whom the ancients claimed invented a working globe, hypothesized four continents in four sections of the earth. Indeed, many geographers argued for the presence of geographic sections; Pythagoras believed that an opposite realm, the antipodes, existed far to the south, mirroring the north. Others, such as Pomponius Mela, author of the *Chorographica,* conceived of the existence of a Southern Hemisphere that was impassable because of the ring of fire. He called such a land the *other world*, "Ora Australis." Cicero's imaginative "Scipio's Dream," in the *Republic,* presented a similar idea. Still others argued that the earth's landmass and peoples lived only in the Northern Hemisphere, the hypothetical south being uninhabitable. Obscure geographers such as Geminus, Cleomedes, and Achilles Tatius spent much time and thought on these theories. Another, Macrobius, in the fifth century CE, called the four continents "terra quadrifiga"; the north and south were separated by an equatorial sea, he believed, agreeing with Crates.

CLAUDIUS PTOLEMY (100–170)

The greatest geographer of the Roman Empire was Claudius Ptolemy, who lived during the second century CE in Alexandria. Ptolemy was an astronomer as well as a geographer. Ptolemy worked in the shadow of Eratosthenes and other Alexandrian geographers to develop the most sophisticated system of geography in the ancient world. He developed standard directions (north, south, east, west). His ideas were, however, erroneous in several respects. He overestimated the size of Asia and underestimated the circumference of Earth by eighty degrees or two-ninths of the earth's surface. He conceived of the earth as being largely composed of land rather than water. Not knowing of the Pacific Ocean or the Americas, Ptolemy's conception of the earth was of three continents (Europe, Asia, Africa), a world-encircling ocean (the Atlantic), and a landlocked Indian Ocean. He assumed that Asia and Africa were connected, with Asia dipping farther south and west than in reality, and he conceived of a vast terra incognita in the Southern Hemisphere, making the Indian Ocean a landlocked sea.

Ptolemy's *Geography* became the standard during the Later Roman Empire and throughout the European Middle Ages. The Renaissance explorer Christopher Columbus relied on Ptolemy's geographical scheme to plan his "Enterprise of the Indies." Ptolemy's errors helped Columbus convince himself that a voyage from Spain west across the Atlantic to China was possible. He did not realize that America, unknown to the ancients, lay between.

ANCIENT MOUNTAINEERING

Among ancient peoples, mountains were places of terror, dread, and mystery, in part because of their forbidding distances, shrouded summits, and thunderous noises but also because mountains were the abodes of the gods. It is rare to find in ancient literature accounts of humans ascending mountains, except out of necessity. But mountains so fascinated humans that they sought to explain the strange phenomena associated with grand peaks. Early attempts at explanation were mythological and fantastic. But with time, ancient thinkers began to ask sophisticated questions and seek realistic answers.

Mythical accounts of mountains are found in many world cultures. Ancient Hindus were in awe of Meru, a legendary mountain north of the Himalaya range. Hindus and Buddhists throughout southern Asia built artificial mountains in imitation of Meru. Likewise, ancient Mesopotamians, Egyptians, and Americans built artificial mountains—ziggurats and pyramids—in a symbolic attempt to approach the divine. Ancient societies of the Near East also had experiences with real mountains, which is reflected in some of the world's oldest literature. The *Epic of Gilgamesh*, composed at the end of the third millennium BCE, describes the adventures of the hero Gilgamesh, who confronted the realm of the divine—mountains—and had the courage to ascend them in pursuit of glory. Mountains were associated with some of the most important gods of the Mesopotamian pantheon. Enlil, whom the *Epic* calls "the father of the gods," took control of a great heavenly mountain that contained the earth and air. A monster protected Mashu, twin peaks where the sun descended and ascended and within which a hellish darkness reigned. Cedar Mountain, guarded by the monster Humbaba, was sacred to Ishtar, a fertility goddess associated with love and lust. Gilgamesh and his friend Enkidu ascended the Cedar Mountain, where they saw "wild barley" growing, blown by the wind, and experienced strange transcendental dreams indicating future events (trans. Sandars).

The Hebrews also had miraculous experiences upon mountains. The Old Testament book of Exodus, composed during the second millennium BCE, describes the prophet Moses receiving from God (Yahweh) the Decalogue, the Ten Commandments, on Mount Sinai, a 7,500-foot peak on the Sinai Peninsula. Sinai was terrifying, encompassed by "thunders and lightnings, and a thick cloud upon the mountain, and a very loud trumpet blast, so that all the people who were in the camp trembled" (Exodus 19:16 RSV). Peals of thunder echoing from a mountain summit covered in dark clouds pierced by frequent strokes of lightning can sound like massive blasts of some otherworldly trumpet.

Greece has a landscape of undulating terrain—mountains and valleys, but few plains. This rocky peninsula jutting into the Mediterranean Sea, between the Ionian and Aegean Seas, hosted the hunter and shepherd—and the poet as well. Particularly the mountains of Greece astonished and inspired the Greeks to attempt explanations of their sublimity and a search to penetrate their mystery.

The great mountains in the neighborhood of Greece that held the Greek imagination for centuries were Parnassus, Nysa, Kyllene, Ida, Dindymum, and Olympus. The Greek world was hemmed in by two Titans suffering the consequences of disobedience to Zeus. Atlas in the west held up the heavens, his shoulders appearing like massive peaks. In the east, at the extreme of the world were the Caucasus Mountains, where Prometheus was chained to rock, daily enduring the torture of a vulture gnawing at his liver. Aeschylus referred to the Caucasus as "star-neighboring mountain tops" (trans. Warner). Closer to home, Parnassus, at over 8,000 feet, loomed over Delphi, the oracle sacred to the archer god Apollo. Parnassus was also the home to the twelve Muses, the daughters of Zeus. Mount Nysa in Thrace was sacred to the god of wine Dionysus. At Nysa, mountain nymphs raised the young god. The ancient *Hymn to Aphrodite* had perhaps Mount Nysa in mind in proclaiming that the oaks and pines of mountains are sacred groves where the fairylike nymphs played. Mount Kyllene in Arcadia, the Greek Peloponnesus, was the birthplace of Hermes, the winged son of Zeus and Maia, the daughter of Atlas. On Mount Kyllene Hermes invented the lyre from a tortoise shell. Mount Ida in Crete was sacred to Zeus, the king of the gods and wielder of the thunderbolt. Mount Ida in northwest Asia Minor was famous for hosting the rivers that watered the Troad, where stood ancient Troy and Ilios. Also on Mount Ida, Aphrodite conceived Aeneas, the great Trojan warrior. Mount Dindymum, in Cyzicus, was sacred to an early fertility goddess, Rhea, mother of Zeus and wife of Cronos. Hence Mopsus the seer, in Apollonius's *Argonautica*, declared that "Rhea's dominion covers the winds, the sea, the whole earth, and the gods' home on snow-capped Olympus. Zeus himself, the Son of Cronos, gives place to her when she leaves her mountain haunts [at Dindymum] and rises into the broad sky" (trans. Rieu).

The most famous mountain was Olympus, which as the highest mountain in Greece, at 9,794 feet, was perfectly suited to be the home of the gods. Greeks refused to climb the mountain because of its height, forbidding countenance, and holy stature. Clouds often hid the distant summit, a perfect cover for the god's secret eternal lives. The frequent storms could be none other than Zeus nodding his great thunderous head and deities, such as his daughter Athena, darting to earth lightning fast to spread word of the father's will. From the perspective of Olympus, the gods observed human behavior, heard their prayers and pleas, and made judgments based upon their awareness of the future in light of the present and the past. The mountain symbolically was raised above human ignorance and time, to provide the bright rays of timeless truth.

The seafaring Greeks were more familiar with the sea than with the mountains. Both phenomena were astonishing and terrifying, but mountains were all the more so because they were nearby, always in sight, yet forbidding and ultimately unknown. Because of their grand appearance, they were a wonderful complement to the inquisitiveness of the Greeks. Mountains served to explain perplexing phenomena, such as lightning and thunder; the mysterious forested slopes; the quake of the earth; the frequent meteorological changes and displays on mountain summits; the sublime sense of the divine apparent in the unapproachable parts of nature; and the relationship of earth and sky. Homer's *Odyssey*, for example, explains the latter phenomenon according to the Titan Atlas, "who holds the columns that bear from land the great thrust of the sky" (trans. Fitzgerald).

Even during subsequent centuries, when Mediterranean societies became more sophisticated and the Greek scientific mind awoke from slumber, mountains were still distant, daunting, and unapproachable. Polybius, who traveled through the Alps and saw Mount Atlas at a distance, allowed his amazement to take hold of his senses: he described the extent of the Alps as being over a million feet! The few recorded ascents of mountains were for military reasons. Thus Alexander of Macedon in the fourth century BCE crossed the Taurus Mountains of southeastern Turkey and the Hindu Kush of Afghanistan. Hannibal of Carthage attacked Rome in 218 BCE after having crossed the Pyrenees, followed by the Alps. The Roman historian Livy (59 BCE–17 CE) recorded the ascent of Mount Hebrus in Thrace at the beginning of the second century BCE. King Philip V of Macedon (r. 221–179), at war with Rome, made the ascent to spy on Roman troop movements. It took three days for the mountaineers to journey through the foothills and ascend the summit: "As they reached the high levels they were increasingly faced with wooded and often impassable ground. Eventually they came upon a track so shaded that it was scarcely possible to see the sky for the density of the trees and the interlacing branches. And then, when they got near to the crest, everything was so covered with mist . . . that they were slowed down as much as if they were on a night march. At last, on the third day, they reached the summit." The descent took two days. The suffering the men experienced was immense, particularly because of the cold; the third night on the summit was awful in this regard. According to Livy, who obviously knew little about mountain climbing, the thick fog that enveloped Philip and his men at the summit was "a rare thing in high places" (trans. Bettenson).

The Greeks and, after them, the Romans rarely attempted to explain mountain phenomena. Science requires not only observation but analysis based on direct experience and experiment. Lacking the will to ascend lofty peaks, one can hardly expect that clear understanding of mountains would exist. This is particularly the case when mountains are considered sacred, associated with the supernatural and transcendent. Lucretius the Epicurean, who refused to believe in anything that could not be explained according to matter in motion, the perpetual movement of invisible atoms, not surprisingly was able to view mountain phenomena from an empirical point of view. Mountains were hollow, Lucretius believed, and volcanic eruptions occurred when atoms of fire were forced out of the cone. Closer to the truth was his observation that clouds develop on mountain peaks because of warm air rushing up the slopes to the cool air at the summit.

The most famous Roman to investigate mountains was Gaius Plinius Secundus, the Elder Pliny (23–79). When in 79 CE, Mount Vesuvius erupted, blasting ash and fire throughout the beautiful region of Campania, Pliny, who could see the volcano from his house at the Bay of Naples, ordered a boat so to investigate the black plume of smoke rising from Vesuvius. He took notes of his observations as the boat reached the shores south of Pompeii and continued to observe the falling ash and pumice until his death from asphyxiation.

At the same time, the first century CE, Christians continued the Jewish fascination with mountains. Jesus of Nazareth, as described in the Gospels of the New Testament, found meaning and transcendence on the small mountains surrounding Jerusalem. Aurelius Augustine, however, several centuries later decried the human fascination with mountains at the expense of self-awareness. The subsequent medieval European attitude toward such physical monuments to the Creator was to eschew concern for the sake of the incorporeal and spiritual. It was left to the Renaissance humanist and mountaineer Francesco Petrarca, in the fourteenth century, to call attention to the possibilities of self-discovery in the experience of ascending the distant peak.

SCIENCE, TRADE, AND CULTURAL EXCHANGE

Ancient peoples were, contrary to expectation, on the move. Even before the advent of sophisticated technological devices to assist humans in moving about the planet at great speeds in a comparatively short amount of time, ancient peoples traveled, carried, traded, penetrated, and engaged in the same geographic movement as modern peoples. Ancient seafarers, as noted above, crossed the world's oceans, connecting continents. The eastern coast of the Atlantic Ocean was busy with shipping, as was the western coast of the Pacific Ocean, and the northern coast of the Indian Ocean. The Mediterranean Sea was busy with sea traffic thousands of years before the birth of Christ. The Greeks colonized North Africa and western Asia. The Romans christened the Mediterranean, *mare nostrum*, "our sea," because Roman shipping and marine transportation dominated the coasts of North Africa, the Levant, and southern Europe for over a thousand years. The Romans developed a sophisticated system of roads that connected the vast distances of the empire, from Britain to the Rhine River to Gaul to Spain to North Africa to Egypt to the Levant to the Black Sea to the Danube River. By the mid-first century CE, there were over three hundred separate roads covering over fifty thousand miles. Roads connected the empire. Dozens of roads spread from Italy to the Alps, connecting Rome with Ravenna, Marseilles, Lyons, Milan, Naples, and Syracuse. Roads followed conquests. They extended into Spain, Gaul, and England, connecting the cities of Toledo, Cadiz, Tarragona, Barcelona, Bordeaux, Paris, Cologne, London, and York. Under Hadrian (r. 117–138), roads in England ended at a seventy-six-mile wall built to hold back the Picts and Scots to the north. Hadrian's Wall was built of stone and cement and was fortified every mile with a small fort or castle. Roads later extended beyond Hadrian's Wall to the Antonine

Wall, built under his successor Antoninus Pius (r. 138–161). To the south, eight roads entered London. Eighteen entered Rome. Throughout the empire there was a sophisticated post system with hostels every eighteen miles where travelers could find refreshment and a place to stay. Perhaps the Romans borrowed the idea from the Persians, who once had the great Royal Road spanning their Asian empire, with hostels every fifteen miles.

The Roman system was called, during the later Roman Empire, the *cursus publicus.* Beginning during the second century BCE, Romans set up milestones every "mile" (4,800 feet) to indicate distance. Cartographers drew early road maps to gauge distance and direction, especially for imperial and military purposes. Of these maps, called *itineraria,* which were drawn on goat's skin or papyrus, only one survives, the Peutinger Table, a copy of an *itinerarium* of the Late Roman Empire.

East of the Roman Empire, the Persian Empire also developed a vast road system to enable communication and trade. Darius I (r. 522–486 BCE) began construction of this extensive network that at its height connected the Indus River to the Kabul River, west to Persepolis then to Susa, west from Susa to Sardis in Anatolia, and then west to Ephesus. The road also connected the Tigris and Euphrates River valleys with the Nile. Like the Roman network, the royal road was maintained extensively and enabled caravans to travel from the Indus to the Aegean in ninety days. Couriers could make the journey in a week. The Persians also accentuated water transport by means of sophisticated ships that could hold up to two hundred tons of cargo. Quays for loading and unloading were found at ports along the sea- and freshwater ports; canals were also constructed. After the conquests of Alexander, the Greek empires of his successors expanded roads, trade, learning, and communication of ideas. After the birth of Christ, the Parthians kept this system of roads and stable communications.

It was inevitable that such large empires, such as the Roman, Persian, Seleucid, Parthian, and Han, would forge links of communication and trade. Caravan routes were forged from the eastern regions of the Seleucid and Parthian empires, Bactria and Sogdiana, northeast across the desert region of Xinjiang to the Han capital of Chang'an. The first explorer to accomplish this was Zhang Qian (164–113), who journeyed to Bactria in 128 BCE and wrote accounts of the regions of the Asian steppes, Bactria, Sogdiana, Parthia, India, and Mesopotamia.

In the *Epitome*, the Roman historian Florus (c. 74–130) wrote of contact between the Han Dynasty and the Emperor Augustus:

> Even the rest of the nations of the world which were not subject to the imperial sway were sensible of its grandeur, and looked with reverence to the Roman people, the great conqueror of nations. Thus even Scythians and Sarmatians sent envoys to seek the friendship of Rome. Nay the Seres [Chinese] came likewise, and the Indians who dwelt beneath the vertical sun [the region of the tropics], bringing presents of precious stones and pearls and elephants, but thinking all of less moment than the vastness of the journey which they had undertaken, and which they said had occupied four years. In truth it needed but to look at their complexion to see that they were people of another world than ours. The Parthians also, as if repenting for their presumption in defeating the Romans, spontaneously brought back the standards

which they had captured in the catastrophe of Crassus. Thus all round the inhabited earth there was an unbroken circle of peace or at least of armistice. (Trans. Yule)

In the wake of the journey to Parthia of Gan Ying at the end of the first century (97) CE, the Chinese and Romans exchanged ambassadors, and mutual knowledge grew of the Romans and Parthians of western Asia and the Chinese of eastern Asia. In the third century CE, historian Yu Huan wrote an account of what he had learned from Roman traders: he mentioned the Nile River, the city of Alexandria, and the Mediterranean Sea. In the several centuries after the birth of Christ, because the Romans had acquired control of Egypt and therefore gained access to the Red Sea and Indian Ocean, and because the Chinese had extended their control to Indochina, there was increasing Roman traffic east and Chinese traffic west via Indonesia and the Indian Ocean. Chinese, Indonesian, Indian, and Roman traders used the winds and currents of the Indian Ocean, which changed according to the seasons, to travel back and forth in their sailing vessels. Arab, Polynesian, and Malay sailors had developed significant ships, rigged with lateen sails for maneuverability in contrary winds, for trading. Indonesian sailors used outrigger canoes to traverse the Indian Ocean to Africa. Overall, there was brisk trade in the ancient world both overland and by sea connecting western, central, and eastern Asia and including Europe and Africa. This trade involved spices, silk, porcelain, medicines, and knowledge.

The journeys of ancient humans into the spatial unknown yielded more questions than answers. What is Earth's circumference, shape, and limits? Can Africa be circumnavigated? How vast is the ocean? Although ancient geographers did not find exact answers, they set the stage for those who would.

FURTHER READING

Aeschylus, *Prometheus Bound*, trans Rex Warner, in *Ten Greek Plays* (Boston: Houghton Mifflin, 1957).

Al-Biruni, *The Chronology of Ancient Nations*, trans. C. Edward Sachau (London: William H. Allen and Co., 1879).

Apollonius of Rhodes, *The Voyage of Argo,* trans. E. V. Rieu (Harmondsworth, England: Penguin Books, 1971).

Arrian, *Anabasis* and *Indica,* trans P. A. Brunt and E. Iliff Robson 2 vols. (Cambridge, MA: Harvard University Press, 1933, 1976).

Jonathan Barnes, *Early Greek Philosophy* (London: Penguin Books, 1987).

Daniel Boorstin, *The Discoverers* (New York: Random House, 1983).

M. Cary, *The Geographical Background of Greek and Roman History* (Oxford: Clarendon Press, 1949).

Chinese Accounts of Rome, Byzantium and the Middle East, c. 91 B.C.E.–1643 C.E., East Asian History Sourcebook, https://sourcebooks.fordham.edu/eastasia/romchin1.asp.

Victor Ehrenberg, *Alexander and the Greeks* (Oxford: Oxford University Press, 1938).

The Epic of Gilgamesh, trans. N. K. Sandars (London: Penguin Books, 1972).

Robin Lane Fox, *The Search for Alexander* (Boston: Little, Brown, 1979).

Homer, *Odyssey,* trans. Robert Fitzgerald (New York: Random House, 1990).

Kathleen Freeman, *Greek City-States* (New York: W. W. Norton, 1950).

R. Ghirshman, *Iran: From the Earliest Times to the Islamic Conquest* (Harmondsworth, England: Penguin Books, 1954).

Thomas Heath, *Aristarchus of Samos* (New York: Dover Books, 1981).

Thor Heyerdahl, *Early Man and the Ocean: A Search for the Beginning of Navigation and Seaborne Civilizations* (New York: Vintage Books, 1980).

John E. Hill, trans., *The Peoples of the West, from the Weilue,* by Yu Huan (2004), http://depts.washington.edu/silkroad/texts/weilue/weilue.html

Livy, *Rome and the Mediterranean,* trans. Henry Bettenson (Harmondsworth, England: Penguin Books, 1976).

J. W. McCrindle, trans., *Ancient India as Described by Megasthenes and Arrian* (London: Trubner & Co., 1877).

The New Oxford Annotated Bible with the Apocrypha, Revised Standard Version (Oxford: Oxford University Press, 1977).

Pausanias, *Description of Greece,* vol. 1, trans. W. H. S. Jones (New York: G. P. Putnam's Sons, 1918).

Pliny the Elder, *Natural History,* 2 vols., trans. H. Rackham (Cambridge, MA: Harvard University Press, 1938, 1947).

Pliny the Younger, *The Letters of the Younger Pliny,* trans. Betty Radice (Harmondsworth, England: Penguin Books, 1963).

Plutarch, "Alexander," in *The Lives of the Noble Grecians and Romans,* trans. John Dryden (New York: Random House, 1992).

Plutarch, *Life of Alexander,* trans. Bernadotte Perrin (Cambridge, MA: Harvard University Press, 1986).

Plutarch, *Moralia,* vol. 12, trans. Harold Cherniss and William Helmbold, Loeb Classical Library (Cambridge, MA: Harvard University Press, 1968).

Polybius, *The Histories,* 6 vols., trans. W. R. Paton (Cambridge, MA: Harvard University Press, 1922–1927).

F. E. Romer, *Pomponius Mela's Description of the World* (Ann Arbor: University of Michigan Press, 1998).

Leonhard Schmitz, *A Manual of Ancient Geography* (Philadelphia: Blanchard and Lea, 1859).

Strabo, *Geography,* trans. H. L. Jones (Cambridge, MA: Harvard University Press, 1917, 1923).

William H. Tillinghast, "The Geographical Knowledge of the Ancients Considered in Relation to the Discovery of America," in *Narrative and Critical History of America,* vol. 1, ed. Justin Winsor (Boston: Houghton, Mifflin, 1889).

E. H. Warmington, *Greek Geography* (London: J. M. Dent, 1934).

Henry Yule, trans., *Cathay and the Way Thither; Being a Collection of Medieval Notices of China,* vol. 1 (London: Hakluyt Society, 1866).

13

Phoenician and Carthaginian Science (1000–200 BCE)

The Phoenicians contributed much to ancient knowledge of geography and navigation. The Phoenicians were the great mariners of antiquity, sailing on trading expeditions out of their port cities of Tyre, Sidon, and Byblos, in the eastern Mediterranean. They opened the entire Mediterranean region to trade, sailing beyond the Strait of Gibraltar (the Pillars of Heracles) at the end of the second millennium BCE. They founded the port cities of Carthage and Utica in North Africa, Cadiz in Spain, and were thought by the Greeks to have founded the trading center of Tartessus, on the Atlantic coast of Spain. The Phoenicians of Carthage were responsible for creating a great trading empire that extended from North Africa to Sicily to Spain, and in the Atlantic from the British Isles to the western coast of Africa. Homer's somewhat erratic understanding of geography in the *Odyssey* owes much to Phoenician exploration. The Greeks, emerging from their Dark Ages in the eighth century, borrowed the Phoenician alphabet. Ironically, however, few records survive from Phoenician city-states, even Carthage. The sources for Phoenician exploration and science come from Greek and Roman writers.

Ancient Greek and Roman thinkers were impressed by Phoenician accomplishments. Pliny the Elder wrote, "The Phoenician people enjoy the glory of having been the inventors of letters, and the first discoverers of the sciences of astronomy, navigation, and the art of war" (trans. Bostock and Riley). Perhaps he meant by this general statement *celestial, marine* navigation and the art of *naval* war. The geographer Pomponius Mela, in *Description of the World*, wrote, "The Phoenicians are a clever branch of the human race and exceptional in regard to the obligations of war and peace, and they made Phoenicia famous. They devised the alphabet, literary pursuits, and other arts too; they figured out how to win access to the sea by ship, how to conduct battle with a navy, and how to rule over other peoples; and they developed the power of sovereignty and the art of battle" (trans. Romer). Pomponius's sweeping statement, like Pliny's, exaggerates their significance but also reveals the vague, legendary history of the Phoenician people.

The Late Roman Platonist Porphyry, among others, argued that the great mathematician and philosopher Pythagoras descended from the Phoenicians:

> Many think that Pythagoras was the son of Mnesarchus, but they differ as to the latter's race; some thinking him a Samian, while Neanthes, in the fifth book of his *Fables* states he was a Syrian, from the city of Tyre. As a famine had arisen in Samos, Mnesarchus went thither to trade, and was naturalized there. There also was born his son Pythagoras, who early manifested studiousness, but was later taken to Tyre, and there entrusted to the Chaldeans, whose doctrines he imbibed. Thence he returned to Ionia, where he first studied under the Syrian Pherecydes, then also under Hermodamas the Creophylian who at that time was an old man residing in Samos. (Trans. Guthrie)

The semimythical Phoenician Sanchoniatho, as interpreted and translated by Philo of Byblos, wrote an account of creation, as recorded by the Christian historian Eusebius of Caesarea:

> He supposes that the beginning of all things was a dark and condensed windy air, or a breeze of thick air and a Chaos turbid and black as Erebus: and that these were unbounded, and for a long series of ages destitute of form. But when this wind became enamoured of its own first principles (the Chaos), and an intimate union took place, that connexion was called Pothos: and it was the beginning of the creation of all things. And it (the Chaos) knew not its own production; but from its embrace with the wind was generated Môt; which some call Ilus (Mud), but others the putrefaction of a watery mixture. And from this sprung all the seed of the creation, and the generation of the universe. And there were certain animals without sensation, from which intelligent animals were produced, and these were called Zophasemin, that is, the overseers of the heavens; and they were formed in the shape of an egg: and from Môt shone forth the sun, and the moon, the less and the greater stars. And when the air began to send forth light, by its fiery influence on the sea and earth, winds were produced, and clouds, and very great defluxions and torrents of the heavenly waters. And when they were thus separated, and carried out of their proper places by the heat of the sun, and all met again in the air, and were dashed against each other, thunder and lightnings were the result: and at the sound of the thunder, the before-mentioned intelligent animals were aroused, and startled by the noise, and moved upon the earth and in the sea, male and female. (After this our author proceeds to say:) These things were found written in the Cosmogony of Taautus [another name for Thoth, an Egyptian god associated with magic, writing, wisdom, and sometimes the moon], and in his commentaries, and were drawn from his observations and the natural signs which by his penetration he perceived and discovered, and with which he has enlightened us. (Afterwards, declaring the names of the winds Notus, Boreas, and the rest, he makes this epilogue:)—But these first men consecrated the productions of the earth, and judged them gods, and worshipped those things, upon which they themselves lived, and all their posterity, and all before them; to these they made libations and sacrifices. (Then he proceeds:—Such were the devices of their worship in accordance with the imbecility and narrowness of their souls.) (Trans. Cory)

Phoenician cities by the end of the second millennium BCE had become powerful and wealthy because of their trading prowess. One famous king of Phoenicia, specifically of Tyre, recorded in the Old Testament, was Hirom (Hiram), who reigned around the same time as the Hebrew kings David and Solomon. According to the Old Testament, Hirom and Solomon were friends, and Hirom sent cedar wood as well as artisans to help build Solomon's temple. The Jewish historian Josephus, in *Against Apion*, provided information from lost sources on the Phoenicians. Josephus wrote,

> I will produce for a witness Dius, one that is believed to have written the Phoenician History after an accurate manner. This Dius, therefore, writes thus, in his *Histories of the Phoenicians*: "Upon the death of Abibalus, his son Hirom took the kingdom. This king raised banks at the eastern parts of the city, and enlarged it; he also joined the temple of Jupiter Olympius, which stood before in an island by itself, to the city, by raising a causeway between them, and adorned that temple with donations of gold. He moreover went up to [Mount] Libanus, and had timber cut down for the building of temples. They say further, that Solomon, when he was king of Jerusalem, sent problems to Hirom to be solved, and desired he would send others back for him to solve, and that he who could not solve the problems proposed to him should pay money to him that solved them. And when Hirom had agreed to the proposals, but was not able to solve the problems, he was obliged to pay a great deal of money, as a penalty for the same. As also they relate, that one OEabdemon, a man of Tyre, did solve the problems, and propose others which Solomon could not solve, upon which he was obliged to repay a great deal of money to Hirom." These things are attested to by Dius, and confirm what we have said upon the same subjects before. (Trans. Whiston)

This anecdote from Josephus and the anecdote about Pythagoras from Porphyry imply that the Phoenicians were adept at mathematics. According to the early twentieth-century historian of mathematics, W. W. Rouse Ball,

> So far as the acquirements of the Phoenicians on this subject are concerned it is impossible to speak with certainty. The magnitude of the commercial transactions of Tyre and Sidon necessitated a considerable development of arithmetic, to which it is probable the name of science might be properly applied. A Babylonian table of the numerical value of the squares of a series of consecutive integers has been found, and this would seem to indicate that properties of numbers were studied. According to Strabo the Tyrians paid particular attention to the sciences of numbers, navigation, and astronomy; they had, we know, considerable commerce with their neighbours and kinsmen the Chaldaeans; and [August] Bockh says that they regularly supplied the weights and measures used in Babylon. Now the Chaldaeans had certainly paid some attention to arithmetic and geometry, as is shown by their astronomical calculations; and, whatever was the extent of their attainments in arithmetic, it is almost certain that the Phoenicians were equally proficient, while it is likely that the knowledge of the latter, such as it was, was communicated to the Greeks. On the whole it seems probable that the early Greeks were largely indebted to the Phoenicians for their knowledge of practical arithmetic or the art of calculation, and perhaps also learnt from them a few properties of numbers. It may be worthy of note that Pythagoras was a Phoenician; and according to Herodotus, but this is more doubtful, Thales was also of that race. (Ball, pp. 2–3)

HERODOTUS ON THE PHOENICIANS

The Phoenicians, like the Greeks, used keel and ribs to design their ships, which relied heavily on two banks of oars: the bireme. Although the Phoenicians originally invented the bireme, they expanded their technology and the size of their ships (as did the Greeks and Romans), to produce the trireme, quadrireme, and quinquereme, which often required hundreds of rowers to staff the huge oars extended from different levels along the sides of the ship. Phoenician cedar was in demand for shipbuilding throughout the Mediterranean world. Herodotus, writing in the mid-fifth century (*The Histories*, book 7), described in detail the peoples of the ancient Near East, recording stories in particular that he had heard about the Phoenicians. He was told, for example, that when the Persian king Xerxes was marching against the Greeks in 480, he ordered the Phoenicians to construct a bridge across the Hellespont (Dardanelles), the strait that separated Asia from Europe. Phoenician bridge builders lashed boats together side by side across the strait, using thick ropes made of flax. They tied massive boards to the decks of the ships to allow the hundreds of thousands of Persian troops, camp followers, and animals to cross.

Herodotus also described one of the great exploring expeditions of all time. He learned from Egyptians, when he traveled to Egypt doing research for his book, that the pharaoh Necho II (r. 610–595) had, around 600 BCE, ordered Phoenician sailors to discover the extent and nature of Africa, called "Libya" by the Greeks. In book 4 of *The Histories*, Herodotus narrated the voyage of the Phoenicians. They set sail from the northern tip of the Red Sea, exited into the Indian Ocean, and proceeded along the eastern coast of Africa. Their small wooden ships were sufficiently seaworthy to navigate the shoal waters along the coast. Herodotus claimed that they patiently made port in autumn, sowed seed, waited, harvested the crop, and then pursued the journey rested and well supplied with food. It took them over two years to make the voyage, during which they observed and recorded their findings. Upon circumnavigating the continent, rounding the Cape of Good Hope sailing east to west, they entered Atlantic waters and sailed up the coast of Africa to the Gulf of Guinea. They then rounded the far western coast of Africa, battling contrary winds and currents, ultimately reaching the Pillars of Heracles and the Mediterranean. Once back to the Nile River, they reported to Necho a strange phenomenon. Sailing west from the Indian to Atlantic Oceans, rounding the Cape of Good Hope, they noted the sun on the port (right-hand) side of their ships. When Herodotus heard this, he was incredulous, knowing from his experience that ships sailing west in the Mediterranean always had the sun on their starboard (left-hand) side. He recorded the dubious story anyway, in so doing providing later observers with clear evidence that the Phoenicians had indeed crossed the Tropic of Capricorn into the Southern Hemisphere, where for travelers going east to west the rays of the sun are always to the north.

DIODORUS SICULUS'S ACCOUNT OF THE PHOENICIANS

The speculative Greeks, and later the Romans, wondered what lay to the west, across the Atlantic. Herodotus's report that the Phoenicians circumnavigated Africa circa 600 BCE, sailing from the Indian to the Atlantic Oceans, was believed by few, who were beholden to the myth of the ring of fire, the Antipodes, encircling Earth's equator. There circulated other vague reports that the Carthaginians explored the North Atlantic to the British Isles, the Azores, and the West African coastline almost to the equator. Diodorus Siculus, the Greek historian of the first century BCE and author of *Universal History*, reported that the Carthaginians discovered a vast, fruitful, and temperate island with large rivers in the Atlantic Ocean. He wrote:

> But now that we have discussed what relates to the islands which lie within the Pillars of Heracles, we shall give an account of those which are in the ocean. For there lies out in the deep off Libya an island of considerable size, and situated as it is in the ocean it is distant from Libya a voyage of a number of days to the west. Its land is fruitful, much of it being mountainous and not a little being a level plain of surpassing beauty. Through it flow navigable rivers which are used for irrigation, and the island contains many parks planted with trees of every variety and gardens in great multitudes which are traversed by streams of sweet water; on it also are private villas of costly construction, and throughout the gardens banqueting houses have been constructed in a setting of flowers, and in them the inhabitants pass their time during the summer season, since the land supplies in abundance everything which contributes to enjoyment and luxury. The mountainous part of the island is covered with dense thickets of great extent and with fruit-trees of every variety, and, inviting men to life among the mountains, it has cozy glens and springs in great number. In a word, this island is well supplied with springs of sweet water which not only makes the use of it enjoyable for those who pass their life there but also contribute to the health and vigour of their bodies. There is also excellent hunting of every manner of beast and wild animal, and the inhabitants, being well supplied with this game at their feasts, lack of nothing which pertains to luxury and extravagance; for in fact the sea which washes the shore of the island contains a multitude of fish, since the character of the ocean is such that it abounds throughout its extent with fish of every variety. And, speaking generally, the climate of the island is so altogether mild that it produces in abundance the fruits of the trees and the other seasonal fruits for the larger part of the year, so that it would appear that the island, because of its exceptional felicity, were a dwelling-place of a race of gods and not of men. In ancient times this island remained undiscovered because of its distance from the entire inhabited world, but it was discovered at a later period for the following reason. The Phoenicians, who from ancient times on made voyages continually for purposes of trade, planted many colonies throughout Libya and not a few as well in the western parts of Europe. And since their ventures turned out according to their expectations, they amassed great wealth and essayed to voyage beyond the Pillars of Heracles into the sea which men call the ocean. And, first of all, upon the Strait itself by the Pillars they founded a city on the shores of Europe, and since the land formed a peninsula they called the city Gadeira; in the city they built many works appropriate to the nature of the region, and among them a costly temple of Heracles, and they instituted magnificent sacrifices which were conducted after the manner of the Phoenicians. And it has come to pass that this shrine has been held in an honour beyond the ordinary, both at the time of its building and in comparatively recent days down even to our own lifetime. Also many Romans, distinguished men who have performed great deeds, have offered vows to this god, and these vows they have performed after the completion of their successes. The Phoenicians, then, while exploring the coast outside the Pillars for the reasons we have stated and while sailing along the shore of Libya, were driven by strong winds a great distance out into the ocean. And after being storm-tossed for many days they were carried ashore on the island we mentioned above, and when they had observed

> its felicity and nature they caused it to be known to all men. Consequently the Tyrrhenians [of Italy], at the time when they were masters of the sea, purposed to dispatch a colony to it; but the Carthaginians prevented their doing so, partly out of concern lest many inhabitants of Carthage should remove there because of the excellence of the island, and partly in order to have ready in it a place in which to seek refuge against an incalculable turn of fortune, in case some total disaster should overtake Carthage. For it was their thought that, since they were masters of the sea, they would thus be able to move, households and all, to an island which was unknown to their conquerors. (Trans. Oldfather)

Some imaginative thinkers have asserted, based on this account from Diodorus Siculus, that the Phoenicians/Carthaginians *discovered* America. Thor Heyerdahl, in *Early Man and the Ocean*, strongly suggests that the Phoenician trading outpost Lixus, in northwestern Africa, was a dynamic port city as early as 1200 BCE. The Phoenicians of Lixus worshipped the sun, built gigantic megaliths, and were excellent astronomers, navigators, and mariners. At the same time, west along the Canary current, where the prevailing winds and currents take mariners directly to Central America, the Olmec civilization appeared at La Venta. Were the Olmecs actually Phoenician colonists? Perhaps.

THE CARTHAGINIANS

The Phoenician colony of Carthage in North Africa became the dominant city of the western Mediterranean. Carthage controlled the trade of the region; dominated Africa, Sicily, Sardinia, southern France, and Spain; and explored the Atlantic coast of western Europe and western Africa. Because of the three Punic Wars between Rome and Carthage, in which Rome was victorious, and particularly the last war, in which Rome completely destroyed Carthage, few records of Carthaginian civilization survived. One of the few is the *Periplus* of Hanno.

The *Periplus* describes a Carthaginian expedition of the sixth century BCE led by Hanno, a navigator of Carthage. Hanno led a fleet west through the Strait of Gibraltar and then south down the coast of Africa past the Tropic of Cancer to the region of the Cape Verde Islands, modern Senegal. The Carthaginians explored, recorded their observations, and took samples. Elephants, crocodiles, and hippopotamuses fascinated them. The native inhabitants were fascinating, too, and a bit terrifying. The Carthaginians made contact with a people they called the Troglodytes, "men of strange appearance," and a savage people whose women had shaggy bodies. The Carthaginians pursued the men without luck, so they contented themselves with flaying the women captives, the skins of which they took back to Carthage as a specimen of the inhabitants and strange lands.

The *Periplus*, based on the Greek, is as follows:

> It was decreed by the Carthaginians that Hanno should sail beyond the Pillars of Hercules and found cities of the Liby-Phenicians. Accordingly he sailed with sixty ships of fifty oars each, and a multitude of men and women to the number of thirty thousand, and provisions and other equipment. When we had set sail and passed the Pillars, after two days' voyage, we founded the first city and named it Thymiaterium. Below this city lay a great plain. Sailing thence westward we came to Soloeis, a

promontory of Libya, thickly covered with trees. Here we built a temple to Poseidon; and proceeded thence half-a-day's journey eastward, till we reached a lake lying not far from the sea, and filled with abundance of great reeds. Here were feeding elephants and a great number of other wild animals. After we had gone a day's sail beyond the lakes we founded cities near to the sea, of which the names were the Fort of Caricon, Gytta, Acra, Melita, and Arambys. Sailing thence we came to Lixus, a great river which flows from Libya. On its banks the Lixitae, a wandering tribe, were feeding their flocks. With these we made friendship, and remained among them certain days. Beyond these dwell the Inhospitable Ethiopians, inhabiting a country that abounds in wild beasts and is divided by high mountains, from which mountains flows, it is said, the river Lixus. About these mountains dwell the Troglodyte, men of strange aspect. Of these the Lixitae said that they could run swifter than horses. Having procured interpreters from these same Lixitas, we coasted for two days along an uninhabited country, going southwards. Thence again we sailed a day's journey eastward. Here in the recess of a certain bay we found a small island, about five furlongs in circumference. In this we made a settlement, and called its name Cerne. We judged from our voyage that this place lay right opposite to Carthage, for the voyage from Carthage to the Pillars was equal to the voyage from the Pillars to Cerne. After this, sailing up a great river which is called Chretes, we came to a lake, in which are three islands greater than Cerne. Proceeding thence a day's sail, we came to the furthest shore of the lake. Here it is overhung by great mountains, in which dwell savage men clothed with the skins of beasts. These drove us away, pelting us with stones, so that we could not land. Sailing thence, we came to another river, great and broad, and full of crocodiles and river-horses [hippopotamuses]. Thence returning back we came again to Cerne; and from Cerne we sailed again towards the south for twelve days, coasting along the land. The whole of this land is inhabited by Ethiopians. These would not await our approach, but fled from us; and their tongue could not be understood even by the Lixitae that were with us. On the last day, we came near to certain large mountains covered with trees, and the wood of these trees was sweet-scented and of divers colours. Sailing by these mountains for the space of two days, we came to a great opening of the sea; and on either side of this sea was a great plain, from which at night we saw fire arising in all directions. Here we watered, and afterwards sailed for five days, until we came to a great bay, which the interpreters told us was called the Western Horn. In this bay was a large island, and in this island a lake of salt water, and again in this lake another island. Here we landed; and in the daytime we could find nothing, but saw wood ashes; but in the night we saw many fires burning, and heard the sound of flutes and cymbals and drums and the noise of confused shouts. Great fear then came upon us, and the prophet bade us leave this place. We sailed therefore quickly thence, being much terrified; and passing on for four days found at night a country full of fire. In the middle was a lofty fire, greater than all the rest, so that it seemed to touch the stars. When day came we found that this was a great mountain which they call the Chariot of the Gods. On the third day of our departure thence, having sailed by streams of fire, we came to a bay which is called the Southern Horn. At the end of this bay lay an island like to that which has been before described. This island had a lake, and in this lake another island, full of savage people, of whom the greater part were women. Their bodies were covered with hair, and our interpreters called them Gorillas. We pursued them, but the men we were not able to catch; for being able to climb the precipices and defending themselves with stones, these all escaped. But we caught three women. But when these, biting and tearing those that led them, would not follow us, we slew them, and flaying off their skins, carried these to Carthage. Further we did not sail, for our food failed us. (Trans. Church)

Hanno's voyage was apparently just one of many carried out by the Carthaginians. Roman and Greek writers periodically referred to Carthaginian exploits, such as the voyage of Himilco, a Carthaginian captain who possibly explored the North Atlantic to the British Isles and west, perhaps even the Sargasso Sea. The account of his voyage does not survive, though a fourth-century poet, Avienus, wrote an account based on the lost account. In Avienus's poem, we find an account of the Scilly Isles off the coast of British Cornwall, based on the eyewitness account of Himilco:

> Beneath this promontory spreads the vast Oestrymnian gulf, in which rise out of the sea the islands Oestrymnides [Scilly Isles], scattered with wide intervals, rich in metal of tin and lead. The people are proud, clever and active, and all engaged in incessant cares of commerce. They furrow the wide rough strait, and the ocean abounding in sea-monsters, with a new species of boat. For they know not how to frame keels with pine or maple, as others use, not to construct their curved barks with fir; but strange to tell, they always equip their vessels with skins joined together, and often travers the salt sea in a hide of leather. It is two days' sail from hence to the Sacred Island [Ireland], as the ancients called it, which spreads a wide space of turf in the midst of the waters, and is inhabited by the Hibernian people. Near to this again is the broad island of Albion [England]. (Trans. Kenrick)

In short, the Phoenicians and Carthaginians were an active, enterprising people, pragmatic explorers and observers of nature who were content less with grand theories than with finding out how the environment worked, in order to build powerful, long-lasting communities.

FURTHER READING

Apollonius of Rhodes, *The Voyage of Argo*, trans. E. V. Rieu (Harmondsworth, England: Penguin Books, 1971).

W. W. Rouse Ball, *A Short History of Mathematics*, 4th ed. (London: Macmillan and Co., 1908).

George F. Bass, "Sea and River Craft in the Ancient Near East," in *Civilizations of the Ancient Near East*, vol. 3, 4 vols., ed. Jack Sasson et al. (New York: Charles Scribner's Sons, 1995).

John Bostock and H. T. Riley, trans., *The Natural History of Pliny*, vol. 1 (London: Henry G. Bohn, 1855).

M. Cary and E. H. Warmington, *The Ancient Explorers* (Harmondsworth, England: Penguin Books, 1963).

Lionel Casson, *Ships and Seamanship in the Ancient World* (Princeton, NJ: Princeton University Press, 1971).

Alfred J. Church, *Carthage, Or the Empire of Africa* (London: T. Fisher Unwin, 1886).

I. P. Cory, trans., *Ancient Fragments* (1832), https://www.sacred-texts.com/cla/af/af01.htm.

Diodorus Siculus, *The Library of History of Diodorus Siculus*, trans. C. H. Oldfather, vol. 3 (London: Loeb Classical Library, 1939).

Herodotus, *The Histories*, trans. Aubrey de Selincourt (Harmondsworth, England: Penguin Books, 1972).

Thor Heyerdahl, *Early Man and the Ocean: A Search for the Beginning of Navigation and Seaborne Civilizations* (New York: Vintage Books, 1980).

Flavius Josephus, *Against Apion*, trans. William Whiston, http://www.gutenberg.org/files/2849/2849.txt.

John Kenrick, *Phoenicia* (London: B. Fellowes, 1855).

Edward Lipinski, "The Phoenicians," in *Civilizations of the Ancient Near East*, 4 vols. (New York: Charles Scribner's Sons, 1995).

Porphyry, *Life of Pythagoras*, trans. Kenneth S. Guthrie, 1920, http://www.tertullian.org/fathers/porphyry_life_of_pythagoras_02_text.htm.

F. E. Romer, trans., *Pomponius Mela's Description of the World* (Ann Arbor: University of Michigan Press, 1998).

14

Astronomy in the Ancient and Medieval Worlds (4000 BCE–1500 CE)

Astronomy developed during the fourth, third, and second millennia BCE in Europe, Africa, Asia, and America, with the erection of megaliths to track astronomical phenomena, the development of lunar and solar calendars and star catalogs and charts, the recording of astronomical observations, the creation of sundials, and the creation of the pseudoscience astrology.

Astronomy was a natural consequence of humans seeking to understand changes in the natural environment that they observed by day and night. The movements of the sun, moon, stars, and constellations fool the generally accurate human ability to observe natural phenomena. On a daily basis, the sun appears to move out from darkness into light and then proceeds to journey across the sky toward its setting and disappearance. Many nights, the moon rises above the eastern horizon in various shapes and then proceeds across the night sky, disappearing in the west near daybreak. The stars, too, and recognizable constellations move across the night sky from east to west. Clearly, it appears that the earth is still while these celestial phenomena are moving from the eastern to the western horizon. Such became the argument for the geocentric, or earth-centered, universe.

The Arab astronomer Al-Biruni (971–1048) provided a fairly simple description of what a human observes:

> Day and night . . . are one revolution of the sun in the rotation of the universe, starting from and returning to a circle, which has been assumed as the beginning of this same Nychthemeron [24 hours]. Whichsoever circle it may be, it being determined by general consent. The circle is a "great" circle; for each great circle is dynamically an horizon. By "dynamically," . . . I mean that it (this circle) may be the horizon of any place on the earth. By the "rotation of the universe," I mean the motion of the celestial sphere, and of all that is in it, which we observe going round on its two poles from east to west. . . . [Further, a] year means one revolution of the sun in the ecliptic . . . and returning to the same point which has been assumed as the starting-point of *his* [the sun's] motion, whichsoever point this may be. In this way the sun includes in his course the four seasons, spring, summer, autumn, winter, and their four different natures: and returns always to the point whence he commenced. (Trans. Sachau)

Al-Biruni described what humans observed over the course of 365¼ days, that the sun would apparently move across the horizon, but each day, its position on the horizon was slightly south or north. If one lived in the Northern Hemisphere, the sun's position would be moving daily south from the summer solstice to the winter solstice and north from the winter solstice to the summer solstice. The precisely opposite observation would take place for a person in the Southern Hemisphere. The "movement" of the sun over north and south over the course of the year is called the ecliptic, and it is caused by the earth's rotation on its axis as it orbits the sun. But to the ancient and medieval mind, the sun was what moved east to west and over the course of the ecliptic.

Ancient peoples, living close to nature and observing natural phenomena, tried to trace the astronomical phenomena that they witnessed. Stonehenge, for example, is the most famous of the hundreds of megalithic sites throughout the world. Megaliths are huge stone structures set up by Neolithic peoples for religious and scientific purposes. Stonehenge, a series of megalithic structures conceived and built over the space of 1,300 years (2800–1500 BCE), served as the means for these ancient peoples to identify and predict solar and lunar phenomena. The debate about exactly what the purpose of Stonehenge was and how precise and intentional its astronomical observations were continues to rage among scholars who call themselves archaeoastronomers. It appears that the ancient astronomers at Stonehenge aligned the massive stones with moonrise and sunrise at various times of the year. Possibly, these Neolithic astronomers also predicted or at least traced eclipses using the arrangement of the monoliths.

Ancient peoples assigned special qualities to the planets, stars, moon, and sun. The sun as the source of warmth and light gained more significance than the other bodies early on. Ancient humans were naturally sun worshippers, whether it was Ra and Amen of the Egyptians, Helios and Apollo of the Greeks, or the One of the Neoplatonists. The moon, imitating the female reproductive cycle of 29½ days, appeared to be involved in fertility and feminine matters. Likewise, the morning star was seen by many ancient cultures as representing a feminine, reproductive power—Ishtar, Aphrodite, and Venus personified the morning star. The other morning star, Mercury, was the herald of day: Hermes, the messenger. The red planet Mars heralded conflict and war. Planets represented metals too; Mars was iron; Mercury, tin; and Venus, copper. The sun, of course, was gold.

Ancient humans as agrarian peoples were weather watchers, concerned about climate and rainfall, when to plant, and when to harvest. From the beginning of writing in Mesopotamia, weather forecasting was part of the creation of the almanac. Myth and superstition, of course, formed the bases of early meteorology. Ancient peoples personalized meteorological phenomena by means of the gods of winds, rain, and thunderstorms. The anger of storm gods such as Enlil and Zeus seemed the only explanation for the violence of a sudden thunderstorm.

BABYLONIAN ASTRONOMY

Sumerian astronomy was vague and inconclusive except for the examination of stars and planets as representing deities. Babylonian astronomy was theoretical rather than empirical; many Babylonian astronomical observations were anecdotal and not based on actual events. Chaldean wisdom was proverbial but often not based on real observations. After the eighth century BCE, Babylonian astronomers recorded lunar and solar eclipses, but they were not able to predict them, and the story that Thales learned how to predict eclipses (such as the one of 585 BCE) from the Babylonians is probably a fable. Yet Diogenes Laertius, the Late Roman commentator, ascribed to Thales a sophisticated knowledge of astronomy: he "discovered the path of the sun from one end of the ecliptic to the other; and . . . , as one account tells us, defined the magnitude of the sun as being seven hundred and twenty times as great as that of the moon. He was also the first person who called the last day of the month the thirtieth" (trans. Yonge).

The Greek astronomer Ptolemy used Babylonian records of eclipses in his astronomy. The Babylonians discovered that solar eclipses occurred at the beginning of the lunar month, at the new moon, whereas lunar eclipses occurred during the middle of the lunar month at full moon. The universe comprised separate spheres for planets and stars. Babylonians also knew of the ecliptic, the path by which the sun makes its daily progress across the heavens. They kept ephemerides for planet positions. They also had a geocentric planetary theory that was remarkably like that of Ptolemy, who must have borrowed from it. The Babylonians conceived of the sun orbiting the earth and the planets orbiting the sun. Their observations were particularly directed toward trying to determine the position of planets relative to fixed stars and especially the rise of the star Sirius. A later contemporary of Ptolemy, Vettius Valens (c. 120–175) of Antioch, wrote the most comprehensive study of astrology, in part based on Babylonian records.

EGYPTIAN ASTRONOMY

Egyptian astronomy relied very little on sophisticated mathematics and was rather practical; it was used to develop agricultural solar calendars. Egyptian astronomers kept track of the heliacal rising of Sirius, the appearance of which indicated the rising of the Nile River. Hence Egyptian priests, the astronomers in ancient Egypt, could always accurately predict the rising of the Nile by means of astronomical observations. Egyptians used a solar calendar. Al-Biruni cited Claudius Ptolemy's *Canon* as the source for his comment that the Egyptians, like other ancient peoples,

> reckon their years as 365 days, and add the quarters of a day in every fourth years as one complete day, when it has summed up thereto. This year they call an intercalary year, because the quarters are intercalated therein. The ancient Egyptians followed the same practice, but with this difference, that they neglected the quarters of a day till they had summed up to the number of days of one complete year, which took place in 1,460 years; then they intercalated one year, and agreed with the people of Alexandria and Constantinople as to the beginning of the year. (Trans. Wright)

Diogenes Laertius reported that Egyptian astronomers kept track of lunar and solar eclipses, recording 373 solar and 832 lunar eclipses. They believed that the gods manifested themselves in the heavenly bodies. Hence the sun was, at various times in Egypt's long history, the gods Ra, Amun, Aten, and Osiris. The goddess Isis manifested herself as the moon. In addition, the Egyptians conceived of a spherical universe. The stars, flames in the heavens, determined human destiny. Herodotus, the Greek traveler, claimed that the Egyptians were the first astrologers. Herodotus also learned what Egyptians knew about the meteorology of the Nile valley. They believed that the strong African winds blew the sun off course along the upper Nile Valley, making it very dry and affecting the level of the Nile downstream in Egypt. According to Diogenes Laertius, Egyptian meteorologists understood the cause of rain to be an atmospheric change.

GREEK ASTRONOMY

The greatest astronomers of antiquity were the Greeks. Early Greek scientists developed the idea of the heavens as a great sphere arching over the earth. Astronomers observed the continually changing horizon and tracked the appearance of stars in the direction of the earth's rotation. Greek astronomers recognized the eastward rotation of the earth; the ecliptic, which is the path of the sun across the horizon over the course of a year; the zodiac, the path of the sun and planets on the ecliptic; and scores of constellations, which they named according to their own myths and legends.

Greek astronomy began during the eighth century BCE. The earliest surviving sundial of antiquity dates from that century. The Greek poet Hesiod, in *Works and Days*, wrote that the wise husbandman knows when to engage in agricultural tasks according to the appearance and movements of the Pleiades, Sirius, and Orion. Homer's poems identified similar celestial phenomena as well. Homer argued, in Plutarch's words, that the sun "has an orbicular energy sometimes appearing over the earth, sometimes going under it." Homer also argued that "the sun is not a fire, but some more potent being, as Aristotle conjectured" (revised, Goodwin). The Greeks, like the Babylonians and Egyptians, brought their mythology to the heavens, naming groups of stars according to great heroes, legendary figures, and animals. The fixed stars formed the background for the wandering heavenly bodies, the planets, which move on the same plane in the band of the zodiac.

The climate of the Aegean Sea influenced Greek views of the gods and nature. Boreas was the source of the prevailing north winds—the Etesian winds—that blew from May to October. The Zephyr, the west wind that blew in like a stampeding horse, was husband to the goddess Iris, the rainbow. The Greek western shores of the Aegean were drier than the Turkish eastern shores of the Aegean; the North Aegean was cooler than the South Aegean. The first Greeks to try to explain meteorological phenomena included the Ionians Anaximander, Anaximenes, and Anaxagoras. Anaximander believed that wind blowing against clouds caused lightning; the winds themselves derived from air. Rainbows, notwithstanding that they were caused by vapor in sunlight, were considered a clear sign of a coming storm. Anaximander's student Anaximenes had a more advanced view: He explained hail and snow as rainwater cooling and solidifying as it fell to earth and rainbows as caused by sunlight hitting water vapor. Anaxagoras said that Egyptians and Greeks planted according to the appearance of the Dog Star, Sirius, in the early spring.

Ionian Greeks such as Thales and Pythagoras formed the first sophisticated school of Greek astronomy. Pythagoras reputedly developed the idea of separate perfect spheres that defined the orbits of the planets, moon, and earth. Planets rubbing against each other were said to cause a sound: the harmony of the spheres. Pythagoras or his followers realized that the moon reflected the light of the sun and that the varying shapes of the moon proved its sphericity, which led to the deduction of the earth's similarity in this regard. Lunar eclipses revealed the curved earth surface as well. Philolaus (470–385 BCE), a follower of Pythagoras, hypothesized the moving earth. He conceived that the earth could not be the center of the universe but rather there must be a central fire around which all the celestial bodies orbited. The sun, moon, and seven planets made nine spheres orbiting around the central fire. Nine is not such a significant number, however, compared to 10 (the sum of 1, 2, 3, and 4). There must be a tenth sphere, which Philolaus imagined to be the antichthon, the "counter earth," which always stays hidden from humans—being precisely on the other side of the sun, it is never seen.

Eudoxus (408–352 BCE), Pythagorean and student of Plato, was an astronomer, mathematician, and physicist. He wrote *Phaenomena,* one of the seminal statements from the ancient world on astronomy. Although the book was filled with erroneous information, arguing for a geocentric universe composed of twenty-seven spheres. A mathematician of note, he anticipated some of Euclid's ideas in geometry, in particular the theory of proportion. He anticipated Archimedes in his development of the method of exhaustion.

According to Strabo, Eudoxus established an astronomical observatory at Heliopolis, Egypt; he also built one at his hometown of Cnidus. From these observatories he tracked stellar phenomena and was particularly intrigued by the star Canopus. His observations and theories were recorded in two books, *Mirror* and *Phaenomena*; the latter book was partially preserved in verse in Aratus's *Phaenomena.*

Eudoxus developed the theory of concentric spheres to attempt to fit together his assumptions of the spherical nature of the earth and the planets, sun and moon, and heavens, as well as his own observations on the movements of the night sky. The theory postulates a series of spheres of different sizes, all perfect circles, moving about the earth. The fixed stars have a sphere by which they move. The sun and moon each require three spheres; each planet has a sphere, but also a larger sphere centered at its poles. Eudoxus even required two more spheres for each planet in order to exactly replicate what he observed in the night sky. So it was that twenty-seven spheres of varying sizes and movements made up Eudoxus's complicated universe. He did not worry about the composition of the spheres nor even their reality. His was a theoretical construct only, intended to predict astronomical phenomena and nothing more.

Eudoxus's theories were carried forth by his students Menaechmus (380–320), Polemarchus (fourth century), and particularly Callippus of Cyzicus (c. 370–310).

Aristotle relied on the work of Eudoxus and Callippus in creating his own scheme of the heavenly spheres. Eventually Ptolemy would take up the cause of Eudoxus's concentric spheres and try to correct some of the theory's errors while still attempting to preserve the conceptual whole.

Aristotle (384–322) and his student Theophrastus (370–286) wrote seminal accounts on meteorology in which they ascribed to astronomical phenomena the causes of meteorological phenomena. They argued that the rising of the Pleiades, the phases of the moon, the appearance of the moon upon rising, and the appearance of the horizon upon sunrise and sunset all have an impact on, and give warning about, the weather.

Heraclides of Pontus (387–312) was a fourth-century BCE astronomer of note and possibly the first to speculate that at least some planets orbit the sun. He was a student of the Academy and was influenced by both Plato and Aristotle. A polymath, he wrote on a variety of subjects. Ancient commentators claimed that Heraclides wrote books on history, playwrights, ethics, the mind, music, poetry, grammar, and nature. He made his mark in astronomy with two startling theories. First, Heraclides argued that the earth spun on its own axis. Second, he argued, according to some commentators, that in analyzing the phenomena of Mercury and Venus never traversing the zodiac, as did the other planets, but rather remaining near the horizon as morning and evening planets, Mercury and Venus orbited the sun rather than the earth. He combined, therefore, the geocentric and heliocentric theories of the solar system, anticipating the theory of Tycho Brahe some 1,800 years later. (Some modern scholars, however, doubt that Heraclides made such a claim, which, they say, is based on a misinterpretation of ancient sources.) Heraclides also believed that the heavenly bodies were deities, that comets were on fire, and that each star was its own universe, with planets and moons.

The Pythagorean concept of the central fire was a heliocentric theory based upon philosophy rather than observations and mathematics. Aristotle, the leading scientist of his age, rejected such a scheme for the geocentric universe. Besides the obvious experience of the sun passing daily from east to west, which according to common sense reveals that it moves while the earth is still, Aristotle argued that for the earth to be orbiting the sun, the stars would pass, rise, and set just as the sun and moon did, which was not the case. Aristotle was too much the practical observer to realize that the distance of the stars from the earth prevented any significant movement in the night sky. In other words, there was very little parallax, stellar movement respective to the earth, a phenomenon that escaped Aristotle.

HELLENISTIC ASTRONOMY

An understanding of parallax did not, however, escape Aristarchus of Samos (310–230 BCE), a mathematician who was the first human to argue that the sun, not the earth, was the center of the universe. Aristarchus assumed that the stars were at such a distance from the earth that their movements respective to the earth were scarcely noticed. He used geometry to try to discover the relative distances and sizes of the sun and moon to each other and to the earth. His estimates were erroneous, yet he was still revolutionary in the attempt.

Aristarchus was a theorist more than an empiricist. His treatise, "On the Sizes and Distances of the Sun and Moon" provided the theoretical, mathematical basis for measurements of the heavenly bodies and calculations of their distances. Aristarchus was a Peripatetic philosopher, a student of Strato (335–269), who was himself the student of Theophrastus.

"On the Sizes and Distances of the Sun and Moon" suggests the respective size of the moon and sun based on measuring their radii and, using geometric and trigonometric techniques based on angles of observation, estimates of the relative distances of the sun and moon from the earth. Aristarchus's estimate that the sun is almost twenty times the distance from the earth as the moon is an underestimate of 50 percent. His estimate that the sun's diameter is seven times that of the earth is a dramatic underestimate (109 times). Aristarchus's hypothesis that the earth moves about the sun, which is the true center of the universe, is known from Archimedes (287–212) and Plutarch. Archimedes reported that Aristarchus conceived of the earth's orbit as circular and the distance of the fixed stars from the center of the universe (sun) to be much greater than hitherto thought. This awareness of the astonishing distances of the stars allowed him to realize that the real movements of the stars respective to the earth (parallax) cannot be observed (without precise scientific instruments). Following Heraclides of Pontus, Aristarchus also argued that the earth orbited on its own axis.

How Aristarchus conceived of this theory is subject to speculation. One possibility is that he assumed that the sun's greater size necessitated that the earth orbit it rather than vice versa. He could also have been influenced by Pythagorean philosophers such as Philolaus (470–385), who suggested that all the heavenly bodies orbited about a central fire.

Autolycus of Pitane (c. 360–290) wrote treatises on the heavenly spheres and the movement of the stars. His books, the earliest surviving Greek mathematical/astronomical texts, were *On Risings and Settings* and *On the Moving Sphere*. He relied heavily on Eudoxus's (408–355) theory of concentric spheres and was the teacher of Arcesilaus (315–240), who was the head of the Academy of Athens in the third century and one of the first Skeptics.

Hipparchus of Nicaea (190–120) was the most important astronomer of the Hellenistic Age after Aristarchus of Samos. Unlike Aristarchus, Hipparchus believed in the spherical, geocentric universe of Aristotle. A noted mathematician, Hipparchus relied on the work of his predecessor, Apollonius of Perga (262–190). Hipparchus, for example, gauged the movement of the sun and the moon using a deferent-epicycle system devised by Apollonius. In his geographical work, Hipparchus was influenced by Eratosthenes (276–195). Hipparchus worked at Alexandria and Rhodes; at the latter city, he created something like an astronomical research center in which he cataloged the stars according to brightness, which indicated to him relative distance from the earth and location. He ended up with a star chart of about 850 stars, tracking their movements and relative brightness. Hipparchus made more accurate measurements than Aristarchus did of the distance of the moon and sun from the earth. Using observations of a solar eclipse taken from Alexandria, where the eclipse was partial, and the northern Aegean, where it was total, Hipparchus worked out an estimate of the distance of the moon from the earth. He discovered that the earth's axis changed over time; he studied the precession of the equinoxes, the circuit that the earth makes over thousands of years as it rotates on its axis (a length of time known as the Great Year). And to try to account for the orbit of the sun around the earth he developed the idea of the eccentric, which Ptolemy of Alexandria would expand upon.

Geminus was an astronomer and mathematician of the first century BCE known for *Introduction to Astronomy* (*Isagoge*) and *Theory of Mathematics.* A Stoic influenced by Posidonius of Rhodes (135–50), Geminus possibly lived in Rhodes as well, an important center of science during the Hellenistic Age. Geminus focused on solar and lunar events and phases, the coordination of solar and lunar calendars, eclipses, the distances of the stars from the earth, constellations, and the zodiac. In *Introduction to Astronomy* as well as in *Theory of Mathematics*, Geminus provided an interesting historical perspective, telling his readers that mathematics had changed over time: Pythagoras's concern for theory and for mathematics as an abstract source of truth gave way to a more practical application of mathematics, represented by Geminus himself.

Geminus continues to be an important source for understanding the development of ancient calendars and dating systems. His *Introduction to Astronomy* tells us of the *octaëteris,* an eight-year cycle at the end of which three intercalary months were added in the ancient Greek lunar calendar to make months and dates conform to the phenomena of the seasons.

CLAUDIUS PTOLEMAEUS (PTOLEMY) (100–170 CE)

One of Alexandrian astronomer, geographer, and mathematician Claudius Ptolemy's accomplishments was a reasonable explanation for the problem of the wandering planets—retrograde motion—which occurs when a planet farther from the sun than earth is "passed" by earth in their respective orbits. The position of the planet in the night sky alters: it seemingly reverses and then resumes its previous course. Retrograde motion of outer planets was confusing to the ancients until Ptolemy came up with his explanation of epicycles, equants, eccentrics, and deferents. Ptolemy's scheme was, of course, based on an Earth-centered universe. He imagined that as planets orbit the earth, they also orbit about a point called the epicycle. The deferent is the orbital pattern around the earth. The eccentric is the center point of the orbital scheme. Earth is not directly in the center of the orbiting planet's deferent. The equant is the true point about which the planet orbits. If this sounds confusing, it was, and it did not reflect reality in the least. But Ptolemy was not concerned about reality; he was concerned about a mathematical scheme that could predict the motions of the planets. In this goal he succeeded brilliantly.

THE ROMANS

The Elder Pliny's (23–79 CE) *Natural History* provides a good summary of how Roman scientists absorbed Greek astronomy. Pliny assumed that the universe was godlike—possibly infinite yet unknowable. The spherical earth, revolving at a tremendous rate of speed, was at the center of the universe. For Pliny, the universe is harmonious and regular, the product of a divine mind utterly beyond human understanding. Perhaps the harmony of the planets produces a beautiful sound, although we earthlings cannot hear it. The planets are distant to the earth in the sense of their respective elevation approaching the starry vault. Each revolves around the earth in a set period; each has a certain character that gives a particular identity. Venus and Mercury, unlike Mars, Jupiter, and Saturn, rarely extend much above the horizon, in morning or in evening. Venus's light is third only to that of the sun and moon—and thus, like those two bodies, has a clear impact on human events. The moon is easily as large as the earth, though the sun dwarfs them both. Pliny discussed the phases of the moon, the solstices and equinoxes of the sun, and the varied forms of eclipse. He discussed comets and their influence on earth events. Indeed, Pliny discussed meteorology in light of astronomy, believing that storms, thunderbolts, climatic change, and the like were influenced by the sun, moon, planets, and stars. He was willing to believe in astronomical portents of changes in human affairs, yet such credulity was countered by some interesting discussions, such as his account of rainbows as being the contact of sunlight with clouds, which was a close approximation to the modern view that molecules of light were refracted by water vapor.

Plutarch (46–120), in his *Natural Questions,* had many common questions about meteorological events, such as what produces thunder and lightning, which he explained according to the combination of cold and heat. He pondered the causes of dew and the differences between seawater and fresh water and their impact on plants and animals, including humans and wellness or sickness. He posited that lightning evaporated sweet, clear water, producing salty scum where it struck the sea. Following Homer, Plutarch considered the cold west wind the swiftest of the winds. In fact, he believed that the first meteorologist of note among the Greeks was Homer, who (he wrote in *The Life and Poetry of Homer*)

> is not ignorant of the causes of disturbances to the elements as earthquakes and eclipses, since the whole earth shares in itself air, fire, and water, by which it is surrounded. Reasonably, in its depths are found vapors full of spirit, which they say being borne outward move the air; when they are restrained, they swell up and break violently forth. That the spirit is held within the earth they consider is caused by the sea, which sometimes obstructs the channels going outward, and sometimes by withdrawing, overturns parts of the earth. This Homer knew, laying the cause of earthquakes on Poseidon, calling him Earth Container and Earth Shaker. (Revised, Goodwin)

Homer "closely observed the nature of the winds, how they arise from the moist element. For the water transformed goes into air. The wind is air in motion." He realized "that the generation of rains comes from the evaporation of the humid" and that "as in the autumnal season when the earth with weight of rain is saturate,—for then the sun on account of the dryness of the ground draws out humors from below and brings from above terrestrial disturbances. The humid exhalations produce rains, the dry ones, winds. When the wind is in impact with a cloud and by its force rends the cloud, it generates thunder and lightning. If the lightning falls, it sends a thunderbolt" (revised, Goodwin).

ASTROLOGY

For most periods in the history of science, astrology and astronomy were scarcely distinct studies. In the ancient world, for example, astrology provided a big impetus to the development of astronomy, as astrologers wanted to know the positions of planets, the moon, and the sun at various times. Astrologers believed that the movement and position of the planets had a determining effect on the future. Mesopotamian, Egyptian, Greek, and Roman astronomers practiced astrology. Theophrastus, the Peripatetic scientist, and Diodorus Siculus (c. 90–30) thought the Chaldeans (Babylonians), and Herodotus thought the Egyptians, could predict the future through astrology. Diodorus, a first century BCE polymath who clearly was impressed by Chaldean astrology, wrote,

> Now, as the Chaldaeans say, the world is by its nature eternal, and neither had a first beginning nor will at a later time suffer destruction; furthermore, both the disposition and the orderly arrangement of the universe have come about by virtue of a divine providence, and to-day whatever takes place in the heavens is in every instance brought to pass, not at haphazard nor by virtue of any spontaneous action, but by some fixed and firmly determined divine decision. And since they have observed the stars over a long period of time and have noted both the movements and the influences of each of them with greater precision than any other men, they foretell to mankind many things that will take place in the future. But above all in importance, they say, is the study of the influence of the five stars known as planets, which they call "Interpreters" when speaking of them as a group, but if referring to them singly, the one named Cronus [Saturn] by the Greeks, which is the most conspicuous and presages more events and such as are of greater importance than the others, they call the star of Helius, whereas the other four they designate as the stars of Ares [Mars], Aphrodite [Venus], Hermes [Mercury], and Zeus [Jupiter], as do our astrologers. The reason why they call them "Interpreters" is that whereas all the other stars are fixed and follow a singular circuit in a regular course, these alone, by virtue of following each its own course, point out future events, thus interpreting to mankind the design of the gods. For sometimes by their risings, sometimes by their settings, and again by their colour, the Chaldaeans say, they give signs of coming events to such as are willing to observe them closely; for at one time they show forth mighty storms of winds, at another excessive rains or heat, at times the appearance of comets, also eclipses of both sun and moon, and earthquakes, and in a word all the conditions which owe their origin to the atmosphere and work both benefits and harm, not only to whole peoples or regions, but also to kings and to persons of private station. Under the course in which these planets move are situated, according to them, thirty stars, which they designate as "counselling gods"; of these one half oversee the regions above the earth and the other half those beneath the earth, having under their purview the affairs of mankind and likewise those of the heavens; and every ten days one of the stars above is sent as a messenger, so to speak, to the

> stars below, and again in like manner one of the stars below the earth to those above, and this movement of theirs is fixed and determined by means of an orbit which is unchanging forever. Twelve of these gods, they say, hold chief authority, and to each of these the Chaldaeans assign a month and one of the signs of the zodiac, as they are called. And through the midst of these signs, they say, both the sun and moon and the five planets make their course, the sun completing his cycle in a year and the moon traversing her circuit in a month. Each of the planets, according to them, has its own particular course, and its velocities and periods of time are subject to change and variation. These stars it is which exert the greatest influence for both good and evil upon the nativity of men; and it is chiefly from the nature of these planets and the study of them that they know what is in store for mankind. And they have made predictions, they say, not only to numerous other kings, but also to Alexander, who defeated Darius, and to Antigonus and Seleucus Nicator who afterwards became kings, and in all their prophecies they are thought to have hit the truth. . . . Moreover, they also foretell to men in private station what will befall them, and with such accuracy that those who have made trial of them marvel at the feat and believe that it transcends the power of man. Beyond the circle of the zodiac they designate twenty-four other stars, of which one half, they say, are situated in the northern parts and one half in the southern, and of these those which are visible they assign to the world of the living, allow those which are invisible they regard as being adjacent to the dead, and so they call them "Judges of the Universe." And under all the stars hitherto mentioned the moon, according to them, takes her way, being nearest the earth because of her weight and completing her course in a very brief period of time, not by reason of her great velocity, but because her orbit is so short. They also agree with the Greeks in saying that her light is reflected and that her eclipses are due to the shadow of the earth. Regarding the eclipse of the sun, however, they offer the weakest kind of explanation, and do not presume to predict it or to define the times of its occurrence with any precision. Again, in connection with the earth they make assertions entirely peculiar to themselves, saying that it is shaped like a boat and hollow, and they offer many plausible arguments about both the earth and all other bodies in the firmament. (Trans. Oldfather)

Under such influence, astrology became widespread in the ancient world and personalized during the Greek Classical and Hellenistic periods. Even so great a scientist as Claudius Ptolemy wrote a treatise on astrology, the *Tetrabiblos*. His book was theoretical, focusing on the physical causes of astrology. In theory, each individual has a particular destiny according to the position of the heavenly bodies at the individual's birth. A horoscope, determining the position of the planets and stars at one's birth, indicates the future. The astrologer determines the position of the planets on the day of an individual's birth to see what zodiacal sign the individual was born under. Each of the twelve signs of the zodiac determines a person's character and destiny. The zodiac is made up of twelve planetary regions set within the ecliptic, the thirty-degree path of the fixed stars. The earliest recorded horoscope is from 410 BCE.

A near contemporary of Ptolemy, Vettius Valens (c. 120–175), wrote one of the most extensive treatises of astrology, the *Anthologies*. He had a fervent, mystical belief in the accuracy of horoscopes, and the book provides details on individual horoscopes of specific people of the second century CE. He believed that the movement of the heavenly bodies—the sun and moon, planets and stars—were the bases for reality, for all human events, for truth itself. The astrologer discovers the truth, both of the now and the future. In the *Anthologies*, Valens provided a list of his sources, which represented a compilation of information on ancient astrologers and astronomers. These include Aristarchus; Asklation, otherwise unknown; Critodemus (first century CE), the author of a book titled *Vision*; Hipparchus, whom Valens used for information on the sun; Hypsicles (190–120) the geometrician; Kidenas (Kidinnu), a fourth-century Babylonian astronomer; Meton, of fifth-century Athens; and the shadowy Zoroaster.

A significant Greek astrologer was Aratus of Soli (315–240 BCE), who penned *Phaenomena,* a treatise in verse on constellations and stars. This was used as a school text in both Greek and Latin translations. The *Phaenomena* opens with a hymn to Zeus, who determines human events. It includes a star map of the constellations, though Aratus did not focus on the role of the stars but, rather, the role of Zeus in human affairs. The fixed stars are images in the sky. He included vague descriptions of the planets, a discussion of the nature of time, and a discussion of meteorological events. The *Phaenomena* influenced later astronomers and astrologers of antiquity and beyond.

Roman astrologers included Manilius (fl. 30 CE), an astrologer and geographer during the early Roman Principate. He wrote *Astronomica*, a poem detailing aspects of the earth and cosmos from an astrological point of view. Manilius relied heavily on Aratus's *Phaenomena. Astronomica*, dedicated to Hermes, opens with a discussion of the origins of the universe and provides a Stoic view of atoms as the basic particles of all matter. The poet discussed the basic elements of the universe as well, such as fire and water. He provided a map of the fixed stars, the constellations, and their role in human affairs. Manilius focused quite a bit on the constellation Orion. He discussed the planets, which occupied set paths, celestial spheres in an organized framework of the universe. He dwelt at length on the celestial spheres of the zodiac as seen from the earth and especially discussed the movement of the Milky Way. Little is known of Manilius, though his work heavily influenced later astronomers and astrologers.

MEDIEVAL ASTROLOGY

Astrology continued to be studied during the medieval period. The Persian Al-Biruni, in *The Chronology of Ancient Nations*, wrote extensively of non-Western astrological interests. Astrologers believed, he wrote, that

> Cancer is the horoscope of the world. For in the first cycle of Sindhind the sun stands in the beginning of Aries above the middle between the two ends of the inhabitable world. In that case, Cancer is the horoscope, which sign according to their tenets, as we have mentioned, signifies the commencement of rotation and growth. Others say, that Cancer was called the horoscope of the world, because of all the zodiacal signs, it stands nearest to the zenith of the inhabitable world, and because in the same sign is . . . Jupiter, which is a star of moderate nature; and as no growth is possible, except when moderate heat acts upon moist substances, it (i.e., Cancer) is fit to be the horoscope of the growth of the world. According to a third view, Cancer was called so, because by its creation the creation of the four elements became complete, and by their becoming complete all growth became complete. (Trans. Sachau)

He commented on the work of twelfth-century astrologer Abraham Ibn Ezra (1089–1167), who assigned length of life to the position of the planets at birth (ibid.).

Medieval astrologers like Al-Biruni were interested in forecasting meteorological events. The creation of almanacs for such purposes goes back to the ancient Mesopotamians and Egyptians. One such almanac writer and meteorological forecaster was the tenth-century Arab physician Sinan ibn Thabit. Al-Biruni relied heavily upon Thabit's work as well as the writings of a variety of especially Greek astronomers to compose an astrological almanac for the year from October to September. The forecast, drawn from a variety of ancient and medieval sources, listed parenthetically by Al-Biruni, for the days of October was as follows:

> 1. People expect rain (Euctemon and Philippus); turbid air (Egyptians and Callippus). 2. Turbid winterly air (Callippus, Egyptians, and Euctemon); rain, (Eudoxus and Metrodorus). 3. Nothing mentioned. 4. Wearing wind (Eudoxus); winterly air (Egyptians). 5. Winterly air (Democritus); beginning of the time of sowing. 6. North wind (Egyptians). 7. South wind (Hipparchus). 8. Nothing mentioned. Winterly air, according to Sinan. 9. . . . East wind (Hipparchus); west wind (Egyptians). 10. Nothing mentioned. 11. Episemasia (Eudoxus and Dositheus). 12. Rain (the Egyptians). 13. Unsteady wind, Episemasia, thunder, and rain (Callippus); north wind or south wind (Eudoxus and Dositheus). Sinan attests that this is frequently true. On this day the waves of the sea are sure to be in great commotion. 14. Episemasia and north wind (Eudoius). 15. Change of the winds (Eudoxus). 16. Nothing mentioned. 17. Bain and Episemasia (Dositheus); west wind or south wind (Egyptians). 18. Nothing mentioned. 19. Bain and Episemasia (Dositheus); west wind or south wind (Egyptians). 20. 21. Nothing mentioned. 22. Unsteady, changing winds (Egyptians). On this day the air begins to get cold. It is no longer time for drinking medicine and for phlebotomy except in case of need. For the Favourable Time for such things are always then, when you intend thereby to preserve the health of the body. For if you are compelled to use such means, you cannot wait for a night or day, for heat or cold, for a lucky or unlucky day. On the contrary, you use it as soon as possible, before the evil takes root, when it would be difficult to eradicate it. 23. Episemasia (Eudorus); north wind or south wind (Caesar). 24. Episemasia (Callippus and Egyptians). 25. Episemasia (Metrodorus); change in the air (Callippus and Euctemon). 26. Nothing mentioned. 27. Winterly air (Egyptians). 28. Nothing mentioned. It is a favourable day for taking a warm bath and for eating things that are of a sharp, biting taste, nothing that is salt or bitter. 29. Hail or frost (Democritus); continual south wind (Hipparchus); tempest and winterly air (Egyptians). 30. Heavy wind (Euctemon and Philippus). The kites, the white carrion-vultures [*vultur perenopterus*], and the swallows migrate to the lowlands, and the ants go into their nest. 31. Violent winds (Callippus and Euctemon); wind and winterly air (Metrodorus and Caesar); south wind (Egyptians). (Trans. Sachau)

On November 7, according to the same source, "astrologers take the horoscope of this time and derive there from an indication as to whether the year will have much rain or little. Herein they rely upon the condition of Venus at the times of her rising and setting." On November 13,

> ships that are at sea on this day put in to shore, and navigation to Persia and Alexandria is suspended. For the sea has certain days when it is in uproar, when the air is turbid, the waves roll, and thick darkness lies over it. Therefore navigation is impracticable. People say that at this time there arises the wind at the bottom of the

> sea that puts the sea in motion. This they conclude from the appearance of a certain sort of fishes which then swim in the upper regions of the sea and on its surface, showing thereby that this storm is blowing at the bottom. Frequently, people say, this submarine storm rises a day earlier. Every sailor recognizes this by certain marks in his special sea. For instance, in the Chinese sea this submarine storm is recognized by the fishing-nets rising of themselves from the bottom of the sea to its surface. On the contrary, they conclude that the sea bottom is quiet if a certain bird sits hatching her eggs—for they hatch in a bundle of chips and wood on the sea, if they do not go on land nor sit down there. They lay their eggs only at that time when the sea is quiet.

In December, at the time of the solstice, which "people call the 'Great Birth', . . . light leaves those limits within which it decreases, and enters those limits within which it increases, that human beings begin growing and increasing, whilst the demons begin withering and perishing." On December 29, it is advisable not to drink cold water upon awakening because "demons vomit into the water, and . . . therefore he who drinks of it is affected by stupidity and phlegm." February 21, "the heat from the interior of the earth [rises] to the surface," and warms the earth. "According to those who hold the opposite view, this change is brought about by the air's receiving heat instead of cold from the body of the sun, for the body of the sun and the near approach of a column of rays are the first cause of the heat." From the end of February to the beginning of March are the "*Days of the Old Woman*, i.e., seven consecutive days beginning with the 26th" during which the days are cold and windy. Syrians, on April 23rd, hold a fair because of the reappearance in the night sky of the Pleiades after forty days. On the 24th, in Mesopotamia, "the Euphrates begins to rise." On May 6th, he wrote, "is the time when the sun passes the (first) 20 degrees of Leo. In this respect the matter stands as we have explained it at the beginning of the rainy season, when the sun moves in Cancer." At the summer solstice, "light subdues darkness." July 18th, the northern, Etesian, winds begin to blow and the seven "dog-days" of summer are experienced. These seven days are critical for acute diseases, the ancients thought. At this time Sirius, the Dog Star, rises. By August 11th, the cool air of autumn is felt. September 3rd, the Etesian winds cease. September 5th, "midsummer ends, and a time comes which is good for bleeding." By the time of the equinox, "the winds, now, blowing . . . are said to be of a psychical nature. To look towards the clouds that rise on this day emaciates the body and affects the soul with disease. I think the reason of this is that people conceive fear on account of the cold and the disappearance of the agreeable time of the year." Al-Biruni noted by way of conclusion, "Astrologers are obliged to know dates . . . , and must date from such an epoch by means of their knowledge of the Permutations, Terminations, Cycles, and Directions, until they find the horoscopes of those people who were born at those times" (trans. Sachau).

Al-Biruni is noteworthy as the first astrologer to identify the lunar mansions, or stations, which are still in use by astrologers. The mansions are based on the lunar zodiac, that is, the path of the moon every night through the constellations of the night sky. The 27 or 28 lunar mansions identified by ancient and medieval cultures allowed astrologers to trace the path of the moon over the course of one lunar month. Astrologers believed that where the moon was in relation to the zodiac on a particular date, and a particular time, could provide information on meteorological and other natural events, including events affecting humans in general and in particular. Al-Biruni discussed the lunar mansions in both *The Book of Instruction in the Elements of the Art of Astrology* and *The Chronology of Ancient Nations*.

Europeans were as equally credulous as the Arabs, Persians, and Chinese in the belief in the wonders of astrology. So much were astrology and alchemy fixed in the medieval mind that we can find references to the two in the great books of the era. Take Geoffrey Chaucer's *Canterbury Tales*, a whimsical account of stories told by various characters on the road to Canterbury, in England. Chaucer (1342–1400) described a physician in the *Prologue* thus: "A Doctor of Physic was with us; in all this world there was none like him for surgery and physic, for he was well grounded in astrology. He watched well times and season for his patient by his natural magic; well could he choose a fortunate ascendant [astrological sign] for his images. He knew the cause of every ailment, were it of hot humour or cold, most of dry, and where it was engendered, and of what humour." In the *Wife of Bath's Prologue*, as she describes her various amours, she claimed, "I had the print of Saint Venus' seal. So God help me, I was fair and rich, a lusty one, young and joyous. For certes in feeling I am all Venerian, and mine heart is Marian. Venus gave me my jollity and my wantonness, and Mars my sturdy hardihood. Mine ascendant was Taurus, and Mars in it. Alas! Alas! That ever love was sin! I followed ay mine inclination by virtue of my stars; this caused that the Venus in me could never resist a good fellow. Yet I have Mars' mark upon my face, for, so God save me! I never loved by discretion, but ever followed by desire." Chaucer also described the alliance of science, alchemy, and astrology in the *Canon's Yeoman's Tale*: "That slippery science," alchemy,

> hath made me so bare that I have naught left wheresoever I go. And thereby am I so deep in debt for gold that I have borrowed. . . . When we be where we shall practice our elvish craft, we seem wondrous wise, our terms be so clerkly and strange. I blow the fire till mine heart fainteth. . . . For all our sleights, we cannot attain unto our end. Our orpiment and our sublimed mercury, our litharge eke ground on a porphyry slab—to use of these a certain number of ounces of each helpeth us naught; our labour is in vain. . . . I will tell you, as was also taught me, the four spirit and the seven bodies in their order, as oft I have heard my lord name them. The first spirit is called quicksilver, the second orpiment, the third sal ammoniac and the fourth brimstone. The seven bodies, lo! Here are they eke: Sol [sun] is gold, Luna [moon] we call silver, Mars iron, Mercury we name quicksilver, Saturn lead, Jupiter tin and Venus copper, by the souls of my forefathers! . . . Ah, nay! Let be! We seek eagerly each and all for the philosopher's stone, called Elixir; for if we had him then were we secure enough; but I make mine avow unto God in heaven, for all our craft and sleight, when we have done our all, he will not come to us. (Trans. MacKaye)

Astrology without the nonsense is based on astronomical principles and examines the impact of heavenly bodies on each other and the earth, which is exactly what modern science does. The difference, of course, is that ancient astrologers gave the heavenly bodies a significance that modern astronomers do not accept. The ancient universe was one of mysterious divine forces; the planets were conceived of as being perfect spherical bodies associated with the divine. Modern astronomers have determined that the planets are made of particles that have no spiritual basis or connection. Dead planets of gas and rock can hardly have an impact on human destiny.

CALENDARS AND DATING SYSTEMS

The daily movement of the sun and nightly phases of the moon gave early humans a sense of passing, of human affairs being somehow linked to celestial phenomena. The female reproductive cycle was remarkably similar to the period from new moon to new moon, and thus ancient societies believed that the moon must have feminine characteristics. Religious festivals in honor of the moon goddess came to be as important as festivals tracing the solar year and the passing seasons. As agriculture developed, calendars became extremely important for tracking the time for planting and harvesting. Festivals built around the equinoxes and solstices were a natural consequence of coordinating humans affairs with solar and lunar movements. The Mesopotamians and Egyptians, the first two civilizations to develop agriculture and a sophisticated society based on the repetitive phenomena of the seasons, created the first calendars. The Babylonians had lunar calendars in which the month began on the evening of the appearance of the crescent moon. The Babylonian calendar required thirteen months in a year. Egyptians had lunar calendars only for festivals but otherwise relied on solar calendars. Egyptian solar calendars were based on 360 days, with five days added on at year's end. The Egyptian calendar coordinated the three farm seasons in four-month segments. The Egyptian new year began in July with the rising of the Nile River and the appearance of Sirius, the Dog Star. Chinese astronomers developed a solar calendar also, perhaps as early as 1300 BCE.

The Greeks made the greatest advances in measuring the lunar and solar years. Plutarch wrote that Solon (640–560) introduced to the Athenians a lunar calendar, although it appears that the Greeks recognized the lunar months—and solstices—as early as the eighth century. According to Plutarch's account, the one day per month when the old moon is replaced by the new moon was called "the old and the new." Solon, following Greek habit, began on this day counting days to twenty. The last ten days were counted in descending order, conforming to the waning moon. That the lunar calendar was short of the solar calendar by 11½ days over the course of the year resulted in much confusion for the religious and thoughtful Greeks, which led mathematicians and astronomers to develop a variety of unique calculating techniques (trans. Dryden).

THE *OCTAËTERIS* AND THE METONIC CYCLE

The remarkable lack of coordination of the solar and lunar years occupied the minds of ancient thinkers. What could be done to devise a calendar that would at the same time keep accurate track of the moon's phases and the sun's daily path? How could religious festivals based on the moon and those based on the sun not become completely confusing? The ancients developed the technique of inserting intercalary days to periodically keep the solar calendar reflecting the observed phenomena. The adding of a day every four years, the leap year, is an ancient as well as modern intercalary method. Greek astronomers and mathematicians were very creative in their attempts to use intercalary years with the least inconvenience and most efficiency. Al-Biruni, the Persian astronomer, provided this explanation for the use of February as the month for adding the extra day on leap year: February

> is the leap-month. It appears to me that the following is the reason—but God knows best!—why people have shortened this month in particular so that it has only 28 days, and why it has not had assigned to it 29 or 30 or 31 days: If it were assigned 29 days and were then to be increased by the leap-day, it would have 30 days and would no longer be distinguishable from the other months in a leap-year. The same would be the case if it had 30 days, whether the year be a leap-year or not. Likewise if it had 31 days, the same similarity with the other months in all sorts of years would exist. For this reason the leap-month has been assigned 28 days, that it might be distinguished from the other months both in leap and common years. (Trans. Sachau)

Geminus, the Roman Stoic and astronomer of the first century BCE, wrote in his *Introduction to Astronomy* that the Greeks of the Archaic Age developed the *octaëteris*: an eight-year cycle in which intercalary months could be added in the least confusing way. Since the lunar year lags behind the solar year by 11½ days, the Greeks discovered that the first point at which to find a reasonable method to calculate intercalary months is at eight years (11½ times 8 equals roughly 90 days), when three months can be added to the lunar calendar to make it conform to the solar calendar. Geminus said that the Greeks decided to break this up in the *octaëteris* by inserting one month at intervals during the eight-year period. The *octaëteris* was still off by 1½ days, which led to the calculation of 16- and 160-year cycles, but most significantly, to the development of the Metonic cycle of 19 years.

The Metonic cycle was the creation of Meton of Athens, who lived in the fifth century BCE. He devised a calendar for use at Athens that showed the months, years, festivals, and ephemerides. Meton realized that 19 solar years equaled exactly 235 lunar months. Greeks knew, therefore, that every 19 years the cycle would begin again of the moon repeating the same phases on the same days of the solar calendar. With Meton's scheme, which still relied on intercalary months, mathematicians could coordinate lunar and solar calendar systems. (The Metonic cycle would in time be used to calculate the annual date of Easter. This was advocated by Anatolius, a Christian Aristotelian philosopher, in the third century CE.)

A contemporary of Meton, Oenopides of Chios, discovered the Great Year recurring every 59 years, which is the same as 730 complete lunar months. Oenopides calculated the year to a very accurate 365 days and 9 hours. According to Diodorus Siculus, "Oenopides . . . passed some time with the priests and astrologers [of Egypt] and learned among other things about the orbit of the sun, that it has an oblique course and moves in a direction opposite to that of the other stars" (trans. Oldfather). Al-Biruni explained in *The Chronology of Ancient Nations* that as ancient astronomers tried to date events in the past, they relied on the *octaëteris* and the Metonic cycle. "The first cycle," Al-Biruni wrote, "employed by those who compute the months by the revolution of the moon and the years by the revolution of the sun, was the cycle of eight years, and the second that of nineteen years. Callippus," who was a fourth-century astronomer from Cyzicus, "was of the number of the mathematicians, and one who himself—or whose people—considered the use of this latter cycle as part of their laws. Thereupon, he computed this cycle (of seventy-six years), uniting for that purpose four cycles of nineteen years" (trans. Sachau).

Callippus, a student of Eudoxus and Aristotle, studied the theory of the concentric spheres of the heavens, theorizing one for each planet. He charted and wrote on the rising and the setting of stars. Callippus is also remembered for his *On the System of the Planets.* The astronomer Hipparchus used an extended cycle of 304 years (four of Callippus's cycles) and developed a system that was highly accurate with respect to solar and lunar years, the latter being off only by one second per month. Hipparchus's estimate for the year was 365 days, five hours, fifty-five minutes, and fifteen seconds. Hence, no matter what the Greeks tried to do, the cycles of solar and lunar years would always be slightly off.

The calendar of early Rome, according to tradition and Plutarch, contained ten months, December being the tenth. The months were of varying numbers of days but totaled 360 days. The Romans had adopted this calendar from another source, perhaps the Etruscans, but did not understand its use, at least according to Plutarch. Then Numa Pompilius became king, succeeding Romulus. Plutarch, who had a balanced approach to history and legend, reported that Numa was a Pythagorean and scientist who reformed the calendar, adding two months, January and February, and making the Roman year begin not in March but in January. He was the first to show the Romans that the cycles of the moon and sun are different by eleven days per year, and that an intercalary month, called Mercedinus, must be occasionally added. This same calendar continued during the republic until the time of Julius Caesar. Upon assuming dictatorial powers, Caesar embarked upon a series of reforms, one of which was the calendar. The Alexandrian astronomer Sosigenes influenced Caesar's calendar reform, which introduced the solar year of 365 days rather than the old lunar year of 355 days. Caesar introduced the 365-day year and a leap day every fourth year. He added an initial two months to bring the seasons and calendar together.

TIMEKEEPING

The Babylonians developed the sexagesimal (hexadecimal) system of base 60, so that hours, minutes, and seconds could be determined as a fraction of 60 or its multiples. The Greeks used the Babylonian system combined with the Egyptian twenty-four-hour day. Time was kept in the Greco-Roman world by means of the sundial, or gnomon, and the water clock. The Greeks borrowed the idea of the gnomon from Egyptian and Babylonian scientists. Meton of Athens, the developer of the Metonic cycle, erected a solar clock in fifth-century Athens. Aristarchus of Samos created a sundial that had "a concave hemispherical surface, with a pointer erected vertically in the middle throwing shadows and so enabling the direction of the height of the sun to be read off by means of lines marked on the surface of the hemisphere" (trans. Heath). The Hellenistic engineer Andronicus of Cyrrhus, around 100 BCE, created a marble sundial on the island of Tinos and a water clock in Athens in a large octagonal structure with a bronze wind gauge at the top. Both Pliny the Younger and Julius Caesar described water clocks that marked the passing of time. Caesar remarked in the *Gallic War* that he determined the length of the British day using a water clock.

CHRONOLOGISTS

An essential part of the art of historical investigation is the work of establishing sequence of events over time, that is, the chronology of rulers, laws, governments, wars, events, human and natural phenomena, types of human expression, institutions, and so on. Mediterranean cultures believed that the earliest civilization was the Egyptian. Many peoples of Egypt, Greece, Rome, and the Christian era dated their own times with reference to Egyptian chronology. Other early chronologists used the reigns of kings or similar government officials or significant events for dating.

The Greek geographer Eratosthenes and historian Apollodorus of Athens (180–120) dated events according to Spartan kings, the ephors; Thucydides (460–400) dated events in his *Peloponnesian War* according to Spartan ephors but also Athenian archons. The historian Polybius (208–126) used the Olympic games, occurring every four years beginning in 776 BCE, as his guide to constructing a regular chronology of historical events. Polybius relied on the work of Eratosthenes, who created a systematic rendering of the Olympiads, which was the basis for his *Chronography.* This system used the Olympiads as the means to measure time in the distant past, before 776, as well as for more contemporary events. According to historian Donald Wilcox, Eratosthenes conceived of chronology in abstract terms, measuring events that were no longer felt or seen, events that had no existence except in the minds of those who recalled them and the scratchings on parchment and stone that purported to record them. Plutarch, for one, thought that dating according to Olympiads was uncertain. Perhaps this explains why historians such as Polybius used other dating systems as well, such as the annual term of office of the Roman consul. Nevertheless, Polybius showed the benefits of using one standard dating system to try to order the occurrence of all human events over time. Roman writers, such as the historians Livy (59 BCE–17 CE) and Tacitus (56–227), relied on the tried-and-true Olympiads and consulships but also dated events according to the hypothetical date of the founding of Rome (753 BCE). Hence Julius Caesar's assassination occurred in 709 (meaning 709 years since the founding of the city).

CHRISTIAN CHRONOLOGY

Christian scholars had to rely on the chronologies produced by their forebears. One of the most cited chronologists was the third century BCE Egyptian chronologist

Manetho. His work was used by pagan as well as Christian chronologists to reach back in time to the earliest civilizations. According to the ninth-century Byzantine chronologist George Syncellus,

> Manetho of Sebennytus, chief priest of the accursed temples of Egypt, who lived later than Bêrôssos in the time of Ptolemy Philadelphus, writes to this Ptolemy, with the same utterance of lies as Bêrôssos, concerning six dynasties or six gods who never existed: these, he says, reigned for 11,985 years. The first of them, the god Hêphaestus, was king for 9000 years. Now some of our historians, reckoning these 9000 years as so many lunar months, and dividing the number of days in these 9000 lunar months by the 365 days in a year, find a total of 727¾ years. They imagine that they have attained a striking result, but one must rather say that it is a ludicrous falsehood which they have tried to pit against Truth. (Trans. Waddell)

Syncellus doubted the pagan Manetho yet used him, as did so many other chronologists, such as Julius Africanus (c. 160–240) and Eusebius of Caesarea, to establish the supposed dynasties of Egyptian kings.

Eusebius of Caesarea (260–339 CE), in his chronological pursuits, used Manetho to help understand Hebrew chronology, as presented in the *Book of Genesis*:

> Now, if you care to compare these figures with Hebrew chronology, you will find that they are in perfect harmony. Egypt is called Mestraïm by the Hebrews; and Mestraïm lived [not] long after the Flood. For after the Flood, Cham (or Ham), son of Noah, begat Aegyptus or Mestraïm, who was the first to set out to establish himself in Egypt, at the time when the tribes began to disperse this way and that. Now the whole time from Adam to the Flood was, according to the Hebrews, 2242 years. But, since the Egyptians claim by a sort of prerogative of antiquity that they have, before the Flood, a line of Gods, Demigods, and Spirits of the Dead, who reigned for more than 20,000 years, it clearly follows that these years should be reckoned as the same number of months as the years recorded by the Hebrews: that is, that all the months contained in the Hebrew record of years, should be reckoned as so many lunar years of the Egyptian calculation, in accordance with the total length of time reckoned from the creation of man in the beginning down to Mestraïm. Mestraïm was indeed the founder of the Egyptian race; and from him the first Egyptian dynasty must be held to spring. But if the number of years is still in excess, it must be supposed that perhaps several Egyptian kings ruled at one and the same time; for they say that the rulers were kings of Thebes, of Memphis, of Saïs, of Ethiopia, and of other places at the same time. It seems, moreover, that different kings held sway in different regions, and that each dynasty was confined to its own nome: thus it was not a succession of kings occupying the throne one after the other, but several kings reigning at the same time in different regions. Hence arose the great total number of years. But let us leave this question and take up in detail the chronology of Egyptian history. (Trans. Waddell)

Eusebius's *Ecclesiastical History* dated the birth of Christ according to the year of Augustus's rule, the years since the Battle of Actium, and the reign of the governor of Syria, Quirinius. Eusebius published the *Chronological Canons,* in which he dated Christian, especially Apostolic, events of the first four centuries CE. Eusebius provided chronological tables of pagan and sacred events to provide comparisons with the past. He relied on some of the chronological estimates of the Hebrew historian Josephus (37–100). Eusebius dated Hebrew events from the birth of Abraham. Some Christian writers, such as Gregory of Tours (538–594) and the fifth-century historian Paulus Orosius, used the creation of Adam as the logical starting point for chronicling human affairs. Even so, Orosius found it more convenient to rely for practical purposes on the years since the founding of Rome. St. Augustine (354–430), the greatest classical thinker on the subject of time, dated the passing of events in personal terms, according to the significance of one's life in relation to the coming of Christ, the Incarnation. The fifth-century chronologist Victorius of Aquitaine, meanwhile, devised a table with which to calculate the proper date of Easter, the Christian celebration of Christ's Resurrection. Victorius decided to date his Easter table beginning with the first Easter. Such a chronology, however, was not useful in the Roman world, where the year began not in the spring but after the winter solstice.

EASTER AND DIONYSIUS EXIGUUS (470–540)

Easter, the day of the celebration of Christ's resurrection, occurs every year on the first Sunday after a full moon from March 21 to April 21. The holiday is a holdover from pagan festivals, is named for the ancient Mesopotamian goddess Astarte, and is based on the lunar as well as the solar calendar. The debate over the yearly date of Easter has been long-standing and frequently acrimonious. The Council of Nicaea, a meeting of Christian bishops during the reign of the emperor Constantine, agreed in 325 CE on the formula for determining the date of Easter. It was decided that Easter would always fall on a Sunday in the spring after the vernal (spring) equinox (March 21). It was left to church mathematicians and astronomers to figure the dates of Easter for coming centuries according to the cycles of phases of the moon. One of these scientists was Dionysius Exiguus, who during the sixth century CE, while working out his table, devised a dating system based on the Incarnation, the supposed date of Christ's birth. Although Dionysius made an errant choice for the year, he established the dating system of years proceeding from Christ's birth. This is the modern chronological system of BC, before Christ (which has given way to BCE, before the Common Era), and AD, *Anno Domini*, in the year of our Lord (which has given way to CE, Common Era). The solar Julian calendar was slightly irregular and, over the centuries, altered the calendar date for the vernal equinox; the Gregorian calendar, adopted in Europe in the sixteenth century and in America two hundred years later, corrected the irregularity, making Easter once again in the month after the vernal equinox.

CONSTELLATIONS

The distant, ancient past is seen on every clear night by the person who peers into the starlit sky. The light of stars hundreds and thousands of light-years away has arrived at the very moment one sees it. The position of the stars has hardly changed in the last several millennia; hence the patterns we see today are very much like those that Homer or Archimedes or Ptolemy saw. The wandering planets in the ancient night sky twinkled to the backdrop of the starry vault, the realm of fixed stars that changed slightly month by month but not with the dramatic movement of the sun, moon, and planets. A few astronomers, such as Hipparchus, understood the tremendous distance of the stars from our solar system; and Hipparchus seems also to have realized that, due to parallax, this distance meant that any movement of the stars would go generally undetected by humans on earth.

The seemingly unchanging, *fixed* stars provided a stable, familiar phenomenon that a person could count on, like sunrise and the full moon, during the course of a long life. So familiar were the stars that early civilizations sought to identify them according to some apparent pattern or in honor of a particular deity or hero. Hence the ancients developed constellations, names for star patterns that were recognizable to young and old alike. Our constellations today are the same as they were three thousand years ago. Then as now, constellations provided bearings for sailors at sea, helped astronomers track the movement of planets and peculiar events such as comets and meteors, and helped storytellers along as they spun yarns for listeners over the centuries. Some constellations could be seen night after night throughout the year by ancient Mediterranean observers. Ursa Major and Ursa Minor, the tail of which forms the North Star, Polaris, were seen year round in the northern horizon. Others came and went with the seasons. Orion the Hunter, for example, appeared in late autumn and departed in the spring, spending his entire stay in the southern horizon (from the perspective of an observer in the Northern Hemisphere). Following the same pattern were the twins Castor and Pollux, Gemini. The signs of the zodiac are, of course, twelve famous constellations that stay within the narrow path of the sun, the ecliptic. Other star formations or single stars were well known among the ancients. Sirius, the Dog Star, arrived after the winter solstice and departed after the spring equinox. The Pleiades, a cluster of stars almost directly above the Mediterranean observer, appeared after the autumn equinox and traveled slowly across the sky for the next six months, departing in the spring. The movement of the constellations, like that of the sun, moon, and planets, was east to west, matching the earth's eastward rotation.

Hypsicles, a second-century BCE Alexandrian mathematician and astronomer, was the first to divide the zodiac by 360 degrees in his book *On the Ascension of Stars*. He wrote, "The circle of the zodiac having been divided into 360 equal arcs, let each of the arcs be called a spatial degree, and likewise, if the time taken by the zodiac circle to return from a point to the same point is divided into 360 equal times, let each of the times be called a temporal degree" (quoted in O'Connor and Robertson, "Hypsicles").

The zodiac was held in such esteem by the ancient Romans, as observed by the seventeenth-century translator of Manilius's *Astronomica*, Edward Sherburne, that the Circus Maximus in the city of Rome was rendered according to Roman "astronomical learning":

> The Order and Disposition represented that of the Heavens. The Circus being an Elliptical or Oval Figure; having twelve Gates or Entries resembling the twelve Signs of the Zodiack. In the Midst an Obelisque, as the Sun: On each side thereof three Metae, denoting the other Six Planets, which in their Respective Courses mark out the several Intervals or Spaces, into which the Mundane System is divided. So that the Circensian Games seem not to have been so much, an Exercise of Charioting and

Racing, as an Astronomical Cursus; wherein the People were not only delighted by the Exhibition of corporal Games, but had their Minds also instructed to apprehend the Course and Order of the Celestial Bodies, which in the Great Circus of the World are continually moving.

NON-WESTERN ASTRONOMERS, CALENDARS, DATING SYSTEMS

Chinese astronomers observed comets, sunspots, novas, and meteors, although accurate explanations eluded them. They developed solar and lunar calendars. Babylonian scholars developed a lunar calendar of thirteen months, and Egyptian scientists formed a calendar based on the rising of the Dog Star, Sirius, in the eastern horizon, which conformed to the rising of the Nile River. Ancient Mayans of Central America developed calendars, while ancient Polynesians sailed the Pacific Ocean using nautical astronomy. Ancient astronomy was mostly employed for religious and magical purposes but sometimes to create calendars for society and agriculture.

Whereas during the Christian era, in the centuries after the reign of Constantine (r. 306–337) the dating system of *Ante Christos/Anno Domini* was established in Europe, the Muslims of North Africa and Asia derived a different dating system, based on the life of the most important prophet of Islam, Muhammad (c. 570–632). "The era of the Flight of the Prophet Muhammad from Makka [Mecca] to Madina [Medina]," Al-Biruni wrote, "is based upon Lunar years, in which the commencements of the months are determined by the appearance of New Moon, not by calculation." The Higira "is used by the whole Muhammadan world" (trans. Sachau). Al-Biruni explained that the date of the Hegira, 622, was indisputable, unlike the date of the birth of the Prophet.

Arab astronomers used astronomy to determine the time of fasting, Ramadan, according to Al-Biruni. He wrote:

They assert that the month of Ramadan has never less than thirty days. However, astronomers and all those who consider the subject attentively, are well aware that the appearance of new-moon does not proceed regularly according to one and the same rule for several reasons: the motion of the moon varies, being sometimes slower, sometimes faster; she is sometimes near the earth, sometimes far distant; she ascends in north and south, and descends in them; and each single one of these occurrences may take place on every point of the ecliptic. And besides, some sections of the ecliptic sink faster, others slower. All this varies according to the different latitudes of the countries, and according to the difference of the atmosphere. This refers either to different places where the air is either naturally clear or dark, being always mixed up with vapours, and mostly dusty, or it refers to different times, the air being dense at one time, and clear at another. Besides, the power of the sight of the observers varies, some being sharp-sighted, others dim-sighted. And all these circumstances, however different they are, are liable to various kinds of coincidences, which may happen at each beginning of the two months of Ramadan and Shawwal [the tenth month] under innumerable forms and varieties. For these reasons the month Ramadan is sometimes incomplete, sometimes complete, and all this varies according to the greater or less latitude of the countries, so that, e.g., in northern countries the month may be complete, whilst the same month is incomplete in southern countries, and vice versa. Further, also, these differences in the various countries do not follow one and the same rule; on the contrary, one identical circumstance may happen to one month several consecutive times or with interruptions. (Trans. Sachau)

FURTHER READING

Al-Biruni, *The Book of Instruction in the Elements of the Art of Astrology*, trans. R. Ramsay Wright (London: Luzac & Co, 1934; reprint Bel Air, MD: Astrology Classics, 2006).

Al-Biruni, *The Chronology of Ancient Nations*, trans. C. Edward Sachau (London: William H. Allen and Co., 1879).

Geoffrey Chaucer, *The Canterbury Tales of Geoffrey Chaucer*, trans. Percy Mackaye (New York: Duffield & Company, 1914).

Diodorus Siculus, *The Library of History of Diodorus Siculus*, vol. 1, trans. C. H. Oldfather (London: Loeb Classical Library, 1933), http://penelope.uchicago.edu/Thayer/E/Roman/Texts/Diodorus_Siculus/home.html.

Diogenes Laertius, *The Lives and Opinions of Eminent Philosophers*, trans. C. D. Yonge (London: Henry G. Bohn, 1853).

Patrick Glauthier, "Repurposing the Stars: Manilius, Astronomica 1, and the Aratean Tradition," *American Journal of Philology* 138, no. 2 (Summer 2017): 267–303, https://www.academia.edu/34309524/Repurposing_the_Stars_Manilius_Astronomica_1_and_the_Aratean_Tradition.

Robert Hannah, *Greek and Roman Calendars* (London: Bloomsbury, 2005), https://books.google.com/books?id=qTWPAQAAQBAJ.

Thomas Heath, *Aristarchus of Samos* (New York: Dover Books, 1981).

O. Neugebauer, *The Exact Sciences in Antiquity* (New York: Dover Books, 1969).

J. J. O'Connor and E. F. Robertson, "Autolycus of Pitane," http://mathshistory.st-andrews.ac.uk/Biographies/Autolycus.html.

J. J. O'Connor and E. F. Robertson, "Heraclides of Pontus," https://mathshistory.st-andrews.ac.uk/Biographies/Heraclides.

J. J. O'Connor and E. F. Robertson, "Hypsicles," http://mathshistory.st-andrews.ac.uk/Biographies/Hypsicles.html.

Plutarch, *Essays and Miscellanies: The Complete Works*, corrected and revised by William W. Goodwin, vol. 3, https://www.gutenberg.org/files/3052/3052-h/3052-h.htm.

Plutarch, "Life of Solon," in *The Lives of the Noble Grecians and Romans*, trans. John Dryden (New York: Random House, 1992).

Mark Riley, "A Survey of Vettius Valens," https://www.csus.edu/indiv/r/rileymt/PDF_folder/VettiusValens.PDF.

Vivian Robson, *The Fixed Stars and Constellations in Astrology* (Whitefish, MT: Kessinger Publishing, 2003).

The Sphere of Marcus Manilius Made an English Poem with Annotations and an Astronomical Appendix, trans. Edward Sherburne (London: Nathanael Brooke, 1675).

Vettius Valens, *Anthologies*, trans. Mark Riley, https://www.csus.edu/indiv/r/rileymt/Vettius%20Valens%20entire.pdf.

W. G. Waddell, trans., *The Fragments of Manetho* (Cambridge, MA: Harvard University Press, 1940), http://penelope.uchicago.edu/Thayer/E/Roman/Texts/Manetho/home.html.

Christopher Warnock, *The Mansions of the Moon: A Lunar Zodiac for Astrology and Magic* (Morrisville, NC: Lulu Press, 2019).

PART 6

The Expansion of Greek Science

15

Hellenistic Greece (331–30 BCE)

The Hellenistic Age began with the conquests of Alexander the Great. From 334 to 324 BCE, Alexander, son of Philip, King of Macedonia, led a coalition of Greek and Macedonian forces into Asia. During a ten-year period, Alexander conquered Asia Minor, Phoenicia, Palestine, Egypt, Iraq, Iran, Bactria and Sogdiana (Afghanistan), and the Indus River valley (Pakistan). Greek culture and ideas, including Greek religious beliefs and institutions, met and merged with African and Asian culture and religious traditions. Although Alexander died shortly after the conclusion of his military campaign, Greek generals and kings continued to control and influence the vast region spreading from Macedonia to Egypt to the Indus River. Three Greek kingdoms emerged in the wake of Alexander's death: the Macedonian, Ptolemaic, and Seleucid. The resulting interaction of Hellenism—Greek culture, institutions, and government—with the traditions of Persia, Egypt, India, Palestine, Mesopotamia, and Anatolia created a volatile, dynamic, creative period of three centuries that historians have labeled the Hellenistic Age. The Hellenistic Age came to an end with the fall of the three Hellenistic kingdoms to the Romans during the second and first centuries BCE.

ALEXANDER THE GREAT (356–323)

The original thinking, unique personality, and irrepressible will of Alexander of Macedon inaugurated the Hellenistic Age (331–30 BCE). Alexander's conquest of the Persian Empire and his vision to link the Greeks and Macedonians of Europe with the Africans of Egypt and the Asians of Anatolia, Palestine, Mesopotamia, and Iran resulted in a mixture of unique cultures that stimulated a new epoch of intellectual and scientific achievement. The diversity of thinkers brought together at such centers of learning as Alexandria, Egypt, founded by Alexander in 331 BCE, resulted in significant scientific writings and theories that have influenced subsequent scientists and philosophers for centuries even to the present. Alexander himself, as the student of Aristotle, was a scientist, a philosopher-king who appended curiosity and scientific discovery to his principal aims of conquest and military glory.

Alexander was raised in a kingdom that valued brutal strength and military prowess. Fourth-century BCE Macedonians had faith in the traditional Greek gods, who were themselves divine and eternal symbols of raw human emotions and violent confrontations. Not surprisingly, Alexander identified with the great Homeric hero and warrior Achilles and the legendary strongman Heracles. His conquest of Asia was motivated in part by his desire to exceed Achilles and Heracles in glory. His mother, Olympias, reputedly believed that she had conceived Alexander with the seed of Zeus, king of the gods; Alexander grew up thinking of himself as (perhaps) the son of Zeus. Olympias, a princess from the kingdom of Epirus west of Macedonia, was a snake charmer who introduced Alexander to other deities, such as Bacchus, the god of wine, whose worship involved orgiastic rituals brought on by divine intoxication. Alexander learned as well that the gods often revealed truth and destiny by means of oracles. Alexander traveled to the Oracle of Zeus-Ammon at Siwah, in the Sahara Desert, after his conquest of Egypt in 331. Alexander regarded Homer's tale in the *Iliad* to have been an accurate account of valor, conduct, and relations between humans and gods. Throughout his life he tried to live up to the poetic, heroic model wrought by Homer.

Another important influence on Alexander's life was the scientist and philosopher Aristotle and his teachings. Aristotle served as Alexander's tutor for three years, from 343 to 340. Aristotle's approach to the divine (and to every other form of human inquiry) was much more sophisticated than the young Alexander had hitherto experienced. The philosopher helped Alexander to see that besides Homeric valor and beliefs are the values of civilized culture and the understanding of the divine by means of the intellect rather than blind superstition. Aristotle opened up Alexander's mind to the infinite possibilities of life and the potential diversity of human behavior and belief.

No precise record survives to indicate what Aristotle taught and Alexander learned. One can hypothesize that Aristotle brought to Alexander many of the theories and observations that would make up the philosopher's immense corpus of writings. Aristotle rejected the Homeric worldview, arguing instead that truth is metaphysical and that the gods are elementary representations of being (*ousia*), which is the transcendent creative force in the universe. Whereas strength mattered most in Macedonia, Aristotle argued that the best form of government is the state that is run according to virtue and encourages virtue in its citizens. Aristotle taught the young prince techniques of observation, inductive reasoning, and deductive logic. He encouraged Alexander to become more than a king—to be a philosopher as well. Plutarch, in his *Life of Alexander,* wrote that Alexander, upon learning that Aristotle had published his book of *Metaphysics*, wrote the philosopher from Asia and chided him for publishing his teachings, hence allowing others to be privy to the same information that Alexander himself had been taught. What was the point in being a king if one's subjects had the same knowledge? Plutarch added that Alexander was particularly interested in what Aristotle taught him about medicine. The philosopher's influence upon the student was so great that Alexander acted as a physician to his friends, making diagnoses and recommending cures and regimens to restore health.

Alexander never attended Aristotle's school, the Lyceum, at Athens, though the king admired the Athenians for their many scientific and philosophic accomplishments. Perhaps the model of Athens inspired Alexander's vision of a center of learning in his new empire that would eclipse the old—the grand but limited ideas of Athens and its philosophers. When in 331 Alexander saw Pharos Island near the mouth of the Nile River in Egypt, he knew he had found the site for his city. Indeed, within a generation of his death, Alexandria had already emerged as one of the leading centers of trade and science in the Mediterranean world.

Alexander, the student of Greek science, brought on his journey to Asia a number of Sophists, seers, writers, and scientists. Callisthenes of Olynth (360–328), for example, the grandnephew of Aristotle, served as official historian of the expedition. As knowledgeable as Aristotle and his followers were, they nevertheless retained the narrow Greek views toward non-Greeks: "barbarians." Callisthenes agreed with his great-uncle that the Persians were barbarians, fit to be conquered by the Greeks. But when Alexander surprised everyone with his epiphany that the Persians and Greeks were one people, Callisthenes was too stubborn to go along with the king; rather, he was open and arrogant in his disagreement. Alexander grew impatient with the philosopher and was all too ready to believe accusations leveled against Callisthenes that he conspired against the king. After Callisthenes's death, the connection between Alexander and Aristotle ended.

Aristander of Telmessos, born in 380, who was also a soothsayer, and Cleitarch of Colophon, accompanied Alexander, as well as Anaxarchus of Abdera (380–320), who sought to make Alexander more temperate; skeptics such as Pyrrho of Elis (c. 365–279); Indian sages such as Calanus (c. 398–323); and soothsayers such as Aristobulus of Cassandreia (375–301). Alexander's interest in geography impelled him to order soldiers to become scientists, as it were; hence Nearchus the Macedonian warrior became an explorer of the Indus River and Indian Ocean. Alexander sent other soldiers to explore places such as the Arabian Peninsula, the Arabian Sea, the Persian Gulf, the Black Sea, the Red Sea, the Caspian Sea, and the many rivers that fed such bodies of water. In all of this Alexander was driven by the desire to discover the unknown. What is the unknown to a person today? Does the unknown drive scientists in their quest for knowledge? How different would be such a pursuit for a person 2,300 years ago, before the age of modern science, before the great accomplishments in science of the Muslims, Romans, and Hellenistic Greeks? Alexander was not a trained scientist. Nor were the men he sent on exploring expeditions. But there is an aspect of science that involves the adventurer, the person who seeks to penetrate the unknown, what Alexander called, in Greek, "pothos." The unknown pulls at the human mind, who seeks to discover what is mysterious, what is over the next hill, what is the cause of so much that humans fear. It is *unknown*. Alexander pursued this just as any scientist today might pursue it. It is an intellectual quest, a drive, an impulse, to discover what lies outside the self—and what lies within the self. Such a scientist as Alexander is not a trained laboratory scientist, an empiricist, like those today. Alexander was an ad hoc scientist, a scientist out of necessity, made by the actions he took and the places he went. There have been so many such ad hoc scientists throughout history. They are often explorers, conquerors, warriors, merchants, people who are lost, guides, hunters, trappers. Throughout history active people journeying into unknown places became ad hoc scientists.

Alexander's biographer Arrian, in his account of Alexander's journey, the *Anabasis*, had the wisdom to ascertain that part of Alexander's motive that drove him forward was the intellectual and scientific quest to know. Alexander was likewise inspired by another such ad hoc scientific explorer, the student of Socrates and mercenary Xenophon, who led Greeks on a perilous journey out of enemy lands back home, becoming an ad hoc scientist along the way. Xenophon's account was also called *Anabasis*.

Alexander's desire to know, his resulting conquests, and the increasing human awareness that succeeded such knowledge of other places and peoples, inspired in him an expanded vision. This vision, based on his expanded sense of self, was that not only had he the power to conquer other peoples, which was an old idea, but he had the power to unite them as well—to cross structured social/geopolitical boundaries and make half a dozen peoples one. No one before Alexander had had such an idea, and few people since have embraced such an idea. Why did he? Perhaps Alexander listened closely to his teacher Aristotle expand Plato's idea of the unified cosmos, which made the young king want to imitate that unified cosmos on earth. Is this not a natural inclination in humans, to, on the one hand, come to know ourselves, our humanity, but, on the other hand, to come to know our roots, the unity and oneness from which we spring? Perhaps Alexander in his way was performing these roles in a geopolitical setting, by first expressing himself as an explorer and conqueror, and then by seeking to take his knowledge and conquests and make them one, unified like the cosmos.

This idea of Alexander's that peoples of the world were inherently equal (next to Alexander) was his most important legacy. Philosophers of varying backgrounds at the leading Hellenistic capitals celebrated the diversity that Alexander's conquests had inspired: the mixture of peoples, ideas, beliefs, customs, and ways of life. A Semite from Cyprus, Zeno (c. 333–262), who arrived at Athens in 311, started a school of philosophy based on Alexander's concept of universal brotherhood. The Stoics studied natural history and human behavior to discover the presence of a universal reason, the Logos, that ordered all things. The goal of thoughtful humans was to imitate the reason and order of the cosmos so to find contentment within themselves.

PYRRHO OF ELIS (C. 365–279 BCE)

Pyrrho of Elis was one of the philosophers of Alexander's expedition who possibly came into contact with Indian philosophers such as Calanus, who taught an extreme form of asceticism. Diogenes Laertius believed that Pyrrho was a follower of the Sophist Anaxarchus, who joined Alexander on his Persian expedition, "and attended him everywhere; so that he even went as far as the Gymnosophists, in

India, and the Magi" (trans. Yonge). Mediterranean writers associated magi with Chaldeans, who were astrologers and soothsayers of Babylon, and gymnosophists, who were the reputed wise men of India. The fourth-century philosopher Clearchus of Soli, according to Diogenes Laertius, claimed that gymnosophists were latter-day magi. Laertius provided an anecdote of an Indian philosopher reproving Anaxarchus for being too obsequious to princes and not teaching the common person, which led Pyrrho to develop a more solitary, reclusive demeanor. He was by profession a painter before he adopted his philosophic garb. Influenced, apparently, by the gymnosophists, "he seems to have taken a noble line in philosophy," according to Diogenes, "introducing the doctrine of incomprehensibility, and the necessity of suspending one's judgment. . . . For he used to say that nothing was honourable, or disgraceful, or just, or unjust. And on the same principle he asserted that there was no such thing as downright truth; but that men did everything in consequence of custom and law. For that nothing was any more this than that." He was quite aware of the transiency of life, and the overall insignificance of human affairs, even comparing "men to wasps, and flies, and birds" (trans. Yonge).

His followers were called "Pyrrhoneans, . . . doubters, and sceptics, and ephectics, or suspenders of their judgment, and investigators, from their principles." They examined all sides of each issue, doubted but "never finding," and "suspending . . . judgment." Further, "the Sceptics persevered in overthrowing all the dogmas of every sect, while they themselves asserted nothing dogmatically; and contented themselves with expressing the opinions of others, without affirming anything themselves, not even that they did affirm nothing" (trans. Yonge). The third-century Skeptic Timon (320–230) was a student of Pyrrho. The most famous student of Pyrrhonism was Sextus Empiricus (c. 160–210).

ALEXANDRIA

Alexandria, founded and named by Alexander the Great in 331 BCE, became the leading center of science and thought in the ancient world from 300 BCE to 500 CE. As the center of Hellenistic culture after Alexander's death in 323, and supported by Ptolemy I and his successors, Alexandria brought together a host of different scientists, philosophers, cultures, and peoples. This dynamic, cosmopolitan mix ensured that the city would be vibrant, diverse, sometimes chaotic, and calm neither in its political affairs nor in its intellectual debates.

Such was the namesake of Alexander the Great, the enigmatic Macedonian conqueror who founded the city after having recently sacked the Phoenician city of Tyre and the Palestinian city of Gaza. Alexander had arrived at Egypt fresh also from defeating Darius III (r. 336–330) at the Battle of Issus. He came to Egypt as a liberator from Persian control; the Egyptians responded by naming him pharaoh of Egypt. Alexander knew that his emerging empire encompassing eastern Europe, northern Africa, and western Asia would require a center of trade, government, and learning. He had the foresight to see that a city at the mouth of the Nile River would fulfill that role.

The ancient biographers Arrian and Plutarch related that Alexander sailed down the Nile to its mouth and then along the Mediterranean coast a brief distance to a natural harbor formed by a quiet lagoon situated between the mainland and an island called Pharos. The harbor was protected from the fresh westerly breeze, and the island hinted of its importance as a seamark for sailors. Alexander decided that such a location would be perfect for the kind of city he envisioned. He himself eagerly set out the plan of the town using barley meal to set the boundaries. According to Plutarch, Alexander conceived of the city as a semicircular arc proceeding from one point in equal segments. Alexander wanted a city built according to the theories of Hippodamus of Miletus (c. 498–408), whose student, Dinocrates of Rhodes, became the lead architect and planner. The city was wide, with a central boulevard connecting different quarters of peoples and places. A causeway connected the island of Pharos to the mainland, which itself was a vast peninsula practically surrounded by water. The causeway formed two harbors on either side; the harbor to the east was enclosed, able to host scores of ships. The southern part of the city had important harbors on Lake Mareotis.

Alexandria quickly eclipsed Athens and other Hellenistic cities, such as Antioch and Pergamum, as the premier center of science during the Hellenistic Age and afterward. Greco-Egyptian scientists and philosophers under the sponsorship of the Ptolemaic pharaohs working with Greeks throughout the Hellenistic world made significant advances in the study of mathematics, astronomy, and geography. The five-hundred-thousand-volume Library at Alexandria in cooperation with the Alexandrian Museum was a multicultural research center.

Upon Alexander's death one of his successors, Ptolemy (367–283), marched to Egypt with Alexander's embalmed corpse, and set himself up as pharaoh (Ptolemy I Soter, r. 323–283), and built a tomb to Alexander. Ptolemy, having taken control of Egypt in the civil wars that followed Alexander's death, not only established the Ptolemaic Dynasty of Egypt but also conceived of both the idea of the lighthouse on Pharos and the Library of Alexandria, and he opened the Museum at Alexandria. Ptolemy was from Macedon and of noble birth; although much older than Alexander, he was one of his first companions, participating with Alexander in the lectures of the noted savant Aristotle. The teachings of Aristotle had a profound effect on Alexander and apparently on Ptolemy as well: he exhibited a love of learning both as a student and a patron.

Ptolemy inaugurated the idea of a massive lighthouse constructed on Pharos; his successor Ptolemy II Philadelphus (r. 283–246) completed it using the talents of the architect Sostratus of Cnidus. The lighthouse, subsequently called simply Pharos, was justifiably considered one of the seven wonders of the ancient world. It was built in three levels: the first rectangular, the second hexagonal, the third a cylinder culminating into the pinnacle, where stood a massive fire fed by wood and resin. Some scholars have hypothesized the existence of mirrors that illuminated the flame. Its builders dedicated Pharos to Zeus and Poseidon. One inscription states that Sostratos dedicated it "to the gods for the protection of sailors" (Empereur, p. 84). Pharos survived for close to 1,500 years, during which it fascinated scores of travelers, sailors, writers, and scientists. It helped to make Alexandria a center of trade, culture, and learning.

Ptolemy patronized the mathematician Euclid (c. 325–265) and Strato (335–269), who later became the director of the Lyceum in Athens. Strato was the tutor to Ptolemy's son, who succeeded the father as Ptolemy II and was himself a patron of scholarship at the Library and Museum; Aristarchus of Samos (310–230) worked in Alexandria. Ptolemy II himself was a zoologist. Under Ptolemy's grandson, Ptolemy III (r. 246–222), the geographer Eratosthenes (276–195) worked at Alexandria. The Ptolemies provided room, board, funds, and lifelong appointments for the scholars who worked under them at the Museum and who frequented the Library.

The Museum, reputedly sponsored by Ptolemy I working with Aristotle's student Demetrius of Phaleron (350–280), housed scholars who indulged in research, writing, speaking, and symposia, which even included the pharaohs. The Library adjoined the Museum and held up to 700,000 papyrus scrolls. The directors of the Library at Alexandria were themselves scholars. The first, Zenodotus (c. 325–260), employed an Aristotelian technique of organizing the papyrus scrolls (rather than bound books) that made up the volumes of the library. Zenodotus had the volumes arranged by subject and author and stored in specific locations. The librarian Callimachus (c. 310–240), a poet, provided something akin to a card catalog with cross-references to help researchers negotiate the vast corpus of scrolls. After Zenodotus and Callimachus, Eratosthenes the mathematician and geographer became the director of the Library.

Great thinkers lived and worked in Alexandria, such as Euclid the mathematician, author of *Elements*; Ctesibius, a pneumatic engineer (c. 285–222); Hero, a mathematician and engineer (62–152); Potamo, founder of the Eclectic school of philosophy and author of *Elements of Philosophy*, who lived in the first century BCE or CE; and physicians such as Herophilus (325–255), Erasistratus (275–194), and Serapion (third century BCE). The Neoplatonist Porphyry wrote that the first-century-CE director of the Library, Chaeremon, a Stoic philosopher and student of Egyptian religion, astrology, and science, and

> others . . . do not think there is any thing else prior to the visible worlds; but in the beginning of their writings on this subject, admit the existence of the Gods of the Egyptians, but of no others, except what are called the planets, the Gods that give completion to the zodiac, and such as rise together with these; and likewise, the sections into decans, and the horoscopes. They also admit the existence of what are called the powerful leaders, whose names are to be found in the calendars, together with their ministrant offices, their risings and settings, and their significations of future events. For Chaeremon saw that what those who say that the sun is the Demiurgus, and likewise what is asserted concerning Osiris and Isis, and all the sacred fables, may be resolved into the stars and the phases, occultations and risings of these, or into the increments or decrements of the moon, or into the course of the sun, or the nocturnal and diurnal hemisphere, or into the river [Nile]. And, in short, the Egyptians resolve all things into physical, and nothing into incorporeal and living essences. Most of them likewise suspend that which is in our power from the motion of the stars; and bind all things, though I know not how, with the indissoluble bonds of necessity, which they call fate. They also connect fate with the Gods; whom, nevertheless, they worship in temples and statues, and other things, as the only dissolvers of fate. (Trans. Taylor)

One of the great thinkers of the ancient world, Claudius Ptolemaeus, the astronomer, geographer, and mathematician, lived and worked in Alexandria. Such scholars made Alexandria the leading center of thought in the ancient world, eclipsing even Athens, Antioch, Rome, and Constantinople.

HELLENISTIC MATHEMATICS AND ASTRONOMY

Hellenistic mathematicians included Apollonius of Perga, who worked with conic sections, and Eudemus of Rhodes (350–290), a student of Aristotle who opened an Aristotelian school at Rhodes. Eudemus wrote a history of mathematics and treatises on arithmetic, astrology, geometry, physics, logic, and angles. Eudemus also wrote a history of astronomy, according to Diogenes Laertius. He was one of the first historians of science; notably, he organized and edited Aristotle's writings through his own notes, which allowed for the survival of much of Aristotle's works. We know through Eudemus that "Thales was the first to *discover . . . the eclipse of the sun* and the fact that the sun's period with respect to the solstices is not always the same"—that is, that the exact time of the solstices differs by year (trans. Heath). It is through Eudemus that the theories of concentric spheres of the fourth-century Platonist Eudoxus of Cnidus (408–355) and his student Callippus of Cyzicus (370–310) are known. He wrote at length about the astronomical theories of Anaximander. Another Rhodian, Posidonius (135–51), who was born in Apamea, worked as an astronomer and geographer. He measured the circumference of the earth and tried to gauge the distance to the sun; he was very interested in the study of meteorology. Posidonius was the teacher of Geminus (10 BCE–60 CE), who wrote a history of astronomy and treatises on geometry and arithmetic.

At Alexandria we find some of the great scientists of all time, such as Euclid, author of *The Elements of Geometry*, which became the standard work on that branch of mathematics for centuries. Euclid's approach owed much to the Platonic concern to discover the hidden reality of the universe by means of mathematics. Other Alexandrian mathematicians and astronomers included Hypsicles (190–120), a second-century mathematician and astronomer and disciple of Euclid's *Elements*. He calculated the length of each day at Alexandria. In *On the Ascension of Stars*, he calculated the extent of the zodiac, dividing it into 360 spatial and temporal degrees. He wrote a book titled *On Polyhedrons*, another on arithmetic progressions, and another on polygonal numbers. Menelaus of Alexandria (70–130), working in Rome in the late first and early second centuries CE, used trigonometry and the geometry of the sphere. He was an astronomer as well.

CLAUDIUS PTOLEMAEUS (100–170 CE)

The greatest Alexandrian scientist was Claudius Ptolemaeus, Ptolemy, one of the most influential scientists of the ancient world. Ptolemy was the head of the Library of Alexandria. He was a mathematician, geographer, and astronomer of note whose conception of the world and the universe influenced thinkers of medieval and Renaissance Europe and the Islamic East. His works were translated into Arabic as *Almagest* and have been known as such ever since.

Ptolemy's astronomy was informed by his predecessors, especially Hipparchus, as well as by Babylonian astronomy. Because of his position as librarian at Alexandria, Ptolemy had before him the vast corpus of Babylonian observations: ephemerides covering many centuries and lunar and solar eclipse records. Indeed, in the *Almagest,* Ptolemy provided numerous tables to describe astronomical phenomena. Ptolemy's geocentric universe was very similar to the Babylonian universe and theories of Greek astronomers before him. In it, Ptolemy assumed that the earth is the center around which the sun orbits. The planets are generally on the same plane as the sun, and from Earth the observer can see their movement. Ptolemy explained a planet's retrograde motion by assuming that planets rotated on an epicycle around the spherical path that took them around the earth. When a planet is rotating about the earth in the distant semicircle of the epicycle, it appears to move rapidly in the direction of the sun. But when it is rotating about the earth in the semicircle of the epicycle closest to earth it appears to reverse its orbit and then go back again—retrograde motion. Ptolemy's explanation of the presence of the morning and evening stars, that Mercury and Venus are visible for short times in the eastern horizon and western horizon, depended on the daily location of these planets in the direction of the sun. The morning star is visible until it reaches an area where it is blocked by sunlight, but then it reappears again on the opposite horizon when the sun sets in the west.

Ptolemy's concern was to create a theoretical model that could explain for the geometer and astronomer the positions of the planets relative to the fixed stars at a given time. Whereas in another work, *Planetary Hypotheses,* Ptolemy treated the planets as physical phenomena and tried to estimate distances from earth, the *Almagest* is entirely theoretical, and one wonders whether he intended his complex system of epicycles to be real or just a good theoretical plan. In the preface to the *Almagest*, Ptolemy claimed that theoretical rather than practical philosophy was best and would be the approach he would take in his book. He followed Aristotle in defining the three areas of investigation as theology, the study of the First Cause; mathematics, the study of forms and motions of heavenly bodies both physical and immaterial; and physics, the study of the actual physical motions and attributes of material phenomena. Ptolemy concluded that of these three areas, the second, mathematics, was most appropriate for the scientist, as the former is unknowable by humans and the latter is subject to the distressing uncertainty and corruptibility of material objects.

Other Hellenistic astronomers included Aristarchus of Samos (310–230), who discovered the heliocentric (sun-centered) universe, though he was alone in this knowledge, since the idea contradicted sensory experience. Hipparchus of Nicaea (190–120), for example, could not doubt what his senses told him, that the sun circled the earth. Nevertheless, Hipparchus catalogued the stars and discovered the precession of the equinoxes. Also significant was Heraclides of Pontus (387–312), who argued that the earth spun on its own axis, and Aratus of Soli (315–240), who preserved (in *Phaenomena*) the theory of Eudoxus of Cnidus that the earth was spherical because the altitude of the stars changed as the latitude of the observer changed.

HELLENISTIC GEOGRAPHY

Apollonius of Rhodes (fl. third century BCE) was not a scientist but, rather, a poet, yet in his poem the *Argonautica* we find a fascinating account of the Hellenistic view of the *ecumene*, the "world." The *Argonautica* is the story of Jason and other Greek heroes who voyaged on the ship *Argo* in quest of the Golden Fleece. The crew included the likes of Heracles; Peleus, the father of Achilles; Telamon; Zetes and Calais, the sons of the wind, Boreas; Meleager; Argus; Castor; and Polydeuces. In the story we also find tales of others famed in Greek mythology, such as Athena, Hera, Poseidon, and Boreas himself. Apollonius told a very ancient story. There was probably a kernel of fact in the story that the Argonauts sailed across the Aegean Sea to the Hellespont, north through the Bosporus to the Black (Euxine) Sea, east to Colchis, and then back across the Black Sea to the Danube River. They ascended the Danube to its source, then descended the Rhone River to the Mediterranean Sea, and paralleled western Italy to Sicily. Near western Greece they were driven off course to North Africa, but they finally returned to Greece by way of Crete and the islands of the Aegean Sea. It was a fantastic journey told many times over the centuries. Apollonius provided the third-century-BCE Greek version.

Eratosthenes of Cyrene (276–195 BCE) was a librarian of Alexandria and a geographer of note. Author of the *Geographica*, Eratosthenes mapped the world and estimated the earth's circumference to 31,300 miles, which was remarkably accurate. Working from the library at Alexandria, Eratosthenes first used latitude and longitude to chart the globe. Inspired by the voyages of Nearchus, he believed in the possibilities of sailing west across the Atlantic to India. He flourished at a time when Alexandria was becoming the premier center of science in the ancient world. He influenced many other thinkers, such as Polybius, Strabo, and Ptolemy of Alexandria.

Eratosthenes's most famous experiment was his measurement of the shadow at Alexandria on the same date and time of day as when the rays of the sun were (he believed) directly overhead at Syrene, Egypt. According to the story (preserved by the first-century-CE astronomer Cleomedes), he determined this phenomenon at Syrene by observing the rays of the sun shining directly on the bottom of a well. Using a gnomon, he determined the angle of the noon shadow at Alexandria to be 7.2 degrees, one-fiftieth of a circle, the spherical earth. The distance from Syrene to Alexandria was 5,000 stadia (about 2.5 million feet), which if multiplied by 50 gives the result of the circumference of the earth, 250,000 stadia, or 23,300 miles, an estimate that was off by only 2,000 miles. (The way to accurately determine the earth's circumference is to multiply the diameter by *pi*. Part of the significance of Eratosthenes's experiment was his realization that the rays of the sun strike the varied places of the earth in parallel [not divergent] lines.)

HELLENISTIC MEDICINE

During the Hellenistic Age, Alexandria dominated medical studies chiefly because of the work of Herophilus and Erasistratus, both of whom were concerned with discovering the precise structure and function of the organs and circulatory system. They engaged in empirical study and dissection, discovering much about the brain and nervous system, the eyes, stomach, liver, reproductive systems, blood, and heart. The kings of Egypt, the Ptolemies, supported Herophilus of Chalcedon and his students by giving them the corpses of condemned criminals to dissect and study. Herophilus spent his professional life as a physician and anatomist at Alexandria. There he performed experiments and dissections to arrive at specific knowledge about the brain, eye, and other organs. He defended the Hippocratic method, seeing the physician as an artist and health as the foundation for all individual accomplishments. The environment is especially important as a cause of disease, and diet, rest, and exercise are important in its prevention. Herophilus's most famous student was Erasistratus.

Erasistratus worked in Alexandria under his teacher Herophilus, performing observations, experiments, and dissections to uncover some of the secrets of the workings of the human body. He researched the brain, studied the heart, and hypothesized about the veins and arteries. As a practicing physician, Erasistratus believed that holistic living was the means to prevent illness. He discarded the four humors of the Hippocratic school in favor of relying upon the pulse to make diagnoses—his was a methodical approach to medicine, as was that of the Roman Asclepiades (120–40) a few centuries later.

Another important Hellenistic life scientist was Nicander of Colophon, a physician during the second century BCE who wrote on agriculture, beekeeping, and poisonous plants and reptiles. Nicander wrote *Alexipharmaca* and *Georgica*, in which he described various poisonous plants and antidotes. For example, he discussed plants in the family Ranunculaceae, such as Aconitum, known as wolf's-bane, leopard's bane, and monkshood—terribly poisonous plants. He described in detail how poison affects a person—the horrible pain—and possible antidotes, including the use of gypsum. Noxious plants he discussed included coriander, hemlock, meadow saffron, henbane, and poppy. "When men drink the tears of the poppy," he wrote, "whose seeds are in a head, they fall fast asleep; for their extremities are chilled; their eyes do not open but are bound quite motionless by their eyelids. With the exhaustion an odorous sweat bathes all the body, turns the cheeks pale, and causes the lips to swell; the bonds of the jaw are relaxed, and through the throat the laboured breath passes faint and chill. And often either the livid nail or wrinkled nostril is a harbinger of death." The fragments of the *Georgica* are apparently random discussions of plants, animals, flowers, cooking, and gardens. He referred to the mulberry, for example, as "the delight of little boys and . . . the harbinger to man of the pleasant season of fruit" (trans. Gow and Scholfield).

HELLENISTIC ALCHEMY

Ancient humans, sensing the presence of the mysterious in nature, tried by magical means to assert control over the forces inherent in the natural environment. The attempt to understand material substances of the earth so as to contrive to alter their character, producing other, more valuable substances, marked the beginnings of chemistry in the ancient world. The first alchemists were Hellenistic chemists working with substances such as sulfur and metals in attempting to improve the appearance and qualities of metals and transform them into something else. They worked under the influence of Aristotle and the atomists, believing that the elements could be expressed in terms of solids, liquids, gases, and colors. They believed that a spiritual substance (*pneuma*) underlay all material substances, and they thought that by changing a substance, the spiritual would be unleashed or changed as well. This change was called transmutation and involved reducing the substance to a fundamental mass of black material and then acting upon it in successive stages to arrive at a white and then a yellow substance, the latter of which, if the alchemist had performed the work correctly, would be gold. The key to all of this was a manipulation of not only the material aspect of the metal but the spiritual aspect as well. In a sense the alchemists thought they were reducing the material to a state of death and then encouraging the seeds of life to regenerate the material into something else. Sulfur was used as a causal agent of such change. Another important material in the alchemist's repertoire was mercury, which seemed to be the ideal substance of change from liquid to metal and back. Its shiny, silver appearance made it particularly valuable. To produce changes to metals, the alchemist developed a variety of apparatuses with which to heat materials and perform experiments. These included cauldrons, beakers, tubes, stills, furnaces, ovens, and so on.

HELLENISTIC INVENTIVENESS

Leading minds of the Hellenistic Age worked in cities throughout the Mediterranean. Archimedes (287–212), although he studied at Alexandria, moved to Syracuse, in Sicily, where he was an inventor, engineer, and mathematician of note. Much of his engineering talent went to devising means of defense against Roman attacks during the Second Punic War. Archimedes discovered *pi*, invented the "Archimedes's screw," researched the nature of volume, and made advancements in geometry. He was an expert in hydraulics.

The Hellenistic Age set the stage for subsequent centuries of scientific inquiry. The Roman conquest of the Hellenistic world during the last two centuries BCE involved political unification under Roman government and the Roman legions. Yet the Hellenistic Greeks conquered the Romans in terms of science, philosophy, art, culture, and religion. The great thinkers and scientists of the later Roman republic and Roman principate were heavily dependent upon their predecessors, particularly the scientists, astronomers, mathematicians, and geographers of the Greek Hellenistic Age.

FURTHER READING

Apollonius of Rhodes, *The Voyage of Argo*, trans. E. V. Rieu (Harmondsworth, England: Penguin Books, 1971).

Arrian, *Anabasis* and *Indica*, trans. P. A. Brunt and E. Iliff Robson, 2 vols. (Cambridge, MA: Harvard University Press, 1933, 1976).

Daniel J. Boorstin, *The Discoverers* (New York: Random House, 1983).

George Botsford and Charles A. Robinson, *Hellenic History*, revised by Donald Kagan (New York: Macmillan Publishing, 1969).

Michael J. Crowe, *Theories of the World from Antiquity to the Copernican Revolution* (New York: Dover Books, 1990).

Diogenes Laertius, *The Lives and Opinions of Eminent Philosophers*, trans. C. D. Yonge. (London: Henry Bohn, 1853).

Will Durant, *The Life of Greece* (New York: Simon and Schuster, 1939).

Victor Ehrenberg, *Alexander and the Greeks* (Oxford: Oxford University Press, 1938).

Jean-Yves Empereur, *Alexandria Rediscovered* (New York: George Braziller, 1998).

Robin Lane Fox, *The Search for Alexander* (Boston: Little, Brown, 1979).

Michael Grant, *From Alexander to Cleopatra: The Hellenistic World* (New York: History Book Club, 2000).

J. R. Hamilton, *Alexander the Great* (Pittsburgh, PA: University of Pittsburgh Press, 1973).

Thomas Heath, *Aristarchus of Samos: The Ancient Copernicus* (New York: Dover Books, 1981).

Thomas S. Kuhn, *The Copernican Revolution: Planetary Astronomy in the Development of Western Thought* (Cambridge, MA: Harvard University Press, 1957).

Henry M. Leicester, *The Historical Background of Chemistry* (New York: Dover Books, 1971).

O. Neugebauer, *The Exact Sciences in Antiquity* (New York: Dover Books, 1969).

Nicander of Colophon, *Alexipharmac and Georgica*, trans. A. S. F. Gow and A. F. Scholfield, Attalus, 1953, http://www.attalus.org/poetry/nicander.html.

J. J. O'Connor and E. F. Robertson, "Hypsicles of Alexandria," https://www-history.mcs.st-andrews.ac.uk/Biographies/Hypsicles.html.

J. J. O'Connor and E. F. Robertson, "Menelaus of Alexandria," http://www-history.mcs.st-andrews.ac.uk/Biographies/Menelaus.html.

Plutarch, *The Lives of the Noble Grecians and Romans*, trans. John Dryden, revised by Arthur Hugh Clough (1864; reprint, New York: Random House, 1992).

Polybius, *The Histories*, trans. W. R. Paton, 6 vols. (Cambridge, MA: Harvard University Press, 1922–1927).

Porphyry, "Letter to Anebo," trans. Thomas Taylor, Tertullian Project, http://www.tertullian.org/fathers/porphyry_anebo_02_text.htm.

16

Greek Philosophy and the Philosophy of Science (750–100 BCE)

Science and philosophy were more closely associated in the ancient world than at any other time in human history. The Greek word for science, *episteme,* means knowledge, while philosophy means love (*philo*) of wisdom (*sophia*). Both of these words, knowledge and wisdom, are pregnant with possibilities, contradictions, and questions. For example, how does one derive knowledge? Or wisdom? How does one recognize knowledge and wisdom in others? Can a person possess knowledge but not wisdom? Wisdom but not knowledge? Wisdom implies awareness of the timeless, yet science often seems very down to earth. To be wise, one is removed from everyday concerns; a scientist is an expert in the ins and outs of the everyday. Wisdom is a product of age and experience. Is knowledge as well? Can one *know* and still be young and foolish?

The ancients, in particular the Greeks, answered these questions by making science devoted to the acquisition of wisdom and making philosophy rely on knowledge. Knowledge can be abstract as well as concrete. Today's scientist eschews the metaphysical (that which transcends the physical) for the physical and tries as much as possible to divorce science from religion (the supernatural). The ancient scientist often could not distinguish between the physical and metaphysical, the natural and the supernatural. For example, the Greek scientist Aristotle sought the essence (*ousia*) in phenomena. An essence can, of course, be physical and concrete, perceived by the five senses. But Aristotle also believed essence could be ephemeral, transcendent, supernatural, and metaphysical.

The Greeks provided an interesting dichotomy with which to approach this question of science and philosophy. All of science and philosophy can be explained according to *being* and *becoming* (i.e., coming *to be*). *Being* is the principle of staticity, of the unchanging, of existence that is eternal and absolute, neither generated nor regenerated. To be (*esse*) is to exist in a fundamental way that is uninterrupted by outside forces, change, time, anything that is becoming. That which is becoming is undergoing change and movement. It is generated and regenerated, and, if absolute, it is an absolute formed by alteration over time. Being implies the metaphysical; becoming implies the material. Philosophers tend to find *being* more attractive; scientists find *becoming* more attractive. Greek philosophers and scientists who focused on being more than becoming were those who emphasized the mind (*nous*), essence (*ousia*), and thought (*logos*). Those who focused on becoming more than being emphasized the primary substances, the four (or five) elements, the corporeal basis of all things.

Among the Greeks, we might include among the scientists of *being* Thales of Miletus, who was one of the first Greek thinkers to conceive of and seek an ultimate basis of reality. Thales's student Anaximander believed the fundamental reality was the *infinite*. Pythagoras tried to identify the ultimate reality with transcendent *number*. Xenophanes of Colophon and Anaxagoras of Clazomenae hypothesized that *mind* is the basis of all things. Parmenides of Elea, a Pythagorean, theorized about *the one* encompassing all reality.

Greek scientists looking at the basis of reality *coming to be* included Heraclitus of Ephesus, who believed that fire was the essential element. Empedocles of Acragas expanded the idea of basic material elements to include not only fire but air, water, and earth. It was left to Leucippus and Democritus to hypothesize an unending plurality of elements coming to be, atoms of constant movement and change.

Aristotle was the greatest philosopher/scientist of the ancient world because he was able to integrate these two points of view, being and becoming. Aristotle assumed a metaphysical being, an essence (*ousia*) to all things. Yet he was quite adamant that change was also fundamental to the universe. Unlike the pure metaphysician, Aristotle relied on experience, observation, and logic—the methods of science—to arrive at an understanding of being. At the same time, he used the methods of the philosopher, logic and reason, to induce universals from particulars and to deduce particulars from universals.

The Stoics also bridged the two Greek approaches to knowledge. Stoic thinkers believed in a divine fire, a universal being and creator who was unlike *mind* or *number*, unlike a metaphysical ultimate reality. The Stoic *logos* was physical, made of atoms like all else in a material universe, yet one and eternal.

Over the course of Western thought, the distinction between science and philosophy is more apparent than real. The past holds many examples of the commensuration of objective and subjective thought, empiricism and idealism. The ideals and metaphysics of the teacher Plato were followed by the more concrete science of the student Aristotle. The theologian Thomas Aquinas was Aristotelian, the mathematician Copernicus a Platonist. Isaac Newton spent more time studying God and the Creation than he did mass, force, and acceleration. Twentieth-century physicists could at the same time speak of the godless Big Bang and the beauty and wonder of the Singularity that proceeded it. Theorists of science have come to see logical positivism as a distant dream and posit the subjectivity of the most concrete and objective scientific discipline. Einstein reshaped our perception of the universe with mathematics stimulated by a creative imagination that refused to believe that "God plays dice." He told the philosopher Martin Buber in 1911, "What we strive for is just to draw his lines after Him." A year later he said, "I want to know how God created this world. I am not interested in this or that phenomenon, in the spectrum of this or that element. I want to know His thoughts, the rest are details" (quoted by Clark, pp. 37–38). Neither Thales nor Anaximander, nor Anaxagoras nor Parmenides, could have stated the relationship between science and philosopher any better.

HELLENISM

Ideas drive science. The initial hypothesis that leads to investigation, experiment, an attempt to answer, derives from a question, formed by a thoughtful response to nature and experience. Questions and hypotheses are generated by a society's worldview, the collection of fundamental assumptions about existence; answers, scientific knowledge, likewise generate the worldview. The single most important idea that drove science in the ancient Mediterranean world was Hellenism.

Homer, in the *Iliad* and *Odyssey*, referred typically to the Greeks as the Achaeans, Argives, and Danaans. Yet in a few instances, he designated the whole of Greece, *Hellas*, and the Greeks as a people, *Hellenes*. The idea of a common people and land had roots in mythology. The hero Hellen, the son of Deucalion, the Noah of the Greeks and son of Prometheus, and Pyrrha, the daughter of Pandora, was the first Greek, the founder of the Hellenes.

Homer's poems describe the characteristics and beliefs that made the Hellenes unique. The poems portray a single world culture of Achaeans (Europeans) and Trojans (Asians). They speak the same language, worship the same deities, have the same customs, live and die to the same code of honor. The heroes of the two poems are "godlike"; the gods are humanlike. The anthropomorphism of Homer's poems does not so much denigrate the divine as elevate what is human. Human struggles, wars, life, and death are beautiful, universal, divine. Scholars refer to Homer's brilliant portrayal of the common humanness of the Achaeans, Trojans, even the gods, as humanism. But Homer's humanism derived from a more fundamental idea: Hellenism.

The idea of Hellenism was fully developed in fifth-century Athens. Classical Athens was a unique moment in human history. The Greek mind, its creativity, wonder, pursuit of knowledge, expression of truth, soared to a pinnacle of human achievement. Phidias (c. 480–425) designed and built the Parthenon on the Athenian acropolis. The playwrights Aeschylus (c. 525–456), Euripides (c. 484–407), and Sophocles (c. 496–406) wrote masterpieces of tragedy. Thucydides (460–400) wrote one of the great histories of all time. Socrates (469–399) questioned and taught and pursued wisdom. Pericles (495–429) molded Athenian politics into the first and perhaps greatest ever democracy. Pericles referred to Athens as "the school of Hellas," a comment that gave praise not only to the Athenians but to the Hellenes as a whole. Protagoras (485–418), an Athenian philosopher, described Athenian society, as well as the fundamental assumptions of Hellenism, when he proclaimed, "Man is the measure of all things."

During the fourth century BCE, Alexander the Great (356–323) epitomized the ideals and behavior of a Hellenistic thinker. Alexander was born and raised in Macedon, north of Greece, a land that the Hellenes considered a cultural backwater. Alexander, when he became king of Macedon, ruler of Greece, and leader of the Hellenes against the Persian Empire, embraced Hellenic thought and culture. Alexander conquered Persia politically but culturally as well. He purposefully established Greek cities, such as Alexandria in Egypt, and encouraged a mix of Greek culture and ideas with Egyptians, Persians, and Indians. Historians refer to the period of Alexander's conquests and the subsequent three centuries (331–30 BCE) as the Hellenistic Age, designating these three centuries in the Mediterranean and Asia as an era of dynamic growth, cultural accomplishment, and scientific achievement.

Hellenism was the inspiration of the Hellenistic Age. Greek questioning, the pursuit of answers using the mind, the focus on human reason and expression, the emphasis on individual accomplishment, and the stress on a common human culture defined Hellenism. The Greeks spoke of this time as the *ecumene*, the common world culture with a common Greek language: *koine*, "spoken Greek." The ecumene encouraged and hosted not only a common lingua franca but also common beliefs, institutions, and culture. Peripatetic philosophy and science, modeled on Aristotle, was widespread throughout the Mediterranean. Stoicism, a philosophy of reason and human universality, became the dominant philosophy of the Hellenistic Age and afterward, when Greek culture was embraced by the Romans. The Roman Empire made the ecumene, the basic ideas of Hellenism, a concrete reality. During the Pax Romana, from 30 BCE to 180 CE, this common Greco-Roman culture encompassed parts of three continents and included up to 100,000,000 people. Scientists such as Pliny the Elder (23–79) and Strabo (63 BCE–21 CE) wrote massive encyclopedic works of knowledge, befitting an ecumenical society. The Roman orator Cicero (106–43) proclaimed himself a *citizen of the world*; the theologian Augustine (354–430) spoke of a universal *city of man* countered by a transcendent *city of God*. The emperor Julian (331–363), when he came to power in 361, after fifty years of imperial support for Christianity, blamed Rome's increasing troubles on the rejection of Hellenism. Julian believed that a society devoted to the traditional gods and the traditional values of virtue, honor, human experience, and individual greatness, would rectify Rome's downward trend.

Julian enjoyed but a brief reign of less than two years. But his was a lasting legacy. Julian and other Hellenists, particularly Greek philosophers, scientists, artists, and writers, spoke to future generations searching for similar ecumenical, secular, humanistic ideals. Thinkers of the European Renaissance called themselves *humanists*, by which they meant an approach to life based on the thought of the Greeks, on Hellenism.

STOICISM

Stoicism, a Greek philosophy named for the *stoa*, or colonnade, of ancient Athens, was a system of thought based on a conception of a material universe composed exclusively of atoms in constant motion and creating varying combinations that constitute the stuff of matter and spirit. The founder of Stoicism was Zeno of

Citium (333–262), in Cyprus, who according to Diogenes Laertius, argued that there is an active and passive principle, the passive being matter, the active being God. Zeno argued this in his treatise *Essence*. Zeno believed in "the creation, and of the destruction of the world, in his treatise on the Universe." Diogenes continued, "The substance of God is asserted by Zeno to be the universal world, and the heaven." It is "Primary Matter," a "providential cause of things." In astronomy, Zeno believed that

> the sun is eclipsed, when the moon runs in front of it on the side towards us . . . ; for when it comes across it in its passage, it conceals it, and again it reveals it; and this is a phenomenon easily seen in a basin of water. And the moon is eclipsed when it comes below the shadow of the earth, on which account this never happens, except at the time of the full moon; and although it is diametrically opposite to the sun every month, still it is not eclipsed every month, because when its motions are obliquely towards the sun, it does not find itself in the same place as the sun, being either a little more to the north, or a little more to the south. When therefore it is found in the same place with the sun, and with the other intermediate objects, then it takes as it were the diameter of the sun, and is eclipsed. (Trans. Yonge)

In meteorology he argued that "lightning is a kindling of the clouds from their being rubbed together, or else broken asunder by the wind, as Zeno tells us in his treatise on the Universe." Regarding the soul, he argued that "the soul is a warm spirit; for that by it we have our breath, and by it we are moved" (trans. Yonge).

The successor to Zeno in the Stoic school at Athens was Cleanthes (c. 330–230), who wrote on a variety of topics, such as time, natural philosophy, justice, virtue, and dialectics. Cleanthes agreed with Zeno in an active principle, God, and a passive principle, matter. Diogenes Laertius wrote that "Cleanthes . . . asserts that all souls continue to exist till they are burnt up." He also believed in the cycle of the creation and destruction of the universe. Sphaerus (c. 285–210), the student of Cleanthes, who taught at Alexandria, wrote books, according to Diogenes Laertius, on atoms, the senses, ethics, passions, government, law, divination, philosophy, reason, death, and dialectics (trans. Yonge).

Chryssipus (279–206) was also the student of Cleanthes, who according to Diogenes Laertius wrote 705 books; he likewise believed in the division between the active God and passive matter. He believed in an active creator God that regulates the universe. Chrysippus, also argued "that the most subtle portion of the aether, which is also called by the Stoics the first God, is what is infused in a sensible manner into all the beings which are in the air, and through every animal and every plant, and through the earth itself according to a certain habit; and that it is this which communicates to them." He believed in the principle of the vacuum. He also thought "that the world is an animal, and that it is endued with reason, and life, and intellect." Diogenes Laertius wrote that Chrysippus argued this in the "first volume of his treatise on Providence, . . . and that it is an animal in this sense, as being an essence endued with life, and with sensation." Further, "Chrysippus says that it is only the souls of the wise that endure." He also wrote in "the second book of his Natural Philosophy" about sight, arguing that "we see because of a body of luminous air which extends from the organ of sight to the object in a conical form" (trans. Yonge).

Apollodorus of Athens (180–120), who lived in the second century BCE and was the student of Panaetius (185–110), was the leader of the Athenian Stoic school. Apollodorus authored, according to Diogenes Laertius, *Natural Philosophy.* Diogenes wrote, "And by the term, the universe, according to Apollodorus, is understood both the world itself, and also the whole of the world itself, and of the exterior vacuum taken together." He agreed with Panaetius that the universe "is imperishable." Moreover, according to Diogenes, "Now a body, says Apollodorus in his Natural Philosophy, is extended in a threefold manner; in length, in breadth, in depth; and then it is called a solid body; and the superficies is the limit of the body having length and breadth alone, but not depth" (trans. Yonge).

Archedemus of Tarsus was a Stoic who lived in the second century BCE and established a school in Babylon. According to Diogenes Laertius, Archedemus believed in a creative god, as he argued "in some treatise on the elements." Regarding the elements, Diogenes continued,

> an element is that out of which at first all things which are produced, and into which all things are resolved at last. And the four elements are all equally an essence without any distinctive quality, namely, matter; but fire is the hot, water the moist, air the cold, and earth the dry—though this last quality is also common to the air. The fire is the highest, and that is called aether, in which first of all the sphere was generated in which the fixed stars are set, then that in which the planets revolve; after that the air, then the water; and the sediment as it were of all is the earth, which is placed in the centre of the rest. (Trans. Yonge)

Stoics further believed that atoms move in a void of no substance. A material universe of constant movement does not easily accommodate the supernatural. Indeed, the Stoics generally removed divine forces from their system of thought, though not going so far as the Epicureans. Rather, Stoics could not conceive of a completely random origin of the universe, and assumed the presence of some mind or thought (composed of atoms, of course) that conceived of and directed, if passively, the universe. This eternal force at once material and spiritual was the *Logos*, variously translated as "word," "defining principle" (Aristotle), or "Holy Spirit." The Stoics conceived of a reality in which the Logos creates and destroys in an unending cycle of change. What remains the same are the atoms and void, which constitute over the everlasting cycles an infinite number of combinations. This accounts for all beings: plant, animal, and humans. Each human is one of a kind with a unique soul. Upon the completion of life, the atoms of the unique human disintegrate into the collective, anonymous whole to be resurrected, as it were, in completely different forms of matter, life, and spirit. The Logos conceived of and directs the process but is necessarily distant and unapproachable. Stoicism is then a philosophy based on determinism that is impersonal and ultimately unknowable. The best the human can do is try to understand the mind of the Logos reflected in the laws of the universe; in so doing, those who are rational and scientific-minded may comprehend some of the natural laws upon which the universe is based. Such understanding of natural law became an important ingredient in the pursuit of scientific knowledge in subsequent centuries.

Diogenes Laertius, in *The Lives and Opinions of Eminent Philosophers*, provided an extensive overview of Stoic thought:

They divide natural philosophy into the topics of bodies, and of principles, and of elements, and of Gods, and of boundaries, and of place, and of the vacuum. And they make these divisions according to species; but according to genera they divide them into three topics, that of the world, that of the elements, and the third is that which reasons on causes. The topic about the world, they say, is subdivided into two parts. For that in one point of view, the mathematicians also have a share in it; and according to it it is that they prosecute their investigations into the nature of the fixed stars and the planets; as, for instance, whether the sun is of such a size as he appears to be, and similarly, whether the moon is; and in the same way they investigate the question of spherical motion, and others of the same character. The other point of view is that which is reserved exclusively for natural philosophers, according to which it is that the existence and substance of things are examined, for instance, whether the sun and the stars consist of matter and form, and whether the sun is born or not born, whether it is living or lifeless, corruptible or incorruptible, whether it is regulated by Providence, and other questions of this kind. The topic which examines into causes they say is also divisible into two parts; and with reference to one of its considerations, the investigations of physicians partake of it; according to which it is that they investigate the dominant principle of the soul, and the things which exist in the soul, and seeds, and things of this kind. And its other division is claimed as belonging to them also by the mathematicians, as, for instance, how we see, what is the cause of our appearance being reflected in a mirror, how clouds are collected, how thunder is produced, and the rainbow, and the halo, and comets, and things of that kind. They think that there are two general principles in the universe, the active and the passive. That the passive is matter, an existence without any distinctive quality. That the active is the reason which exists in the passive, that is to say, God. For that he, being eternal, and existing throughout all matter, makes everything. . . . But they say that principles and elements differ from one another. For that the one had no generation or beginning, and will have no end; but that the elements may be destroyed by the operation of fire. Also, that the elements are bodies, but principles have no bodies and no forms, and elements too have forms. . . . They also teach that God is unity, and that he is called Mind, and Fate, and Jupiter, and by many other names besides. And that, as he was in the beginning by himself, he turned into water the whole substance which pervaded the air; and as the seed is contained in the produce, so too, he being the seminal principle of the world, remained behind in moisture, making matter fit to be employed by himself in the production of those things which were to come after; and then, first of all, he made the four elements, fire, water, air, and earth. . . . They also speak of the world in a threefold sense; at one time meaning God himself, whom they call a being of a certain quality, having for his peculiar manifestation universal substance, a being imperishable, and who never had any generation, being the maker of the arrangement and order that we see; and who, after certain periods of time, absorbs all substance in himself, and then re-produces it from himself. And this arrangement of the stars they call the world, and so the third sense is one composed of both the preceding ones. . . . And the heaven is the most remote circumference of the World, in which all the Divine Nature is situated. . . , since mind penetrates into every part of the world, just as the soul pervades us; but it is in a greater degree in some parts, and in a less degree in others. For instance, it penetrates as a habit, as, for instance, into the bones and sinews; and into some it penetrates as the mind does, for instance, into the dominant principle. . . . And on the outside there is diffused around it a boundless vacuum, which is incorporeal. And it is incorporeal inasmuch, as it is capable of being contained by bodies, but is not so. And that there is no such thing as a vacuum in the world, but that it is all closely united and compact; for that this condition is necessarily brought about by the concord and harmony which exist between the heavenly bodies and those of the earth. . . . And they say that these things are all incorporeal, and all alike. Moreover, that time is incorporeal, since it is an interval of the motion of the world. And that of time, the past and the future are both illimitable, but the present is limited. And they assert that the world is perishable, inasmuch as it was produced by reason, and is one of the things which are perceptible by the senses; and whatever has its parts perishable, must also be perishable in the whole. And the parts of the world are perishable, for they change into one another. Therefore, the whole world is perishable. And again, if anything admits of a change for the worse it is perishable; therefore, the world is perishable, for it can be dried up, and it can be covered with water. Now the world was created when its substance was changed from fire to moisture, by the action of the air; and then its denser parts coagulated, and so the earth was made, and the thinner portions were evaporated and became air; and this being rarefied more and more, produced fire. And then, by the combination of all these elements, were produced plants and animals, and other kinds of things. . . . For that which is an animal, is better than that which is not an animal. But nothing is better than the world; therefore the world is an animal. And it is endued with life, as is plain from the fact of our own soul being as it were a fragment broken off from it. . . . The world, then, is finite, and the vacuum infinite. Of the stars, those which are fixed are only moved in connection with the movements of the entire heaven; but the planets move according to their own peculiar and separate motions. And the sun takes an oblique path through the circle of the zodiac, and in the same manner also does the moon, which is of a winding form. . . . Therefore [the sun] is fire, because it performs all the functions of fire. And it is larger than the earth, as is proved by the fact of the whole earth being illuminated by it, and also the whole heaven. Also the fact of the earth throwing a conical shadow, proves that the sun is greater than it; and the sun is seen in every part, because of its magnitude. But the moon is of a more earthy nature than the sun, inasmuch as it is nearer the earth. . . . And all the other stars derive their nourishment from the earth. They also consider that the stars are of a spherical figure, and that the earth is immovable. And that the moon has not a light of her own, but that she borrows it from the sun. . . . They also say that God is an animal immortal, rational, perfect, and intellectual in his happiness, unsusceptible of any kind of evil, having a foreknowledge of the world and of all that is in the world; however, that he has not the figure of a man; and that he is the creator of the universe, and as it were, the Father of all things in common, and that a portion of him pervades everything, which is called by different names, according to its powers. . . . And his nature they define to be, that which keeps the world together, and sometimes that which produces the things upon the earth. And nature is a habit which derives its movements from itself, perfecting and holding together all that arises out of it, according to the principles of production, in certain definite periods, and doing the same as the things from which it is separated. And it has for its object, suitableness and pleasure, as is plain from its having created man. . . . And fate . . . is a connected . . . cause of existing things, or the reason according to which the world is regulated. They also say that divination has a universal existence, since Providence has; and they define it as an act on account of certain results. . . . Now matter is that from which anything whatever is produced. And it is called by a twofold appellation, essence and matter; the one as relating to all things taken together, and the other to things in particular and separate. The one which relates to all things taken together, never becomes either greater or less; but the one relating to things in particular, does become greater or less, as the case may be. . . . It is also subject to change . . . ; for if it were immutable, then the things which have been produced out of it would not have been produced; on which account he also says that it is infinitely divisible. . . . They also say that there are some Demons, who have a sympathy with mankind, being surveyors of all human affairs; and that there are heroes, which are the souls of virtuous men, which have left their bodies. Of all the things which take place in the air they say that winter is the effect of the air above the earth being cooled, on account of the retirement of the sun to a greater distance than before; that spring is a good temperature of the air, according to the sun's approach towards us; that summer is the effect of the air above the earth being warmed by the approach of the sun towards the north; that autumn is caused by the retreat of the sun from us. . . . And the cause of the production of the winds is the sun, which evaporates the clouds. . . . And that comets, and bearded stars, and meteors, are fires which have an existence when the density of the air is borne upwards to the regions of the aether. That ray of light is a kindling of sudden fire, borne through the air with great rapidity, and displaying an appearance of length; that rain proceeds from the clouds, being a transformation of them into water, whenever the moisture which is caught up from the earth or from the sea, by the sun, is not able to be otherwise disposed of; for when it is solidified, it is then called hoar-frost. And hail is a cloud congealed, and subsequently dispersed by the wind. . . . Thunder is the noise made by [clouds] on the occasion of their being rubbed together or broken asunder; and the thunderbolt is a sudden kindling which falls with great violence on the earth, from the clouds being rubbed together or broken asunder, or, as others say, it is a conversion of fiery air violently brought down to the earth. A typhon is a vast thunderbolt, violent and full of wind, or a smoky breath of a cloud broken asunder. . . . They also think that the general arrangement of the world is in this fashion; that the earth is in the middle, occupying the place of the centre; next to which comes the water, of a spherical form; and having the same centre as the earth; so that the earth is in the water; and next to the water comes the air, which has also a spherical form. And that there are five circles in the heaven; of which the first is the arctic circle, which is always visible; the second is the tropical summer circle; the third is the equinoctial circle; the fourth, the winter tropical circle; and the fifth the antarctic, which is not visible. And they are called parallel, because they do not incline to one another; they are drawn however around the same centre. But the zodiac is oblique, cutting the parallel circles. There are also five zones on the earth; the first is the northern one. Placed under the arctic circle, uninhabitable by reason of the cold; the second is temperate; the third is uninhabitable because of the heat, and is called the torrid zone; the fourth is a temperate zone, on the other side of the torrid zone; the fifth is the southern zone, being also uninhabitable by reason of the cold. . . . Also, that the soul is sensible, and that it is a spirit which is born with us; consequently it is a body and continues to exist after death; that nevertheless it is perishable. But that the soul of the universe is imperishable, and that the souls which exist in animals are only parts of that of the universe. . . . And they further teach that there are eight parts of the soul; the five senses, and the generative faculties, and voice, and reason. And we see because of a body of luminous air which extends from the organ of sight to the object in a conical form . . . . And the apex of this cone is close to the eye, and its base is formed by the object which is seen; so that that which is seen is as it were reported to the eye by this continuous cone of air extended towards it like a staff. In the same way, we hear because the air between the speaker and the hearer is struck in a spherical manner; and is then agitated in waves, resembling the circular eddies which one sees in a cistern when a stone is dropped into it. . . . These then are the doctrines on the subject of natural philosophy entertained by them, which it seems sufficient for us to detail, having regard to the due proportions of this book. (Trans. Yonge)

In short, according to Diogenes Laertius, the philosophy of the Stoics encompassed metaphysics, on the nature of being; physics and chemistry, on the nature of matter; biology, on the nature of life; astronomy, on the nature of the heavens; geography, on the nature of the earth; meteorology, on the nature of the weather; and the various characteristics of human senses, such as sight and hearing.

THE LOGOS

Logos is one of the few Greek ideas that continues to be argued among intellectuals and scientists in the third millennium CE. The great debate among Greek scientists during the first millennium BCE, involved the question of what the creative force, the underlying reality, of the universe was. Philosophers by the time of Thales and his successors had abandoned their simplistic belief in the Greek gods—but what was there to replace Zeus, the Fates, Hades, the nymphs, the great mother goddess Cybele, and Athena? What else, indeed, but the *word*—the Logos. Logos as the *word* represented to ancient thinkers the universal transcendence, the creative expression of the universal mind.

Logos is based on a Greek verb infinitive, *legein*. It had a variety of different meanings in the ancient world. The first poets and historians, such as Herodotus, referred to the spirit of truth guiding their work as the Logos. Aristotle referred to the "defining principle," the Logos, which can mean a principle of scientific or philosophic language or a principle of existence, of being. This latter view was adopted by the philosopher Zeno, who argued that Logos is the eternal idea or word, the source of all being. Humans derive from the Logos and return to the Logos. Human beings and society that imitate the harmony of the universal Logos will achieve happiness. Said happiness, however, is limited to the duration of human existence. Zeno tried to bridge the gap between the idealist philosophers, such as Plato and Parmenides, who believed that being was a force of mind, of spirit, and the materialists, such as Heraclitus and Democritus, who believed that being was material rather than spiritual—that being is a substance. Zeno's theory was that Logos is material and spiritual, a substance and an idea.

Zeno founded the philosophy of Stoicism, which along with Epicureanism was the route that the Logos generally took from 300 BCE to 200 CE. Stoic and Epicurean philosophers were attracted to a concept that combines, into one phenomenon, thought and action, body and soul, time and eternity, creation and destruction. Marcus Aurelius (121–180), the most famous Stoic philosopher, interpreted the Logos as the holy spirit, the divine fire. Logos created and destroyed, gave life and then took it. Logos made up all existence: each and everything, each individual. The Stoics put a new twist on the dualism of body and soul, matter and spirit, arguing that the soul was invisible, spiritual, yet also material. All ideas, all truth, is material. One's soul exists of atoms, which will be rearranged upon one's death. The soul lasts only as long as the body does. The Logos is in each human even as it transcends each human.

A conception of Logos as nonmaterial, spiritual, was provided by the philosopher Philo of Alexandria and the theologian Jesus of Nazareth and his followers.

Philo argued that the utterance of God, as represented in the Old Testament book of Genesis, was the same as the Greek and Roman Logos as described especially by Platonists and Stoics. Logos is God, God's thought and utterance, God's creative action, God's intermediation between God and God's creation.

JESUS AS LOGOS

The Epicurean and Stoic logos was an impersonal, anonymous force of creation and destruction. Other thinkers of the ancient world sought a more personal force of creation, a being who encompassed and transcended each individual, who brought about life and then embraced humans at death. John, the disciple of Jesus of Nazareth, conceived of Jesus as the Logos, the timeless word, the eternal transcendence that encompassed body and soul, life and death, truth, God. "In the beginning was the word," wrote John, "the word was with God, and the word was God." According to the Gospel of John, "the word was made flesh, and dwelt among us" as Jesus, the Christ. "All things were made by" the word. He is the being that unites and transcends all being (John, chapter 1, RSV).

The concept of Logos, and the debate surrounding it, is not ancient but very modern, very real. Scientists still wonder about the origins of life, of the universe, of being. Philosophers still wonder whether being is material or spiritual. Christians still argue that Christ was and is the Logos, the "defining principle" of Aristotle, the "word" of Zeno, the "holy spirit" of Marcus Aurelius.

TIME

Time—its nature, meaning, and origins—was one of the more difficult objects of philosophical and scientific inquiry tackled by ancient thinkers. Scientists, theologians, philosophers, and historians have for centuries tried to comprehend time. But because humans are so much a part of, so dependent upon, time, a true understanding has been and continues to be elusive. The questions of today were the questions the ancients asked about time. Is time an artificial measure, a tool by which humans trace their own existence? Or does time have an independent existence, separate from human experience? Is it true that there is geologic time, that the universe carves a temporal path from its beginning to its end? Or is time separate from human awareness?—hence without humans to know and trace it, time is meaningless, nonexistent. Is time therefore an absolute, a constant that can be measured with mechanical devices, a certainty that, as Newton believed, has very little fluctuation, hence allows humans the confidence to base their lives upon it? Clocks, chronometers, and calendars help us to safely trace the passing of years, days, months, which gives us meaning, helps us to know ourselves and our world. Perhaps time is relative, as Einstein believed: since it depends upon the individual observer, time is inconstant and fluctuating, governed by outside forces, significant only insofar as it yields for us a sense of uncertainty, inconstancy, meaninglessness, and anomie. Is one human's time the same as another human's time? Is the time of the twenty-first-century United States the same as the time of first-century

Rome? What does it mean to say that we live in the year 2021? How does it help to know one's age? What does it mean to guide one's life by the clock? What does it mean to regulate institutions, government, the most minute human events, according to the passage of seconds, minutes, and hours?

Two of the most fundamental explanations of the nature of time were developed in the ancient Mediterranean. Time as an objective phenomenon is an external, natural force separate from human thought and experience. Time as a component of nature exists even if humans do not—time has an independent existence that humans struggle to perceive and understand. Time as a subjective phenomenon, on the contrary, is an internal, human experience understood best by those who understand self. A personal sense of time is a unique experience enjoyed by each person individually over the course of life.

The earliest Greek writers and philosophers considered time as a divine phenomenon, often subjective and personal yet at times distant and untouchable. Cronos, the son of Uranos (heaven) and Gaia (earth), was, according to Greek mythology, the god of time. Cronos was an odd one, jealous of his own power, and suspicious of his own children: he ate five of his six children. The oldest of the six, Zeus (sky), was hidden from him, however. Zeus conquered Cronos and sent him to the depths of Tartarus but not before Cronos brought back Zeus's sisters Hestia (hearth), Demeter (grain), and Hera (fertility), and brothers Hades (hell) and Poseidon (sea). Perhaps Cronos's confusion about power and responsibility explains why the poets such as Homer ignored the precise movement and logic of time and cast human events in isolated, sporadic moments seemingly unconnected with the continuum of past/present/future. So contradictory was Homer's sense of time that his poems are models of *anachronism*, which means literally "against time," a confusion of chronology. Homer's *Iliad* and *Odyssey* record the people and events of the distant Mycenaean Age (which ended during the twelfth century BCE) joined to a portrait of society, institutions, and culture at the height of the Greek Dark Ages (1000 BCE), while adding random perspectives of the time (eighth century BCE) in which the poems were written down after centuries of oral telling and retelling.

By the end of the Greek Dark Ages in the eighth century BCE, Greek society had become increasingly sophisticated. The city-state (polis) was the center of trade, organized government, and intellectual speculation. The order and structure of a civilization usually requires a concrete perspective on the passing of events and a record of what has happened. Greek thinkers developed a sense of objective time as external to human affairs. The sixth-century thinker Pherecydes of Syros (according to the Neoplatonist writer Damascius) conceived of time as the begetter of "fire and air and water" (trans. Barnes), the primal elements. But time itself is not begotten, therefore is uncreated, eternal, and infinite. Anaximander of Miletus (610–540), according to Aristotle, conceived of the *infinite,* which is the unending of space and time. Hippolytus, in *Refutation of All Heresies*, explained that Anaximander believed that time is the eternal action of "being, existence, and passing away" (trans. Heath). Aristotle claimed that Pythagoras (570–490) believed "that *time* is the *motion of the whole* (universe)" (trans. Heath).

According to the Greek mathematician Eudemus, Pythagoras argued that all phenomena experienced eternal recurrence. The heavenly spheres of the universe move according to mathematical precision, in constant repetition of the same harmonic circles. If the heavenly spheres constantly return to the beginning, then why not humans as well? Melissus (fifth century BCE), the student of Parmenides, stated emphatically that "whatever existed always existed and always will exist." Socrates, in Plato's *Phaedo*, argued something similar, professing his belief in the recollection of ideas and phenomena previous to this life, which implies a fundamental cyclical state of time. Plato (427–346), in the *Timaeus*, explained the nature of time as due to the Creator's desire to have "a moving image of eternity" that can be gauged "according to number." Aristotle agreed with his teacher in the *Physics*, arguing that time depends upon "motion that admits of numeration," in Bertrand Russell's words (Russell, pp. 144, 206).

At the same time, the Greek awareness of their spectacular accomplishments in the Mediterranean world of the first millennium BCE spawned a complex sense of history to record the passage of time in light of political events. Herodotus (490–430), the first Greek historian of note, relied on the works of his predecessors Hecataeus of Miletus (550–476) and Hellanicus of Lesbos (491–405) to develop a subjective sense of time as inquiries, *historia*. The Greek sense of time in historical writing was chronologically imprecise and episodic because dating systems were imprecise, far from standardized. Olympiads were a favorite chronological marker, as were the term of office of archons, the election of priests and priestesses, and the reigns of kings or, under the Roman Empire, the annual terms of consuls. Such points on the human scale of time were overall unsatisfactory. Historians and biographers were less apt to place the lives of humans in a broader context of linear events and more apt to isolate the object of inquiry in a subjective profile of life, character, and personality.

Hellenistic philosophy and historiography continued the balancing of objective and subjective notions of time. Epicurean and Stoic philosophers argued for recurrence and infinite cycles of life and death, creation and destruction. What was indestructible and infinite—the atom—was also uncreated, its numbers forming and re-forming in countless combinations of matter, alive and dead. Lucretius (98–55), the Roman Epicurean, wrote in *On the Nature of the Universe* that time was objective, hence it existed only insofar as an observer might recognize and trace the ongoing transformation of matter. Time was completely merged with natural movement and otherwise had no existence.

One finds in the Stoic philosopher Marcus Aurelius's *Meditations*, however, a strong sense of the subjective human experience of time. Aurelius could make little sense out of life and death because of his materialistic assumptions about the indestructibility of matter, the material basis of the soul, and the insignificance of human existence. He believed in time as an unending series of moments over which the individual had absolutely no control. Life is random; the self is a product of random atoms in constant movement. Each individual's life is a solitary moment in a vast whole. There is no meaning in the duration of time; the only meaning is in the fleeting moment. Ironically, as Aurelius professed the meaninglessness of life, he desperately sought meaning. As he proclaimed the anonymity of time, he sought to personalize it, make it his own. In this he ultimately failed because of his materialistic fatalism. The self, according to Marcus Aurelius, yearns for significance but cannot find it in a purposeless, godless world.

A one-time Stoic turned Neoplatonist and then Christian, Aurelius Augustine (354–430), reassessed ancient theories of time in light of Hebrew and Christian theology to arrive at a completely original and personal view of time. The societies of the ancient Near East, such as the Sumerians, Egyptians, and Hebrews, lacked a clear chronological dating system and, like the Greeks and Romans, dated events according to pharaohs, kings, prophets, and the passage of generations of father and son. Hebrew historical narratives, such as Chronicles and Kings in the Old Testament, are clear narratives of events with but a simple sense of the linear movement of time. The Hebrews lived their lives between the past and future: the previous greatness of their ancestors and the future coming of the Messiah. This obsessive sense of concern for the past and future found its way into Christian theology during and after the first century CE because of the teachings of Jesus of Nazareth, described in the Gospels and the Acts of the Apostles. Luke the physician, who wrote the third Gospel as well as Acts, used the same episodic, subjective time frame by which to couch his narrative of events that were otherwise chronologically imprecise. Christians made their mark in the history of human comprehension of time by seeing the birth of Jesus as the most significant event in human history, one that required a sense of division: before the incarnation of God to man—that is, before Jesus—and after his crucifixion and resurrection. It was not until the sixth century, and the chronology of Dionysus Exiguus, that the dating system of BC (*ante Christos*) and AD (*anno domini*) came into being. More significant than this new dating system was Aurelius Augustine's sophisticated study of the nature of time.

Augustine personalized time in a way no one had before him. His *Confessions* recounts his own life from birth to conversion and baptism at middle age. There are no dates, and the passage of time is not marked by clear references to the reign of kings and the like, the typical Greek and Roman approach to providing an external, objective human or natural events. Yet the narrative is temporal and linear, and the reader does achieve a clear sense of Augustine's path through his own life, in terms of his own time. Augustine's time is his own; no one else experiences his time—all have their own. All experience time according to the pattern of their own life. Augustine's philosophy appears to be an extreme form of relativism, yet he tempers it and provides a universal standard—in other words, he provides an objective time frame from which to gauge one's own time. This objective standard is unexpected and untraditional in terms of Greco-Roman writing. It is the Incarnation—the birth of Jesus of Nazareth. Augustine's life is understood only by means of reference to this event that occurred centuries before, but which according to Christian theology is made personal and real to each individual Christian. Here in the thought of Augustine is the culmination of the ancient approach toward time: the objective and subjective, the external and personal experiences of time, are combined in the life of the Christian who lives in reference to, and fully in, the life of the Son of Man, Jesus of Nazareth.

FURTHER READING

Augustine, *Confessions*, trans. R. S. Pine-Coffin (Harmondsworth, England: Penguin Books, 1961).

Marcus Aurelius, *Meditations*, trans. Maxwell Staniforth (Harmondsworth, England: Penguin Books, 1964).

Jonathan Barnes, trans., *Early Greek Philosophy* (London: Penguin Books, 1987).

John Boardman, Jasper Griffin, and Oswyn Murray, *The Oxford History of the Classical World* (Oxford: Oxford University Press, 1986).

Ronald W. Clark, *Einstein: The Life and Times* (New York: Avon Books, 1984).

Paul Davies, *God and the New Physics* (New York: Simon and Schuster, 1983).

Diogenes Laertius, *The Lives and Opinions of Eminent Philosophers*, trans. C. D. Yonge (London: Henry Bohn, 1853).

M. I. Finley, *The World of Odysseus* (Harmondsworth, England: Penguin Books, 1972).

Moses Hadas, *Humanism: The Greek Ideal and Its Survival* (New York: Mentor Books, 1972).

Thomas Heath, *Aristarchus of Samos* (New York: Dover Books, 1981).

The New Oxford Annotated Bible with the Apocrypha, Revised Standard Version (Oxford: Oxford University Press, 1977).

R. M. Ogilvie, *Roman Literature and Society* (Harmondsworth, England: Penguin Books, 1980).

Bertrand Russell, *A History of Western Philosophy* (New York: Simon and Schuster, 1945).

Philip Wheelwright, ed. and trans., *Aristotle* (New York: Odyssey Press, 1951).

Donald Wilcox, *The Measure of Time's Past: Pre-Newtonian Chronologies and the Rhetoric of Relative Time* (Chicago: University of Chicago Press, 1987).

The Works of the Emperor Julian, trans. W. C. Wright (Cambridge, MA: Harvard University Press, 1962, 1969).

J. E. Zimmerman, *Dictionary of Classical Mythology* (New York: Harper and Row, 1971).

17

Greek Science in the Roman Empire (750 BCE–500 CE)

Greek philosophy and science conquered the Roman Empire intellectually at the same time as Roman armies overran Greece and other areas dominated by Hellenistic thinking. Epicureanism, Stoicism, Platonism, and Peripateticism came to dominate the philosophy and science of the Romans.

EPICUREANISM

Epicurus (341–271) and many of his followers did not stray from philosophy to science except for when, under the influence of Rome, the philosophy acquired the scientific perspective of a materialistic universe. Epicurean philosophy made its way to Rome by means of Philodemus of Gadara (c. 110–30), a writer and philosopher who influenced many first-century Romans such as Julius Caesar, Horace, Virgil, and Lucretius.

Titus Lucretius Carus (98–55) was the most famous disciple, as well as the greatest exponent, of the teachings of the Greek philosopher Epicurus. Lucretius's poem, *On the Nature of Things*, is an account of the structure of the universe, written in verse according to Epicurean philosophy. *On the Nature of Things* is the work of a polymath who used the Epicurean philosophy of materialism to describe and explain matter, motion, mind, soul, body, society, and natural phenomena. The essence of all things, he argued, is found in constantly moving, indestructible, yet invisible objects: atoms.

Lucretius lived during the first half of the first century BCE. Romans of the second and first centuries actively embraced Greek culture and philosophy in the wake of the Roman political conquest of the Greek world in the decades after 200 BCE. Scientists, philosophers, physicians, and teachers were typically Greeks writing in Greek. Lucretius, like his contemporary Cicero, was an exception: Lucretius was a Latin poet and a Roman who wrote Greek philosophy. His philosophy was formed by the third-century-BCE Greek atomist Epicurus who, along with his forebears

Leucippus and Democritus, advocated a materialist philosophy that shunned the spiritual and metaphysical, branding as superstitions such beliefs.

Leucippus and Democritus in the fourth century, and Epicurus in the third century, were atomists who hypothesized that atoms were the fundamental structures of existence. All life, spiritual as well as material, as well as the world, planets, stars, and sun were made of invisible particles constantly in motion. Gods might or might not exist—in any case, they were irrelevant. Perhaps mind, beauty, the good, the word, and being existed; if so, atoms were the components of what other philosophers considered to be grand metaphysical, spiritual forces. Nothing could be further from the thought of Parmenides, Socrates, and Plato than atoms were. The atomists believed that speculation on the spiritual world could be unending and unsubstantiated by the senses and experience. They demanded that explanations of existence be reduced to common sense and the evidence of the senses. This radical empiricism required a disbelief in anything that could not be seen, heard, smelled, touched, or tasted. Everything else must be the products of overactive imaginations. The atomists allowed themselves one general supposition: there must be a force that causes the movement in things detected by humans. This force must be the basis of cause, of *the nature of things,* hence the fundamental causative force in the universe. But cause, like all of existence, must result from a thing—something material, composed of matter—not an idea.

Lucretius, in support of the empirical arguments of the atomists, used the example of wind. The senses allow humans to detect the presence of the hot winds of summer and the cool breezes of winter. The phenomenon itself is invisible to the eye, but humans can trace it in blowing leaves, the dust being hurled about in a storm, the waves being whipped up, and the billowing sails of a ship at sea. Something that is invisible yet sensed by taste, smell, touch, and hearing clearly exists and clearly has a material basis. But that basis is so small that the sense of sight cannot detect it. It is so small it cannot be made smaller, cannot be cut (Greek, *atomos*). The irreducible form of matter that makes up the essence of all things is the atom.

In *On the Nature of Things*, Lucretius took many pains to hypothesize the qualities and characteristics of the atom. Since nothing cannot exist, the atom must be indestructible, eternal, and infinite. Atoms clearly move and cause motion and thus cannot be packed together without space between. There must be vacuity, a vacuum, in which atoms move. Atoms make up all that is. They are of different sizes, colors, combinations, and patterns. One thing differs from another because of atomic structure. Lucretius and the atomists did not hypothesize the presence of the electron, proton, and neutron. That discovery awaited the atomists of the twentieth century. Rather, Lucretius realized that there must be at least a part that was irreducible—that this atom must be infinite in number and composition because otherwise there must exist a limit on things, and he could not envision such a limit. Limitless atoms meant a limitless universe.

This notion of the limitless universe, which had its origins with Greek philosophers such as Parmenides and was later embraced by the atomists, was a true revolution in thought. The limitless universe frees humans from impositions on their power and free will. It frees creation from limits posed on its extent, quantity, and reproductivity. A limitless universe can host limitless time, limitless motion, limitless types of living beings, limitless stars and planets, limitless universes.

Limitless was the anonymous power and impersonal will exercised by atoms in their capacity as building blocks of the four elements. Like Democritus and, before him, Empedocles (495–435), Lucretius assumed that earth, air, fire, and water made up all things—and making up the four elements were atoms. Without mind or will or fate, atoms determined the universe, even if the particulars were never clear. But one knew that from atoms one was born, and to the atoms one would go upon death. There was no escape.

Lucretius believed humans had a limited power in their ability to grasp the infiniteness in scope, power, and duration of atoms. Once one knew the simple truth, everything could be understood. This was clearly the goal of *On the Nature of Things*, which was unfinished at Lucretius's early and untimely death. He was prepared to take on all topics that had occupied the minds of past philosophers. His atomic theory allowed an explanation of reality, of birth and death, of the mind and soul, of the senses, of emotions, of thought and imagination, of sex and conception. Lucretius conceived of the universe evolving over time from a massive disjointed mixture of atoms at the beginning of time to their subsequent coagulation into distinct forms—the stars, planets, sun, moon, and earth. The earth itself was initially water, but the deep was heated by the fire of the sun and slowly dried to reveal earth. Above and surrounding the earth is the air and, beyond that, ether. The heavenly bodies are buoyed by the air in the ether and rotate around the solid earth, the most substantial, heaviest, and central unmoved body. The sun, a ball of fire, daily crosses from horizon to horizon, only to sink below the earth and then be regenerated again in the morning. For regeneration from the countless atoms is the nature of things. More convincing was Lucretius's argument that the moon reflected the light of the sun and became full or waned according to its position with respect to the sun's rays.

When considering the origins of humans, Lucretius found himself in a dilemma. The Greeks and Romans, like most peoples, assumed a divine role in the astonishing creation of humans. Lucretius, however, was obliged by his atomic theory to conclude that humans and animals simply emerged from the atomic mix. Earth, like a young mother, bore her children, but not consciously, not purposefully. In an attempt to reconcile myth and science, Lucretius hypothesized that the earth also generated monsters of every sort. Anticipating Darwin, Lucretius proposed that such monsters became extinct because they were unable to find sufficient food and adapt to the environment. Lucretius stuck to the old assumption of Homer and Hesiod that the first humans were strong and heroic yet primitive and savage. His portrait of early humans had an interesting evolutionary tint to it. The first humans were primitive, without laws or morals, living according to the naked laws of nature. In time, mutual survival brought humans together into society. Fire tamed their savagery. Civilization emerged—social structure, laws, cities, trade, metallurgy, and organized religion. This latter development Lucretius blamed on human ignorance. The human mind assigns to mysterious phenomena the agency of the divine. Lucretius, in short, composed an anthropology of humans struggling to survive, lifting themselves from the simple struggle of life, developing agriculture to sustain a growing population and to generate wealth, and creating the arts and sciences to make life comfortable, to express joy in song, and to recover the past in verse and prose. It was not Prometheus who gave fire and civilizing arts and sciences to humans. Rather, the generations of humans struggling against time and nature developed a unique, sophisticated society.

Lucretius concluded *On the Nature of Things* with a fascinating scientific potpourri. He began with meteorology. Thunder, he argued, occurred because of the clash of clouds, wind blowing through them, and the rupture of clouds, like the burst of a bladder filled with air. Thunder, of course, followed close on the heels of lightning, which was produced with atoms of fire, inherent in clouds, which were forced to emerge from clouds by the wind. Mountaintops were often cloudy due to the force of the wind driving condensed "cloudlets" up craggy slopes to the summits.

The wind, the agent of causation in Lucretius's meteorology, skims the surfaces of rivers and seas, collecting atoms of moisture, which form eventually in clouds. Moisture-swollen clouds release rain. Sunlight reflects on raindrops to produce the rainbow. The wind is also responsible for earthquakes. The largely hollow earth serves as a pathway for violent winds that, along with raging seas and massive caverns, sometimes topple mountains and collapse caves—an earthquake on the surface of the earth is the result. Winds soar through hollow mountains, collecting atoms of fire and producing volcanic eruptions.

Last, Lucretius considered magnets and epidemics. The strange power of attraction and repulsion exhibited by magnetic stones was due, once again, to the unusual behavior of atoms. Air was found among atoms in all things. When the atoms from a magnet stone emerged outward toward a piece of iron, the air was scattered, creating a vacuum, which forced the iron to move toward the magnet. Lucretius concluded *On the Nature of Things* with a description of plague. Some atoms are destructive to humans and animals. The quality of air varies from place to place; the air of one country that is healthy to its inhabitants is dangerous to the foreigner. Unfortunately, one does not have to travel to a foreign country to be infected by the plague-ridden atoms of dangerous air. Air masses move about the world; some arrive in neighborhoods from distant places carrying their dangerous atomic cargo. Thus epidemics and plagues result.

The concluding pages of Lucretius's *On the Nature of Things* provide a brutally real description of plague—the suffering, torture, horror, fear, and death. Strangely, this was a fit conclusion to the book. The Epicurean philosophy had little room for happiness, salvation, grace, and redemption. Life is slow death. Pain is ubiquitous. The universe is morally and emotionally dead, an anonymous and unfeeling force. One must have low expectations, seek the path of least resistance, avoid pain, and accept the few and simple pleasures life has to offer. Death cannot be avoided, nor must it be feared. Death brings annihilation—a lack of knowing, an end of feeling, which, in Lucretius's world, is not altogether bad.

The Epicureans, like the Stoics, conceived of the soul but a material one based on atoms. Plutarch, who was a Platonist, not a materialist, thought that Homer must have influenced the materialists toward their peculiar belief regarding the soul. Plutarch indicated that Homer used the Greek word for soul, *psyche*, literally "breath," as "the vital spirit, being humid, a breath; when it is extinguished he likens it to smoke. And the word 'spirit' he uses for soul." Likewise, "the Stoics define the soul as a cognate spirit, sensible to exhalations. It has its origin from the humid portions of the body. In this they follow Homer" (revised Goodwin).

STOICISM

The great Stoic philosophers were Zeno (333–262), Cleanthes (330–230), Chryssipus (279–206), Cicero (106–43), Seneca (5–65), Epictetus (55–135), and Marcus Aurelius. The former three founders were Greek. Greeks such as Polybius (208–126) and Panaetius (185–110) brought Stoicism to Rome in the second century BCE. Cicero learned of Stoicism by means of Posidonius of Rhodes.

Posidonius of Rhodes (135–51 BCE), originally from Syria, was a Stoic philosopher and Greek thinker who was very influential in first-century Rome. A student of Panaetius, who first introduced Roman intellectuals to Stoic thought, Posidonius continued the Stoic influence upon the Romans particularly because of his friendship with Marcus Tullius Cicero. He was known to be the teacher of Gnaeus Pompey (106–48). Posidonius, influenced by the geographer Eratosthenes, was a student of geodesy, the study of the earth's shape and size. He used mathematics, specifically geometry, as well as scientific instruments such as the gnomon, a sundial that casts a shadow, to estimate the circumference of the earth to 180,000 stades, a Roman measurement. Like Herodotus, Posidonius was a traveler, in his case journeying throughout the Mediterranean and Europe to acquire scientific data. A polymath, Posidonius penned a periplus based on his travels. Posidonius was also a student of geology (volcanoes) and meteorology, studying the tides. Diogenes Laertius stated that Posidonius wrote in *Meteorology* that "the rainbow is the reflexion of the sun's rays from the moist clouds, or, . . . a manifestation of a section of the sun or moon, in a cloud suffused with dew; being hollow and continuous to the sight; so that it is reflected as in a mirror, under the appearance of a circle." He also argued that "snow is moisture from a congealed cloud." In natural philosophy, he wrote a treatise on *Heavenly Phenomena* and discourses on *Natural Philosophy*, and a treatise on *Meteorological Elements*, in which he asserted that "the world is a thing which is peculiarly of such and such a quality consisting of universal substance, . . . being a system compounded of heaven and earth, and all the creatures which exist in them; or it may be called a system compounded of Gods and men, and of the things created on their account." Posidonius also argued "that the world is one and also finite, having a spherical form. For that such a shape is the most convenient for motion." Diogenes further recorded this: "And the sun is pure fire, as Posidonius asserts in the seventh book of his treatise on the Heavenly Bodies, and it is larger than the earth, as the same author informs us, in the sixteenth book of his Disclosures on Natural Philosophy. Also it is spherical, as he says in another place, being made on the same principle as the world is." Further, "all these fiery bodies, and all the other stars, receive nutriment; the sun from the vast sea, being a sort of intellectual appendage; and the moon from the fresh waters, being mingled with the air, and also near the earth." In mathematics, he argued that "a line is the limit of a superficies, or length without breadth, or something which has nothing but length. A point is the boundary of a line, and is the smallest of all symbols" (trans. Yonge).

Antipater of Tyre was a first-century-BCE Stoic who, like Panaetius and Posidonius, influenced Roman Stoics such as Cicero. "And thus the whole world," as Diogenes Laertius wrote, "being a living thing, endowed with a soul and with reason, has the aether as its dominant principle, as Antipater, of Tyre, says in the eighth book of his treatise on the World." Antipater also argued "that the world is one and also finite, having a spherical form." Further, "body [or matter] is . . . a substance and finite; as Antipater says, in the second book of his treatise on Substance; and Apollodorus, in his Natural Philosophy, agrees with him" (trans. Yonge).

Marcus Tullius Cicero was one of the best known of the Roman Stoic philosophers. He lived during the time of internal discord that destroyed the Roman republic. Cicero translated his uncertainty about the state and corporeal matters into uncertainty about the nature and end of life. He was not quite sure what to make of religion, but he certainly doubted the traditional Roman pantheon of gods and goddesses. Cicero was a theorist about government and morality, though his own politics and ethics were sometimes questionable. He wrote a variety of dialogues and orations, most notably *On the Republic*, *On the Laws*, *Divination*, *On the Nature of the Gods*, and *On the Character of the Orator*.

Greek philosophy had a profound impact on Cicero. Two leaders of the Academy, Philo of Larissa (c. 159–84) and Antiochus of Ascalon (125–68), were among his teachers. The Stoics Posidonius and Diodotus (whose dates are uncertain) were his friends and mentors. Philodemus of Gadara and Phaedrus (c. 140–70) were his Epicurean mentors. Indeed, Cicero's *On the Nature of the Gods* is an imaginary dialogue between an Epicurean, a Stoic, and an Academician. This dialogue provides an interesting analysis of the scientific and religious arguments for belief or disbelief in the divine.

Cicero's *On the Character of the Orator* presents the view that the successful orator must have broad knowledge, including of natural history and philosophy and political science, but especially knowledge of humans. The orator understands the individual and group feelings and expectations of the listeners. Cicero's psychology introduces the idea that empathy is required from any orator who aims to reach and influence an audience.

Cicero's imaginative essay "Scipio's Dream," found in book 6 of *Republic*, presents a cosmic, geographical, and psychological worldview. Cicero, like many geographers of his time, conceived of continents in different hemispheres of the earth, separated by the oceans. For example, at the opposite side of the world from Italy were people whose feet imprinted the ground opposite to the Romans—these were the *antipodes.* Cicero agreed with most ancient geographers that there were five zones on the earth, two polar, two temperate, and a zone of fire at the equator.

Cicero is best remembered for his role in trying to halt the disintegration of the Roman republic at the hands of military tyrants such as Catiline, Caesar, and Octavian. Cicero looked to an earlier time when the Senate ruled benevolently over a people content with public service and dedication to the Senate and people of Rome.

Seneca was an extremely influential statesman and Stoic philosopher of the mid-first century CE. He was primarily a moralist, but he also speculated on nature in his *Natural Questions*, and he explored the human psyche in the *Moral Letters*. A native of Spain, Seneca gained a reputation in Rome for his essays and plays. He served under the emperor Claudius and was made tutor to Claudius's adopted son Nero. Upon Nero's accession in 54 CE, Seneca served as the young emperor's chief adviser. After about ten years, Nero grew suspicious of Seneca, who retired to his villas and his books. He was compelled to commit suicide in 65 CE, after a plot to topple Nero was discovered. Nero thought that Seneca was behind it.

Seneca was one of the leading Stoics of his day. Stoicism had as one of its chief tenets that nature was rational and orderly, and the person who imitated such reason and order would enjoy the best sort of existence. Natural history was therefore important to a Stoic such as Seneca. His *Natural Questions*, addressed to Lucilius, like the *Moral Letters*, reveals this concern. The book examines meteorology, astronomy, and geosciences. Seneca explained meteors according to changes in the atmosphere. He described the origin of winds, the varied precipitation falling from clouds, the weather of Egypt compared to Italy, the causes of earthquakes, and the nature of comets.

The *Moral Letters* reveals Seneca to have deeply pondered the nature of happiness and how to achieve it. His psychology focused on the constant fear of death, the ongoing awareness of the passage of time and mortality, and the resulting anxiety that overwhelmed, depressed, imprisoned, and froze human action. Seneca advocated a form of *apatheia*, the ability to withstand change and contingency, which were the forerunners of death. One must treat death, he preached in letters to his friend Lucilius, as another act of life, as something one simply did, like eating, sleeping, and bathing. The key is to avoid a constant focus on the future. One must rid oneself of thoughts or sensations of fear as well as of hope—anything that represents the utter unknown of the future. Seneca's argument regarding the quandary of hope and fear was very similar to that of St. Augustine centuries later. Time involves anticipation of the future, awareness of the present, and recollection of the past. But if memory brings on obsessions about past wrongs and sins, while anticipation of the future suggests all sorts of awful possibilities, and the present moment goes by too quickly to make sense of it, then humans are destined, it seems, to suffer unhappiness.

Seneca's solution to the dilemma of human suffering was to assume that one was in constant recovery from the illness of anxiety: the only medicine that worked was acceptance of life and self. The latter involves knowing oneself, one's faults and virtues, and accepting one's existence and, especially, one's mortality.

Seneca's psychology of self had a profound influence on many subsequent psychologists and philosophers, such as Epictetus, Augustine, Montaigne (1533–1592), and Erasmus (1466–1536), and through them the modern study of the mind, emotions, and behavior. Plutarch, an observer, not an advocate, of Stoicism, argued, in *The Life and Poetry of Homer*, that Stoics such as Seneca, who believed that "virtue by itself is sufficient for happiness," took their "cue from the Homeric poems in which [Homer] has the wisest and most prudent man

[Odysseus] on account of virtue despising trouble and disregarding pleasure" (revised Goodwin).

PLATONISM

Platonic philosophers and scientists during the Roman Empire by way of the Academy in Athens brought the influence of Plato to Roman thinkers. Examples include Philo of Larissa and Antiochus of Ascalon. Antiochus began a new approach to Platonic philosophy known as Middle Platonism. He was an eclectic philosopher who tried to combine Platonic tenets with those of the Stoics and Peripatetics.

The best-known Platonic philosopher of the Roman Empire was Plutarch of Chaeronea (45–120 CE). Plutarch is well known for his moralistic biographies of great Greek and Roman statespersons and warriors. Plutarch was an eclectic, encyclopedic thinker who understood the philosophies and scientific theories of the past. He was willing to employ a given theory, especially from the masters Plato and Aristotle, whenever the biographical, historical, philosophical, or scientific needs of the moment demanded it. Plutarch was in some respects a compiler and commentator, particularly with regard to the writings on the physical sciences contained in his massive *Moralia.* Plutarch was at his best as a biographer and student of human nature and behavior. As the sixteenth-century French philosopher Montaigne realized, Plutarch was particularly gifted in his ability to understand the human psyche.

Plutarch was a native of Chaeronea in Greece, where centuries before, the Macedonian king Philip II had conquered the Greek city-states (338), forever depriving them of their freedom. Plutarch spent his life in Chaeronea, serving in various political capacities. He was for many years a priest of Apollo at the nearby Oracle of Delphi. He traveled to Rome, where he came to know many of the leading intellectuals, and Egypt, where, like his many philosophic predecessors, he became inspired by the mysteries of the past. He was educated at Athens. His mentor was the Pythagorean and Peripatetic philosopher Ammonius of Athens. Plutarch was a prolific writer, turning out not only his extensive *Parallel Lives* but a series of essays, *Moralia*, as well.

Plutarch's psychology derived from his lifelong study of human behavior and morality. This interest in the human motivation to act and the consequences of doing so led Plutarch to write the first great biographical portraits of individual humans. These are the *Parallel Lives*, fifty in all, somewhat incomplete portraits of men of mixed characters and accomplishments. Plutarch's aim, as he wrote in his *Life of Alexander*, was to draw individual portraits of life. He sought to penetrate the human soul, to go beyond the exterior behavior and appearance and find the essence of what is human.

Plutarch's biography of Alexander is a case in point. Modern scholars trained in historicism and the theories of personality development often point out the limitations of Plutarch's psychology, noting that it did not allow for changes in the stages of life. It is true that Plutarch was subject to the assumptions of his own time. Hence, he could blame Alexander's drunken rage that killed his friend Cleitus on demonic possession. Plutarch thought that Alexander's hot personality reflected his hot bodily constitution. Yet where in literature does one find a more penetrating portrait of insecurity propelling a person forward in constant acts of personal validation? Plutarch's Alexander is molded by his youth. Olympias, his mother, estranged from his father Philip, encouraged Alexander's suspicions and fears directed toward his gruff, one-eyed father and suggested to the toddler that his true father was Zeus. Alexander's confusion was accentuated by constant comparisons to Heracles and Achilles. The hatred of his mother and father for each other instilled a deep void in Alexander's life that Alexander filled by continual activity. Childlike feelings of anger became warlike acts of violence against the enemies of Macedonia. As a man he used power and glory to erase the anxieties and fears of childhood. A contradictory self-image gave way to a godlike, great king of Persia. Uncertainty about self was buried in a series of conquests, victories, and unheard of, foolhardy, accomplishments.

Plutarch claimed to write *bios,* stories of human life, discovering that, as he wrote in his *Life of Timoleon,* by studying past humans he came to identify, even empathize with, these people long dead. As Michel de Montaigne, a student of Plutarch's thought, discovered when he perused Plutarch's writings, Plutarch created a personal dialogue with an individual human from the past; this dialogue involved empathetic give-and-take on the part of Plutarch and, vicariously, his object of inquiry. Plutarch had the insight that the only sure way to uncover the deepest emotions in a person is to form a bridge to connect one's emotions and personal past with another. This empathetic tie remains today one of the fundamental approaches of psychoanalysis and counseling.

The individual struggle to sort reality from image and fantasy occurs on a broader human scale, as Plutarch revealed in his *Lives.* Plutarch lived in an age when philosophy struggled to make sense out of the fictions and myths of the past. The Platonic and Peripatetic schools refused to give credence to the Homeric gods and heroes. Epicureans and Stoics discounted any form of anthropomorphic gods. Stoics elevated the divine to the status of a universal Logos, while Epicureans were practical atheists, joined in their sentiments by the Skeptics. Plutarch used the medium of the human past to sort through the rival claims to truth. When approaching the life of the legendary Theseus, for example, Plutarch was duly suspicious of the numerous myths and stories surrounding the life of the Athenian hero; yet at the same time Plutarch sensed that even in myth there is some truth.

One of the most perplexing anthropomorphic devices that Homer used to express the influence of the divine upon human actions was his anticipation of the modern idea of the human conscience in episodes wherein gods whispered direction to humans. In book 1 of the *Iliad,* for example, Achilles prepares to strike Agamemnon in anger but is checked by the goddess Athena, who arrives just in time to forestall this hasty action by wise words of advice. Plutarch's explanation for tales of divine intervention was pragmatic and reasonable. He assumed that there must be a divine presence in some way or form in human events. Plutarch rejected the Epicurean argument that the gods existed but had absolutely no role in human affairs. Nor did he accept the Stoic ideal of a divine force that anonymously set forth all things but then had an utterly passive role in human affairs. Plutarch assumed the existence of the divine and assumed that the divine had a role in human affairs and a concern for individual humans. Homer's portrayal of gods qua conscience fit perfectly well with Plutarch's assumptions that the divine had a subtle influence on human thought, encouraging and suggesting but never demanding and requiring, putting thoughts into one's head or calling out in a distant inner voice. Here, the god is invisible, anonymous, known only to the especially perceptive, sensitive human, who understands the divine as a message of hope.

Plutarch interpreted dreams in this fashion as well. Dreams reveal the inner workings of the mind, upon which the divine might act, again in an ever-so-subtle way. So when Alexander was trying to decide where he should found his city in Egypt, he had a dream in which a wise old man came to him and suggested the Isle of Pharos. Alexander, who read Homer as a person would a Bible, awoke knowing that Homer was leading him to this place, which turned out to be the superb location upon which Alexandria was built. According to the Greeks, the Muses or some other deity inspired and spoke through the poets. On this occasion, Homer was Alexander's "Muse." Plutarch related a similar idea in his essay, "On Socrates' Personal Deity," in which the idea that each human has a deity, one might say conscience, is defended, based in part on Socrates's own experience as Plato described in several of his dialogues, notably *Apology*.

Plutarch was often bold in his interpretations of human psychology. In "On the Use of Reason by 'Irrational' Animals," he broke from the teachings of Plato and Aristotle to suggest the view that animals were naturally intelligent, even more intelligent than their human counterparts. Human reason is acquired by teaching, study, and practice—by artificial means ("art"), in other words. Animals, however, are instinctually wise, hence are less apt than humans to stray from their inherent proclivity toward wise and moral behavior. The context for this strange argument is an even stranger story involving Odysseus, held against his will on the isle inhabited by Circe, the witch, and forced to debate with one of his men in the form of a pig (which was Circe's doing) over the question, Is it better to be a man or a hog? The hog, Gryllus, turns out to be much the better debater. He points out to the king of Ithaca that humans display courage, honor, or responsibility only to protect their reputations or because they are afraid of customs or laws. They are so compelled by these outside forces that they lack free will, which cannot be said of an animal who willingly goes to battle to find food or to protect its young. Similarly, animals, unlike humans, are not governed by greed for possessions, and they exercise more restraint on their desires than humans do. Only humans are omnivorous, bisexual, and pleasure-seeking. In short, Gryllus shows that animals, being close to nature, are therefore more intelligent than humans, who do all that they can to separate themselves from nature by forming artificial environments.

Some scholars doubt that Plutarch really meant what he argued in *On the Use of Reason by "Irrational" Animals*. But Montaigne believed that Plutarch was serious. Montaigne's *Apology for Raimond Sebond*, for example, relies heavily on Plutarch's apparent Skepticism. Montaigne used Plutarch rather like a breviary for twenty years as he fought to control the obsessions and anxieties of his mind, perplexed with illness and images of death. Montaigne finally arrived at the conclusion to which Plutarch had been guiding his readers all along, in his essay "On

Contentment." Plutarch believed that all humans, even the mentally ill, possessed within themselves the basis of happiness, which derived from acceptance of self, position, responsibility, life, and death. The proclivity of humans is to find meanness, evil, suffering, and trouble in life rather than what is good and enjoyable. Obsessions about past actions destroy present moments and cast doom upon the future.

The future, as well as the past and the present, fascinated Plutarch, because he found in the passing of time, moment to moment, the source of discontent. Humans become obsessed with time, Plutarch wrote in "On Contentment." Obsession with the future forces people to ignore the present because they are always in anticipation of what will come next, and to ignore the past as being irrelevant to upcoming moments. Like Augustine, Plutarch argued that time is a continuum of past, present, future—indistinguishable except by means of memory and anticipation in the present moment and held together by a sense of the unity and efficacy of time and life. Fighting depression caused by memory or dread is as simple as pushing the anxiety into the background of one's life portrait as it is being sketched moment by moment. Obsessions cannot be ignored but merely made unimportant. Anticipating modern psychology, Plutarch advocated for an individual suffering from depression to confront the fears, to purposefully obsess and ritualize, to discover that obsessions brought into the light, out of the dark, become less terrifying.

Plutarch directly tied human happiness with the human and natural past. The latter, he believed, was the product of the benevolence of the Good, which provides all that a human needed to survive, thrive, and be content. Plutarch's approach to the study of nature, then, had a therapeutic value to it. Science had the potential to provide the student of nature with insights necessary to understand the whole and one's place in it, which, in short, is the key to happiness. To this end Plutarch refused to limit his studies to any one discipline but, like his hero Plato, tried to understand a variety of topics of inquiry by the use of reason, observation, and analysis.

Plutarch was particularly interested in the principles of heat and cold and the action of heat and cold on the core elements of the earth. In his *On the Principle of Cold*, he argued that cold derived not from air or water but from earth itself. He wondered whether cold had an independent existence apart from heat: Is there something that exists in nature that is simply *cold* rather than dependent upon relationships with other phenomena? In *Life of Camillus*, Plutarch identified fire, or heat, as the basic principle of causation and motion, inaugurating movement in all things, which otherwise lie cold and still.

Fascinated by meteorology, Plutarch like others before him recognized that the meeting of cold and warm air produces violent thunderstorms. He was fascinated by the interaction of cold and warm water and its effect on sea creatures, as seen in *Causes of Natural Phenomena* (*Quaestiones Naturales*). He pondered the causes of dew and the differences between seawater and fresh water and their impact on plants and animals, including humans, with regard to wellness or sickness. He posited that lightning evaporated sweet, clear water, producing salty scum where it struck the sea. Following Homer, Plutarch considered the cold west wind the swiftest of the winds.

Causes of Natural Phenomena is similar to a Peripatetic exercise in asking learned, open-ended questions about nature: plants, animals, meteorology. Indeed, Plutarch relied heavily on Aristotle (384–322) and Theophrastus (370–286) in deriving the various possible answers to the questions that he asked. In other of his essays Plutarch wrote on a multitude of topics, such as the number of worlds in what appeared to be the fixed system of the heavens; the proclivity of fire to grow more bold in cold weather; the phenomenon of demigods in oracles; the close association of science and religion; whether the moon is inhabited and if so, by what sort of beings; and the possibility of settled lands west of Britain in the Atlantic.

For the historian of science and philosophy, Plutarch's writings are a treasure of references and quotes from the great thinkers who preceded him. Many intellectuals of the Pax Romana have important parts in Plutarch's dialogues. Moreover, in his *Lives,* Plutarch included important vignettes of many noteworthy scientists. In *Life of Marcellus,* for example, Plutarch provided a full description of the activities of Archimedes of Syracuse (287–212). We learn much about Plato (427–346) from the *Life of Dion*, about Aristotle from the *Life of Alexander*, about Pythagorean thought from the *Life of Numa Pompilius*, and about the social sciences from the *Life of Solon*.

PERIPATETICISM

The followers of Aristotle continued to have a marked influence on Roman thought before and after the birth of Christ. These included eclectic thinkers such as Porphyry (234–305), who was the disciple of the Neoplatonist Plotinus (205–270); historians, such as Ammianus Marcellinus (c. 325–395); agriculturalists, such as the early fifth-century writer Palladius; and the first-century-BCE commentator Andronicus of Rhodes. Andronicus's student Boethus of Sidon (c. 75–10), "in his treatise on Nature, calls the substance of God the sphere of the fixed stars." According to Diogenes Laertius, Boethus agreed with Chrysippus, Posidonius, and Zeno "that all things are produced by fate." "But Boethus denies that the world is an animal" (trans. Yonge). Among other Peripatetic philosophers was Adrastus of Aphrodisias, who lived in the second century CE (who also wrote commentaries on Plato), and—one of the most significant within the Roman Empire—Alexander of Aphrodisias, who flourished at the end of the second century CE. Arab commentator Ibn Abi Usaibi'ah (1203–1270) provided an extensive list of Alexander of Aphrodisias's works:

(1) A commentary on Aristotle's "Categories."

(2) A commentary on Aristotle's "De Interpretatione."

(3) A commentary on Aristotle's "Analytica I," reaching only to the part on beautiful forms; we possess two commentaries by Alexander on this book, each better than the other.

(4) A commentary on "Analytica II."

(5) A commentary on the "Topics," of which only parts have been found; they relate to the first, fifth, sixth, seventh, and eighth treatises.

(6) A commentary on "Physics."
(7) A commentary on "De Caelo et Mundi," part of the first treatise.
(8) A commentary on "De Generatione et Corruptione."
(9) A commentary on "Meteorologica."
(10) "The Soul," a treatise.
(11) A treatise on the inversion of premises.
(12) A treatise on providence.
(13) A treatise on the difference between matter and species.
(14) A treatise on refuting those who say that there is nothing without a cause.
(15) A treatise refuting the proposition that sight results from rays that are sent out from the eye.
(16) A treatise on color, and what it is according to the philosophical viewpoint.
(17) A treatise on the sentence, with special reference to Aristotle.
(18) A treatise on melancholy.
(19) A treatise on species and genera.
(20) A treatise refuting the eighth treatise of Galen's "Demonstration."
(21) A treatise refuting Galen's criticism of Aristotle, who said that everything moves owing to a motive force.
(22) A treatise refuting Galen on the subject of the measure of the possible.
(23) A treatise on the members into which bodies are divided.
(24) A treatise on the intellect according to Aristotle.
(25) An epistle on the world, and which of its parts depend for their existence and continuance on control by the other parts.
(26) "On Monotheism."
(27) A treatise on the beginnings of the universe according to Aristotle.
(28) "The Philosophers' View of Monotheism."
(29) A treatise on creation from the void.
(30) A treatise on the nature of regularities.
(31) A treatise explaining the Aristotelian view of the Platonic method of division.
(32) A treatise on the proposition that entities are not bodies.
(33) A treatise on potentiality.
(34) A treatise on contrasts, which according to Aristotle, are the origin of everything.
(35) A treatise on time.
(36) A treatise on matter.
(37) A treatise on Aristotle's proposition that one force can absorb all contrasts.
(38) A treatise on the difference between matter and kind.

(39) A treatise on matter, nonexistence and existence, and the solution of the problem of the ancients, who had been led to negate existence on the strength of Aristotle's "Auscultatio Physica."

(40) A treatise on regularities and generalities, demonstrating that no valid principles can be derived from them.

(41) A treatise refuting those who claim that the members of any species are not necessarily to be found in that species alone but may occur in many other species that are not classified in the hierarchy.

(42) A treatise on excerpts from Aristotle's book which in Latin is called "Theologia," i.e., the dogma of the unity of the Omnipotent God.

(43) A treatise on the thesis that every cause is evident in all things and not only in a single thing.

(44) A treatise on the validity of spiritual forms which contain no matter.

(45) A treatise on the afflictions which beset the entrance to the stomach.

(46) A treatise on species.

(47) A treatise including a chapter of the second treatise of Aristotle's "Book on the Soul."

(48) An epistle on the power emanating from the movement of the noble body toward the bodies which are below the level of generation and corruption. (Trans. Kopf)

Ibn Abi Usaibi'ah's list reveals that Alexander was more than just an Aristotelian commentator, as he wrote on medicine, including commentaries on Galen, physics, theology, time, and human emotions.

FURTHER READING

John Boardman, Jasper Griffin, and Oswyn Murray, *The Oxford History of the Classical World* (Oxford: Oxford University Press, 1986).

Cicero, *Basic Works*, ed. Moses Hadas (New York: Modern Library, 1951).

Clarke, John, trans., *Physical Science in the Time of Nero: Being a Translation of the Quaestiones Naturales of Seneca* (New York: Macmillan, 1910).

Diogenes Laertius, *The Lives and Opinions of Eminent Philosophers*, trans. C. D. Yonge (London: Henry Bohn, 1853).

Will Durant, *The Life of Greece* (New York: Simon and Schuster, 1939).

Michael Grant, *From Alexander to Cleopatra* (New York: History Book Club, 2000).

Ibn Abi Usaibi'ah, *History of Physicians*, trans. L. Kopf (Bethesda, MD: National Library of Medicine, 1971).

Lucretius, *The Nature of the Universe*, trans. R. E. Latham (Harmondsworth, England: Penguin Books, 1951).

R. M. Ogilvie, *Roman Literature and Society* (Harmondsworth, England: Penguin Books, 1980).

Pliny the Elder, *Natural History*, trans. John F. Healy (London: Penguin Books, 1991).

Plutarch, *Essays*, trans. Robin Waterfield (London: Penguin Books, 1992).

Plutarch, *Essays and Miscellanies: The Complete Works*, vol. 3, rev. William W. Goodwin, 2009, https://www.gutenberg.org/files/3052/3052-h/3052-h.htm.

Plutarch, *The Lives of the Noble Grecians and Romans*, trans. John Dryden (New York: Random House, 1992).

Plutarch, *Moralia*, 15 vols., Loeb Classical Library (Cambridge, MA: Harvard University Press, 1968–1976).

Seneca, *Letters from a Stoic*, trans. Robin Campbell (London: Penguin Books, 1969).

PART 7

Greco-Roman Science

18

Medical Science in Ancient Greece (750–100 BCE)

The Greeks of the first millennium BCE had, like their predecessors of the Near East, a superstitious approach to healing. Asclepios/Asclepius was the god of healing worshipped by the Greeks as well as the Romans. Homer's *Iliad* first identifies Asclepios, not as a god but as a human, a contemporary of Heracles, Theseus, and Jason. Chiron the Centaur, the teacher of Achilles the warrior and Jason the sailor and adventurer, imparted his knowledge of medicine and surgery to Asclepios. Asclepios, in turn, taught his sons Machaon and Podalirios the art of healing, and they their sons so that as time passed, the Greeks believed that the progeny of Asclepios existed among them, teaching medicine and healing the sick. Asclepios himself was eventually given the patrimony of the original god of healing Apollo. The story goes that Apollo's mortal lover Coronis, pregnant with his child, was killed by the angry god for loving another. Upon her death the babe Asclepios, taken from her womb, was given to Chiron to raise. As the son of Apollo, he was deified by the Greeks to become the god of medicine. The *Homeric Hymns*, composed at some point around 1000 BCE, includes a hymn to the god Asclepios. The story of the mortal Asclepios becoming the deified patron of medicine was perhaps a Greek borrowing of the Egyptian story of the early hero Imhotep, who was deified to become the Egyptian god of healing and magic.

In Plutarch's study of Homer, *The Life and Poetry of Homer*, he argued that Homer's poems were an inspiration to Greek medical theorists and practitioners:

> The study of symptoms [Homer] goes over in the case of Achilles. For he was a disciple of Charon. He first observed, then, the causes of the pestilence which was attacking the Greeks. For he knew that the causes of common diseases were from Apollo, who seems to be the same as the Sun. For he notices the seasons of the year. If these are intemperate, they become the causes of disease. For, in general, the safety and destruction of men are to be ascribed to Apollo, of women to Artemis, i.e., to the Sun and Moon, making them the casters of arrows by reason of the rays they throw out. So dividing the male and female he makes the male of the warmer temperament. On this account, at any rate, he says Telemachus is of this type, "by

> the guidance of Apollo"; but the daughters of Tyndarus grew up, he says, under the protection of Artemis. (Revised Goodwin)

Homer, wrote Plutarch, related that "the rising of the Dog Star" was the "sign and cause of fever and disease." He related food to disease: "For food, whether dry or humid, is generative of blood. And this nourishes the body; if it is excessive or corrupt, it becomes the cause of disease." Moreover, "the practical part of medicine he carefully distinguishes. In this is the dietetic. First, he knew the periods and cures of diseases. It is evident that he thinks a light diet is healthful. For he pictures his heroes making use of cooked food and so removes extravagant attention about things to eat. And since the stomach needs constant repletion, when cooked food, which has the closest relation to the body, is digested in the heart and veins, and the surfeit is cast forth." Homer wrote as well of the efficacy of medicinal wine, exercise, sleep, being moderate in all things, and fresh air, in which he anticipated the Hippocratic school. Regarding knowledge of surgery, "Machaon heals Menelaus by first removing the javelin; then he examines the wound and presses out the blood, and scatters over it dry medicaments. And it is evident that this is done by him in a technical fashion. Eurypalus, who is wounded in the thigh, first treats it with a sharp knife, then he washes it with clear water; afterward to diminish the pain, he employs an herb. For there are many in existence that heal wounds. He knew this, too, that bitter things are suitable; for to dry up wounds requires exsiccation." And regarding materia medica, "This noted, too, in Homer, that he knows the distinction of drugs. Some are to be used as plasters, others as powders. . . . But some are to be drunk, as where Helen mixes a medicine in a bowl" (revised Goodwin).

ASCLEPIADS

The belief in Asclepios, his tradition, progeny, and healing power lasted for millennia. Priests of Asclepios practiced his healing arts; the sick slept in the god's temples to be visited by Asclepios during the night and be able to greet the morning free of illness. Temples hosting cults of Asclepios were found throughout the Greek world. Epidaurus, on the east coast of the Peloponnesian Peninsula, and Cos, in the Aegean Sea, were centers of the worship of Asclepios. Hygeia (health) and Panacea were daughters of Asclepios. Statues of Asclepios typically show the god holding a staff around which the "asclepian snake" is coiled. To honor the god, priests allowed the small, brown, nonpoisonous snake to inhabit the temples of Asclepios. Snakes symbolized regeneration, the hope of many worshippers of Asclepios. Asclepiads were known to use the mistletoe and bark of the willow tree in cures of physical ailments.

The Asclepiads formed a cult at Cos, where priests and physicians who considered themselves descendants of Asclepios practiced medicine. Cos was an island community and, like its Aegean neighbors Miletus, Halicarnassus, and Chios, a center of Ionian culture during the Archaic and Classical Ages of Greek history. The scientific community at Cos featured a clan or guild of physicians that claimed descent either physically or symbolically from the mythical healer and god Asclepios.

The Asclepiads of Cos developed the theory of the four humors that regulate health in the human body. If any of the four—phlegm, blood, yellow bile, and black bile—became more or less dominant than normal, compared to the others, illness was the result. The healthy body rarely had such imbalance. Diet, rest, exercise, and climate helped to prevent imbalance and to correct imbalance in case of illness.

Ionian medicine at Cos was similar to Ionian science in general during the sixth and fifth centuries BCE. Ionian scientists such as Thales, Anaximander, Anaximenes, and Anaxagoras were interested in explaining natural phenomena by physical or natural causes; they observed the effects of nature to help explain its causes and processes; they tried to understand by thought and observation the essence and patterns of nature. Likewise, the Asclepiads were interested in explaining disease according to cause-and-effect relationships, natural rather supernatural forces.

HIPPOCRATES (460–377 BCE)

Hippocrates is generally recognized as the greatest physician in the history of the ancient world. An Asclepiad who lived on the Aegean island of Cos, Hippocrates broke medicine from its reliance upon superstition and faith healing to a more empirical approach toward the study of the human body and illness. How distinct the "Father of Medicine" was from his fellow Asclepiads is not clear, nor is it certain what of the so-called Hippocratic Corpus (see below) came from him as opposed to others of his association.

Hippocrates and his followers believed that disease resulted from an imbalance of nature, a disruption of the normal balance of the four humors. In the natural conflict between health and disease, the illness would typically reach a crisis point at which nature determined the future course of the illness—slow recovery or a critical phase usually resulting in death. Much of the Hippocratic writings that survive deal with observations of illness, the crisis point, and the resolution. In some sense, Hippocrates was more of a medical observer and scientist than a healer. A good many of his patients died. Nevertheless, it was his attempt to explain rationally what disease was, how it was caused, and what course it took that made Hippocrates the most significant physician of all time.

Hippocratic Corpus

By 400 BCE there existed a body of medical writings that modern scholars consider to be by Hippocrates or from the Hippocratic school of thought. The *Hippocratic Corpus* is a collection of writings generally oriented around fifth-century Cos and thought to reflect Hippocrates's thought and teachings, even if it is not clearly known whether the master wrote any of the treatises and case studies. These varied works propound the theory of the four humors, discuss the impact of the environment on disease and health, prescribe remedies for disease, and describe varied diseases using specific case studies. The Hippocratic approach assumed that the body had a precise balance, that disease reflected an imbalance. The physician sought evidence that one of the four humors was dominant in the system. The job of the physician was primarily to diagnose the illness and to observe the progress of the patient as the disease took its course, reached a crisis, and then culminated in death or recovery. A general theme of the writings is that medicine is an art form not distinct from the repertoire of the philosopher. Art and science frequently complement each other. Moreover, the Hippocratic Corpus emphasizes the ability of the human body to heal itself. This holistic approach to medicine, hence the Hippocratic approach to medicine, has seen a resurgence in the Western world in recent decades.

The Hippocratic Corpus emphasizes the rational, scientific mind in the detection and analysis of illness. The *Prenotions of Cos* and *First Prorrhetic* reveal a scientific concern for prognosis. *Regime in Acute Diseases* discusses possible remedies for illness such as purgatives, emetics, baths, and drinking potions of barley water, wine, honey water, or honey vinegar. The treatise *Ancient Medicine* describes the four humors as forces or powers inherent in the human body and not actual physical substances. Nevertheless, the humors represent the contrasts of hot and cold, moist and dry, aridity, fluidity, clarity, and dimness. The diseases catalogued in the Hippocratic Corpus include malaria and its consequences—fever, ague, and delirium; consumption, asthma, and other pulmonary complaints; and diarrhea, dysentery, and other intestinal problems. Climate and meteorological conditions are addressed in the Hippocratic Corpus as well. The change of seasons, particularly at the solstice and equinox, can have an impact on health. The environment of various regions, Hippocrates argued, helped determine the varied physical, mental, and racial characteristics of different people. Asia, which had fewer climatic challenges, produced people who were less aggressive and warlike than Europeans were, with their harsh and varied climate. The latter were hard men and not the soft, effeminate types of Asia. Herodotus learned from Egyptian priests that the weather had a direct connection to health and illness. Hippocrates agreed, discussing meteorology and human health in his treatise *Airs, Waters, and Places.* The winds, climate, and meteorological conditions of a place—the prevalence of heat or cold, sun, moisture or drought, the season of the year—all had an impact on disease, all determined wellness or illness. For example, a community that received the hot winds of the south tended to have inhabitants with excess phlegm dripping from their sinuses, men who were weak and fat, women who were subject to constant bleeding and diarrhea, and children who suffered from asthma.

Epidemics

A large part of the Hippocratic Corpus is a detailed account of various epidemics at obscure times and places. In case after case, the physician observes disease in generally anonymous people. Typically, the disease worsens to a crisis, and then either a slow recovery or degeneration to death ensues. Observations of the patient included bodily heat and moisture, lethargy, thirst, appetite, the color and smell of the urine and feces, and the consistency of the latter. At one point Hippocrates was very explicit as to what he looked for in making a diagnosis. In *Epidemics*, Hippocrates "framed" his "judgments" by

> learning from the common nature of all disease and the particular nature of the individual, from the disease, the patient, the regimen prescribed and the prescriber—for these make a diagnosis more favorable or less; from the constitution, both as a whole and with respect to the parts, of the weather and of each region; from the custom, mode of life, practices and ages of each patient; from talk, manner, silence, thoughts, sleep or absence of sleep, the nature and time of dreams, pluckings, scratchings, tears; from the exacerbations, stools, urine, sputa, vomit, the antecedents and consequents of each member in the successions of diseases, and the abscessions to a fatal issue or a crisis, sweat, rigor, chill, cough, sneezes, hiccoughs, breathing, belchings, flatulence, silent or noisy, hemorrhages, and hemorrhoids. From these things must we consider what their consequences will be. (Trans. Jones)

Sometimes the patient, after struggling through a painful illness, quickly reached a crisis and turned the corner toward health. Meton, for example, "was seized with fever, and painful heaviness in the loins." For several days he experienced "heaviness in the head; stools thin, bilious, rather red" and "urine rather black; had a rather black cloud floating in it, spread out, which did not settle." On the "fifth day" the patient suffered "violent epistaxis of unmixed blood from the left nostril; sweat; crisis. After the crisis sleeplessness; wandering; urine thin and rather black. His head was bathed; sleep; reason restored. The patient suffered no relapse, but after the crisis bled several times from the nose" (trans. Jones).

At other times the crisis was reached only after months. "The wife of Epicrates," for example, "gave birth to a daughter, and the delivery was in every respect normal. On the second day after the delivery," however, "she was seized with acute fever, pain at the stomach and in the genitals. A pessary relieved these symptoms, but there was pain in the head, neck and loins. No sleep. From the bowels passed scanty stools, bilious, thin and unmixed. Urine thin and blackish. Delirium on the night of the sixth day from the day the fever began." Pain, delirium, fever, and abnormal urine and stools continued for weeks. "About the twenty-first day," "heaviness all over the left side, with pain; slight coughing; urine thick, turbid, reddish, no sediment on standing. In other respects easier; no fever. From the beginning she had pain in the throat; redness uvula drawn back; throughout there persisted an acrid flux, smarting, and salt." The new mother continued in this vein for another sixty days. Finally, on the "eightieth day," there was a "complete crisis with cessation of fever" (trans. Jones).

Just as often, however, Hippocrates reported on diseases that were short and terminal. "Crito in Thasos, while walking about, was seized with a violent pain in the great toe. He took to bed the same day with shivering and nausea; regained a little warmth; at night was delirious." The next day there was "swelling of the whole foot, which was rather red about the ankle, and distended; black blisters; acute fever; mad delirium. Alvine discharges unmixed, bilious and rather frequent. He died on the second day from the commencement" (trans. Jones).

Largely because of the influence of Hippocrates, the physician became less of an itinerant healer and more a stable member of a community. Some city-states employed physicians for healing the poor. Women physicians specializing in women's health became more prevalent as well.

Near contemporaries of Hippocrates had different approaches to medicine. Petron of Aegina believed that disease derived from an imbalance of the wet (cold) and dry (hot) in the human body. The Pythagorean Hippo of Croton likewise considered the amount of moisture to be the key to health. Another Pythagorean philosopher, Philolaus, determined that there were three rather than four humors: bile, blood, and phlegm; he also focused on relative heat and cold and believed in the vegetarian diet. Thrasymachus of Sardis focused on the balance of bile, phlegm, and pus. Menecrates sought a balance of blood, bile, phlegm, and breath (*pneuma*) for good health. He argued that black bile was a source of melancholy. The first Greek philosopher-physician who put aside superstition to approach medicine in realistic, if philosophical, terms was Alcmaeon of Croton in Italy. Alcmaeon was a Pythagorean who developed theories on the nature of disease. Like others of his time, he focused on balance as essential to health. Whether he was a practicing physician is not clear. Empedocles also had a reputation as a physician, though he was more the seer than the healer. An important female practitioner of medicine was Aspasia of Athens. Philistion of Locris argued that the soul was located in the heart, and that when the hot and cold elements were imbalanced, diseases resulted. Theophrastus, Aristotle's successor at the Lyceum, wrote an extensive study of plants, including some of their medicinal properties. Bolos of Mende, a contemporary of Herodotus, wrote *On Sympathies and Antipathies*, in which he classified flora and fauna based on the four elements. He also penned *On Natural Drugs*. Syennesis of Cyprus wrote on the nature of the circulation of blood. Hippocrates's student Polybus wrote the medical treatise, *On the Nature of Man*.

Other contemporaries of Hippocrates with different points of view included Herophilus and his student Erasistratus. Herophilus, a native of Chalcedon, worked as a physician and scientist in Alexandria, supported by the Ptolemaic Dynasty. A student of Praxagoras, who identified the pulse as an important diagnostic tool, Herophilus discovered that the pulse was not inherent to the arteries but instead derived from the pumping of the heart. He tried to describe the pulse according to the theory of music. Herophilus's understanding of human physiology was informed by experiments, dissections, and autopsies (of executed criminals). He explored the nervous system and the circulation of the blood. Herophilus discovered the duodenum, which is a part of the small intestine. He believed in the Hippocratic emphasis on the environment. His student Erasistratus also worked and taught in Alexandria. He was similarly focused on the pulse, the heart, and the brain. He believed the environment was important, but he did not abide by Hippocrates's four humors. Influenced by Aristotle and his followers of the Peripatetic school, Erasistratus tried to find cause-and-effect relationships in medicine. Erasistratus was unique in his ability to connect mind and body. He studied the brain and nervous system, identifying the cerebrum and cerebellum and discovering that the nerves were not airy tunnels. He explored illnesses of the mind (*psyche*) and nervous system, thereby beginning the field of study that today we call psychiatry.

Since blindness was one of the chief physical disabilities in the ancient world, physicians and philosophers were concerned with its causes and how to alleviate suffering. This ophthalmological focus was related to the interest among Greek and Roman philosophers in optics. The atomists, for example, such as Leucippus and Democritus, believed that since everything in existence is corporeal and the human eye must receive some sort of stimulus to account for vision, the stimulus must derive from atoms moving from the object perceived to the eye; this is the *intromission* theory of vision. Plato, on the other hand, believed, as he wrote in *Timaeus*, that vision resulted from a light, derived from the essence of fire, which streams into the eye, and is understood by the intellect:

> And so in the vessel of the head, they [the gods] first of all put a face in which they inserted organs to minister in all things to the providence of the soul, and they appointed this part, which has authority, to be by nature the part which is in front. And of the organs they first contrived the eyes to give light, and the principle according to which they were inserted was as follows: So much of fire as would not burn, but gave a gentle light, they formed into a substance akin to the light of every-day life; and the pure fire which is within us and related thereto they made to flow through the eyes in a stream smooth and dense, compressing the whole eye, and especially the centre part, so that it kept out everything of a coarser nature, and allowed to pass only this pure element. When the light of day surrounds the stream of vision, then like falls upon like, and they coalesce, and one body is formed by natural affinity in the line of vision, wherever the light that falls from within meets with an external object. And the whole stream of vision, being similarly affected in virtue of similarity, diffuses the motions of what it touches or what touches it over the whole body, until they reach the soul, causing that perception which we call sight. (Trans. Jowett)

Plato's student Aristotle disagreed with the teacher in some respects; for example, he realized that the liquid in the eye makes it more receptive to light and color, and it then transports images and colors to the intellect. The Stoic philosophers, both Greek and Roman, believed that the pneuma was the medium from the brain to the eye and the air next to the eye, from which objects are perceived. This *extramission* theory of vision would be prevalent for centuries. Hence Platonic and Stoic thought rejected the Atomist notion that invisible particles bombarded the eyeball and caused the effect of vision. The most sophisticated Roman medical thought of the Pax Romana, represented by Galen, embraced the Stoic view.

Legacy of Hippocrates

Hippocrates's reach forward in time was extensive. The Hippocratic method of seeking physical causes to illness, observing symptoms, formulating prognoses, and prescribing medicine, rest, and healthy living continued to be practiced by physicians for centuries. No less a physician than the Roman Galen practiced the Hippocratic approach to medicine. More than anything, however, Hippocrates influenced later medical practice by means of the Hippocratic Oath, which reads:

> I swear by Apollo Physician, by Asclepius, by Health, by Panacea and by all the gods and goddesses, making them my witnesses, that I will carry out, according to my ability and judgment, this oath and this indenture. To hold my teacher in this art equal to my own parents; to make him partner in my livelihood; when he is in need of money to share mine with him; to consider his family as my own brothers, and to teach them this art, if they want to learn it, without fee or indenture; to impart precept, oral instruction, and all other instruction to my own sons, the sons of my teacher, and to indentured pupils who have taken the physician's oath, but to nobody

> else. I will use treatment to help the sick according to my ability and judgment, but never with a view to injury and wrong-doing. Neither will I administer a poison to anybody when asked to do so, nor will I suggest such a course. Similarly I will not give to a woman a pessary to cause abortion. But I will keep pure and holy both my life and my art. I will not use the knife, not even, verily, on sufferers from stone, but I will give place to such as are craftsmen therein. Into whatsoever houses I enter, I will enter to help the sick, and I will abstain from all intentional wrong-doing and harm, especially from abusing the bodies of man or woman, bond or free. And whatsoever I shall see or hear in the course of my profession, as well as outside my profession in my intercourse with men, if it be what should not be published abroad, I will never divulge, holding such things to be holy secrets. Now if I carry out this oath, and break it not, may I gain for ever reputation among all men for my life and for my art; but if I transgress it and forswear myself, may the opposite befall me. (Trans. Jones)

Hippocrates, along with Galen, became the most important source of information for medieval medicine. The Arab commentator Ibn Abi Usaibi'ah, in *History of Physicians*, quoted from the Hippocratic Corpus at length. In *Law of Medicine*, Hippocrates wrote (as quoted in Ibn Abi Usaibi'ah's *History of Physicians*):

> Medicine is the most noble of all the arts; but lack of understanding on the part of those who practice it has caused people to become deprived of it. No fault has been found with it in any city except for the ignorance of those who claim to be masters but who do not deserve to be mentioned in connection with it. They are like those imaginary figures invoked by storytellers to amuse the people—just as these are figures without shapes, so are these physicians—many in name but very few indeed. . . . Whoever wishes to study the medical art ought to have a good and compliant nature, matched by ardent desire and an exemplary studiousness. Now, most important of all is the natural ability; if it is appropriate, one should embark on study and not become impatient, so that [the knowledge] may impress itself on the mind and reap good fruit, as can be observed in terrestrial plants: the natural ability is like the soil, the benefit of instruction like the seed and the cultivation of study like a grain in fertile earth. When a careful study of the medical art proceeds as mentioned, physicians, when coming to the cities, will be physicians not only in name but indeed. The knowledge of medicine, for those who have acquired it, is an excellent treasure and a splendid store filled with secrets and open joy, whereas lack of it in those pretending to possess it turns it into a wicked art and a rotten store, barren of joy and always accompanied with fear and hastiness. Fear is the mark of inability and hastiness, of a want of competence in the art. (Trans. Kopf)

In the *Disposition of Medicine*, according to Ibn Abi Usaibi'ah, Hippocrates wrote:

> A student of medicine should be of free birth, good nature, young age, medium stature, and well-proportioned limbs. He should also be of good understanding, a pleasant speech, sound judgment when consulted, chaste, brave, and not greedy for money, self-controlled in anger, but not repressing it completely, and not dull. He should sympathize with the patient, feel concern for him and be able to keep secrets; for many patients who tell us of their ailments do not want others to know about them. He should also be able to stand abuse, for some of those affected with pleurisy and fixed notions caused by melancholy assail us with it. We should be capable of tolerating them in such a condition, knowing that it does not come from then [sic]

> but originates in a sickness which is extraneous to their nature. Moreover, the physician should clip his hair to a medium, proportionate length, neither crop it close nor leave it like a topknot and should neither pare down his nails completely, nor allow them to grow above the fingertips. His garments should be white, clean and comfortable. In his gait he should never be hasty, for this is a sign of rashness, nor lagging, for this indicates inertia of the soul. When summoned to a patient he should sit down with legs crossed and ask him about his state of health calmly and sedately, without impatience or agitation. This outward demeanor, this garb and this deportment are to my mind more important than other things. (Trans. Kopf)

"As regards Hippocrates' treatment and healing of diseases," Ibn Abi Usaibi'ah continued, "he always took the utmost care to be of service to his patients and to heal them. It is reported that he was the first to introduce, create and establish the hospital. In a certain part of a garden which he owned, near his house, he prepared a special place for patients and installed servants to look after them. That place he named Xenodochium, i.e., a meeting place for patients. The same meaning attaches to the word bimaristan [in Arabic], which derives from Persian: bimar in Persian means 'sick' and istan 'place'—'the place of the sick.'" Throughout his life Hippocrates had, in that way, no preoccupation other than to study the medical craft and to establish its laws, treat the sick, bring them relief, and rid them of their ailments. Many case histories of his patients are reported in his book known as *Epidemiae*. "Epidemiae" means "arriving diseases" (trans. Kopf).

Ibn Abi Usaibi'ah reported a story about Hippocrates as related by the Islamic physician, Ibn Juljul (943–994):

> I have seen an interesting story about Hippocrates, which I find pleasant to relate in order to point out thereby the virtuousness of that man. It is to the effect that Polemon, the author of the book on physiognomy, asserted in this work that he could deduce the psychological characteristics of a person from his physical constitution. Once the disciples of Hippocrates assembled and said to one another: "Do you know anyone in our time who is superior to this excellent man." They replied: "We know of no one!" Then one of them said: "Let us make him a test case for Polemon, with regard to his pretension as to his knowledge of physiognomies." So they drew a portrait of Hippocrates, took it to Polemon and said to him: "You eminent man, look at this person and judge his psychological traits from his physique." He gazed at it, compared the limbs with one another, and gave his judgment as follows: "This is a man who is fond of debauchery." "You are lying," they said, "This is the portrait of the wise Hippocrates." "It is impossible that my science should be wrong," he replied. "Ask him himself, for this man in not prone to lie." So they went back to Hippocrates and told him the whole story and what Polemon had said to them. Hippocrates said: "Polemon is right. I am fond of debauchery, but I restrain myself." This proves Hippocrates' noble character, his self-control and self-discipline by virtuousness. (Trans. Kopf)

The thirteenth-century Egyptian scientist and compiler Al-Mubashshir ibn Fatik, in *Choice Maxims and Best Sayings*, listed the corpus of Hippocratic works as they had come down to medieval Arab science:

(1) "The Book of Fetuses," in three treatises, the first comprising the theory of the formation of sperm, the second the theory of the formation of the embryo and the third the theory of the formation of the limbs.

(2) "The Book of the Nature of Man." This consists of two treatises, and treats of the characteristics of the body and its components.

(3) "The Book of Airs, Waters and Places," also consisting of three treatises. In the first Hippocrates offers information on the climates of various lands and the local diseases they cause; in the second he treats of various lands and the local diseases they cause; in the second he treats of the characteristics of the drinking waters and of the seasons of the year and the local diseases they cause; in the third he discusses those things which by their very existence cause local diseases.

(4) "The Book of Aphorisms," in seven chapters. In it he sets forth the generalia of medicine so that they may serve as laws for the physician's benefit—directives as to how to understand matters of medical practice with which he may be confronted. This book contains summaries of what he laid down in his other works. This is evident to anyone who examines it attentively, for the aphorisms are arranged as condensations of his "Prognostics," "On Airs, Waters and Places" and "On Acute Diseases," as well as highlights and leading points from his book "Epidemiae," which means "arriving diseases," and finally aphorisms from his book "On Female Complaints" and from all his other works.

(5) "The Book of Prognostics," in three chapters. Here he sets forth the characteristics on the basis of which the physician can recognize the state of every disease in all the three phases—past, present and future. He also explains that if the patient is told of what happened in the past, he will trust his physician and willingly surrender to his treatment, so that he will be able to treat him according to the precepts of the art. If he recognizes the present condition he can then apply the proper medicines and devices; and if he knows the prospective developments of the disease he can prepare himself with all that may check them before they assail the patient so vehemently that no time is left to counteract them appropriately.

(6) "The Book of Acute Diseases," in three chapters: the first contains the theory of the regulation of diet and evacuation in acute diseases; the second deals with treatment by hot packs and bloodletting and with the composition of laxatives and the like; the third contains the theory of the medicinal use of wine, honey-water, oxymel, cold water and hot baths.

(7) "The Book of Female Complaints," in two chapters. At first he sets forth the complaints caused by a retention or excessive flow of the menses, and then he mentions diseases common during and after pregnancy.

(8) "Epidemiae," in seven chapters. In it he explains the epidemic diseases and the regimen and treatment they call for. He says that they fall into two categories—the first involving a single disease and the other a deadly disease called the double-death, affecting both men and animals. [He makes this distinction] in order that the physician may deal with each category by the appropriate means. Here he gives excerpts from his diaries. Galen says: "I and other commentators know that the fourth, fifth and seventh chapters of this book are wrongly attributed to Hippocrates." He further points out

that the first and third chapter treat of epidemic diseases and that the second and sixth contain Hippocrates' diaries—some of which he himself had written, while others were recorded by his son for his own use from what he had heard from his father; these contain some of his best thoughts and explanations. Galen asserts that people abandoned the study of the fourth, fifth and seventh chapters of the book, so that they became extinct.

(9) "The Book of Humors," in three chapters. This book discloses the condition of the humors, namely, their quantity and quality, the prognosis by their inherent symptoms and the method and cautious approach demanded in treating each of them.

(10) "The Book of Diet," in four chapters. From this one can learn about the causes and means by which food accumulates in the body, makes it grow and replaces in it what becomes dissolved.

(11) "The Book of the Physician's Workshop," in three chapters. A description of the necessary medical operations that are carried out with the hands only, such as bandaging, constriction setting of broken bones, suturing, replacement of dislocations, fomentation, hot packs, etc. Galen says: "Hippocrates proceeded from the assumption that this book would be the first of his works to be read, a view followed by all the commentators, among whom I am included." He called it, "The Shop in Which the Physician Sits to Cure the Sick." A better rendering, however, would be "The Book of the Operations Performed in the Physician's Shop."

(12) "The Book of Fractures and Dislocations," in three chapters. It contains everything a physician ought to know on this subject.

Hippocrates has other books—some of which are merely attributed to him. These include:

(13) "On the Complaints of Virgins."

(14) "On the Areas of the Body."

(15) "On the Heart."

(16) "On Dentition."

(17) "On the Eye."

(18) A letter to Thessalus.

(19) "On the Flow of the Blood."

(20) "On Inflations."

(21) "On Burning Fever."

(22) "On Glands."

(23) An epistle to King Demetrius, known as "The Healing Treatise."

(24) "On the Use of Humids."

(25) "The Precepts."

(26) "The Covenant," also known as "The Book of Oaths." Hippocrates wrote this for his pupils and also for those they would treat, in order that they might

be guided by it and not offend against the stipulation he therein imposed on them and in order to dispel by his statements the odium he incurred for transferring this art from hereditary transmission to free dissemination.

(27) "The Law of Medicine."

(28) "The Testament," known as "The Etiquette of Medicine," in which he outlines the rules to be followed by a physician with regard to outward appearance, style of dress and deportment.

(29) "On Luxations."

(30) "On Head Injuries."

(31) "On Meats."

(32) "On the Prognosis of Diseases Caused by a Change of Air."

(33) "On the Natures of Animals."

(34) "The Book of Symptoms"—expounding the symptoms of the twenty-five cases, which presage death.

(35) "On the Indications of Crisis."

(36) "On Superfetation."

(37) "Introduction to Medicine."

(38) "On Children Born in the Seventh Month."

(39) "On Wounds."

(40) "On Weeks."

(41) "On Madness."

(42) "On Pustules."

(43) "On Children Born in the Eighth Month."

(44) "On Bloodletting and Scarification."

(45) "On the Basilic Vein."

(46) "A Little on Plato's Invectives against Eros."

(47) "On Urine."

(48) "On Colors."

(49) A letter to Antigonus on the preservation of health.

(50) "On Diseases."

(51) "On Juveniles."

(52) "On the Divine Disease" [epilepsy]. In the first chapter of his commentary on the "Prognostics" Galen says with regard to this book that Hippocrates refutes here those supposing that God, the Blessed and Exalted, may be the cause of some diseases.

(53) A letter to Caesar, King of the Romans, on the fortunes of men according to the temperament of the year.

(54) "On Medicine by Inspiration"; it is reported that this book contains all that entered his heart and, when applied, the result was in accordance with his anticipations.

(55) An epistle to Artaxerxes the Great, King of Persia, on the occasion of the plague which visited the Persians during his reign.
(56) An epistle to a Group of the People of Abdera, the City of Democritus the savant, in reply to their epistle to him in which they invited him to come and treat Democritus.
(57) "On the Difference of Seasons and the Proper Preparation of Foods."
(58) "On Man's Constitution."
(59) "On the Extraction of Arrowheads."
(60) "On Predictions," Book I.
(61) "On Predictions," Book II. (Trans. Kopf)

The eleventh-century polymath Al-Biruni wrote that "Hippocrates in his book of the seasons, forbids taking hot drugs and bleeding twenty days before and after the rising of [the Dog] star [Sirius], because it is the hottest time of summer and the heat reaches its maximum, and because summer time by itself warms, dissolves, and takes away all moist substances. However, Hippocrates does not forbid those things if you take but very little of them. Afterwards, when autumn comes with its cold and dryness, you cannot be sure whether the natural warmth may not be entirely extinguished" (trans. Sachau).

FURTHER READING

Al-Biruni, *The Chronology of Ancient Nations*, trans. C. Edward Sachau (London: William H. Allen and Co., 1879).

Galen, *On the Natural Faculties*, trans. A. J. Brock, Loeb Classical Library (Cambridge, MA: Harvard University Press, 1916).

Robert Graves, *The Greek Myths*, vol. 1 (Harmondsworth, England: Penguin Books, 1960).

Hesiod, *The Homeric Hymns and Homerica*, trans. Hugh G. Evelyn-White, Loeb Classical Library (Cambridge, MA: Harvard University Press, 1982).

Ibn Abi Usaibi'ah, *History of Physicians*, trans. L. Kopf (Bethesda, MD: National Library of Medicine, 1971).

W. H. S. Jones, trans., *Hippocrates*, vol. 1, Loeb Classical Library (Cambridge, MA: Harvard University Press, 1923).

David C. Lindberg, *Theories of Vision from Al-kindi to Kepler* (Chicago: University of Chicago Press, 1981).

G. E. R. Lloyd, *Methods and Problems in Greek Science: Selected Papers* (Cambridge: Cambridge University Press, 1991).

Vivian Nutton, *Ancient Medicine*, 2nd ed. (New York: Routledge, 2013).

Plato, *Timaeus*, trans. Benjamin Jowett, http://www.gutenberg.org/files/1572/1572-h/1572-h.htm.

Plutarch, *Essays and Miscellanies: The Complete Works*, vol. 3, corrected and revised by William W. Goodwin, https://www.gutenberg.org/files/3052/3052-h/3052-h.htm.

Betty Radice, *Who's Who in the Ancient World* (Harmondsworth, England: Penguin Books, 1973).

19

Medical Science in the Roman Empire (300 BCE–500 CE)

The cult of Asclepius grew during the Hellenistic Age and afterward, as Romans adopted the worship of the healing god (known to the Greeks as Asclepios). The writings of Aelius Aristides during the second century CE give a detailed description of Aristides's personal relationship with Asclepius. In his *Sacred Teachings*, Aristides recorded his dreams in which the god advised him on medical concerns. Aristides believed that he and Asclepius experienced a healing union initiated by the god himself. Aristides had numerous health problems—real or imagined—that drove him to the worship of Asclepius. The god prescribed in dreams ritual acts of healing and atonement. Aristides was to go barefoot in winter, take mud baths, and bathe in cold water during all seasons. Aristides was clearly attempting to expiate guilt by such extraordinary measures. His gastrointestinal and nervous illnesses demanded the healing of his mind and soul more than anything else.

The cult of Asclepius thrived during the principate into the Later Roman Empire. The pagan writer Celsus claimed that "a great multitude of men, both Greeks and barbarians," enjoyed a healing relationship with the god (quoted in Dodds, p. 45). Marcus Aurelius, perhaps encouraged by his physician Galen, benefited from the counsel of Asclepius. During the third century, however, the cult of Asclepius declined. Devotees to Asclepius, such as the pagan Neoplatonist Porphyry, in *Against the Christians*, claimed that Christian opposition, in particular the counter claims of the healing presence of Christ, led to a decline in the worship of Asclepius and increased sickness in the empire. Christians found in their dreams the answers to physical, mental, and emotional complaints. And the expiation of guilt and illness required nothing so harsh as a mud bath in winter!

PRAGMATIC MEDICINE

On the other hand, most Roman physicians embraced the practical, commonsense Roman worldview. Asclepiades (120–40), for example, was a Greek from

Bithynia who embraced the doctrines of the atomists and materialists, such as Democritus and Epicurus. Asclepiades adapted Greek medicine to Roman society and thought. As an Epicurean, Asclepiades was a thoroughgoing materialist in science and medicine, believing that there was nothing spiritual or holistic in healing, nor was he concerned with the impact of environment and the particulars of the individual case. There was no "art" in Asclepiades's healing. Breaking, then, from the Hippocratic school, Asclepiades and other such "Methodists" argued that disease was the result of constriction or dilation of tissues by which fluids flowed with ease or not. The Methodists believed that treatment must abide by opposites, that is, either a reversal of the constriction or of the dilation.

According to the Elder Pliny, Asclepiades corresponded with King Mithridates of Pontus (r. 120–63), advising him on the king's materia medica, especially his obsession with antidotes. Mithridates reputedly took small amounts of poison to inure his body to it and to test various antidotes. Pliny claimed that after Gnaeus Pompey defeated Mithridates in 63 BCE, he discovered a corpus of medical notes taken and kept by the king. Asclepiades advised medicated wine to treat various illnesses and believed that onion juice was an important remedy for digestive complaints and illness of the eyes. He also used "hydrotherapy" in various ways—baths of varying temperature and the plentiful use of cold water applied to the body and, in copious amounts—as a remedy for various ailments.

CELSUS (C. 25 BCE–50 CE)

Aurelius Cornelius Celsus was a Roman physician from Spain who lived and worked at the beginning of the Roman principate. Celsus was a polymath who wrote a variety of works on topics such as agriculture, oratory, warfare, and medical science. His greatest work was *On Medicine,* written in Latin, in eight books. Celsus relied on his predecessors Hippocrates, Erasistratus, and Asclepiades. He was particularly indebted to Greek medical science. His was an encyclopedic work that encompassed all aspects of medicine, ranging from surgery to materia medica, to hearsay and superstition, to concrete discoveries in medicine and practical advice on good diet and health. Book 1, for example, of *On Medicine,* discusses in detail the daily regimen of diet, exercise, purging the system, and so on to arrive at good health. Book 2 discusses in Hippocratic fashion the impact of the environment upon good health. Scholars disagree on whether Celsus was an interested observer or a practicing physician. He did refer to actual patients and seems to have practiced much of what he preached. *On Medicine,* on the other hand, is a descriptive work focusing on observation, symptoms, diagnosis, and prevention rather than direct intervention to cure disease. Celsus's comments on medical practice relied chiefly on the traditional methods of purging the system and bloodletting. At the same time, he was the first physician to describe the means to treat cataracts in the eye. Examining Celsus's achievements, one realizes that Roman medical science in the Augustan Age was sophisticated. Roman physicians and surgeons knew about dentistry, urology, obstetrics, ophthalmology, anesthesia, plastic surgery, and diseases of the ears and throat.

DIOSCORIDES (C. 40–90 CE)

Dioscorides was an expert on pharmacology who either traded in drugs, practiced medicine, or served as a physician to Roman soldiers. He apparently traveled widely during the reigns of Claudius (41–54) and Nero (54–68). He was born in southeastern Anatolia and traveled throughout the eastern portion of the Roman Empire. Many physicians were itinerants; probably Dioscorides was too. During the reign of Nero, Dioscorides published *De Materia Medica,* a collection of herbs and their remedies that he had collected and observed over the course of his many travels. *De Materia Medica* classified drugs according to their respective properties and became a fundamental source during the Byzantine and Roman Empires for knowledge about herbal remedies. Dioscorides discussed how concoctions should be prepared and their efficacy in providing relief for various ailments. *De Materia Medica* "is an assemblage of data about the medicinal properties of slightly more than 1,000 natural products drugs, mostly from the plant kingdom." The book "also included drugs derived from animals and minerals. . . . Dioscorides related 4,740 medicinal actions, such as stimulants, antiseptics, and anti-inflammatory agents" (quoted in Riddle, p. xviii).

The Arab commentator Ibn Abi Usaibi'ah, author of *History of Physicians,* wrote of Dioscorides:

> Dioscorides of Anazarba, the man pure of soul, who brought great benefit to the people, the Arabized, the victorious, who toured the land, adopted the science of simple drugs from deserts, islands and seas; who described them, experimented with them and detailed their uses before being questioned about their effect, so that when experiments succeeded and he found that when put to the test they came out with the same results, he proved that and illustrated them according to the pattern. He was the pioneer of all the simples, which were taken up by those who came after him. From him they learned about all they needed to know about these drugs. Blessed is that good soul which strove ceaselessly for the good of all mankind. (Trans. Kopf)

Ibn Abi Usaibi'ah provided a fascinating example of Dioscorides's observational abilities (or credulity): "Dioscorides reports in his book that when arrows shot at them remain in their bodies the wild goats of Crete eat the plant called diftany, which is a kind of mint; whereupon the arrows fall out without causing the animals any injury" (trans. Kopf).

Ibn Abi Usaibi'ah quoted another commentator, Hunayn Ibn Ishaq, as saying of Dioscorides that he

> was called by his people by the name Azdash Niadish, which means "defecter from us." All this was because he did not mix with his people but kept to the mountains and forests, where he used to abide all the time. He never conformed to his people, never asked for their advice or judgment or accepted their rules. Because of that, his people called him by this name. "Discori" means in Greek trees, "Dos" means God, in other words, "God inspired him with trees and wild flowers." The author says: The fact that Dioscorides was in the habit of traveling from one country to another in search of wild flowers in order to observe them in their places of growth is proved by his own words at the opening of his book, when he addresses the person to whom he dedicated the work: "As for us, we have had since childhood, as you know, a

> boundless thirst to know the basics of treatment, for which purpose we have traveled to many countries. Our life has been that of one who did not stay in one place." (Trans. Kopf)

Ibn Abi Usaibi'ah listed the works of Dioscorides:

> The first treatise deals with a collection of medicines which have a good scent: potions, ointments, resins and big trees. . . . The second speaks of animals, animal juices, seeds, pills, pulse seeds, edible seeds, hot seeds and hot medicines. The third deals with the origin of plants, the cactus, seeds, resins and flowerless weeds. The fourth details drugs, most of which are cold weeds, hot weeds, some of which are laxatives and some emetics. It concludes with a discussion of weeds useful for the treatment of poisons. The fifth speaks of the vine and the kinds of drinks and mineral drugs, Galen says about this book: "I have studied 14 volumes on simple drugs for many communities but I could not find a more complete book than the one by Dioscorides from Anazarba!" (Trans. Kopf)

ELDER PLINY (23–79 CE)

During the first century, the Elder Pliny created a materia medica based on his study on the writings of physicians and other scientists as well as on his own experience. He cited Cato the Elder (234–149) as a source for cabbage being used as a poultice on wounds and Asclepiades for using onion juice as a cure for ailments of the eyes and digestive tract. Pliny believed that vinegar used as a salve and rinse, as well as when gargled and ingested, was efficacious in curing and reducing pain in a variety of diseases and ailments. Vinegar was useful for asthma and other respiratory ailments and was a surprisingly good antidote for snakebite. Indeed, Pliny recorded a variety of antidotes for snakebite and poison. His impact on medicine in the Later Roman Empire was significant. A variety of commentators wrote on Pliny's science, including his medicine. For example, Quintus Serenus Sammonicus was a third- or fourth-century compiler of Pliny's medicine, publishing the *Book of Medicine* on Pliny's work. His work is filled with prescriptions—some realistic, others anecdotal, many based on superstition. Another third-century commentator, Gargilius Martialis, included material that he acquired from Pliny's work.

SORANUS OF EPHESUS (C. 98–138 CE)

Soranus of Ephesus was a physician around the time of the reigns of the emperors Trajan (98–117) and Hadrian (117–138). An advocate of the Methodist school of healing, he practiced gynecology in Alexandria and Rome, perhaps at Ephesus as well. In the treatise, *Gynecology*, he discussed midwifery and obstetrics as well as infant health and pediatrics. He provided practical information for midwives in delivering babies. He broke from some of the teachings of the Hippocratic school regarding childbirth, such as the idea of the wandering womb, arguing that the womb is instead stationary inside the woman. A student of women's health and mental health, Soranus was known for his compassion toward patients. *On Acute and Chronic Diseases* was partially preserved by the Late Roman commentator and physician Caelius Aurelianus, who lived in the fifth century. Galen, clearly influenced by Soranus, preserved some of his teachings, such as in *On Sustaining Causes*.

RUFUS OF EPHESUS (C. 80–150 CE)

A contemporary of Soranus who likewise influenced Galen was Rufus of Ephesus, a student of the work and writings of Hippocrates; his teachings and writings reflected a pragmatic approach to medicine. His work on human anatomy was particularly comprehensive. He had great impact on anatomical nomenclature. Little is known of his life, though he probably lived in Anatolia, perhaps also in Alexandria. He was influenced by Dioscorides, and he himself influenced Galen and Oribasius. The Arab commentator on medicine Ibn Abi Usaibi'ah wrote:

> Rufus the Great . . . was from the city of Ephesus and was unrivaled his time in the medical art. Galen mentioned him in some of his books, holding him in high esteem, and copying from him. These are Rufus' books.

(1) "On Melancholy," in two chapters; it is one of his best works.
(2) The Book of the Forty Chapters.
(3) "Nomenclature of the Parts of Man."
(4) "On the Disease with Which Hydrophobia appears."
(5) "On Jaundice and Afflictions of the Gallbladder."
(6) "On the Diseases that Affect the Joints."
(7) "On Emaciation."
(8) "The Regimen of Him Who Is Not Attended by a Physician," in two chapters.
(9) "On Angina."
(10) "On the Medicine of Hippocrates."
(11) "On the Use of Wine."
(12) "On the Treatment of Women Who Do Not Conceive."
(13) "On the Rules for the Preservation of Health."
(14) "On Epilepsy."
(15) "On Quartan Fever."
(16) "On Pleurisy and Pneumonia."
(17) "Regimen," in two chapters.
(18) "On Sexual Potency," in one chapter.
(19) "On Medicine."
(20) A treatise on the work done in hospitals.
(21) "On Milk."
(22) "On the Agonies of Death."
(23) "On Virgins."
(24) "On Figs."

(25) "On Foul Breath."
(26) "On Vomiting."
(27) "On Deadly Drugs."
(28) "On the Medicines for Kidney and Bladder Diseases."
(29) "On Whether Much Drinking of Water with Meals Is Beneficial."
(30) "On Solid Tumors."
(31) "On Memory."
(32) "On Dionydes' Disease, Which Is Suppuration."
(33) "On Injuries."
(34) "On Regimen in Old Age."
(35) "On the Prescriptions of Physicians."
(36) "On Clysters."
(37) "On Parturition."
(38) "On Luxations."
(39) "On the Treatment of the Retention of the Menses."
(40) "On Chronic Diseases according to the View of Hippocrates."
(41) "On the Degrees of Drugs."
(42) "On What the Physician Should Ask the Patient."
(43) "On Bringing up Children."
(44) "On Vertigo."
(45) "On Urine."
(46) "On the Drug called Licorice."
(47) "On Catarrh of the Lungs."
(48) "On Chronic Diseases of the Liver."
(49) "On the Fact that Men May Become Affected with a Stoppage of Breathing."
(50) "On Purchasing Slaves."
(51) "On the Treatment of an Epileptic Boy."
(52) "On the Regimen of Pregnant Women."
(53) "On Indigestion."
(54) "On the Wild Rue."
(55) A treatise on perspiration.
(56) "On Constipation." (Trans. Kopf)

As Ibn Abi Usaibi'ah noted, one of Rufus's most significant works was *On Melancholy*, in which he sought an explanation for depression, mental malaise, and extreme moods based on the imbalance of the humors in the body. Al-Rāzī (Rhazes; 854–925), who was interested in the science of the mind, analyzed Rufus's comments on melancholy found in his book, *On Black Bile*: "The beginning of melancholy is indicated by fear, anxiety and suspicion aimed at one particular thing whilst no disease is present in any other respect. . . . Examples of their imaginations include the following. Some are afraid of thunder; others ardently desire to discuss death; others want to wash themselves, or hate a particular food, a particular drink or a particular kind of animal; or they imagine that they have swallowed a viper or something similar" (ed. by Pormann).

A contemporary of Rufus, Athenaeus of Attalia, who flourished in the first century CE, was the founder of the Pneumatic school of medicine. Galen provided an account of this school of thought in *On Sustaining Causes*:

> Athenaeus' three types are as follows: the first consists of containing causes, the second of preceding causes, and the third of the matter of antecedent causes: for this is what they call everything external to the body which harms it and produces disease in it. If what is produced in the body belongs to the class of what causes disease, then, while it has not actually brought the disease about, it is called the preceding cause. Alterations are produced in the natural *pneuma* by these causes together with those which are external, and with the body moistened or desiccated, chilled or heated, these are said to be the containing causes of diseases. (Trans. Hankinson)

The key word here, *pneuma*, air or spirit, implies a spiritual rather than material cause of disease.

GALEN (130–200 CE)

Galen was a Greek physician from Pergamum, Asia Minor, who was a master of medical science and physician to the Roman emperor Marcus Aurelius (r. 161–180). Galen's first love was philosophy; the ongoing influence of Aristotle is clear in his writings. His father made sure that Galen had the best education, involving the trivium and quadrivium, with a special focus on mathematics. Galen was an able rhetorician. While still in his teens, he began the study of medicine, traveling to many of the major centers of learning, particularly Alexandria, where he came to know the writings of Asclepiades, Erasistratus, Herophilus, and Hippocrates. Accepting the latter as his literary and scientific mentor, Galen was a physician to gladiators and then had a successful medical practice at Rome. There was much vitriol in Galen's writings and apparently in his lectures and conversation as well. He was unafraid of offending others by challenging their beliefs and advocating his own as the only true way. He was forced to flee Rome for a time but returned to be physician and tutor to Commodus, the son of Marcus Aurelius and future emperor. His literary output was immense. In his works we find a physician and scientist who studied the brain and nervous system, the digestive and circulatory systems, anatomy, skeletal structure, the function of the organs, and materia medica. Galen, like other Roman scientists, experimented on animals as a substitute for humans so he could learn how to perform surgery on human patients. Ibn Abi Usaibi'ah wrote that "Galen, in his book 'On Clysters,' says on the authority of Herodotus that is . . . a bird called ibis which led to the knowledge of enemas. Galen states that this bird is voracious, leaving no flesh whatsoever that it would not eat. As a result, its bowels become constipated because noxious humors collect in them and increase in quantity. When suffering greatly in consequence, the bird repairs to the sea, picks up some water in its beak and introduces it into its anus. Through this water the humors chained in its belly are driven out. Thereafter, the bird reverts to its accustomed diet." Galen and others made advances in anatomical research. Although a rationalist, Galen still sometimes relied on the workings of the divine to understand the intricate ways of the human body. Nature, he believed, was the wise creator of the human body, which was an image of the rational and good rather than a mere organism or machinelike substance. He believed dreams could be used in prognosis and that the moon could affect the human body.

Ibn Abi Usaibi'ah wrote that Galen found healing by visiting the Temple of Asclepius, and that "Galen further repeats several times in his writings that Asclepius' medicine had a divine character. He remarks that his kind of medicine bears to our own the same type of resemblance that our medicine bears to that of quacks." Ibn Abi Usaibi'ah quoted (and edited) Galen at length about his views of Asclepius:

> It is said that Asclepius was the son of Apollo and a descendant of Phlegyas and Coronis, his [Apollo's] spouse, and that he was composed of both mortal and immortal elements. This suggests that he took an interest in human beings because they were of his kind and that he possessed an immortal nature which was superior to the human variety. It was from the effects of medicine that the poet derived this name, i.e., Asclepius. The report that Asclepius descended from Phlegyas originates from the fact that the latter name is derived from the word for flame [plegma], that is to say, [he was] the son of that animal force which generates heat. That he [Asclepius] was the son of Apollo is a record stemming from the belief that a physician should in a certain measure possess the faculty of foretelling the future; for it would be improper if a good physician were not to know certain things that are to happen later on. His portrayal was that of a bearded man adorned with long locks. As to why he was presented bearded and his father [Apollo] beardless, some say that he was sculpted thus because he was so when God made him ascend to his presence; others say the reason was that his profession calls for probity and a venerable age, while still others hold that it was because Asclepius surpassed his father in medical skill. If you look at him attentively you will find that he stands erect as if about to set to work, his garments tucked up, such an appearance indicating that it befits physicians to philosophize at every moment. You also see that those parts of the body which are shameful to expose are covered and those required for exercising the art are bared. He is represented holding in his hand a crooked staff with shoots of the marsh-mallow tree, which is to indicate that a person who avails himself of the benefits of the art of medicine may attain such a great age that he will need a stick to lean upon, or else that a man upon whom God, the Blessed and Exalted, has bestowed certain favors is found deserving of being presented with a staff, just as one was given to Hephaestus, Zeus and Hermes. We find that with that staff Zeus gladdens the people he holds dear and also arouses the sleeping. Asclepius' staff was represented as being of the marsh-mallow tree because it banishes any sickness. As to the crookedness and numerous shoots of the staff, these signify the multiple subdivisions of medicine. We also see that they did not leave the staff without ornamentation and elaboration, but carved it with the image of a long-lived animal, the serpent, which coils around it. This animal is comparable to Asclepius in many respects e. g. it has keen eyesight, is very wakeful and in fact never sleeps; it becomes a man aspiring to the medical art, so that he be not detained from it by sleep and that he be most sagacious, so as to make good progress and inform his

> patients of the present situation and of what may come to pass. For Hippocrates exhorted thus: 'I hold it excellent that the physician exercise foresight'. . . . The serpent on Asclepius' staff has had yet another explanation, namely: that animal, the serpent, lives very long, so that it is said to be eternal; and those availing themselves of medicine may also enjoy a long life. We find, for example, that Democritus and Herodotus, lived to a very great age by following the rules of medicine. In addition, just as that animal sloughs off its skin, which the Greeks call "old age" [geres] so man, by exploiting the art of medicine, may rid himself of "old age," i.e., disease, and gain health. . . . Portraits show Asclepius with a laurel wreath on his head, for this tree banishes grief. For the same reason, since it befits physicians to keep grief at bay, we find Hermes, when called the awe-inspiring, crowned with a similar garland. Since physicians are to keep grief away, Asclepius was supplied with such a wreath; or perhaps people felt that since a wreath served as a symbol of both medicine and divination, the one worn by physicians and diviners should be the same; another possible explanation is that this tree also has the faculty of healing diseases. You find, for example, that if laurel is put in a certain place, poisonous creeping things will flee from there. The same effect is produced by the plant called Conyta [fleabane]. The fruit of the laurel, which is called the grain, when rubbed over the body, produces an effect which is similar to that of castor. When depicting that serpent they put in Asclepius' hand an egg, thereby alluding to the fact that the entire world is in need of medicine, the egg symbolizing the universe. . . . We should also speak of the slaughters performed in the name of Asclepius for the purpose of finding favor with God, the Blessed and Exalted, through him. We may say, however, that there has never been anyone who used a goat for making a sacrifice to God in the name of Asclepius, for the hair of that animal is not as easy to spin as wool, and he who frequently partakes of goat meat is apt to become afflicted with epilepsy. The reason is that the food prepared from this animal produces bad chyme, has a drying effect, is coarse and acrid-smelling and has properties similar to those of atrabilious blood. We find, rather, that people use cocks for making sacrifices to God, the Blessed and Exalted, in the name of Asclepius. They also have a tradition that Socrates, too, offered him such sacrifice. Such was the condition of this divine man, who taught mankind that the medical art is a permanent possession much more valuable than the inventions of Dionysus and Demeter. (Trans. Kopf)

Generally, however, Galen believed in experiment and observation by which to arrive at medical theories and diagnoses. He was a methodical, logical thinker, who, like Aristotle, could overpower his readers and encourage belief by the relentless force and sophistication of his arguments.

Influenced by Hippocrates, the great medical thinker of the fifth century BCE, Galen argued that the human body is made up of four humors: black bile, yellow bile, phlegm, and blood. Health depends on maintaining a balance of the humors by means of proper food, regular exercise, adequate sleep—in short, a temperate lifestyle. An overabundance of black bile, for example, can lead to cancer. Galen accepted the Hippocratic view that one must treat illness with opposites to reestablish the humoral balance. Hence, wetness must act upon dryness; what is hot must be cooled. Blood is warm and moist, which is countered by black bile, which is cold and dry. Phlegm, which is cold and moist, is countered by yellow bile, which is warm and dry. Galen argued that drugs and foods influence the humoral balance in the body: too much food can produce sickness, followed by disease.

Galen's writings on food and nutrition include *On the Powers of Foods,* in which he experimented with the impact of fruit, such as apples and pears, on bowel movements, *On Barley Soup*, and *On the Causes of Disease.* Ibn Abi Usaibi'ah quoted Galen's opinion of the importance of good health:

> Is not all this helpful and pleasurable only by reason of health, which is a blessing truly deserving of that name? The reason is that health is a most perfect blessing which holds no intermediate position between good and bad, nor ranks second among the kinds of good, as was maintained by certain philosophers, namely those known as Peripatetics and Stoics. All the noble virtues for which men strive so eagerly during their whole life depend only upon health. For instance, if a person wishes to show courage, vigor and spirit in combat, defending his people by going to war on their behalf, he must make use of his bodily strength. A man who desires to be just, give everyone his due, do all that should be done, observe the law and be right in his opinions and deeds, will not achieve these aims without being healthy. With regard to salvation, too, it is held that perfection can be attained only through good health, for it is like something born from health. In short, if anyone wishes to say, out of deference to some belief or in order to convince others of an affected lie, that he does not aspire to health, such a statement would be mere talk. If he were to admit the truth he would say: Health, in reality, is the most perfect blessing. (Trans. Kopf)

In the five centuries that separated the work of Hippocrates and the work of Galen, medicine had been heavily influenced by the materialist strain of Epicurean writers who believed that the mechanistic movement of atoms in the body has the greatest impact upon health. Galen was particularly incensed at Asclepiades, who refused to believe that a holistic approach to medicine was important but, rather, that most illness was due to the constriction of tissues and vessels. One of Galen's extant works, *On the Natural Faculties*, is a sustained attack against the teachings of his predecessors except Hippocrates. Galen thought it absurd to assume that the human body has no living, spiritual component—that the body is but atoms in motion in varied combinations. How can one diagnose illness without considering the will and vitality of the human being?

The power of Galen's thought and his many writings influenced medical thought and practice during subsequent centuries. Galen was read by Christians and Muslims alike, who saw in him an Aristotelian thinker who also believed in the "vital spirit" present in each individual. The pagan Galen could therefore be adopted by monotheists looking for a physician who accepted the will and presence of the divine in human existence. Galen also argued that the human body acts like a system in which all parts work toward the functioning of the whole. Each part is intertwined, attracting and expelling like positive and negative forces—not in isolation but in unison.

ROMAN MEDICINE AFTER GALEN

The most noteworthy physicians in the centuries after Galen appear to have lived and practiced in Alexandria. Zeno of Cyprus founded a school in Alexandria at the beginning of the fourth century that produced a number of physicians, such as Oribasius (c. 320–400). Eunapius, the fourth-century author of *Lives of the Sophists*, wrote,

> In those days many famous physicians flourished, among whom was Zeno of Cyprus, who established a celebrated school of medicine. Nay, he survived down to the time of Julian the sophist, and after him there were contemporaries of Prohaeresius who were the successors of Zeno. He had trained himself in oratory as well as in the practice of medicine. Of his famous pupils some took up one or other of these professions, thus dividing among them what they had learned from him; others again took up both; but whether they inherited his medical practice or his oratory, every one of them prospered mightily. (Trans. Wright)

Thanks in part to his relationship with the emperor Julian, Oribasius was a dominant physician of the fourth century. A compiler and commentator, he published a seventy-book *Synopsis*, in which he examined the work of previous Greco-Roman physicians such as Celsus, Asclepiades, Erasistratus, Herophilus, Hippocrates, and Galen. A leading physician at the end of the fourth and beginning of the fifth century was Theodorus Priscianus, who wrote a book of easy remedies, *Euporiston*, and another of household remedies, *Physica*. He was a student of Vindicianus Afer, a late-fourth-century physician and North African statesman praised by St. Augustine. Theodorus's *Euporiston* discussed external, internal, and gynecological diseases. Like Hippocrates, he believed that natural remedies were best for most ailments. Theodorus's works were written for people who had little access to physicians or were inherent do-it-yourselfers. Other writers of the time wrote similar treatises. For example, Marcellus Empiricus wrote a book of medical remedies in the early fifth century CE.

FURTHER READING

M. Bujalkova, "Rufus of Ephesus and His Contribution to the Development of Anatomical Nomenclature," *Acta Medico-Historica Adriatica* 9, no. 1 (2011): 89–100.

A. Cornelius Celsus, *On Medicine*, trans. W. G. Spencer (Cambridge, MA: Harvard University Press, 1938).

E. R. Dodds, *Pagan and Christian in an Age of Anxiety* (New York: W. W. Norton, 1965).

Will Durant, *Caesar and Christ* (New York: Simon and Schuster, 1944).

Eunapius, *Lives of the Philosophers and Sophists*, trans. Wilmer C. Wright, 1921, http://www.tertullian.org/fathers/eunapius_02_text.htm.

Galen, *On the Natural Faculties*, trans. A. J. Brock (Cambridge, MA: Harvard University Press, 1916).

Galen, "On Sustaining Causes," trans. R. J. Hankinson, in *The Cambridge Companion to the Stoics*, ed. Brad Inwood (Cambridge: Cambridge University Press, 2003).

Mark Grant, ed. and trans., *Galen on Food and Diet* (New York: Routledge, 2000).

Ibn Abi Usaibi'ah, *History of Physicians*, trans. L. Kopf (Bethesda, MD: National Library of Medicine, 1971).

Marianna Karamanou, Gregory Tsoucalas, George Creatas, and George Androutsos, "The Effect of Soranus of Ephesus (98–139) on the Work of Midwives," *Women and Birth* 26 (2013): 226–28.

R. M. Ogilvie, *Roman Literature and Society* (Harmondsworth, England: Penguin Books, 1980).

Plinio Prioreschi, *Roman Medicine* (Omaha, NE: Horatius Press, 1998).

John M. Riddle, *Dioscorides on Pharmacy and Medicine* (Austin: University of Texas Press, 1985).

Rufus of Ephesus, *On Melancholy*, ed. Peter E. Pormann (Tuebingen, Germany: Mohr Siebeck Publishing Company, 2008).

20

Mathematical Science in Ancient Greece and Rome (750 BCE–200 CE)

Mathematics, the systematic analysis of numerical symbols that represent reality, which began in third- and second-millennia-BCE Mesopotamia, was brought to fruition by first millennium BCE Greek thinkers. The great mathematicians of ancient Greece—Pythagoras, Aristarchus, Archimedes, Eudoxus, Euclid, Apollonius, Eudemus, Geminus, Nichomachus—made advances in arithmetic, geometry, algebra, trigonometry, astronomy, and engineering. The accomplishments of Greek mathematicians were not surpassed until the Renaissance and Enlightenment in Europe, and the work of Copernicus, Leibnitz, Pascal, and Newton.

Mathematics as an activity of abstract science began with the early human fascination for number—not only as the amount of items but as an abstract concept as well. The ancients conceived of *one*, a singularity, the whole of all things; *two*, a dualism, the counterpart of one thing with another, such as good and evil; *three*, the additional substance or idea that includes what is not either the one or the other but is in between, a third. Ancient mathematicians also arrived at the concept of place value (of tens, hundreds, thousands, etc.) in denoting numbers. As one historian of mathematics has written,

> The mathematics of Egypt, Babylon, China, Greece, and India was developing from within. Algebra is not, as often assumed, an artificial effort of human ingenuity, but rather the natural expression of man's interest in the numerical side of the universe of thought. Tables of square and cubic numbers in Babylon; geometric progressions, involving the idea of powers, together with linear and quadratic equations in Egypt; the so-called Pythagorean theorem in India, and possibly in China, before the time of Pythagoras; and the geometrical solution of quadratic equations even before Euclid in Greece, are not isolated facts of the history of mathematics. While they do indeed mark stages in the development of pure mathematics, this is only a small part of their significance. More vital is the implication that the algebraical side of mathematics has an intrinsic interest for the human mind not conditioned upon time or place, but dependent simply upon the development of the reasoning faculty. We may say that the study of powers of numbers, and the related study of quadratic equations, were an evolution out of a natural interest in numbers. (Trans. Karpinski)

The Mesopotamians and Egyptians had some influence on the Greeks, though the accounts are mostly anecdotal. The first great Greek scientist, Thales of Miletus (625–545), reputedly studied with Egyptian savants. But Thales and his successors diverged from the Egyptians and Mesopotamians, providing a highly abstract foundation for Greek mathematics. The polymath Plutarch (46–120) wrote at the beginning of the second century CE that "all the so-called mathematical sciences are like smooth flat mirrors in which traces and images of intelligible truth are reflected" (trans. Barnes). Plutarch's comment reflects the impact of the Ionian Greek philosophers of the seventh and sixth centuries, Pythagoras (570–490) and his followers of the sixth and fifth centuries, the Athenian Plato (427–347), and the Alexandrian mathematician Euclid (325–265) on Greek mathematical thought.

The late Roman commentator Proclus (412–485) said of Thales that he was a geometrician who, regarding triangles, was the first to comprehend that the angles at the base of an isosceles triangle are equal, and that two "triangles with one equal side and two equal angles are equal" (trans. Barnes). Thales knew that the vertex of intersecting lines have equal angles. Proclus said of Thales that he proved that the diameter of a circle divides it into two equal halves. Thales used his mathematical mind in astronomy, to study eclipses and navigation, and in engineering, to divert a river from its course. But Thales was primarily a philosopher who sought the nature of things, the ultimate causes of the phenomena of life that he found around him.

Thales imparted his philosophical and mathematical interests to his student Anaximander (610–540). According to the philosopher and scientist Aristotle (384–322), Anaximander believed that the *infinite* was the basic principle of all things. Time, that is temporal movement, is infinite because it has no limit to its duration. Magnitude, the extent of a thing, is infinite in that it can be divided over and over and never reach an indivisible number. Being, that which *is*, is infinite by the very fact of never changing or moving. Multitude, the numbering of things, is infinite, for the summation and multiplication of things will never end. For Anaximander, number, the symbol of reality, is made up of multitude, being, magnitude, and time. His impact on subsequent thinkers was significant. Leucippus (fl. early fifth century), for example, conceived of his materialistic philosophy of atoms as the basis of all reality upon the principle of infinity; atoms exist at all times and all places, always have, and always will.

Pythagoras was the intellectual heir to Anaximander and Thales. For the Pythagoreans, number is found in all aspects of life; it is the harmonic basis of reality. Number is reflected in the changing of days, the passing from day to night, the sounds of nature, and the motions of the planets and stars. The changes in observed phenomena, the passages of time and movement through life, are not real but shadows of what is real, which can be expressed mathematically. Life is full of similarity (even numbers) and contrasts (odd numbers), individual beings or entities, such as a person (whole numbers, integers); and parts of things, such as petals on a stem (fractions).

Pythagoras launched a school of thought that influenced mathematics in Greco-Roman culture for centuries. Pythagorean mathematicians included his daughter,

Theono of Thurii (fl. sixth century BCE), who, after Pythagoras's death, continued his work on astronomy and numbers. Another of Pythagoras's disciples, Philolaus of Croton (470–385), used the theory of number in the science of astronomy. Philolaus argued that there must be an additional planet in the solar system, a "counter-earth," because it was impossible to have only nine heavenly bodies (sun, moon, six planets, fixed stars); there must be ten because of the tetractys, the numbers 1, 2, 3, and 4, ordered in separate rows: "Ten is the perfect number: it contains the important musical rations, and it can be arranged to form a perfect triangle." In *On the World*, Philolaus wrote that "all things that are known have a number," for otherwise "nothing could be thought of or known." Number, he wrote, "has two proper forms, odd and even (and a third, even-odd, mixed from both); and of each form there are many shapes which each thing in itself signifies" (trans. Barnes). The Pythagoreans had a reputation for being mystics; the fifth-century commentator Proclus, who was sympathetic to Pythagorean ideas, wrote that Philolaus believed that angles of geometric shapes were dear to gods—in particular Cronos (time), Hades (death), Ares (war), and Dionysus (wine). Proclus also claimed that Pythagoreans discovered the theorem "that every triangle has internal angles equal to two right angles" (trans. Barnes). Plato wrote that the Pythagorean Hippasus (fl. fifth century BCE) discovered the mathematical/musical ratios of 4:3, 3:2, 2:1. Indeed, the Pythagoreans believed in a harmony of the universe, the "music of the spheres," reflected in perfect, unchanging numbers. As the Pythagorean Nicomachus of Gerasa (60–120) believed, "Lines, numbers, harmonic intervals, and the revolutions of circles bear aid to the learning of the doctrines of wisdom" (trans. D'Ooge). He further claimed, in *Introduction to Arithmetic*, that even and odd numbers always resolve into a harmony of the whole, that numbers reveal the mind and motives of God.

The impact of Pythagoras and his followers was extensive, encompassing a thousand years of antiquity and stretching into medieval and Renaissance Europe as well. Socrates (469–399), his student Plato, Plato's student Aristotle, Plato's Academy, and Aristotle's Lyceum were beholden to the Pythagoreans for the repeated focus on the metaphysical nature of mathematics. Although mathematics was clearly a practical science for business, navigation, warfare, city planning, construction, engineering, and agriculture, nevertheless Socrates (in Plato's *Republic*) concluded that the ultimate application of mathematics is "knowledge of the eternal, and not of aught perishing and transient" (trans. Jowett). To gain this knowledge, one must study arithmetic, geometry, astronomy, and harmonics. In geometry, for example, Plato conceived that the symmetry of the universe is based in five solids formed from equilateral triangles and squares: the tetrahedron (four equilateral triangles), octahedron (faces of eight equilateral triangles forming two opposing pyramidal shapes), icosahedron (faces of twenty triangles formed from combination of five equilateral triangles), hexahedron (the cube), and dodecahedron (formed of twelve pentagons). Aristotle, more the scientist than Plato, was skeptical of some of the Pythagorean claims and, although he accepted the reality of the immaterial and transcendent truths, questioned whether mathematics could reveal the nature of these truths. Aristotle more than any other ancient scientist established the analytical and methodical foundations for mathematical investigations. It was to Aristotle, for example, that Euclid turned to formulate the basic principles of the definitions and elements upon which his book, *The Elements*, was based. Aristotle argued that there were a few truths that were not provable and must be assumed in mathematics. In arithmetic, one must assume the unit before proceeding to an understanding of even and odd numbers, which require proof. In geometry, one must assume the point and the line before proceeding to an understanding of geometric figures, which require proof. Aristotle discussed the basic axioms and definitions of mathematics in *Metaphysics* and *Posterior Analytics*.

Euclid, relying upon Aristotle, wrote one of the seminal works, *The Elements*, in the history of mathematics. Euclid was the first great thinker of the Hellenistic Age, when Greek culture spread to North Africa and western and central Asia in the wake of the conquests of Alexander the Great (356–323). Euclid lived in Alexandria, Egypt. According to the Late Roman commentator Proclus, Euclid had been a student at Plato's Academy, which in subsequent centuries required incoming students to know, above all else, Euclid's *Elements*. Euclid was a prolific writer, penning books on geometry, mathematical propositions, astronomy, music, optics, and *Conics*, to which his disciple Apollonius of Perga (262–190) added. Besides Apollonius, Euclid influenced Archimedes (287–212) and a host of subsequent ancient, medieval, and modern mathematicians and scientists. *The Elements*, Euclid's most famous work, provides a series of a priori truths, elements that cannot be doubted, from which further axioms, postulates, and definitions are derived. *The Elements* begins with the statement: "1. A **point** is that which has no part", meaning, following Aristotle, that a point is a position or place that is indivisible. Euclid's next definition is: "2. A **line** is breadthless length", which refers to the first dimension of reality, not encompassing the second (breadth) or the third (depth). Aristotle summed up these first two definitions as "a line by its motion produces a surface, and a point by its motion a line" (trans. Heath), the point itself being motionless.

HELLENISTIC MATHEMATICS

Pythagorean, Platonic, Aristotelian, and Euclidean mathematics set the foundation for subsequent developments during the Hellenistic Age (331–30). Greeks journeying to Iraq in the wake of Alexander's conquests made contact with Babylonian mathematicians. They adopted the use of zero as a place notation from Babylonians, although the Greeks were the first to use the sign "0," the Babylonians having only a cuneiform symbol. Hellenistic mathematicians discovered irrational numbers and geometric algebra. Greeks combined Babylonian quadratic equations with irrational numbers to form, in Otto Neugebauer's words, a geometry of algebra.

An example of a Hellenistic mathematician was Archimedes of Syracuse, a disciple of Euclid, who spent time in Alexandria studying with Euclideans. Archimedes published such works as *On the Sphere and Cylinder*, *Measurement of the Circle*, *On Conoids and Spheroids*, *On the Equilibrium of Planes*, *On Floating Bodies*, and *The Method Treating of Mechanical Problems*. He was one of the first mathematicians to employ his skill in the fashioning of technology and inventions. Archimedes used mechanics to invent engines of war, such as cranes and catapults, which helped the Syracusans fight off attackers. He is famous for using hydrostatics to invent a screw used for irrigation and water displacement. He discovered, according to the later commentator Pappus of Alexandria (fl. third century CE), thirteen solids. Archimedes approximated π (pi). He also used mathematics in astronomy, as did many other Hellenistic mathematicians, such as Eratosthenes of Alexandria (276–195), one of Archimedes's friends and correspondents, and Aristarchus of Samos (310–230), who was the first to hypothesize the heliocentric (sun-centered) universe; Aristarchus was a theorist more than an empiricist. His treatise, "On the Sizes and Distances of the Sun and Moon," provided the theoretical, mathematical basis for his heliocentric argument.

Hellenistic mathematicians made some of the great discoveries in mathematics. Apollonius of Perga, whose work on epicycles anticipated the theories of epicycles and eccentrics used by Ptolemy in the *Almagest*, wrote one of the most definitive works on geometry during the ancient world. His *Conics* was a masterpiece of examining the intersection of the cone and plane. He coined the terms for many geometrical shapes, such as the hyperbola, ellipse, and parabola. He studied optics, reputedly invented a "burning-glass," and he invented a sundial and experimented with hydraulics. Other important Hellenistic mathematicians included Posidonius of Rhodes (135–50), an astronomer and student of geodesy who made, for his time, the most accurate estimate of the earth's circumference; and Hipparchus of Rhodes (190–120), one of the inventors of trigonometry, who was an astronomer and student of geodesy who provided a very accurate estimation of the length of the year. Eudemus of Rhodes (350–290), a student of Aristotle who edited the teacher's works, had an Aristotelian school at Rhodes and wrote histories of mathematics and astronomy; through him the astronomical ideas of early Ionian thinkers are known. Another Rhodian, Geminus, living in the first century BCE, was heavily influenced by Posidonius's work at Rhodes and wrote a history of mathematics and treatises on geometry and arithmetic. Hypsicles of Alexandria (190–120), who was a mathematician and astronomer, wrote in the wake of Euclid a book *On Polyhedrons*, examining solids in spheres. He reputedly defined polygonal numbers thus: "If there are as many numbers as we please beginning from 1 and increasing by the same common difference, then, when the common difference is 1, the sum of all the numbers is a triangular number; when 2, a square; when 3, a pentagonal number [and so on]. And the number of angles is called after the number which exceeds the common difference by 2, and the side after the number of terms including 1" (quoted in O'Connor and Robertson, "Hypsicles"). Menelaus (70–130), also of Alexandria, used trigonometry and geometry in astronomy and to make detailed studies of the sphere. Theodosius of Tripolis (160–90), an astrologer and astronomer, authored a work on spherical geometry, *Sphaerica*, composed in three parts. He wrote, "If in a sphere a great circle cut another circle at right angles, it bisects it and passes through its poles." And "if in a sphere a great circle touch another (second) circle and cut a third which is parallel to the second and lies between it and the centre, and if the pole of the great circle lies between the two parallel circles, then any great circles which touch the third will be inclined to the (first) great circle" (quoted in Gow, p. 288). According to the historian of mathematics Reviel Netz, Greek mathematics used very esoteric yet not symbolic language to express mathematical concepts.

Mathematics during the subsequent 1,500 years was in many respects a footnote to ancient Greek mathematics. One of the most important mathematicians of the Roman Empire, Claudius Ptolemy (100–170), worked in the shadow of Eratosthenes and Hipparchus. Ptolemy, in *Almagest*, defined mathematics as "the kind of science which shows up quality with respect to forms and local motions, seeking figure, number, and magnitude, and also place, time, and similar things." Mathematics falls between physics and theology, he argued, "not only because it can be conceived both through the sense and without the senses, but also because it is an accident in absolutely all beings both mortal and immortal, changing with those things that ever change, according to their inseparable form, and preserving unchangeable the changelessness of form in things eternal and of an ethereal nature." Ptolemy believed that mathematics was on a surer foundation for providing knowledge than either physics or theology was, because arithmetic and geometry involve "indisputable procedures" (trans. Taliaferro). Mathematics assists in the understanding of theology because mathematics identifies those unchanging, eternal, truths upon which theology is based. Likewise, mathematics assists in understanding movement and change, the basis for physics. Mathematics was, for Ptolemy, the fundamental science.

The fourth-century-CE geometer Pappus of Alexandria (290–350) wrote *Mathematical Collection*, a commentary on ancient mathematicians such as Ptolemy as well as Pythagoras, Archimedes, Euclid, and Apollonius of Perga. His *Collection* included discussions on plane geometry, solid geometry, squaring the circle, the Pythagorean Theorem, volume, astronomy, conics (commenting on Apollonius of Perga), analytical geometry, and mechanics. Pappus preserved a discussion of the thirteen solids discovered by Archimedes. Proclus of Athens, also a geometrician, wrote *Commentary on Euclid*, and commentaries on Aristotle and ancient astronomers. One of the first female mathematicians was Hypatia of Alexander (370–415), the daughter of Theon of Alexandria (335–405), a noted mathematician and commentator on Ptolemy and Euclid in his own right. Hypatia wrote on spheres and cones, commenting on Archimedes and Apollonius. Hypatia and Theon were influenced by another Alexandrian, Diophantus (200–284), author of *Arithmetica*; he was an early algebraist and used quadratic equations; he was perhaps influenced by Hindu mathematicians.

FURTHER READING

Jonathan Barnes, trans., *Early Greek Philosophy* (London: Penguin Books, 1987).

Euclid, *The Thirteen Books of the Elements*, trans. Thomas Heath (New York: Dover Books, 1925).

David Fowler, *The Mathematics of Plato's Academy: A New Reconstruction* (Oxford: Oxford University Press, 1999).

James Gow, *A Short History of Greek Mathematics* (Cambridge: Cambridge University Press, 2010).

Greek Mathematical Works, vol. 2, trans. Ivor Thomas (Cambridge, MA: Harvard University Press, 1941).

Thomas Heath, *Aristarchus of Samos* (New York: Dover Books, 1981).

Reviel Netz, *The Shaping of Deduction in Greek Mathematics: A Study in Cognitive History* (Cambridge: Cambridge University Press, 1999).

O. Neugebauer, *The Exact Sciences in Antiquity* (New York: Dover Books, 1969).

Nichomachus of Gerasa, *Introduction to Arithmetic*, trans. Martin Luther D'Ooge (New York: Macmillan, 1926).

J. J. O'Connor and E. F. Robertson. *History of Mathematics*, School of Mathematics and Statistics, University of St. Andrew's, Scotland, http://www-history.mcs.st-andrews.ac.uk/history/Indexes/HistoryTopics.html.

J. J. O'Connor and E. F. Robertson, "Hypsicles of Alexandria," https://www-history.mcs.st-andrews.ac.uk/Biographies/Hypsicles.html.

Plato, *The Dialogues*, trans. Benjamin Jowett (Chicago: Encyclopædia Britannica, 1952).

Ptolemy, *The Almagest*, trans. R. Catesby Taliaferro (Chicago: Encyclopedia Britannica, 1952).

Robert of Chester's Latin Translation of the Algebra of Al-Khowarizmi, trans. and ed. Louis C. Karpinski (New York: Macmillan, 1915).

21

Science during the Pax Romana (31 BCE–180 CE)

The Roman principate is named for the preferred title of Augustus (Octavian) Caesar (63 BCE–14 CE), who was only eighteen years old when his great-uncle and adoptive father Julius Caesar was assassinated by senators at Rome. Yet Octavian quickly gained the loyalty of Caesar's troops and supporters, made war against the Brutus and Cassius, the leaders of the conspiracy, and then slowly became the most powerful man in the empire. In 31 BCE he defeated his rival Antony at the Battle of Actium and assumed control of the Roman Empire. Octavian ostensibly returned power to the "Senate and People of Rome" but was given the special powers of the proconsul, commander of Roman military forces, and tribune, defender of the people, with veto power. Octavian also had personal control over the province of Egypt, which gave him unlimited wealth and power over a primary source of food in Rome. In 27, the Senate conferred upon Octavian the titles of Augustus (honored, revered) and Imperator (conqueror, one holding imperium, power). Augustus preferred, however, the title Princeps, which Roman leaders in good standing in the republic had long used; it meant simply "first citizen." Henceforth Augustus and his successors during the two-hundred-year principate would rule Rome with almost total power but would style themselves as the princeps of Rome.

JULIO-CLAUDIANS

Augustus, after years of civil conflict, tried to restore the empire to health and prosperity. He initiated a campaign to bring back the old morality of the Romans, and he sponsored the work of many poets and historians, such as Virgil (70–19) and Horace (65–8). This atmosphere of prosperity and achievement inspired others to engage in science and medicine. Augustus was undoubtedly influenced by his adoptive father Julius, who supported teachers, writers, and researchers in the liberal arts. Three dynasties of Roman emperors oversaw a relatively peaceful time in the history of Rome that encouraged thought and speculation into nature. The three dynasties were the *Julio-Claudian* Dynasty (27 BCE–68 CE), under the emperors Augustus (27 BCE–14 CE), Tiberius (14–37), Caligula (37–41), Claudius (41–54), and Nero (54–68); the *Flavian* Dynasty (69–96), under the emperors Vespasian (69–79), Titus (79–81), and Domitian (81–96); and the *Antonine* Dynasty, under the emperors Nerva (96–98), Trajan (98–117), Hadrian (117–138), Antoninus Pius (138–161), and Marcus Aurelius (161–180).

As Rome moved toward an autocratic society, the state became more important in sustaining learning and science. Roman emperors granted special benefits to people involved in the liberal arts and supported scientific research in a variety of fields, such as medicine and engineering.

The two-hundred-year period of the Pax Romana was a time when the Romans achieved an extremely sophisticated culture based on wealth from conquests and borrowings from the Greeks. There are many examples of Roman literary and scientific achievement; yet, even during this time, few Romans were educated, and even fewer had anything like a scientific education. Among the population, philosophers and scientists were a very small minority. Rome was an aristocratic society, and as in most societies, education followed upon wealth. Romans practiced slavery, and slaves, of course, would be uneducated (though some educated people taken in war became slaves). Women were typically uneducated; there were no legal prohibitions to their receiving an education, but the Romans were patriarchal, and the society and culture were male-dominated. As among the Greeks, there were particular exceptional women who became well educated and dabbled in the sciences. Epicurus had a lover, Leontion, who practiced Epicurean philosophy. Galen had a female friend, Arria, who was a brilliant Platonic philosopher. In medicine, there were female midwives and a few female physicians. There were several noteworthy female mathematicians who resided in Alexandria. Some Romans, such as Plutarch, believed that women were as capable as men and could be educated accordingly.

The Romans, like the Greeks, embraced education as a means to benefit society—a learned citizen was one who contributed to the republic or democracy. This ideal, developed first in Greece, was the *enkyklios paideia*, which was embraced by Romans throughout the centuries and then became the basis for medieval European education: the liberal arts. The term *liberal arts* refers to the study of subjects for the purpose of acquiring the habits and personality of a person who lives a free life, that is, who lives in such a way as to be free from outside influences and to know precisely what one's personal beliefs, derived from personal experience, are. The Greeks provided the foundation for the trivium and quadrivium, as they were defined later, in the Middle Ages, because of their focus on philosophy, literature, rhetoric, history, art and architecture, mathematics, and the life and physical sciences. The Roman Empire encompassed the learning of the ancient Greeks and brought such learning forward into the centuries after the birth of Christ.

Education during the Roman Empire, as in Greece in previous centuries, focused especially on grammar, logic, and rhetoric—the trivium. Students—mostly wealthy boys from good families—attended private schools where the curriculum focused on great literature, such as the works of Homer, and how to read, write, and speak. The focus on the trivium could still involve scientific study (aspects of the quadrivium). The great orator Quintilian (35–100 CE), who was a teacher and author of *Institutes of Oratory*, wrote this of geometry: "Elements of this science are of value for the instruction of children; for admittedly it exercises their minds, sharpens their wits and generates quickness of perception. But it is considered that the value of geometry resides in the process of learning, and not as with other sciences in the knowledge thus acquired" (Carrier, pp 80–81). Scientific study was not as important as rhetoric but could still help a student learn how to think, especially since the topics of philosophy included highly advanced concepts. Orators were expected to be cognizant of all aspects of learning. The physician Galen argued that the best physicians were those with a well-rounded education that did not ignore grammar, logic, rhetoric, and mathematics. Since the physician must know how the environment affects diseases, Galen believed physicians should have a deep understanding of astronomy and meteorology. Vitruvius, an architect and engineer, believed that the best engineers had a complete understanding of all aspects of the trivium and quadrivium. Students who pursued their studies beyond a basic primary or secondary education would typically study under a great master of oratory, astronomy, mathematics, medicine, or engineering.

During the Pax Romana, there was a variety of ways that the common people—those who did not enjoy the formal education of the *enkyklios paideia*—could enjoy some of the crumbs from the table of scientific discourse. There were public lectures on science that anyone could attend, and there were numerous libraries in Rome and other cities where people could examine scientific works.

ROMAN PRAGMATISM

The Roman mindset was less speculative and metaphysical than that of the Greeks, who excelled at philosophy and the search for *being*. The Roman scientist was interested in knowledge that could be usefully applied to problems of everyday life as well as issues of order, mobility, structure, power, and majesty. Roman scientists were polymaths, like the Elder Pliny (23–79). Another was Plutarch (46–120), whose vast corpus of writings included the *Natural Questions*, in which he explored remarkable phenomena, such as the changing color of the octopus to deceive prey and predators.

Roman roadbuilding is a good example of Roman pragmatism. Roads are useful for travel and trade, of course, but initially the Romans constructed roads as a consequence of, or to complement, military activity. Logistics, unimpeded communications and supply lines, were keys to Roman military success. Roman surveyors solved the problem of drainage and picked the best routes with the least fluctuation in elevation. Surveyors also solved the problem of elevation and gravity in the construction of aqueducts, some of which still exist, which brought fresh water to towns using the force of nature: water flows as long as there exists an uninterrupted declivity. The aqueducts were astonishing structures of elegance and utility that relied on sophisticated engineering techniques. Roman engineers learned how to distribute the material forces of weight and gravity by means of the vaulted arch. The emperor Hadrian showed his ability as an engineer in the design of the domed Pantheon, which became the design of choice for builders of public buildings for centuries to come. The Roman Colosseum not only was a magnificent engineering achievement but also featured an intricate design for a partial cloth covering, or shade, to be used on sunny days. Many of the ideas and techniques of Roman construction, engineering, and architecture are preserved in Vitruvius's *On Architecture.*

VITRUVIUS (90–20 BCE)

Marcus Vitruvius Pollio was a Roman architect during the age of Augustus and the learned author of *On Architecture.* Little is known of his life except for his book, in which he addressed the emperor Augustus and his interest in the Roman infrastructure of public buildings and forums, hydraulics, and engineering. He clearly benefited from a complete education in the trivium and quadrivium, which were supplemented by learning about engineering and hydraulics. Rhetoric served him well in his book. The study of astronomy was essential for building according to set environments. He studied music to understand acoustics. Vitruvius discussed in detail building materials, the impact of climate on building, the functionality of rooms in a dwelling, interior design and wall paintings, and the dangers of lead pipes. Vitruvius argued for the active sharing of scientific knowledge. An engineer, he examined aqueducts, Archimedes's discoveries on the displacement of water, the water screw, the lever, rotary power, and the use of machinery in making weapons of war.

On Architecture describes in detail the science of acoustics. Vitruvius compared the human voice in speech or song to the waves spreading out from a stone tossed into still water; the voice moves both vertically and horizontally. The human voice, he argued, has three types of modulation in tone. An advocate for using the pure shapes of the universe—the knowledge of geometry—in building, Vitruvius was a student of the Pythagorean theorem and the works of Archimedes (287–212). He argued that the Roman theater should be built with four equilateral triangles coming to a point in the center. He hoped that the theater's acoustics would rival the music of the heavenly spheres. Vitruvius also discussed the building of a hydraulic organ.

Vitruvius was also interested in astronomy and its practical uses in timekeeping. He discussed sundials, water clocks, and how the gnomon can measure the sun's shadow at the winter solstice to get an idea of the extent of the world. A long section of *On Architecture* focused on the geocentric view of the planets, moon, and sun and the revolution of the heavens about the two poles on the earth. Fascinated by astrology, Vitruvius discussed in detail the zodiac and constellations. He claimed that the greatest astrologists were the Chaldeans, in particular Berossus (a Babylonian, fl. beginning of the third century), who had a school in Cos, in the Aegean. Vitruvius is often remembered for his conception of the perfect dimensions of man, his hands and legs outstretched within a square and circle.

AGRICULTURE

As the Roman Empire was initially built on successful cultivation of the Italian soil by hardy farmers and shepherds, agriculture was a favorite research topic for

Roman scientists. Roman writers, such as the historian Livy (59–17 CE), eulogized the sturdy farmers, such as Cincinnatus (c. 519–430), who were loyal to Rome and formed the basis for middle-class armies of freeholders who could defeat any army of the Mediterranean world. The practical Roman mindset produced quite a few agricultural writers, such as the Elder Cato, Varro, the physician Celsus (25 BCE–50 CE), Columella, the Elder Pliny, and Palladius (fl. 500 CE). These writers emphasized the pastoral, the valuable peace that accompanied the farm life. Some, such as Cato, wrote to the increasing number of wealthy plantation owners who rarely got their hands dirty but, rather, utilized armies of slaves, captured in war, who performed the manual labor. Roman politics of the second century BCE involved conflict over land, as small freeholders were forced to sell to the grasping owners of slave plantations, *latifundia*.

Cato, Marcus Porcius (234–149 BCE)

The Elder Cato was a philosopher, statesman, historian, and agricultural writer. His *On Agriculture* was written in Latin, since he was very opposed to the growing Greek intellectual influence in Roman republican society. Cato's audience consisted of the large landowners who created the *latifundia*, the infamous plantations dedicated to wine and olive production, rather than the small freeholders who were by the second century BCE becoming a thing of the past.

Cato influenced Columella, Varro, Palladius, and the Elder Pliny, among many other subsequent writers on agricultural topics. Pliny praised Cato for his expertise in viticulture, olive growing and the production of oil, the purchase of farm property, deciding what the land was best suited for, and particular crops and their medical efficacy. Cato, for example, praised cabbage as important in combating hypochondria, sleeplessness, and nightmares. Eating hare was also a sure remedy for insomnia. Pliny also relied on Cato for his comments on materia medica and veterinary medicine.

Cato believed that the Roman moral strength could be sapped by too heavy reliance upon other cultures. He wrote his son the following advice respecting Greek physicians:

> I shall speak about Greek doctors at the appropriate point, Marcus, my son. I shall tell you what I discovered at Athens and convince you that it is good to skim through their literature, but not to study it in depth. They are a worthless and intractable lot--in this area accept me as a prophet. For when the Greeks give us their literature it will undermine our whole way of life, and even more so if they send us their doctors. They have sworn among themselves to kill all foreigners with their medicine, but they do so for a fee, to win credit and destroy us easily. They also persistently call us foreigners and treat us with less consideration than others and belittle us as country bumpkins. I have forbidden you to have anything to do with doctors! (Trans. Healy)

Varro (116–27 BCE)

Marcus Terrentius Varro was one of the great scholars of the Roman republic. His many works included *On Agriculture*, *Antiquities*, and *On the Latin Language*, which provided a scientific account of Roman agriculture, an analysis of the human and natural past, and a scholarly study of Latin. Varro wrote *On Agriculture* during the period of civil conflict after the assassination of Caesar. He believed in practical experience gained by years working with the land as well as experimentation with crops and soils. Aurelius Augustine (354–430), who used *On the Latin Language* as well as Varro's histories, thought Varro was one of the most important thinkers of the past. Varro's agricultural writings influenced others such as Columella and Palladius. Varro wrote on farming and ranching, advocating an empirical approach to the study of plants and animals. Varro's *On the Latin Language* argued the belief, especially prevalent in first-century-BCE Rome, that truth is dependent upon how it is presented in speaking and writing. The *Antiquities* consisted of forty-one books devoted to anthropology and theology. Augustine despised Varro's paganism but approved his rational approach toward the Roman pantheon of gods. Varro described the varied approaches of Greek philosophers to the divine but reserved judgment as to which theory was most accurate. In the end, Augustine relied on Varro's natural theology to condemn the paganism of the Romans and support his own views on the supremacy of Christianity. Varro also influenced the *Etymologies* of Isidore of Seville (560–636). Isidore claimed that "Varro says there are four kinds of divination, namely, by earth, air, water, fire; hence geomancy, hydromancy, aeromancy, pyromancy" (trans. Brehaut).

Columella (5 BCE–60 CE)

Lucius Junius Moderatus Columella, a native of Cadiz, Spain, was the author of *On Agriculture, On Trees,* and the lost *Against the Astrologers.* His books reveal the Roman concern for practical techniques of applied science. Columella wrote on aviculture (raising birds), leaning on the work of Hyginus in this regard, and viticulture (cultivation of grapes, making of wine), so important to the Roman economy. He particularly emphasized viticulture as the best means to make land profitable. Columella's practical advice included the idea that wheat would do well in a field that lay fallow for a year and that, since birds bathed in dust and ash, the poultry man should line henhouses with such matter. Columella also advised crop rotation. He believed that agriculturists should be cognizant of other sciences, such as astronomy and meteorology. Columella's writings had a significant impact on the work of later Latin scientists and agriculturalists, such as Pliny and Palladius.

HISTORY AND GEOGRAPHY

The Roman focus on practical knowledge and empire building meant that the science of geography was especially important. Roman geographers included Gaius Julius Hyginus (64 BCE–17 CE), a native of Spain and a freedman, who was the librarian of the Palatine library and a writer of biographies and histories as well as treatises on geography, agriculture, aviculture, and religion. Augustus sponsored the work of Hyginus on the geography of Italy. Suetonius (c. 69–130), the Roman biographer, wrote of Hyginus:

> Gaius Julius Hyginus, a freedman of Augustus and a Spaniard by birth (some think that he was a native of Alexandria and was brought to Rome when a boy by Caesar after the capture of the city), was a zealous pupil and imitator of the Greek grammarian Cornelius Alexander [105–35], whom many called "Polyhistor" because of his knowledge of the past, and some "History." Hyginus was in charge of the Palatine Library, but nevertheless took many pupils. He was an intimate friend of the poet Ovid and of Clodius Licinus the ex-consul and historian, being supported as long as he lived by the writer's generosity. He had a freedman Julius Modestus, who followed in his patron's footsteps as student and scholar. (Trans. Rolfe)

The Greek Strabo (63 BCE–21 CE) produced a massive compendium of geographical facts and observations about nature ranging from Europe to Africa to Asia. The cartographer Pomponius Mela of Spain (fl. 45 CE) produced a multivolume *Chorographia* during the reign of the emperor Claudius wherein he described the many lands subjugated by the Romans.

The works of Livy (59 BCE–17 CE), Tacitus (56–117), and Plutarch (46–120) reveal that historical writing requires the complement of geographical investigation. Historical writing was the art and science par excellence during the principate. Roman historians, as students of the entire experience of humankind, of necessity engaged in speculation on human nature and natural history. During Augustus's reign, Titus Livius produced a massive history of republican Rome that included wonderful descriptions of the imposing forces of nature, such as the Alps. The polymath Cornelius Tacitus, who wrote during the reign of Trajan, explored the human psyche (in his study of Tiberius, for example) and described the culture of the Germanic tribes of northern Europe in the *Germania.* Plutarch, a Greek writing during the principate, composed not only *Lives* but *Moral Essays* as well, in which he explored a variety of scientific issues ranging from human behavior to world geography to astronomy. Plutarch was a polymath, as was his predecessor in the art of knowing a little about a lot: the Elder Pliny. Pliny was a historian and geographer, a tirelessly inquisitive man who lost his life when Mount Vesuvius erupted in 79 CE. Pliny's *Natural History,* a wonderful catalogue of hearsay and observation, was published posthumously.

The Eruption of Vesuvius

Mount Vesuvius has been an active volcano for thousands of years. The region of western Italy dominated by the four-thousand-foot volcano experienced many warnings of the mountain's volatility before the massive explosion of 79 CE. The historian Dio Cassius (163–225) described how for years Vesuvius had erupted, sending up flames, smoke, and ash. The curious had climbed to the crater at its summit, which looked like an amphitheater. But quite often "the normal state of affairs" was the rumbling of the mountain, the frequent orange glow at night, the warnings of what might occur (trans. Cary). Dio implied that the inhabitants of Pompeii and Herculaneum were so familiar with the mountain's habits that the roars, quakes, and fires preceding the August 24 eruption were ignored.

Strabo, a scientist and geographer, described in volume 5 of his *Geography* the volcano at one of its peaceful moments: "Mt. Vesuvius dominates this region. All but its summit is clad in exceptionally fine fields. The summit itself is mostly flat, and entirely barren. The soil looks like ash, and there are cave-like pits of blackened rock, looking gnawed by fire. This area appears to have been on fire in the past and to have had craters of flame which were subsequently extinguished by a lack of fuel. No doubt this is the reason for the fertility of the surrounding area" (trans. Jones).

Seneca (5–65), the Roman stoic and naturalist, described the earthquake of 62 that presaged the total destruction of 79. The earthquake hit in February, and "devastated Campania. The area is never safe from this sort of danger, but it had escaped damage and outlived the scare many times before" (trans. Clarke). Herculaneum was partially destroyed, Pompeii almost totally. But the people immediately set to rebuilding and were still doing so in August 79.

Pliny the Elder, in his *Natural History*, also described the region, its fertility and beauty. He knew Campania well, had observed Vesuvius and surrounding mountains repeatedly, and, as a commander of the Roman navy, he was stationed at a nearby port, Misenum, on August 24 of 79. Pliny's nephew, the letter-writing naturalist Pliny the Younger, then seventeen years old, was present that morning. He recalled later, in a letter to the historian Cornelius Tacitus, that his uncle had determined to investigate the reason for the large dark cloud rising from Vesuvius and called for a boat to be made ready. Science then had to make way for official duties. Having received word that the people of the coast at Pompeii and Herculaneum were in danger, he ordered the galleys to be launched to rescue the inhabitants. As they sailed across the Bay of Naples directly toward the massive cloud looming above the mountain, spreading outward, Pliny took notes of his observations. The boats were prevented by the number of volcanic rocks falling from the sky, making hazardous the shallow water along the eastern shore of the bay. Pliny ordered the ships south to the town of Stabiae, where the wind, current, and quakes made escape impossible. Pliny ordered a bath and dinner and slept. Meanwhile pumice stones and ash fell, darkening the sky. Buildings were rocking, threatening to collapse. Pliny and his followers rushed to the seashore but found there was no escape. The stench of sulfur was overwhelming. Poisonous gas, heat, and ash finally brought death.

Pliny the Elder (23–79 CE)

Pliny the Elder, Gaius Plinius Secundus, was a Roman polymath and author of *Natural History*, which is a diverse collection of anecdotes, history, geography, medical information of varying worth, discussions of astronomy and earth science, and a catalog of Roman knowledge on botany and zoology. Pliny's information varies in its quality; sometimes he seems amazingly credulous; at other times he appears a worthwhile scientist. *Natural History* is filled with interesting information and useful facts, which may explain why it was one of the most widely read books during the Late Roman Empire, Middle Ages, and Renaissance.

Pliny was born during the reign of the second Roman emperor Tiberius (14–37 CE); he died in the same year that the emperor Vespasian died and his son Titus,

Pliny's friend, assumed the throne (79). Pliny was of the equestrian class, which the Roman emperors relied on for political support and to fill the many administrative positions in the far-flung Roman Empire. Pliny served in the military as a provincial governor (procurator) and as head of the Roman fleet, the latter a position he held at his death. Indeed, as noted above, the fleet was moored at Misenum on the Bay of Naples, up the coast from Pompeii and Herculaneum, when Mount Vesuvius erupted in August 79. Pliny died when he responded to the eruption with a sense of duty to help those who were trying to flee the volcano as well as with a sense of curiosity about the phenomenon of a volcanic eruption. A Stoic, he had a no-nonsense approach to life, approaching death fearlessly, doing his duty to the last, and not allowing emotions to overwhelm reason. Pliny believed in the Stoic concept of afterlife: that whatever it is, it will not involve conscious awareness, hence this life is all a person has. Pliny filled every moment with activity and study. He slept little, preferring to write his many books at night rather than to repose. Besides *Natural History*, he wrote books on military affairs and the German people. It is *Natural History*, of course, for which he is best known, because of the book's influence for centuries after his death, because of its eclectic and universal approach to learning, and because it is accessible to a variety of readers of different skills and interests.

Pliny's ultimate concern was to make the sciences—human knowledge—accessible to all people. His attempt at universal knowledge is of mixed quality. *Natural History*'s presentation of astronomy is elementary, useful perhaps for the farmer but not the savant. Pliny, likewise, was not an expert in geodesy, the study of the shape of the earth. His approach to meteorology was also simple—but then a simple explication of natural events was his goal.

Pliny's encyclopedic approach is clearly seen in his discussion of astronomy and meteorology. Pliny accepted contemporary Greek astronomical theories about a geocentric universe: the heavenly bodies of sun, moon, Mercury, Venus, Mars, Jupiter, Saturn, and stars revolving around the earth, which itself rotates at a remarkably high speed. He provided full descriptions of eclipses, the movement of the planets, constellations, comets, and other astronomical phenomena. Yet he was much too absorbed, as were his contemporaries, in the belief that the movements of the heavenly bodies affected human affairs, even natural phenomena. A case in point was the impact of planets and stars on meteorological events, that thunderstorms and other weather changes were caused by the influence of planets and stars in the heavenly sphere above earth. Even if somewhat absurd, Pliny's discussion of some meteorological phenomena, such as rainbows, earthquakes, and the nature and force of winds, is remarkable.

Pliny's discussion of metals and metallurgy contained much common sense, being largely based on his own observations and analysis. He discussed the nature and uses of gold, silver, iron, lead, tin, copper, mercury, and bronze, providing some fascinating insights with little accompanying fantasy. He provided an extensive description of mining practices and the dangers miners endured in their quest for metals. He was fascinated with precious metals, yet he knew their limitations and also knew the negative impact that gold and silver would have on the greedy Roman character. One is surprised to find Pliny condemning the Roman obsession with gold and silver and its mining and wondering why it is necessary to disembowel the earth for the sake of metals used to satisfy greed and vanity. Pliny, always interested in the practical side of science, discussed a few of the apparent medicinal benefits of metals, such as the application of silver or lead on bodily parts needing cooling and passivity.

As one might expect from a compiler, Pliny's life and human sciences were rudimentary, based largely on hearsay. The anthropological discussion in *Natural History* reveals Pliny's utter credulity when it comes to world peoples. Likewise, his zoology was marred by a willingness to accept strange stories: he uncritically accepted remarkable tales about humanlike behavior, characteristics, emotions, and intelligence of animals such as elephants and lions. Pliny recorded accounts of mythical animals such as werewolves, although in this case he doubted their reality. On the other hand, Pliny's account of bees and their culture was a complete study based on close observation. He believed that the honeybee had social and behavioral characteristics similar to Rome under the republic. Pliny, a representative Roman thinker, echoed Aristotle's comments comparing beasts and humans: humans are at the pinnacle of the animal scale of being, being the only creature to exhibit consistent rational thought and behavior. Pliny did, however, conceive of animals showing remarkable humanlike characteristics, such as pity, logical (if rudimentary) thought, piety, empathy, and a sense of duty. Pliny was credulous in many respects respecting human and animal life, assuming that in the far reaches of Asia and Africa, strange creatures and humans with odd characteristics existed.

Pliny's discussion of flora was erudite and exhaustive although generally derived from other botanical writers, such as the student of Aristotle, Theophrastus (370–286). His focus, as throughout *Natural History,* was on useful knowledge. Hence he provided a long discussion on papyrus, spice trees, viticulture, olive trees, the benefits of forest trees, and agriculture. Pliny saw enough of the effects of wine on Romans to condemn wine-drinking out of hand as a terrible vice. Pliny developed an extensive materia medica that collected observations and folklore involving concoctions useful in treating illness.

Although Pliny was not an original scientist, he was nevertheless significant as a compiler. His encyclopedic approach to learning had a profound effect on subsequent compilers, particularly of the Late Roman Empire and European Middle Ages. More significant is Pliny's willingness to cite other writers and to catalog systematically their writings. Modern students of the ancient world and science know of ancient writers and writings that would otherwise be unknown, save for Pliny's efforts.

Pliny the Younger (61–113 CE)

Gaius Plinius Luci was the nephew and heir of Pliny the Elder. He was, like his uncle, interested in a variety of topics of inquiry, some of them scientific, which are revealed in his *Letters.* Pliny was a consul, senator, and lawyer. He was a warm supporter of the emperor Trajan (r. 98–117 CE), who rewarded Pliny with the governorship of Bithynia and Pontus in the eastern Mediterranean. Pliny's intellectual interests were focused mostly on literature and rhetoric. He was not a scientist but rather an educated Roman. Yet his *Letters* contain important portraits of the culture and science of his time.

It is through the Younger Pliny that we obtain most of our information on the eruption of Mount Vesuvius in August 79, which resulted in the death of his uncle and adoptive father, Pliny the Elder. His description of the eruption of Mount Vesuvius was recorded at the bidding of Cornelius Tacitus (56–117), who was writing his history of the time and sought to record all remarkable phenomena. Pliny the Younger described the eruption as causing a massive cone of smoke and ash—white, yet spotted with a dark mixture—soaring above the mountain and then spreading out in all directions like an umbrella. As night fell, the mountain vomited unremitting fire. Daybreak on the morning of August 25 at Stabiae near Pompeii was indistinguishable because of the intense darkness. The nephew, meanwhile, and his mother remained in Misenum, across the Bay of Naples from Pompeii. The eruption of Vesuvius caused everything to shake in a continual earthquake. The sea churned as well, and the tide receded to such an extent that marine animals were thrust upon the shore and stranded. Pliny and his mother joined an exodus from Misenum, the people wandering blindly in the utter darkness. Ash fell like heavy rain and covered everything. People cried out in fear of the end of the world and their abandonment by the gods.

Pliny claimed that he was terrified but less superstitious than others about the eruption of Vesuvius. Nevertheless, he painted an interesting portrait of the leading Roman intellectuals of his age, for example, his friends Cornelius Tacitus, the historian, and Suetonius Tranquillus (69–130), the biographer, as believers in the traditional Roman gods and goddesses and the prophetic nature of dreams and animal sacrifice.

Pliny's letters reveal the Roman fascination with water. Like many of his contemporaries, Pliny believed that bathing, particularly in cold water, was a guarantor of good health. He enjoyed watching natural hydraulics at work, the fall of springs and filling of ponds. As governor of Bithynia, he oversaw the building and maintenance of several public works, particularly aqueducts and canals.

Scientists such as Pliny the Elder and his nephew took a rational, scientific approach to the eruption and earthquakes of Vesuvius. The credulous, however, interpreted the calamity of the eruption of Vesuvius as the work of the gods and demons. Dio Cassius, writing years later in his *History of Rome*, recorded uncritically the perceptions of many survivors, that

> numbers of huge men appeared, but bigger than any human. . . . They were seen on the mountain, in the surrounding countryside, and in the cities, wandering over the earth day and night, and also journeying through the air. Then came a terrible dryness, and sudden violent earthquakes, so that the whole plain seethed and the heights leaped into the air. There were frequent rumblings, some underground, sounding like thunder, others on the surface, making a bellowing sound. The sea joined in the roar, and the sky added its peal. Then suddenly a dreadful crash was heard, as if the mountains were collapsing in on themselves. First huge stones flew up as high as the mountain top, then came a great quantity of fire and endless smoke, so that the air was darkened and the sun entirely hidden, as if eclipsed. Thus day turned into night and darkness came out of the light. Some thought that the Giants were rising

> again in revolt (for many of their forms could still be discerned in the smoke, and a sound as of trumpets were heard), others believed that the whole universe was being resolved into chaos or fire. (Trans. Cary)

And that "chaos or fire" reflected how Stoic philosophers and scientists interpreted the end of time.

STOICS

Roman science during the final centuries BCE had become heavily dependent upon the Greeks. The Epicurean view of a material universe was eloquently expressed by Lucretius (98–55) in *On the Nature of Things*. Stoicism, founded by Zeno, a Greek, was the most popular philosophy among Roman intellectuals. Its leading advocate during the late republic was Cicero (106–43) who, although he was primarily an orator and statesmen, produced works such as *On the Nature of the Gods* and *On Divination*. A century later, during the reign of Nero, another great Stoic thinker of the principate, Seneca, by his daily practice of philosophy engaged in spirited inquiries into the nature of things. His surviving *Epistles* reveal his preoccupation with morality, nature, and science. Seneca wrote other works, too, such as *Problems in Natural Science*. A half century later still, during the reign of Trajan the Younger, Pliny revealed his scientific inclinations in his correspondence. The emperor Marcus Aurelius was a fine example of a Stoic philosopher. His *Meditations* reveal his inquisitiveness toward history, nature, and the origin of things and their final end. Other emperors likewise considered themselves to be philosophers: Nero engaged in verse and architecture; Hadrian loved Greek philosophy and also delved into architecture; Claudius was a biographer, philologist, and historian.

Epictetus (c. 50–135 CE)

Epictetus was a Greek slave who became a leading voice of Stoicism during the Roman principate. He was a student of Musonius Rufus (c. 30–101), a Roman Stoic philosopher. After being freed by his master Epaphroditus and then banished from Rome by the emperor Domitian, Epictetus set up a school in western Greece, on the Adriatic at Nicopolis. There he had great influence on Greek and Roman Stoics, his most famous student being the historian and biographer of Alexander the Great, Flavius Arrianus, Arrian (88–175). Arrian recorded Epictetus's teachings in the *Discourses,* which he claimed were his own notes taken during the philosopher's lectures. Arrian also published the *Enchiridion,* a handbook of Epictetus's Stoic teachings. Epictetus also influenced the Roman emperor Marcus Aurelius.

Epictetus contributed to the development of the science of psychology in antiquity. Much of his moral philosophy dealt with the search for happiness and contentment; the path to coming to know oneself; and the ways to cope with anxiety. Epictetus's answer to all of these issues was to accept oneself, one's situation in life, one's surroundings, and the workings of time. He believed correct thinking about life and self could help in the healing process of body and mind. His focus on reason led him to discount the typical Roman reliance upon soothsayers and diviners; one should divine one's future based on awareness of self, Epictetus argued. He counseled acceptance of sickness rather than resistance and the endless search for doctors and cures. Ultimately the individual has free choice to respond appropriately or not to the contingencies of life. At all times, one's "moral purpose" (*prohairesis*) must guide the response to peculiar situations and the manifold events of life.

Epictetus's *Discourses* were so curt and direct that his readers, particularly those of a similar mindset, could scarcely disagree with his approach to life and thought. He deeply influenced Marcus Aurelius and, much later, Michel de Montaigne, who saw in Epictetus a philosopher who could address humans' most essential limitations: fear of the future and fear of death.

Marcus Aurelius (121–180 CE)

One of the great exponents of Zeno's philosophy of Stoicism was the Roman emperor Marcus Aurelius, the author of *Meditations*. By the time of his reign in the mid- to late second century CE, Stoicism was firmly entrenched as the most popular philosophy among Romans. Marcus Aurelius had trained himself as a philosopher even before he was adopted by the emperor Antoninus Pius. The Stoics believed that fate directed all human affairs. Aurelius felt a need to console himself when in 161 he became Roman emperor. His *Meditations* was a diary meant only for himself, written sporadically in consolation, expression, rumination, joy, and sorrow. And yet *Meditations* was also an account of the scientific worldview of the ancient Stoic philosopher.

The study of *physics*, that is, the natural world, its causes and effects, was one of the chief topics of Stoic inquiry. Aurelius studied human behavior and natural phenomena to arrive at the notion that there was a universal law of nature that governed and ordered all things. Believing that said law was not the product of chance in a universe completely oriented around fate, Aurelius, like other Stoics, assumed the existence of a universal mind, the holy spirit, the Logos. The Logos was a creative yet impersonal force of reason that erects, controls, and destroys successive universes in infinite time. Aurelius believed that the Logos was a thrifty creator who allowed only such beings that were necessary in an economical universe. A materialist, Aurelius believed that all things, even being, the soul, the Logos, are physical phenomena. "I consist of a formal element and a material," he wrote in *Meditations*. "Neither of these can ever pass away into nothing, any more than either of them came into being from nothing. Consequently every part of me will one day be re-fashioned, by a process of transition, into some other portion of the universe" (trans. Staniforth).

Marcus Aurelius was also an important philosopher on the nature of time. "To see the things of the present moment is to see all that is now, all that has been since time began, and all that shall be unto the world's end; for all things are of one kind and one form" (trans. Staniforth). Time, like the universe, like being, is interconnected, a unity embraced in its totality by the Logos.

The *Meditations* is hardly the work of a man content with himself, his life, his destiny, and the cosmos. Stoicism in theory taught the acceptance, hence the conquest, of death. Aurelius repeatedly tried to convince himself that death was nothing to fear—and yet the fear crept in daily, weekly, yearly, recurrently. The Stoic promise, based on materialism, that one's destiny lies in the return of one's atoms to the universe, never to know or to love again, led to despair rather than happiness in the life and writings of Marcus Aurelius.

SKEPTICS

Sextus Empiricus (160–210 CE) was a disciple of the Skeptic philosopher Pyrrho (365–279). Sextus wrote the *Outlines of Pyrrhonism* in defense of Pyrrho's thought and *Against the Mathematicians* to pour out his doubt and ire against scientists, mathematicians, and philosophers. Sextus Empiricus argued that most philosophers of his time were Dogmatists, meaning that they believed they had discovered what was truth; he counted the Academics, Peripatetics, Epicureans, and Skeptics among these Dogmatists. But a Skeptic, according to Sextus, was a philosopher who did not know and who therefore reserved the ability to doubt until truth could be discovered. Diogenes Laertius described Pyrrhonists, therefore Skeptics, thus: "The Pyrrhonean system, then, is a simple explanation of appearances, or of notions of every kind, by means of which, comparing one thing with another, one arrives at the conclusion, that there is nothing in all these notions, but contradiction and confusion" (trans. Yonge).

ROMAN MEDICINE

The *Pax Romana* also featured Roman physicians working in the shadow of Greeks such as Hippocrates. Asclepiades in the first century BCE and Celsus in the first century CE were the leading Roman physicians before Galen. The Epicurean Asclepiades believed that physical health and disease were the product of the movement and combinations of invisible atoms; his was a "mechanistic" view of medicine. He was nevertheless tireless in treating malarial patients, the mentally disturbed, and those with arthritic complaints. Celsus, on the other hand, was a polymath who wrote on medicine, a compendium of Hippocratic teachings and Greek medical techniques and terminology. By the beginning of the second century CE, Roman medicine, based on Greek models, particularly Hippocrates's, became specialized into varying fields of expertise: gynecology, urology, ophthalmology, surgery, and dentistry. Women physicians specialized in obstetrics and herbal medicine, including a variety of herbs used in birth control and abortion.

Pliny's contemporary Dioscorides was an herbalist and botanist who created an extensive pharmacology. Hailing from Anatolia, he was a traveling physician or apothecary, either independently or attached to the Roman legions; he collected specimens of flora and fauna, and he published a complete materia medica. In some respects, Dioscorides was one of the first real Roman pharmacists—he wrote a pharmacology that listed hundreds of floral remedies. He was preceded in his work by a century by the shadowy Greek thinker Crateus. Dioscorides's impact on subsequent scientists was profound.

Galen of Pergamum (130–200) was a follower of Hippocrates, yet his originality and medical expertise made him much more than an imitator. Galen was physician to gladiators as well as to emperors over the course of his long life. He believed in the unity of the living organism, in the wonderful creativity of nature (*techne*) in forming the human, and in the vital principle (*pneuma*) that is the spiritual center of the human being. Galen summarized his thought in *On the Natural Faculties*, in which he attacked his opponents past and present and discussed his theories of the operation of the kidneys in producing urine, the role of the liver in storing food, and the action of the veins in the flow of blood. Building on the theories of Hippocrates, Galen's forceful approach to medicine had an overwhelming impact on subsequent centuries until the dawn of modern medical science after 1500 CE. During the Later Roman Empire, Galen's followers, such as Oribasius (c. 320–400), recorded the master's teachings in books such as the *Synopsis*, thereby allowing the continued study of Galen's theories and observations in the Eastern Roman Empire, in the Byzantine Empire, and among Arab scholars of the Middle Ages.

FURTHER READING

Marcus Aurelius, *Meditations*, trans. Maxwell Staniforth (Harmondsworth, England: Penguin Books, 1964).

Jonathan Barnes, trans., *Early Greek Philosophy* (London: Penguin Books, 1987).

Anthony Birley, *Marcus Aurelius* (Boston: Little, Brown, 1966).

Ernest Brehaut, trans., *An Encyclopedist of the Dark Ages: Isidore of Seville* (New York: Columbia University, 1912).

Richard Carrier, *Science Education in the Early Roman Empire* (Durham, NC: Pitchstone Publishing, 2016).

Cassius Dio, *Roman History*, trans. Earnest Cary (Cambridge, MA: Harvard University Press, 1914–1927).

John Clarke, trans., *Physical Science in the Time of Nero: Being a Translation of the Quaestiones Naturales of Seneca* (New York: Macmillan, 1910).

Columella, *On Agriculture*, trans. Edward H. Heffner (Cambridge, MA: Harvard University Press, 1989).

Diogenes Laertius, *The Lives and Opinions of Eminent Philosophers*, trans. C. D. Yonge (London: Henry Bohn, 1853).

Epictetus, *The Discourses*, ed. Christopher Gill, Everyman's Library (Rutland, VT: Tuttle Publishing, 2001).

Galen, *On the Natural Faculties*, trans. A. J. Brock, Loeb Classical Library (Cambridge, MA: Harvard University Press, 1916).

R. M. Ogilvie, *Roman Literature and Society* (Harmondsworth, England: Penguin Books, 1980).

Pliny the Elder, *Natural History*, trans John F. Healy (London: Penguin Books, 1991).

Pliny the Elder, *Natural History*, 2 vols., trans H. Rackham (Cambridge, MA: Harvard University Press, 1938, 1947).

Pliny the Younger, *The Letters of the Younger Pliny*, trans. Betty Radice (Harmondsworth, England: Penguin Books, 1963).

F. H. Sandbach, *The Stoics* (New York: W. W. Norton, 1975).

Seneca, *Letters from a Stoic*, trans. Robin Campbell (London: Penguin Books, 1969).

Sextus Empiricus, *Outlines of Scepticism*, ed. Julia Annas and Jonathan Barnes (Cambridge: University of Cambridge Press, 2000).

Strabo, *Geography*, trans. H. L. Jones (Cambridge, MA: Harvard University Press, 1917, 1923).

Suetonius, *On Grammarians*, trans. J. C. Rolfe, 1914, http://penelope.uchicago.edu/Thayer/E/Roman/Texts/Suetonius/de_Grammaticis*.html.

Vitruvius, *The Ten Books on Architecture*, trans. Morris H. Morgan (Cambridge, MA: Harvard University Press, 1914).

PART 8

Applied Science in the Ancient and Medieval Worlds

22

Engineering and Technology in the Ancient and Medieval Worlds (4000 BCE–1500 CE)

Early humans discovered hundreds of centuries ago that the key to survival in an environment filled with natural dangers and predators was to use reason as the means to develop tools to alter the environment. Such were the beginnings of human technology. Humans fashioned tools by which to dig, cut, lance, bolster, barricade, and change the natural environment according to a preconceived vision of how something might work. When one human using one tool was insufficient, humans developed systems to organize dozens, scores, hundreds, or thousands of workers working simultaneously to achieve a change in the natural environment. Mesopotamia (Iraq), the Tigris and Euphrates River valleys, was the location of some of the earliest technological discoveries by humans. The Neolithic Revolution after 10,000 BCE led to signal developments in agriculture and early attempts to construct shelters of mud brick as well as walled towns and monuments.

The Mesopotamians, followed by other ancient civilizations, such as in Egypt, India, and China, developed a systematic structure and organization by which to channel the water from rivers, in Mesopotamia the Tigris and Euphrates, by which to irrigate crops and control flooding. The ancient geographer Strabo wrote of the canal system,

> The Euphrates rises to flood-tide at the beginning of summer, beginning first to rise in the spring when the snows in Armenia melt; so that of necessity it forms lakes and deluges the ploughed lands, unless the excess of the stream, or the surface water, is distributed by means of trenches and canals, as is the case with the Nile in Aegypt. Now this is the origin of the canals; but there is need of much labour to keep them up, for the soil is so deep and soft and yielding that it is easily swept out by the streams, and the plains are laid bare, and the canals are easily filled, and their mouths choked, by the silt; and thus it results again that the overflow of the waters, emptying into the plains near the sea, forms lakes and marshes and reed-beds. (Trans. Jones)

The Mesopotamians were the first builders of monumental architecture: the ziggurat or pyramid, an idea that was shared with other ancient cultures over time.

Ancient observers such as Strabo were duly impressed by the accomplishments of ancient Mesopotamians. He wrote, "Here too is the tomb of Belus, now in ruins, having been demolished by Xerxes, as it is said. It was a quadrangular pyramid of baked brick, not only being a stadium in height, but also having sides a stadium in length. Alexander intended to repair this pyramid; but it would have been a large task and would have required a long time (for merely the clearing away of the mound was a task for ten thousand men for two months), so that he could not finish what he had attempted" (trans. Jones). This ziggurat was known in Sumerian as *Etemenanki* and is called today the Tower of Babel.

The Mesopotamians were the first to develop symbols, markings, to indicate natural, human, and physical names that earlier humans had devised in oral communication. Cuneiform symbols made with a stylus in wet clay developed in third-millennium Mesopotamia in response to the need to record agricultural surplus. In Egypt, pictograph and ideogram hieroglyphics similarly allowed for human communication, an analysis of the divine, and stories of humans and gods to be not just told but also recorded. Toward the end of the second millennium BCE, the Phoenicians of the eastern Mediterranean developed a simple, twenty-two-letter alphabet that became the basis for the Greek and Latin alphabets. The Phoenician alphabet allowed Greek philosophers and scientists to begin to develop their conceptions of existence and analyses of reality.

PYRAMIDS

Scholars are divided about the chronology of pyramid building in world history: Who were the first to build pyramids, the Sumerians or the Egyptians? The Egyptians most certainly were the master pyramid builders. They constructed pyramids in several stages, using ramps on which the large limestone blocks would be dragged up to the level of building activity. The higher the pyramid got, the wider the ramp had to be and the greater the number of ramps needed to accommodate the increasing structural pressure. Ramps would doubtless have been built along the sides of the pyramid. A step pyramid structure would have formed the inside of the outer pyramid to bolster it. Pyramids like the Great Pyramid of Cheops were built in the summer with the rising of the Nile so that water would bring barges close to the construction site. Some scholars hypothesize that the Egyptians developed a working knowledge of pi (π) in building the pyramids.

A number of distinct civilizations, worldwide, constructed pyramids: Egyptians, Mesopotamians, Harrapan peoples of the Indus River, ancient Peruvians, ancient Mayans, ancient Olmec peoples all built pyramids of differing sizes and shapes. Some scholars who believe in the diffusion of culture from one source throughout the world believe that ancient mariners sailed reed ships across the Atlantic and Indian Oceans, bringing the idea of an artificial mountain, or pyramid; hence we find similar constructions throughout the world.

The Egyptians also solved problems in hydraulics, the science of the movement and force of water. Dependent on the rising and falling of the Nile River, Egyptians had to organize their activities to utilize most efficiently the land's water resources. The Nile floods in the spring when the snow of the highlands of Ethiopia melts and the Nile grows big, overflowing its low banks to inundate the land of upper (southern) and lower (northern) Egypt. The spring rising of the Nile inundates the parched land of Egypt, actually swelling the soil and making drainage difficult but agriculture successful. Land of higher elevation surrounding the Nile required artificial watering, which the Egyptian farmer accomplished by using a tool called the shadoof. The shadoof had a pail tied to a rotating pole, connected to a stationary pole, that acted as a lever to raise the pail filled with water, rotate it to a waiting pool or channel, empty it, then return it to the river to retrieve more water. Similar was the *sakia*, a rotating waterwheel fitted with cups that would scoop up water as the wheel rotated and then, upon reaching the pinnacle and beginning the descent, deposit the water in a waiting canal. The Egyptians of the Old Dynasty of the third millennium BCE had basins made of stone and lavatories and copper pipes to eliminate water and waste. Egyptian metallurgists and craftspersons even devised copper pipes and pipe fittings, plugs, and drains.

THE BRONZE AGE

Metallurgy, the science of metals, is one of the oldest forms of inquiry and applied science in human history. Precivilized societies used metals such as gold, silver, and copper for decorative purposes. Metalworkers used stone tools to hammer metal into such items as necklaces and bracelets. During the fifth millennium BCE, humans began to fashion more utilitarian implements, tools and weapons, from copper, but it was not until the fourth millennium BCE that copper was smelted and combined with arsenic, lead, and tin to form various alloys of bronze. The Bronze Age existed for several millennia, during which bronze was the most effective metal used in tools and weapons. During the first millennium BCE, iron and its derivative, steel, became the material of choice for tools and weapons. Copper, bronze, wrought iron, lead, and steel were henceforth the most useful metals, supplemented in more recent times by, for example, aluminum, nickel, silicon, and titanium.

The importance of technology in the development of civilization is such that scholars have long identified stages in the advance of human society according to tool use and technological development. The Stone Age prior to the fifth millennium BCE gave way to the Copper Age, followed shortly by the Bronze Age, which lasted for over three thousand years. During the fifth millennium BCE, smiths learned how to mine copper stone (chalcopyrite) and extract copper by means of primitive smelting techniques, which allowed for the creation of copper tools and weapons. Copper, which is malleable, was superior to stone and wood and could be repeatedly heated and reused, and the edges of copper weapons could be repeatedly sharpened. In the fourth millennium BCE, however, smiths discovered that bronze tools and weapons were stronger, and lasted longer, than comparable copper tools and weapons. The smelting of copper and tin to produce bronze required very high temperatures (over one thousand degrees centigrade), which demanded forges and bellows. Bronze produced by smelting copper (which contains a small amount of arsenic), however, could be done at lower temperatures in a large kiln with a blowpipe. As society grew in sophistication, in Mesopotamia and Egypt at the beginning of the third millennium BCE, the forges of metalworkers allowed for a superior alloy of bronze to be produced. A problem for early metallurgists was the difficulty in extracting tin from cassiterite. At the same time, few places in the ancient Near East had deposits of copper stone and cassiterite in the same locales. The demand for tin, therefore, stimulated trade between societies.

Initially bronze was fashioned by combining copper with arsenic, which yielded a stronger material than copper alone. The smelting process, however, produced dangerous fumes from the arsenic, disabling or killing the smith. By the end of the fourth millennium, bronze workers had substituted tin for arsenic. Bronze was henceforth the tool and weapon material of choice, the production of which encouraged war and trade and the expansion of Bronze Age civilization.

Smiths were sufficiently important in ancient societies that they were represented by (and anthropomorphized as) smith gods, such as the Mesopotamian Gibil, the Egyptian Ptah, the Greek Hephaistos, and Latin Hephaestus. Hephaestus was different from the other Olympian deities as he was misshapen and crippled. Was it perhaps because Hephaestus, or his earthly counterparts, smelted copper and arsenic to produce bronze? Hephaestus's deformity has some similarity to arsenic poisoning, which doubtless plagued the earliest smiths, until someone over five thousand years ago realized the connection between arsenic and illness and substituted tin for arsenic. Since arsenic is naturally occurring in copper in small amounts, whereas tin is difficult to extract from the mineral cassiterite, the switch was difficult yet worth the effort, as it preserved the health and lives of craftspersons and produced a harder bronze for tools and weapons.

Bronze Age civilizations included Sumeria in Mesopotamia; the Minoan civilization of Crete; the Mycenaean civilization of Greece; Egypt along the Nile River; the Harappan civilization of the Indus River valley; and the Shang civilization of China. Anatolia (modern Turkey) and Mesopotamia had plentiful deposits of copper and tin, as well as iron.

Change wrought by bronze was a slow revolution, rather like the domestication of agriculture thousands of years earlier, which transformed society and culture. Historians have long referred to this time of the development of human technology as the Bronze Age. Bronze, in societies such as ancient Mesopotamia and ancient Egypt, was the catalyst for change and progress. A simple farming, Copper Age society did not have the vision and drive of Bronze Age society, which featured a burgeoning population, agricultural surplus, specialized occupations, an expanding class structure, a growing economy, and greater social organization and governmental control. The development of bronze was at the heart of this transforming society.

The foundation of Bronze Age society was the increasing agricultural surplus of fourth-millennium Mesopotamia and Egypt. This surplus resulted from leadership and organization, in which a structured government organized people to labor and build—for example, in Mesopotamia—canals to channel floodwaters to control flood damage and to irrigate parched earth. Such organization paid off in increasing agricultural yields. Those who had control over the surplus food were the powerful and wealthy, the upper echelons of society—landowners, priests, merchants, warriors and kings, who controlled the craftspersons, shopkeepers, teamsters, farmers, laborers, and slaves. People who lived in the same area, who worked together, forged a common identity, a sense of place. Thriving communities with good harvests clearly had the protection of the divine; as a result, localized deities were worshipped, led by a class of priests, and monuments were built to the gods. Cultivated lands, surplus food, growing wealth, walled towns, and burgeoning cities with divine patrons: all contributed to Bronze Age culture. At its heart were the tools used to plant and build and the weapons used to protect and conquer. Agricultural surplus allowed trade caravans to bring from other places—such as the mountains of northern Mesopotamia—the tin needed for the forging of bronze tools and weapons. Smiths became respected craftspersons able to know the various combinations of bronze and tin to produce soft or hard bronze: less tin (one part tin to fifteen or sixteen parts copper) produced a malleable bronze, whereas more tin (one part tin to nine or ten parts copper) produced bronze for casting.

Bronze for several thousand years was the most effective metal for making swords, shields, spear tips, and daggers. Bronze does not rust (oxidize) like iron and was more efficient and potentially lethal than any number of wood, bone, stone, copper, or iron weapons. This superior form of military technology allowed for the first empires in world history, such as that of Sargon of Akkad at the end of the third millennium BCE. Warfare and trade led to the diffusion of bronze weapons and tools, as well as of knowledge of how to make bronze. Bronze Age cultures were spread throughout the globe. The earliest, besides the Mesopotamian and Egyptian, included the Phoenicians of Tyre, Sidon, Byblos, Carthage, and Cadiz, who garnered control of the trade in tin during the first millennium BCE. Other Bronze Age cultures included the Harappan of the third millennium BCE Indus River valley and the Shang culture of second millennium BCE China.

One of the most fascinating and least known of Bronze Age societies was centered on the island of Crete during the third and second millennia BCE. Sir Arthur Evans, who performed the first archaeological digs at Knossos and other Cretan cities, designated the forgotten civilization Minoan after Minos, the mythical king of Crete and son of Zeus. Minoan culture was sophisticated for its time, partly derivative from Near East civilizations. There was a distinct social structure that included a royal house, aristocratic priests, middle-class craftspersons, merchants, sailors, and farmers and slaves. The remains of the palace at Knossos shows an intricate, well-decorated structure with enough rooms and halls to make it seem like a labyrinth to later generations. The Minoans had a thalassocracy, an empire built on the sea, on trade, and on a fleet of wooden triremes imposing the will of King Minos and his successors on surrounding subject states. They were devoted to the worship of the mother goddess, and the bull was sacred. Sculpture, art, metallurgy, and pottery were well developed at Crete. The highpoint of Minoan civilization was the development of a written script, which scholars call Linear A.

Archaeologists have discovered a similar culture at Santorini, a small island sixty miles north of Crete. A massive volcanic eruption in the fifteenth century BCE partially destroyed the island, called Thera in antiquity, and ended a beautiful culture of sophisticated stonework, metallurgy, multistoried buildings, trade, and art. The ancient Therans built a town (Akrotiri) that included colonnaded homes decorated with bright colors and beautiful murals of sea creatures and simple pleasures. Lead pipes brought water to some homes and served as part of an elaborate sewage system that included drains below slabs of stone that made up roads and alleys.

Greece and the islands of the Aegean were initially subject to the empire of Crete before gaining independence by the mid-second millennium BCE. This civilization, centered at the Greek Peloponnesus, was unknown until its discovery by Heinrich Schliemann in the nineteenth century. Schliemann had already discovered the ancient site of Troy, in northwest Turkey. He claimed to have discovered the homeland of King Priam and Prince Hector of Troy and the kingdoms of Menelaos and Agamemnon, brothers and kings of Sparta and Mycenae and Argos. Schliemann referred to the Bronze Age cultures of the Aegean and Greece as the Mycenaean civilization, whereas modern scholars usually refer to it as Helladic.

The Mycenaeans were similar to the Minoans in their sophisticated metallurgy, shipbuilding, trade, military, and fortresses. They possessed a form of writing, perhaps derived from the Minoan civilization, which scholars call Linear B. Kings and aristocrats ruled, and most people were farmers and herders, but there was also a middle class of craftspersons, scribes, physicians, and merchants. The Mycenaean civilization was patriarchal, as reflected in its deities, dominated by the male god Zeus. The people of Greece, like those of Crete, tried to explain the workings of nature but could do no better than to imagine supernatural forces at work behind the immensity of the sky, the depths of the sea, the thunderbolt, the fertility of soil and humans, human emotions, illness, and death. The Greek gods, ruling from Mount Olympus, personified the forces of nature for these Bronze Age people.

THE IRON AGE

The Bronze Age came to an end with the emergence of the Iron Age at the end of the second millennium BCE. The use of iron, like bronze, was spread through war. Invaders from northern Europe wielding iron weapons migrated into the Balkans and Asia Minor and traveled south to Palestine and Egypt. Nomadic and primitive, illiterate and violent, but somehow knowledgeable of the secret to smelting iron, they were largely unstoppable. The Greeks called the invaders Dorians; they destroyed the Mycenaean civilization and took control especially of western Greece and the Peloponnesus. Perhaps they had a role in the fall of Troy. These same invaders conquered the Hittite Empire of Asia Minor, threatened (as the Philistines) the Hebrews in Palestine, and attacked the Egyptians. Egypt repelled them, but the Egyptians were nevertheless terrified of these people from the sea. The invaders are sometimes referred to as the *sea peoples*.

This revolution in metallurgy that led to the dissemination of power and wealth and the rise of more widespread tools for farmers and weapons for soldiers occurred at the end of the second millennium BCE in Asia Minor. This was the development of the process of digging iron ore from the ground and heating it to such high temperatures that finished iron could be wrought, using the anvil and hammer; sufficient *carburization*, melding a small portion of carbon with the iron, produced steel.

The first iron users were the Hittites of eastern Anatolia, who relied on the expertise of iron-working Armenians. These people had apparently discovered the means to heat iron in the same fashion as copper was heated to drive out the impurities. Smiths heated iron ore, removed it, beat it, heated it again, and repeated the process, a method that produced wrought iron, which lacked carbon and was more malleable even than bronze. Wrought iron was not useful as a weapon and was essentially decorative. The Egyptians—who borrowed the use of iron from the Hittites, at first using meteoric iron, calling it the "iron of heaven"—used wrought iron; it was easier to work with, hence more available. The process of turning iron into steel probably occurred for the first time toward the end of the second millennium BCE in Anatolia—the Hittite Empire. It involved the use of the forge and bellows. Charcoal was heated to extreme temperatures—white hot, over two thousand degrees Fahrenheit—and then the iron was smelted to a point that impurities were removed, and ultimately steel was produced. Steel weapons, more rare in the ancient world, were superior to bronze weapons. The iron-using cultures were less sophisticated than the bronze cultures were, but as time passed during the first millennium BCE, even the bronze cultures adopted the use of iron. Bronze tools and weapons were expensive and difficult to make; thus they were limited to aristocrats. Iron signaled the coming power of the middle-class merchant and farmer of the first millennium BCE. The spread of literacy and coinage also helped in the movement of the masses toward economic and social power and influence compared to the old, landed aristocracy. And then, increasingly, philosophy and science became activities in which people besides the privileged few could participate.

The first peoples to use iron weapons with a steel edge—the Hittites of Turkey, the Dorians who invaded Greece and destroyed the Mycenaean civilization, and the sea peoples, such as the Philistines, who swept through the eastern Mediterranean ravaging and conquering—were not noted for their cultural advances. Yet these seemingly primitive peoples conquered more sophisticated peoples who still used bronze weapons. Particularly after 1000 BCE in the Mediterranean region, warriors turned to iron as the weapon of choice. During the first millennium BCE, bronze weapons were still used, but iron was preferred. Hittite civilization collapsed because of the invasions from northern Europe, which also destroyed the Mycenaean civilization. Mycenaean peoples who fled before the attacks of iron-wielding Dorians were the Ionians, people who inhabited during historical times cities ranging from Athens in eastern Greece to Halicarnassus in southwest Turkey to Ephesus on the west coast of Anatolia and the various island city-states of the Aegean Sea. The Greeks during the Archaic and Classical Ages knew that the people of the Greek Peloponnesus differed from the people of the Aegean and Anatolia. The Ionians were cultured and scientific while the Dorians were warlike and aggressive. Even so, they shared the Greek language, similar customs, the same gods and goddesses and mythology, and called themselves Hellenes. They were descended, they thought, from the mythical hero Hellen, the son of

Deucalion. Even so, people such as the Spartans were rigorous in the training for war, while the Athenians were rigorous in the training for art and oratory. The Spartans and Corinthians lauded people of action, willing to die for the cause of their people. Athenians and Milesians lauded those such as Demosthenes who could make words flow from the mouth like honey.

GREEK AND ROMAN METALLURGY

The ancient science of metallurgy was influenced by philosophic conceptions of the elements: earth, air, fire, water, and, according to the Greek polymath Aristotle, aether. The earliest treatise on metallurgy, by Aristotle in his *Meteorology*, hypothesized that metals and stones are produced by the exhalations of the earth. "The vaporous exhalation is the cause of all metals," he wrote, "those bodies which are either fusible or malleable such as iron, copper, gold. All these originate from the imprisonment of the vaporous exhalation in the earth" (trans. Webster). Aristotle's notion that metals were a product of the congealing of earth's vapors into a solid mass was echoed by his student Theophrastus, in *On Stones*. Other ancient metallurgical writers included Posidonius of Rhodes, who believed some materials were formed by frozen vapor, and Pliny the Elder, who argued that metals can change, grow, and regenerate. For ancient scientists, the division between the spiritual and physical was very narrow, even nonexistent. Greco-Roman Stoic philosophers, for example, believed that spirit or soul (*pneuma*) was found in all things, even stones and metals. Scientists and philosophers in China at the same time had remarkably similar ideas. The five elements of fire, water, wood, earth, and metal, Chinese philosophers of the late first millennium argued, were represented by two complementary primary spiritual and physical phenomena, yin and yang. The combination of mystical and spiritual beliefs with physical science in the Greco-Roman and Asian cultures resulted in the pseudoscience of alchemy.

During the Hellenistic Age of Greco-Roman science, scientists began experimenting with the possibilities of transforming one substance into another by a metamorphosis of the material and spiritual qualities of the substance. Using sulfur and mercury, for example, the alchemist sought by heating and cooling to transmute substances from black or dark to white or light, and eventually to gold. In the first century CE, Pliny the Elder believed that mixing mercury and gold, unheated, resulted in all impurities in the gold being removed. Chinese alchemists at the same time worked with similar materials, including cinnabar, to perform the same hopeful result.

Metallurgy and alchemy resulted from humans attempting to apply knowledge to the problems of existence; for Greek philosophers and scientists of western Anatolia—Ionian Greece—philosophy, science, and mathematics, understanding natural phenomena, were the means to the end of human control.

Even after the introduction of iron, bronze was used as readily for ornamental purposes and statues as for tools and weapons. There was an extensive bronze industry throughout the Greco-Roman world in the centuries before and after the birth of Christ. According to the Roman scientist and historian Pliny the Elder, Greco-Roman smiths turned out bronze that was comprised of one part tin to eight parts copper. The best bronze for use in molds to create bronze statues, according to Pliny, also included small amounts of lead and silver.

Bronze continued to be employed in art, sculpture, and weaponry in ensuing centuries. Notwithstanding the growing sophistication in the manufacture of steel, bronze is still used to produce materials that can resist corrosion and wear and provide for more flexibility than steel or other metals can. Bronze was for centuries the material used in cannons on land and sea, partly because of its resistance to highly corrosive salt water.

GREEK ENGINEERING AND TECHNOLOGY

The Greeks were responsible for some wonderful inventions, buildings, and cities. The most famous—and still awe-inspiring to look at—is the Parthenon, named for the maiden or virgin (*parthenos*) goddess Athena, designed by Ictinus and Callicrates, and built from 447 to 438 BCE under the authority of Pericles. Strabo wrote of Athens and the Parthenon that "the city itself is a rock situated in a plain and surrounded by dwellings. On the rock is the sacred precinct of Athena, comprising both the old temple of Athena Polias, in which is the lamp that is never quenched, and the Parthenon built by Ictinus, in which is the work in ivory by Pheidias, the Athena" (trans. Jones). Phidias was in charge of the artistic rendering of the Parthenon as well as other temples on the Athenian Acropolis.

ALEXANDRIA AND DINOCRATES OF RHODES

The Hellenistic city of Alexandria was a center of research into hydraulics, engineering, and inventiveness. Alexandria was planned by Alexander when he entered Egypt as conqueror and was named pharaoh of Egypt in 332 BCE. Among Alexander's entourage was an urban designer and architect, Dinocrates (Deinocrates) of Rhodes (fl. end of the fourth century BCE). According to an anecdote preserved by Vitruvius, Dinocrates came to Alexander's attention when he proposed converting Mount Athos, which is situated on a peninsula jutting into the Aegean Sea in northern Greece, into a huge statue of Alexander. Dinocrates reputedly introduced himself to Alexander as a Macedonian architect: "I have made a design for the shaping of Mount Athos into the statue of a man, in whose left hand I have represented a very spacious fortified city, and in his right a bowl to receive the water of all the streams which are in that mountain, so that it may pour from the bowl into the sea." Alexander liked the idea, though he showed himself wiser than the architect: he

> inquired whether there were any fields in the neighbourhood that could maintain the city in corn. On finding that this was impossible without transport from beyond the sea, "Dinocrates," quoth he, "I appreciate your design as excellent in composition, and I am delighted with it, but I apprehend that anybody who should found a city in that spot would be censured for bad judgement. For as a newborn babe cannot be

> nourished without the nurse's milk, nor conducted to the approaches that lead to growth in life, so a city cannot thrive without fields and the fruits thereof pouring into its walls, nor have a large population without plenty of food, nor maintain its population without a supply of it. Therefore, while thinking that your design is commendable, I consider the site as not commendable; but I would have you stay with me, because I mean to make use of your services." (Trans. Morgan)

This he did, appointing Dinocrates as lead architect and planner for his conception of a new city at the mouth of the Nile. "Alexander, observing a harbour rendered safe by nature, an excellent centre for trade, cornfields throughout all Egypt, and the great usefulness of the mighty river Nile, ordered [Dinocrates] to build the city of Alexandria, named after the king" (Morgan). Dinocrates was a student of Hippodamus of Miletus (498–408), who believed in an ordered city of rectangular streets and buildings where the predominant shape would be the right angle.

Strabo described the layout of Alexandria in the first century BCE:

> The shape of the area of the city is like a chlamys; the long sides of it are those that are washed by the two waters, having a diameter of about thirty stadia, and the short sides are the isthmuses, each being seven or eight stadia wide and pinched in on one side by the sea and on the other by the lake. The city as a whole is intersected by streets practicable for horse-riding and chariot-driving, and by two that are very broad, extending to more than a plethrum in breadth, which cut one another into two sections and at right angles. And the city contains most beautiful public precincts and also the royal palaces, which constitute one-fourth or even one-third of the whole circuit of the city; for just as each of the kings, from love of splendour, was wont to add some adornment to the public monuments, so also he would invest himself at his own expense with a residence, in addition to those already built, so that now, to quote the words of the poet, 'there is building upon building.' All, however, are connected with one another and the harbour, even those that lie outside the harbour. (Trans. Jones)

Dinocrates worked alongside the hydrologist or mining engineer Crates of Olynthus, whose labors in Boeotia working with flooding lakes came to Alexander's attention. Crates designed the intricate waterworks and canals of Alexandria.

Dinocrates was well known for other architectural works, too, such as the funeral pyre for Hephaestion, Alexander's boyhood friend and closest of his generals, who preceded Alexander in death by three months. According to Diodorus Siculus,

> Alexander collected artisans and an army of workmen and tore down the city wall to a distance of ten furlongs. He collected the baked tiles and levelled off the place which was to receive the pyre, and then constructed this square in shape, each side being a furlong in length. He divided up the area into thirty compartments and laying out the roofs upon the trunks of palm trees wrought the whole structure into a square shape. Then he decorated all the exterior walls. Upon the foundation course were golden prows of quinqueremes in close order, two hundred and forty in all. Upon the cat-heads each carried two kneeling archers four cubits in height, and (on the deck) armed male figures five cubits high, while the intervening spaces were occupied by red banners fashioned out of felt. Above these, on the second level, stood torches fifteen cubits high with golden wreaths about their handles. At their flaming ends perched eagles with outspread wings looking downward, while about

> their bases were serpents looking up at the eagles. On the third level were carved a multitude of wild animals being pursued by hunters. The fourth level carried a centauromachy rendered in gold, while the fifth showed lions and bulls alternating, also in gold. The next higher level was covered with Macedonian and Persian arms, testifying to the prowess of the one people and to the defeats of the other. On top of all stood Sirens, hollowed out and able to conceal within them persons who sang a lament in mourning for the dead. The total height of the pyre was more than one hundred and thirty cubits. (Trans. Oldfather)

Dinocrates also worked on temples, including at Delphi and Delos, and he was involved in the restoration of the Temple of Artemis of Ephesus.

PHILO OF BYZANTIUM (280–220 BCE)

Philo was the student of (or otherwise affiliated with) Ctesibius of Alexandria (285–222), an inventor and engineer who made advances in understanding air pressure and hydraulics. An engineer and inventor as well, Philo's specialty was pneumatics, and he wrote a book of that title, in which he developed a number of theories and a multitude of devices such as an air pump that operated bellows, a chain pump, whistles and sirens, and pistons, all of which used principles of air expansion and contraction due to heat. His work, like that of Ctesibius and the later scientist Hero of Alexandria, foreshadowed modern steam power. Like Hero, however, Philo's inventions were used more for show or leisure than for performing work. Philo is best known for identifying and writing about the Seven Wonders of the World.

Philo's *Seven Wonders of the Ancient World* discusses those astonishing examples of human creativity and construction of ancient culture that the modern world continues to look upon in astonishment. The seven wonders were the Great Pyramid of Egypt, the Colossus of Rhodes, the Lighthouse at Alexandria, the Mausoleum at Halicarnassus, the Hanging Gardens of Babylon, the Temple of Artemis at Ephesus, and the Temple of Zeus at Olympia.

The Great Pyramid of Cheops, built about 2650 BCE, was observed by Herodotus on his visit to Egypt during the fifth century. Herodotus learned from Egyptian priests that the pyramid was built with the labor of one hundred thousand men who dragged massive limestone blocks from barges that had sailed up canals filled with water from the flooding Nile. Levers were used to raise the blocks from each level or step to the next. The height of the pyramid was the same as each side of its base. Pliny the Elder considered the pyramids unforgivable examples of hubris, built by peasant-slaves in massive building projects. The largest, he said, covered five acres and was 725 feet high. The ancients were uncertain how they were built—Pliny thought perhaps by means of mud brick ramps or heaps of salt. Pliny doubted the theory that the huge limestone blocks were brought to the building site during the Nile's flooding.

The Colossus of Rhodes, designed and constructed by Chares of Lindos on the island of Rhodes, was a massive bronze statue, over one hundred feet high, of the sun god Apollo. The statue took twelve years to build, from 292 to 280 BCE. Bronze covered an iron structure that was built by piling up hills of sand around the statue as it rose in height. Pliny the Elder recorded that the Colossus stood for sixty-six years, falling to pieces during an earthquake.

The Lighthouse of Alexandria at Pharos was designed and built by Sostratos of Cnidus. The tower was built in three levels and was four hundred feet high. Pliny the Elder claimed that the lighthouse cost eight hundred talents—a talent was a unit of measure of silver or gold that equaled roughly six thousand drachmas in Classical Athens. Strabo the geographer provided a description of Pharos and the lighthouse:

> Pharos is an oblong isle, is very close to the mainland, and forms with it a harbour with two mouths; the shore of the mainland forms a bay, since it thrusts two promontories into the open sea, and between these is situated the island, which closes the bay, for it lies lengthwise parallel to the shore. Of the extremities of Pharos, the eastern one lies closer to the mainland and to the promontory opposite it (the promontory called Lochias), and thus makes the harbour narrow at the mouth; and in addition to the narrowness of the intervening passage there are also rocks, some under the water, and others projecting out of it, which at all hours roughen the waves that strike them from the open sea. And likewise the extremity of the isle is a rock, which is washed all round by the sea and has upon it a tower that is admirably constructed of white marble with many stories and bears the same name as the island. This was an offering made by Sostratus of Cnidus, a friend of the kings, for the safety of mariners, as the inscription says: for since the coast was harbourless and low on either side, and also had reefs and shallows, those who were sailing from the open sea thither needed some lofty and conspicuous sign to enable them to direct their course aright to the entrance of the harbour. (Trans. Jones)

The Mausoleum of Halicarnassus was designed by Pythius of Halicarnassus from 355 to 350 BCE as a marble monument to the king of Caria Mausolus (r. 377–353). It was fifty-five meters high, was supported by thirty-two columns, and had a step-pyramid at the top. According to Vitruvius, Pythius believed that education in the liberal arts was key to such brilliant architecture, writing in his "Commentaries that an architect ought to be able to accomplish much more in all the arts and sciences than the men who, by their own particular kinds of work and the practice of it, have brought each a single subject to the highest perfection" (trans. Morgan).

According to legend, King Nebuchadnezzar II (r. 604–562) built gardens for his wife, who missed the forested mountains of her native land. Ancient authorities described the gardens from hearsay. Reputedly an elaborate system of irrigation kept the trees and shrubs blooming in constant dazzling colors. The ruins of the Hanging Gardens of Babylon at the southern palace of King Nebuchadnezzar II suggest layered walls that hosted a diverse flora. Diodorus Siculus described the gardens this way:

> There was also, because the acropolis [of Babylon], the Hanging Garden, as it is called, which was built . . . by a later Syrian king to please one of his concubines; for she, they say, being a Persian by race and longing for the meadows of her mountains, asked the king to imitate, through the artifice of a planted garden, the distinctive landscape of Persia. The park extended four plethra on each side, and since the approach to the garden sloped like a hillside and the several parts of the structure rose from one another tier on tier, the appearance of the whole resembled that of a

> theatre. When the ascending terraces had been built, there had been constructed beneath them galleries which carried the entire weight of the planted garden and rose little by little one above the other along the approach; and the uppermost gallery, which was fifty cubits high, bore the highest surface of the park, which was made level with the circuit wall of the battlements of the city. Furthermore, the walls, which had been constructed at great expense, were twenty-two feet thick, while the passage-way between each two walls was ten feet wide. The roofs of the galleries were covered over with beams of stone sixteen feet long, inclusive of the overlap, and four feet wide. The roof above these beams had first a layer of reeds laid in great quantities of bitumen, over this two courses of baked brick bonded by cement, and as a third layer a covering of lead, to the end that the moisture from the soil might not penetrate beneath. On all this again earth had been piled to a depth sufficient for the roots of the largest trees; and the ground, which was levelled off, was thickly planted with trees of every kind that, by their great size or any other charm, could give pleasure to beholder. And since the galleries, each projecting beyond another, all received the light, they contained many royal lodgings of every description; and there was one gallery which contained openings leading from the topmost surface and machines for supplying the garden with water, the machines raising the water in great abundance from the river, although no one outside could see it being done. (Trans. Oldfather)

The Temple of Artemis (Roman Diana) at Ephesus was, according to Pliny the Elder, 425 long and 225 feet wide, with 127 columns of up to 60 feet in height. The architect, Chersiphron of Knossos, received, it was thought, guidance in building the temple from Artemis herself, in dreams. The temple was built on low-lying, marshy ground to escape the impact of earthquakes. Sandbags, stacked one upon another, supported the weighty stone; once the capital and lintel were in place, Chersiphron had the sand released from the bags, one after another, so gently easing the massive stones into place. Fire destroyed the Temple of Artemis in 356 BCE, the same day, it was claimed, as the birth of Alexander of Macedon. Soothsayers interpreted the omen as indicating that Alexander would scorch all of Asia. The temple was rebuilt in 334 BCE by Dinocrates of Rhodes, the urban planner of the city of Alexandria.

The Temple of Zeus at Olympia was designed by Libon of Ellis and was completed in 456 BCE. The wonder, a statue of Zeus designed and sculpted by the art director of the Parthenon, Phidias, was forty-four feet tall and twenty-two feet wide.

ARCHIMEDES (287–212 BCE)

Meanwhile, in the western Mediterranean, scientist Archimedes of Syracuse worked as a mathematician, hydrologist, engineer, and designer of military machines. In mathematics, his work on geometry, particularly cones, spheres, and cylinders, was unsurpassed. He anticipated calculus, and studied in depth hydrostatics, mechanics, matter, and force. He invented the screw used in irrigation and solved many engineering problems associated with the use of the pulley, wedge, and lever.

Archimedes developed several basic propositions of hydraulics that became a basis for many of his inventions. His studies of fluid displacement showed that when pressure is applied to a given volume of water the part under pressure will necessarily displace another part not under the same pressure. He examined the weight of objects relative to volume of water: a solid of the same weight as a proportional amount of fluid will float at just below the surface. A solid of lesser weight than a proportional amount of fluid will extend part of its surface above the water. The amount of water displaced by the solid is of the same weight as the water. If a solid is forced into water, the reciprocal force is proportional to the weight of the water and weight of the water displaced. Archimedes's screw was a simple machine that resembled a large screw enclosed by a cylindrical frame that, when set in the water and rotated manually, would wind about, bringing water to the top, which was above ground, where it would be emptied into a waiting pool.

THE SIEGE OF SYRACUSE

During the Second Punic War and the battle for Sicily in 212 BCE, the Romans laid siege to the Greek city of Syracuse. The rulers of the city compelled Archimedes to employ his inventions based on research into the principles of mechanics to help defend the city. Plutarch wrote:

> When, therefore, the Romans assaulted the walls in two places at once, fear and consternation stupefied the Syracusans, believing that nothing was able to resist that violence and those forces. But when Archimedes began to ply his engines, he at once shot against the land forces all sorts of missile weapons, and immense masses of stone that came down with incredible noise and violence; against which no man could stand; for they knocked down those upon whom they fell in heaps, breaking all their ranks and files. In the meantime huge poles thrust out from the walls over the ships sunk some by the great weights which they led down from on high upon them; others they lifted up into the air by an iron hand or beak like a crane's beak and, when they had drawn them up by the prow, and set them on end upon the poop, they plunged them to the bottom of the sea; or else the ships, drawn by engines within, and whirled about, were dashed against steep rocks that stood jutting out under the walls, with great destruction of the soldiers that were aboard them. A ship was frequently lifted up to a great height in the air (a dreadful thing to behold), and was rolled to and fro, and kept swinging, until the mariners were all thrown out, when at length it was dashed against the rocks, or let fall. (Trans. Dryden)

According to Plutarch, Archimedes died during the Roman siege of Syracuse, having incurred the wrath of Roman soldiers with his military machines.

MILITARY SCIENCE

Archimedes exemplified Greek achievements in military technology and organization. Greek city-states of the Classical Age developed the best fighting force of the time, centered on hoplite foot soldiers. The hoplites were heavily armed infantrymen who marched in a phalanx formation. They were highly maneuverable and, when they locked shields together, appeared impregnable to their enemies. Greek hoplite armies successfully defeated Persian armies in the fifth and fourth centuries BCE and had some success against the Romans in the third century.

Vitruvius claimed that the Carthaginians were the first to develop the battering ram to destroy city walls. He wrote,

> A carpenter from Tyre, Bright by name and by nature, was led by this invention into setting up a mast from which he hung another crosswise like a steelyard, and so, by swinging it vigorously to and fro, he threw down the wall of Cadiz. Geras of Chalcedon was the first to make a wooden platform with wheels under it, upon which he constructed a frame work of uprights and crosspieces, and within it he hung the ram, and covered it with oxhide for the better protection of the men who were stationed in the machine to batter the wall. As the machine made but slow progress, he first gave it the name of the tortoise of the ram. (Trans. Morgan)

Demetrius (c. 337–283), a king of Macedon and son of Antigonus I, one of Alexander's generals, enlarged and made more grand siege engines such as the turtle, a protective battering ram, and a large drill for boring. Similar kinds of military inventions came from Diodes of Pella, who accompanied Alexander on his Asian campaigns. Alexander's conquests introduced Greek scientists to the use of petroleum, sulfur, and arsenic in formulating weapons and poisons in warfare.

The Greeks, specifically the historian Thucydides, claimed that the first fighting warship, the trireme, was an invention of the Peloponnesian city-state of Corinth. Ameinocles was a Corinthian shipwright who was the inventor of the warship, Thucydides wrote. With these triremes the Corinthians defeated the Corcyraeans, reputedly in the seventh century. Notwithstanding Thucydides's Hellenic patriotism, the Phoenicians might have been the first to invent the trireme, but the Greeks were certainly the first to use fleets of such deadly ships in warfare. In particular, Thucydides's fellow Athenians built a huge empire based in part on their naval prowess. The trireme evolved from earlier Greek ships, the penteconter and the bireme, the former using fifty oarsmen, the latter having two banks of oars. Warships had hoplite marines aboard that would fight opponents on board or at shore. The trireme boasted a bronze-sheathed ram that could penetrate the hull of enemy ships, sending their crews quickly to the bottom of the sea. The Romans borrowed the Greek trireme, and after the First Punic War, began to dominate the Mediterranean with their versatile warships.

ROMAN ENGINEERING AND SCIENCE

The great builders of antiquity were the Romans. The Roman scientist was interested in knowledge that could be usefully applied to problems of everyday life as well as issues of order, mobility, structure, power, and majesty. The Roman practical genius combined with military necessity, the riches of empire, and soldiers who knew how to work and to build to create lasting roads and bridges that have withstood the passing millennia. Roads are useful for travel and trade, of course, but initially the Romans constructed roads as a consequence of, or to facilitate, military activity. Logistics, unimpeded communications and supply lines, were keys to Roman military success. The Roman legions built as they marched, confident that they would be returning the same way again and again once enemy territory was pacified and brought into the Roman Empire. The empire, as it expanded throughout the Mediterranean world, required good roads for communications. Often the roads traversed hilly, wet land that necessitated numerous bridges. The demands of civilization meant that the Roman builders of roads and bridges were constantly at work.

The first Roman road is in many ways still the most grand. The Via Appia, or Appian Way, connects the city of Rome with Capua to the south. It was built at the insistence of the Roman senator Appius Claudius Crassus in 312 BCE. The Via Appia, like subsequent roads, was built by Roman engineers using soldiers to do the labor. They dug a running pit three feet deep and fifteen feet wide and layered it with gravel and stone. Roman engineers did not use cement in building the long-lasting roads of the empire. Rather, basalt or limestone laid into a bed that was built using gravel and heated sand formed the road. The stones, basalt and silex, were quarried from ancient volcanic sites, brought to the construction site, and carefully cut. Workers placed the octagonal stones in a bed of gravel and heated sand. Workers wedged the stones together so tight as to not allow even a knife's edge to penetrate. Roman surveyors used the *groma*, a surveying instrument that relied on plumb lines, to build the road with a slight incline from the center to allow for good drainage. Low-lying, marshy regions required additional care and the building of causeways. The Via Appia traversed the Pontine Marshes and was forced to end at Lake Pontia in Campania at the Forum Appii. Beyond the lake, the Via Appia continued to Capua. Along the road, numerous tombs were built, beginning with that of its creator, Appius, who died shortly after the completion of the road that bears his name.

As the centuries passed, Roman roads changed to meet the conditions of different lands. Some roads were built with curbs. Lands such as Africa did not have volcanic stone, so other rock, such as limestone, was cut by masons into rectangles and laid closely together. Roman surveyors solved the problem of drainage and picked the best routes with the least fluctuation in elevation.

The Romans were also the greatest bridge builders in antiquity. Few people before the Romans understood the art and science of bridging rivers and streams. One reads about pontoon bridges and bridges formed by lashing boats together, but no country except Rome made bridges of stone that spanned rivers and time. Some of the most famous Roman bridges included the Milvian Bridge that crossed the Tiber, at the place where Constantine (r. 306–337) defeated Maxentius in 311 CE, and the mile-long bridge spanning the Danube, built by the emperor Trajan (r. 98–117) under the direction of the engineer Apollodorus of Damascus. The Romans were also the greatest hydraulic engineers of antiquity. Aqueducts were not unknown to other cultures—reputedly one Eupalinus built an aqueduct through a mountain on the island of Samos in the sixth century BCE. But the Romans made the building of aqueducts a science and an art. Appius Claudius Crassus, the creator of the Via Appia, was also the force behind the construction of the first aqueduct built to provide fresh water for the city. Marcus Agrippa (c. 63–12), the friend and associate of Octavian (Augustus) Caesar (r. 31 BCE–14 CE), brought Rome's hydraulics to their most efficient level; he was responsible for the Pont du Gard, a bridge and aqueduct that still stand. Agrippa oversaw constructing and repairing aqueducts, cleaning sewers, and ensuring a continual flow of water for private and public use by means of public fountains, public and private baths, and water to flush latrines and sewers. Wealthy Romans could have water brought to their homes by means of lead pipes. There was much illegal tapping of water, which the authorities attempted to stop, usually in vain.

By the first century CE, there were eleven aqueducts bringing water to the city. Roman aqueducts were astonishing structures of elegance and utility that relied on sophisticated engineering techniques. Apollodorus of Damascus designed the Alcántara in Spain. Another noteworthy aqueduct was built at Segovia, Spain, during the reign of Augustus. Valen's aqueduct was built during the late fourth century; it lasted 800 years and spanned the 150-mile breadth of the Balkans to Constantinople. Some of these imposing works were built with stone blocks arranged together with such precision as to withstand gravity, time, and geologic events. Others used cement, which the Romans developed, combining volcanic rock with lime, sand, and water. Bridges required a firm base in the marsh or river; engineers dug through silt to rock, to which they cemented ashlars of stone or pylons kept upright with huge boulders. Surveyors solved the problem of elevation and gravity in the construction of aqueducts—some of which still exist—which brought fresh water to towns using the force of gravity. Water flows as long as there exists an uninterrupted declivity.

FRONTINUS (C. 40–103 CE)

Sextus Julius Frontinus was the water commissioner in charge of the upkeep of the aqueducts that brought fresh water to the city of Rome during the reigns of Nerva (96–98) and Trajan. His treatise *On the Water-Management of the City of Rome* provides a detailed description of the origins, number, placement, upkeep, and utility of the Roman aqueducts. Rome's first aqueduct was the Aqua Appian, also named for Appius Claudius Crassus. Frontinus systematically described such aqueducts as the Old Anio, Marcia, Tepula, Julia, Virgo, Augusta, Claudia, and New Anio. He laid out their extent in paces and distinguished those parts that were subterranean or aboveground upon a series of arches. Frontinus carefully described the bases of measurements (in digits and inches) of the varied pipes used to convey water. He estimated the amount of water delivered by each aqueduct in *quinaria,* a unit of measure developed by the Romans that was 5/4 of a digit (which was 1/16 of a foot). The *quinaria* measured not only length but volume as well. Frontinus proclaimed that the aqueducts brought health to the city: fresh, pure water for drinking and bathing and water to cleanse the city streets and sewers. Work gangs of slaves maintained the aqueducts. Frontinus thought that the Roman aqueducts were Rome's great symbol, as the pyramids were to Egypt; yet the pyramids accomplished nothing, while the Roman aqueducts were not only awesome and beautiful but pragmatic as well.

Roman engineers learned how to distribute the material forces of weight and gravity by means of the vaulted arch, created by building a brick structural frame into which concrete was poured. The thrust of huge structures was balanced so that no buttresses were needed on walls. The emperor Hadrian

(117–138) showed his ability as an engineer in the design of the domed Pantheon, which became the form of choice for builders of public buildings for centuries to come. The Pantheon was built to honor all the world's deities. The diameter and height of the Pantheon are precisely the same, forming a perfect circle. The cupola at the extreme height of the ceiling is the only source of light. The dome relies on panels of concrete and other materials, built lighter at the top, heavier in descent. The panels of concrete on the dome also grow larger in descent, each supporting the one above.

The Roman Colosseum, begun during the reign of Nero (54–68), was built of stone without the use of cement. Upon completion, during the reign of Titus (79–81), it could seat as many as fifty thousand spectators, who watched gladiatorial contests, mock battles, and wild animal hunts. Of particular note was the velarium, a partial covering to keep the audience protected from sun and the elements. The velarium was vastly complicated, requiring hundreds of seamen to operate the miles of rope, pulleys, and winches.

MARCUS VITRUVIUS POLLIO (90–20 BCE)

Many of the ideas and techniques of Roman construction, engineering, and architecture are preserved in Vitruvius's *On Architecture.* Vitruvius's book, dedicated to Augustus Caesar, was a basic manual on the nature of architecture, including the characteristics of the architect, and the various fields of learning that the architect must study. "Let him be educated," he wrote, "skilful with the pencil, instructed in geometry, know much history, have followed the philosophers with attention, understand music, have some knowledge of medicine, know the opinions of the jurists, and be acquainted with astronomy and the theory of the heavens." He explained that the architect "ought to be an educated man so as to leave a more lasting remembrance in his treatises. Secondly, he must have a knowledge of drawing so that he can readily make sketches to show the appearance of the work which he proposes. Geometry, also, is of much assistance in architecture, and in particular it teaches us the use of the rule and compasses, by which especially we acquire readiness in making plans for buildings in their grounds, and rightly apply the square, the level, and the plummet. By means of optics, again, the light in buildings can be drawn from fixed quarters of the sky. It is true that it is by arithmetic that the total cost of buildings is calculated and measurements are computed, but difficult questions involving symmetry are solved by means of geometrical theories and methods." Furthermore, he wrote,

> Philosophy treats of physics . . . where a more careful knowledge is required because the problems which come under this head are numerous and of very different kinds; as, for example, in the case of the conducting of water. For at points of intake and at curves, and at places where it is raised to a level, currents of air naturally form in one way or another; and nobody who has not learned the fundamental principles of physics from philosophy will be able to provide against the damage which they do. So the reader of Ctesibius or Archimedes and the other writers of treatises of the same class will not be able to appreciate them unless he has been trained in these subjects by the philosophers.

He believed that the science of harmonics was important in constructing theaters: "In theatres, likewise, there are the bronze vessels which are placed in niches under the seats in accordance with the musical intervals on mathematical principles. These vessels are arranged with a view to musical concords or harmony, and apportioned in the compass of the fourth, the fifth, and the octave, and so on up to the double octave, in such a way that when the voice of an actor falls in unison with any of them its power is increased, and it reaches the ears of the audience with greater clearness and sweetness" (trans. Morgan).

The art and science of architecture, according to Vitruvius, involved order, arrangement, eurythmy or harmony, symmetry, propriety, and economy. These qualities go into buildings, "time-pieces," and machinery. Buildings are constructed for "defensive, . . . religious, and . . . utilitarian purposes." Under the former, "comes the planning of walls, towers, and gates, [and] permanent devices for resistance against hostile attacks." Vitruvius believed in a science to urban planning:

> For fortified towns the following general principles are to be observed. First comes the choice of a very healthy site. Such a site will be high, neither misty nor frosty, and in a climate neither hot nor cold, but temperate; further, without marshes in the neighbourhood. For when the morning breezes blow toward the town at sunrise, if they bring with them mists from marshes and, mingled with the mist, the poisonous breath of the creatures of the marshes to be wafted into the bodies of the inhabitants, they will make the site unhealthy. Again, if the town is on the coast with a southern or western exposure, it will not be healthy, because in summer the southern sky grows hot at sunrise and is fiery at noon, while a western exposure grows warm after sunrise, is hot at noon, and at evening all aglow. (Trans. Morgan)

As the heat of the sun drains humans of their energy, so heat itself is "a universal solvent, melting out of things their power of resistance, and sucking away and removing their natural strength with its fiery exhalations so that they grow soft, and hence weak, under its glow. We see this in the case of iron which, however hard it may naturally be, yet when heated thoroughly in a furnace fire can be easily worked into any kind of shape, and still, if cooled while it is soft and white hot, it hardens again with a mere dip into cold water and takes on its former quality." Expanding on this principle of heat, he wrote,

> For while all bodies are composed of the four elements, that is, of heat, moisture, the earthy, and air, yet there are mixtures according to natural temperament which make up the natures of all the different animals of the world, each after its kind. Therefore, if one of these elements, heat, becomes predominant in any body whatsoever, it destroys and dissolves all the others with its violence. This defect may be due to violent heat from certain quarters of the sky, pouring into the open pores in too great proportion to admit of a mixture suited to the natural temperament of the body in question. Again, if too much moisture enters the channels of a body, and thus introduces disproportion, the other elements, adulterated by the liquid, are impaired, and the virtues of the mixture dissolved. This defect, in turn, may arise from the cooling properties of moist winds and breezes blowing upon the body. In the same way, increase or diminution of the proportion of air or of the earthy which is natural to the body may enfeeble the other elements; the predominance of the earthy being due to overmuch food, that of air to a heavy atmosphere. (Trans. Morgan)

In devising the streets of the town, Vitruvius argued that it was important to take account of the winds: "cold winds are disagreeable, hot winds enervating, most winds unhealthy" (trans. Morgan). He described a scientific approach to designing city streets in accord with the different directions of the winds to achieve the most healthful experience of the inhabitants. Vitruvius's *On Architecture* describes building materials; temples; the nature of symmetry; public places; acoustics (as in a theater); private places, such as homes; hydrology, the principles and methods of constructing aqueducts; sundials, waterclocks, and astronomy; machines for building, hoisting, climbing, and accomplished by pneumatic principles; and waterwheels and water screws.

ROMAN MILITARY SCIENCE AND TECHNOLOGY

Vitruvius believed that there was a science to military exercises, in his case involving the creation of weapons of war. He believed that musical theory aided in the construction of siege engines:

> Music, also, the architect ought to understand so that he may have knowledge of the canonical and mathematical theory, and besides be able to tune ballistae, catapultae, and scorpiones to the proper key. For to the right and left in the beams are the holes in the frames through which the strings of twisted sinew are stretched by means of windlasses and bars, and these strings must not be clamped and made fast until they give the same correct note to the ear of the skilled workman. For the arms thrust through those stretched strings must, on being let go, strike their blow together at the same moment; but if they are not in unison, they will prevent the course of projectiles from being straight. (Trans. Morgan)

Of ballistics in warfare, he wrote, "Ballistae are constructed on varying principles to produce an identical result. Some are worked by handspikes and windlasses, some by blocks and pulleys, others by capstans, others again by means of drums. No ballista, however, is made without regard to the given amount of weight of the stone which the engine is intended to throw" (trans. Morgan). Vitruvius wrote of the science of devising successful fortifications. He discussed the ways to build walls to be impregnable to attack, including towers atop that provided the means of attacking those who tried to scale the walls; he advised against square towers, preferring round or polygonal, as being able to resist the force of the battering ram. Further,

> special pains should be taken that there be no easy avenue by which to storm the wall. The roads should be encompassed at steep points, and planned so as to approach the gates, not in a straight line, but from the right to the left; for as a result of this, the right hand side of the assailants, unprotected by their shields, will be next the wall. Towns should be laid out not as an exact square nor with salient angles, but in circular form, to give a view of the enemy from many points. Defence is difficult where there are salient angles, because the angle protects the enemy rather than the inhabitants. (Trans. Morgan)

HERON OF ALEXANDRIA (C. 10–75 CE)

The Alexandrian inventor, physicist, and mathematician Heron was a creative mind of the first century CE. Influenced by Aristotle and the atomists, he also built on the work of the mechanical engineer Ctesibius of Alexandria. He wrote *Pneumatica* and *Automatapoeica*, which described his ideas on physical forces and mechanisms to displace weight, water, and air. He was interested in land measurement, wrote on surveying, and invented a forerunner of the theodolite, a surveyor's instrument for measuring angles; he also invented an odometer. Heron, like most ancient engineers, also turned his skills to military science, working on siege engines, slings, missiles, and other ballistics.

Heron was fascinated by actions upon air and water. He argued that air is a material substance that exists within an apparently empty container. He analyzed the displacement of air by pouring water into a jar. He experimented with compression and argued for the presence of vacuums in nature. He was one of the founders of theories of kinetic energy. He explained the action of fire on substances according to the Aristotelian theory that heavy objects fall toward the center while lighter objects ascend toward the heavens. Heron invented a steam mechanism that featured a cauldron of boiling water that released steam through a small tube entering a sphere with two pipes at right angles. As the steam was forced through the pipes, the sphere rotated. Another device heated air that filled a container of oil; the oil was forced by the air into tubes held within statues; the oil dripped from the stone hands holding cups for libations. Heron also experimented with pistons, valves, pneumatics, and hydraulics. Heron's devices were built at a time of slavery when there was no demand for such labor-saving machines. Most of his inventions were used as toys or for tricks to amuse the rich. For example, Heron contrived a device, "An Altar Organ blown by the agency of a Wind-mill" that, by forcing air through small valves, would produce the appearance and sound of bird's singing. Another device used principles of heat and air pressure: it was a device made of an iron cauldron filled with water that was heated by a fire. Steam was forced through a small opening at the top of the device, which provided sufficient force to cause a small ball to hang and dance just above the opening.

The Romans of the first millennium BCE used military science to build a vast empire that surrounded the Mediterranean Sea and encompassed three continents: Europe, western Asia, and North Africa. Roman soldiers were superb engineers. They knew the basic techniques of surveying, built stone roads that lasted for centuries, erected walls that tourists still explore, and devised a system of military camps that were impregnable to enemy attack. The Roman army evolved over centuries as the Romans constantly observed weaknesses in army units, learned from the techniques of their opponents, and implemented necessary changes. Commanders kept up with military technology as it developed in the ancient world and outfitted the Roman legions with the best siege engines, catapults, pikes, spears, helmets, shields, and swords. The Romans adopted Greek hoplite fighting techniques and improved upon them. They also developed a superior system of logistics to maintain communications and ensure supplies. Yet the greatest accomplishment of the Roman army was, ironically, their system of encampment and defense.

MILITARY DISCIPLINE

Roman military success was marked by superior organization, rigorous training, and attention to detail. The Roman legion formed the core of the army. The Greek historian Polybius (208–126) reported that during his time (second century BCE), 4,200 men made up one legion. Each legion had ten maniples of 420 men that acted as a single unit. The maniples were grouped on the battlefield in a checkerboard fashion, to allow for utmost maneuverability in all directions to respond to superior enemy forces on the wings, front, or rear. Maniples were further divided into cohorts and centuries, commanded by a centurion. Centurions unceasingly drilled new recruits and veteran legionnaires not just in war but in fortifications and encampment as well. Discipline was a high priority. Death was often the punishment for faltering at one's post. According to the historian Josephus (37–100), each soldier carried, in addition to his weapons, "a saw, a basket, a pick and an axe, not to mention a strap, a bill-hook, a chain and three day's rations" (trans. Williamson). In truth, the legionnaire spent more time shoveling and carrying dirt, building and taking down, than he did fighting.

VEGETIUS

The fifth-century-CE Roman military scientist Flavius Vegetius Renatus, author of *Epitoma Rei Militari*, *Of Military Matters*, wrote of the Roman marching camp: "For nothing else is found to be so useful, so necessary in time of war. If a camp is properly built, the men spend their days and nights safe and sound inside the rampart, even if the enemy is besieging them. It is as though they carried around with them a fortified city wherever they go" (trans. Milner). Vegetius's rules for building a camp were as follows: If there was no immediate threat, the Romans constructed a ditch that was three feet deep and four feet wide; a three-foot earthen wall stood behind the ditch. Soldiers placed sharpened wooden stakes into the wall to repel possible enemy attack. If in enemy territory, where attack seemed inevitable, the ditch was expanded to nine feet wide and seven feet deep. If under attack, the legionnaires would nevertheless set to their task of constructing the camp, building a ditch twelve feet wide and nine feet deep, backed by a wall that was twelve feet high with stakes mounted on top.

THE *CASTRA*

Polybius, a Greek captured by the Romans who recognized the military genius of his captors, penned a multivolume history of the Roman rise to power that provides the best description of the Roman military camp, the *castra*. Whenever marching, at day's end or under attack requiring defense, officers and surveyors reconnoitered a likely position for the camp. The ground had to be level and dry, an open space with a fresh water supply and good drainage; preferably the camp would not be surrounded by hills and thick woods. Surveyors, led by a master builder/surveyor, quickly formed the plan for the camp, measuring a forty-thousand-square-foot area formed as a square, everything at right angles, with fortifications, gates, and roads. Flags of varying colors designated the location of the officer's tents, the tents of foot-soldiers, the depots for supplies or booty, latrines, pits for rubbish, and corrals for pack animals and horses. Surveyors ensured that the tents were set back two hundred feet from the fortifications. Streets were generally one hundred feet long. Three streets connected the four gates of the camp and met at the central headquarters. One road was perpendicular to two parallel roads. Temporary camps were made of turf and timber; soldiers quickly erected skin tents that sheltered eight men. Permanent winter camps used stone; the soldier's quarters were small snug huts with thatched roofs. Permanent camps included granaries, a hospital, a training area, and barracks. According to Polybius, the Roman legionnaires were so well trained that the layout of the camp and their own duties were obvious. Their camp routine "resembles the return of an army to its native city"—each man goes to the location of his tent as he would go to the location of his home (trans. Paton).

Josephus, who, like Polybius, was a Roman prisoner turned supporter and historian who had personal acquaintance with the Roman camp, reported that while on military campaign, dawn signified the time to break camp and march, and each soldier had specific duties in breaking camp. Heralds blowing horns would signal tasks to accomplish in precise order: first, to bring all tents down; second, to load baggage on the mules; third, to set fire to the encampment to prevent its use by the enemy. If they were under attack, the herald asked three times if the men were "ready for war," to which the men shouted three times with right arms raised, "We are ready." Josephus characterized the camp as "an improvised city" that "springs up with its market-place, its artisan quarter, its seats of judgement." The permanent camps he saw commanded by Vespasian and his son Titus in the 60s CE had towers on the walls at regular intervals and catapults positioned to hurl objects at the enemy. The Romans, unlike their enemies who built fortified camps, took the time to level the ground to ensure order and regularity. Josephus was impressed by the Roman discipline requiring that "whatever hostile territory they may invade, they engage in no battle until they have fortified their camp" (trans. Williamson). This explains why the Romans experienced few major military disasters caused by ambush. The most famous, the destruction of Publius Quinctilius Varus's legions in 9 CE, was due to a lack of discipline and the inability of the commanders to organize the troops into forming the emergency camp.

The decline of the Roman Empire, or transition from the Roman Empire into the Western Roman Empire followed by independent kingdoms and the Eastern Roman Empire into the Byzantine Empire, resulted in part from the breakdown of military forces, at least in the west. The Roman legions were no longer able to prevent the "barbarian" forces of the Goths, Vandals, Huns, Franks, and other *Germanic* tribes from crossing the Rhine and Danube Rivers, which led to the collapse of the Roman Empire in the West in the fifth century. An era of kingdoms in control of the varied parts of the old Roman Empire followed, with frequent chaotic battle and a decline in urban culture and social structure. It was not until the eighth century that there was something of a revival in Europe.

MEDIEVAL TECHNOLOGICAL DEVELOPMENTS

The Carolingian Renaissance beginning in the eighth century in Gaul/France was fueled in part by agricultural and economic advances that resulted from technological developments beginning in the sixth century. A series of technological advances in farming resulted in increased agricultural yield. These developments included a heavy plow with coulter, ploughshare, and mouldboard, which resulted in the ability to cut deep, horizontally and vertically, into thick European soil and turn the soil over in preparation for planting. A related technological development was the invention and adoption of the collar harness for horses, allowing the pulling of heavier plows. The old yoke harness choked horses, but the collar harness more evenly distributed the force of the plow on the horse's shoulders. Also, medieval agriculturalists increasingly used horseshoes to enable more efficient plowing by horses. The heavy plow and collar harness led to advances in agriculture, more food resulting in better nutrition and longer lives. Added to these technological developments were new techniques in agriculture, namely, the threefold rotation system. The principle of this system, adopted throughout Europe over time in the Middle Ages, was to alternate crops in fields yearly, allowing one in three fields to lie fallow, which resulted in greater productivity of the soil.

At the same time, the founder of the Carolingian Dynasty, Charles Martel (c. 688–741), introduced a new technological development in mounted combat. The stirrup, which allows a rider to focus less on using leg strength to ride the horse, was developed in Asia before the Common Era. The stirrup came to Europe by the early Middle Ages and was adopted by Charles Martel to arm his knights in battle. The introduction of the stirrup meant that a heavily armed soldier, with armor, sword, and lance, could put more power into his thrust against the enemy, which turned the tide in warfare. Historians such as Lynn White Jr. believe that this "mounted shock combat" enabled Martel and his knights to turn back the Muslim advance into France at the Battle of Poitiers in 732. Charles's army of knights with armor, weapons, and horses became the norm in medieval Europe, and helped bring about the dominance of the institution of feudalism, in which a knight was granted land by a lord to sustain his fighting prowess in return for fealty to the lord.

After the turn of the millennium, the medieval mind increasingly came to see the possibilities of machine power to manipulate the natural environment. Medieval artisans developed, and farmers used wind and water mills for grinding grain. Rotating millstones gave forth the idea of rotating machines, which led to the crank, where one arm in rotation was connected at a ninety-degree angle to another arm moving in reciprocal motion. Flywheels were developed to equalize continuous rotating motion. The treadle, moving a machine by foot power, was devised for the spinning wheel. Such devices symbolized the new physical ideas of "regularity, mathematically predictable relationships, [and] facts quantitatively measurable" (White, p. 125). All of these developments, the gear, crank, flywheel, treadle, spring, mill, cam, and rod "facilitated the conversion of reciprocating into continuous rotary motion" (White, p. 129). The medieval mind was fascinated by the idea of perpetual motion and looked forward to machines that could help harness the power inherent in nature.

GUNPOWDER AND OTHER ANCIENT INCENDIARY WEAPONS

The invention of fire—that is, the human ability to contrive it by artificial means—in the distant past, and the consequent interest that humans had about fire—its aid to survival as well as destruction—fascinated humans long after the Paleolithic and Neolithic periods had ended in most parts of the world and continued to fascinate civilized humans. Ancient humans used fire for a variety of means, in part to create, as in metallurgy, and to destroy, as in war. Ancient humans discovered incendiary materials such as petroleum, tar, sulfur—what Homer's *Odyssey* calls *brimstone*—that could be used for useful and destructive purposes. The Byzantines developed during the seventh century an incendiary device called Greek fire, perhaps made of a mixture of sulfur and other ingredients, such as naphtha, which could be launched against enemy ships, causing great destruction and terror. The Byzantine empress Anna Comnena in *The Alexiad* provided a description of the use of Greek fire at the end of the twelfth century: "On the prow of each vessel [were] the heads of lions and other land animals affixed; they were made of bronze or iron, and the mouths were open; the thin layer of gold with which they were covered made the very sight of them terrifying. The fire to be hurled at the enemy though tubes was made to issue from the mouths of these figure-heads in such a way that they appeared to be belching out the fire." The Greek fire destroyed the enemy, who "were frightened out of their wits, partly because of the fire being directed at them (they were unaccustomed to such equipment: fire naturally rises upwards, but this was being shot in whatever direction the Romans [Byzantines] wished, often downwards and sideways, to port or starboard)" (trans. Sewter).

The Chinese were the first to develop gunpowder, sometime during the beginning centuries of the Common Era. They called gunpowder "fire-chemical" and "fire-drug," which betrayed in part the alchemical origins of gunpowder. Gunpowder might be made with different ingredients, such as saltpeter, sulfur, nitrates, and charcoal. The Chinese knew how to develop fireworks with low nitrates and small-scale rockets. The higher the nitrate content, the greater the explosion and propulsion of an object. Because of its alchemical origins, gunpowder was often considered medicinal for sores and worms. By the ninth and tenth centuries, the Chinese had developed gunpowder weapons that could be propelled, causing an explosion or a fire, and protochemical weapons that gave off poisonous gas. The Chinese used bamboo as the casing for the prototype of a gun as well as for fireworks. In time, with increasing explosive power, metal replaced bamboo. By the fourteenth century, cannon prototypes were used in Southeast Asia. Also by the fourteenth century, trade between East and West led to the adoption of gunpower, guns, and cannons in Europe.

FURTHER READING

Aristotle, *Meteorology*, trans. Webster, in *The Works of Aristotle*, vol. 1 (Chicago: Encyclopedia Britannica, 1952).

K. Aterman, "From Horus the Child to Hephaestus Who Limps: A Romp through History," *American Journal of Medical Genetics* 5 (1999), http://www.ncbi.nlm.nih.gov/pubmed/10076885.

Paul Budd, "Recasting the Bronze Age," *New Scientist* (October 1993), https://www.newscientist.com/article/mg14018964-100-recasting-the-bronze-age.

Lionel Casson, *Libraries in the Ancient World* (New Haven, CT: Yale University Press, 2001).

Gordon Childe, *What Happened in History* (New York: Penguin Books, 1946).

Anna Comnena, *The Alexiad*, trans. E. R. A. Sewter (Harmondsworth, England: Penguin Books, 1969).

Diodorus Siculus, *The Library of History of Diodorus Siculus*, trans. C. H. Oldfather, vol. 1 (London: Loeb Classical Library, 1933), http://penelope.uchicago.edu/Thayer/E/Roman/Texts/Diodorus_Siculus/home.html.

Will Durant, *The Life of Greece* (New York: Simon and Schuster, 1939).

Adolf Erman, *Life in Ancient Egypt*, trans. H. M. Tirard (New York: Dover Books, 1894).

Michael Grant, *From Alexander to Cleopatra* (New York: History Book Club, 1982).

Thomas Heath, *Aristarchus of Samos* (New York: Dover Books, 1981).

"Irrigation," *Encyclopædia Britannica*, vol. 12 (1962).

Josephus, *The Jewish War*, trans. G. A. Williamson (Harmondsworth, England: Penguin Books, 1969).

Gerda de Kleijn, *The Water Supply of Ancient Rome: City Area, Water, and Population* (Amsterdam: Gieben, 2001).

Henry M. Leicester, *The Historical Background of Chemistry* (New York: Dover Books, 1971).

Joseph Needham, *Science & Civilisation in China*, vol. 5, *Chemistry and Chemical Technology*, part 7, *Military Technology; The Gunpowder Epic* (Cambridge: Cambridge University Press, 1986).

Pliny the Elder, *Natural History: A Selection*, trans John F. Healy (London: Penguin Books, 1991).

Plutarch, *The Lives of the Noble Grecians and Romans*, trans. John Dryden (New York: Random House, 1992).

Polybius, *The Histories*, trans. W. R. Paton, 6 vols. (Cambridge, MA: Harvard University Press, 1922–1927).

Strabo, *Geography*, trans. H. L. Jones, 8 vols. (Cambridge, MA: Harvard University Press, 1917–1932), http://penelope.uchicago.edu/Thayer/E/Roman/Texts/Strabo/home.html.

Thucydides, *The Peloponnesian War*, trans. Rex Warner (Harmondsworth, England: Penguin Books, 1972).

Vegetius, *Epitome of Military Science*, trans. N. P. Milner (Liverpool: Liverpool University Press, 1993).

Vitruvius, *The Ten Books on Architecture*, trans. Morris H. Morgan (Cambridge, MA: Harvard University Press, 1914).

Lynn White Jr., *Medieval Technology and Social Change* (Oxford: Clarendon Press, 1962).

23

Social and Behavioral Sciences in the Ancient and Medieval Worlds (500 BCE–1500 CE)

Social science is literally the scientific study of society in all its forms: family, community, kinship, nation, institutions, laws and norms, ethnic groups, and human culture. Although the social sciences are modern scientific disciplines of study, one finds the roots of the formal study of sociology, anthropology, ethnology, geography, economics, and political science in the ancient world, particularly among the Greeks. Social science is, of course, vastly different in scope and method from the physical and life sciences. The latter scientists manipulate the objects of study in ways that social scientists refuse to do. There are few laboratory experiments in the study of society. Ancient scientists, however, performed few experiments, hence physical and life sciences were less empirical, more reliant upon reason, logic, observation, and analysis. In short, social science was scarcely different from the physical and life sciences in the ancient world.

Sixth- and fifth-century-BCE Athens was the place and time during which the basic ideas of social science were developed, primarily by such thinkers as Solon, Pericles, Thucydides, Socrates, and Plato. The lawgiver Solon made the Athenian law code more equitable than before, and more fitting to the people and place of ancient Attica. Pericles oversaw the implementation of the basic core values and institutions of democracy—trial by jury, equality among citizens, public responsibility and service, and the secret ballot. Thucydides the historian used the Peloponnesian War (431–404) to make lasting observations about human behavior. Socrates, Plato, and Aristotle, as revealed in Plato's *Dialogues* and such works by Aristotle as *Politics*, developed such core ideas of political and social science as the *social compact*, wherein government is a voluntary association among free humans who join together to promote their own survival, as well as the idea that civilized society forms a *division of labor*, wherein each human performs the task that most benefits the public good.

HISTORICAL STUDY

History was science in the ancient world. Aristotle, in *Poetics*, contrasted poetry's concern for the subjective and intuitive with the objective and factual account of past events found in the written discourse of history. Homer and Hesiod were poets, Aristotle argued, hence they described the general truth of human experience; historians such as Thucydides described the factual basis of particular human events.

Historical study as much as scientific analysis is an attribute of civilization. The Sumerians and Egyptians, upon inventing cuneiform and hieroglyphic writing, used their invention to record crops and livestock, thus preserving a written record of past experience. The Hebrews of Israel and Judah early in the first millennium BCE wrote narratives of creation, the first humans, the migration of the Hebrews, the first prophets and lawgivers, the chronologies of kings and wars. Some of these narratives verge on myth; others are quite concrete and factually explicit. The Hebrew sense of history, as revealed in the Old Testament, derived from their incredible sense of consciousness as a chosen people, which required an understanding of the path of the chosen through time. Knowing from the Pentateuch and the first prophet, Moses, what God (Yahweh) intended for them, they looked to the known past for guidance—indeed, the past was the future in a sense.

HERODOTUS

If the Hebrews were certain of their past and future, the Greeks of the first millennium BCE were not. Questioning all things, especially the past, characterized such Greeks as Herodotus, who referred to historical study as *historia*, which meant an "arbiter" to Homer, and "researches" or "inquiries" into human experience to Herodotus. Herodotus of Halicarnassus (485–425), the first great Greek practitioner of this search for knowledge of the past, wrote his "Researches . . . to Preserve the Memory of the Past" (Selincourt, *Histories*). Herodotus's inquiries were not exactly what Aristotle had in mind when he defined history in *Poetics*. Herodotus's *Histories* are wandering and anecdotal, recording stories of varying worth and historical accuracy based on his sources. Herodotus wandered the eastern Mediterranean during the mid-fifth century BCE, searching for answers to explain why the Persians under Darius I in 490 and Xerxes in 480 attacked the city-states of Greece and how, against all apparent reason, the Greeks won these two contests of the Persian Wars. Herodotus included ruminations on human nature, discussions of Asian and African culture, and detailed descriptions of natural phenomena, such as the flooding of the Nile, and that odd creature, the hippopotamus. He sought information on the extent of continents and seas, the origins of rivers, and the differences in human customs. Herodotus was not a precise recorder of events, but he was an inquisitive wanderer, a first-class geographer, and a tireless researcher.

THUCYDIDES (460–400 BCE)

Herodotus hardly fit the ideal of the objective inquirer, but the Athenian historian Thucydides, who lived a generation later, did. Thucydides was an Athenian who served in the Peloponnesian War, the great war between the Dorians of southern and western Greece and the western Mediterranean, led by Sparta, and the Ionians of eastern Greece, the Aegean, and western Turkey, led by Athens. An Athenian general who lost his commission early in the war, he decided to stay involved by observing events, talking to participants, and recording his own assessments of the war. He lived in Thrace for most of the war and returned to Athens at its conclusion in 404. Thucydides wrote *The Peloponnesian War* based on the assumption that it would be one of the greatest conflicts in history. He used the pathos and drama of war to portray constant human emotions and characteristics. Thucydides was a precursor to modern social and behavioral scientists and their search for laws of human behavior. Thucydides, therefore, had an objective ideal. He criticized poets and storytellers, claiming "the subjects they treat of being out of the reach of evidence, and time having robbed most of them of historical value by enthroning them in the region of legend" (trans. Warner). Like Herodotus, Thucydides filled his account with numerous digressions. Unlike Herodotus, these digressions were not to entertain the reader with fascinating stories and legends. Rather, Thucydides digressed to correct false assumptions and incorrect accounts of past events. He was in other respects very much like the modern scientific scholar, following a strict chronological narrative and rarely allowing himself to speculate on the events of the war. He did, to be sure, put speculative speeches into the mouths of the Dorian and Ionian statesmen. Clearly based on reports and rumors, these speeches nevertheless had the look and feel of authenticity. Thucydides knew to use a precise form—thesis, arguments, exhortations, and conclusion—for these speeches and used this familiar pattern to re-create the essential characteristic of Greek political culture based on rhetoric.

Thucydides purposely set forth to treat historical inquiry like a science by asking questions, collecting available facts, and setting forth answers based on reason, analysis, and an objective mindset. In the first book of his history, Thucydides, seeking an accurate picture of the Trojan War and the Mycenaean kings, engaged in deductive analysis based on reason and experience to try to reconstruct in general terms the Greek past. He set himself apart from his predecessors Homer, Hecataeus, Hellanicus, and Herodotus. He condemned their uncritical use of sources and reliance upon myth and poetry. History has little to do with poetry, Thucydides argued. Aristotle, in the *Poetics* (book 9), would echo this same sentiment half a century later. Aristotle compared history unfavorably to poetry, in that the former refers to particulars, the latter to universals, hence to the truth. Thucydides, however, believed the exact opposite, seeing in the particulars of human experience the true kernels of universal human truth.

Thucydides argued that by examining an event such as a war we might see how human behavior will repeat itself, not in the particulars of time and place but in the general behavioral response of humans to certain conditions. In this way history can become a science of human behavior, wherein patterns will be found to occur again and again. His portrait of the whims of public opinion and the power of demagoguery at Athens will be seen again and again over the course of centuries in similar democracies and republics. The unexpected and irrational in human affairs, so clearly seen in the case of Alciabiades, accused of impiety and driven to the Spartan side, is a constant feature of Thucydides's work. His account of the plague that raged through Athens early in the war is a realistic portrait of the impact of fear and disaster upon human institutions, morality, and decency. Thucydides was impatient with the credulous beliefs among the Greeks that healing could come from the gods, particularly in the sanctuary of a temple of Asclepius or Apollo. The gods, he believed, had no apparent impact on the prevention of illness or recovery. This rational approach to medicine was being activated by a contemporary of Thucydides, Hippocrates of Cos. Indeed, Thucydides's description of the plague in *The Peloponnesian War* shows clear indications that he was influenced by Hippocrates.

Many were the imitators of Thucydides in subsequent centuries; yet few could match his tight narrative and seemingly scientific precision. True, Thucydides began the practice imitated by other Greek and Roman historians and biographers of re-creating speeches based on his understanding of human character and the particulars of time and place. This merely showed his genius for interpretation, for portraying the generalities of human existence from sporadic, uncertain, or nonexistent sources. Thucydides's greatness lay in his ability to organize a clear narrative of sequential events over time, to see time as one episode after another, moving from future possibilities to present existence and awareness to past recollection. Humans are typically blinded by the present moment, which is fleeting (while the future is unknown and the past unclear). Through the chronological historical narrative, patterned by Thucydides, the reader can escape from the dependence on the present moment, to see human existence and behavior over a broad expanse of time, thereby gaining an appreciation of what it means to be human.

XENOPHON (430–355 BCE)

Xenophon was a native Athenian writer, philosopher, historian, adventurer, and soldier. He is famous for his *Anabasis,* an account of a dangerous journey into the heart of Persia and a successful retreat back to Greece led by Xenophon himself. He also wrote *Hellenica*, a continuation of Thucydides's brilliant historical narrative; *Oeconomicus*, about running one's personal household (*oikos*); and *Symposium* and *Memorabilia*, both of which are reminiscences of former times when he was a student of the great Athenian philosopher Socrates. Xenophon's philosophy was not as good as Plato's. His history was not as good as Thucydides's. And his *Oeconomicus* had limited value. And Xenophon was not much of a scientist. Nevertheless, it was through the *Anabasis* that Xenophon made his mark on the history of literature, of thought, and even of science.

Xenophon was a young Athenian nobleman, a veteran of the Peloponnesian War, and a student of Socrates when a friend asked him to join a contingent of

Greek hoplites (heavily armed infantrymen) who were to march into Asia Minor on an expedition against the Persians. They were to be led by Cyrus, a brother of the king of Persia Artaxerxes II (r. 404–358). Xenophon agreed but claimed later that he did not realize Cyrus's real intent, to topple his brother. They marched, however, in 401, into the heart of Persia, where Cyrus was killed in an engagement with the Persian army under Artaxerxes. The Greeks, leaderless and surrounded in hostile territory, began a dramatic journey amid sporadic fighting to return to Asia Minor, the Black Sea, and eventually Greece. During this journey, Xenophon rose to be one of the commanders. He and most of the soldiers survived, after which Xenophon used his sword to fight for the Spartans. He lived in the Peloponnesus for many years, returning near the end of his life to Athens.

The *Anabasis* is a wonderful adventure story that provides a fairly good historical narrative filled with action, speeches, and descriptions of Asia. Xenophon imitated Thucydides's style of writing, presenting a no-nonsense narrative that provides a chronological viewpoint following a clear temporal order. Xenophon rarely digressed but stayed with his story. Although he clearly believed in the Olympian gods, there is little that is ridiculous in the *Anabasis.* From a historical point of view, it is unique as a firsthand account of historical events—history as personal experience. Xenophon was observant and intelligent, and he provided some good descriptions of the Asian peoples, cultures, and institutions and the landscapes of Mesopotamia, Armenia, and Asia Minor. Xenophon was not trained in the study of human culture, but he became a reasonably good anthropologist out of the necessity of escaping Persian troops through a foreign and dangerous country.

Xenophon's significance from a scientific standpoint was limited. He influenced others who sought to discover more about Asia's geography and people. Notably he inspired Aristotle's interest in the Persian Empire, which was passed on to Aristotle's student Alexander, who used the *Anabasis* as a guide to Persia, its peoples and way of life. That Alexander became an ad hoc geographer and anthropologist like Xenophon was not a coincidence. Out of Alexander's anabasis came some other good narrative adventures filled with geographical, anthropological, and natural insights—for example, Nearchus's *Indica.*

ROMAN GOVERNMENT AND HISTORIOGRAPHY

The Romans, ever practical and concerned with building successful communities, government, and empire, developed social and political systems based on observations of human nature and experience. Polybius, the Greek who wrote a history of Rome in the second century BCE, argued that only by means of history, an examination of human experience, might correct behavior be molded and successful institutions be adopted. Livy echoed Polybius in his history of Rome, declaring that morality is decided by a reading of history and an imitation of good behavior and an avoidance of the bad.

The Roman Constitution as presented by Polybius was a masterful experiment in political science. Polybius argued that Rome balanced power by combining the rule of one person (kingship) with the rule of many people (aristocracy and oligarchy) with the rule of the people en masse (democracy). Rome's mixed constitution gave executive power to the consuls, deliberative power to the Senate, and legislative power to the people in Assembly. The two consuls served annually and could not be reelected. They were elected by the people: one came from the lower and middle classes, the plebeians; the other from the aristocracy, the patricians. Each had a veto over the other's actions; the Senate exercised a powerful influence over both consuls. The numbers in the Senate constantly changed as its membership was not dependent upon election; rather, consuls immediately joined the Senate after their one-year term expired. Senators served for life. It was their job to advise the consuls and recommend legislation for the Assembly to consider. The Assembly was comprised of the male citizens. It passed the laws and elected the consuls. The rights of the Assembly were also represented by the Tribune, who had the veto power and was bound to exercise it in the protection of the people from wanton abuse of power.

POLYBIUS (208–126 BCE)

Greek power was eclipsed first by the Macedonians in the fourth century and then by the Romans in the second century. Polybius was a Greek historian captured by the Romans. During the second century BCE, the Greek Polybius was sent to Rome as a prisoner, where he became the client of Scipio Amelianus. Polybius along with Panaetius (185–110) brought his Stoic beliefs in reason and traditional morality to Rome, which had a profound impact on the Roman worldview. Polybius came to admire the Romans, who had already embraced a lifestyle very much like the Stoics; he brought his Stoic didacticism to a universal history of his time. He took an interest in Rome's rise to power and so traced Roman history in his multivolume *historia*. Polybius's *Histories* narrates the rise of Roman power throughout the Mediterranean from the third to the second centuries BCE. His work was pragmatic and ecumenical, examining all affairs relating to Rome and the Mediterranean during this time. Polybius merged the apparently competing techniques of Herodotus's storytelling with Thucydides's infatuation with a strict chronological narrative, seeking to describe the truth of human events and not to sacrifice truth to rhetorical flourish. He condemned the historian Timaeus (c. 345–250) in this regard, who used high rhetoric to mask the fact that he wrote about events of which he had no part and of places that he had not seen. Polybius wrote a precise account of history following a chronological narrative, using Olympiads as the basic framework of dates; yet he digressed and theorized, related fascinating stories, and described the geography, culture, customs, flora, and fauna of regions conquered by the Romans, such as North Africa. Polybius journeyed to the Alps, the great barrier to Italy that the Carthaginian general Hannibal crossed. Polybius believed that history was an expression of the truth of human experience. As in book 12 of his *Histories*, he constantly raised the banner of an objective account of the past using a clear analysis of all available documents and firsthand knowledge of peoples and places. Polybius's book contains, in parts, a history of science and is based on a scientific approach to history as well.

Geographer

Polybius enlivened his history with firsthand accounts of the landscape and peoples of the Mediterranean. He described Italy as a triangle with the Apennine Mountains forming the center from which land and waters descended to the Adriatic Sea in the east and the Tyrrhenian Sea in the west. The Po drained the northern region of Italy, its waters reaching flood stage about the time of the rising of the Dog Star, that is, mid-July. Likewise, the Euphrates in Mesopotamia rises at the same time. The Lotus of North Africa is useful in producing fruit from which wine can be fermented. Polybius traveled to Africa and wrote of its geographical features, such as the behavior of wild animals, the habits of the Egyptians, and the environs of Mount Atlas. Unlike some ancient geographers, Polybius argued (in a lost treatise, *On the Parts of the Globe under the Celestial Equator*) that the equatorial zone of Earth is a delightful place to live. Heat is concentrated in the tropic zones (Tropic of Cancer and Tropic of Capricorn) because the sun's rays settle dreadfully upon these spots, at the solstices, for about forty days. The geographer Strabo (63 BCE–21 CE) wrote that Polybius believed the equatorial zone to be higher in elevation, with more rainfall than the tropics. In lost parts of his *Histories*, Polybius discussed the supposed travels of Odysseus, making realistic assessments of actual locales to match the strange monsters and mythical forces described in Homer's poem *The Odyssey*. Polybius also described in detail but with much skepticism the supposed voyage of Pytheas of Massilia. After visiting the Alps of Europe, Polybius declared them uninhabitable because of snow and height. Astonished by the immensity of the Alps, Polybius imagined their extent to be "two thousand, two hundred stades" (trans. Paton), which would equal over a million feet (a *stadium* being equal to about six hundred feet)! Indeed, all of the distances in Polybius's *Histories* are grossly overestimated.

Political Scientist

One of the more famous parts of Polybius's massive history is book 6, which describes in detail the Roman system of government. Polybius argued that a mixed government like that of the Roman republic was the best form of government because it combined democracy, aristocracy, and kingship, any one of which was inadequate on its own. Polybius formed his political science on a cyclical view of history. He believed that humans congregated together in herds, as did animals, but unlike animals, humans used reason to form ideas of duty and justice. The initial form of government in human society is kingship, which is simply leadership by the best warrior and most courageous leader. As a hereditary principle worms its way into this natural form of leadership, oppression often results, because inadequate leaders must use force and intimidation to secure their rule. Eventually the people rebel, led by wealthy aristocrats, who take charge of government upon the toppling of the king. But this leads to a new form of oppression, the oligarchy, which again is toppled by the will of the people, who form themselves into a body politic, a democracy. For a generation or so, Polybius argued, democracy works, but eventually the people grow lazily content with their freedom and let it slowly slip away into civil war and chaos, bringing the society around once again to primitive kingship.

DIODORUS SICULUS (C. 90–30 BCE)

Diodorus was a Greek Sicilian who assigned himself the task of writing a universal history. Like Herodotus three centuries earlier, Diodorus traveled to locate information for his history; and like Herodotus, he referred to his book as *historia*, "inquiries"; and like Herodotus, he went to Egypt. He claimed that he had "visited a large portion of both Asia and Europe," which was an exaggeration, and the project had taken him thirty years to complete. Like Livy, who was a Latin writer of the next generation, Diodorus saw history as having a didactic purpose to teach behavior "to use the ignorant mistakes of others as warning examples for the correction of error." He wrote,

> It is fitting that all men should ever accord great gratitude to those writers who have composed universal histories, since they have aspired to help by their individual labours human society as a whole; for by offering a schooling, which entails no danger, in what is advantageous they prove their readers, through such a presentation of events, with a most excellent kind of experience. For although the learning which is acquired by experience in each separate case, with all the attendant toils and dangers, does indeed enable a man to discern in each instance where utility lies—and this is the reason why the most widely experienced of our heroes suffered great misfortunes . . . yet the understanding of the failures and successes of other men, which is acquired by the study of history, affords a schooling that is free from actual experience of ills. (Trans. Oldfather)

Diodorus believed that his universal history was a record of humans acting in accord with Providence:

> And such historians have therein shown themselves to be, as it were, ministers of Divine Providence. For just as Providence, having brought the orderly arrangement of the visible stars and the natures of men together into one common relationship, continually directs their courses through all eternity, apportioning to each that which falls to it by the direction of fate, so likewise the historians, in recording the common affairs of the inhabited world as though they were those of a single state, have made of their treatises a single reckoning of past events and a common clearing-house of knowledge concerning them. (Trans. Oldfather)

He thought of history as the original study of wisdom, the means to know the truth. Although human existence is brief in time, history, the story of great deeds, is a "voice most divine," a herald of time (ibid.).

Diodorus aimed to write a history covering all known peoples in all recorded times. He dated events according to the chronology of Apollodorus of Athens (180–120), who had dated the end of the Trojan War to 1184 BCE. Diodorus's account went into his own time of the first century BCE. Diodorus followed the Stoics (rather than the Peripatetics) in believing that the world had a definite origin and would at some point come to an end. His account, following the Stoic philosophy, of the origins of the universe and life, is worth noting:

> When in the beginning, as their account runs, the universe was being formed, both heaven and earth were indistinguishable in appearance, since their elements were intermingled: then, when their bodies separated from one another, the universe took on in all its parts the ordered form in which it is now seen; the air set up a continual motion, and the fiery element in it gathered into the highest regions, since anything of such a nature moves upward by reason of its lightness (and it is for this reason that the sun and the multitude of other stars became involved in the universal whirl); while all that was mud-like and thick and contained an admixture of moisture sank because of its weight into one place; and as this continually turned about upon itself and became compressed, out of the wet it formed the sea, and out of what was firmer, the land, which was like potter's clay and entirely soft. But as the sun's fire shone upon the land, it first of all became firm, and then, since its surface was in a ferment because of the warmth, portions of the wet swelled up in masses in many places, and in these pustules covered with delicate membranes made their appearance. Such a phenomenon can be seen even yet in swamps and marshy places whenever, the ground having become cold, the air suddenly and without any gradual change becomes intensely warm. And while the wet was being impregnated with life by reason of the warmth in the manner described, by night the living things forthwith received their nourishment from the mist that fell from the enveloping air, and by day were made solid by the intense heat; and finally, when the embryos had attained their full development and the membranes had been thoroughly heated and broken open, there was produced every form of animal life. These, such as had partaken of the most warmth set off to the higher regions, having become winged, and such as retained an earthy consistency came to be numbered in the class of creeping things and of the other land animals, while those whose composition partook the most of the wet element gathered into the region congenial to them, receiving the name of water animals. And since the earth constantly grew more solid through the action of the sun's fire and of the winds, it was finally no longer able to generate any of the larger animals, but each kind of living creatures was now begotten by breeding with one another. (Trans. Oldfather)

Diodorus believed, based on his experiences in Egypt, that the Nile River valley was the most likely place for human and animal life to have begun; besides, the earliest notion of the deities seemed to have originated in Egypt. The Egyptians taught Diodorus that the first deities were Osiris the sun and Isis the moon, and

> practically all the physical matter which is essential to the generation of all things is furnished by these gods, the sun contributing the fiery element and the spirit, the moon the wet and the dry, and both together the air; and it is through these elements that all things are engendered and nourished. And so it is out of the sun and moon that the whole physical body of the universe is made complete; and as for the five parts just named of these bodies—the spirit, the fire, the dry, as well as the wet, and, lastly, the air-like—just as in the case of a man we enumerate head and hands and feet and the other parts, so in the same way the body of the universe is composed in its entirety of these parts. (Trans. Oldfather)

TITUS LIVIUS (59 BCE–17 CE)

Rome eventually came to dominate up to one hundred million people stretching from England to Mesopotamia to Morocco. The drama and grandeur of Rome's rise to power encouraged the thoughtful, the analytical, and the didactic to record Rome's history. Because of the historical writings of Sallust (86–35), Caesar (100–44), Livy, Appian (95–165), Tacitus, Suetonius (69–130), Plutarch (46–120), Dio Cassius (163–225), and Ammianus Marcellinus, the great Roman contribution to the history of thought was through the writing of history. A pervasive historical perspective and sense of time pervaded Rome's politics, customs, and beliefs. The Roman genus, the spirit of Rome, united past and present by means of the tremendous awareness of ancestors and descendants possessed by each Roman. The Roman worldview was molded by a state that was completely temporal: the account of conquest and conquerors over time fascinated Romans. The greatest Roman narrator of historical events was the historian Titus Livius, Livy, whose account of the origins and rise to power of the city on the Tiber towered above other towering historical writers. Yet Livy claimed that his history had less to do with temporal affairs and more to do with transcendent morals: "The study of history is the best medicine for a sick mind; for in history you have a record of the infinite variety of human experience plainly set out for all to see; and in that record you can find for yourself and your country both examples and warnings; fine things to take as models, base things, rotten through and through, to avoid" (trans. Selincourt). Livy's was didactic history, a narrative of past events to prepare those of the future for similar events. Livy was a masterful storyteller who sacrificed a strict accounting of detailed fact for a sometimes heavy reliance on legend and anecdote.

TACITUS (56–117 CE)

If Livy was not inclined toward a scientific viewpoint in his historical writing, the same cannot be said about Cornelius Tacitus, the best Roman historian of the centuries after the birth of Christ.

Tacitus was a historian, biographer, and ethnographer. He was a senator during the Flavian and Antonine dynasties of the Roman principate. His works included the *Annals*, *Histories*, *Agricola*, and *Germania*. He was a friend and correspondent with Pliny the Elder (23–79) and Pliny the Younger (61–113). Tacitus was one of the most sophisticated Roman authors, a polymath of note, and, as an ethnologist, he penned an enduring portrait of the peoples of Germany. Even more than Thucydides, Tacitus sought legitimacy in his terse historical writing. To write Rome's history, Tacitus used whatever valid historical documents he could find, including memoirs, Senate records, eyewitness accounts, letters from observant friends (such as Pliny the Younger), contemporary letters and speeches, historical narratives, and official journals. His accounts also included perceptive assessments of the natural environment. And as if this were not enough, Tacitus was the first historian to write from what historians today would call a behavioral science approach. Tacitus sought to re-create the personalities, motives, emotional distresses, and bouts of insanity of the objects of his inquiry, particularly the emperors Tiberius (r. 14–37), Caligula (r. 37–41), and Nero (r. 54–68).

Tacitus's *Annals* has often been praised for its penetrating psychological analyses of emperors such as Tiberius and for its ability to portray collective behavior.

Tacitus was keenly aware of the abuse of power by the princeps and the army during the age of the Julio-Claudian emperors (31 BCE–68 CE). Tacitus yearned for the republican past when Romans worked together as a supportive community and selfishness and lust for power—the products of inequality and increasing wealth—had not yet set in.

Generally Tacitus, as an objective thinker, leaves a lot to be desired. He believed completely in Roman polytheism and the pantheon of gods, Stoic and Epicurean skepticism having not, apparently, caused sufficient doubts. Like many historians, Tacitus recorded miracles, wonders, omens, and prophecy. Tacitus's *Histories* relates an incident involving the future emperor Vespasian, who received clear messages from the gods and even restored sight to a blind man (81–82). The healing was not science, but miracle.

Germania

The best example of Tacitus as a social and behavioral scientist is *Germania*. Tacitus relied on a variety of Greek and Roman sources, such as Pliny the Elder, who had written histories of the wars with the Germans, and Posidonius of Rhodes (135–50), who had lived during the previous century. Tacitus was also the son-in-law of the Roman commander Agricola, who had served as governor of Britain. In his biography of Agricola, Tacitus described the customs and lifestyle of the Britons and provided an interesting account of Roman explorations around the British Isles, approaching almost to Thule. The *Germania*, however, provided a fuller and more sensitive account of the Germanic peoples. The Romans considered the Germans to be barbarians, and Tacitus clearly echoed this point of view. Yet the *Germania* is less a condemnation of the German peoples and more a full description of the land in which they lived and their physical characteristics, customs, institutions, way of life, and communities. The loyalty, constancy, and self-sacrifice of the Germans toward one another impressed Tacitus. He believed that if they were uncivilized compared to the Romans, the Germans were morally superior to the Romans: "Good morality is more effective in Germany than good laws are elsewhere," he said tellingly (trans. Mattingly).

ARRIAN (89–180 CE)

Flavius Arrianus Xenophon, Arrian, is important to the history of science as a compiler and commentator on the works of others. Arrian wrote the *Anabasis*, one of the most important sources for the reign of Alexander the Great, in which Alexander's scientific interests are discussed at length. Arrian's *Indica*, a literary and scientific appendage to the *Anabasis*, is the account of Nearchus's voyage from the Indus River to the Tigris River in 325 BCE. In both of these works Arrian discussed Greek exploration, the geography and anthropology of Asia, and the scientific inclinations of those violent, warlike men, the Macedonians. Arrian's third major contribution to the literature of ancient science was his compilation of the writings of the Stoic philosopher Epictetus (50–135).

Arrian was a Greek, a native of Nicomedia in western Turkey, who rose to prominence among the Romans in the second century CE. He was a student of the philosopher Epictetus in the Epiran town of Nicopolis. Subsequently Arrian served as a soldier, became a consul of Rome, was a priest in his native Bithynia, and crowned his career as governor of Cappadocia in Turkey. Cappadocia was one of Rome's frontier provinces; Arrian was involved in many battles against outside aggressors. For a Greek to become governor of an important province was at this time exceptional. Reasons for his success include his birth into a wealthy family and the fact that his father was already a citizen of Rome; hence Arrian became a citizen at birth. Stoic philosophy, learned from Epictetus, gave Arrian the training and mindset of successful Roman men of action. Also, Arrian was a student of the fourth-century-BCE historian, philosopher, and adventurer Xenophon, whose interest in military affairs, horsemanship, and the hunt became passions for Arrian as well. Xenophon had written an account, the *Anabasis*, of the Greek incursion into Persia that inspired Alexander's own expedition into the Persian Empire. Arrian modeled himself after Xenophon such that he named his account of Alexander's journey *Anabasis* as well.

Historian of Science

One would hardly claim Arrian to have been a scientist, but in his interests and writings he was clearly one of a growing number of historians of science of the Pax Romana and the Late Roman Empire. In his *Anabasis* Arrian described the scientific interests of Alexander and the activities of the numerous philosophers and scientists that the conqueror brought with him. Arrian's *Indica* is an account of the voyage of exploration and science of Alexander's lieutenant Nearchus. Arrian also wrote a fascinating treatise of geographic exploration, the *Circumnavigation of the Black Sea,* in which he described a journey taken in the early 130s CE to explore the entire coastal reaches of the Black (Euxine) Sea.

Arrian's writings are typically modeled on Attic and Ionic Greek of centuries past. An exception is his compilation of Epictetus's works. As a young man, Arrian attended the lectures of this ex-slave and formidable teacher. Epictetus did not record his lectures or write anything else; the student Arrian decided to record the teachings of the master. He did so in the spoken Greek of the period, the Koine. Arrian claimed, in an opening letter to a friend that introduces the *Discourses* of Epictetus, that he took notes of the lectures not intending them to be published.

In the *Anabasis* Arrian relied on the extant sources of his time, many of which were by scientists or written in a scientific style. Of the latter is Nearchus's *Indica,* an account of the coastal wastelands from the Indus to the Euphrates Rivers. Arrian also used the incomplete account of Alexander's conquests by Callisthenes (c. 360–328), who was a nephew of Aristotle and a scientist and philosopher. Also of importance was the account of Alexander written by Aristobulus (375–301), a scientist who accompanied the expedition and described his experiences shortly before his death a generation later.

AMMIANUS MARCELLINUS (325–395 CE)

Centuries passed, the Roman Empire went through political, social, economic, and moral crises, and the scope of history became confusing, not at all like the five centuries BCE of Rome's rise to power. Nevertheless, in the fourth century CE, one man attempted to write the history of the Roman people, taking up where Cornelius Tacitus had left off. Ammianus Marcellinus's *History* was a fourteen-volume work that encompassed the three centuries from 100 to 400 CE. Marcellinus modeled himself on Tacitus's style, method, and approach. He recorded the political and military events of the Later Roman Empire but also digressed to describe the customs, beliefs, institutions, and geography of surrounding peoples—the Gauls, Goths, Parthians, and others—as well as other aspects of natural history. A pagan, he held the beliefs in the supernatural common to his time. Marcellinus's history is a narrative of cause and effect, with arbitrary actions of Adrastia (Nemesis) underlying this. As the universal spiritual power pervades the earth and cosmos, some humans, exercising prescience, gain knowledge of the future with the help of Themis, who makes known what fate has determined. The gods send birds, whose flight, noises, and path indicate the future for the clever augur. Another source of future knowledge comes from the sun, Helios, the life and breath of the universe who sends thought (divine sparks) to the human seeking mental awareness.

Ancient historical writing did not, of course, involve experiments, lab reports, and mathematics. But science involves more than the manipulation of nature in a controlled environment. Greeks and Romans such as Ammianus Marcellinus believed science to be the systematic acquisition of knowledge. The great historians among the Greeks and Romans believed historical study, along with geographical and anthropological study, was as much a science as the study of natural philosophy and natural history.

MEDIEVAL HISTORIOGRAPHY

During the process of the transition of the Western Roman Empire to separate kingdoms and the Eastern Roman Empire to the Byzantine Empire, the standards and style of Greco-Roman historiography continued to be modeled by Western European, Eastern European, and Islamic historians, with some changes according to the fundamental Christian or Islamic worldview. Some medieval writers continued in the path of Greco-Roman historians in largely dealing with secular human affairs without an undue reliance on superstition or hagiography. Others, however, were smitten by the supernatural, with a fierce desire to defend the religious view of the times in which they wrote. The former included Procopius of Caesarea (c. 500–565), who wrote *History of the Wars*, an account of the wars waged by the Byzantines against the Sassanid Persians to the east and the Vandals and Goths to the west. Procopius was the secretary to the emperor Justinian's General Belisarius, and much of his account was based on eyewitness. His was a straightforward narrative style similar to that of his predecessors, such as Ammianus Marcellinus. Procopius is best known for his *Secret History*, an appendage to his *History of the Wars*, which describes the corruption of Justinian's court, especially the wily designs and character of the empress Theodora. A contemporary of Procopius writing about the history of western Europe, Gregory of Tours (538–594), wrote a similar account of events in the sixth century; Gregory's work, *The History of the Franks*, discussed the Merovingian rulers of France but also provided an overview of the history of the world from the creation to Gregory's own time. A similar straightforward literary account of contemporary history came from the pen of Einhard (c. 770–840) in *The Life of Charlemagne*, the king of the Franks and patron of the liberal arts. Ibn Khurradadhbih (c. 820–911) wrote the *Book of Roads and Kingdoms*, a geographical text based on his knowledge gained as postal director for the Abbasids. His work became the basis for many subsequent medieval geographers. Medieval Muslim historians included Al Mas'udi (896–946), a historian and geographer from Baghdad who wrote *The Meadows of Gold and Mines of Precious Gems*, a universal history. Al-Biruni (971–1048) wrote the exhaustive historical and scientific chronology, *The Chronology of Ancient Nations*. Ibn Hayyan (987–1076) wrote a history of Muslim Spain, relying on the accounts of earlier Muslim Andalusian historians. Ahmad ibn Rusta, a tenth-century Persian, wrote a multivolume encyclopedia of geographic and historical knowledge. His source, in part, was the *Book of Roads and Kingdoms* of Ibn Khurradadhbih. The tenth-century Persian Istakhri wrote a comparable *Book of Roads and Kingdoms*. Ibn Rusta wrote important accounts of the peoples of western and central Asia, such as the Magyars, Khazars, and Rus. The medieval Mesopotamian physician Ibn Abi Usaibi'ah (1203–1270) wrote an extensive *History of Physicians*. Ibn Khaldun (1332–1405) of northwestern Africa was a historian and social scientist who wrote *Kitāb al-'Ibār*, or *Book of Lessons*, a book of vast learning focusing on geography, economics, history, and philosophy. Another fourteenth-century North African scholar and explorer, Ibn Battuta (c. 1304–1369), wrote *A Gift to Those Who Contemplate the Wonder of Cities and the Marvels of Traveling*, based on his many travels through Europe, Asia, and Africa.

A chronological account of the times of late medieval Europe was provided by Geoffroy de Villehardouin (c. 1150–1218), who wrote *The Conquest of Constantinople*, an account of the Fourth Crusade. Other sober accounts of historical events in eastern Europe included *Chronographia* by Michael Psellus (c. 1018–1078), an account of political and military events of the Byzantine Empire in the tenth and eleventh centuries. Anna Comnena (1083–1153), an Aristotelian scholar, wrote a similar narrative of Byzantine affairs focusing on the reign of her father Alexius I (r. 1081–1118)—*The Alexiad*.

Famous examples of historical writing that bordered on the mythical and fabulous were by two medieval British historians. Bede (673–735), a Roman Catholic monk, wrote several religious histories of England, such as *A History of the English Church and People*. In his narrative, Bede discussed at length the various miracles and prodigies consistent with his Christian worldview. Bede focused on the wonders and miraculous cures of the saints of the English church. Comparable in terms of challenging the credulity of readers was *The History of the Kings of Britain* by Geoffrey of Monmouth (c. 1095–1155), who provided a mythical account of the founding of England by Brutus, who was descended from Aeneas, the Trojan who journeyed to Italy, where he became the progenitor of the Roman people. Geoffrey also discussed the myth of King Arthur and wrote at length on the prodigies of the sorcerer Merlin. From one of two medieval mapmakers who challenged the credulity of their viewers came *Map of the World*, drawn by Richard of Haldingham, a thirteenth-century Roman Catholic cleric living in England, who allowed faith to direct his cartography into making Jerusalem the center of the world. Muhammad al-Idrisi (1099–1165), meanwhile, in the twelfth century, composed a complex map of the world with the Southern Hemisphere on top, and the Northern Hemisphere on the bottom.

CHINESE HISTORICAL WRITING

Chinese historical writing began with Confucius (K'ung fu-tzu, 551–479), who traditionally wrote books such as the *Spring and Autumn Annals*. Confucianism went wonderfully well with historical thought, as it was focused on tradition and used the past for guidance, filial piety, order (derived from past tradition), and good behavior (based on past morals). Confucian historians were didactic, teaching morality. A good example is Sima Qian (145–90), who wrote *Records of the Historian*. His book is a semireligious moral tract of events in the first century millennium. But Qian did not sacrifice rationality in his moral treatment of past events. Another example is *Tso Chuan*, *The Commentary of Tso*, written perhaps in the fourth century. This moralistic handbook focuses on the application of virtue to social and political life. History is driven by human will and morality: humans are free moral agents in historical events. *Kuo yü, Conversations from the States* and *Chan-kuo ts'e, Intrigues of the Warring States* are didactic histories written in the Confucian tradition.

During the Han Dynasty, the Confucian focus on the past, on historical tradition, guided Chinese government, society, and learning. Chinese historians followed in the same pattern established by Confucian historians such as Sima Qian. The Chinese conceived of five elements, or agents, involved in natural and human history. Each dynasty was identified with one of the five agents (wood, fire, earth, metal, water). Ban Gu (c. 32–92) wrote *Han Shu, History of the Former Han*, covering the Han Empire from 209 BCE to 25 CE. Several centuries later, Fan Ye (398–445) wrote *Records of the Former Han*, covering the period of the first and second centuries CE of the Han Dynasty. Liu Chih-Chi (661–721) wrote *Generalities of History* (*Shih-T'ung*), which was didactic history written to inform the reader about political systems. Liu was noteworthy for practicing internal source criticism.

BEHAVIORAL SCIENCE

One of the greatest accomplishments of ancient science was the discovery of the human psyche. Psychology, the study of the *psyche* (soul), was largely brought about by Greek philosophers, in particular the Platonists, Peripatetics, Stoics, Epicureans, Gnostics, and Christians. The Greeks were able to at once apply analysis, imagination, speculation, logic, and intuition to a thorny, mysterious, hidden object of inquiry—the mind. During the span of a millennium from the appearance of Homer's *Iliad* and *Odyssey* to Augustine's conversion experience, ancient thinkers were able to discover the existence of the individual, the singular mind tied to unique experiences; the emergence of the self from a sea of others, the collective mass of humanity; the identification of the self with the Other, the *numinous*, the sum and total of all existence; and how the individual's momentary awareness fits in the overall fluidity of time.

Greek psychology did not develop in a vacuum. The Greeks built upon the insights of their predecessors, the Mesopotamians, Egyptians, and Hebrews. Mesopotamia, where the first civilization developed, hosted a people, the Sumerians, who engaged in speculation into the relation of humans to the divine, the ultimate purpose of existence, and the role of the individual in the whole. The human population increased dramatically during the third millennium BCE in Mesopotamia, and the Sumerians were able to isolate individual human achievements and personality traits. The *Epic of Gilgamesh,* for example, describes an individual human who is complex, unique, and tragic. Gilgamesh's search for eternal life and happiness is a general human search made singular to the life of one man. Gilgamesh, moreover, is part god, and he can compete on the gods' level, which indicates a growing awareness of the significance and value of the individual person.

Homeric Psychology

A similar approach to human psychology is found in the *Iliad* and *Odyssey* of the Greek poet Homer. Homer's tale is a timeless portrait of human struggle against fate. Unlike the Hebrews, fate in Homer's poems is anonymous and unknowable, and tremendous will and strength of character are required for individuals to forge their own life in time. Homer put a premium on the characteristics of the individual hero. In so doing he created a personality standard of the heroic individual who strives against fate by means of the peculiar human trait of *arete*—manly courage. The hidden presence of gods poetically describes the unconscious mind of individuals who typically heed the advice of this inner voice and act accordingly. Athena is almost like an extension of Odysseus—an alter ego. Likewise, Telemachus is a reflection of Odysseus. His story is a story of Odysseus at an early age: the relations with Athena, his bearing and nascent courage, his thoughtfulness, his uncertainty but willingness to search for answers and to confront himself, others, and his destiny.

The first time we meet Odysseus is at Calypso's island in the *Odyssey,* he is not some great hero but, rather, a man sunk to the bottom, weeping, in despair. Is this the hero we have come to expect? Why doesn't he fight, do something? Things get even worse for him. How will he respond? Here, Homer brilliantly portrays humankind. He introduces us to a human in deepest despair. When humans are down, almost defeated yet struggling to live, that is when humans are best known. This is a portrait of naked humanness, more revealing than later, when Odysseus is in all his glory. At this point, when Odysseus arrives at Ithaka, Homer presents the suitors, a group of self-perceived heroes in self-denial, fooling themselves, forgetting what is true, what will happen, denying the future. Earlier in the poem, the story of Polyphemus shows the contrast of a more settled society, with customs such as laws reflecting the will of the gods, versus barbaric people who respect no such customs, do not live in settled community with others. Perhaps Odysseus is revealing a step toward a more civilized society in his use of analytical thought and self-awareness. He is also an individual struggling with his own destiny: a portrait of individualism.

Dreams

Much of the writings of ancient peoples involved stories of a dreamlike quality. Indeed, dreams were considered supernatural phenomena. Until the work of Sigmund Freud, Carl Jung, and their successors, the scientist would scarcely give attention to dreams as a proper object of inquiry. Dreams were stuff for the mystic, perhaps, but not for the scientist. The ancients, of course, could not analyze dreams as the working out of daily experience, as a necessary and healthy activity during sleep or, for the psychoanalyst, the reflections of the unconscious mind. Ancient thinkers did, however, recognize the significance of dreams, even if understanding was elusive. Ancient humans could not conceive of dreams as anything other than a divine message hinting of what will be. There came to be a class of pseudoscientists who made their living predicting the future by means of the interpretation of dreams. Ancient literature is filled with such stories. In Genesis, in the Old Testament, Joseph interprets the dreams of the pharaoh and so gains a preeminent position in the kingdom. In the book of Daniel, the Hebrew prophet Daniel interprets the dreams of the Babylonian king Nebuchadnezzar (r. 604–562). Homer's poems are filled with accounts of dreams sent by the gods to inform humans of future possibilities. Typically, the Homeric world personified a natural occurrence by deifying it; hence Oneiros was the god of dreams. Common to all of these peoples of the ancient Mediterranean was the belief that dreams reflect natural or supernatural phenomena, and it takes reason and the analytical mind to correctly interpret them.

Aristotle (384–322) was one of the greatest students of human psychology. He made extensive studies of sleep, discovering that human mental activity continues as during wakefulness. He examined dreams and declared them to be natural rather than supernatural and subject to scientific analysis rather than religious faith. In *On Memory and Reminiscence*, he studied how the human mind responded to time and experience. He understood the role of sense perception in observation and analysis, the chief means by which an Aristotelian scientist acquired knowledge. Aristotle referred to the *psyche*, or soul, as the force of being in the human upon which bodily actions are dependent.

The Italian school of ancient Greece had a different interpretation of dreams. The fifth-century-BCE materialists Leucippus and Democritus held the view that invisible atoms composed all existence, including even the mind, thoughts, dreams, the soul, and being: all were composed of atoms constantly in movement; but identity was restricted to the here and now, its future limited by death.

There were several significant students of dreams during the Roman Empire. The most famous was the physician Galen (130–200), who believed that dreams helped direct him in diagnosis and healing in general. Galen, who was physician to the emperor Marcus Aurelius (121–180), imparted this respect for the teaching significance of dreams to the Stoic emperor. Ibn Abi Usaibi'ah wrote in his *History of the Physicians* about Galen's belief in dreams:

> An example is provided by Galen's report in his book on phlebotomy concerning an order he received to perform a bloodletting on an artery. He says: "In a dream I was twice ordered to open the artery between the index and the thumb of the right hand. When I awoke I opened that artery and let the blood flow until it stopped of itself; for so I had been commanded in my dream. The amount of blood withdrawn was less than one liter, but in that way I was immediately relieved of a pain I had long been suffering from at the spot where the liver touches the diaphragm. At the time I was first afflicted with that pain I was a youngster." He goes on to say: "A man I know, in the city of Pergamum, was cured by God, the Exalted, of a chronic pain in his side through bloodletting on the artery in the palm of his hand. What induced him to undergo that bloodletting was a dream he had had." (Trans. Kopf)

Ibn Abi Usaibi'ah wrote further:

> In the fourteenth chapter of his book "The Stratagem of Healing" Galen says: "I once saw a tongue so enlarged and swollen that the mouth could not contain it. The man affected with this complaint had never undergone a bloodletting. He was sixty years old. The time I first saw him was the tenth hour of the day [i.e., our 4th hour after noon], I felt I should purge his bowels with the pills usually employed in such cases; they are composed of aloe scammony and the pulp of coloquintida. I administered the drug toward evening and enjoined the patient to place something cooling on the diseased organ. I said to him: 'Do so in order that I may be able to observe the effect and to determine the treatment accordingly.' Another physician whom he had asked to come did not agree with me on that and for this reason, after the patient had taken the pills, the deliberation as to the method of treating the organ itself was deferred to the following day. All of us had been hoping that the substance employed for treatment would show its beneficial effect, and so we would try it out on him, because the entire body would by then have been purged and the substance which had been pouring forth to the organ would have flowed downward. During that night, in a dream, the patient was visited by a clear and unequivocal vision, in consequence of which he lauded my advice and took what I had recommended to him as stuff for that remedy. Namely, in his dream, he saw a person who ordered him to keep some sap of lettuce in his mouth. He did as he was ordered and recovered completely, requiring no other treatment" (Trans. Kopf)

And further:

> In his commentary on Hippocrates' "Oaths" Galen says: "The great majority of people bear witness that it was God, the Blessed and Exalted, who through dreams and visions inspired them with medical knowledge which rescued them from severe diseases. We find, for example, countless people who were healed by God, the Blessed and Exalted—some with the aid of Serapis, others with the aid of Asclepius. In the cities of Epidaurus, Cos and Pergamum, the last being my native town, in short, in all the temples of the Greeks and of other nations there are cases of serious diseases cured through dreams and visions." (Trans. Kopf)

Ibn Abi Usaibi'ah related a similar account involving the fourth-century physician Oribasius:

> A man had a large stone in his bladder. He says: "I treated him with every drug believed to be effective in crushing stones, but it was of no avail—the man was on the point of death. Then, in a dream, he saw someone approaching him, holding a small bird in his hand. This person said to him: 'Take this bird, which is called the wren and which lives in swamps and thickets, burn it and eat of its ashes so that you may be relieved of your sickness.' Upon waking, the man did as had been told and thereby caused the stone to come out of his bladder, crumbling like ashes. His recovery was complete." (Trans. Kopf)

A fourth-century-CE physician, Gennadius, learned of the soul's immortality by means of a dream. The most significant students of dreams were Artemidorus of Daldis and Aelius Aristides (117–181), both of the second century CE. Artemidorus wrote *Oneirocritica*, a book on dreams and their meaning. He took a scientific approach to dream prognostication. He assiduously recorded every aspect of dreams, compiling a precise record of this peculiar human activity. Similar was Aelius Aristides, a priest of Asclepius, the Greek god of healing. Aristides kept a full account of his many dreams over the space of several decades. He believed that Asclepius healed or gave advice on healing through dreams.

Ibn Abi Usaibi'ah recorded a comment from Ali ibn Ridwan (c. 998–1061), the Egyptian astrologer, astronomer, and commentator on Galen:

> [In] Alī ibn Ridwān's commentary on Galen's book "On the Sects in Medicine" [*De sectis*] I have copied the following: "For many years I had been suffering from a severe headache due to an overfilling of the blood vessels of the head. I performed a venesection, but the pain did not cease; I repeated it several times but the pain continued as before. Then I saw Galen in a dream. He ordered me to read to him his 'Stratagem of Healing,' and I read seven chapters of that work. When I reached the end of the seventh chapter, he said: 'Have you now forgotten the headache you are suffering from?' And he ordered me to rub the occipital protuberance. After that I woke up, rubbed it and was cured of my headache on the spot." (Trans. Kopf)

Plutarch (46–120)

Other students of psychology during and after the Pax Romana included Plutarch, the Greek author of *Moralia* and *Parallel Lives*. Plutarch made human (and animal) psychology one of his favorite topics. Plutarch employed an empathetic approach to converse, as it were, with past humans such as Alexander the Great, to discover fresh insights into their personalities. Plutarch made a study of depression and its causes; he believed depression was the result of obsessive thoughts about the momentary present and the approaching future. One must enjoy the present by using the past, which is an extensive repository of pleasant memories, an encyclopedia of wisdom never to be equaled by oneself, and a peaceful balance to the irritations and fears of the present moment.

Plutarch believed that the mind was a powerful agent in arriving at contentment. Reason was not a foil but rather an aid to happiness. One must exercise control over depressing images and obsessive thoughts. Thoughts cannot always be avoided. In Plutarch's essay *On Contentment*, he recommended that the anxious and depressed person purposely confront the fears and obsessions so as to make them less powerful. Modern psychologists often recommend the same technique to their patients.

Plutarch was not as original in his animal psychology as he was in human psychology. In *Whether Land or Sea Animals Are Cleverer* and *On the Use of Reason by "Irrational" Animals*, he argued that animals did possess reason. Nature itself provides animals with courage, acceptance, and patience, which are all too often lacking in humans. Plutarch's arguments are not too different from the Pythagorean belief in the transmigration of souls. *On the Use of Reason by "Irrational" Animals*, for example, pits the wise Odysseus against a philosophic hog, who seems to get the better of the famous Ithacan. Plutarch as a young man went so far as to advocate vegetarianism in his essay *On the Eating of Flesh.*

Porphyry (234–305)

Another student of Plato, Porphyry, was also intrigued by various psychological questions, especially with regard to the ability of soothsayers to predict the future and whether dreams are reflections of a future reality. He wrote, in a letter to an Egyptian priest:

> What that is which is effected in divination? For we frequently obtain a knowledge of future events through dreams, when we are asleep; not being, at that time, in a tumultuous ecstasy, for the body is then quiescent; but we do not apprehend what then takes place, in the same manner as when we are awake. But many, through enthusiasm and divine inspiration, predict future events, and are then in so wakeful a state, as even to energize according to sense, and yet they are not conscious of the state they are in, or at least, not so much as they were before. Some also of those who suffer a mental alienation, energize enthusiastically on hearing cymbals or drums, or a certain modulated sound, such as those who are Corybantically inspired, those who are possessed by Sabazius, and those who are inspired by the mother of the Gods. But some energize enthusiastically by drinking water, as the priest of Clarius, in Colophon; others, by being seated at the mouth of a cavern, as those who prophesy at Delphi; and others by imbibing the vapour from water, as the prophetesses in Branchidse. Some also become enthusiastic by standing on characters, as those that are filled from the intromission of spirits. Others, who are conscious what they are doing in other respects, are divinely inspired according to the phantastic part; some, indeed, receiving darkness for a cooperator, others certain potions, but others incantations and compositions: and some energize, according to the imagination, through water; others in a wall, others in the open air, and others in the sun, or in some other of the celestial bodies. Some also establish the art of the investigation of futurity through the viscera, through birds, and through the stars. I likewise ask concerning the mode of divination, what it is, and what the quality by which it is distinguished? All diviners, indeed, assert, that they obtain a foreknowledge of future events through Gods or daemons, and that it is not possible for any others to know that which is future, than those who are the lords of futurity. I doubt, therefore, whether divinity is so far subservient to men, as not to be averse to some becoming diviners from meal. But, concerning the causes of divination, it is dubious whether a God, an angel, or a daemon, or some other power, is present in

> manifestations, or divinations, or certain other sacred energies, as is the case with those powers that are drawn down through you [priests] by the necessities with which invocation is attended. Or does the soul assert and imagine these things, and are they, as some think, the passions of the soul, excited from small incentives? Or is a certain mixed form of subsistence produced from our soul, and divine inspiration externally derived? Hence it must be said, that the soul generates the power which has an imaginative perception of futurity, through motions of this kind, or that the things which are adduced from matter constitute daemons, through the powers that are inherent in them, and especially things adduced from the matter which is taken from animals. For in sleep, when we are not employed about anything, we sometimes obtain a knowledge of the future. But that a passion of the soul is the cause of divination, is indicated by this, that the senses are occupied, that fumigations are introduced, and that invocations are employed; and likewise, that not all men, but those that are more simple and young, are more adapted to prediction. (Trans. Taylor)

Augustine (354–430)

One of the great psychology treatises of the ancient world was Augustine's *Confessions.* Few people have documented so completely depression and recovery, the obsessive-compulsive personality, and the sources and resolution of anxiety and the identity crisis. Martin Luther, who suffered from many of the same problems as Augustine, was an Augustinian monk deeply influenced by the *Confessions* and *City of God,* Augustine's other grand treatise dealing with the duality of mind and body. Jonathan Edwards, eighteenth-century America's great theologian, was a student of Augustine's psychology. Sigmund Freud and Carl Jung were also similarly influenced.

Asian Psycho-Philosophical Treatises

Asian philosophical treatises in Confucianism, Taoism, Hinduism, and Buddhism penetrated deeply into an examination of the relationship between each individual human and the natural world and supernatural experiences.

Confucianism typically focused on order and normative behavior, though there was a sense of a cosmic force in the universe that held everything together, as signified by the yin-yang duality. The fourth-century Confucian Mencius (372–289) had a mystical aspect to his thought: "A conception of reality as an extension of the mind, and . . . the gnostic desirability of achieving some kind of awareness of oneness of the knower and the known." Mencius thought that the human mind reflects the cosmic order: "When man develops his mind, he knows nature, and when he knows his own nature, he knows Heaven" (Mote, p. 60). Chinese Taoists could agree with Mencius's comments. Taoism tended to be more naturalistic, less focused on society and institutions. Taoism sought the essence of life, as present in the universe, the way or *tao.* Both philosophies had a similar approach to how the individual thinker confronted the cosmos. They "display the same predilection for concepts by intuition rather than by postulation, for suggestive rather than explicit language, for similitude rather than syllogism" (Mote, p. 70). The *Tao Te*

Ching, the search for the way, written (traditionally) by Lao Tze, reveals an obscure and remote approach to finding individual truth in the mystery of nature. If this truth is found, then pure happiness will result. The fourth-century-BCE thinker Chuang Tzu (Zhuang Zhou, c. 369–286) elaborated on Taoism and downplayed somewhat this search for personal happiness, arguing that humans could only hope for relative, not absolute, happiness.

Buddhism, a derivative of Hinduism in India that spread to China after the Han Dynasty, challenged traditional Confucian and Taoist ideas. Buddhism was primarily psychological rather than religious or philosophical in focus. Like Hindus, Buddhists conceived of endless cycles of wandering and suffering in life; after one life comes another, as determined by karma, one's actions in the current life. Buddhists believed that a person could escape from evil, that is, time and the endless experiences of bodily life, by purification over multiple lifetimes that would result, eventually, in Nirvana (liberation from self, liberation of consciousness of self). To achieve this, Siddhartha Gautama (563–483), the Buddha, taught that the liberation of self was achieved by means of the Four Noble Truths and the Eightfold Path.

The *Sermon at Benares* provides succinctly the Buddha's approach to salvation:

> The Tathagata [Buddha] does not seek salvation in austerities, but neither does he for that reason indulge in worldly pleasures, nor live in abundance. The Tathagata has found the middle path. There are two extremes, O bhikkhus [Hindu monks], which the man who has given up the world ought not to follow—the habitual practice, on the one hand, of self-indulgence which is unworthy, vain and fit only for the worldly-minded and the habitual practice, on the other hand, of self-mortification, which is painful, useless and unprofitable. Neither abstinence from fish or flesh, nor going naked, nor shaving the head, nor wearing matted hair, nor dressing in a rough garment, nor covering oneself with dirt, nor sacrificing to Agni [Hindu god of fire], will cleanse a man who is not free from delusions. Reading the Vedas, making offerings to priests, or sacrifices to the gods, self-mortification by heat or cold, and many such penances performed for the sake of immortality, these do not cleanse the man who is not free from delusions. Anger, drunkenness, obstinacy, bigotry, deception, envy, self-praise, disparaging others, superciliousness and evil intentions constitute uncleanness; not verily the eating of flesh. A middle path, O bhikkhus, avoiding the two extremes, has been discovered by the Tathagata—a path which opens the eyes, and bestows understanding,—which leads to peace of mind, to the higher wisdom, to full enlightenment, to Nirvana! . . . How can any one be free from self by leading a wretched life, if he does not succeed in quenching the fires of lust, if he still hankers after either worldly or heavenly pleasures. But he in whom self has become extinct is free from lust; he will desire neither worldly nor heavenly pleasures, and the satisfaction of his natural wants will not defile him. . . . He who recognizes the existence of suffering, its cause, its remedy, and its cessation has fathomed the four noble truths. He will walk in the right path. Right views will be the torch to light his way. Right aspirations will be his guide. Right speech will be his dwelling-place on the road. His gait will be straight, for it is right behavior. His refreshments will be the right way of earning his livelihood. Right efforts will be his steps: right thoughts his breath; and right contemplation will give him the peace that follows in his footprints. . . . Birth is attended with pain, decay is painful, disease is painful, death is painful. Union with the unpleasant is painful, painful is separation from the pleasant; and any craving that is unsatisfied, that too is painful.

> In brief, bodily conditions which spring from attachment are painful. . . . Verily, it is that craving which causes the renewal of existence, accompanied by sensual delight, seeking satisfaction now here, now there, the craving for the gratification of the passions, the craving for a future life, and the craving for happiness in this life. . . . Verily, it is the destruction, in which no passion remains, of this very thirst; it is the laying aside of, the being free from, the dwelling no longer upon this thirst. . . . Now this, O bhikkhus, is the noble truth concerning the way which leads to the destruction of sorrow. Verily! it is this noble eightfold path; that is to say: Right views; right aspirations; right speech; right behavior; right livelihood; right effort; right thoughts; and right contemplation. (Trans. Carus)

Religious treatises such as Buddha's sermon, Jesus's parables, the Psalms of David, or the Hindu text Bhagavad Gita are avenues that allow a person to arrive at an understanding of human psychology. The *Sermon at Benares* addresses the inner conflict that all humans experience between immediate desire based on fear in the moment and the realization that not giving into impulsive feelings and thoughts will result in contentment. Jesus's parables are directed at the same conflict humans experience in wanting to believe, to have faith in the Other, at the same time that fear and selfishness drive people into narcissism. The Psalms of David in the Old Testament likewise present the quandary of the Psalmist, usually David, and his belief in God, desire to surrender to God, yet selfish need for honor, glory, riches, and power. In the Bhagavad Gita, further, the reader is presented with another quandary that all humans face at some point in life: the expectations of society and family, religious and social institutions, that is, a sense of *duty*, in contrast to the feelings of the self, the feelings of personal morality, of what an individual seeks notwithstanding the expectations of others. The warrior Arjuna is faced with a commitment and duty to fight, but the enemy represents people that he knows, even distant relatives, and he is overwhelmed with horror at the prospect of killing those whom he would in another place and time sit with, converse with, and befriend. Why must he kill just because of societal obligations? Isn't it best to love, not hate, and to embrace others, not attack and kill them? The god Krishna, who appears in the Bhagavad Gita as Arjuna's chariot driver, confronts the misgivings of Arjuna, and convinces him that sometimes religious and social duty is more important than individual desire and will. This inner confrontation and dialogue that all humans engage in is wonderfully, poetically expressed in the Bhagavad Gita.

FURTHER READING

Al-Biruni, *The Chronology of Ancient Nations*, trans. C. Edward Sachau (London: William H. Allen and Co., 1879).

Ammianus Marcellinus, trans. John C. Rolfe, 2 vols. (Cambridge, MA: Harvard University Press, 1950).

Arrian, *Anabasis* and *Indica*, trans. P. A. Brunt and E. Cliff Robson, 2 vols. (Cambridge, MA: Harvard University Press, 1933, 1976).

Arrian, *The Campaigns of Alexander*, trans. Aubrey de Selincourt (Harmondsworth, England: Penguin Books, 1971).

Bede, *A History of the English Church and People*, trans. Leo Sherley-Price (Harmondsworth, England: Penguin Books, 1968).

The Bhagavad Gita, or the Message of the Master, trans. Yogi Ramacharaka (Chicago: Yogi Publication Society, 1907).

Ernst Breisach, *Historiography: Ancient, Medieval and Modern* (Chicago: University of Chicago Press, 1983).

Jacques Brunschwig and Geoffrey Lloyd, *Greek Thought: A Guide to Classical Knowledge* (Cambridge, MA: Harvard University Press, 2000).

Anna Comnena, *The Alexiad*, trans. E. R. A. Sewter (Harmondsworth, England: Penguin Books, 1969).

William Theodore De Bary, trans., *Sources of Chinese Tradition*, vol. 1 (New York: Columbia University Press, 1960).

Diodorus Siculus, *The Library of History of Diodorus Siculus*, trans. C. H. Oldfather, vol. 1. (London: Loeb Classical Library, 1933).

Einhard, *The Life of Charlemagne*, trans. Lewis Thorpe (Harmondsworth, England: Penguin Books, 1969).

Epictetus, *The Discourses*, ed. Christopher Gill (Rutland, VT: Tuttle Publishing, 2001).

M. I. Finley, *Aspects of Antiquity: Discoveries and Controversies* (Harmondsworth, England: Penguin Books, 1977).

Geoffrey of Monmouth, *History of the Kings of Britain*, trans. Lewis Thorpe (Harmondsworth, England: Penguin Books, 1966).

The Gospel of Buddha: Compiled from Ancient Records, trans. Paul Carus (Chicago: Open Court Publishing, 1915).

Michael Grant, *Readings in the Classical Historians* (New York: Scribner's, 1992).

Gregory of Tours, *The History of the Franks*, trans. Lewis Thorpe (Harmondsworth, England: Penguin Books, 1974).

J. R. Hamilton, *Alexander the Great* (Pittsburgh: University of Pittsburgh Press, 1973).

Herodotus, *The Histories*, trans. Aubrey De Selincourt (Harmondsworth, England: Penguin Books, 1972).

Ibn Abi Usaibi'ah, *History of Physicians*, trans. L. Kopf (Bethesda, MD: National Library of Medicine, 1971).

Ibn Fadlān, *Ibn Fadlān and the Land of Darkness: Arab Travellers in the Far North*, trans. Paul Lunde and Caroline Stone (London: Penguin Books, 2012).

Abd Ar Rahman bin Muhammed ibn Khaldun, *The Muqaddimah*, trans. Franz Rosenthal, https://asadullahali.files.wordpress.com/2012/10/ibn_khaldun-al_muqaddimah.pdf.

Joinville and Villehardouin, *Chronicles of the Crusades*, trans. M. R. B. Shaw (Harmondsworth, England: Penguin Books, 1963).

Livy, *The Early History of Rome*, trans. Aubrey de Selincourt (Harmondsworth, England: Penguin Books, 1971).

Frederick W. Mote, *Intellectual Foundations of China* (New York: Knopf, 1971).

Polybius, *The Histories*, trans. W. R. Paton, 6 vols. (Cambridge, MA: Harvard University Press, 1922–1927).

Porphyry, "Letter to Anebo," in *Iamblichus on the Mysteries of the Egyptians, Chaldeans, and Assyrians*, trans. Thomas Taylor (Chiswick, England: C. Whittingham, 1821).

Procopius, *The Secret History*, trans. G. A. Williamson (London: Penguin Books, 1981).

Michael Psellus, *Fourteen Byzantine Rulers: The Chronographia of Michael Psellus*, trans. E. R. A. Sewter (Harmondsworth, England: Penguin Books, 1966).

Henry Rowell, *Ammianus Marcellinus, Soldier Historian of the Late Roman Empire* (Cincinnati, OH: University of Cincinnati Press, 1964).

Chester Starr, *The Awakening of the Greek Historical Spirit* (New York: Knopf, 1968).

Tacitus, *The Agricola and the Germania*, trans. H. Mattingly and S. A. Handford (Harmondsworth, England: Penguin Books, 1970).

Tacitus, *The Histories*, trans. Kenneth Wellesley (Harmondsworth, England: Penguin Books, 1972).

E. A. Thompson, *The Historical Work of Ammianus Marcellinus* (Cambridge: Cambridge University Press, 1947).

Thucydides, *The Peloponnesian War*, trans. Rex Warner (Harmondsworth, England: Penguin Books, 1972).

Xenophon, *A History of My Times*, trans. Rex Warner (Harmondsworth, England: Penguin Books, 1979).

Xenophon, *The Persian Expedition*, trans. Rex Warner (Harmondsworth, England: Penguin Books, 1972).

PART 9

Science in the Later Roman and Byzantine Empires

24

Late Roman Science (180–500 CE)

ROME'S DECLINE

After 200 CE, there was little scientific creativity in the ancient Mediterranean world. The achievements of Aristotle, Ptolemy, Euclid, and Galen would set the standard for the understanding of the universe, the world, mathematics, and medicine in the coming centuries. Not until 1500, during the Renaissance in Europe, would there be similar accomplishments fundamental to scientific theory and practice. Scientists from the third to the sixth centuries and afterward tended to be synthesizers and proselytizers, pedagogues repeating the same theories to students, commentators making the same comments in books and lectures. Of course, the years from 200 to 500 are traditionally categorized as a long process in the fall of Rome. Edward Gibbon, the great historian of Rome's fall, traced the decline of the ancient civilization, the loss of past greatness, the military, political, and intellectual weakness of the time. Some scholars argue that the principal cause of the decline of the Roman Empire, particularly in the Western Roman Empire (Britain, Spain, Gaul, Italy, North Africa), was that agriculture, the dominant feature in the Roman economy, declined in the third and fourth centuries. Declining agricultural productivity and food surplus spelled disaster for the lower classes, who frequently were victims of famine, malnutrition, and disease. The population of the empire was dramatically reduced during the third century. More and more thousands of acres went untilled. Declining food reserves encouraged declining population, leading to less acreage under cultivation. The problem did not go away for centuries; in the meantime, the Western Empire fell to Germanic tribes who knew little about agriculture, which helped to bring about the infamous Dark Ages of Europe.

Theurgists, magicians, and Sophists dominated what passed for science. Neoplatonists and Aristotelians subjected science to the realm of philosophy. The practice of medicine became too often the realm of the charlatan and astrologer. The great debate between pagan humanists and Christian intellectuals distracted thinkers from the general decline in political stability, economic health, artistic and literary creativity, and scientific accomplishment. By 500, scientific thought and practice in the Western Roman Empire was buried under the suffering, poverty, famine, and sickness of petty Germanic warlords competing for control of what was left of Rome. In the Eastern Roman (Byzantine) Empire, it was a different story, not so much decline as business as usual, and students of Aristotle, Euclid, Ptolemy, and Galen carried upon their intellectual shoulders the scientific legacy of the ancient world.

Superstition has often been a strange bedfellow of science over the course of human history. The ancient world, notwithstanding its huge accomplishments in science, hosted an amazing variety of magicians, soothsayers, prognosticators, astrologers, theurgists, and the like. Even during the principate of the first and second centuries CE, when Roman culture was at its height, superstition typically dwarfed scientific inquiry. This imbalance of fantasy and reality grew during the third century and afterward, as the empire fell victim to civil conflict and its usual consequences of moral and intellectual malaise.

Thinkers during the Later Roman Empire continued the tendency begun during the Hellenistic Age and continuing into the Roman principate of dividing scientific and philosophic theories and methods into two schools of thought: the Platonic and the Aristotelian. The former, with its focus on the metaphysical, gained new strength with the teachings of Plotinus and his followers Porphyry and Iamblichus. The scientific thought of these Neoplatonists was limited by their distrust of the senses and belief that reality is rarely seen or experienced in human existence. Yet their counterparts, the Aristotelians (Peripatetic philosophers), were hardly more successful at science, being so heavily dependent upon the teachings of Aristotle and expending most of their intellectual energy on the task of passing on what the Master said.

NEOPLATONISM

The Neoplatonic philosophy of the Later Roman Empire explained physical reality according to transcendent universals. Neoplatonists used reason, abstract thinking, mathematics, and intuitive thought to identify the intelligible world of the hidden essences of the divine, the intellectual world of the visible manifestations of the divine, the human realm of soul and body, and the material world of nature. *Neoplatonism* is a modern term used to distinguish followers of the philosophy of Plato during the Later Roman Empire from his earlier followers of the Academy. The leading Neoplatonists were Plotinus, Porphyry, Iamblichus, Maximus of Ephesus, and the Roman emperor Julian. Their thinking was inspired not only by Plato but by other Greek philosophers such as Pythagoras, Parmenides, Anaxagoras, and Aristotle.

Plotinus (205–270 CE)

Plotinus sought to adopt Plato's philosophy of the intelligible world of ideal forms to a more universal conception of the absolute source of all things, the One, and the human ability to approach the One through intensive self-examination leading to a mystical union. Plotinus wrote the *Enneads* (literally, the *nine* tracts in each of six books), which outlined his philosophy of the realms of existence in order, from the intelligible world of the One; the intellectual world of the manifestation of the One through the Logos, the word; the world of the soul, both a universal world soul and the individual soul that makes each human unique; and the world of the body, of nature, and of demons, which is the counterpart of the Good, and the multiplicity that becomes somehow swept up in the encompassing One. Like Plato, Plotinus believed that a mystical union with the transcendent One could be achieved by intuition combined with reason built on a solid foundation of training in mathematics and logic.

Plotinus was a native Egyptian who taught at Alexandria and Rome. He lived during a time of civil unrest, economic dislocation, and cultural malaise. Plotinus compared his world to a Greek drama—the reality of the actors, the stage, and sets being highly questionable. Greek philosophy involved an unending repetition of the same theories, a continuing discussion of the same books by the great thinkers of the past. Plotinus sought to reinvigorate thought through his teachings and writings. Plato inspired him more than any other philosopher, and Plotinus adopted Plato's theories of ideal forms and methods to achieve knowledge and to approach the Good. Eunapius (c. 345–414), writing about a century after Plotinus's death, claimed that "altars in honor of Plotinus are still warm, and his books are in the hands of educated men, more so than the dialogues of Plato" (trans. Wright, *Lives of the Sophists*).

In *Enneads*, Plotinus argued that the intelligible world is a trinity of sorts: One, Intelligence, and Soul. The One is the source of all being, the unmoved mover, the Good. The One encompasses all; it is the unity, the source of all reality, seen and unseen. Intelligence, begotten of the One, is the Logos, the creative word that encompasses all things. The Logos has its visible counterpart in light, the tremendous power and warmth of the sun. The Soul is a world soul, for all existence is spiritual, and soul imbues all things. The planets and the stars have being, as do humans. Notwithstanding the body, the soul makes the human. The manifestations in the intellectual world of human existence that conform to the intelligible world of ideal forms are the individual sense of a unified reality, rational thought, and one's feeling of an individual unique soul.

Plotinus distinguished between his philosophy based on transcendent ideals that accommodate human freedom and the philosophies of his rivals. To the materialists, such as the Epicureans, and their theories of invisible atoms combining and recombining in chance arrangements, Plotinus argued that "to derive everything from physical causes, whether atoms or elements, and to assume that their unregulated motion will create order, reason, and the motive power of the soul, is absurd and impossible" (Katz, p. 61). Plotinus questioned as well how science could be erected from such uncertain foundations. Philosophers such as the Stoics argued that all events were fated by the creative principle, the divine fire—that there was a necessity to all things. Plotinus, on the contrary, was a believer in the rational human acting according to free will. Other philosophers—astrologers in particular—believed in necessity and determination according to the motion of the stars and the arrangement of the planets. Plotinus believed in the Platonic notion that the heavenly bodies moved in perfect spheres and were themselves divine and ethereal. He did not deny the influence of the heavens upon humankind and nature, nor did he disapprove of astrologers involved in prophesizing the future, but he was adamant that "we should not be denied our individuality" (Katz, p. 64). The anonymous and passive One gave humans, collectively and individually, power over many of the details of corporeal, temporal existence.

Like Plato, Plotinus taught that the philosopher can achieve awareness of one's inner being, the world soul, universal intelligence, and ultimately the One by means of the study of mathematics and dialectical philosophy, the ability to synthesize through logic what appear to be contradictions (the thesis and antithesis). The Neoplatonist, like Pythagoras (570–490), believed that "Number" encompassed the universe, beginning, of course, with 1, the One. Geometry, in particular, was the route to understand the incorporeal forms that make up existence. Neoplatonist psychology involved the recognition of the duality of human existence, that the individual human is at the same time good and evil, soul and body, incorporeal and corporeal. The individual must recognize that bodily impulses drag one down to earth, as it were, preventing one from achieving awareness of the Good within oneself. The goal of philosophy is the greater recognition of the truth within oneself, which can yield to an understanding of reality and possible union with the One.

Plotinus is notable for his penetrating study of the human psyche, its relationship to nature and the ultimate reality, and the deep well of the self. Anticipating modern psychologists such as Carl Jung, Plotinus believed that all reality is joined spiritually, that all humans commune with each other and with the intellectual world and the intelligible world. The One encompasses all, even each individual human. This is the Greek concept of *sympatheia*. Plotinus was one of the first philosophers to explore the self through his writings. In *Ennead* IV, Plotinus wrote: "Sometimes I wake from the slumber of the body to return to myself; and turning my attention for external things to what is within me, I behold the most marvelous beauty. I then fully believe that I have a superior destiny. I live the highest life and am at one with the divinity" (Katz, pp. 101–2). Porphyry, his disciple, who wrote a biography of his teacher, described several moments of such spiritual union during Plotinus's life. Plotinus compared the longing of the individual soul with the unified world soul as the extended drink after an endless thirst. At times, he felt divided between his soul surrounded by the things of this world and the universal soul that has nothing to do with what is corporeal and transient. The individual longs for the ultimate reunification, coming home to the One, but until then must suffer through a divided life, being simultaneously pulled in two directions, to the transcendent and the transient, good and evil. Plotinus believed that said union relies upon the human rather than the One. It is natural, in the words of E. R. Dodds "that all things tend to revert to their source" (p. 88), hence the individual soul cannot help but return to the world soul.

Plotinus's philosophy emerged during the third century, when the Roman Empire was wracked by civil unrest, war, poverty, and disease—in short, chaos and anarchy, which bred anxiety, alienation, despair, and a loss of hope.

Philosophy was the means to rise above the conflicts and contradictions of earthly existence. Other philosophies and religions, too, gained strength during such turbulent times. Christianity was growing during the third century, and it competed with Neoplatonism for adherents. The commensurability of the Neoplatonist triad to the Christian concept of trinity (Godhead, Logos, Holy Spirit) is remarkable. Many Christians, such as the Roman emperor Julian (330–363), became Neoplatonists, and Neoplatonists such as Aurelius Augustine (354–430) often became Christians. Then there were those such as the theologian Origen (184–254) who combined Neoplatonic and Christian thinking to form a belief based on reason seeking union with the logos, the word that is the expression of reason.

Porphyry (234–305 CE)

Fourth-century-CE Neoplatonism succumbed to declining leadership and the weakening of Plotinus's original philosophical standards with a turn toward mysticism and fancy. After Plotinus's death in 270, his disciple Porphyry wrote of the master's life and teachings. Porphyry was one of the great Neoplatonic thinkers of the Later Roman Empire. He adopted Plotinus's Platonic beliefs in the division of existence into the intelligible, intellectual, and natural worlds. He agreed with Plotinus that the essence of all things is the anonymous One. Porphyry also held the Platonic notions of the importance of human free will in dominating the corporeal passions and temporal limitations so to gain reunification with the One, and the transcendency of the human soul, encompassing a unified world soul.

Porphyry lived at a time of civil unrest and social despair followed by the autocratic reign of the emperor Diocletian (r. 284–305). He was, perhaps, a Phoenician by birth, though he studied under the philosopher Longinus (213–273) at Athens and under Plotinus at Rome. The biographer Eunapius claimed (probably in error) that Porphyry immigrated to Sicily and contemplated suicide because Plotinus taught him to hate his body. Porphyry claimed that Plotinus experienced a mystical union with the One on several occasions, while he only accomplished it once or twice. He was at least human enough to marry and have children. Porphyry was a polymath who mastered a variety of topics ranging from Pythagorean philosophy to mathematics and logic to rhetoric and astrology. He wrote biographies of Plotinus and Pythagoras and commentaries on Homer, and he enjoyed quoting philosophical maxims to make a point. He agreed with Heraclitus (540–480) that what is good and just to the gods is incomprehensible to humans, and with Pythagoras that eating any kind of flesh was sinful (because of the transmigration of souls).

Porphyry conceived of the Greek pantheon of gods as representing manifestations of the One, which is "of the nature of light" who "dwells in an atmosphere of ethereal fire." In *On Images*, Porphyry conceived of the One as being manifested in the intellectual world as the Sun (Helios or Apollo), although to the Greeks the One was worshipped as "Zeus alone first cause of all." Hera, Zeus's wife in Greek mythology, is "the ethereal and aerial power" while the son Apollo represents the sun, and Zeus's daughters, the Muses, are the seven heavenly spheres and the starry vault (trans. Gifford). Fertility goddesses—Hestia, Rhea, Demeter, Kore—represent the productive and regenerative natural world. Asclepius symbolizes the healing power of the sun. Hecate, Artemis, and Athena are like the moon, with its nocturnal light giving wisdom and blessing women with fertile wombs. Meanwhile Aphrodite is the morning and evening stars, Kronos is time, and Hermes is the source of rational thought. Porphyry believed that the pantheons of gods in different cultures represented the same natural and transcendent powers of nature, the mind, and the universe.

His inquisitiveness into the nature of the divine is revealed in a letter he wrote to an Egyptian priest/philosopher, Anebo, in which Porphyry asked, "Why, since [all] the Gods dwell in the heavens, theurgists only invoke the terrestrial and subterranean Gods? Likewise, how some of the Gods are said to be aquatic and aerial? And how different Gods are allotted different places, and the parts of bodies according to circumscription, though they have an infinite, impartible, and incomprehensible power? How there will be a union of them with each other, if they are separated by the divisible circumscriptions of parts, and by the difference of places and subject bodies?" (trans. Taylor).

Iamblichus (250–325 CE)

Porphyry himself had disciples, such as Iamblichus, who became the leading Neoplatonist of the fourth century CE. A native of Syria, Iamblichus was the student of Porphyry and follower of Plotinus. Iamblichus broke from the ascetic, deeply intuitive, and spiritual approach of his forebears to embrace magic and theurgy. He was closer in spirit to Pythagoras, whom Iamblichus admired sufficiently to write the treatise *On the Pythagorean Way of Life*. As a theurgist, Iamblichus used magic to call upon the power of the gods. He could wow his audience and disciples, some of whom were credulous enough to believe in such magic; others were incredulous but recognized the power that such trickery could have over the simpleminded. Iamblichus died during the reign of Constantine, yet his teachings, writings, and example continued to influence later philosophers and theurgists such as Maximus of Ephesus and Julian the Roman emperor, who referred to Iamblichus as "that truly godlike man, who ranks next to Pythagoras and Plato" (trans. Wright, *Works*).

Iamblichus was one of quite a few holy men of the third and fourth centuries who attracted disciples and sycophants because of their unique skills at calling upon the divine. Eunapius the biographer recorded one instance where Iamblichus was able to produce two youthful deities before his astonished (and credulous) followers. In *On the Mysteries* Iamblichus wrote in detail about the varied techniques to call upon the gods and enjoy a mystical union. Gods might so change you as to withstand fire and knife wounds and other such pain. Or gods might inhabit your body so as to allow elevation and other such gravity-defying tricks. There was a hierarchy of divine possession. The elite holy man such as Iamblichus claimed not only to be able to give himself to the god, or absorb the god in him, but also—the pinnacle of mystical joining—become equal to the god and merge together in a dual union. Mysticism required not just the willing mind but spells and incantations, even the power of numbers. He lauded Pythagoras for giving to others the source of knowledge and power found in the "tetrad," described by Julian as "the mystery of the Four," the numbers 1, 2, 3, 4 that, when arranged in accumulating order, added to 10, which was "symbolical of all proportion and perfection." The tetrad (or tetractys) appears as follows (as described in Barnes, *Early Greek Philosophy*, p. 212):

```
   *
  * *
 * * *
* * * *
```

Besides *On the Mysteries* and *On the Pythagorean Way of Life*, Iamblichus wrote *On the Soul, Theological Arithmetic*, and *Commentary to Nichomachus' Introduction to Arithmetic*, from all of which fragments survive.

Iamblichus made the unfortunate decision to see the Asian savior god Mithras as the manifestation of the Logos, the source of the intellectual world. Mithras's followers identified their savior with the Good and the sun. There was, in fact, a cult of the sun at the end of the third and beginning of the fourth centuries. *Sol Invictus*, the "Unconquered Sun," was popular with Roman soldiers and emperors, such as Aurelian (r. 270–275) and Constantine (r. 306–337). Mithras became identified with the Unconquered Sun and with other cults as well. The ancient cult to the mother goddess Cybele featured a savior god, Attis, who became associated with Mithras and the sun. Out of this confusing mess emerged Helios.

COMMENTATORS

During late antiquity, the Roman Empire was disintegrating, the barbarian kingdoms of western Europe were being created, and the Byzantine Empire was emerging from the Eastern Roman Empire. There were few original thinkers and scientists during these centuries—most surviving works of science and philosophy were by commentators who examined the thinking of the greats of the past, such as Aristotle (384–322), and wrote extensive commentaries to preserve their teachings. Much of what we know about early Greek scientists comes from the work of the commentators of late antiquity. The most important of these commentators included Philostratus, Diogenes Laertius, Eunapius, and Proclus.

Philostratus (170–250 CE)

As the empire began to crumble in the wake of the Pax Romana of the first and second centuries, the scientific centers of the Roman Empire continued to be Athens, Alexandria, Pergamum, and Antioch. The city of Rome was sometimes important, too, especially under the patronage of the emperors and their families. Philostratus, who wrote the *Life of Apollonius* and *Lives of the Sophists*, was active at the court of Septimius Severus (r. 193–211) and patronized by the empress Julia Domna (160–217). The consort of the emperor Septimius Severus and mother to the emperor Caracalla (r. 198–217), she was reputedly foretold by a horoscope that she would be an empress, which compelled the astrologist Septimius Severus to marry her. She cultivated the pagan arts and literature, bringing together Sophists and astrologers and others involved in pseudoscience. Philostratus wrote *Lives of the Sophists* and *Life of Apollonius of Tyana,* at her bidding. Aelian, the author of *On the Characteristics of Animals,* was also patronized by Julia.

Philostratus was himself a Sophist, a paid teacher of "wisdom": rhetoric and philosophy. It should not surprise us that compilations of past science, books that told the story of great philosophers and Sophists of the past, should be popular and widespread during years when there was little intellectual creativity. The *Life of Apollonius of Tyana* is a somewhat imaginary account of the life of the worker of wonders Apollonius of Tyana, who practiced his miraculous arts at Byzantium in the second century CE. Besides Philostratus's *Lives of the Sophists*, there were Eunapius's *Lives of the Sophists* and Diogenes Laertius's *Lives of the Philosophers.* These works, typical for a time of intellectual malaise, tended to be anecdotal accounts of great thinkers of mythic quality. Philostratus's *Lives of the Sophists* included brief biographies of such famous figures as Eudoxus of Cnidus (408–355), a scientist; Favorinus (c. 85–155), a geographer and historian; Protagoras (485–415), the Sophist made famous by Plato in his *Dialogues*; Hippias of Elis (c. 460–400), a geographer and astronomer; Aelius Aristides (117–181), an analyzer of dreams and devotee of Asclepius; and Aelian (175–235), a zoologist. Diogenes wrote of the great ones of the past, the early Ionian masters of science such as Thales (626–545) and Anaximander (610–540), and the Athenians Plato (427–347) and Aristotle. Eunapius, on the other hand, centered his efforts on the Neoplatonic heroes of the fourth century, his own time, and their champion, the emperor Julian.

Claudius Aelian (175–235 CE)

Aelian was a Roman who wrote in Greek during the first half of the third century. He was a Sophist connected, perhaps, to the court of Julia Domna, the empress and patroness of philosophers. Philostratus, who knew him, wrote a brief life of Aelian in his *Lives of the Sophists.* Aelian's *On the Characteristics of Animals* is an eclectic compilation of facts about animals, purporting to illustrate their moral (and immoral) behavior. Aelian's work is not an original contribution to science; rather, it is heavily reliant upon earlier authors, and as such, it furnishes a varied catalogue of ancient writers and commentators. Aelian took as his model such writers as Herodotus (490–430), who wrote of places in Europe, Asia, and Africa and presented uncritically fact and fancy. Aelian, however, never took the time to travel the Mediterranean to learn firsthand the facts and stories upon which he wrote. Aelian's work proceeds from one topic to another in an apparent random fashion, the order being circumscribed only to specific animals under discussion according to hearsay, legend, and myth.

On the Characteristics of Animals is anecdotal, filled with comparisons to current practices, beliefs, and verbal expressions. For example, Aelian described the Egyptian veneration for lions, which came to people in their dreams and gave them a sense of the future. He assigned, after Democritus (460–370), hot weather and warm south winds for producing a more rapid birth in animals because the organs and tissues are warm and fluid. Bitches have many babies in a litter because they have many wombs. Ancient myth informs the credulous (such as Aelian) that some animals are particularly loved by the gods, who use animals to send their messages and do their will. Many animals have human characteristics. Some have nurtured and raised infant humans (one thinks of the she-wolf raising Romulus and Remus). Others come to the aid of humans, as when the dolphin saved Arion. Some, such as dolphins, mares, and stingrays, enjoy human music, and others enjoy human dance; Indian elephants enjoy the scent of flowers, drink wine, and are gratified when the forerunners of veterinarians apply salves and other concoctions to heal their wounds. Aelian reported on a thieving octopus, a male hare that bore baby rabbits, tritons (half human, half fish) seen at sea, and the two hearts of the elephant, one good and one bad. Interested in medicine, Aelian described sovereign remedies for physical ailments derived from various animal parts. The sea urchin is good for stomach problems; the ashes of the hedgehog mixed with pitch is good for hair loss; hedgehog ashes and wine help purify the kidneys.

If Aelian was not a discriminating scientist, he was at least a writer interested in compiling all that he had learned on natural history. *On the Characteristics of Animals* helps scholars supplement surviving fragments of earlier philosophers. Most importantly, Aelian's work reveals the amazing degree to which people of the Late Roman Empire bought into the claims and stories of pseudoscientists.

Diogenes Laertius (Third Century CE)

Diogenes Laertius is one of the more important sources for uncovering the history of ancient science. The *Lives and Opinions of Eminent Philosophers in Ten Books* examines the lives of ancient thinkers beginning with the Ionian Thales and ending with the Skeptic Saturninus (fl. c. 200). Diogenes divided Greek philosophy into the Ionian and Italian schools; the former was highlighted by Anaximander, Socrates (469–399), Plato, Aristotle, and Zeno (333–262), the founders of the Academy, Peripatetics, Stoics, and Cynics; the latter featured Pythagoras, Xenophanes (570–478), Parmenides (fl. 485), and Epicurus (341–271), the founders of the Pythagorean and Epicurean schools.

Little is known of Diogenes's life. Besides the *Lives*, he wrote *Epigrams in Various Metres*, known today only in fragments. The selection of his topics in the *Lives* indicates he wrote after 200 CE. Diogenes was a Greek philosopher interested in the development of Hellenic thought as it expanded throughout the Mediterranean world. It is possible that he was one of the many philosophers who were sponsored by the empress Julia Domna in the early third century.

Greek thought by 200 lacked creativity and dynamism; Diogenes's *Lives* reflects this malaise. Diogenes was a storyteller who delighted in the well-placed anecdote. He was not usually critical of his sources and perceived his role as an antiquarian collector of the relics of thought. Diogenes provided little historical, political, and geographic scope with which to place the lives of his heroes. His are temporally disjointed lives (*bios*) in the same style as other biographers, such as Plutarch. He was rarely skeptical, being generally credulous about gods, prophecy, and magic. Diogenes was also not too particular about whom he classified as a philosopher. In his *Lives*, we find a wide range of thinkers, natural scientists, mathematicians, metaphysicians, and biologists, all collectively considered lovers of wisdom (*sophia*), seekers of knowledge (*episteme*), hence scientists.

The Seven Sages were legendary thinkers identified by ancients of the Greco-Roman world to explain the origins of thought, literature, and science. Diogenes's *Lives of the Philosophers* lists the seven as Thales, Solon (640–560), Periander (c. 627–587), Cleobulus (fl. sixth century), Chilon (c. 620–520), Bias (fl. sixth century), and Pittaeus, although Diogenes noted that sometimes these were replaced by others: Anacharsis the Scythian (fl. sixth century), Myson of Chen, Pherecydes of Syros (fl. sixth century), Epimenides the Cretan, and Pisistratus (600–527), tyrant of Athens. Thales of Miletus was, perhaps, the only physical scientist among the lot. He was an astronomer, engineer, and metaphysician who was clearly influenced by Egyptian and Mesopotamian scientists. The Athenian Solon was known for his wisdom and statesmanship in creating the Athenian state. Besides Thales and Solon, most of the other Seven Sages are known only by anecdotal information. Diogenes Laertius called Chilon, Pittaeus, Bias, and Cleobulus moral philosophers, although the latter was familiar with Egyptian philosophy. Anacharsis was reputed by some to be the inventor of the anchor and the potter's wheel. Diogenes claimed that Myson argued for a concrete approach to acquiring knowledge based on accumulating facts with which to support arguments. Epimenides, according to Diogenes, saved Athens from the ravages of plague. According to Pliny the Elder, he lived for 157 years, having spent 57 of those asleep in a cave. Plutarch claimed he was a soothsayer. Pherecydes, whom some call the teacher of Pythagoras, was, perhaps, a philosopher and scientist of time, inventing the sundial. Pliny the Elder claimed (without giving details) that Pherecydes was able to predict earthquakes.

Diogenes Laertius's *Lives of the Philosophers* is valuable in several ways. His is one of the few relatively complete collections of ancient thinkers written by an ancient, if mediocre, thinker. Many of his "facts," anecdotes, conversations, and letters are unique to his book, not found elsewhere in surviving sources. Diogenes, moreover, provided an undiscriminating collection of authors and commentators that are otherwise unknown, or whose works are lost save for the bits and pieces that he and other literary collectors have passed down to modern times.

Proclus (412–485 CE)

Proclus of Lydia was a Late Roman mathematician and Neoplatonist philosopher who studied at Athens and Alexandria, and was the longtime head of a Platonic school in Athens. His disciple Marinus of Samaria (fl. fifth century) wrote *Life of Proclus*, which includes an account of Proclus's character:

> The primary elements of the soul were innate in him, and he had no need of learning them, and even so they were highly developed in him. His was a great memory, an intelligence suited to all kinds of studies; he was liberal, affable, loving, and fraternal to truth, justice, courage and temperance. Never had he voluntarily told a lie; lies he abhorred, and he cherished sincerity and veracity. What else could be expected from a man who was to achieve the presence of True Being? Since youth, he was impassioned for truth, for truth is the source of all goods, among gods as

among men. His profound scorn for sensuality, and his inclination to temperance was well illustrated by his extreme ardor and overwhelming leaning towards science, and all kinds of sciences, which do not even allow a first start to the pleasures of gross and animal life, and, on the contrary have the power to impress us with the pure and unmingled joys of the soul. (Trans. Guthrie, *Life of Proclus*)

Marinus wrote further of Proclus's Neoplatonic beliefs: "He is the author of many hitherto unknown theories, that were physical, intellectual, or still more divine. For he was the first to assert the existence of a kind of souls that are capable of simultaneously seeing several Ideas. He had very properly postulated their existence as intermediate between the Mind *(Nous)* which embraces all things together by a single intuition, and the souls whose discursive thoughts pass, and who are unable to conceive more than a single idea at one time" (trans. Guthrie, *Life of Proclus*).

Proclus was a student of Zoroastrian texts, the *Chaldean Oracles*, which examined levels of supernatural, transcendent beings according to ancient Persian theology. Marinus, and no doubt other disciples as well, believed that Proclus had special powers over natural elements, such as meteorological events; he was also a dream interpreter and healer and experienced visions of deities such as the goddess Hecate and the healing god Asclepius. An astrologist, he knew the precise situation of the planets and constellations at the moment of his birth, as recorded by Marinus:

The Sun was in Aries, at 16 degrees 26 minutes
The Moon was in Gemini, at 17 degrees 29 minutes
Saturn in Taurus, at 24 degrees 23 minutes
Jupiter in Taurus, at 24 degrees 41 minutes
Mars in Sagittarius at 29 degrees 50 minutes
Venus in Pisces, at 23 degrees
Mercury in Aquarius at 4 degrees 42 minutes
The horoscope was taken in Aries at 8 degrees 19 minutes
The meridian in Capricorn at 4 degrees 42 minutes
The ascendant at 24 degrees 33 minutes
The preceding New Moon in Aquarius at 8 degrees 51 minutes.
(Marinus of Samaria, Trans. Guthrie)

Proclus was a mathematician and astronomer of note. He wrote the *Commentary on Euclid* and *On Motion*. This latter work reads remarkably like Euclid's *Elements*:

BOOK I

1. CONTINUOUS are things whose term is one.
2. CONTIGUOUS are things whose terms join.
3. CONSEQUENT are things between whom is nothing of the same kind.
4. The first moment of motion is the one than whom there is nothing longer nor shorter.
5. The ORIGINAL PLACE is the one that is neither larger nor smaller than the contained body.
6. That is at rest which, before and after, both itself and its parts remain in one and the same place.

PROPOSITION

1. Two indivisibles do not touch.
2. From two indivisibles no continuum can be formed.
3. What is interposed between indivisibles must be continuous.
4. Two indivisibles cannot be consequent.
5. Every continuum is divisible, and ever divisible.
6. If a quantity is composed of indivisibles, the motion made thereon will also be indivisible,
7. If the motion is composed of indivisibles, its time will also be composed of indivisibles.
8. Amidst things moved unequally the swiftest, is the one traversing the greater space.
9. Things moved unequally,—if one takes more time than is required by the slowest, but less than the swiftest in this medium space of time the swiftest will cover the most ground and the slowest less.
10. Of things moved with unequal swiftness an equal apace is traversed by the swiftest in less time.
11. All time, magnitude and motion are divisible in infinity.
12. It is impossible to traverse an infinite magnitude in finite time.
13. No finite magnitude is traversed in infinite time.
14. If the swiftest bears to the slowest a sesquialter ratio, the lines were not indivisible.
15. A moment remains the same either in the past or future.
16. A moment is indivisible.
17. Everything that moves, moves in time.
18. Everything that rests, rests in time.
19. Everything that moves is indivisible.
20. If the parts of some motion correspond to the parts of some continuum, certainly all the motion will be of all.
21. All that is moved primarily is moved in that in which it is moved.
22. All that is moved, is primarily moved in an indivisible.
23. No mutation has any beginning from the moved, which is motion; no mutation has a mutation as beginning, at the time of which that which is moved may be said to be moved primarily.
24. If mutation is of something which possesses quantity, the first of this quantity will be incomprehensible.
25. If the first time of the mutation, whatever it be, is taken in any part of time, there will be a part of motion
26. All that is moved is moved primarily.
27. All that is moved was moved primarily.
28. If that which is moved is infinite, a finite quantity will not pass infinite time.
29. If what is moved is infinite, an infinite magnitude will not pass in finite time.
30. So motion is not infinite, because it is repeated.
31. All that is moved in a place is all in the moment according to the first place.
32. Everything indivisible in quantity is immovable by itself.

BOOK SECOND

POSTULATES

1. Every natural body is movable according to the place.
2. All motion which is made according to the place is either circular, straight, or mixed.

3. Every natural body is moved by a single one of the above notions.
4. Every natural body is either simple or composite.
5. Every simple movement is of a simple body.
6. Every simple body is moved by a single movement according to its nature.

DEFINITIONS

1. Past intervals of movable things are in a ratio to their speed.
2. That is heavy which is borne to its centre.
3. That is light which is moved from its centre.
4. Circular notion consists in being borne from one point to the same.
5. Contrary movements are those borne from contraries to contraries.
6. Time is the number of the notion of celestial bodies.
7. Motion is single when it does not differ in kind, is in one subject, and occurs in continuous time.

PROPOSITIONS

1. Things which by nature are borne circularly are simple.
2. Things which are naturally borne circularly are not identical with those naturally borne straight or mixed.
3. Things naturally borne circularly have neither weight nor lightness.
4. Nothing is contrary to circular motion
5. Things naturally borne circularly are not subject to generation or decay.
6. Every thing naturally borne circularly is finite.
7. Bodies of infinite magnitude have infinite powers.
8. Bodies of finite magnitude do not have infinite powers.
9. Powers of things moved in equal speed are in alternate ratio to the times of movement.
10. No weight nor lightness is infinite.
11. Nothing infinite can suffer from the finite.
12. Nothing finite can suffer from the infinite.
13. Nothing infinite can suffer from the infinite.
14. Simple bodies are finite in kind.
15. No sensible body is infinite.
16. Time is continuous and also perpetual.
17. Circular movement is perpetual.
18. That which is the cause of perpetual movement is perpetual.
19. That which is immovable precedes the movable things, and the things which move; it claims the principal rank.
20. All that is moved is moved by somebody.
21. As the First moves the circular conversion, it has no parts. (Trans. Guthrie, *On Motion*)

Proclus attempted to measure the sun's diameter using a water clock. He wrote commentaries on Aristotle and on astronomers such as Hipparchus (190–120). He studied eclipses, spheres of planets, and the distances between planets.

ARISTOTELIAN INFLUENCES

Platonic (Neoplatonic) thinkers believed in the contemplative life removed from the demands of public service, which fit their view that the activities, sights, and sounds of daily life are in the end insignificant, mere shadows of what is true and real. Aristotelian (Peripatetic) thinkers, on the other hand, embraced the apparent reality of everyday life, believing that in the everyday lies the hints of truth to the observant philosopher and scientist. In the Eastern Roman (Byzantine) Empire, there were thinkers influenced by Aristotle such as Themistius (317–388) and Ammianus Marcellinus (325–395). In the Western Roman Empire, Aristotle influenced a wide range of thinkers, from Porphyry to Palladius (fl. 500) to Aurelius Augustine. A Rhodian, Andronicus, during the first century BCE published an edition of Aristotle's works, generating renewed interest in the philosopher among Romans. In the first century CE, Aristocles of Messene produced commentaries on Aristotle's work. A century later, Alexander of Aphrodisias used Andronicus's edition of Aristotle's work to provide extensive commentaries completely in the Peripatetic tradition, unalloyed by Platonic thought.

Another Aristotelian, Anatolius, lived in the third century; he was the Christian bishop of Laodicea, and author of *Elements of Arithmetic.* He advocated the use of the Metonic cycle in figuring Easter. Porphyry, though a student of Plotinus, hence Plato, saw the significance of Aristotle as well and wrote a host of commentaries on Aristotle from the Neoplatonic position. He was interested in Aristotle's *Physics* and *Categories* and wrote *Isagoge*, a textbook introduction to Aristotle's principles of logic that was a basic source for the next millennium. Gaius Marcus Victorinus (c. 290–364), a Roman African philosopher and rhetorician of the mid-fourth century, teaching in Rome, who converted from paganism to Christianity, made Latin translations of the works of Plotinus and Porphyry and also wrote commentaries on, and translated into Latin, Aristotle's *Categories* and *On Interpretation.*

Palladius (fl. 500 CE)

The Peripatetic influence in philosophy and science had a long reach throughout antiquity, influencing Greeks and Romans, pagans and Christians. The Peripatetics emphasized observation, empiricism, and concrete results. An example is Rutilius Taurus Aemilianus Palladius, of Gaul, the author of the book (in Latin) *On Husbandry.* Very little is known about his life. He came from an important Roman family, the Aemiliani, and was probably a wealthy landowner. He lived sometime during the fourth or fifth centuries, when the Roman economy had not yet ground to a halt in the western half of the empire but agriculture was clearly on the decline. Palladius appears to have been a pagan rather than a Christian. His fourteen-volume *On Husbandry* became an important resource for landowners during the European Middle Ages. Palladius's sources included the early Roman agricultural writers Columella (5–60) and Cato (234–149).

On Husbandry is divided into books based on the seasons of the year. It reads rather like an almanac—it discusses the best times to engage in the various activities of husbandry. Palladius followed a lunar cycle to direct his agricultural planning. *On Husbandry* gives advice on the use of manure—of different animals for different crops. He discussed ways to construct a warm stable for farm animals. He gave advice on beekeeping; on the proper tools to use for pruning, digging, plowing, and cutting; on when to plant asparagus, wheat, beets, lettuce, and radishes; on the planting of nuts and orchards; and on the use of prayers and the value of keeping faith in preparing for a good harvest. Palladius wrote in prose, though he had a poetic side as well. His writing is reminiscent of the pastoral literature of the Augustan Age.

Palladius, writing in the fourth to fifth centuries, indicated that Roman agricultural methods were the same throughout the thousand-year history of the Roman Empire. The sandy Mediterranean soil required simple tools; the heavy plough of the Middle Ages was developed in response to the thick and rocky soil of northern Europe. The Romans practiced an elementary form of crop rotation involving leaving land fallow in intermittent years. The major change in Roman agriculture from the first millennium BCE to the first millennium CE was the increasing number of slaves and peasants involved in cultivation of land that they did not own. The inequality of Late Roman agriculture meant that there was a small class of the very rich who were looked to for leadership in war and local politics and a huge dependent class of agriculture laborers, the *coloni*.

CHRISTIANITY AND THE LATER ROMAN EMPIRE

Christianity, as it developed during the first few centuries CE, brought together into one system of thought the psychology of antiquity as it was developed over the course of several millennia by the Mesopotamians, Egyptians, Hebrews, and Greeks. The Gospels and Epistles of the Greek New Testament present an approach to the human psyche that is at the same time spiritual and materialist, based on surrender as well as freedom. The crucial development in ancient Christian psychology revolves around the life and teachings of Jesus of Nazareth combined with the belief by the early Christian church that he was the Christ, the Logos.

The fourth Gospel, the Gospel of John, is the crucial text in the emergence of Christianity from the psychology of the ancient Near East and ancient Greeks. John, son of Zebedee, who along with his brother James was a fisherman and one of Christ's first disciples, was the "disciple whom Jesus loved," the author of the fourth Gospel, three Epistles, and perhaps Revelation. John was a teacher of many second-century apostles, such as Papias, Polycarp, and Ignatius. John interpreted Jesus as the Greek Logos, the creative force of the universe, the source of reason, knowledge, and being. The Jesus of the Gospel of John is as well the begotten, coeternal *Word* of the Hebrew conception of God, Yahweh. What was completely new and unique amid the religious and philosophical history of the ancient Mediterranean was that Jesus, besides being the Logos, the *nous* (mind), and the *ousia* (being), was also a simple peasant born to a carpenter and his wife in the poor rural society at the close of first-century-BCE Palestine.

That Jesus was a man or, as he called himself, the Son of Man, had a profound effect on the development of Christian psychology. It is altogether a different experience to find within oneself a sense of knowing, being, feeling, and thinking that is human rather than eternal, infinite, transcendent, and anonymous. A Christian knew the knowing that Jesus knew. A Christian felt the knowing that Jesus felt. A

Christian thought the knowing that Jesus thought. Christian psychology is so overwhelmed with human feelings and experiences of love, joy, faith, hope, sorrow, suffering, and pain that merely by believing in Jesus as the Son of Man and Christ, the believer felt less alone, a part of something encompassing the whole of human experience, akin to a man who was god incarnate who suffered and died, the fear of which is precisely the source of anxiety, obsession, and depression among humans. Christ is the means of contentment for the Christian. Christ is the means of achieving a satisfactory sense of identity, of having a meaningful understanding of life, of accepting oneself, life and death, and the universe itself.

Harkening back to Yahweh's proclamation to Moses that "I am that I am," the Gospel of John quotes Jesus proclaiming to his disciples and the Jews, "I am" (*ego eimi*). Jesus, a man, proclaimed himself the transcendent Logos. But he also set forth an invitation to the human who seeks to know the transcendent, that to simply proclaim "I am" is to know the truth, the source of freedom.

In addition to the Gospels, the New Testament, the Nag Hammadi Scrolls, and the works of other philosophers and theologians of the first several centuries CE reveal a high level of sophistication in the study of human psychology and criticism of Greco-Roman philosophy and religion.

Paul of Tarsus (fl. First Century CE)

Paul of Tarsus was a Roman citizen, a tent maker, a Pharisee, and, after his conversion to Christianity, a sophisticated student of theology and human psychology. Paul, in his Epistles, which form a good part of the New Testament, portrayed himself as a man of great passions of body and mind. He believed fervently and acted on his beliefs. He felt the needs of the flesh and acted on those needs. He felt the needs of the mind and acted on those needs. His conversion to Christianity on the road to Damascus, as described in the Acts of the Apostles of the New Testament, gave him a new understanding of life, of his own desires and needs, though it did not eradicate those desires of the body and mind.

Before his conversion, Paul tried to channel the desires of his body and passion of his mind through the Mosaic law, as described in the Old Testament, but he was never quite successful. As he wrote in the Letter to the Romans 7:15 (RSV): "I do not understand my own actions. For I do not do what I want, but I do the very thing I hate." His conversion and understanding of the significance of the Messiah gave him the realization that the Old Testament Law is limited, and cannot yield salvation, cannot bring peace and truth. Rather, the law made him realize how sinful he was. "Wretched man that I am! Who will deliver me from this body of death?" (RSV).

Living under the Old Testament law, there was a dualism in Paul's thought. Law was truth, law was of God, but Paul lived in sin; Paul did not live according to God. The separation between Paul and God was vast. But then the *incarnation*, Paul believed, brought God to live on earth, to assume the "likeness of sinful flesh," to experience all that Paul experienced: a body, pain, suffering, fear, sin, death. God brought himself to Paul, he believed, to experience what Paul experienced. Paul, knowing that Jesus Christ had experienced humanity as Paul did, realized that all of his sins had been experienced by Christ, had been assumed by Christ, and there was no longer a duality of law and Paul but a unity of sinful, human Paul and Christ who experienced human sin and knew what Paul experienced. So even though Paul continued to sin, because, after all, he had a body with passions and needs and a mind that continued to seek, he realized that Christ himself experienced this body with passions and needs and a mind that sought, and therefore it was acceptable for Paul. The guilt of the law was lessened if not eradicated. The fruitless attempts by Paul not to sin continued, but the psychic impact was not as terrible, and Paul did not feel so lonely, because he felt Christ within him. Indeed, he argued, Christ was with him even at his most horrible moments, the moments when the body plagued him most, when he was feeling lust or experiencing horrible pain, or when he died. At these moments Paul felt no longer alone confronting an awful and terrible punishing God; rather he experienced a God of passion, and like a child, Paul cried out, "Abba, father," and felt the loving embrace of the arms of the Father.

Pagan and Christian in an Age of Anxiety

Christianity combined with other religious and philosophical sects flooding the Roman Empire of the second and third centuries CE to produce one of the most religiously intellectual dynamic times in human history. Gnostics such as Valentinus (c. 100–160), Stoics such as Marcus Aurelius (121–180), Neoplatonists such as Plotinus, Christians such as Origen, and philosophers such as Aristides and Hermias tried to find meaning at a time when the ancient world was coming to an end, the Roman Empire was showing signs of collapse, the world seemed old and decayed, and humans were weighed down with sin and suffering. For many, the mind was the best escape to a new and better reality. This was certainly true of Plotinus, who believed that within the self was the means of achieving the divine, to unite with the ultimate reality, the One, in an ecstatic transformation. Plotinus symbolized, in the words of E. R. Dodds (p. 37), "the progressive withdrawal of divinity from the material world" in late antiquity. Gnostic philosophers, Christians, and Platonic philosophers believed that reality was nonmaterial, noncorporeal, hence had to be ignored, neglected, or punished. This period from the third to the fifth centuries was the time of the great desert hermits and self-mutilators such as Simon Stylites (c. 390–459). The body had nothing to do with personal identity; indeed, it hampered human experience with the divine world. Total sexual abstinence was extremely popular; some fanatics castrated themselves; others refused to bathe; some loaded chains on their bodies to purge it of sin; others practically starved themselves in the name of purity. Such actions imply tremendous guilt, which appears to have been a common experience as a response by individuals and groups to the growing problems of the Roman Empire, such as civil war, famine, epidemics, violence, apathy, and so on. Amid such problems, moral and natural philosophers sought to find what was true and real amid the shadows and fancies of falsehood.

Justin Martyr (100–165)

Justin Martyr, a Platonist philosopher who converted to Christianity at age thirty, became an important apologist for the emerging second-century church. A pagan Gentile living in Syria, he was well educated and learned about Pythagoreanism, Stoicism, Aristotelianism, and Platonism before his conversion to Christianity. He was martyred by the Romans for refusing to recant his beliefs in 165.

In his many writings, Justin described his conversion, criticized pagan religion, and provided support for his own beliefs. In *Dialogue with Trypho*, he described his education and his dedication to Platonic thinking before he met an old man, by the seashore, with whom he had an extensive dialogue; the man convinced Justin that Plato alone without the prophets of Scripture is insufficient by which to understand God. In *The Discourse to the Greeks*, he attacked Greek mythology and the Greek pantheon of deities as contradicting common sense, reason, and the Greco-Roman Logos, which Justin, like John before him, interpreted to be Jesus of Nazareth. Like his contemporary Aristides of Athens, Justin addressed an *Apology* in support of Christianity to the Roman emperor, at the time Antoninus Pius (r. 138–161). Justin's *Horatory Address to the Greeks* provides an extensive discussion, and refutation, of Greek philosophy. He first discussed the Ionians, then the Italians of Magna Graecia, and then the materialists, noting that the inability of Thales, Anaximander, Anaximenes, Heraclitus, Hippasus, Anaxagoras, Pythagoras, Epicurus, and Empedocles to agree on the fundamental principle revealed their ignorance. Next, he addressed the teachings of Plato and Aristotle, noting how teacher and student contradicted each other, "so that one can see that they not only are unable to understand our earthly matters, but also, being at variance among themselves regarding these things, they will appear unworthy of credit when they treat of things heavenly" (trans. Dods). Justin defended the authority of Moses and provided evidence from ancient tradition, such as a poem by Orpheus, in support of monotheism. Justin went on to argue that Plato was actually influenced by the Hebrew prophets so that he had a monotheistic view. Justin argued as well that the poet Homer had been in Egypt, where he had been influenced by the writings of Moses. If this was not enough, Justin also cited the Roman Sibyl, arguing that her prophecies were akin to the writings of the Hebrew prophets!

Marcianus Aristides of Athens (?–134)

Aristides was an Athenian philosopher who embraced Christianity and had the courage to address the emperor Hadrian (r. 117–138) when the pagan emperor visited Athens in 125 CE. Previous emperors, such as Hadrian's successor Trajan, had persecuted Christians: Pope Clement I, the fourth pope, was martyred under the emperor Trajan in 99. Ignatius of Antioch (50–110?), the third bishop of Antioch, was a disciple (perhaps) of St. John and was martyred at Rome (also under Trajan). Aristides therefore took a risk when addressing the pagan emperor Hadrian, who was pontifex maximus of Rome, dedicated to the cult of emperor worship, and a believer in the traditional gods, for which he rebuilt and dedicated new temples of worship. *The Apology* was a sophisticated comparison of the tenets of Christianity with Judaism, traditional Greek religion, and the polytheism of other peoples of the Mediterranean world. In *The Apology*, Aristides expressed subtly how he came to embrace the Christian god:

> I, O king, by the grace of God came into this world; and having contemplated the heavens and the earth and the seas, and beheld the sun and the rest of the orderly creation, I was amazed at the arrangement of the world; and I comprehended that the world and all that is therein are moved by the impulse of another, and I understood that he that moveth them is God, who is hidden in them and concealed from them: and this is well known, that that which moveth is more powerful than that which is moved. And that I should investigate concerning this Mover of all, as to how He exists—for this is evident to me, for He is incomprehensible in His nature—and that I should dispute concerning the stedfastness of His government, so as to comprehend it fully, is not profitable for me; for no one is able perfectly to comprehend it. But I say concerning the Mover of the world, that He is God of all, who made all for the sake of man; and it is evident to me that this is expedient, that one should fear God, and not grieve man. (Trans. Kay)

Unlike what pagan polytheists believed, Aristides argued, the Christian god is neither male nor female, nor substance, nor jealous, nor filled with wrath, nor subject to time, nor subject to emotions or any limitations whatsoever. He pointed out that polytheistic peoples conceive of the universe according to the four elements—earth, air, fire, and water—and conform their deities to these elements, making the uncreated creator a creature bound to the substances of the universe. He argued as well that the anthropomorphism of humans, making the divine conform to human thoughts and feelings, is illogical and inconsistent, and it leads to all sorts of barbarity and immorality among humans.

Aristides was well versed in the Torah and the writings of the Gospels, he appears to have been familiar with the Epistles of Paul, and he accommodated these writings to his understanding of Greek philosophy, particularly of the Athenians, such as Plato and Aristotle. Aristides anticipated the work of others of his ilk, such as the philosopher Hermias, who later, in the second or third century, wrote a condemnation of Greek philosophy.

Hermias the Philosopher (fl. Second to Third Century)

Hermias, about whom little is known, was a Greek philosopher who embraced Christianity, during the course of which, it appears, he struggled with the competing ideas of the Ionian and Italian schools of philosophy, the idealists, empiricists, and atomists, so that he realized the apparent truth of the comment of Paul of Tarsus: "The wisdom of the world is folly in the sight of God" (1 Corinthians 3:19 RSV). Hermias had become a Skeptic regarding the philosophic and scientific ruminations of the Greeks, which turned him to Christianity. He wrote,

> The philosophers put forth their doctrines, saying things that neither sound the same, nor mean the same as one another. For some of them say that the soul is fire, like Democritus; air, like the Stoics; some say it is the mind; and some say it is motion, as Heraclitus; some say it is exhalation; some an influence flowing from the stars; some say it is number in motion, as Pythagoras; some say it is generative water, as Hippo; some say an element from elements; some say it is harmony, as Dinarchus; some say

the blood, as Critias; some the breath; some say unity, as Pythagoras; and so the ancients say contrary things. How many statements are there about these things! how many attempts! how many also of sophists who carry on a strife rather than seek the truth! (Trans. Giles)

Hermias's writing is satirical yet revealing of his misgivings as a philosopher with the varied contradictions of the most profound Greek philosophers over the nature of reality. He wrote that first Anaxagoras (610–540), the Ionian philosopher, "catches me, he teaches me thus: The beginning of all things is mind, and this is the cause and regulator of all things, and gives arrangement to things unarranged, and motion to things unmoved, and distinction to things mixed, and order to things disordered." Then, however, Hermias confronted the writing of Parmenides of the fifth-century Italian school of Greek philosophers: "Being is one, and everlasting, and endless, and immoveable, and in every way alike." The power of this idea captivated Hermias, and "I know not why I change to this doctrine: Parmenides has driven Anaxagoras out of my mind. But when I am on the point of thinking that I have now a firm doctrine," Anaximenes (586–525), "the Ionian philosopher, catching hold of me, cries out, 'But I tell you, everything is air, and this air, thickening and settling, becomes water and air; rarefying and spreading, it becomes aether and fire: but returning into its own nature, it becomes thin air: but if also it becomes condensed, (says he) it is changed.' And thus again I pass over to this opinion of his, and cherish Anaximenes." But then Hermias found the writings of the materialist philosopher Empedocles (495–435), followed by the humanist Protagoras (485–415), then the Ionian Thales (625–545), followed by another Ionian, Anaximander (610–540); then he found Archelaus of Athens of the fifth century, the teacher of Socrates, followed by the student of Socrates, Plato, and then Aristotle the empiricist, followed by the fifth-century atomist Leucippus. The arguments of other philosophers follow: the materialists Democritus (460–370) and Heraclitus (540–480), the Stoic Cleanthes (330–230), the Platonist Carneades (214–139), and then the great mathematician Pythagoras, whose mathematical conceptions of reality intrigued Hermias until he confronted the materialism of Epicurus (341–271), and it seemed that in this endless circle of speculation, Hermias found no end. "For all things already are the darkness of ignorance to me," he wrote, "and black error, and endless wandering, and unprofitable fancy, and ignorance not to be comprehended: unless else I intend to number the very atoms also, out of which such great worlds have arisen, that I may leave nothing unexamined, especially of things so necessary and useful, from which both houses and cities prosper. These things have I gone through, wishing to point out the opposition which is in their doctrines, and how their examination of things will go on to infinity and no limit, for their end is inexplicable and useless, being confirmed neither by one manifest fact, nor by one sound argument." Hence, unconvinced by the speculations of Greek philosophers and scientists, he turned to the teachings of Paul of Tarsus, and Christianity.

Clement of Alexandria (c. 150–215 CE)

Clement of Alexandria was a Christian and author of *Miscellanies*, in which he provided commentary on ancient Greek and Roman philosophers and scientists. A convert to Christianity from paganism, Clement had studied Greek and Roman philosophers, such as Plato; after conversion, he used his knowledge of Greek philosophy in support of Christianity. He was one of the ancient sources for information on the Seven Sages of antiquity. Clement lived and taught in the cosmopolitan city of Alexandria.

Clement wrote, in *Miscellanies*, "Philosophy does not ruin life by being the originator of false practices and base deeds, although some have calumniated it, though it be the clear image of truth, a divine gift to the Greeks; nor does it drag us away from the faith, as if we were bewitched by some delusive art, but rather, so to speak, by the use of an ampler circuit, obtains a common exercise demonstrative of the faith. Further, juxtaposition of doctrines, by comparison, saves the truth, from which follows knowledge." Clement believed that all knowledge, pagan and Christian, science and philosophy, has a supernatural origin: "Scripture calls every secular science or art by the one name wisdom (there are other arts and sciences invented over and above by human reason), and that artistic and skillful invention is from God." Clement argued that prior to the coming of Christ, Greek philosophers and scientists provided the basis, the "handmaid," for the coming of Christianity. He argued that science must be pious, engaged in discovering the ways of God (trans. Wilson).

Clement was a historian of Greek philosophy and science as well, providing a brief summary of the origins and extent of science and philosophy, beginning with the legendary Seven Sages:

> The Greeks say, that after Orpheus and Linus, and the most ancient of the poets that appeared among them, the seven, called wise, were the first that were admired for their wisdom. Of whom four were of Asia—Thales of Miletus, and Bias of Priene, Pittacus of Mitylene, and Cleobulus of Lindos; and two of Europe. Solon the Athenian, and Chilon the Lacedæmonian; and the seventh, some say, was Periander of Corinth; others, Anacharsis the Scythian; others, Epimenides the Cretan. . . . Others have enumerated Acusilaus the Argive among the seven wise men; and others, Pherecydes of Syros. And Plato substitutes Myso the Chenian for Periander, whom he deemed unworthy of wisdom, on account of his having reigned as a tyrant. That the wise men among the Greeks flourished after the age of Moses, will, a little after, be shown. But the style of philosophy among them, as Hebraic and enigmatical, is now to be considered. They adopted brevity, as suited for exhortation, and most useful. Even Plato says, that of old this mode was purposely in vogue among all the Greeks, especially the Lacedæmonians and Cretans, who enjoyed the best laws.
>
> The expression, Know yourself, some supposed to be Chilon's. But Chamæleon, in his book *About the Gods*, ascribes it to Thales; Aristotle to the Pythian. It may be an injunction to the pursuit of knowledge. For it is not possible to know the parts without the essence of the whole; and one must study the genesis of the universe, that thereby we may be able to learn the nature of man. Again, to Chilon the Lacedæmonian they attribute, Let nothing be too much. Strato, in his book *Of Inventions*, ascribes the apophthegm to Stratodemus of Tegea. Didymus assigns it to Solon; as also to Cleobulus the saying, A middle course is best. . . . And the aphorism, Practice conquers everything, they will have it to be Periander's; and likewise the advice, Know the opportunity, to have been a saying of Pittacus. Solon made laws for the Athenians, Pittacus for the Mitylenians. And at a late date, Pythagoras, the pupil of Pherecydes, first called himself a philosopher. Accordingly, after the fore-mentioned three men, there were three schools of philosophy, named

after the places where they lived: the Italic from Pythagoras, the Ionic from Thales, the Eleatic from Xenophanes. Pythagoras was a Samian, the son of Mnesarchus, as Hippobotus says: according to Aristoxenus, in his life of Pythagoras and Aristarchus and Theopompus, he was a Tuscan; and according to Neanthes, a Syrian or a Tyrian. So that Pythagoras was, according to the most, of barbarian extraction. Thales, too, as Leander and Herodotus relate, was a Phœnician; as some suppose, a Milesian. He alone seems to have met the prophets of the Egyptians. But no one is described as his teacher, nor is any one mentioned as the teacher of Pherecydes of Syros, who had Pythagoras as his pupil. But the Italic philosophy, that of Pythagoras, grew old in Metapontum in Italy. Anaximander of Miletus, the son of Praxiades, succeeded Thales; and was himself succeeded by Anaximenes of Miletus, the son of Eurustratus; after whom came Anaxagoras of Clazomenæ, the son of Hegesibulus. He transferred his school from Ionia to Athens. He was succeeded by Archelaus, whose pupil Socrates was. . . . Antisthenes, after being a pupil of Socrates, introduced the Cynic philosophy, and Plato withdrew to the Academy. Aristotle, after studying philosophy under Plato, withdrew to the Lyceum, and founded the Peripatetic sect. He was succeeded by Theophrastus, who was succeeded by Strato, and he by Lycon, then Critolaus, and then Diodorus. Speusippus was the successor of Plato; his successor was Xenocrates; and the successor of the latter, Polemo. And the disciples of Polemo were Crates and Crantor, in whom the old Academy founded by Plato ceased. Arcesilaus was the associate of Crantor; from whom, down to Hegesilaus, the Middle Academy flourished. Then Carneades succeeded Hegesilaus, and others came in succession. The disciple of Crates was Zeno of Citium, the founder of the Stoic sect. He was succeeded by Cleanthes; and the latter by Chrysippus, and others after him. Xenophanes of Colophon was the founder of the Eleatic school, who, Timæus says, lived in the time of Hiero, lord of Sicily, and Epicharmus the poet; and Apollodorus says that he was born in the fortieth Olympiad, and reached to the times of Darius and Cyrus. Parmenides, accordingly, was the disciple of Xenophanes, and Zeno of him; then came Leucippus, and then Democritus. Disciples of Democritus were Protagoras of Abdera, and Metrodorus of Chios, whose pupil was Diogenes of Smyrna; and his again Anaxarchus, and his Pyrrho, and his Nausiphanes. Some say that Epicurus was a scholar of his. (Trans. Wilson)

Hippolytus (c. 180–235 CE)

Hippolytus was a Christian theologian who wrote *Refutation of All Heresies*, in which he condemned pagan philosophy and science while offering commentary and excerpts. Little is known about his life. But his book, the *Refutation*, was much studied and became an important source into the writings and thought of the "natural philosophers" Thales, Pythagoras, Empedocles, Heraclitus, Anaximander, Anaximenes, Xenophanes, Parmenides, Leucippus, Democritus, Ecphantus, Hippos, and Archelaus; the "moral philosophers" Socrates and Plato; the "logicians," most especially Aristotle but also the Stoics Chrysippus, Zeno, and Epicurus; Skeptics such as Pyrrho (365–279); and other philosophers as well, such as the Indian Brahmins, the Celtic Druids, and the poet Hesiod. In short, Hippolytus provided an extensive commentary on the principal philosophers and scientists of ancient Greece.

Origen (184–253)

The Christian theologian Origen, born in Alexandria, founded a school in Caesarea, where he taught theology as well as the subjects that became in time the trivium and quadrivium: logic, mathematics, astronomy, and philosophy. Origen was one of the great theologians of the Christian church but also a Platonic scholar, thereby encompassing both Greco-Roman and Christian thought in the third century. He believed that the ancient *paideia* trained students to be able to approach, study, and comprehend Christian scripture and theology. He believed that the study of nature would allow a person to put behind material things to focus upon God. He disagreed with Stoic and Aristotelian teachings but found Plato's emphasis on ideal forms to be consistent with Christianity, hence he advocated the Platonic approach to studying nature and God. The Christian theologian Jerome (340–420), in *On Illustrious Men*, wrote that Origen "understood dialectics, as well as geometry, arithmetic, music, grammar, and rhetoric, and taught all the schools of philosophers, in such a way that he had also diligent students in secular literature, and lectured to them daily, and the crowds that flocked to him were marvelous. These he received in the hope that through the instrumentality of this secular literature, he might establish them in the faith of Christ" (quoted in Carrier, p. 154).

Origen's thoughts on the relationship between Greek philosophy and science and Christian theology are recorded in the *Philocalia*, in which Gregory Nazianzus (329–390) and Basil of Caesarea (330–379) recorded excerpts from the thoughts and writings of Origen. For example, the *Philocalia* cites the following as Origen's view on Greek philosophy:

> Natural ability, as you know, if properly trained, may be of the utmost possible service in promoting what I may call the "object" of a man's training. You, for instance, have ability enough to make you an expert in Roman law, or a philosopher in one of the Greek schools held in high esteem. I should like you, however, to make Christianity your 'object,' and to bring the whole force of your ability to bear upon it, with good effect. I am therefore very desirous that you should accept such parts even of Greek philosophy as may serve for the ordinary elementary instruction of our schools, and be a kind of preparation for Christianity: also those portions of geometry and astronomy likely to be of use in the interpretation of the sacred Scriptures, so that, what the pupils of the philosophers say about geometry and music, grammar, rhetoric, and astronomy, viz. that they are the handmaidens of philosophy, we may say of philosophy itself in relation to Christianity. (Trans. Lewis)

Origen was a student of logic, and believed that theological interpretation could benefit from its application: "If any one doubts the soundness of this reasoning, let him consider whether a problem in ethics, or physics, or theology, can be properly conceived without accurately finding the meaning, and without close regard to the clear rules of logic. What absurdity is there in listening to those who determine the exact meaning of words in different languages, and in carefully attending to things signified? And we sometimes through ignorance of logic fall into great errors, because we do not clear up the equivocal senses, ambiguities, misapplications, literal meanings, and distinctions" (trans. Lewis).

Aurelius Augustine (354–430 CE)

Aurelius Augustine was a theologian and philosopher who was one of the great thinkers of all time in the fields of psychology, the theory of knowledge, and the nature of time. Augustine was a prolific writer best known for his *Confessions* and *City of God*. The former is an autobiographical portrait of his struggle with sin and redemption through conversion and faith in God. The *City of God* is a massive treatise on history and the human relationship with God. Both works became fundamental sources of Christian theology, spirituality, and faith. Ironically, they are also sophisticated analyses of human nature. Augustine's works also allow us to see the intellectual perceptions of the Late Roman Empire. Augustine was at various times in his life a student of Stoicism, Neoplatonism, Manicheism, and, of course, Christianity. Before he turned to Christianity. Augustine was the typical obsessive individual who was overwhelmed with feelings of guilt that resulted in depression even as he was becoming rich and famous in fourth-century Italy. Having tried to find happiness by exorcising his guilt by means of self-examination, self-control, and the use of reason, and finding himself even more miserable, Augustine turned to Christianity in response to a religious experience in which he heard the call of a child telling him to "take and read" the Bible. Augustine did so, and he became a completely new man, born again, free from oppressive feelings of guilt, confident that his life was valid. Augustine discovered that he could not think or will himself happy, he had to surrender, accept, and embrace something larger than himself to find himself. He wrote the *Confessions* to describe his journey to contentment and the connection of his psyche to the Logos, Jesus Christ. The *Confessions* is in many ways the beginning of modern psychology.

Augustine represents the transition from the ancient to the medieval worlds. He used the best of ancient philosophy and advanced it further under the umbrella of Christianity. He was the last great thinker of antiquity and the first great thinker of medieval Europe. His ideas on human psychology and time have never really been surpassed, and they set the stage for later philosophers and scientists: Francesco Petrarca, Martin Luther, Michel de Montaigne, René Descartes, Jonathan Edwards, Immanuel Kant, Sigmund Freud, William James, and Albert Einstein.

FURTHER READING

Aelian, *On the Characteristics of Animals*, 3 vols., trans. A. F. Scholfield (Cambridge, MA: Harvard University Press, 1971).

Aristides, *The Apology of Aristides, Translated from the Syriac*, trans. D. M. Kay, in *Ante-Nicene Fathers*, vol. 9 (Buffalo, NY: Christian Literature Publishing, 1896), http://www.newadvent.org/fathers/1012.htm.

Aristotle, *On Sleep and Sleeplessness, On Prophesying by Dreams, On Memory and Reminiscence*, trans. J. I. Beare, in *The Parva Naturalia* (Oxford: Clarendon Press, 1908).

Augustine, *Confessions*, trans. R. S. Pine-Coffin (Harmondsworth, England: Penguin Books, 1961).

Jonathan Barnes, *Early Greek Philosophy* (London: Penguin Books, 1987).

Daniel Boorstin, *The Discoverers* (New York: Random House, 1983).

Glen Bowersock, *Julian the Apostate* (Cambridge, MA: Harvard University Press, 1978).

Peter Brown, *The Making of Late Antiquity* (Cambridge, MA: Harvard University Press, 1978).

Robert Browning, *The Emperor Julian* (Berkeley: University of California Press, 1976).

Richard Carrier, *Science Education in the Early Roman Empire* (Durham, NC: Pitchstone Publishing, 2016).

Henry Chadwick, *Augustine* (Oxford: Oxford University Press, 1986).

Clement of Alexandria, *The Stromata*, trans. William Wilson, from *Ante-Nicene Fathers*, vol. 2 (Buffalo, NY: Christian Literature Publishing, 1885), http://www.newadvent.org/fathers/02101.htm.

Diogenes Laertius, *Lives of Eminent Philosophers*, trans. R. D. Hicks, 2 vols. (Cambridge, MA: Harvard University Press, 1931, 1938).

E. R. Dodds, *Pagan and Christian in an Age of Anxiety* (New York: Norton, 1965).

Glanville Downey, "*Philanthropia* in Religion and Statecraft in the Fourth Century after Christ," *Historia* 4 (1955): 199–208.

Eunapius, *Lives of the Sophists*, trans. W. C. Wright (Cambridge, MA: Harvard University Press, 1968).

Galen, *On the Natural Faculties*, trans. A. J. Brock (Cambridge, MA: Harvard University Press, 1916).

Michael Grant, *The Climax of Rome* (Boston: Little, Brown, 1968).

Hermias the Philosopher, *Derision of Gentile Philosophers*, trans. J. A. Giles, in *The Writings of the Early Christians of the Second Century, namely Athenagoras, . . . Hermias . . .* (1857), http://www.tertullian.org/fathers/hermias_1_satire.htm.

Hippolytus, *Refutation of All Heresies*, trans. J. H. MacMahon, in *Ante-Nicene Fathers*, vol. 5, ed. Alexander Roberts, James Donaldson, and A. Cleveland Coxe (Buffalo, NY: Christian Literature Publishing, 1886), http://www.newadvent.org/fathers/050101.htm.

Tom B. Jones, *In the Twilight of Antiquity* (Minneapolis: University of Minnesota Press, 1978).

Julian, Emperor of Rome, *The Works of the Emperor Julian*, trans. W. C. Wright (Cambridge, MA: Harvard University Press, 1962, 1969).

Justin Martyr, *Horatory Address to the Greeks*, trans. Marcus Dods, in *Ante-Nicene Fathers*, vol. 1 (Buffalo, NY: Christian Literature Publishing, 1885), http://www.newadvent.org/fathers/0129.htm.

Joseph Katz, ed., *The Philosophy of Plotinus* (New York: Appleton Century Crofts, 1950).

Marinus of Samaria, *Life of Proclus, or Concerning Happiness*, trans. Kenneth S. Guthrie (1925), http://www.tertullian.org/fathers/marinus_01_life_of_proclus.htm.

The Philocalia of Origen: A Compilation of Selected Passages from Origen's Works Made by St. Gregory of Nazianzus and St. Basil of Caesarea, trans. George Lewis (Edinburgh: T. & T. Clark, 1911).

Philostratus, *Life of Apollonius of Tyana*, trans. F. C. Conybeare (Cambridge, MA: Harvard University Press, 1912).

Philostratus, *Lives of the Sophists*, trans. W. C. Wright (Cambridge, MA: Harvard University Press, 1921).

Plutarch, *Essays*, trans. Robin Waterfield (London: Penguin Books, 1992).

Plutarch, *The Lives of the Noble Grecians and Romans*, trans. John Dryden (New York: Random House, 1992).

Porphyry, "Letter to Anebo," trans. Thomas Taylor, in *Iamblichus on the Mysteries of the Egyptians, Chaldeans, and Assyrians* (Chiswick, England: C. Whittingham, 1821), http://www.tertullian.org/fathers/porphyry_anebo_02_text.htm.

Porphyry, *On Images*, trans. Edwin H. Gifford, in Eusebius, *Preparation for the Gospel.* (Oxford: Clarendon Press, 1903), http://classics.mit.edu/Porphyry/images.html.

Proclus, *On Motion*, trans. Kenneth S. Guthrie, (1925), http://www.tertullian.org/fathers/proclus_on_motion.htm.

Henri de Valesius, trans., *The History of the Church . . . Written in Greek by Eusebius Pamphilus, Socrates Scholasticus, Evagrius* (1692; reprint ed., London: Henry Bohn, 1853).

Donald Wilcox, *The Measure of Times Past: Pre-Newtonian Chronologies and the Rhetoric of Relative Time* (Chicago: University of Chicago Press, 1987).

John Wippel and Allan Wolter, eds. *Medieval Philosophy* (New York: Free Press, 1969).

25

Byzantine Science (300–1450 CE)

Upon the collapse of the Roman Empire in the West during the fifth century, most intellectual concerns, including science, fell subject to the violence and chaos of war and the primitive urge to survive in a time of declining trade and agriculture. The Eastern Roman Empire, subsequently called the Byzantine Empire, continued to thrive under energetic emperors as well as thinkers trying to carry the Aristotelian and Platonic torches forward into the future. The legacy of ancient science was seen most dramatically during the period of Byzantine dominance in Eastern Europe during the Middle Ages.

The Byzantine Empire, which lasted for over 1,100 years, began when the Roman Emperor Constantine (r. 306–337) determined to focus the attention of his rule upon the Eastern Roman Empire. Constantine's predecessor, Diocletian (r. 284–305), to prevent the disintegration of the empire occurring during the third century, had organized the empire into two halves, Western Roman Empire and Eastern Roman Empire, and four prefectures, over which was an emperor, titled Augustus, or a vice emperor, titled Caesar. Diocletian made his capital Nicomedia in Turkey. Constantine followed suit, realizing that the Eastern Roman Empire was wealthier and easier to defend. He decided in 325 to construct a New Rome at the site of an ancient Greek city, Byzantion, which had been founded as a colony in the eighth century BCE by Dorian settlers. Byzantion possessed a wonderful site situated on a peninsula jutting into deep water overlooking the Bosporus and Sea of Marmara. The city lay astride sea lanes passing from the Black (Euxine) to the Aegean Seas, and land trade routes from western Asia to eastern Europe. Byzantion thrived as a Greek polis and then, after the birth of Christ, as the Greco-Roman city of Byzantium. The city, as a crossroads of trade, was necessarily a cosmopolitan mixing place of cultures, institutions, and ideas. As a Greek and Roman city, it was a well-known center for astrologers, seers, and prognosticators.

Constantine realized its strategic location at the junction of two continents, Europe and Asia, and waterways, the Black Sea, Sea of Marmara, Aegean Sea, and Mediterranean Sea. The city featured a deep harbor and was surrounded by water on three sides, making it easily defensible. The city was christened Constantinople in 330; here, where Constantine ruled, was the palace of the emperor, the forum, the Hippodrome (where games were held), and the church of the holy wisdom, Hagia Sophia. Other churches, buildings, and statues surrounded the city's monumental core, which was located at the eastern end of the city, overlooking the Sea of Marmara. High cliffs and walls guarded the city on three sides; south was the Sea of Marmara, east was the Bosporus, and north was the Golden Horn, a secure, deep harbor that was the basis for the city's extensive trade. The city could be accessed by land only to the European west, which Constantine fortified by a massive wall; his successors Theodosius (r. 379–395) and Justinian (r. 527–565) built walls too.

Constantinople quickly became a leading city of the Roman Empire and the leading city of the East. Under Constantine, Constantinople became the premier center of learning in the empire (next to Alexandria, in Egypt). The city was wealthy and cosmopolitan, and it featured an interesting culture combining pagan and Christian traditions and thought. Constantine had been a pagan when he experienced a sudden conversion to Christianity in 312. Ostensibly a Christian, the emperor retained many of his pagan beliefs. A soldier rather than an intellectual, he nevertheless sponsored philosophers and theologians in their endeavors to understand the natural and supernatural worlds.

A Christian emperor, Constantine threw his autocratic weight in terms of wealth and power behind the church, providing exemptions of service and taxation for the clergy, granting the church tremendous wealth, and allowing bishops significant influence over his decisions. Ancient sources indicate that Constantine supported literature, philosophy, and science, though this was often a challenge for him. Constantine's Christianity was of the most elementary kind, the intricacies and subtleties of his adopted religion escaping the emperor, making him a pawn for the many bishops who had political and theological agenda.

In the year 325, at Nicaea in western Turkey, bishops from throughout the Roman Empire came by order of Constantine to reach a unified position on the many divisive issues facing the church, chief of which was the question of the role of Jesus Christ in the church, and whether Jesus was the Logos, eternally at one with God the Father, or the Created Son, not coeternal but, rather, subordinate to the Father. One of the most significant documents in the history of Christianity emerged from the Council of Nicaea: the Nicene Creed, which proclaimed the Triune God, which became the basis for the Christianity of western and eastern Europe. Even today millions of Christians proclaim the Nicene Creed as their statement of faith.

A year later, however, Constantine accused Crispus, his eldest son, and Fausta, his wife and the daughter of Maximian, former emperor with Diocletian, of unknown conspiracies; both were executed. The executions revealed Constantine's credulous and suspicious personality as well as his autocratic conceptions of his own power. Constantine's family had divergent loyalties, a house of intrigue and violence—and yet it was the leading Christian family of the empire.

Initially Constantinople was a military, administrative, and trade center rather than an intellectual center. Athens, Antioch, and Alexandria continued to take the lead in scientific and philosophical matters. Constantine, however, set the precedent for the emperor being the head of the church, which meant that he was the focal point of the ongoing theological debates centering upon the Trinity, leadership in the church, and the accepted dogmas and canons of ecclesiastical rule. Christian philosophers and theologians therefore made Constantinople their home so that they could be near the emperor and, if possible, influence his decisions.

Amid personal disaster and the challenges to his autocratic authority from the church, Constantine advocated learning and education in his huge empire. He supported philosophers such as Sopater of Apamea (c. 242–325), a Neoplatonist, who was a student of Iamblichus (250–325). Little is known of Sopater, though Eunapius (345–414) told an anecdote, in *Lives of the Sophists*, that when a grain shipment was late arriving to Constantinople, Sopater's enemies accused him of using magic to change the winds, thereby delaying the shipment; the credulous Constantine believed the accusations and had the philosopher executed. Perhaps the story suggests more simply that pagan philosophy was under attack during the last years of Constantine's reign. Another philosopher at the court of Constantine was Eugenius, from Paphlagonia, the father of the philosopher and rhetorician Themistius (317–388), who was active under Constantine's successors. Constantine encouraged the work of philosophers and teachers with financial support. Along with releasing clergy from obligations of taxes and civic service, Constantine also included professors of grammar, rhetoric, and medicine. The Theodosian Code provides examples of laws promulgated under Constantine to this effect. A law of 321, reissued in 324, exempted physicians, grammarians, and professors of literature from municipal duties. Another law from 333 also exempted the children of such thinkers. A law of 326 exempted chief physicians from taxation. In 334, a law exempted architects from taxation.

One problem with such exemptions was that cities were already struggling to meet the various obligations imposed by the autocrats Diocletian and Constantine. Increasingly in the fourth century, the urban, *curial*, class in the Greek cities had financial obligations that often could not be paid. Decurions, the members of the curial class that were compelled by their status to serve on city councils, were incredibly important to the success of the empire. The cities were the financial and administrative backbone of the empire, and the decurions were the financial and administrative backbone of the cities. Even so, autocrats imposed severe tax and financial obligations on cities, including land confiscation to pay for imperial needs. The empire, as well as each city, was reliant upon teachers, mechanics, geometricians, and architects. Not only were decurions responsible for meeting tax quotas but they also had to pay for city services, such as maintenance of public buildings, roads, and aqueducts, which included the support of "salaried professors of rhetoric and literature" as well as physicians (trans. Pharr).

Constantine died in 337; he left the empire to his three surviving sons, Constantine II, Constans, and Constantius II, co-emperors of Rome. Predictably, upon Constantine's death, the many stored-up rivalries and ambitions among his brothers, nephews, cousins, and sons came to a head. A bloodbath ensued, and many members of the House of Constantine were eliminated. In time, the three sons of Constantine became rivals for power, and periodic civil conflict continued.

Of the four sons of Constantine, the best known was the longest survivor of the four, Constantius II. Constantius ruled the Roman Empire either jointly or alone from 337 to 361, when he died unexpectedly while on route to face his cousin

Julian, who had been proclaimed Augustus (emperor) by the Roman legions in Gaul. The Christian and pagan sources of Constantius's life and work provide a tragic picture of a hardworking, dedicated Roman attempting to uphold the struggling empire of the present and valiant empire of the past, a grand empire of memory and tradition, who at the same time was fighting the irrevocable forces of history, leading both himself and the Roman Empire to ruin and transformation.

Constantius's rule sat astride the crossroads of time and place. Constantius ruled from the New Rome, Constantinople. He was raised to respect the Latin heritage of Rome, but his was a Greek life; the Balkans and Anatolia formed his character and personality, even if his tutors had him reading Cicero and Tacitus. Moreover, Constantius was the second Christian emperor, who guaranteed the success of Christianity in the Roman and Byzantine Empires. Yet while his father presided over the Council of Nicaea, which proclaimed the Triune God, Constantius himself turned to Arianism, believing that Jesus the Son was not equal to God the Father. Constantius ruled according to the ongoing imperial facade of the princeps even as he turned increasingly to Eastern imperial traditions of autocracy.

A well-educated philosopher-prince attracted to the literary and scientific traditions of the Greco-Roman world, the Christian emperor Constantius sponsored rhetoricians, philosophers, and scientists. The patron of Julian, his cousin, Constantius had him raised by the best Greco-Roman teachers. Constantius was friends with the rhetorician and philosopher Themistius, who became a leader of the Constantinople intellectual scene. Themistius, under Constantius's patronage, was proconsul (urban prefect) of Constantinople; he helped to build up the city and Senate. Like his father, Constantius passed laws to bolster education, rhetoric, and philosophy, and he attacked soothsayers, magicians, and astrologers. Claudius Mamertinus, a pagan supporter of Julian, stated, perhaps exaggeratedly, that astrology was abandoned during Constantius's reign because of his laws against astrology and divination.

THEMISTIUS (317–388 CE)

Themistius was one of the greatest commentators on the science of Aristotle during the Later Roman Empire. He was a pagan philosopher whose tact, learning, and approach to philosophy earned him the admiration of pagan as well as Christian emperors. Themistius was born in 317, just a few years after the emperor Constantine had converted to Christianity. Themistius's father Eugenius was a Greek Sophist at Constantinople. Themistius followed the path of his father, operating a school at Constantinople from 345 to 355. During this time, he attracted the attention of the student and future emperor Julian, who was delighted to find a pagan philosopher who was the equal to any Christian counterpart. Themistius served as an orator, panegyrist, tutor, and adviser to six emperors until his death in 388.

Themistius was one of the many philosophers of the Later Roman Empire who tried to reconcile the apparent disparity between the teachings of Plato and Aristotle. Themistius believed in personal enlightenment used in public service. For example, in his orations to the emperor Constantius II, Themistius argued that *philanthropia* is the most important quality for a ruler to have. *Philanthropia* meant to the ancient world love, clemency, mildness, justice, and humanity. Themistius hoped that the emperor, the ruler of the Roman world, would imitate God, the ruler of the universe. Here is the Platonic concern for the awareness of truth combined with the statesman's concern for its proper application in government and society.

Themistius's contribution to the study of Aristotle was his *Paraphrase*, which featured important discussions of Aristotle's *On the Soul, Metaphysics, Posterior Analytics*, and *On Heaven*. Themistius's work was translated into Arabic, Hebrew, and Latin. It was through the work of such Aristotelian commentators that Aristotle's science came to be known to the Muslim world as well as to late medieval thinkers such as Thomas Aquinas.

THE EMPEROR JULIAN (331–363 CE)

Julian assumed the role of emperor in 361 upon the unexpected death of his cousin Constantius. He ruled the Roman Empire from 361 to 363. A self-perceived philosopher-king and student of ancient science, Julian sought to promote *Hellenism*, Greek thought and learning, during a time dominated by Christianity. Julian was the nephew of the emperor Constantine. When Constantine's sons came to power upon the father's death in 337, their supporters eliminated rivals to their power, including Julian's family. Julian himself was spared because of his youth; but he was subsequently kept under a sort of house arrest. He rejected the Christian teachings of his tutors by the time he was in his teens and sought to study under the great philosophers and scientists of the Greeks. Constantius allowed this, seeing nothing very harmful in Julian's search for learning. Julian became a studious and intelligent neophyte philosopher hungry for knowledge of Greek philosophy and science. He became attracted to Neoplatonism and Stoicism and became friends with clear-headed thinkers such as Themistius, Libanius the orator (314–393), and Oribasius the physician. Julian also succumbed to the fascination of his age with magic and superstition; he eulogized the Neoplatonist Iamblichus and became a disciple of the theurgist Maximus. Nevertheless, Julian was an important patron of science and philosophy during his brief reign. His court at Constantinople was staffed with Greeks who believed that the emperor was leading the empire to a new Hellenistic Age.

Julian's surviving letters reveal the scope of his scientific interests and the network of fourth-century science and philosophy. Julian surrounded himself with Neoplatonists and Aristotelians, pagans as well as Christians. He was a student while at Athens of the Christian philosopher Prohaeresius (c. 276–368) and had a long-standing friendship with the Aristotelian Aetius, who became a Christian bishop. Julian thanked the philosopher Priscus (c. 305–395) for introducing him to the study of Aristotle. Indeed, Julian claimed that Aristotle and Plato formed the basis for all inquiry. His correspondence with the Aristotelian commentator Themistius focused on the contemplative side of Aristotle's philosophy. But Julian was clearly more attracted to Platonism, particularly the watered-down version practiced by Iamblichus and his followers. In one letter Julian referred to Iamblichus, the disciple of the great Neoplatonists Plotinus and Porphyry and leader of fourth-century Neoplatonism, as "that truly godlike man, who ranks next to Pythagoras and Plato" (trans. Wright, *Works*). Julian sacrificed reason for fantasy when he became a follower of Maximus, the theurgist and miracle worker.

Julian saw Helios, the sun, as the answer to the strange mixture of Greek and Asian philosophy and mysticism that was fourth-century Neoplatonism. In his hymn to Helios, Julian explained that Helios was the mediator, the Logos, between the Intelligible and Intellectual worlds. Helios is the visible manifestation of the One, hence Helios is approachable, can be seen, felt, and prayed to, unlike the unfathomable, distant, anonymous One. To Julian, Helios is the light, the truth, the father of humankind. Seeing Greek myth as allegorical stories meant to represent a deeper truth, Julian saw this truth as Helios, who in a way encompassed the Greek and Roman pantheon of gods and goddesses

Julian was interested in serious science. The "science of music" fascinated him, and he enjoyed studying the purported "inventor of ancient music," the philosopher Amphion, whom Julian claimed was responsible for developing the science of harmony (trans. Wright, *Works*). In letters to Ecdicius, an official in Egypt, Julian requested that as the temple to the Egyptian god Serapis was being restored, that the "Nilometer," a device used to measure the rising and lowering of the Nile River, be activated again (trans. Wright, *Works*). Perhaps because of the influence of his friend Oribasius, Julian was very interested in the study of medicine. He wrote to Zeno, an Alexandrian physician, that "Homer was right when he said 'One physician is worth many other men'" (trans. Wright, *Works*). Julian supported the study of medicine and work of physicians by decree in 362: "That the science of medicine is salutary for mankind is plainly testified by experience. Hence the sons of the philosophers are right in proclaiming that this science also is descended from heaven. For by its means the infirmity of our nature and the disorders that attack us are corrected. Therefore, in accordance with reason and justice, we decree what is in harmony with the acts of former Emperors, and of our benevolence ordain that for the future ye may live free from the burdens attaching to senators" (trans. Wright, *Works*). This exemption applied to military and civil service. Indeed, one might conclude that the most important legacy of Julian's reign as emperor of Rome was his encouragement of the study of science and medicine, especially his support of his friend Oribasius, who produced an important *Synopsis* of the work of Galen (130–200).

ORIBASIUS OF PERGAMON (320–400 CE)

Oribasius was a Greek physician during the Later Roman Empire, a confidant to the pagan emperor Julian, and a synthesizer of the writings of the great Roman physician Galen. Galen's work had a tremendous impact on the European Middle Ages and Renaissance and among Arab scholars partly because of Oribasius's *Synopsis*, which made the vast output of Galen understandable to scholars and practitioners of medicine. Oribasius was born a little over a century after Galen's death in Galen's hometown, Pergamon, in Asia Minor. Pergamon was a center of

Greek learning, culture, and science; because of Galen and Oribasius, it was the leading center of medicine in late antiquity as well. According to the ancient biographer Eunapius, Oribasius was a precocious learner when growing up at Pergamon toward the end of Constantine's reign and the beginning of the reign of his son and successor, Constantius II. Oribasius studied with the "healing sophist" Zeno of Cyprus at Alexandria. At some point Oribasius was studying or practicing medicine at Athens when he met the future emperor Julian, himself a student at Athens in the early 350s. The two became friends; Oribasius earned the right, even after Julian became Caesar and Augustus, to counsel and speak frankly to the emperor. Eunapius claimed that Oribasius taught the neophyte philosopher the bearing and patience necessary in an emperor. Julian in turn encouraged Oribasius to prepare the seventy-book *Synopsis* of Celsus (25–50), Asclepiades (120–40), Erasistratus (275–194), Herophilus (325–255), Hippocrates (460–377), and especially Galen. Oribasius was a compiler and synthesizer, but he also had some original thought, which was connected to his Hellenism. Oribasius supported and contributed to Julian's program of Hellenistic reform, which included not only the reopening of pagan temples after years of suppression by Christian emperors but also a new emphasis on the Greek tradition of rational, humanistic inquiry. Oribasius shared Julian's absorption with Greek Neoplatonism, a philosophy that focused on a spiritual, transcendent reality. At the same time, Oribasius was schooled in the practical medical philosophy of Galen. Oribasius made an important contribution to the history of medicine and science by combining the rational pragmatism of the Roman mindset with the metaphysical approach of the Neoplatonic philosopher.

After Julian died fighting against the Parthians in 363, subsequent Christian emperors exiled the pagan Oribasius. According to Eunapius, Oribasius became legendary even among the "barbarians" because of his healing arts. Eventually he was recalled from exile, and was still living when Eunapius wrote his *Lives of the Philosophers* at the end of the fourth and beginning of the fifth centuries. Eunapius, also a pagan, called Oribasius friend and relied on him heavily in the preparation of the *Lives*.

As a physician Oribasius modeled Galen's approach to medicine. In discussing the illness and death of the aged Chrysanthius (c. 310–390), a fellow pagan Sophist, Eunapius described how initially physicians bled the old man, sick with a terrible stomachache; but the bleeding weakened and brought him near death. Oribasius finally arrived to help his friend Chrysanthius. Oribasius, like, Galen was devoted to the theories of Hippocrates. The Hippocratic physician believed that nature is a whole, as is the human organism. One must treat the whole organism rather than focus on isolated parts. The well person is a balanced whole, the four *humors* of the body—blood, phlegm, black bile, yellow bile—in perfect accord. The sick person has an imbalance of the humors. The way to bring the body back in balance is to treat an excess of a humor with its opposite. Bring heat to cold, and vice versa. Oribasius diagnosed Chrysanthius with a chill in his stomach and joints, an excess of phlegm, so he applied heat to the affected areas. Chrysanthius, an old man weakened by the bleeding, never recovered. Oribasius no doubt did not despair on the loss of his patient, because a Hippocratic physician knew that death is part of the whole, in balance with the constant generation of new life.

MAXIMUS OF EPHESUS (C. 310–372)

Although Julian worked with genuine philosophers and scientists such as Themistius and Oribasius, at the same time he relied too heavily on pseudoscientists. Maximus is a good example of the degeneration of Greco-Roman thought during the fourth century. He was a native of Ephesus in western Turkey, a follower of the Neoplatonist Plotinus and his disciple Iamblichus. Plotinus's esoteric and metaphysical Neoplatonic thought had been watered down a century after his death to become the province of magicians and astrologers. One of these was Maximus, who used magic and divination, and the formulaic incantations of the *Chaldaean Oracles*, to convince the credulous of his power and influence over nature and the gods. Maximus's most famous pupil was the emperor Julian.

With the decline of civilization in western Europe after 200, there was very little good philosophy and even less good science. Yet men still yearned to observe nature and exercise power over their environment. Unwilling or unable to engage in clear observation of natural phenomena and the systematic ordering of data, intellectuals of the third and fourth centuries turned more to pseudoscience, superstition, and the fabulous and remarkable. Julian not only became friends with clear-headed thinkers such as Themistius the philosopher, Libanius the orator, and Oribasius the physician, but something in Julian, something about his culture, demanded more than the simple tasks of contemplation and recording of observations. Julian eventually attached himself to a group of Sophists who were disciples of the aged philosopher Aidesios. Julian studied under one of these men, Eusebius, who was "passionately absorbed in working marvels, and . . . the study of the science of divination." One anecdote, described by the biographer Eunapius, has Eusebius criticizing Maximus of Ephesus for his tricks, which astonished his observers: he could make the statue of the goddess Hekate smile, and the torches that she held, as well as the head on her head, would burst into flames at the commands of Maximus. Julian, rather than being incited against Maximus for his theatrics, declared to Eusebius, "You have shown me the man I was in search of" (trans. Wright, *Lives of the Sophists*).

Maximus taught a corrupt form of Neoplatonism. The philosophy of Plotinus focused on the unity of all reality in the One, which was too great for human observation or comprehension. Hence Neoplatonists such as Plotinus taught that there were sometimes visible manifestations of the One. Maximus taught that the Greek goddess Hekate was a manifestation of the great mother of nature, the warmth and love of the One. Maximus was a theurgist, part theologian, part philosopher, part charlatan, and part magician. Perhaps he believed that his magic was the go-between between humans and the divine. Quite a few scientists of the past and present have believed as much. He and his intellectual milieu are good examples of a time when magic was considered science, and science was thought to be magic.

Neoplatonism at the hands of Julian and Maximus was the means to explain what otherwise eluded explanation. The Neoplatonists sought the origins of nature, an explanation of the universe, an understanding of human personality and behavior. Their search for answers took them into the realm of philosophy and mathematics, which were the foremost outlets available to ancient scientists seeking the ultimate truth. In time, however, reflecting the overall degeneration of culture in the fourth century, Neoplatonism became a simplistic search for easy answers, a new form of sun worship, which symbolized the descent of science to the realm of magic during the Later Roman Empire.

There were some important centers of philosophy and science as the Roman world was going through a transition to the medieval West and Byzantine East. Sardis, once a capital of the Midean empire, by the Later Roman Empire a small city, boasted as a native the historian of philosophy Eunapius (345–414). Eunapius was the pagan author of *Lives of the Sophists*, an important source of the lives of Late Roman philosophers and scientists of the third and fourth centuries. Eunapius's *Lives* is not a critical source, as much of his work deals with Neoplatonism and its fourth-century expression dominated by theurgists such as Iamblichus and Maximus of Ephesus. This period was a time of confusion and anxiety, the decline of paganism and rise of Christianity. Christians were confused as they sought to explain their basic doctrines, such as of the Trinity. Pagans were similarly uncertain about what role the ancient beliefs played in society. Pythagorean and Platonic philosophers tried to bring back the Hellenic past. Eunapius was a faithful follower of these efforts, though they were ultimately doomed to failure.

GREGORY OF NYSSA (335–395)

Another center of thought was the province of Cappadocia, in Anatolia, where the philosophers and theologians Basil the Great, Gregory of Nyssa, and Gregory of Nazianzus lived and worked at the end of the fourth century. Gregory of Nyssa (335–395), for example, was a Christian theologian who appreciated Neoplatonic philosophy; he studied what he called "profane" learning—the Greek and Roman heritage—as a means to understanding the theology of Christianity. In this, he claimed he was like Abraham, who used the philosophy of the Chaldeans to bolster his understanding of the god of the Hebrews. It was especially Moses, however, whom Gregory thought was the greatest philosopher and theologian of the past. In *The Life of Moses*, Gregory advocated Greek and Roman learning as a means to acquire virtue. Such learning included "moral and natural philosophy, geometry, astronomy, dialectic," and other pagan subjects. He realized the benefits and limitations of pagan science and philosophy for the Christian, when he wrote, "Pagan philosophy says that the soul is immortal. This is a pious offspring. But it also says that souls pass from bodies to bodies and are changed from a rational to an irrational nature. This is a fleshly and alien foreskin. . . . It says there is a God, but it thinks of him as material. It acknowledges him as Creator, but says he needed matter for creation. It affirms that he is both good and powerful, but that in all things he submits to the necessity of fate" (trans. Malherbe and Ferguson).

Gregory believed that the Old Testament was a great metaphor for the human understanding of the transcendent God.

AMMIANUS MARCELLINUS (C. 325–395)

The city of Antioch, found in the northeastern corner of the Mediterranean region on the Orontes River, was another important center of Byzantine learning. Here, intellectuals such as the pagan orator and friend of Julian, Libanius, taught; his pupils included the great theologian John Chrysostom (c. 349–407). Another Antioch native and theologian was Theodoret (c. 393–457). The most famous scientific mind to be produced by Antioch was Ammianus Marcellinus.

Ammianus Marcellinus was a soldier turned historian, a Greek from an aristocratic family who served in the Roman military under the emperors Constantius II and Julian. As a soldier, he traveled throughout the Roman Empire, from Gaul (France) in the west to the eastern extreme of the empire at Mesopotamia (Tigris and Euphrates River valleys). Marcellinus was a well-educated man who, in his surviving *History*, was willing to discuss and speculate upon a variety of natural and human phenomena. In the role of a man of affairs and historian who occasionally dipped into the world of the scientist, he mirrored his hero Cornelius Tacitus (56–117), the great historian of the Roman principate. Indeed, Ammianus Marcellinus conceived his massive *History* to be a continuation of the *Histories* of Tacitus. The first fourteen books of Marcellinus's *History* are lost. In the surviving books, however, we have a full narrative account of many of the events of fourth-century Rome as well as plentiful discussions of the lands and peoples of Rome and its enemies. Marcellinus was a pagan during a time when Christianity was becoming more widespread throughout the Roman Empire and most emperors (except Marcellinus's hero Julian) were Christian. Marcellinus had a classical education and was familiar with the best writers and thinkers of the Greco-Roman world.

Because of the many years in which Ammianus Marcellinus traveled about the Roman Empire as a soldier and the wide variety of peoples and places he saw, his *History* is particularly useful in its geographic and ethnographic descriptions. In book 14, for example, Marcellinus described the Saracens, a nomadic people living on the fringes of the empire near Arabia. The men were great horsemen but not agriculturalists. Plowing the soil was eschewed by men who lived on horseback under the great canopy of the open sky and twinkling stars of night. Of the Gauls, whom Marcellinus had come to know in war, he wrote of their irrepressible courage, particularly exhibited by Gallic matrons who fought with the same fury and strength as their husbands. Marcellinus was particularly good at describing the Alps, the Rhine River, the Rhone River, Mesopotamia, and other places that he had seen on his travels.

At other times, eschewing experience for the wisdom of other sources, Marcellinus relied on authors whom he considered like-minded, such as Strabo (63 BCE–21 CE) the geographer and Pliny (23–79) the natural historian. For his long and fascinating discussion of the Euxine (Black) Sea, in which he mixed legend, myth, and natural history, Marcellinus used Pliny and Strabo extensively. From Strabo, Marcellinus learned of the Druids, whom he compared to

Pythagorean philosophers in their attempts to uncover the essence of things, and the Euhages, also of Britain, who pried into the mysteries of nature. For his description of Egyptian hieroglyphics, he relied on Plutarch (46–120), Seneca (5–65), and Diodorus Siculus (90–30 BCE). Pliny the Elder's *Natural History* assisted Marcellinus in wading through the varied theories to account for the rising of the Nile River in Egypt. Marcellinus knew from Eratosthenes (276–195) that the sun was directly overhead at the summer solstice at Syrene. He identified the source of the word *pyramid* from the Greek *pyre,* or fire, the flames of which were seen to form a point as they rose. Marcellinus was heavily dependent upon Herodotus (490–430) for his discussion of Persia. However, when he took time to consider the origins of pearls, that they are formed by the rays of the moon, he relied upon his own singular imagination.

As an astronomer Marcellinus was not, of course, original, relying heavily on his forebears, particularly Claudius Ptolemy (100–170). The universe of the Later Roman Empire was geocentric, the moon closest to the earth, the sun in the third spot between Venus and Mars. The inaccuracies of such a system were daunting for an Aristotle or a Ptolemy and simply overwhelming for a lesser thinker such as Marcellinus. His account of a solar eclipse in 360 CE is fascinating for its description of the darkness of the day such that stars could be seen. Marcellinus knew that a lunar eclipse occurred when the moon stood at the opposite extreme of the spherical path or elliptic of the sun. Marcellinus's explanation of rainbows was accurate that the light of the sun confronting mist in the atmosphere forms the rainbow. He thought, however, that the rainbow is opposite in respect to the sun, rather like what happens in his theory of the lunar eclipse, and it extends across the entire starry vault of the heavens. Marcellinus's theory of meteors fit perfectly the ancient conception of the spiritual nature of the heavenly bodies. He condemned the idea that meteors were physical bodies—rather, he believed them to be sparks of heaven or at least rays of light produced by the confrontation of sunlight and dense clouds.

Northwestern Turkey, even today subject to frequent tremors of the earth, experienced an earthquake in August 358; Marcellinus was (perhaps) an eyewitness. The city of Nicomedia was flattened by the earthquake, which indicated its coming by the darkness of the sky and terrible thunderstorms that revealed the anger of the gods. Marcellinus duly presented an extensive discussion on the cause of earthquakes, describing the view of Aristotle (384–322), that water surging through fissures in the earth was the cause; of Anaxagoras (510–428), that it was subterranean winds; and of Anaximander (610–540), that it was an earth either too dry or soaked by rains that succumbed to the force of wind.

Ammianus Marcellinus was a supporter of Julian, the emperor and Neoplatonist. Marcellinus's conception of human and natural history was that of a Neoplatonist—believing that the source of all being exuded from the invisible One. Like Julian as well, he had the Stoic belief in the order of things and in the importance of human experience in the acquisition of knowledge.

Nearby Antioch was the Syrian city of Beirut, which featured a highly recognized law school. Further south, Gaza, at the southeastern end of the Mediterranean, was a center of learning in part because the scholars in the city were influenced by nearby Alexandria and in part because Gaza was seen as an alternative Christian learning center to pagan Athens. Gaza's school of learning produced scholars such as the historian Procopius of Caesarea (500–565). One of the leading Gaza intellectuals was Procopius of Gaza (c. 470–528), a polymath who wrote on mechanical devices and theology. He was influenced by, among others, Philo of Alexandria (20 BCE–40 CE), Gregory Nyssa, and Origen (184–254). Timothy of Gaza (fl. 500 CE), a student of the fifth-century Alexandrian savant Horapollon, wrote a zoological poem, *On Animals*, which was very influential among Arab scholars; the book is a series of observations and anecdotes about animals, such as the giraffe, which according to Timothy was generated by the coupling of other animal species. Aeneas of Gaza (c. 450–518) wrote a clever treatise, *Theophrastus*, imagining how Aristotle's famous student was so overwhelmed with the evidence of Christianity as to relinquish his pagan beliefs.

HYPATIA OF ALEXANDRIA (370–415 CE)

South and west of Gaza was the center of learning of the late ancient world, Alexandria, which featured a long tradition of study, teaching, and writing in the sciences, mathematics, and philosophy. A notable Alexandrian scholar of the early fifth century was Hierocles (c. 400–460), a pagan philosopher and Neoplatonist—heavily influenced by Plato and Pythagoras—and famous as a teacher in Alexandria. Preceding him in time but surpassing him in learning was Hypatia, one of the few ancient female scientists for which we have clear information, owing to the tragic nature of her death, her accomplishments as a mathematician, and her many students. Hypatia was a pagan Neoplatonist in Christian Alexandria. She was an astronomer and mathematician and wrote several commentaries on her precursors, such as the third century, BCE, mathematician Apollonius of Perga (262–190), the second-century CE astronomer Ptolemy of Alexandria and the third-century CE mathematicians Diophantus of Alexandria (200–284). Indeed, third- and fourth-century Alexandria was the most important center of science in the Later Roman Empire. Hypatia's predecessors in Alexandria included Diophantus, who wrote the *Arithmetica*; a female mathematician called Pandrosion; and Hypatia's father, Theon (335–405), a scholar of Archimedes (287–212) the mathematician and engineer, a commentator on Euclid (325–265), and an extensive scholar and commentator on Ptolemy. Theon wrote a commentary on the chronological canon of Ptolemy. The Persian scholar Al-Biruni, in *The Chronology of Ancient Nations*, wrote, "According to the statement of Theon, in his Canon, the people of Constantinople, and of Alexandria, and the other Greeks, the Syrians and Chaldaeans, [and] the Egyptians of our time, . . . all use the solar years, which consists of nearly 365¼ days" (trans. Wright, *Book of Instruction*).

Hypatia was the teacher of the famous Christian bishop Synesius of Cyrene (c. 370–413), who was a mathematician in his own right. Palladas, the fourth-century grammarian, wrote of Hypatia, "Revered Hypatia, ornament of learning, stainless star of wise teaching, when I see you and listen to your discourse, I worship you, looking on the starry house of the virgin; for your business is in heaven" (Jones, p. 102).

As Hypatia's writings are lost, it is difficult to reconstruct her work and its significance. She doubtless was a student of Ptolemy's *Almagest*, as was her father. She also was a commentator on geometry, especially Archimedes's work on the circle and Apollonius's work on conic sections. She was interested in dreams, like her student Synesius. Hypatia and Synesius also corresponded about scientific apparati such as the astrolabe, used for measuring the positions of the stars in the heavens. Synesius was the designer, reputedly, of an astrolabe.

Alexandria was famous for its religious conflicts, during one of which Hypatia ultimately fell victim. Christians accused Hypatia of practicing magic and astrology, which in late-fourth- and early fifth-century Rome were illegal, if widespread among the populace. Apparently this undeserved reputation as a sorceress led to her terrible death at the hands of an Alexandrian mob. The Christian historian Socrates Scholasticus (c. 380–450) in the *Ecclesiastical History* (7:5), noted that Hypatia was very confident in her bearing, wore the garb of a Greek philosopher, and associated with men as an equal. Hypatia, reputedly a beautiful woman, was stripped, tortured, dismembered, and burnt at the hands of angry Christians in 415.

Christian Alexandria continued, however, to sponsor pagan scholars, notably the fifth-century philosopher Hermias, his son Ammonius, and his student John Philoponus. Ammonius (440–520), the student of Proclus of Athens (412–485), was the head of the school founded by Horapollo at Alexandria, where he wrote and delivered commentaries on Aristotle and Plato. John Philoponus (490–570), who taught at Alexandria before moving to Constantinople during the reign of Justinian, wrote commentaries on Aristotelian science and philosophy as well. Philiponus wrote of Asclepius (as related by Ibn Abi Usaibi'ah in *History of Physicians*):

> Asclepius lived for ninety years, spending fifty of them as a youth and man, prior to being filled with divine powers, and forty as a savant and teacher. He left two sons, both skilled in the art of medicine, whom he enjoined to teach this lore only to their sons and his relatives, but not to strangers. His instructions to his successors were the same, and these he ordered to do two things in addition: first, to settle in the center of the inhabited holy places of the Greeks lands, namely on three islands, one of which was Cos, the island of Hippocrates, and secondly, not to divulge the medical art to strangers but to transmit it from father to son. The two sons of Asclepius accompanied Agamemnon when he set out to conquer Troy. Agamemnon held them in great esteem and loved them solely because of their high rank in science. (Trans. Kopf)

Ibn Abi Usaibi'ah recorded a delightful anecdote about John Philiponus's conversion to a life of science:

> Yahyā the Grammarian [John Philiponus] had a thorough knowledge of grammar, logic and philosophy, but he also wrote commentaries on numerous medical writings. He was regarded as a philosopher, being one of the outstanding scholars of that science in his time. He acquired his proficiency as follows: at first he was a boatman carrying passengers on his ferry. He thirsted for knowledge, and when he was ferrying people from the Academy and the science teacher on Alexandria Island who were discussing subjects they had just been studying, he listened enthusiastically. When he had gained some insight into the science, he began to consider his own

condition and said to himself: "I have passed the age of forty without training myself in anything. Knowing nothing but a boatman's trade, how can I master any of the sciences?" While thus immersed in thought, he suddenly observed an ant which was attempting to drag a date-stone uphill. Whenever it had made some progress, it slipped back again, but it never gave up and each time reached further up. This went on all day and he continued to watch it, until it had brought the date-stone to its destination. Seeing this, Yahyā concluded: "If this weak animal has achieved its purpose by exerting himself, I should be much better equipped to reach my goal by strenuous effort." He immediately went and sold his boat and joined the Academy. Studying grammar, lexicology and logic, he made excellent progress, and as he had begun by studying grammar, he was surnamed accordingly and became renowned in this field. He wrote numerous books, commentaries, etc. (Trans. Kopf)

Ibn Abi Usaibi'ah listed the works by John Philoponus as:

(1) A commentary on Aristotle's "Categories."

(2) A commentary on Aristotle's "Prior Analytics," which, however, does not cover the work fully.

(3) A commentary on Aristotles's "Posterior Analytics."

(4) A commentary on Aristotle's "Topics."

(5) A commentary on Aristotle's "Physics."

(6) A commentary on Aristotle's "De Generatione et Corruptione."

(7) A commentary on Aristotle's "Mayal" [?].

(8) A commentary on Galen's "Book of Differences."

(9) A commentary on Galen's "Smaller Art."

(10) A commentary on Galen's "Lesser Book of the Pulse."

(11) A commentary on Galen's "To Glaucon."

(12) A commentary on Galen's "Book of Elements."

(13) A commentary on Galen's "Book of Temperaments."

(14) A commentary on Galen's "Book of Natural Forces."

(15) A commentary on Galen's "Lesser Book of Anatomy."

(16) A commentary on Galen's "Book of Causes and Symptoms."

(17) A commentary on Galen's "Diagnostics of Internal Diseases."

(18) A commentary on Galen's "Greater Book of the Pulse."

(19) A commentary on Galen's "Book of Fevers."

(20) A commentary on Galen's "Book of Crisis."

(21) A commentary on Galen's "Book of the Days of Crisis."

(22) A commentary on Galen's "Stratagem of Healing."

(23) A commentary on Galen's "Regimen of the Healthy."

(24) A commentary on Galen's "Uses of the Parts of Animals."

(25) An abstract of Galen's "Book of Theriac."

(26) An abstract of Galen's "On Phlebotomy."

(27) "The Refutation of Proclus," in 18 discourses.
(28) A book on the proposition that the force of every finite body is finite.
(29) "The Refutation of Aristotle," in discourses.
(30) "A Discourse in Refutation of Nestorius."
(31) A book refuting the opinions of the agnostics, in 2 discourses.
(32) Another tract, refuting the opinions of a different school of thought.
(33) A treatise on the pulse.
(34) A refutation of the "Eighteen Problems" of Proclus the Successor, the Neoplatonist.
(35) A commentary on Porphyry's "Isagoge."

One of Ammonius's students was Simplicius (500–540 CE), author of *Commentary on the Physics* and *Commentary on the Heavens*. Like other commentators, he provided extensive quotes from early philosophers and scientists, many of whom are otherwise scarcely known, their works lost. Simplicius's writings provide much information on otherwise unknown works by sixth- and fifth-century-BCE scientists and philosophers such as Hippo, Hippasus, Diogenes of Apollonia, Zeno of Elea, Parmenides, Melissus, Xenophanes, and Thales.

THE AGE OF JUSTINIAN

The Byzantine emperor Justinian, who ruled from 527 to 565, sought not only to restore the Roman Empire to its ancient grandeur but also to build Constantinople into a center of beauty, learning, and Christianity. Justinian considered himself a Roman emperor; the goal of his long reign was to make concrete this perception by conquering the old Roman West, then in the hands of Germanic kings, and by resurrecting all things Roman, including thought, culture, and science. Justinian, like Constantine, was not an intellectual. Yet he patronized a host of thinkers—Aristotelian commentators, Christian theologians, Platonic thinkers, and some of the greatest legal minds of all time.

The Roman Empire's greatest accomplishment, perhaps, was the creation of Roman law. The worldview of the Romans was practical, focused on living, building, and growing; they were accomplished in empire building and creating successful government. Law was the basis. The Romans created a republican form of government in the sixth century BCE. According to tradition, they revolted against Etruscan kings who ruled the small city on the Tiber River. The Romans, like the Americans centuries later, grew to distrust the wielding of power by individuals. They believed that power, *imperium*, must be shared and distributed. The Roman system divided power between the magistrates, Senate, and Assembly. The magistrates included consuls, praetors, quaestors, and censors. The Romans elected two consuls annually, each of whom had a veto power over the other. Praetors were judicial officials that made decisions on everyday matters affecting Romans. After the republic collapsed, replaced by the rule of one man, the princeps or emperor, the praetors retained the authority to issue decisions on matters of law. During the principate of the Pax Romana, the first several centuries CE, these praetors, or *jurists*, often wrote extensive briefs discussing their opinions of various matters of Roman law. Often they were trained in formal schools of law. The most famous of these jurists were Gaius, Papinian (140–212), Julius Paulus (fl. third century CE), Salvius Julianus (c. 110–170), and Ulpian (c. 170–228). The writings of these five, and a few others, formed the basis for legal opinion in subsequent centuries. Aurelius Hermogenianus, a legal expert during the time of Diocletian, used the writings of these jurists to form a collection of legal pronouncements that served those supporting Roman law during the fourth century. In the fifth century, Theodosius II (r. 379–395) ordered a compilation of Roman law and opinions focusing on imperial decisions beginning with Constantine. This resulted in the Theodosian Code, published in 438.

About ninety years later, the emperor Justinian ordered a remake and expansion of earlier law codes. He assigned a Commission of Ten experts, headed by Tribonian (c. 500–547), to form the code, which was published in 529. Tribonian earned high praise from Edward Gibbon in *The Decline and Fall of the Roman Empire*: "This extraordinary man, the object of so much praise and censure, was a native of Side in Pamphylia; and his genius . . . embraced . . . all the business and knowledge of the age. Tribonian composed, both in prose and verse, on a strange diversity of curious and abstruse subjects: a double panegyric of Justinian and the life of the philosopher Theodosius; the nature of happiness and the duties of government; Homer's catalogue and the four and twenty sorts of metre; the astronomical canon of Ptolemy; the changes of the months; the houses of the planets and the harmonic system of the world" (p. 396).

Justinian ordered Tribonian to lead a larger group of sixteen lawyers and legal experts, a few from the universities at Constantinople and Beirut, to compile a digest of the opinions of jurists collected over the past six centuries. The *Digest* was published in 533. It contained the legal opinions of thirty-nine jurists going back to Quintus Mucius Scaevola in the first century BCE. The vast majority of the opinions derived from Gaius, Papinian, Pomponius, Julian, and Ulpian. A summary and brief text based on the *Digest* was published by Tribonian, Theophilus, and Dorotheus as well—the *Institutes*. These were scientists of law and government, forerunners of today's political scientists. The introduction of the *Institutes* reads: "Imperial majesty should not only be adorned with military might but also graced by laws, so that in times of peace and war alike the state may be governed aright and so that the Emperor of Rome may not only shine forth victorious on the battlefield, but may also by every legal means cast out the wickedness of the perverters of justice, and thus at one and the same time prove as assiduous in upholding the law as he is triumphant over his vanquished foes" (trans. Kolbert). The aim of the *Institutes* was to serve as a sort of text for law students. Divided into four sections, the *Institutes* contained "the first elements of the whole science of law."

Much of the *Digest* of Justinian comprises laws related to people, property, and actions. A good part of it is the Roman law of *Delicts*, which concerns the relationships among people based on what Ulpian declared to be "the maxims of the law . . . : to live honestly, to harm no one, and to give everyone his due" (trans. Kolbert). The *Digest* is systematic, based on extensive logic and analysis of all possibilities under consideration relating to specific instances of human relationships. The legal and political scholars involved in its creation analyzed thousands of laws, hypothesized past situations, and deduced and induced the meaning and intention of past laws and their promulgators to gather the most significant laws and legal decisions and to create new standards of the Roman legal system that would last for centuries. In this sense the *Digest* is science.

The emperor Justinian, like his predecessors and successors, ruled from Constantinople, but considered himself the legitimate emperor of Rome. To prove his point, Justinian reunited much of the western empire under his rule, though this brief unification hardly survived him. He believed in the majesty of Rome, the *genius* of Rome, and sought to awe enemies and inhabitants alike with its power and grandeur. Although Justinian was unquestionably Christian, he still recognized the legacy of the Greek and Roman literature and science and allowed education to still be based on the trivium and quadrivium, which formed the bases for more advanced studies at schools and universities throughout the Eastern Roman, or Byzantine, Empire, at Gaza, Beirut, Antioch, Alexandria, and Constantinople. The state sponsored the education and salaries of a variety of scientific minds: physicians, engineers, and architects.

One of the most dynamic cities of the Byzantine Empire under Justinian was Gaza, an ancient Philistine city at the southern edge of Palestine on the roads to and from Phoenicia, Petra, Egypt, and Alexandria. A Semitic city for most of its past, Gaza was nominally Phoenician in the fourth century BCE, when Alexander the Great laid siege to it. In his biography of Alexander, Arrian (88–175) described the city thus: "Gaza is about two and a half miles from the sea; the approach to it from the coast is over deep sand, and the sea off-shore is all encumbered with shoals. It was a large town, standing high on an eminence, and encircled by a strongly built wall—the last town, on the edge of the desert, as one travels south from Phoenicia to Egypt" (trans. Selincourt). Hellenistic and Roman influence made Gaza during the Later Roman Empire a Semitic city with a predominant Greek culture; the inhabitants spoke Greek and Aramaic. During the fifth century CE, Gaza became a Christian city. Like any Greek city, Gaza had a hippodrome and amphitheater, temples, and churches. The city, with streets in a grid dominated by right angles, was patterned after the ideas of the Milesian city designer Hippodamus (498–408). In the centuries after the foundation of Constantinople, Gaza increased in importance as one of the intellectual centers of the Eastern Roman (Byzantine) Empire along with the imperial city, Antioch, Athens, and Alexandria.

Sixth-century Gaza looked less to Constantinople than to the Greek city of Alexandria in Egypt for scientific and intellectual leadership. Procopius of Gaza, for example, one of the well-known intellectuals of sixth-century Gaza, studied at Alexandria; he was famous for his panegyrics and literary descriptions. One, of a complicated clock, still survives. Another sixth-century writer, John of Gaza, produced a literary account of the universe based on a painting on display at the city. Aeneas of Gaza, at the same time, wrote an imaginary account of the Peripatetic school and its leader, Theophrastus. In good Aristotelian fashion, Timothy of Gaza penned a scientific description of the animal kingdom. The schools of Gaza were considered of the same rank as those of Alexandria and Constantinople.

Students at Gaza received a traditional Greek education (*paideia*) along with a sufficient grounding in Christian theology. The two approaches combined to produce "Christian scholars," in Glanville Downey's words (*Gaza*, p. 115).

Choricius, one of the leading intellectuals of early sixth-century Gaza, is a good example of the Christian scholar. He combined Christian theology with allusions to classical mythology in his writings, the most famous of which were detailed portrayals of Gaza churches. His description of the Basilica of St. Stephen, for example, combines myth, faith, and mathematics in one literary tour de force:

> One of the girdles of the apse—the highest one, I mean—there rests a novel shape. Geometrical terminology, I understand, calls this a half-cone, the term receiving its origins as follows: you have perhaps seen in your country the pine tree, and if this was originally a maiden—for there are some who tell this story, how Boreas, smitten with amorous jealousy, was about to slay her, when Earth, deeply pitying her plight, set up a tree of the same name as the maiden. I neither believe the people who tell this nor is it my intention to relate it, but only to say that it bears a fruit which is called the cone. This is the origin of the term applied to the form. This much I can describe to you by a graphic image. But if you wish to her a full description, it is like this. A carpenter, cutting circles, or ribs of the framework, five in number, from the material of his craft, wood, and cutting each of them equally in two, and joining nine of these slices to each other by their tips, and also joining them by their middles to the girdle which, as I just now said, was the highest course of the church, sets upon them panels of wood, which he hollows out to the required pine-cone shape, equal in number to the ribs, which begin in broad fashion from below and gradually become narrower and rise up to a sharp point, so as to fit the concavity to the wall; and drawing together all the tips into one, and bending them gently in a gradual curve, he produced a most pleasing spectacle. (Quoted in Downey, *Gaza*, pp. 137–38)

Science at the end of the ancient world, in the sixth century CE, was elementary compared to the great accomplishments of the past, during the time of Aristotle and Ptolemy. Yet cities such as Gaza, by relating the concepts and writings of past thinkers, were able to bring them forward to later ages, to the Renaissance, and to the present.

During Justinian's long reign, many public buildings and churches were built, the most notable being a new church dedicated to Holy Wisdom. Constantine's Hagia Sophia had burned early in Justinian's reign, and the emperor decided to build a new one on a much grander scale. He employed two of the most famous scientists of the day, Anthemius of Tralles (474–534) and Isidorus of Miletus (c. 442–537). These two mathematicians served as architects for the church. They planned and created a massive church with a central nave (for worship services) over which was a giant dome—107 feet in diameter and 180 feet tall—that Procopius, the historian of Justinian's reign, thought hung suspended from heaven. Indeed, the natural forces weighing upon the dome seemed insignificant compared to the immaculate rising stone formed into countless arches, hemicycles, and smaller domes that countered the constant downward thrust. The two engineers and mathematicians were also the architects of the Church of the Holy Apostles in Constantinople.

Justinian sponsored other Byzantine scholars and scientists, especially those of the University of Constantinople, such as Cosmas Indicopleustes. Cosmas was a traveler and perceptive geographer, author of *The Christian Topography*, in which he described many of the lands he had visited, such as Abyssinia and India. Strangely, Cosmas also believed that the earth (indeed, even the entire universe) was like a massive tabernacle, square rather than round. Hierocles's *A Fellow-Traveler of Hierocles* is a less ambitious yet more descriptive and concrete analysis of the Byzantine Empire. Hesychius of Miletus wrote commentaries on ancient writings, including the *Life of Plato*.

Justinian envisioned Constantinople as a promulgator of Greek culture in a thriving Roman Empire. If the latter goal was not entirely successful, the former goal was triumphantly accomplished. For centuries after Justinian's death, Constantinople served as a beacon of Greek culture and a repository of ancient documents, unsurpassed anywhere else in the world. Byzantines and Muslims alike shared an interest in the writings of such greats as Aristotle. It was through Arab translations of many of Aristotle's works that Western Europeans in the thirteenth century became reacquainted with Aristotle, leading to an Aristotelian, scientific revival that culminated in the scientific achievements of the European Renaissance.

After the age of Justinian, thought, including science, went through a precipitous decline during the seventh century, as the Byzantine Empire was attacked on all sides by various enemy forces, especially the Muslims. Exceptions included Theophylat Simocatta (c. 580–640), Egyptian natural philosopher and historian of Byzantine emperor Maurice (r. 582–602). During the eighth century, the empire was racked by the controversy over images—the iconoclastic controversy. Still, during the reign of Michael III (884–867), education in the palace school at Constantinople focused increasingly upon the trivium and quadrivium under the foremost scientist of the time, Photius (c. 810–893). Photius produced the *Bibliotheca*, a book of excerpts from and commentaries on the great writers, including philosophers and scientists, of the past, particularly the Byzantine past. The palace school at Constantinople continued in subsequent centuries, under the emperors who formed the Macedonian Dynasty, to produce excellent thinkers, especially Michael Psellus. One of the emperors, Constantine VII Porphyrogenitus (r. 913–959), was himself a scholar, a historian and geographer who, in his learning and patronage, sponsored intense intellectual activity in Constantinople. After a respite of intellectual decline, under Constantine IX Monomachus (r. 1042–1059), there was a revival of learning, again at the school at Constantinople, inspired by Michael Psellus. Constantinople at this time was alive with the study of law and philosophy.

MICHAEL PSELLUS (1018–1096)

One of the pinnacles of the study of ancient Greek science, philosophy, and mathematics during the European Middle Ages occurred during the eleventh century under the leadership of Michael Psellus, Byzantine historian, philosopher, and scientist. Psellus was extremely gifted in many ways, serving as adviser to several Byzantine emperors and holding government posts of importance. He was a Christian who believed that he could acquire knowledge through deductive and inductive thinking, mathematics, and the study of Aristotle, Plato, and Neoplatonists such as Plotinus and Porphyry. He also studied ancient texts on medicine and astronomy. Michael, though a philosopher who focused on reason and logic, also accepted astrology as a proper science but was not willing to assume that the movement of the stars and planets governed human affairs.

Michael's masterpiece was the *Chronographia*, an account of the reign of Byzantine emperors from Basil II to Michael VII (976–1078). The book was partly autobiographical. Psellus described his role in the creation of the intellectual ferment under Constantine IX. "I applied myself," he wrote, "to the study of philosophy, and having acquainted myself thoroughly with the art of reasoning, both *deductive*, from cause to immediate effect, and *inductive*, tracing causes from all manner of effects, I turned to natural science and aspired to a knowledge of the fundamental principles of philosophy through mathematics" (trans. Sewter, *Fourteen Byzantine Rulers*). As a young man, Psellus discovered the wonders of Greek philosophy, especially the work of the Platonists, and read Plotinus, Porphyry, Iamblichus, and Proclus. Proclus inspired Psellus to study "abstract conceptions in . . . mathematics, which hold a position midway between the science of corporeal nature, with the external apprehension of these bodies, and the ideas themselves, the object of pure thought" (trans. Sewter, *Fourteen Byzantine Rulers*). He intensely studied the quadrivium and, seeking even higher, deeper knowledge, turned to mystical knowledge. An adviser to several emperors, Psellus turned to the study of military tactics in order to advise the emperor Romanus IV (r. 1068–1071).

Thanks in part to the efforts of Michael Psellus, there was a Renaissance in Byzantine intellectual activity in the eleventh and twelfth centuries. Scholars focused on the quadrivium. Excellent histories were written by Nicephorus Bryennius (c. 1062–1137) and Anna Comnena (1083–1153), who wrote the *Alexiad* about the era of her father, Alexius I Comnenus (r. 1081–1118). Anna was well educated in the Greek and Roman classics, including history and philosophy, such as Aristotle. The emperor Manuel I Comnenus (r. 1143–1180) studied Ptolemy's *Almagest* and wrote a treatise on astrology. The sack of Constantinople during the Fourth Crusade led to the decline of Byzantine power and intellectual influence.

FURTHER READING

Al-Biruni, *The Book of Instruction in the Elements of the Art of Astrology*, trans. R. Ramsay Wright (London: Luzac & Co, 1934; reprint Bel Air, MD: Astrology Classics, 2006).

Arrian, *The Campaigns of Alexander*, trans. Aubrey de Selincourt (Harmondsworth, England: Penguin Books, 1971).

Diana Bowder, *The Age of Constantine and Julian* (London: Paul Elek, 1978).

The Christian Topography of Cosmas, an Egyptian Monk, trans. J. W. McCrindle (London: Hakluyt Society, 1897).

Fatih Cimok, *Istanbul* (Istanbul: A Turizm Yayinlari, 1989).

Anna Comnena, *The Alexiad*, trans. E. R. A. Sewter (Harmondsworth, England: Penguin Books, 1969).

Dieting for an Emperor: A Translation of Books 1 and 4 of Oribasius' Medical Compilations with an Introduction and Commentary, trans. Mark David Grant (Leiden, Netherlands: Brill, 1997).

Glanville Downey, *Constantinople in the Age of Justinian* (Norman: University of Oklahoma Press, 1960).

Glanville Downey, "Education in the Christian Roman Empire: Christian and Pagan Theories under Constantine and His Successors," *Speculum* 32 (1957): 48–61.

Glanville Downey, *Gaza in the Early Sixth Century* (Norman: University of Oklahoma Press, 1963).

Eunapius, *Lives of the Sophists*, trans. W. C. Wright (Cambridge, MA: Harvard University Press, 1968).

Edward Gibbon, *The Decline and Fall of the Roman Empire*, vol. 4 (New York: E. P. Dutton & Co., 1910).

Gregory of Nyssa, *The Life of Moses*, trans. Abraham J. Malherbe and Everett Ferguson (New York: Paulist Press, 1978).

Ibn Abi Usaibi'ah, *History of Physicians*, trans. L. Kopf (Bethesda, MD: National Library of Medicine, 1971).

A. H. M. Jones, *The Greek City from Alexander to Justinian* (Oxford: Clarendon Press, 1940).

Justinian, *The Digest of Roman Law: Theft, Rapine, Damage and Insult*, trans. C. F. Kolbert (Harmondsworth, England: Penguin Books, 1979).

Ramsey MacMullen, *Constantine* (New York: Harper and Row, 1971).

H. Muller, *Christian and Pagan in the Fourth Century A. D.* (Pretoria, South Africa: Union Booksellers, 1946).

The Oxford Handbook of Byzantine Studies, ed. Elizabeth Jeffreys, John Haldon, and Robin Cormak (Oxford: University Press, 2008).

John G. Pedley, *Sardis in the Age of Croesus* (Norman: University of Oklahoma Press, 1968).

Clyde Pharr, trans., *The Theodosian Code and Novels* (Princeton, NJ: Princeton University Press, 1952).

Michael Psellus, *Fourteen Byzantine Rulers: The Chronographia of Michael Psellus*, trans. E. R. A. Sewter (Harmondsworth, England: Penguin Books, 1966).

John C. Rolfe, trans., *Ammianus Marcellinus*, 2 vols. (Cambridge, MA: Harvard University Press, 1950).

H. Rowell, *Ammianus Marcellinus, Soldier Historian of the Late Roman Empire* (Cincinnati, OH: University of Cincinnati Press, 1964).

John H. Smith, *Constantine the Great* (New York: Scribner, 1971).

The Stanford Encyclopedia of Philosophy, https://plato.stanford.edu.

Louis J. Swift and James H. Oliver, "Constantius II on Flavius Philippus," *American Journal of Philology* 83 (1962): 247–64.

E. A. Thompson, *The Historical Work of Ammianus Marcellinus* (Cambridge: Cambridge University Press, 1947).

Robert Todd, trans., *Themistius: On Aristotle's On the Soul* (London: Bloomsbury, 1996).

A. A. Vasiliev, *History of the Byzantine Empire*, 2 vols. (Madison: University of Wisconsin, 1958).

The Works of the Emperor Julian, trans. W. C. Wright (Cambridge, MA: Harvard University Press, 1962, 1969).

PART 10

The Impact of Ancient Science upon Medieval Asia, Africa, and Europe

PART 10

The Impact of Ancient Science upon Medieval Asia, Africa, and Europe

26

Science in Medieval Mesopotamia and the Levant (600–1500 CE)

A historical, philosophical, political, and religious event that had a profound impact on the preservation, development, and spread of the scientific legacy of the Greco-Roman world was the emergence of Islam in the seventh century CE. Islam is a monotheistic religion based on the teachings of Muhammad (570–632). Followers of Islam are Muslims. The religion originated in Arabia, from which it spread throughout Africa, Asia, and Europe (Spain and southeastern Europe) to other places throughout the world, such as Indonesia. The holy book of Islam is the Qur'an, believed to be the revelation of God through the Angel Gabriel to the Prophet Muhammad who recorded (orally) God's Holy Word. Parts of the Qur'an are based on the Torah (Old Testament), and parts of the Qur'an recall the life and teachings of Jesus, who to Muslims is not the Messiah but rather a great prophet preceding Muhammad. Muslims, Hebrews, and Christians are joined historically through Abraham, father of the Hebrews through his son Isaac, father of the Muslims (Arabs) through his son Ishmael, and part of the genealogy of Jesus of Nazareth (Christians believe) through the descent of Joseph from Abraham's line.

The origins of Islam occur because of the experiences of Muhammad, an illiterate merchant in Mecca, which at the end of the sixth century was a polytheistic Arabian city, a crossroads of trade from Asia to Europe. Christians, Jews, Zoroastrians, and pagans visited the city; hence, there were monotheistic traditions mixing with polytheistic traditions there. Muslims believe that Muhammad had a revelation from the Angel Gabriel that he must submit to Allah, the one God. Muhammad then began to recite what Gabriel told him to his followers (c. 610). There was much resistance to his teachings in Mecca, which forced Muhammad to flee (the Hegira) to Medina. Muhammad and his teachings came to control Medina, from which he launched an attack on Mecca and captured the city. In control, Muhammad cleansed Mecca of all idols, all polytheism. There was controversy over who were the rightful successors upon Muhammad's death in 632. Abu Bakr (father-in-law, father of Muhammad's wife Aisha) claimed to be the rightful heir.

His followers, the Sunni, are the majority Muslims today. Ali (son-in-law, husband to Fatimah) claimed to be the heir as well. His followers are the Shi'ites, who are non-Arab. Islam, like all religions at the time, was aggressive and expansionistic. Muhammad controlled Arabia during his lifetime; subsequently, Islam spread to Egypt, Syria, and Persia under the first four caliphs (632–661) and then North Africa, Spain, Turkey, and Asia to the Indus River under the Umayyad Caliphate (661–750), the capital at Damascus. The capital of the Abbasid Dynasty (750–1258) was at Baghdad; from there, Islam spread to Africa and southern and eastern Asia.

Islam has a strong scientific and medical tradition going back to Muhammad, who first developed the religion in Mecca, a cosmopolitan crossroads of varying religions and philosophies. The Qur'an posits the Abrahamic god who is omniscient and omnipotent. It is God, according to Sura 10, "who hath appointed the sun for brightness, and the moon for a light, and hath ordained her stations that ye may learn the number of years and the reckoning of time. God hath not created all this but for truth. He maketh his signs clear to those who understand" (trans. Rodwell). This is the same theme found in the New Testament, such as Paul's Letter to the Romans, that one can use the marks of the creation as the means of discovering God and can use one's understanding of the Creator to contemplate the Creation. Muhammad was clearly influenced by Jewish, Christian, Byzantine, and Zoroastrian traditions. Upon his death, Islam spread from the Arabian Peninsula north and west, and with it the Arabic language, which became a medium for intellectual discussion and scientific learning. Moreover, teaching the Qur'an required a literate people, and writing, calligraphy, and papermaking flourished, impacting intellectual treatises. Damascus and Baghdad, in particular, became centers of learning, and Arabic science and medicine flourished.

The beginnings of Muslim interest in Greek science occurred during the Abbasid Dynasty beginning in the eighth century. Abbasid scholars were heavily influenced by the work of Nestorian and Monophysite Christian exiles of the Byzantium Empire living in Syria who had for several centuries translated and studied many Greek scientific works into Syriac.

The intellectual center of the Abbasid Dynasty in Baghdad was the House of Wisdom, a library and learning center where the examination and transcription of Greek, Roman, and Byzantine philosophic and scientific manuscripts were ongoing for centuries. The Persian mathematician Al-Khwarizmi (c. 790–850) praised Caliph Al-Ma'mun (r. 813–833) for his "fondness of science" and for "affability and condescension which he shows to the learned, that promptitude with which he protects and supports them in the elucidation of obscurities and in the removal of difficulties" (trans. Rosen). Medical practice and theory at the House of Wisdom in particular and Baghdad in general during the eighth, ninth, and tenth centuries derived from the efforts of one family, Syriac Christians who hailed from Gondishapur in Iran: the Bukhtishu family. Upon the conquest of Khuzistan in the seventh century, including the Academy of Gondishapur (Jundishapur), the academy came under the authority of the Abbasid Dynasty. A century later, the caliph Al-Mansur (r. 754–775) recruited Jibril ibn Bukhtishu (d. 828), a physician at Gondishapur, to come to Baghdad to be in charge of a new medical school. Jibril ibn Bukhtishu's descendants, also physicians at Gondishapur, continued working in

Baghdad in subsequent centuries. The Bukhtishu family taught the works and practices of the ancient Roman physician Galen (130–200). About the same that Jibril ibn Bukhtishu came to Baghdad, a hospital was established as well. In time, many more were built. These were teaching hospitals that employed dozens of physicians. The historian of Hindu science Benoy Kumar Sarkar claimed that Hindu physicians worked at Baghdad hospitals.

HUNAYN IBN ISHAQ (809–873)

It was partly by means of Arab scholars that the West was generally reintroduced to the writings of Hippocrates and other Greek thinkers. One of the most prolific of these translators was Hunayn ibn Ishaq, a Nestorian Christian who lived in Baghdad and was connected to the House of Wisdom. He translated a variety of Greek works into Arabic. His books included *Anecdotes of Philosophers and Savants*, from which subsequent Arab scholars quoted extensively. He wrote, for example, a saying of Hippocrates (460–377), that "a patient who feels a desire has, in my opinion, better chances than a healthy person who feels no desire for anything" (trans. Kopf). A physician and ophthalmologist, Hunayn ibn Ishaq taught his students in Hippocratic fashion, according to Ibn Abi Usaibi'ah (1203–1270):

> (1) Generally speaking, the body is treated in five ways: What is in the head by gargling; what is in the stomach, by vomiting; what is in the belly, by purging the bowels; what is between the two skins, by sweating; and what is in the depth and in the veins by bloodletting. (2) The yellow bile has its seat in the gallbladder, and it controls the liver; the phlegm has its seat in the stomach, and it controls the chest; the black bile has its seat in the spleen, and it controls the heart; the blood has its seat in the heart, and it controls the head. (3) To a pupil of his he said: "The best way for you to approach people is to give them love, to care for their needs, and to be informed about their condition and be ready to help them." (Trans. Kopf)

Hunayn ibn Ishaq was a prolific writer of varied subjects in medicine. He wrote *The Book of Questions*, introducing his students to the basics of medicine. He also wrote *The Ten Treatises on the Eye*, which consisted, according to Ibn Abi Usaibi'ah, of the following:

> (I) The first treatise discusses the nature and structure of the eye. (II) The second treatise examines the nature and uses of the brain. (III) The third treatise studies the optic nerve, sight and how vision comes about. (IV) The fourth treatise gives a general account of all that is indispensable for the preservation and restoration of health and various kinds of treatment. (V) The fifth treatise discusses the causes of the accidents to which the eye is prone. (VI) The sixth treatise deals with the symptoms of eye diseases. (VII) The seventh treatise discusses the properties of remedies in general. (VIII) The eighth treatise discusses the remedies applicable to the eye in particular. (IX) The ninth treatise discusses the medicinal treatment of eye diseases. (X) The tenth treatise discusses the compound drugs suitable for eye diseases. (Trans. Kopf)

This work on ophthalmology far surpassed what the ancients had understood about the eye. Ibn Ishaq also included the first diagram of the anatomy of the eye. A serious student and writer on Galen, he wrote a variety of treatises analyzing

Galen's work on pharmacology, obesity, sperm, prognostics, and extensive works on Galen's analyses of Hippocrates's work. Hunayn ibn Ishaq wrote original works on logic, symptoms of disease, the pulse, fevers, urination, stomach problems, dehydration, dentistry, pediatrics, melancholy, epilepsy, kidney stones, asthma, Aristotle, veterinary medicine, religion, astronomy, agriculture, meteorology, rainbows, and human history. Hunayn ibn Ishaq had a number of pupils who followed in his path of learning, including his son, Ishaq ibn Hunayn, who was similarly a commentator and historian of medicine, writing on Ptolemy (100–170), Euclid (325–265), pharmacology, logic, Galen, and Hippocrates.

Many Arabs of western Asia were students of Aristotle (384–322), whose writings were preserved and translated into Arabic and hungrily read by Arab scientists and physicians. Aristotle's *Logic* particularly fascinated the medieval world and elicited such comments as that recorded by Ibn Abi Usaibi'ah:

> These are all the parts of Aristotle's "Logic" and all that is contained in each. The fourth part is the most important, for it deals with the foremost purpose of the science of logic. The other parts are subsidiary. The first three are introductions, and the last four have two purposes, the first being to supplement the fourth part—only that some do it more and some less—and the second being to circumscribe its sphere more clearly by setting forth the peculiarities of each of them. A man seeking truth and justice should not use what appears to him to be a syllogism without knowing it to be such, for otherwise he may be diverted from a certainty to what is merely a strong suspicion and thus, unwittingly, pass on to matters of eloquence or even be satisfied with errors. He may think that something is true when it is not, or use poetical expressions without knowing that they are poetical, and adopt a faulty line of reasoning, believing that in all this he is following the path of truth and reaching his goal, while in fact he is very far from it. He might be compared to the man who is ignorant of spices and drugs and cannot distinguish between them and poisons and who, being unable to tell the cause of a disease from its remedy, brings about his own death. As for the second purpose, Aristotle worked out the details of each of the four sciences so that, if a man wanted to become an expert logician, he might know how many things he has to learn and how he can verify whether he or another speaker has followed the path of eloquence or of another discipline; the same applies to the one who wants to become a good poet: he is told how many things he has to learn and how he may verify whether he or another has followed the path of poetry or has missed it or confused it with another discipline. It is also useful to him who wants to be able to mislead others without anyone being able to mislead him. He also has to know how many things he has to learn and how to test every utterance and opinion as to whether he has misled or been misled, and in what respect. (Trans. Kopf)

AL-KINDI (C. 801–873)

One of the first great Aristotelian scholars associated with the House of Wisdom in Baghdad was Ya`qub ibn Ishāq al-Kindī. A tutor to the Abbasid caliphs, his list of his scientific and philosophical works is astounding and included books on natural philosophy, religion, logic, mathematics, arithmetic, the sphere, geometry, the sextant, astrolabes, sundials, clocks, "burning glasses," music, astronomy, the motions of the planets, the sun and stars, the atmosphere, history, time and the nature of the infinite, cause and effect, physics, and books on Aristotle, Porphyry, Ptolemy, Plato, Euclid, and Manicheism. He studied the constellations and the zodiac, and he wrote on astrology. The North African historian Ibn Khaldun (1332–1406) wrote of Al-Kindi's astrology: "Ya'qub b. Ishaq al-Kindi, astrologer to ar-Rashid and alMa'mun, composed a book on the conjunctions affecting Islam. The Shi'ah called the book al-Jafr, after the name of their own book, which is attributed to Ja'far as-Sadiq. In his book, al-Kindi is said to have made complete forecasts concerning the 'Abbasid dynasty. He indicated that the destruction of (the 'Abbasid dynasty) and the fall of Baghdad would take place in the middle of the seventh [thirteenth] century and that its destruction would result from the destruction of Islam" (trans. Rosenthal). Al-Kindi expressed the view common to late ancient, medieval, and Arabic thinkers and commentators respecting the importance of ancient thought to their own understanding: "It is fitting them to remain faithful to the principle which we have followed in all our works, which is first to record in complete quotations all that the Ancients have said on the subject, secondly to complete what the Ancients have not fully expressed, and this according to the usage of our Arabic language, the customs of our age and our own ability" (quoted in Lindberg, p. 18). Al-Kindi understood all of nature as being linked by rays of power; everything is therefore connected, even the human mind to outside natural phenomena. He found it easy to accept, as a result, the extramission theory of vision, popular with ancient Greek Stoics and philosophers such as Euclid, which viewed vision as the result of rays emanating from the eyes onto the air surrounding objects in the nearby field of vision.

Al-Kindi's student, Ahmad ibn al-Tayyib al-Sarakhsī, Al-Sarakhsi (c. 835–899) followed his teacher as a tutor of Abbasid caliphs and participant in the House of Wisdom, as well as a writer of books examining Porphyry, Pythagoras, Socrates, Galen, Aristotle (especially his *Logic*), astrology, music, arithmetic, medicine, diseases, sleep, meteorology, dreams, chess, omens, and the soul.

AL-KHWARIZMI (780–850)

Mohammed Ben Musa Al-Khwarizmi was a ninth-century Persian scholar at the House of Wisdom in Baghdad. A mathematician, astronomer, and geographer, he wrote *The Compendious Book on Calculation by Completion and Balancing*, a book on algebra dealing with polynomial equations. He is often called the "Father of Algebra" for his work. Algebra literally meant *completion* by reducing equations to their most elementary form so as to solve them. His book also focused on *balancing*, meaning "the process of reducing positive terms of the same power when they occur on both sides of an equation" (quoted in J. J. O'Connor and E. F. Robertson, "Al-Khwarizmi"). Al-Khwarizmi was a commentator on Ptolemy, and he developed tables for sine and cosine in trigonometry. He dedicated his algebra to Al-Ma'mun, his patron. He prefaced the book stating it was "a short work on Calculating by (the rules of) Completion and Reduction, confining it to what is easiest and most useful in arithmetic, such as men constantly require in cases of inheritance, legacies, partition, law-suits, and trade, and in all their dealings with one another, or where the measuring of lands, the digging of canals, geometrical computation, and other objects of various sorts and kinds of concerned" (trans. Rosen).

Al-Khwarizmi was aware of the vast tradition of learning toward which he contributed his understanding of mathematics. He wrote,

> The learned in times which have passed away, and among nations which have ceased to exist, were constantly employed in writing books on the several departments of science and on the various branches of knowledge, bearing in mind those that were to come after them, and hoping for a reward proportionate to their ability, and trusting that their endeavours would meet with acknowledgment, attention, and remembrance—content as they were even with a small degree of praise; small, if compared with the pains which they had undergone, and the difficulties which they had encountered in revealing the secrets and obscurities of science. Some applied themselves to obtain information which was not known before them, and left it to posterity; others commented upon the difficulties in the works left by their predecessors, and defined the best method (of study), or rendered the access (to science) easier or placed it more within reach; others again discovered mistakes in preceding works, and arranged that which was confused, or adjusted what was irregular, and corrected the faults of their fellow-labourers, without arrogance towards them, or taking pride in what they did themselves. (Trans. Rosen)

Al-Khwarizmi was influenced by a variety of earlier sources, including Hindu mathematicians. Indeed, he gained fame during his time for his work on Hindu astronomical tables, the Zij al-Sindhind. He wrote books on these astronomical tables, the sundial, the astrolabe, the *Book of Chronology*, arithmetic, the quadrivium, and Ptolemy's *Geography*. He studied geometry and music as well. His numerical analysis was on a par with geometric analysis.

Al-Khwarizmi's work on arithmetic undoubtedly preceded his work on algebra, and he introduced his algebra with the basics of simple arithmetic. *The Compendious Book on Calculation by Completion and Balancing* is remarkable for its approach of trying to make mathematics simple, balanced, and complete. He argued that completion and reduction of numbers involve three types: "roots, squares, and simple numbers relative to neither root nor square," from which he derived the various ways of using equations involving various variables: linear and quadratic equations, for example. He wrote,

> When I considered what people generally want in calculating, I found that it always is a number. I also observed that every number is composed of units, and that any number may be divided into units. Moreover, I found that every number, which may be expressed from one to ten, surpasses the preceding by one unit: afterwards the ten is doubled or tripled, just as before the units were: thus arise twenty, thirty, &c., until a hundred; then the hundred is doubled and tripled in the same manner as the units and the tens, up to a thousand; then the thousand can be thus repeated at any complex number; and so forth to the utmost limit of numeration. (Trans. Rosen)

Al-Khwarizmi's work was first translated into Spanish by John of Seville (c. 1100–1180); his astronomical tables and trigonometrical works were translated by Adelard of Bath (c. 1080–1152). Gerard of Cremona (1114–1187) translated *The Compendious Book on Calculation by Completion and Balancing* into Latin. Robert of Chester, who lived in Segovia in Spain in the twelfth century and who first translated the Qur'an into Latin, also translated some of Al-Khwarizmi's works, such as *The Compendious Book* as well as other Arabic scientific works.

Following Al-Khwarizmi, important Mesopotamian mathematicians included Al-Battani, known also as Albategnius (c. 853–929), a mathematician, astrologer, and astronomer. He was born to the Sabian sect of star worshippers, which led to his fascination with astronomy and astrology. He was a commentator on Ptolemy, catalogued hundreds of stars, wrote on the zodiac, estimated the year at 365 days, 5 hours, 46 minutes, and 24 seconds, was a significant founder of trigonometry, and published tables on the motions of the sun, moon, and planets. Scientists in ninth-century Baghdad also included Thabit ibn Qurra (836–901). Ibn Abi Usaibi'ah wrote that Thabit Ibn Qurra "carried out remarkable observations of the sun. He recorded them in a book, which contains his views as to the solar year, the sun at its apogee, the length of the solar year and the measuring and modes of deviation of the sun's movement. In addition, he was adept at translating into Arabic, displaying an elegant style. He also had an excellent knowledge of Syriac and other languages" (trans. Kopf). The same source provides an astonishing list of Thabit ibn Qurra's literary output, which included works on medicine, including ophthalmology, pharmacology, and anatomy; geometry; astronomy such as on the moon, stars, eclipses, and planets; astrology; commentaries on Aristotle's logic; and commentaries on Galen, Hippocrates, Ptolemy, Euclid, Socrates, and Plato. His work on smallpox and measles was well known in Europe; it later had an impact on the development of smallpox inoculation during the European and American Enlightenment. Another mathematician, Sharaf al-Din al-Tusi (1135–1213), born in Iran, moved to Damascus and taught mathematics. He was also an astronomer and astrologer, a student of algebra and geometry. He wrote on the astrolabe. Al-Tusi's most famous disciple, Kamal al-Din ibn Yunus (1156–1242), was the teacher of the Persian scholar Nasir al-Din al-Tusi (1201–1274).

AL-FARABI (872–950)

Al-Farabi was a Damascus mathematician, cosmologist, commentator on Aristotle, and medical thinker. According to Ibn Abi Usaibi'ah,

> Al-Farabi had at first been a watchman in an orchard in Damascus, but that despite this occupation, he had constantly applied himself to philosophy and to the thoughts of the ancients and the interpretation of their meaning. He was so poor that when staying up at night in order to read and write, he used his watchman's lamp as a light. He remained in this condition for some time. Then his affairs took a rapid turn for the better: his learning became known, his writings won fame, he acquired many pupils and became the greatest authority of his age. . . . He felt no inclination for any worldly pursuit. It is said that he would leave his house at night, and go to the watchmen to read by the light of their lamps. In music he reached an unsurpassable standard in both theoretical knowledge and practical skill. It is reported that he devised a marvelous instrument with which he produced wonderful soul-stirring melodies. . . . According to one report, he was induced to study philosophy by the fact that a man deposited with him some of Aristotle's works, which, upon inspection, he found to his liking, so that he felt impelled to study them. He did not desist until he understood them thoroughly and became a full-fledged philosopher. (trans. Kopf)

Al-Farabi was a commentator on Aristotle's *Logic*, dividing his commentary into eight books:

(1) Laws of the different intelligibles and the terms leading up to them, contained in the book called "al-Maqūlāt" in Arabic and "Kategorias" in Greek.

(2) Laws of the compound terms, which are intelligibles consisting of two different concepts (with the terms leading up to them, consisting of two different notions), contained in the book called "al-`Ibāra" in Arabic and "Peri Hermeneias" in Greek.

(3) Propositions relating to the analogies common to the five sciences, contained in the book called "al-Qīyās" in Arabic and "Analytica I" in Greek.

(4) Laws for verifying logical demonstrations and laws governing philosophy and making all its processes more complete, perfect and exact, contained in the book called "al-Burhān" in Arabic and "Analytica II" in Greek.

(5) Laws for verifying statements for testing the soundness of questions and answers; in short, laws governing the science of polemics, by which the latter becomes more perfect, correct and efficient, contained in the book "Topics," i.e., "The Rules of Polemics."

(6) Laws as to things liable to distort and obscure the truth. This book mentions all the things used in order to create confusion and error in the sciences and in debate and all that is necessary to contradict inconsistent utterances when made by the distorter or his hearers: how to start and how to finish, how to avoid errors and whence these derive. All this is to be found in the book called "Sophistics," i.e., "the misleading philosophy."

(7) Laws for testing exhortative utterances and the different kinds of discourses: rhetorical as well as homiletic utterances, whether rhetorical or not. The book mentions everything pertaining to speeches, describes the way to compose them in each branch of the arts, and makes suggestions for rendering them more beautiful and persuasive. The book is called "Rhetorics."

(8) Laws governing poetry and the kinds of poetical utterances that are composed in each branch of the arts. The book mentions everything pertaining to poetry: how many kinds there are of poetry and poetical sayings, how and of what elements to compose each of them and how to harmonize the elements so as to make the result more beautiful, easier to understand and clearer in meaning; also, what is necessary to make poetry more eloquent and lofty. The work is entitled "Poetics." (Trans. Kopf)

Al-Farabi, as quoted by Ibn Abi Usaibi'ah, provided a short history describing how Greek philosophy survived to the time of Islam. He argued that Alexandria was the center of philosophy up to the time of Augustus, who made Andronicus of Rhodes, who lived in the first century BCE and was a famous Aristotelian commentator, in charge of the works of Aristotle. Augustus, Al-Farabi claimed,

> inspected the libraries and supervised the output of books and found that they housed copies of the works of Aristotle, which had been written in the days of the author and at the time of Theophrastus. He also found that the teachers and philosophers had

> written books on the same subjects as Aristotle. He gave orders to copy those works which had been copied in the days of Aristotle and his disciples and to make them the basis of teaching while discarding the others. Placing Andronicus in charge of that scheme, he ordered him to make copies which he would take to Rome with him and others which he would leave at the Academy of Alexandria. He also ordered him to appoint a successor, who would take up teaching in Alexandria in his stead, and to come with him to Rome. (Trans. Kopf)

Subsequently, when Christianity dominated Rome under Constantine and his successors, Christian bishops refused to allow much of the work of Aristotle to be studied, as it would threaten Christianity. Not until the rise of Islam were Aristotle's works rescued by the more openminded Muslims.

One of these, Ibn Abi Usaibi'ah claimed, was Al-Farabi himself, who

> surpassed all Muslim scholars in that art, in the probing of its depths, the elucidation of its obscurities, the exploration of its secrets and his ready understanding of it. He assembled the essentials of that art in books, accurately and lucidly written, . . . and clearly setting forth the five methods of logic, describing the modes of their application and the use of analogy with regard to each of them. So his books on that subject turned out to be highly satisfactory and extremely erudite. In addition, he wrote an excellent and original book entitled "An Enumeration of the Sciences and a Definition of their Aims." Students of all sciences must have recourse to it and study it right from the start. He also wrote a book on the aims of Platonic and Aristotelian philosophy, which testifies to his proficiency in philosophy and his accurate knowledge of the various scientific disciplines. It is the best guide to the speculative method and to procedures of investigations. He reveals in it the secrets and achievements of the sciences, one by one, and explains how to proceed, gradually, from one to another. He starts with the philosophy of Plato, explains the aim which Plato pursued by it and enumerates his writings in this field. He then goes on to the philosophy of Aristotle, introducing it with remarkable discourse, in which he sets forth how he himself gradually comprehended Aristotle's philosophy. Then he describes the aims pursued by Aristotle in his writings on logic and physics, book by book, until, according to the copy at my disposal, he winds up with the beginning of metaphysics and the method of deriving proofs on it from physics. I know of no work more useful to the student of philosophy, for it explains the concepts common to all sciences and those specific to each of them. Only through that book can we understand the concepts of the categories and the premises underlying the various sciences. (Trans. Kopf)

Al-Farabi's literary production was vast. Besides his many commentaries on Aristotle, he wrote on Ptolemy's *Almagest*, works on logic, a commentary on Porphyry's *Isagoge*, history, music, philosophy, Galen, Hippocrates, Plato, politics, religion, law, rhetoric, military science, meteorology, astrology, mechanical devices, dreams, metaphysics, zoology, astronomy, language, alchemy, geometry, the soul, and physics.

ABU AL-FARAJ IBN AL-TAYYIB (980–1043)

Ibn al-Tayyib, of Baghdad, was a prolific scholar who wrote on Aristotle's logic, Hippocrates's works, and extensively on Galen's works, as well as on drugs, dreams, and ophthalmology. And being a Christian, he wrote a commentary on the New Testament Gospels. Ibn Abi Usaibi'ah, in *History of Physicians*, wrote, "Abū al-Faraj was famous for his medical knowledge. He was an eminent personality, held in great esteem, a man of vast learning, a prolific writer and an expert in philosophy, to which he applied himself with zeal. He wrote commentaries on many of Aristotle's philosophical works and of Hippocrates' and Galen's books on medicine. He possessed enormous talent for composition; most of his extant works were dictated by him" (trans. Kopf).

IBN BUTLAN (1001–1038)

One of Ibn al-Tayyib's disciples, Ibn Butlan, also a Christian scholar in eleventh-century Baghdad, traveled widely and corresponded with other significant scientists, such as the Egyptian physician Ali ibn Ridwan (988–1061). He was a satirist as well, writing *The Physician's Dinner Party.* He died in a monastery in Antioch. Ibn Abi Usaibi'ah provided a transcript of comments made by Ibn Butlan when he traveled through Constantinople, stayed in the city for a year, and experienced an outbreak of the plague. Ibn Butlan wrote,

> The most widely known outbreak of the plague in our time was the one in the year 446/1054 [Islamic/Christian time], when Sirius appeared in the Gemini. In the autumn of that year, fourteen thousand dead were buried in St. Luke's Church after all the cemeteries of Constantinople had been filled. In midsummer 447/1055, when the Nile did not rise as usual, most of the inhabitants of Fustāt and Damascus died, and so did all outsiders, except those whom it pleased Allāh to spare. The plague spread to Irāq, where it destroyed most of the population, and the country was thereupon laid waste by invasions of hostile armies. This situation continued until 454/1062. In most countries people became affected with ulcers caused by black bile and with swellings of the spleen; the sequence of the attacks of fever changed and the usual order of crises was upset, so that the science of prognostics had to follow a different line. . . . Since the Sirius star appeared in the year 445/1053 in the sign of the Gemini, which is the ascendant of Egypt, the plague in Fustāt was caused by the Nile's failure to rise. So Ptolemy's prediction—Woe to the people of Egypt when one of the meteors causing melting establishes itself in the Gemini—came true. And when Saturn entered the sign of the Scorpion, the devastation of Iraq, Mosul and al-Jazirah became complete, habitations in Bakr, Rabī'ah, Mudar, Fāris, Kirmān, the Maghrib, Yemen, Fustāt and Syria became deserted, the position of the kings of the earth became precarious and wars, death and plagues abounded. Ptolemy's statement that if Saturn and Mars came into conjunction in the sign of the Scorpion the world would be wrecked had thus come true. (Trans. L. Kopf)

Ibn Abi Usaibi'ah, in *History of Physicians*, told also of Qusta ibn Luqa al-Ba'lbakki (c. 820–912), who was a Christian of Greek heritage who lived in Armenia, where he translated many Greek works of science into Arabic. He was himself a mathematician and physician and wrote numerous books on infection, gout, the elements, acute diseases, the four humors, the liver, animals, dreams, geometry, algebra, astronomy, logic, and Euclid. Another featured medical thinker and polymath discussed by Ibn Abi Usaibi'ah was Abd al-Latif al-Baghdadi (1162–1231), who "hailed from Mosul, but he himself was born in Baghdad. He became renowned for his knowledge of various sciences and his scholarship. He had a pleasant diction and wrote copiously, his special field being Arabic grammar and lexicology. He was also well-versed in Muslim theology and in medicine. While in Damascus, he devoted much of his time to the medical art and became renowned for his mastery of its theory. Numerous students and even physicians frequented him to study under his guidance." He wrote a vast number of works, such as commentaries on Aristotle, Hippocrates, Dioscorides, Galen, and Ibn Sina (980–1037); books on mathematics, logic, religion, theriac, plants, medicine, diseases, pharmacology, fever, zoology, Egyptian history, physics, metaphysics, alchemy, languages, and predestination; and "The Amazing Book on the History of Animals." He was a significant student of human anatomy and observed and wrote on the human skeletal structure (trans. Kopf).

IBN ABI USAIBI'AH (1203–1270 CE)

Ibn Abi Usaibi'ah was a Syrian Arab physician of the thirteenth century who practiced medicine in Cairo and Syria, near Damascus. The student of Al-Dakhwar (1170–1230), he worked briefly with his colleague Ibn al-Nafis (1213–1288) in Cairo before returning to Damascus, where he practiced medicine and wrote the *History of Physicians*, tracing them from Asclepius, the legendary founder; to the Greeks, such as Hippocrates; to the Romans, such as Galen; to medieval Asian and African physicians up to his own time. The *History of Physicians* relies heavily on other commentators of his own and earlier times. It is filled with philosophical discussion mixed with anecdotal reflections. The author traced the generations of physicians in almost a biblical way, with centuries separating various legendary physicians of the past. Many pre-Socratic and post-Socratic philosophers, considered by Ibn Abi Usaibi'ah as physicians, are mentioned, such as Pythagoras, Parmenides, Plato, and Aristotle, among others.

Ibn Abi Usaibi'ah assigned to Allah the role of bringer and healer of disease: "The creator of the spirit of life and the healer of sickness, who bestows abundant favors upon him whom He prefers and threatens painful punishment and affliction to him who disobeys Him; He Who, by His wondrous deeds, caused creatures to come into being from the void and Who, by His most perfect acts and with gravest wisdom, decrees maladies and reveals the remedy" (trans. Kopf).

Ibn Abi Usaibi'ah prefaced his *History of Physicians* with the basic truth that human "aspirations are two: [heavenly] bliss and [sensual] pleasure. And these two aims cannot be attained by man except in a state of health, for the pleasure to be derived from this world and the bliss hoped for in that to come can only be gained through permanent good health and bodily vigor; these, moreover, can be secured only thanks to the art of medicine, which nurtures existing health and restores the lost state" (trans. Kopf).

Like many physicians of antiquity, Ibn Abi Usaibi'ah argued that animals developed a wide range of self-healing techniques. He explained: "Examples of this kind are numerous; and if animals, which possess no intellect, know instinctively what benefits them and promotes their welfare, human beings, who are endowed with reason, judgment and a sense of responsibility, in short, the paragon of animals, should be much better equipped with such knowledge. This is the strongest argument of those who believe that medicine stems from the inspiration and guidance which God, glory be to Him, bestows on His creatures" (trans. Kopf).

Asclepius, according to Ibn Abi Usaibi'ah, was the founder of medicine. He relied on many sources, mostly fanciful, to make this argument. Abu Ma`shar al-Balkhi (787–886), for example, a ninth-century astrologer in Baghdad, wrote "in his 'Book of the Thousands' that . . . Asclepius was not the first to be deified in medicine nor did he inaugurate it but learnt it from others and followed in their footsteps." He also says that Asclepius was the disciple of Hermes the Egyptian and continues by saying,

> There were three Hermes. Hermes the First, the one of threefold grace, lived before the Flood. . . . He was the first to talk about lofty things, such as the movements of the stars, and his grandfather Kayōmarth, who is Adam, taught him [the hours of the] day and night. He was also the first who built temples and praised God in them, and it was he who initiated the study of medicine and discoursed upon it. For his contemporaries he composed many books of rhythmic poems on terrestrial and celestial matters, rhymes that were habitual in their language. He was the first to give warning of the Flood, foreseeing that affliction, consisting of water and fire, would rain down on the earth from the heavens. The domicile which he chose for himself was Upper Egypt [Sā`id], where he built pyramids and towns of mud. Fearing that knowledge might be destroyed by the Flood, he erected the temples in Akhmim [Panopolis] and had their walls engraved with pictures of all the techniques and technicians and of all the tools employed by them. The sciences were also depicted in drawings for the benefit of future generations, since he was anxious to preserve them for posterity and feared that all trace of them might vanish from the earth. . . . Hermes the Second hailed from Babylonia. He inhabited the city of the Chaldeans, that is Babylon, and lived after the Flood, in the time of Nazīr Bālī, who first built the city of Babylon after Nimrod, the son of Kush. He excelled in medicine and philosophy and was well-versed in the properties of numbers. Pythagoras the arithmetician was his disciple. In the domains of medicine, philosophy and arithmetic this Hermes revived what had been obliterated by the Flood in Babylonia. The said city of the Chaldeans was the city of Eastern philosophers, who first established laws and fixed rules. Finally, Hermes the Third lived after the Flood and inhabited the city of Miṣr [Egypt]. The author of a book on injurious animals, he was a physician, a philosopher and a student of the properties of deadly drugs and noxious beasts. He frequently roamed around the country, knowing the layout and natural features of cities and the character of their inhabitants. Fine and valuable sayings on alchemy are ascribed to him which also have a bearing on many of the arts, such as [the manufacture of] glass, pearls, earthenware and the like. He had a pupil known as Asclepius, who resided in Syria. (Trans. Kopf)

The first volume of Ibn Abi Usaibi'ah's *History of Physicians* focuses predominantly on Hippocrates and Galen. Indeed, the writings of Galen are one of his chief sources for the *History of Physicians.* Ibn Abi Usaibi'ah wrote, for example,

> Galen, in the third chapter of his book on the characteristics of the soul [De moribus], says: "Besides his knowledge of medicine, Hippocrates was so well-versed in astrology that none of his contemporaries came near him. He was also familiar with the elements of which the bodies of living creatures are composed and from which

generation and corruption of all bodies subject to generation and corruption derive. He was the first to demonstrate conclusively the things which we have mentioned and to demonstrate how sickness and health come about in all living creatures and in plants. It was he who discovered the various kinds of diseases and the methods for curing them." (Trans. Kopf)

Of Hippocrates's wisdom, Ibn Abi Usaibi'ah recorded such wise sayings as these:

(1) Medicine is both reasoning and experience.

(2) If all men had been created of one "nature," no one would fall ill; for there would be nothing that would not lie in keeping with that nature and thus causing illness.

(3) A habit, if inveterate, becomes second nature.

(4) The more expert a man is in astrology, the better he knows the stars and their natures and the more proficient he is in imitating [them].

(5) As long as a human being finds himself in the sensual world, he must inevitably accept a share—small or large—of sensual [affections].

(6) Every disease whose cause is known has its remedy.

(7) People, when in a state of health, ate the food of wild beasts, and that made them ill; when we gave them the food of birds they recovered.

(8) We eat to live, we do not live to eat.

(9) Do not eat for the sake of eating.

(10) Every sick person should be treated with the medicines of his own country; for nature resorts to that to which it is accustomed.

(11) Wine is the friend of the body; the apple is the friend of the soul.

(12) When asked why it is that the body becomes stimulated most strongly when one has taken a medicine, Hippocrates said: "Because a house is most full of dust while it is being swept."

(13) Do not take medicine unless you need it; for if you take it needlessly and it does not find an illness to act upon, it will act upon health and cause illness.

(14) The sperm in the loins is like the water in a well; if you drain it, it gushes forth and if you leave it, it oozes away.

(15) He who performs coitus [strikes] the water of life.

(16) When asked how often a man should copulate, he said: "Once a year."—"And if he cannot [abide by this]?"—"Once a month."—"And if he cannot?" "Once a week."—"And if he cannot?"—"It is his soul, he may liberate it when he likes."

(17) The principal pleasures of this world are four: the pleasure of eating, the pleasure of drinking, the pleasure of sexual intercourse, and the pleasure of hearing. The first three pleasures cannot be attained, even in the smallest measure, without toil and labor, and are harmful if over-indulged in, whereas the pleasure of hearing, whether sparse or abundant, is free of toil and exempt from effort.

(18) When treachery becomes nature with man, indiscriminate faith in men becomes a failure; and when the means of subsistence are distributed equitably, greed becomes senseless.

(19) Having few dependents is one of the sources of prosperity.

(20) Health is a hidden possession; only he who lacks it knows how to appreciate it.

(21) Asked what kind of life was best, he said: "Safety with poverty is better than wealth with fear."

(22) Seeing people burying a woman he said: "An excellent husband married you."

(23) While teaching Hippocrates is said to have turned to a youngster among his pupils. When the adults reproached him for preferring the youth to them, he said: "Don't you know why I prefer him to you?" and they said that they did not. He said: "What is the most wondrous thing in this world?" One of them replied: "The heaven, the spheres and the stars." Another said: "The earth and the animals and plants on it." A third said: "Man and his constitution." Each of them, in turn, said something, and in every instance Hippocrates said "No." Then he asked the youth: "What is the most wondrous thing in this world?" and he answered: "O learned one, if everything in this world is wondrous, there is no wonder." Whereupon the savant said: "This is why I preferred him."

(24) To fight passion is easier than to cure an illness.

(25) To get rid of serious ailments is a great art.

(26) Once, on visiting a patient, Hippocrates said to him: "I, the disease and you are three. If you help me to overcome it by taking my advice, we shall be two and the disease will be isolated. Then we shall overpower it, for if two join hands against one, they will overcome him."

(27) When he was on his deathbed, he said: "Learn from me the quintessence of the science: He who sleeps much, has a soft nature and a moist skin will live long." (Trans. Kopf)

Ibn Abi Usaibi'ah's work shows the staggering influence of ancient Greek medicine, especially the Hippocratic corpus, on Roman medical thought, as represented by Galen. The *History of Physicians* is also testament to the influence that Galen had over the development of medieval Arab medicine. Ibn Abi Usaibi'ah provides an astonishing list of Galen's works, many of which cover the thought and technique of Hippocrates, that had been translated into Arabic. The list, with Ibn Abi Usaibi'ah's commentary, follows:

(1) "The Memorandum," i.e., the catalogue indicating all the books he had written, his purpose in writing them, the persons to whom they are dedicated and his age at the time of writing them. There are two parts to it, one mentioning his medical books, the other his works on philosophy, logic, rhetoric, grammar, etc.

(2) "On the Order of Reading His Books," in one chapter; intended to inform his readers of the most useful order of his books, from the first to the last.

(3) "The Clear Explanation," in one chapter. Galen said that this was the first book to be read by one who desires to study medicine, for in it he describes the view of each group of physicians as to experiments, analogy and

procedure, their arguments, their means of verifying their claims and refuting their opponents, and their method of distinguishing truth from falsehood. He wrote this treatise when he was thirty or so, when he first visited Rome.

(4) "The Smaller Book of the Pulse," in one chapter, dedicated to Tothors. Galen says that it contains summaries of other, more detailed books, and thus serves as an epitome of them.

(5) "The Smaller Book of the Pulse," in one chapter, dedicated to Tothors and other students and purporting to describe what scholars need to know about the pulse. It first mentions different kinds of pulse—not all of them, but only those which are easy to understand. It then lists the causes of changes of the pulse, natural, unnatural, and supernatural. Galen wrote this treatise at the same time as "The Clear Explanation."

(6) "To Glaucon—on the Way to Cure the Sick." Glaucon in Greek means blue. Glaucon was a philosopher who, aware of Galen's great achievements in medicine, asked him to write this book for him. Since the physician cannot succeed in treating an illness before diagnosing it, Galen, before dealing with therapy, describes the symptoms of different diseases. The first chapter mentions the symptoms and cures of fevers—not of all of them, but only those which are very common. This chapter is divided into two parts: the first describes fevers which have no special symptoms, the second those which are accompanied by peculiar phenomena. The second chapter describes the symptoms and treatment of swellings. This book was written at the same time as "The Clear Explanation."

(7) "On the Bones," in one chapter, subtitled "for students," because Galen wanted students to study anatomy before all the other branches of medicine: he who does not know it can study methodology. In this book Galen describes each bone first separately, and then its junction with other bones. He wrote it together with other books for students.

(8) "On the Muscles," in one chapter, Galen did not destine this book for students, but the Alexandrians included it in a series of textbooks compiled by them by combining the two above-mentioned treatises with three others that Galen had written for students: "On the Dissection of Nerves," "On the Dissection of Nonpulsating Arteries," and "On the Dissection of Pulsating Arteries." They combined that series into one book of five chapters, entitled "Anatomy for Students." In "On the Muscles" Galen very concisely describes each of the muscles of each member and indicates their number, types, points of origin and functions.

(9) "On the Nerves." Also a treatise for students, indicating how many pairs of nerves derive from the lower brain, of what type they are, how and where they are divided, and what functions they perform.

(10) "On the Arteries," in one chapter. In this book written for students and dedicated to Autisthenes, Galen describes the pulsating and non-pulsating arteries. The Alexandrians divided it into two chapters, one on the veins and the other on the arteries. Galen's purpose here was to indicate how many veins

originate in the liver, of what type they are and how and where they subdivide and how many arteries originate in the heart, their types and mode of subdivision.

(11) "On the Elements according to Hippocrates," in one chapter. This book explains that all things which are capable of existence and destruction, viz, animals, plants and minerals, consist of four elements: fire, air, water and earth; these are also the primary very indirect constituents of the human body. The secondary, more immediate constituents of the human body and the bodies of the other creatures that have blood are the four humors: blood, phlegm and the two biles.

(12) "On the Humors," in three chapters. The first two chapters indicate the humors in the animal body—how many they are, of what kinds, and the characteristics of each—and the third chapter describes the humors in drugs—how they are to be distinguished and defined.

(13) "On the Natural Powers," in three chapters. This book purports to explain that the body functions by means of three natural powers: regeneration, sleep and nutrition. The faculty of regeneration consists of two powers, one of which acts on the blood until it forms the organs which have parallel parts, while the other determines the shape of those organs. Galen describes the position, size and proportions of each compound organ. Nutrition is subdivided into four secondary powers: ingestion, retention, conversion and excretion.

(14) "On Diseases and Afflictions," in five chapters composed separately, but combined into one book by the Alexandrians. Galen entitled the first chapter "On the Kinds of Disease"; he states in it how many types of disease there are, subdividing each into its minutest varieties. The second chapter, "On the Causes of Diseases" states how many causes there are of each disease, what they are, etc. The third chapter, "On the Kinds of Afflictions," describes how many and what type of causes there are to each affliction.

(15) "Diagnostics of Internal Diseases." This book is known also as "The Painful Spots," in six chapters. This book describes the symptoms of internal diseases and the diseases themselves. The first chapter and part of the second present the general methods by which diseases and their locations can be determined. The second chapter also points out Archigenes' errors in this field. The last part of the second chapter and the remaining four chapters systematically describe all the internal organs and their diseases, beginning with the brain, indicating the symptoms and how to diagnose the disease by them.

(16) "The Greater Book of the Pulse," in sixteen chapters grouped into four parts of four chapters each. The first part is entitled "On the Varieties of the Pulse"; it describes how many and what kinds of primary pulse there are, and how each of them is subdivided. The first chapter of this part gives a complete description of the varieties and subdivisions of the pulse, by way of an epitane of the other parts. The remaining three chapters are devoted to a demonstration and a discussion of the same and their scope. The second part is entitled "On the Determination of the Pulse"; it describes how each kind of pulse can

be determined by feeling the arteries. The third part, "On the Causes of the Pulse," indicates from what each kind of pulse arises. The fourth part, "Prolegomena to the Knowledge of the Pulse," shows how the preceding data in respect of each kind of pulse are obtained.

(17) "On Fevers," in two chapters. This book describes the principal kinds of fever, their varieties and symptoms. The first chapter presents two types of fever, one of the spirit, the other of the principal organs; the second chapter describes a third kind of fever, which afflicts the humors.

(18) "On Crisis," in three chapters. This book indicates how we can diagnose delirium, when it occurs and why, and to what it may lead.

(19) "On the Days of Crisis," three chapters. The first two chapters indicate how the degree of resistance changes during illness, on which days the crisis occurs and on which it cannot occur, which are the days on which the delirium is benign and on which it is malignant, etc. The third chapter states the reasons why the days of illness differ as to the degree of resistance.

(20) "The Road to Health," in fourteen chapters. This book shows how every disease can be cured on the strength of analogy. It concentrates upon the general circumstances according to which the treatment of diseases should be determined. In this connection some specific examples are given. Six chapters are dedicated to a man called Hieron. The first and second explain the correct principles on which therapeutics should be based and refute the wrong ones propounded by Erasistratus and his companions. The other four describe the treatment specific to each organ. When Hieron died, Galen abandoned the book until he was asked by Eugenianus to finish it, whereupon he dedicated the remaining eight chapters to him. The first six of these describe the treatment of diseases which afflict the parallel organs, and the last two the treatment of those attacking the compound organs. The first of the six mentions how to treat each kind of bad humor when occurring in one organ only and explains it by way of analogy to what takes place in the stomach. The next chapter, the eighth, expounds the treatment of the daily recurrent kinds of fever which afflict the spirit. The ninth chapter describes the treatment of raging fevers, the tenth the treatment of the fever which afflicts the principal organs i.e., hectic fever. Here, Galen covers all aspects of the use of hot baths. The eleventh and twelfth chapters describe fevers resulting from putrefaction of the humors—the eleventh fevers not accompanied by peculiar phenomena, the twelfth fevers which are so accompanied.

(21) "Therapeutics Based on Dissection," known also as "The Great Book of Anatomy," in fifteen chapters. Galen notes that this book is a comprehensive encyclopedia of anatomy. The first chapter describes the muscles and joints of the arms; the second the muscles and joints of the legs; the third the nerves and arteries of the arms and legs; the fourth, the muscles moving the cheeks and lips and those moving the lower jaw up and down; the fifth the muscles of the chest, abdomen, hips, and loins; the sixth the digestive system, including the stomach, intestines, liver, spleen, kidneys, bladder, etc.; the seventh

and eighth the dissection of the respiratory system. (The seventh states the results ensuing from dissection of the heart, lungs and arteries of an animal which just died but whose body is still partly functioning, while the eighth sets out the results following dissection of the whole chest). The ninth chapter is wholly devoted to the brain and spinal cord. The tenth describes the eyes, tongue, esophagus and the like; the eleventh the larynx, the bone which resembles a Greek λ and the nerves connected with them; the twelfth the sex organs; the thirteenth the arteries and veins; the fourteenth the nerves originating in the brain; the fifteenth the nerves arising from the spinal cord. "This book," says Galen, "is necessary to the study of anatomy. I have written others which are not indispensable, but are also useful."

(22) A summary of Marinus' book on anatomy. Marinus' book comprised twenty chapters, and Galen condensed them into four.

(23) A summary of Lucus' book on anatomy. The original seventeen chapters are condensed into two.

(24) "On Dissension among the Ancients Concerning Anatomy," in two chapters. This book discusses disagreements on anatomical points among Galen's predecessors, explaining how they arose and which of them are merely verbal and which are substantive.

(25) "On the Dissection of Dead Bodies," in one chapter, describing what can be learnt from the dissection of animal corpses.

(26) "On Vivisection," in two chapters, showing what can be learnt from the dissection of live animals.

(27) "On Hippocrates' Knowledge of Anatomy." This book in five chapters, was written when Galen was still a youth. Its purpose was to prove, by quoting examples from his Works, that Hippocrates was reliable in matters of anatomy.

(28) "Erasistratus' Views on Anatomy," in three chapters, also written in Galen's youth and dedicated to Boethus. Its purpose is to note everything concerning anatomy in Erasistratus' books and to point out his correct observations and his errors.

(29) "On the Unknown," to Lucus, on anatomy, in four chapters.

(30) "On Discussions with Lucus Concerning Anatomy," in two chapters.

(31) "On the Anatomy of the Womb," in one short chapter, written in Galen's youth for a pregnant woman. It is a comprehensive anatomical description of the womb and explains the changes occurring in it during pregnancy.

(32) "On the Articulation of the First Cervical Vertebra," in one chapter.

(33) "On the Differences between Symmetrical Organs."

(34) "On the Anatomy of the Vocal Apparatus," in one chapter. . . .

(35) "On the Anatomy of the Eye," in one chapter. . . .

(36) "On the Movement of the Chest and Lungs," in three chapters, written in Galen's youth, after his first return from Rome, when he was staying with

Valvus in Smyrna. One of his schoolmates there had commissioned it. The first two chapters and the beginning of the third describe what he learnt from his master Valvus; the remainder presents his own discoveries.

(37) "On the Diseases of the Respiratory Organs" in two chapters, written for Boethus during Galen's first journey to Rome. Its purpose is to explain in which organs breathing is voluntary and in which it is automatic.

(38) "On the Voice," in four chapters, written after the preceding book and intended to explain what the voice is, how it is brought about, from what, which organs produce it or aid in its production, and finally, how voices differ.

(39) "On the Movements of the Muscles," in two chapters, intended to explain these movements—how they are produced and how a muscle performing a certain movement elicits different movements This book also discusses respiration—whether it is voluntary or automatic—and examines many fine points pertaining to it.

(40) A treatise entitled "Criticism of the Errors of Those Who Distinguish between the Urine and the Blood."

(41) A treatise entitled "On the Function of the Pulse."

(42) A treatise entitled "On the Function of Respiration."

(43) A treatise entitled "Pulsating Arteries—Whether the Blood Runs in Them Automatically or Not."

(44) "On the Power of Laxatives," in one chapter, explaining that the action of these drugs is not identical. Not all laxatives reduce whatever they find in the stomach to its natural consistency and then expel it, but each attracts a humor which is congenial to it.

(45) "On Habits," in one chapter, explaining that habit is a factor to be taken into account. Appended to this book is a commentary by Erophilos on Galen's citations from Plato on this subject, and a commentary by Galen himself on his citations from Hippocrates.

(46) "On the Opinions of Plato and Hippocrates," in ten chapters, showing that Plato agreed with most of Hippocrates' views even before he borrowed them from him, and that Aristotle was wrong on all his points of disagreement with them. This book covers the entire range of the mind's faculties, including thought, will and memory, the three principles in which the powers of the body originate, and other miscellaneous matters.

(47) "On Peculiar Movements," in one chapter, dealing with certain movements which had been incomprehensible to Galen and his predecessors and which he subsequently came to understand.

(48) "On the Olfactory Organ," in one chapter.

(49) "On the Functions of the Parts of the Body," seventeen chapters, the first and second of which show the wisdom of the Creator, may He be glorified, in creating the hand. The third chapter shows His wisdom in creating the foot; the fourth and fifth His wisdom manifested in the digestive organs. The sixth and seventh chapters deal with the respiratory organs, the eighth and ninth with

the contents of the skull; the tenth with the eyes, the eleventh with the other components of the face; the twelfth with the parts that belong to both the head and neck; the thirteenth with the region of the shoulders and spine. The next two chapters reveal the Creator's wisdom as expressed in the reproductive organs. The sixteenth deals with the organs extending throughout the body, such as the blood vessels, nerves, etc. The seventeenth describes the location and size of all the organs and stresses the advantages of the book as a whole.

(50) "A Treatise on the Original Construction of the Body," a sequel to the first two chapters of "On the Humors." Its purpose is apparent from the title.

(51) A treatise on the fertility of the body, a minor work, the subject of which is evident from the title.

(52) A treatise on the evils of changing humors, the aim of the work being obvious from the title. He mentions the kind of bad humor which affects the whole body and the condition of the body thus affected and the kinds of humor which are apt to change in the different parts of the body.

(53) "Simple Drugs," in eleven chapters. The first two expose the error of those who followed wrong methods to determine the power of drugs. The third chapter provides a solid basis for the whole study of the primary qualities of drugs. The fourth chapter explains the secondary properties of drugs, such as taste and smell and how they yield information concerning the primary properties. The fifth chapter explains the actions of drugs in the body, viz, desiccation, calefaction, refrigeration and humectation. The next three chapters describe the qualities of each vegetable. The ninth indicates the qualities of drugs which are derived from the such as dust, mud, stones and minerals, the tenth the qualities of animal drugs, and the eleventh the qualities of drugs which originate in the sea and brackish waters.

(54) A treatise on the symptoms of eye diseases, written in his youth for a young oculist. It enumerates the afflictions liable to occur to each coat of the eye and describes their symptoms.

(55) A treatise on the phases of diseases, describing four phases, i.e., outbreak, progress, climax, and subsidence.

(56) "On Plethora," also known as "On Abundance," in one chapter, describing superabundance of humors and the symptoms of each of its varieties.

(57) A treatise on swellings. Galen describes swellings as kinds of roughness caused by nature. In this work he deals with each different type of swelling and discusses the respective symptoms.

(58) A treatise on the apparent causes of swellings on the exterior of the body. Here, Galen explains the real causes and refutes those who negate them.

(59) A treatise on the causes of diseases, an etiological study.

(60) A treatise on tremors, ague, palsy and spasms.

(61) A treatise on the branches of medicine, these being classified according to various principles.

(62) "The Lymph," in two chapters, intended to show that blood is not, as Aristotle thinks, the substance from which all the parts of the body originate, but that the lymph is the source of all the principal white parts of the body, while it is only the red flesh which is derived from the menstrual blood.

(63) A treatise on the birth of a seven-month infant.

(64) A treatise on black bile, its various types and symptoms.

(65) A work on attacks and characteristics of fevers in one chapter, refuting misstatements made in this connection. Galen entitled this book "The Refutation of Those Who Have Discussed Impressions." . . .

(66) A summary of the "Greater Book of the Pulse," in one chapter, claiming to deal exhaustively with the subject. . . .

(67) "On the Pulse," against Archigenes. Galen refers to it as consisting of eight chapters.

(68) "On Wrong Breathing," in three chapters, dealing with the kinds, causes and symptoms of faulty respiration. The first chapter describes the kinds and causes of wrong breathing, the second the respective symptoms, and the third adduces proofs from Hippocrates.

(69) "The Treasures of Advanced Knowledge" in one chapter. This book advocates the pursuit of advanced knowledge and indicates ingenious methods of acquiring it. It also describes and explains astonishing examples of Galen's therapeutics. An excellent work.

(70) A summary of "The Road to Health," in two chapters.

(71) "On Phlebotomy," in three chapters, the first of which refutes Aristotle, who opposed phlebotomy; the second is an attack on Aristotle's followers in Rome, with exactly the same arguments, the third chapter describes the therapeutic advantages of phlebotomy.

(72) "On Lassitude," in one chapter, explaining the nature, types and treatment of this condition.

(73) A treatise on the characteristics of a person prone to epilepsy.

(74) "On the Properties of Foods," in three chapters, listing all nourishing foods and beverages and the value of each.

(75) "On Gentle Treatment," in one chapter; the title is self-explanatory.

(76) A summary of the above, in one chapter.

(77) "On Good and Bad Chyme," in one chapter, describing different foods and explaining which of them produce good and which bad chyme.

(78) "On Erasistratus' Opinions Concerning the Treatment of the Sick," in eight chapters, tracing Erasistratus' therapeutical method and determining its good points and shortcomings.

(79) "On the Treatment of Acute Diseases," according to Hippocrates, in one chapter.

(80) "The Composition of Drugs," in seventeen chapters. Seven chapters deal with each type of compound medicine separately. For instance the drug which

builds the flesh in an acute ulcer is said to be one kind, that which alleviates its acuteness another kind, that which heals it yet another, and so on. The purpose here is to describe the method of classifying drugs, which is why these seven chapters are subtitled "The Composition of Drugs According to Groups and Types." The other ten chapters are subtitled "The Composition of Drugs, according to the Site," meaning that the composition of drugs is discussed here not theoretically, according to the action of each upon a certain ailment but practically, according to the site, or the organ afflicted. The discussion begins with the head and goes on through all the parts of the body. This book no longer exists as a whole, but has been split into two separate parts. . . .

(81) "On Drugs Which Are Easy to Find," i. e., drugs available everywhere; two chapters. . . .

(82) "On Homeopathic Drugs," in two chapters, the first of which describes theriac and the second other unguents.

(83) "Theriac," in one short chapter dedicated to Magelianus.

(84) "How to Remain Healthy," in six chapters, explaining how robust people—both those who are in perfect health and those whose health is less than perfect, both freemen and slaves—may preserve their well-being.

(85) "To Aspolus," in one chapter, a study of whether healthy people remain so thanks to medicine or to asceticism. This is the treatise alluded to at the beginning of Galen's book on the regimen of healthy men, where he says: "The art which achieves the preservation of the body is one, and only one as I have shown elsewhere."

(86) "On Playing with a Little Ball," in one chapter, praising this activity and the game of cricket, which the author prefers to all other sports.

(87) A commentary on Hippocrates' "Belief," in one chapter.

(88) A commentary on Hippocrates' "Book of Members," in seven chapters.

(89) A commentary on Hippocrates' "Book of Fractions," in three chapters.

(90) A commentary on Hippocrates' "Book of the Reduction of Luxations," in four chapters.

(91) A commentary on Hippocrates' "Book of Advanced Knowledge," in three chapters.

(92) A commentary on Hippocrates' "Treatment of Acute Diseases," . . . three chapters exist of this commentary, while in the index of his books Galen says that he wrote five. The three extant chapters deal, respectively, with the authentic part and the doubtful parts of the work in question.

(93) A commentary on Hippocrates' "Book of Ulcers," in one chapter.

(94) A commentary on Hippocrates' "Wounds in the Head," in one chapter.

(95) A commentary chapter on Hippocrates' "Epidemiae"; the first chapter is discussed in three chapters, the second in six, the third in three and the sixth in eight. Galen did not write a commentary on the fourth, fifth and seventh chapters, holding that they were wrongly attributed to Hippocrates.

(96) A commentary on Hippocrates' "Book of Humors," in three chapters.

(97) A commentary on Hippocrates' "Book of Increased Caution." . . .

(98) A commentary on Hippocrates' "Categories," in three chapters.

(99) A commentary on Hippocrates' "Airs, Waters, Places," in three chapters. Some copies contain four chapters, but only the three-chapter version is authentic.

(100) A commentary on Hippocrates' "Book of Foods," in four chapters.

(101) A commentary on Hippocrates' "Book of the Nature of the Fetus." . . .

(102) A commentary on Hippocrates' "Book of Human Nature," in two chapters.

(103) "On the Identity of the Views Expressed by Hippocrates in his 'Book of Human Nature' and in the Remainder of his Books," in three chapters. Galen mentions that he wrote it after commenting upon the "Book of Human Nature," upon hearing that some people found fault with this book and claimed it was not by Hippocrates.

(104) A treatise claiming that the perfect physician should be a philosopher.

(105) "On the Authentic Books of Hippocrates and Those Wrongly Ascribed to Him," in one chapter.

(106) "On Quintus's Arguments against the Hippocratic Theory of the Four Elements," in one chapter. . . .

(107) "On Labored Slumber, according to Hippocrates." . . .

(108) "On Hippocrates' Phraseology." . . .

(109) "On the Essence of the Soul," according to Asclepiades; one chapter.

(110) "On Medical Experiments," in one chapter, accurately stating the arguments of the experimentalists against those of the analogists.

(111) "Encouragement of the Study of Medicine," in one chapter. . . .

(112) "Instances of Experiments," in one chapter.

(113) "On the Test of the Virtuous Physician," in one chapter.

(114) "Credo" in one chapter, in which he states what he knows and what he does not know.

(115) "On Medical Terms," explaining terminology used by physicians, in five chapters. . . .

(116) "Demonstration," in fifteen chapters, intended to show a method of making things evident, as Aristotle tried to do in his "Fourth Book on Logic." . . .

(117) "On Faulty Analogies," in one chapter.

(118) "On the Structure of the Sciences." . . .

(119) "On the Knowledge of One's Own Vices," in two chapters. . . .

(120) "On Morals," in four chapters, describing different moral defects, their causes, symptoms and ways to counteract them.

(121) A treatise on how to banish grief, written for a man who had asked Galen why he was not at all grieved when everything he had stored in the vast

treasure-houses in Rome was lost in the great fire. He told him and explained what one should grieve about and what not.

(122) A treatise demonstrating that good people can benefit by their enemies.

(123) "On Medical Subjects Touched upon in Plato's 'Timaeus,'" in four chapters.

(124) "On the Fact that the Mental Faculties are Determined by the Humor of the Body," in one chapter, the subject being evident from the title.

(125) An anthology of Plato's writings. . . .

(126) "On the Fact that the Prime Mover Does Not Move," in one chapter.

(127) "Introduction to Logic," in one chapter, containing what students should know of the science of demonstration.

(128) A treatise on the number of analogies.

(129) A commentary on Aristotle's second book, known as "Peri Hermeneias," in three chapters. . . .

(130) "On What He Who Modulates His Voice in Speaking Should Do," in seven chapters. (Trans. Kopf)

The *History of Physicians* provides details on many of the physicians and scientists so influenced by Galen's writings. These include Oribasius (320–400), the physician and friend of the emperor Julian, who, according to Ibn Abi Usaibi'ah, wrote extensively on Late Roman materia medica. A near contemporary of Oribasius, living at the beginning of the fourth century, Philagrius of Epirus, similarly influenced by Galen, wrote extensively on diseases and their remedies. Ibn Abi Usaibi'ah provided an extensive list of and commentary on translators of Greek and Roman scientific works into Arabic.

ALCHEMY

Closely allied to medical sciences was alchemy, an especially popular topic of study in Islamic medieval thought. Influenced in particular by the Hellenistic alchemists of centuries before, perhaps even Chinese alchemists, Islamic alchemists focused on numerology, mysticism, magic, astrology, and alchemy. The most famous Arabic alchemist, Jabir ibn Hayyan (c. 721–815) (known as Geber by later Europeans), a purported astrologer, alchemist, numerologist, physician, magician, and chemist of the eighth century, is a shadowy figure who represents more a school of thought than an actual person. The Jabir Corpus of alchemical writings was heavily influenced by Aristotelian scientific principles. Metals had external and internal qualities of heat and cold, moistness and dryness; alchemists believed that they could transform metals by bringing forth the internal qualities of the metal. Here they relied on a complex numerology. The qualities of the metals were connected to the four primal elements of earth, air, fire, and water. The Arab alchemists used distillation apparatuses to bring forth an elixir used in the transformation of metals.

FURTHER READING

Peter Adamson, "Al-Sarakhsi, Ahmad ibn al-Tayyib," https://www.academia.edu/36578109/Encyclopedia_of_Medieval_Philosophy_-_Sarakhsi.

The Algebra of Mohammed Ben Musa, trans. Frederic Rosen (London: Oriental Translation Fund, 1831).

S. M. Deen, *Science under Islam: Rise, Decline and Revival* (Morrisville, NC: Lulu, 2010).

Ibn Abi Usaibi'ah, *History of Physicians*, trans. L. Kopf (Bethesda, MD: National Library of Medicine, 1971).

Abd Ar Rahman bin Muhammed ibn Khaldun, *The Muqaddimah*, trans. Franz Rosenthal, https://asadullahali.files.wordpress.com/2012/10/ibn_khaldun-al_muqaddimah.pdf.

The Koran, trans. J. M. Rodwell (London: J. M. Dent & Sons, 1909).

Henry Leicester, *The Historical Background of Chemistry* (New York: Dover Publications, 1971).

Daren Lin, "A Foundation of Western Ophthalmology in Medieval Islamic Medicine," *University of Western Ontario Medical Journal* 78 (2008): 41–45, http://www.uwomj.com/wp-content/uploads/2013/06/v78n1.41-45.pdf.

David C. Lindberg, *Theories of Vision from Al-Kindi to Kepler* (Chicago: University of Chicago Press, 1981).

J. J. O'Connor and E. F. Robertson, "Abu Abdallah Mohammad ibn Jabir Al-Battani," http://mathshistory.st-andrews.ac.uk/Biographies/Al-Battani.html.

J. J. O'Connor and E. F. Robertson, "Abu Ja'far Muhammad ibn Musa al-Khwarizmi," http://mathshistory.st-andrews.ac.uk/Biographies/Al-Khwarizmi.html.

J. J. O'Connor and E. F. Robertson, "Sharaf al-Din al-Muzaffar al-Tusi," http://mathshistory.st-andrews.ac.uk/Biographies/Al-Tusi_Sharaf.html.

Robert of Chester's Latin Translation of the Algebra of Al-Khowarizmi, trans. and ed. Louis C. Karpinski (New York: Macmillan, 1915).

Benoy Kumar Sarkar, *Hindu Achievements in Exact Science: A Study in the History of Scientific Development* (London: Longmans, Green and Co., 1918).

Emilie Savage-Smith, *Islamic Culture and the Medical Arts*, National Library of Medicine (1994), https://www.nlm.nih.gov/exhibition/islamic_medical/index.html#toc.

27

Science in Moorish Spain (700–1450 CE)

The expansion of Islam west across Africa and north spanning the Strait of Gibraltar into Spain from the seventh into the eighth century inaugurated Islamic rule for centuries in the southern Mediterranean. In Spain, the period of Muslim rule, under the Umayyad Dynasty, resulted in an interesting combination of European and North African cultural and social characteristics known as Moorish Spain. Intellectual ferment in Spain during the Middle Ages was characterized by the development of Islamic scholarship combined with Greek and Latin elements. A centerpiece for this scientific and philosophical renaissance was Andalusia.

Physicians and scientists in Andalusia during the Middle Ages included Ibn al-Samina, a physician and native of Cordoba. The Arab historian and physician Ibn Abi Usaibi'ah wrote: "He was well versed in arithmetic, astrology and medicine, adept in the sciences and interested in various spheres of knowledge; he was a distinguished scholar of grammar, lexicography, metrics and poetics, Muslim law, Hadīth, history and dialectics. . . . He traveled East, but returned to Spain and died there in the year 315/927" (trans. Kopf).

Al-Majriti (950–1007) was a native of Cordoba, and "the foremost Andalusian mathematician of his time," according to Ibn Abi Usaibi'ah. He "had a better knowledge of astronomy than his predecessors, engaged in the observation of stars and ardently applied himself to the study of Ptolemy's 'Almagest.' He wrote a book on the improvement of what we call commercial arithmetic and another epitomizing that part of al-Battānī's 'Astronomical Tables,' which deals with the equation of planets" (trans. Kopf). Al-Majriti had a good understanding of chemistry and practiced alchemy. Al-Majriti's students included Ibn al-Saffar, who wrote on astronomy and the uses of the astrolabe in the eleventh century, and Abu 'l-Hasan Ali ibn Sulayman al-Zahrawi, a mathematician and physician.

IBN JULJUL (943–994 CE)

Ibn Juljul was a tenth-century physician and pharmacologist who wrote on the history of medicine and philosophy. He was a commentator on Plato, Hippocrates, and Dioscorides (30–90). According to Ibn Abi Usaibi'ah, he "was a distinguished physician. An expert therapist, he displayed remarkable versatility in his profession. He lived in the days of Hishām [II] whom he served as physician. He was well acquainted with the properties of simple drugs. He explained the names of the simple drugs occurring in the book of Dioscorides of Anazarba."

Indeed, Ibn Juljul became quite the expert on the first-century Roman physician Dioscorides. In *Interpretation of the Names of Simple Drugs Occurring in the Book of Dioscorides*, Ibn Juljul provided a fascinating discussion illustrating precisely how the Arabic world preserved the writings of ancient scientists that were then brought to the attention of western Europeans:

> Dioscorides' book [Ibn Juljul wrote] "was translated from Greek into Arabic in Baghdād in the Abbasid era during the reign of Ja`far al-Mutawakkil" [Caliph of Baghdad during the mid-ninth century] by Stephen, the son of Basil, the interpreter. Hunayn ibn Ishāq, the translator, corrected the text and approved it. Stephen translated into Arabic those Greek names for which he knew an Arabic equivalent, while leaving in the Greek original those for which he did not know an Arabic term, hoping that Allāh would later send someone equipped with the necessary knowledge to be able to translate them into Arabic, for nomenclature is but a matter of convention among the people of the country concerned, who agree to denote the different kinds of drugs as seems best to them—either by a derivative word or by any other means fixed by common consent. Stephen trusted that he would be succeeded by others who would know the drugs for which he did not know a name and would name them in accordance with the state of knowledge at their time, so that the names would become generally known. That book was circulated in Andalusia as translated by Stephen, containing the Arabic names which the translator knew and the original Greek names for which he knew no corresponding Arabic terms. It was utilized as far as the terms employed were intelligible according to the usage prevalent in the East and in Andalusia. This state of affairs continued until al-Nāsir `Abd al-Rahmān became ruler of Andalusia [after 929 CE]. Romanus, Emperor of Constantinople, sent him a message—I think it was in the year 337/948 and presented him with gifts of great value, including the book of Dioscorides, with pictures of herbs in the marvellous Byzantine style and written in the Greek language. . . . Romanus wrote in his letter to al-Nāsir: "Dioscorides' book cannot be utilized except with the help of a person who knows Greek well and is acquainted with the drugs concerned. If there is someone in your country equipped with the necessary knowledge, you will, O king, derive great profit from the book." . . . At that time, no Andalusian Christian in Cordoba was able to read Greek. So the Greek original of Dioscorides' book remained in `Abd al Rahman al-Nāsir's library without being translated into Arabic. The people of Andalusia continued to use Stephen's translation, which had been brought from Baghdad. In his reply al-Nāsir asked the Emperor Romanus to send him a man who knew Greek and Latin and who might teach some of his slaves, who in turn would become translators. So the Emperor sent a monk named Nicholas, who arrived in Cordoba in the year 340/951. At that time, some physicians in Cordoba were engaged in painstaking research with a view to translating into Arabic the names of those drugs occurring in Dioscorides' book, which had not been understood. The one who most eagerly applied himself to this pursuit, with a view to ingratiating himself with King `Abd al-Rahmān al-Nāsir, was Hasday ibn Shaprut [sic], the Israelite. Nicholas the Monk, who became his favorite and close friend, explained to him those obscure names of drugs. Moreover, Nicholas was the first in Cordoba to prepare the theriac called al-Fārūq from purely vegetal ingredients. The

> physicians who, at that time, tried to establish the meanings of the names used by Dioscorides and to identify the drugs themselves included Muhammad, known as al-Shajjār [the herbalist]; a man known as al-Basbāsī; Abū `Uthmān al-Hazzāz, surnamed al-Yābisī; Muhammad ibn Sa`īd, the physician; `Abd al-Rahmān ibn Ishāq ibn Haitham, and Abū `Abd Allāh al-Sikillī [the Sicilian], who spoke Greek and was acquainted with the drugs from personal observation. All these were contemporaries of Nicholas the Monk, and I saw them as well as Nicholas and became acquainted with them in the days of al-Mustansir al-Hakam. At the beginning of the reign of this ruler [ca 961 CE], Nicholas died. Thanks to the efforts of those persons, it became possible, in Cordoba alone of the whole of Andalusia, to identify those drugs in a way precluding all doubt, resulting from direct knowledge of the drugs themselves. The correct pronunciation of the names was also established. There remained only a few drugs—about ten—of minor importance that could not be identified. (Trans. Kopf)

Ibn Juljul concluded his account stating, "I tried hard to ascertain the primary medicinal substance, the substance which is the basis of compound drugs. At last Allāh, in His mercy, granted me such insight that I hit on it. It was my intention to preserve what I feared might become obliterated and the benefit of which might be lost to mankind; for Allāh has created healing, spreading it in what the land brings forth in the animals that live on the earth—whether walking, swimming or creeping—and in the minerals hidden underground. In all these there is healing, divine grace and benevolence" (trans. Kopf).

Ibn Juljul wrote of Plato (427–346), according to Ibn Abi Usaibi'ah:

> The master Plato was one of the people of Athens, a Greek philosopher and physician, with a knowledge of geometry and the nature of numbers. He wrote a book on medicine which he dedicated to his disciple Timaeus. He composed many prose and poetic works on philosophical topics. His style was unique; in this field of composition and style he invented the art of prefacing, which is the relation of all sayings to the five uniting principles, other than which must not be found in any of the compound things in existence. Having perfected his knowledge of the nature of numbers and the five uniting principles, he proceeded to the science of the whole cosmos. He came to know the obstacles to the formation of compound and harmonized particles, their different colors and variations, their composition according to their proportion. Thus he reached to the science of drawing. He found the first movement, which is the summation of all other movements, divided it according to numerical proportion and fixed the compound particles on this basis. He then turned to the science of drawing images by which time he was a master in the science of composition and each one of its components (he wrote a book on this). His philosophical sayings are extraordinary. He was also among those who set down the laws and ordinances of the day, which he included in his books on politics and laws. (Trans. Kopf)

One of the leading Andalusian pharmacologists after Ibn Juljul was Ibn al-Wafid (997–1074) (Abenguefit), who lived and ruled in Toledo. He authored *Book of Simple Medicines*, as well as other treatises on agriculture, medical experiments, and eye diseases. He was a student of Aristotle and Galen. According to Sa'id al-Andalusi,

> He specialized in the field of simple drugs until he had a more detailed and exact knowledge than any of his contemporaries. He composed a remarkably well-ordered

book on this subject in which he compiled the relevant information contained in Dioscorides and Galen. He organized his work in the best way possible. He told me it had taken him about twenty years to collect the material and put it in order, to verify the names and qualities of the medicaments, including details of their powers and strength, before he himself considered it satisfactorily and complete. He had a subtle conception and a sound technique in medicine, both of which were expressed in his belief that drugs should not be used when it is sufficient to have recourse to an alimentary regime or the like. If it was absolutely necessary to employ drugs, he did not use compound ones before trying the simples. If he was forced to use the compound drugs, he restricted himself to as small dosages as possible. His way of curing serious maladies and dread diseases by the simplest and gentlest treatments was renowned as extraordinary, and long remembered as miraculous. (Trans. Kopf)

SA'ID AL-ANDALUSI (1029–1070)

Sa'id al-Andalusi, often called Judge Sa'id, lived in Toledo, Spain, in the eleventh century, where he served in jurisprudence, teaching, writing, and science. His most famous work was *Book of the Categories of Nations*. An encyclopedist and polymath, he wrote about most aspects of knowledge: physical, mathematical, and social sciences. He wrote about most of the Mediterranean peoples, focusing extensively on the Greeks, Romans, Hebrews, Egyptians, Chaldeans, and Spanish as well as other peoples to the east, the Arabians, Indians, Persians, and Chinese.

Sa'id al-Andalusi recorded the sayings of philosophers and their general significance and personality. He wrote of Socrates (469–399), "Socrates was one of Pythagoras' disciples. From all the branches of philosophy, he devoted himself solely to the divine sciences, casting aside worldly pleasures. He openly opposed the Greek idolatry and confronted the rulers with dispute and arguments." Of Pythagoras (570–490), he wrote,

Pythagoras came some time after Pendacles. He learned wisdom from the followers of Solomon the son of David, peace be upon them, when they came to Egypt from Damascus. Prior to that he learned geometry from the Egyptians. Then he returned to Greece, where he introduced the sciences of geometry, natural science, and theology. On his own initiative he founded the science of musicology and composition, in accordance with numerical measurements, claiming that he attained this by prophetic inspiration. He devised strange symbols and far-reaching notions on the formation of the world and its order and on the properties of numbers and their degrees. Concerning the Resurrection he had theories which come close to those of Pendacles in that beyond the domain of nature there is a celestial, spiritual world whose beauty and wonder are unfathomable by the human mind and which pure souls yearn toward; moreover, every person who has succeeded in improving himself by avoiding vanity, tyranny, deceit, envy and other physical lusts will become eligible to enter the spiritual world and to discover what he wishes to reap from its pearls of heavenly wisdom. Then, what the soul longs for will penetrate him, just like musical tunes alighting on the ear, and he will not have to make any effort to gain his desires. Pythagoras wrote books on the rules of arithmetic, music, etc. (Trans. Kopf)

Pendacles was a mythical character that Judge Sa'id believed was one of the first Greek philosophers, who "are the noblest of people and the greatest of the learned because of the real interest they showed in the arts of wisdom, in the logical and mathematical sciences, natural and divine sciences and domestic and civilian policies." Pendacles, according to Sa'id, combined Greek with Islamic philosophy. He was reputedly educated by Luqman, a sage mentioned in the Qur'an, who lived, Sa'id believed, in Damascus. Pendacles traveled from Damascus to Greece, where he had an influence on philosophers such as Pythagoras. "Pendacles was the first who united the meanings of the epithets of Allah the Almighty, saying that they all mean one thing, that if He is described as knowledge, magnanimity and power. He has no specific definition as characterized by these different names. He is the One in reality Who does not generate in any way, as compare with all other beings. Particles of the universe are subject to increase, either in part or in meaning or in kind, but the nature of the Creator transcends all this." Pendacles wrote on metaphysics and medicine. "Aristotle," Judge Sa'id wrote, "was the acme of Greek philosophy, the keystone of Greek medicine and the chief of the Greek scholars. He was the first to isolate the science of demonstration from the rest of the logical sciences, and he was the one to establish it in its three forms. He then made it the instrument of all the philosophical sciences, so that he was nicknamed the Master of Logic. He wrote excellent books, both general and special, in all the philosophical branches" (trans. Kopf).

Judge Sa'id also commented on Muslim scientists of his own or earlier times. He wrote of the Baghdad scientist associated with the House of Wisdom, Abu Yusuf Ya'qub ibn Ishaq Al-Kindi (801–873): his writings

> include books on the science of logic which have a wide circulation, but are of little didactic value because they lack the analytic element that is indispensable for distinguishing truth from untruth in any sphere of research. It was the art of synthesis that Ya`qub intended to discuss in those books, but they can be utilized only by those having some preliminary knowledge of the subject because the fundamentals of any topic of study can be acquired only by the art of analytics. I do not know what induced Ya`qub to omit this essential art, whether ignorance of its importance or a miserly reluctance to divulge it to the public. Whatever it was, his exposition is defective. In addition, al-Kindi wrote a great number of epistles on various sciences, in which he set forth fallacious doctrines and mistaken opinions. (Trans. Kopf)

Of the Persian, Rhazes (Al-Razi), Judge Sa'id wrote that Rhazes "did not go deeply into theology and could not grasp its ultimate significance. This warped his mind, and he adopted hateful opinions and followed wicked paths. He censured people whom he could not understand and whose ways he could not learn" (trans. Kopf).

Judge Sa'id also commented on his fellow Andalusian philosophers and scientists, one of whom was the tenth-century Cordoban Ibn al-Samina: "He was well versed in arithmetic, astrology and medicine, adept in the sciences and interested in various spheres of knowledge; he was a distinguished scholar of grammar, lexicography, metrics and poetics, Muslim law, Hadīth, history and dialectics." Judge Sa'id commented on the Cordoban physical scientist Al-Majriti (950–1007), saying that he was "the foremost Andalusian mathematician of his time," studying and commenting on Ptolemy's *Almagest*, and on the astronomical tables of Al-Battani (851–929) as well as the tables of Al-Khwarizmi (790–850). He said of the eleventh-century Cordoban mathematician Ibn al-Samh (979–1035) that he "had a profound knowledge of arithmetic and geometry, was well-versed in astronomy and also applied himself to medicine." Al-Samh wrote on geometry, numbers, astronomy, and the astrolabe. Another eleventh-century mathematician, astrology, and physician, Al-Kirmani (996–1021), lived in Saragossa, Spain, having traveled to Mesopotamia, where he learned of and brought to Spain the *Epistles of the Brethren of Purity*, a book of Platonic and Islamic philosophy. Judge Sa'id commented on Al-Kirmani's "valuable observations in this field and . . . remarkable skill in cauterization, amputation, incisions, ablations and other operations." A tenth-century physician and architect, Muhammad ibn Tamlih, assisted in the construction of the Great Mosque of Cordoba, according to Judge Sa'id (trans. Kopf).

IBN BAJJA (1085–1138)

Other significant scientists in medieval Andalusia included Ibn Bajja (Avempace), who was, according to Ibn Abi Usaibi'ah, "a guiding light in his time in the philosophical sciences, although his life was full of difficult moments, . . . He distinguished himself in Arabic and literature, knew the Qur'ān by heart, and was considered one of the best physicians. He was proficient in music and playing the lute." He influenced Ibn Rushd, Albertus Magnus (c. 1200–1280), and Maimonides (1138–1204). His body of work included commentaries on Aristotle's *Physics*, a "Discourse on Certain Parts of Aristotle's Book of Meteors," as well as other commentaries on Aristotle's zoological and botanical works; and writings on astronomy, and geometry, philosophy, and medicine.

Another contemporary, Ibn Zuhr (Avenzoar) (1094–1162), was "an expert in simple and compound medicines and a perfect practitioner. His fame blazed in Andalusia and other countries. Medical men studied his books since none of his contemporaries equaled him in the medical theory and practice. There are many anecdotes about how he diagnosed and treated maladies in a way never known before. He served the Almoravids, who showered favors and money upon him." His works included *The Simplification of Treatments and Regiments*, *The Book of Nutrition*, and other works on medical science.

IBN RUSHD (1126–1198)

Averroes, Ibn Rushd, was known as the *commentator*, a title that he earned from his numerous works of study on Aristotle and from using Aristotle to tackle metaphysical topics. "He who studies anatomy increases his belief in God," he once said.

Ibn Rushd studied under Abu Ja'far Ibn Harun, who lived in the twelfth century and who, according to Ibn Abi Usaibi'ah,

> was one of the leading personalities in Seville, adept in the philosophical sciences, who studied Aristotle and other ancients. He was an upstanding physician, expert in the general and specialized aspects of the medical art, and an excellent practitioner with a remarkable technique. He served Abū Ya'qūb, the father of al-Mansūr. He

> was a disciple of the jurist Abū Bakr ibn al-`Arabī and studied the Hadīth for a time under his guidance. Abū Ja`far ibn Hārūn himself became a master of the Hadīth and the teacher of Abū al-Walid ibn Rushd in medicine and the different sciences. He was born in Trujillo, a frontier town of Andalusia, which al-Mansūr found deserted by its fleeing inhabitants and which was thereupon repopulated by Muslims. Abū Ja`far was also an expert ophthalmologist and left excellent writings on therapeutics. (Trans. Kopf)

Ibn Rushd was a diverse scholar who mastered many subjects. He wrote on Islam, jurisprudence, the *General Principles* of medicine, zoology, philosophy, time, astronomy, fever, and theriac. He was a particularly well-known Aristotelian commentator, who wrote commentaries on Aristotle's logic, ethics, metaphysics, physics, and syllogism; a treatise on *A Refutation of Doubts Raised with Regard to the Philosophers and His Proof of the Existence of Primary Matter, and an Inquiry Showing that Aristotle's Demonstration is the Evident Truth*; another treatise on *The Beliefs of the Peripatetics and the Muslim Theologians on the Mode of the World's Existence, in Which They Approach Agreement*; and another on *The Book of Inquiry into the Question "Is it Possible for Our Intellect, Called the Material One, to Conceive Images Distinct from It?" This Is the Subject Which Aristotle Promised to Discuss in His "Book of the Soul."* He also wrote at length on Galen (130–200), including commentaries on Galen's "Book of Principles," his "Book of Temperaments," his "Book of Natural Forces," his "Book of Maladies and Accidents," his "Book of Acquaintance," his "Book of Fevers," his "Book of Simple Drugs," and his "The Stratagem of Healing." He also wrote commentaries on Ibn Sina's (980–1037) *Book of Medicine* and Ibn Sina's "Classification of Beings as 'Absolutely Possible', 'Possible by Itself,' 'Extrinsically Necessary' and 'Necessary by Itself.'"

Ibn Rushd wrote of his *Book of General Principles*:

> This book, concerned with the treatment of diseases in general, is written in the most precise and lucid manner we have been able to contrive. It now remains for us to write about the cure of the afflictions of the various parts of the body, though it is not absolutely necessary for us to do so because the subject matter is implicitly contained in the aforementioned "General Principles," and its presentation is thus in fact a repetition. We shall discuss the treatment of diseases member by member, in the manner commonly adopted by the authors of books on therapeutics, so as to add detailed information to the general conclusions. This is the best procedure, for it results in a maximum of detail. However, we shall defer this task until such time we shall have more leisure for it, as at present we are preoccupied with other important matters. Anyone who, having read this book without the other part would like to continue his study of therapeutics, had best turn to Abū Marwān ibn Zuhr's book called al-Taisīr,' which was written in our time and at our request and which we ourselves copied so that it might be published. This is the aforementioned "Book of Details," closely linked with my "Book of General Principles." However, Ibn Zuhr, in the fashion of the authors of books on therapeutics, discusses the symptoms and causes of diseases along with their treatment. Whoever has read our book need not trouble with the former; it will be sufficient for him to study the treatment. To sum up, whoever has taken advantage of our "Book of General Principles" is able to distinguish between what is right and what is wrong in the treatments described by the authors of books on therapeutics and in their explanations of the way to apply them. (Trans. Kopf)

In the period after Ibn Rushd's death, Christian kingdoms increasingly took control of Muslim holdings in Spain; the caliphate of Cordova declined in power so that by the end of the fifteenth century, the Iberian Peninsula was completely in Christian Europe hands.

FURTHER READING

Sa'id al-Andalusi, *Science in the Medieval World: "Book of the Categories of Nations"*, trans. Sema'an I. Salem and Alok Kumar (Austin: University of Texas Press, 1991).

Ibn Abi Usaibi'ah, *History of Physicians*, trans. L. Kopf (Bethesda, MD: National Library of Medicine, 1971).

Henry Leicester, *The Historical Background of Chemistry* (New York: Dover Books, 1971).

John Wippel and Allan Wolter, eds., *Medieval Philosophy* (New York: Free Press, 1969).

28

Science in Medieval Africa (500–1500 CE)

At the time of the death of Aurelius Augustine (354–430), author of the *City of God*, the Vandals, a Germanic people of north central Europe who had migrated to the Western Roman Empire during the fifth century, had crossed into the Maghreb, northwestern Africa, in 429, and were threatening Hippo, Augustine's bishopric. The Vandals ruled the region for a century, falling to Justinian's general Belisarius in 534. Subsequently the Byzantine Empire ruled North Africa until the seventh century, when the Muslims conquered the region. Henceforth North Africa, including the Maghreb, was a mixture of Greco-Roman, Coptic, Muslim, and Berber culture. During the period of Muslim rule, Greco-Roman philosophy and science merged with, and had a large influence upon, Muslim science and thought. Examples of Muslim scholars influenced by Greco-Roman thought include Ishaq ibn `Imran, Ishaq ibn Sulayman, Ibn al-Jazzar, Ibn al-Haytham, Ali ibn Ridwan, Ibn Fatik, Ibn Jumay‘, Maimonides, Al-Nafis, Al-Baitar, and Ibn Khaldun.

Tenth-century Iraqi Arab physician Ishaq ibn `Imran, who flourished around 900 and lived in what is now Kairouan, Tunisia, was, according to Ibn Abi Usabai'ah, "a skillful physician, an expert in the preparation of compound drugs and a competent diagnostician. In his knowledge and outstanding talents he resembled the ancients" (trans. Kopf). He lived at the time of the Aghlabid Dynasty, when Kairouan was a center of Islamic scholarship. He was the author of books on pharmacology; medical techniques, including bloodletting; the causes of colic; commentaries on Hippocrates (460–377) and Galen (130–200); analyses of pus, urine and sperm; dropsy; melancholy; and the soul. His student, Ishaq ibn Sulayman, an Egyptian who traveled to Kairouan, "was an outstanding physician, a savant renowned for his skill and knowledge, an excellent writer and a man of far-reaching aspirations. In addition to having an excellent knowledge of medicine, he was well-versed in logic and a number of other disciplines. He lived for more than a hundred years, but never took a wife or begot offspring." His books included *The Book of Fevers*, about which Ibn Abi Usabai'ah wrote that "no finer work on this subject can be found," and about which the Egyptian physician Ali ibn Ridwan wrote, "This book is a useful and excellent compendium. I have applied a great number of (the cures) indicated therein and found them unsurpassable." Ishaq ibn Sulayman also wrote books on simple drugs, urine, elements, metaphysics, logic, medicine, the pulse, and theriac, an antidote to snake venom. Ishaq ibn Sulayman's student living in Kairouan was Ibn al-Jazzar (895–979), a physician and pharmacologist who reputedly would dispense medicines for free based on his inspection of his patients' urine samples. He wrote books on disease, medicines, compound drugs, how to achieve long life, biographies and histories, the stomach, medicines for the impoverished, the etiology of diseases, bloodletting, nasal inflammation, leprosy, plague, the importance of hot baths, and a book of aphorisms. One of his famous works was *Medicine for the Poor and the Destitute*. Another was a book known throughout Europe in Latin translation, *Provisions for the Traveler* (*Viaticum peregrinantis*).

IBN AL-HAYTHAM (965–1040)

Ibn al-Haytham (Alhazen) was born in Mesopotamia but spent his life as a scientist and teacher in Cairo, the capital of the Fatimid Dynasty (909–1171). He was a polymath, studying optics, physics, mathematics, zoology, astronomy, geography, and theology. He believed in using the experimental method. He was a student of Aristotle (384–322) and Ptolemy (100–170), upon whom he relied heavily in his optics and physics, though he did have a critical viewpoint. In astronomy, Alhazen studied and wrote on the stars, moon, eclipses, the Milky Way, and rainbows.

Alhazen, a student of optics and ophthalmology, wrote *De Aspectibus*, in which he expanded upon Greek and Roman observations on vision and the nature of the eye to show that vision results from light from the object reflected upon the eye. He opened his treatise examining in detail the phenomenon of light in nature, as well as the relationship of light to color. Then, he analyzed the anatomy of the eye and argued that nerves connect the eye to the brain. He argued that perpendicular rays from objects were the means by which sight was accomplished.

ALI IBN RIDWAN (998–1062)

Ali ibn Ridwan was a physician, astrologer, astronomer, and a commentator on Galen (130–200) and Ptolemy. He was born in Giza but moved to and studied and lived in Cairo, where he was chief physician to the caliph. He left behind a brief autobiography, recorded by Ibn Abi Usaibi'ah:

> Since everyone should choose the profession most suitable for him, since medicine is next door to philosophy with regard to obeying God, glory and power are His, since my horoscope pointed to medicine as my calling and since a livelihood derived from a learned profession seemed more desirable to me than any other, I began to study medicine at the age of fifteen. But the best will be for me to tell you all about myself. I was born in Egypt, at a place situated 30° latitude and 55° longitude, under the sign of Aries according to the astronomical tables of Yahyā ibn Abī Mansūr. When I was six years old, I began to devote myself to study, and when I was ten, I moved to the capital and exerted myself to acquire knowledge. In my fifteenth year, I began to study medicine and philosophy, but I had no money for my upkeep, so

that I experienced great hardship with my studies. At one time, I earned my livelihood by astrological fortune-telling, at another by medical work and occasionally by teaching. This situation lasted until my thirty-second year. At that time, I became known as a physician, and what I then earned by practicing medicine was not only sufficient for my support but left me a surplus which I have retained until now, i.e., up to the end of my fifty-ninth year. With my extra income, I bought real estate in this city which, if Allāh decrees that she remain safe and if He allows me to attain old age, will ensure my subsistence. Since the age of thirty-two I have lived according to a plan, which I modified every year until I crystalized it in the shape in which I am applying it now, on the threshold of my sixtieth year. It provides, inter alia, that I so exert myself in performing my daily professional duties as to make up for the lack of health-preserving physical exercise. After resting from these exertions, I partake of such food as is calculated to keep me healthy. In performing my work, I always endeavor to be modest and kind, to help the troubled, raise the spirits of the worried and assist the needy. In doing all this, it is my aim to derive pleasure from work and beautiful experiences; in addition, of course, there is some material reward, which I spend in part for the well-being of my body and the upkeep of my home, in such a way as amounts neither to extravagance nor to stinginess, but always strikes a happy balance, as prescribed by common sense. In looking after my household, I repair and replace as is necessary; and I store in my house such commodities as food and drink, honey, olive-oil and firewood, and also clothing. The money left over after all these expenditures is spent on various kinds of charity and useful investments, such as gifts to relatives, friends and neighbors and the improvement of my house. The revenue yielded by my real estate is assigned to its repair and embellishment and similar investments when the occasion arises. When I am faced with some new project, in commerce, building or any other field, and I find that it is very likely to succeed, I hasten to carry it out; but if its prospects are meager, I put it aside, awaiting further developments and making the necessary preparations. I take care that my dress conforms to that of people of distinction and that it is clean and fragrant. I refrain from offensive talk, saying only what is fitting. I avoid swearing and base thoughts and eschew self-conceit over-assertiveness. I am inaccessible to both covetousness and despair, and if a calamity takes me unaware, I put my trust in Allāh the Exalted, and face it, as befits a sound mind, without cowardice or rashness. When effecting any commercial transaction, I settle the account in due time, neither advancing money nor incurring debt, except when the need arises. When somebody asks me for a loan, I give it to him without asking anything in return. The hours of the day left after work I spend in worshiping Allāh, praised be He, I take delight in contemplating the kingdom of heaven and earth and glorifying its perfection. I meditate on Aristotle's treatise on "Regimen" and vow to follow its instructions morning and evening. In my leisure time I review my doings and experiences of the preceding day, and what was good, pleasant or useful I rejoice about and what was bad, ugly or harmful I regret and resolve never to repeat. As to the things I take pleasure in, they are the invocation of Allāh, power and glory are His, and by contemplating the kingdom of heaven and earth. The ancients and savants have written a great number of books on that subject, of which I see fit to refer to the following: five books on literature, ten on religious law, the books of Hippocrates and Galen on medicine; some books on allied disciplines, such as Dioscorides' "Book of Herbs," the works of Rufus, Oribasius and Paulus, and al-Rāzī's "al-Hawi"; four books on agriculture and pharmacology. (Trans. Kopf)

His literary output was astonishing and included many commentaries on Galen's work, *The Book of Principles*, covering medicine; a book on leprosy; a work on pharmacology; commentaries on Hippocrates; a commentary on Pythagoras (570–490); medical treatises on purgatives, fevers, humors, sexual potency, reproduction, pulse, respiration, and tumors; a book on ornithology; commentaries on Plato (427–347) and Aristotle; treatises on logic; and treatises on natural history.

IBN FATIK (C. 1019–1097)

Emir Abu al-Wafa' al-Mubashshir ibn Fatik, an eleventh-century Egyptian mathematician, logician, and physician, wrote *Choice Maxims and Best Sayings*, *The Book of Beginning* (on logic), and a book on medicine. He quoted widely from ancient philosophers. Ibn Abi Usaibi'ah wrote that Ibn Fatik "was the author of excellent works on logic and other philosophical disciplines, which have become renowned among specialists. He also engaged assiduously in copying books; I have seen numerous volumes in his handwriting, containing works by ancient authors. He acquired a huge number of books, many of which are still extant" (trans. Kopf).

Ibn Fātik's *Choice Maxims and Best Sayings* provided a list of the sayings and maxims of Hippocrates, such as

> Lasting health is gained by not being too lazy to work and by refraining from stuffing oneself with food and drink. . . . If you do what should be done the way it should be done and the result is not what it should be, do not change your course of action as long as your original intention remains unaltered. . . . Diminishing what is harmful is better than augmenting what is useful. . . . I possess nothing of the virtue of knowledge except my knowing that I do not know. . . . The owner of a thing is its master. But he who wishes to be free should not covet what does not belong to him; he should shun it, lest he become its slave. . . . This world does not last; so, if you can, do good, but if not, pretend to—and seek the best of reputations. . . . Knowledge is the soul and action the body. Knowledge is the root and action the branch. Knowledge is the procreator and action the offspring. Action originates in knowledge, not knowledge in action. . . . Action is the servant of knowledge, and knowledge is an end [in itself]. Knowledge is the pioneer and action its agent. (Trans. Kopf)

IBN JUMAY' (C. TWELFTH CENTURY, D. 1198)

Ibn Jumay' was a Jewish physician living in Egypt. An associate of Maimonides, Ibn Jumay' was known for identifying the cataleptic fit, a neurological disorder of rigidity of the body that can make a person appear dead. Ibn Abi Usaibi'ah recorded an anecdote describing this discovery:

> One day, when Ibn Jumai` was sitting in his shop near the Candle Market in Fustāt, a funeral procession passed by. After looking at it, he shouted to the kinsfolk of the dead, that their beloved was not dead and that if they interred him, they would be burying a living person. The people stood open-mouthed at this, astonished at his words, and refused to believe him. Then they said to one another: "It will not harm us to check what he says. If he is right, so much the better; and if he is wrong, nothing will have changed." So they called to him to approach and said to him: "Prove what you have told us." He instructed them to return home, remove the shrouds

from the body and carry it into the bathroom. There he poured some hot water on it to warm it up, and treated it with aromatic substances, which had resulted in his sneezing. So the people were . . . able to note some sign of sensation in it; when it made some slight movements, Ibn Jumai` said: "Rejoice at his return to life." He then continued treating him until he recovered consciousness and felt well. This was the beginning of Ibn Jumai`'s fame as a practitioner and scientist. It seemed as if he had performed a miracle. Later, he was asked how he had known that the body carried on a bier and covered with shrouds still had some life in it, and he replied: "I looked at his feet and saw that they were turned upward, whereas the feet of those who are dead are stretched out flat. So I guessed that he was alive, and my guess proved correct."

Ibn Jumay', a believer in clinical practice in medicine, wrote the following works:

(1) "A Guide to the Welfare of the Soul and Body," in four chapters.
(2) "The Explanation of the Hidden Meaning," being a revision of the "Qānūn" (of Ibn Sīnā).
(3) An epistle on Alexandria, the condition of its air, its water and the like, and the ways of life of its inhabitants.
(4) An epistle to Qadī al-Mākin Abū'l-Qāsim Alī ibn al-Husayn on what he should do in a place where he does not find a physician.
(5) A treatise on the lemon, the use of its juice as a beverage and its beneficial properties.
(6) A treatise on rhubarb and its beneficial properties.
(7) A treatise on humpbacks.
(8) A treatise on the treatment of colic, entitled "The Epistle Dedicated to Saif on Royal Medicines." (Trans. Kopf)

MAIMONIDES (1138–1204)

A Jew born in Muslim Spain who moved to Egypt where he wrote medical and scientific treatises in Arabic, Maimonides (Abu 'Imran Musa ibn 'Ubayd Allah ibn Maymum al-Qurtubi), was an Aristotelian commentator and student of a wide range of Peripatetic writings. Maimonides was by profession a physician, by avocation a philosopher, and in the Jewish community a great commentator on the Talmud and Torah. In his medical training and writing, he was heavily influenced by Islamic writers such as Rhazes (854–925) and ancient writers such as Hippocrates and Galen. He served as a court physician in Cairo to Saladin (r. 1174–1193), who employed a huge staff of physicians, including the expert on eye diseases, Al Qaysi. Maimonides's medical practice was such as to leave him practically exhausted, but somehow he found time to write. He wrote *Abstracts on Galen*, *Aphorisms of Hippocrates*, *Medical Aphorisms of Moses*, *Treatise on Hemorrhoids*, *Treatise on Cohabitation*, *Treatise on Asthma*, *Treatise on Poisons and their Antidotes*, *Regimen of Health*, *Discourse on the Explanation of Fits*, and *Glossary of Drug Names*. Maimonides mastered a variety of medical techniques and approaches focusing on the sources of disease, how to alleviate disease, how to prevent disease through lifestyle, how to diagnose, symptoms, pharmacology, and gynecology. He analyzed pneumonia, hepatitis, hemorrhoids, asthma, snakebite, and mental health.

As a philosopher, Maimonides was an Aristotelian who in his *Guide of the Perplexed* provided a logical proof for the existence of God, which had a profound impact on the development of Scholasticism in medieval Europe. Using in part Aristotle's *Metaphysics* and *Physics*, Maimonides composed twenty-five premises on the nature of existence, movement, infinity, motion, cause and effect, the first cause, the nature of time, and the relationship between act and potential.

Other physicians in addition to Maimonides in medieval Egypt included the pharmacologist and botanist Ibn al-Baitar (1197–1248), who was born in Andalusia, Spain, but spent his life traveling in the eastern Mediterranean and worked in Cairo. He wrote *Book of Simple Drugs and Food.* His work, based in part on the experimental method, involved the collection of plants and determining their efficacy for medical treatments. He studied essential oils, such as orange water and rosewater.

In addition, medieval Egypt under Islamic rule contributed the idea of the hospital to the modern world. Saladin (1137–1193) founded the Nasiri hospital in Cairo, which was led by the physician and anatomist Ibn al-Nafis (1213–1288). Al-Nafis wrote *The Comprehensive Book of Medicine*, a commentary on Hippocrates's *On the Nature of Man*, and was the reputed discoverer of the circulatory system. He was also a student of ophthalmology, authoring *Perfected Book of Ophthalmology.* At his death in 1288 the Mansuri Hospital, named for Mansur Qalaun, which served Cairo for centuries, was being built. The Mansuri Hospital had specialized wings for treatment, whether it was gastrointestinal complaints, mental complaints, or ophthalmological issues; it included a surgical center, a pharmacy, and a library, and it was also a teaching hospital. The hospital had a kitchen, living quarters, and clean water sources. Like other Islamic hospitals, the Mansuri Hospital served all people regardless of wealth, charged nothing to the poor, and relied on bequests from charitable Muslims.

IBN KHALDUN (1332–1406)

Ibn Khaldun was a native of northwest Africa who spent many years in Morocco, Tunisia, and Egypt. He was a polymath, an encyclopedist, who wrote a universal history, *Kitāb al-'Ibār*, or *Book of Lessons*, of which the introduction, *The Muqaddimah*, or *Prolegomena*, was particularly famous.

Ibn Khaldun's historical work covered social and natural sciences. He viewed history as a science, the scientific study of humanity, and believed that there was a providential role in history, that God was active in human affairs. He conceived of history as involving the critical study of human relationships, which required the historian to wade through stories and anecdotes and all kinds of untruth in the pursuit of truth. This required critical thought from the historian. The historian must as well be a student of cause and effect in human and natural affairs. The historian must be a student of human civilization, that structure of institutions, organizations, activities, and thought that theoretically elevates humans above animals. "The inner meaning of history," he wrote, "involves speculation and an attempt to get at the truth, subtle explanation of the causes and origins of existing things, and deep knowledge of the how and why of events." History "is firmly rooted in philosophy," and is therefore a science (trans. Rosenthal).

Ibn Khaldun was a social-scientific historian. Human structures and institutions, the accoutrements of civilization, fascinated him. He studied the basis of human social organization that went into settled communities. He differentiated between civilized societies, with sedentary cities filled with luxuries, and nomadic peoples such as North African Bedouins who were less prone to luxury and laziness and hence more courageous. Ibn Khaldun argued that the civilizing influence of laws has a downside, that is, they make people reliant upon government rather than upon themselves; he praised self-reliance. He studied the nature and type of civilizations that existed worldwide. He expounded on the importance of practical town planning, finding the best places in terms of healthy water and air. He was a student of the law. He wrote at length on scholarship, education, linguistics, poetry, and prose.

Geography fascinated Ibn Khaldun, and he wrote at length on different peoples in the different zones of the earth. He drew a map to illustrate the globe. Ibn Khaldun described seven zones of a spherical earth that is covered more by water than by land. He followed ancient and earlier medieval geographers in conceiving of the habitable parts of the earth being north of the equator and south of the arctic. Ibn Khaldun described the "surrounding sea" encircling the lands joined together that represented the three known continents Africa, Asia, and Europe. He described the Mediterranean, Red, Indian, and Caspian Seas and the Persian Gulf. He discussed significant rivers such as the Nile, the greatest, as well as the Tigris and Euphrates. Relying heavily on ancient sources such as Claudius Ptolemy as well as more contemporary Arabic accounts, he recorded fascinating details such as the European exploration and conquest in the fourteenth century of the Canary Islands. Ibn Khaldun, however, claimed that sailors in general eschewed exploring the surrounding sea. He was in general well informed about southern Europe, western Asia, Egypt, Abyssinia, Arabia, and the northern parts of the Indian Ocean, and he had heard of Ceylon, India, Tibet, Java, China, and Korea.

Ibn Khaldun, despite his learning and social-scientific logic, was still a believer in the supernatural, the importance of dreams, soothsaying, and the predictions of astrologers. He was fascinated by the many astrologers who predicted the duration of Muslim dynasties, and of Islam itself. His studies of the work of astrologers included that of Theophilus (695–785), an astrologer to the Abbasids in Mesopotamia, who "said that the Muslim dynasty would have the duration of the great conjunction, that is, 960 years. When the conjunction occurs again in the sign of Scorpio, as it had at the beginning of Islam, and when the position of the stars in the conjunction that dominates Islam has changed, it will be less effective, or there will be new judgments that will make a change of opinion necessary" (trans. Rosenthal). He argued that dream interpretation was a science.

Ibn Khaldun was a writer on economics: he studied prices, labor, profit, capitalism, servitude, the nature of wealth, the importance of agriculture, and commerce. Of agriculture, he wrote, "Agriculture is the oldest of all crafts, in as much as it provides the food that is the main factor in perfecting human life, since man can exist without anything else but not without food." Of commerce, he wrote,

"Commerce is a natural way of making profits. However, most of its practices and methods are tricky and designed to obtain the (profit) margin between purchase prices and sales prices. This surplus makes it possible to earn a profit. Therefore, the law permits cunning in commerce, since (commerce) contains an element of gambling. It does not, however, mean taking away the property of others without giving anything in return. Therefore, it is legal," though, he added, not necessarily *right* (trans. Rosenthal).

Ibn Khaldun's *Book of Lessons* was much more than a social-scientific text. It included a variety of other topics, such as midwifery, of which he wrote:

> When the embryo has gone through all its stages and is completely and perfectly formed in the womb—the period God determined for its remaining in the womb is as a rule nine months—it seeks to come out, because God implanted such a desire in (unborn children). But the opening is too narrow for it, and it is difficult for (the embryo to come out). It often splits one of the walls of the vagina by its pressure, and often the close connection and attachment of (its) covering membranes with the uterus are ruptured. All this is painful and hurts very much. This is the meaning of labor pains. In this connection, the midwife may offer some succor by massaging the back, the buttocks, and the lower extremities adjacent to the uterus. She thus stimulates the activity of the (force) pushing the embryo out, and facilitates the difficulties encountered in this connection as much as she can. She uses as much strength as she thinks is required by the difficulty of (the process). When the embryo has come out, it remains connected with the uterus by the umbilical cord at its stomach, through which it was fed. That cord is a superfluous special limb for feeding the child. The midwife cuts it but so that she does not go beyond the place where (it starts to be) superfluous and does not harm the stomach of the child or the uterus of the mother. She then treats the place of the operation with cauterization or whatever other treatment she sees fit. When the embryo comes out of that narrow opening with its humid bones that can easily be bent and curved, it may happen that its limbs and joints change their shape, because they were only recently formed and because the substances (of which it consists) are humid. Therefore, the midwife undertakes to massage and correct (the newborn child), until every limb has resumed its natural shape and the position destined for it, and (the child) has again its normal form. After that, she goes back to the woman in labor and massages and kneads her, so that the membranes of the embryo may come out. They are sometimes somewhat late in coming out. On such an occasion, it is feared that the constricting power (muscle) might resume its natural position before all the membranes are brought out. They are superfluities. They might become putrid, and their putridity might enter the uterus, which could be fatal. The midwife takes precautions against that. She tries to stimulate the ejection, until the membranes which are late in coming out come out, too. (Trans. Rosenthal)

He wrote as well about medicine, arguing that fevers derive generally from bad digestion of food.

Ibn Khaldun believed that humans are the ultimate thinking creatures who achieve knowledge, advance learning, record thoughts, calculate figures, form conceptualizations, and organize their thoughts. "Man," he wrote, "belongs to the genus of animals and that God distinguished him from them by the ability to think, which He gave man and through which man is able to arrange his actions in an orderly manner. This is the discerning intellect. Or, when it helps him to acquire from his fellow men a knowledge of ideas and of the things that are useful or detrimental to him, it is the experimental intellect. Or, when it helps him to obtain perception of the existent things as they are, whether they are absent or present, it is the speculative intellect" (trans. Rosenthal). He believed that theology and the Qur'an were subject to human scientific speculation. Other sciences were logic, physics, metaphysics, mathematics—such as geometry and arithmetic—music, astronomy, and alchemy.

FURTHER READING

Al-Biruni, *The Chronology of Ancient Nations*, trans. C. Edward Sachau (London: William H. Allen and Co., 1879).

Alhacen's Theory of Visual Perception: A Critical Edition, with English Translation and Commentary, of the First Three Books of Alhacen's De Aspectibus, the Medieval Latin Version of Ibn al-Haytham's Kitab al-Manazir, trans. A. Mark Smith, vol. 2 (Philadelphia: American Philosophical Society, 2001).

Nor Afifah Borhan, Nuramirah Mohd Nor, and Aminuddin Ruskam, "Ibn Al-Baitar: The Pioneer of [*sic*] Botanist and Pharmacist," http://eprints.utm.my/id/eprint/60916/1/AminuddinRuskam2014_IbnAlBaitarthePioneerofBotanist.pdf.

Hans Daiber, Anna Akasoy, Emilie Savage-Smith, eds., *Islamic Philosophy, Theology and Science*, vol. 83 (Leiden, Netherlands: Koninklijke Brill, NV, 2012).

Mohammad Abdullah Enan, *Ibn Khaldun: His Life and Works* (Kuala Lumpur, Malaysia: The Other Press, 2007).

Nahyan Fancy, *Science and Religion in Mamluk Egypt: Ibn al-Nafis, Pulmonary Transit and Bodily Resurrection* (London: Routledge, 2013).

Ibn Abi Usaibi'ah, *History of Physicians*, trans. L. Kopf (Bethesda, MD: National Library of Medicine, 1971).

Abd Ar Rahman bin Muhammed ibn Khaldun, *The Muqaddimah*, trans. Franz Rosenthal, https://asadullahali.files.wordpress.com/2012/10/ibn_khaldun-al_muqaddimah.pdf.

Joel L. Kraemer, *Maimonides: The Life and World of One of Civilization's Greatest Minds* (New York: Doubleday, 2008).

J. J. O'Connor and E. F. Robertson, "Abu Ali al-Hasan ibn al-Haytham," http://mathshistory.st-andrews.ac.uk/Biographies/Al-Haytham.html.

Fred Rosner, "The Life of Moses Maimonides, a Prominent Medieval Physician," *Einstein Journal of Biology and Medicine* 19 (125–128), http://www.einstein.yu.edu/uploadedFiles/EJBM/19Rosner125.pdf.

Emilie Savage-Smith, *Islamic Culture and the Medical Arts* (Bethesda, MD: National Library of Medicine, 1994), https://www.nlm.nih.gov/exhibition/islamic_medical/index.html#toc.

29

Science in Medieval Central and Southern Asia (500–1500 CE)

CENTRAL ASIA/PERSIA

The Sasanian Dynasty, dominating central Asia from 224 to 651 CE, fell to the Rashidun caliphate of Arab Islamic conquerors in the mid-seventh century. The Umayyad caliphate dominated central Asia during the next century, followed in subsequent centuries by the Abbasid caliphate. Hereafter Islam became the dominant religion in culture, including science and philosophy, in central Asia. Due to the influence of the Byzantine Empire on the expanding Islamic culture, Islamic science in central Asia continued to be devoted to commentaries and interpretations of ancient Greek scientists such as Aristotle and Ptolemy and ancient Roman scientists such as Galen. The influence of ancient science on medieval Persia is exemplified by Abu Maʿshar, Rhazes, Al Majusi, Ibn Sina, Al-Biruni, Omar Khayyam, and Al Tusi.

ABU MAʿSHAR (787–886)

A native of Afghanistan, the Persian Abbasid astrologer and mathematician Abu Maʿshar, known in Europe as Albumasar, tried to estimate chronologies of distant periods, for example, the deluge. He migrated to Baghdad during the first part of the ninth century, where he became the most notable astrologer of his time. Many ancient and medieval scholars believed that such a deluge, or worldwide flood, had occurred, as described in *The Epic of Gilgamesh* and Genesis of the Old Testament. However, according to Al-Biruni,

> [most] Persians, and the great mass of the Magians, deny the Deluge altogether; they believe that the rule (of the world) has remained with them without any interruption ever since Gayomarth Gilshah, who was, according to them, the first man. In denying the Deluge, the Indians, Chinese, and the various nations of the east, concur with them. Some, however, of the Persians admit the fact of the Deluge, but they describe it in a different way from what it is described in the books of the

> prophets. They say, a partial deluge occurred in Syria and the west at the time of Tahmurath, but it did not extend over the whole of the then civilized world, and only few nations were drowned in it. (Trans. Sachau)

Abu Ma'Shar was one of the Persians who agreed that the deluge had occurred, but he provided a novel theory as to precisely when. Al-Biruni wrote,

> He supposed that the Deluge had taken place at the conjunction of the stars in the last part of Pisces, and the first part of Aries, and he tried to compute their places for that time. Then he found, that they—all of them—stood in conjunction in the space between the twenty-seventh 10 degree of Pisces, and the end of the first degree of Aries. Further, he supposed that between that time and the epoch of the AEra. Alexandri [conquest of Alexander the Great], there is an interval of 2,790 intercalated years 7 months and 26 days. This computation comes near to that of the Christians, being 249 years and 3 months less than the estimate of the astronomers. Now, when he thought that he had well established the computation of this sum according to the method, which he has explained, and when he had arrived at the result, that the duration of those periods, which astronomers call "star-cycles," was 360,000 years, the beginning of which was to precede the time of the Deluge by 180,000 years, he drew the inconsiderate conclusion, that the Deluge had occurred once in every 180,000 years, and that it would again occur in future at similar intervals. This man, who is so proud of his ingenuity, had computed these starcycles only from the motions of the stars, as they had been fixed by the observations of the Persians; but they (the cycles) differ from the cycles, which have been based upon the observations of the Indians, known as the "cycles of Sindhind." (Trans. Sachau)

Al-Biruni, in providing arguments to contradict the theories of Abu Ma'Shar, engaged in a speculative discourse in which he supposed that such erroneous astronomers as Abu Ma'Shar could hypothetically assume the extent of the temporal nature of the universe to extend into billions of years. He wrote:

> For it is quite possible that these (celestial) bodies were scattered, not united at the time when the Creator designed and created them, they having these motions, by which—as calculation shows—they must meet each other in one point in such a time (as above mentioned). It would be the same, as if we, e.g., supposed a circle, in different separate places of which we put living beings, of whom some move fast, others slowly, each of them, however, being carried on in equal motions—of its peculiar sort of motion—in equal times; further, suppose that we knew their distances and places at a certain time, and the measure of the distance over which each of them travels in one Nychthemeron [24 hours]. If you then ask the mathematician as to the length of time, after which they would meet each other in a certain point, or before which they had met each other in that identical point, no blame attaches to him, if he speaks of billions of years. Nor does it follow from his account that those beings existed at that (past) time (when they met each other), or that they would still exist at that (future) time (when they are to meet again); but this only follows from his account, if it is properly explained, that, if these beings really existed (in the past), or would still exist (in future) in that same condition, the result (as to their conjunctions) could be no other but that one at which he had arrived by calculation. But then the verification of this subject is the task of a science which was not the science of Abu-Ma'shar. (Trans. Sachau)

Abu Ma'Shar's work on astrology long outlived him and had an influence on medieval Byzantine and European astrologers and philosophers, such as the thirteenth-century thinker Albertus Magnus (1200–1280) and the English philosopher Roger

Bacon (1220–1292). He was instrumental in the rediscovery of Aristotle (384–322) by medieval Europeans.

RHAZES (C. 854–925)

Abū Bakr Muhammad ibn Zakariyyā al-Rāzī, Rhazes, was the "Arab Galen" according to contemporaries. A Persian physician and pharmacologist, Rhazes lived for a while in Baghdad and was founder and head of the Baghdad Hospital. He also founded a hospital in his hometown of Rayy. His interests were broad and his writing prolific. He wrote treatises and books on dreams, logic, chemistry, diseases such as scarlet fever and measles, kidney stones, diseases of the eyes, human anatomy, pharmacopoeia, mathematics, astronomy, philosophy and theology, magnets, laxatives, and purgatives; he kept casebooks in which he described various diseases, such as to the eye; he was a commentator on Galen (130–200), Plutarch (46–120), and Hippocrates (460–377). He wrote, "When Galen and Aristotle agree upon something, it is the truth; when they disagree, it would be very difficult for us to decide where the truth lies."

Specific titles of Rhazes's works, according to the *History of Physicians* by Ibn Abi Usaibi'ah, include:

> Al-Hāwi [the Collection], which is his best and most comprehensive medical work, as he had gathered in it everything he found separately concerning maladies and their treatment in all the other books, from those of the ancients up to his contemporaries. . . . "Spiritual Medicine," known also as "Psychiatry," the aim of which is the treatment of the spirit's composition; twenty chapters; . . . a book proving that man was fashioned by a wise and capable creator, giving proofs from dissection and the functions of the members which show that the creation of Man could not have been by chance. . . . "The Fame of Nature," in which he intended to give an introduction to the natural sciences and an easy guide for the student to the understanding of the many various meanings in the books on that subject. . . . "Isagoge," an introduction to logic. . . . "The Shape of the World," in which he aimed to clarify that the earth is round and situated at the middle of the celestial sphere, which has two poles on which it revolves; that the sun is bigger than the earth and the moon smaller than it, and so on. . . . A book for those who cannot reach a doctor, aiming to define the various maladies; here he discourses at length, mentioning one malady after the other, and how each can be treated by common drugs; this book is known as "The Medical Book of the Poor." . . . "The Drugs Found Everywhere," in which he mentions drugs which spare the skilled physician the need of any others, if he adds to them what is found in every house and kitchen. . . . "The Reason for Which the Earth is Fixed in the Middle of the Celestial Sphere in Spite of its Revolving." . . . A book opposing Ibn al-Yamān's refutation of psychiatry. . . . A book on the fact that the world cannot be the loftiest thing we see. . . . A study on movement, which is not imaginary but real. . . . A treatise concerning the fact that the body has its own movement and that movement is a natural principle. . . . Ode on Logic. . . . Ode on Theology. . . . "The Spheres and a Short Account of the Fates." . . . "On Periods, Which are Time, and on Emptiness and Fullness, which are Space." . . . "Metaphysics." . . . "Matter, Absolute and Partial." . . . A book on metaphysics, following Plato. . . . A collection, known as "The Stronghold of Medicine," in which his aim was to collect everything which he could find in any medical book, old or new, and bring it under the same heading in one book. It is divided into twelve parts:

a. The preservation of health, the treatment of maladies, dislocations, surgery and drugs
b. The quality of foods and drugs and what has to be known for therapeutics
c. Compound medicines and what must be known about them, in the form of a Pharmacopoeia
d. Necessary data on the pulverization of drugs, their burning, melting, washing, extracting of their powers and preservation; how long each drug can be kept, etc.
e. The chemistry of medicine, describing the various drugs, their color, taste, odor, components, what is good and what is bad in this domain, etc.
f. On permutation, mentioning a replacement for each food or drug when it is not available
g. Explanation of names, weights and measures which are used by druggists, and the names of members and maladies in Greek, Syriac, Persian, Hindu, and Arabic, in the form of books called "Shaqshamāhī"
h. Anatomy and the functions of the members
i. The natural factors in the art of medicine, aiming to show the natural causes of illnesses
j. Introduction to medicine; two essays—one on natural phenomena, the other on the origins of medicine
k. Collection of case histories, prescriptions, etc.
l. Data he found in Galen's books which are not mentioned by Hunayn or in Galen's Index. . . .

An epistle proving that the rising and setting of the sun and stars which we imagine is caused not by the revolutions of the earth but by those of the celestial sphere. . . . On the fact that even the expert physician cannot cure all illnesses, for this would be an impossibility even for one with the skill of Hippocrates; but the physician deserves to be praised and thanked and the art of medicine glorified and honored, even though the physician cannot do that [cure all illnesses] even after he has risen above his contemporaries. (Trans. Kopf)

Rhazes's *Book of Secret of Secrets* is an alchemical recipe book, filled with information on alkalies, what he called "sharp waters," such as vinegar and lemon juice, which he believed were solvents to use with metals. He believed in the alchemist idea of transmutation of substances and tried to discover recipes to encourage such change.

Rhazes was a medical practitioner as well as a writer. He observed countless diseases in the hospitals at Rayy and Baghdad that allowed him to have a critical view toward the writings of ancient medicine—he could be critical even toward Galen. His work was focused on finding therapies for illnesses ranging from urologic diseases to diabetes to headaches to dysentery to toothache. A prolific writer on medicine and philosophy, his works became famous among both Arabs and Europeans. His *Book of Medicine Dedicated to Mansur* was translated into Latin in the twelfth century, and *The Comprehensive Book on Medicine* was translated into Latin in the thirteenth century.

AL-MAJUSI (930–994)

Ali ibn al-'Abbas al-Majusi, known in Europe as Haly Abbas, was a Zoroastrian Persian physician and writer who wrote *The Complete Book of the Medical Art* based on his understanding of the work of Hippocrates and Galen. He had a critical viewpoint toward ancient medicine as well as toward his near contemporary Rhazes. He lived in Bagdad, serving the ruler Al Dawlah, to whom he dedicated *The Complete Book of the Medical Art*, known in Europe as the *Royal Book, Liber Regius*. Al-Majusi was an encyclopedic writer of medicine who wrote on theory and practice. His work, in several volumes, focused on human anatomy, causes of disease, the impact of the environment on disease, the Hippocratic and Galenic approaches to disease, and the various therapies used to relieve symptoms.

IBN SINA (980–1037)

Ibn Sina (Avicenna) was, at the beginning of the eleventh century, the greatest living Aristotelian scholar. He had an encyclopedic mind, wrote many commentaries, and was best known for his studies of medicine. European scientists relied on his works for centuries. He was a religious thinker as well as a logician, a philosopher who studied nature and the author of nature, God. In this he was like many of the Greek and Roman philosophers, as well as most contemporary medieval philosophers in Asia, Africa, and Europe, who believed wholeheartedly in the omnipotence and omniscience of a divine power the characteristics of which through reason and science could be discovered by humans.

Ibn Abi Usaibi'ah, *History of Physicians*, recorded an autobiographical account penned by Ibn Sina:

> My father was a native of Balkh, but moved from that city to Bukhārā . . . and . . . was appointed governor of the village of Kharmaithan, a government estate in the Bukhārā region and a most excellent locality. Near it is another village, Afshana, where my father married my mother and settled down. In that village I, and later my brother, was born. Then we moved to Bukhārā itself. I became a teacher of the Qur'ān and of literature. At the age of ten, I had already mastered the whole of the Qur'ān and an abundance of literature, so that I was held in great esteem. Later, Abū `Abdallāh al-Nātilī, called the philosopher, came to Bukhara, and my father lodged him in our house in the hope that he would teach me. Prior to his arrival I had applied myself to the study of law. . . . I was a very good debater, since I had become familiar with the customary ways of asking questions and raising objections to the answers. Thereafter I began to study [Porphyry's] "Isagoge" under the tuition of al-Nātilī, and when, upon my question, he told me the definition of a genus, saying it was a category comprising many different species. I began to verify this definition in a way he had never heard of. Full of admiration, he warned my father not to let me occupy myself with anything but science. Whatever problem he set me I solved better than he. I studied the rudiments of logic under his tuition, but as to its intricacies, he was not acquainted with them. I then started reading books by myself with the help of commentaries, until I achieved complete mastery of the art of logic. The same procedure was applied to Euclid's *Book* [*of Elements*]. I studied the first five or six chapters under al-Nātilī and then undertook to solve [the problems of] the remaining parts by myself. Next I turned to the [Ptolemy's] *Almagest*, and when I had finished the introductory parts and reached the chapter on geometry, al-Nātilī said to me: "Try to study and solve them by yourself and then report to me what you have found so that I may tell you what is right and what is wrong." He had not mastered the book and after I had worked out the solutions, [it became clear] how many [difficult] passages he had not understood until I expounded them to him. Later,

al-Nātilī left me and went to Kurkānj, and I applied myself to the study of natural history and metaphysics, reading both original texts and commentaries. The gates of knowledge opened to me. I then felt an inclination for medicine and set myself to perusing books on this subject. Medicine is not one of the difficult sciences; no wonder, therefore that I had soon made such progress in it that distinguished physicians came to study under me. I also tended the sick and thus gained an indescribable amount of practical knowledge of methods of treatment. In addition, I continued studying law and participating in discussions on this subject. At that time I was sixteen years old. Thereafter I devoted a year and a half entirely to study and reading, revising [texts on] logic and all the branches of philosophy. During this period I never slept a whole night, nor did I use the daytime for anything but study. I kept some blank sheets before me, and whenever I came across a difficult problem, I set it down and then considered what conclusions might be drawn from it so that, in the end, the solution came to me. Whenever I was baffled by a problem and could not determine the answer, I repaired to the great mosque and humbly prayed to the Creator until the obscure became clear and the difficult easy. In the evening I returned home, placed a lamp in front of me and applied myself to reading and writing. When sleep threatened to overpower me or I felt weak, I drank a cup of sharāb in order that my strength might be restored and I might be able to go on reading. Whenever I actually did doze off, 1 would dream of the very problem which had been occupying my mind, so that I solved many questions while asleep. I persevered in this way until I had mastered all the sciences and knew as much of them as is within the power of man to know. Everything I learnt at that time is as fresh in my memory as if I had learnt it now, and I have added nothing. So I mastered logic, the natural sciences and mathematics. Then I turned to metaphysics and read the [Aristotle's] "Book of Metaphysics," but I did not understand it, and the purpose of the author remained obscure to me. I reread it forty times, so that I came to know it by heart, but I still did not understand it nor the aim it served. I despaired of myself and said: This is a book which is impossible to comprehend. One day, at the time of the afternoon prayer, I happened to pass through the booksellers' [quarter] when I saw in the hand of an auctioneer a volume that he was advertising and offering to me. I gruffly declined the offer, convinced that such a purchase would be of no advantage. Whereupon he said to me: "Buy it for it is cheap. I am selling it to you for three dirhams because its owner is in need of money." So I bought it and behold, it was a book by Abū Nasr al-Farabī on the purpose of the "Book of Metaphysics." I went home, hastened to read it, and presently understood the "Book of Metaphysics," because I remembered its contents. I was delighted and the next day spent a considerable amount in alms to the poor as a thanksgiving to God the Exalted. At that time, Nūh ibn Mansūr was Sultan of Bukhara. He contracted a disease which the physicians vainly tried to cure. As my name was well known among the medical men by reason of my assiduous studies, they mentioned me to the Sultan and advised him to send for me. I came and joined them in attending the Sultan, and as a consequence, I entered his service. One day, I asked his permission to enter the court library, examine it and read its store of medical books. He granted my request, and I entered a building of many rooms, each of which were piled high with bookshelves. In one room were books on the Arabic language and on poetry, in another jurisprudence, so on, each room containing books on a specific subject. I perused the catalogue of the ancients and picked out those books that I needed. I saw books the titles of which many had never heard of and which I had never seen before or did see afterward. I read those books, profited by their contents and came to know the position occupied by each author in his discipline. When I was eighteen years of

> age, I was already familiar with all these sciences. At that time my memory was better able to absorb facts, but my knowledge is now more mature. At any rate, it is the same knowledge, for nothing new has accrued to me since. There lived in my neighborhood a man named Abū'l-Husayn al-`Arūdī. He asked me to write a compendium of science for him, which I did and which I named after him. It deals with all the sciences except mathematics. At the time I was twenty-one years old. Another of my neighbors was a man named Abū Bakr al-Barqī, who was born in Khwārizm. He was a renowned jurist, devoting himself entirely to jurisprudence, Qur'anic exegesis and ascetic practices. But he felt an inclination for the sciences and asked me to write commentaries on certain books for him. I wrote for him "al-Hāsil wal-Mahsūl," comprising nearly twenty volumes, and also—in the field of ethics—a book entitled "Virtue and Sin." These two books are to be found only with him for he never lent them to anyone for copying. Then my father died and life underwent frequent changes. After serving in the administration of the Sultan, I was obliged to leave Bukhārā and to proceed to Kurkānj, where Abū'l-Husayn al-Sahlī, a lover of the sciences, held the office of vizier. I was introduced to the Emir Alī ibn Ma`mūn, dressed in the garb of a jurist with the tailasān and the taht al-hanak. I was granted a monthly allowance sufficient for the maintenance of a person of my standing. (Trans. Kopf)

Ibn Sina relied on Aristotle's theory on the nature of causes as applied to medicine. Aristotle argued that causation is based on material, efficient, formal, and final causes. How do these apply to medicine? Ibn Sina asked. Science, he argued, including the science of medicine, requires investigation into how things come about, in nature, in the human body. Some causes are visible, obvious, others hidden, subtle. Effects in nature, effects in the human body, derive from causes, and until a person, a scientist or physician, understands these causes, the effects will be misunderstood, and a possible cure will be lost. In medicine, the material cause involves the affected person and the causes of disease, which Ibn Sina understood to be the four Hippocratic humors: blood, black bile, yellow bile, phlegm. Next are the efficient causes: the place in which a person lives, the atmosphere and environment, meteorological conditions, the type of food ingested and drink imbibed. The formal causes are the basic fundamental constitution, or health, bodily strength or weakness, of the person under consideration. The final causes are how the physician seeks to approach the disease present in the patient's body in order to seek a cure.

Ibn Sina's major work, modeled on Aristotle, *The Cure*, was an encyclopedic work on logic, natural science, mathematics, and philosophy. It was highly influential among thinkers of medieval Europe, such as Thomas Aquinas (1225–1274). His major medical work, also extremely influential, was *The Canon*. This encyclopedic book discussed diseases, remedies, and pharmacopeia. He believed in the power of ice and heat in healing and in the use of various analgesics such as opium and mandrake. Ibn Sina wrote on human psychology, in *Treatise on the Soul*, in which he argued for a mind-body duality, that the soul is independent of the body and self-aware. His writings on logic were heavily influenced by Aristotle and Porphyry (234–305) the Neoplatonist. In the *Book of the Remedy*, he hypothesized on metals and other substances in an alchemical way. Nevertheless, he did not believe in the transmutation of metals and did believe that all that an alchemist could accomplish was a change in the appearance of materials.

AL-BIRUNI (C. 971–1048)

One of Ibn Sina's friends was Abū 'l-Raihān al-Bīrūnī, who was a physician, astronomer, astrologer, geologist, and historian. He wrote works on the past, on the astrolabe, on Ptolemy (130–200), and a book, as described by Ibn Abi Usaibi'ah, "on medicine (materia medica); in it he gave an exhaustive account of the nature of different drugs, their names, the different opinions held about them by his predecessors and statements made in their regard by physicians and others. He arranged it in alphabetical order" (trans. Kopf). Al-Biruni combined astronomy, chronology, and history in *The Chronology of Ancient Nations*, which examined the ways different historical peoples used the motions of the sun and moon to gauge time and create calendars. This was an exhaustive study; Al-Biruni was particularly strong on his discussions of Arab, Muslim, and Persian astronomers and chronologists, though he was quite aware of Western scientists as well. He was critical of astrologers such as Abraham ibn Ezra (c. 1093–1167), the twelfth-century author of *Navivitatibus*, in which he quantified human lifespan according to the position of the planets at birth, but Al-Biruni could be credulous toward stories of giants and incredibly long human life spans as recorded in the Torah. He was fascinated by years and sequence and provided complex chronological tables on, for example, the reigns of kings: Assyrian, Babylonian, Chaldean, Roman, Byzantine, and Persian kings, among others. He was fascinated by, and wrote on, the game of chess. His anecdotes and commentaries on other scientists provide interesting insights into ancient and medieval beliefs. For example, in discussing the heat of the sun, he provided a perspective of the different beliefs people held during his time:

> Regarding the rays of the sun many theories have been brought forward. Some say that they are fiery particles similar to the essence of the sun, going out from his body. Others say that the air is getting warm by its being situated opposite to the sun, in the same way as the air is getting warm by being opposite to the fire. This is the theory of those who maintain that the sun is a hot, fiery substance. Others, again, say that the air is getting warm by the rapid motion of the rays in the air, which is so rapid as to seem timeless. . . . This is the theory of those who maintain that the nature of the sun has nothing in common with the natures of the four elements. Further, there is a difference of opinion regarding the motion of the rays. Some say this motion is timeless, since the rays are not bodies. Others say this motion proceeds in very short time; that, however, there is nothing more rapid in existence by which you might measure the degree of its rapidity. E.g., the motion of the sound in the air is not so fast as the motion of the rays; therefore the former has been compared with the latter, and thereby its time (i.e., the degree of its rapidity) has been determined. As the reason of the heat which exists in the rays of the sun, people assign the acuteness of the angles of their reflexion. This, however, is not the case. On the contrary, the heat exists in the rays (is inherent in them). (Trans. Sachau)

He wrote of Muslim astronomers, that they named the days of the week, even the hours of the day, according to the celestial bodies that orbited the earth. "The sun is the dominus of the first day, and at the same time the dominus of the first hour. The second hour is ruled by the planet of the sphere next under the sphere of the sun, i.e. Venus. . . . According to this system, the dominus of the twenty-fifth hour is the moon, and this is the first hour of Monday" (trans. Sachau).

An astrologer and meteorologist, Al-Biruni provided an extensive almanac in *The Chronology of Ancient Nations*, reporting on weather changes and supernatural events in each day of the year. The solar and lunar calendars determined religious festivals, and the months they were held in, for various Asian peoples, especially the Persians. His learning, as reflected in *The Chronology of Ancient Nations*, was astonishing. He was familiar with a plethora of Arab, Greek, and Latin authors—among the latter, Callippus of Cyzicus the astronomer (370–310), Hipparchus of Nicaea the astronomer (190–120), Dositheos the theologian (first century CE), the astronomer Eudoxus of Cnidus (408–355), the atomist philosopher Democritus (460–370), Philippus the writer of *Epigrams* (first century CE), Conon of Samos the astronomer (c. 280–220), Claudius Ptolemy (100–170), Hippocrates (460–377), and Galen (130–200). He used these sources to plot an almanac of meteorological changes throughout the entire year. Al-Biruni was also fascinated by Asian Indian culture, learning, and science. He wrote a *History of India* that explored these many topics.

A polymath, Al-Biruni attempted expertise in most subjects of the physical, earth, biological, and human sciences. He wrote on hydrology, geology, geography, astronomy, chronology, meteorology, biology, medicine, pharmacology, and mathematics. He speculated on the source of the Nile in *The Chronology of Ancient Nations*:

> The Nile, again, has high water when there is low water in both Tigris and Euphrates, because its source lies in the Mons Lunoe, as has been said, beyond the Abyssinian city Assuan in the southern region, coming either exactly from the equator or from countries south of the equator. This is, however, a matter of doubt, because the equatorial zone is not inhabited, as we have before mentioned. It is evident that in those regions there is no freezing of moist substances at all. If, therefore, the high water of the Nile is caused by falling dew, it is evident that the dew does not stay where it has fallen, but that it directly flows off to the Nile. But if the high water is caused by the springs, these also have the most abundant water in spring. Therefore the Nile has high water in summer, for when the sun is near us and our zenith, it is far distant from the zenith of those places whence the Nile originates, and which in consequence have winter. (Trans. Sachau)

An astrologer as well, Al-Biruni wrote *The Book of Instruction in the Elements of the Art of Astrology* in 1029. In this book and *The Chronology of Ancient Nations*, Al-Biruni provided an extensive overview of astrological methods and ideas for a variety of peoples, especially those of western Asia. For example, he wrote of the "illiterate" Arabs, those perhaps before the coming of Islam, that "it was their object to learn . . . all meteorological changes in the seasons of the year." Ancient and medieval peoples used the solar and lunar zodiacs for their astrology. The Arabs used the latter, as they

> could not recognize the Lunar Stations except by certain marks, visible to the eye. Therefore they marked the Stations by those fixed stars which lie within them. And the rising of the fixed stars in the east early after the rise of dawn they considered as a sign of the sun's entering some one of the Stations, and so they could do, since the stars do not recede from their places except after the lapse of long spaces of time, and, besides, the Arabs were not educated enough to notice such a variation.

Further, they composed verses and rhymed poetry, so that these things could easily be remembered by illiterate people, and recorded therein the annual physical influences which, according to their observation and experience, coincided with the rising of each particular Station. (Trans. Sachau)

On science, Al-Biruni wrote:

Now, if we in some places wander about through various branches of science, and plunge into subjects which are not very closely connected with the order of our discussion, we must say that we do not do this because we seek to be lengthy and verbose, but as guided by the desire of preventing the reader from getting tired. For if the mind is continually occupied with the study of one single science, it gets easily tired and impatient; but if the mind wanders from one science to another, it is as if it were wandering about in gardens, where, when it is roving over one, another one already presents itself; in consequence of which, the mind has a longing for them, and enjoys the sight of them; as people say, "Everything that is new offers enjoyment." (Trans. Sachau)

Al-Biruni wondered at the "creative power of nature," which "never drops any material unused." He believed that "none of the existing bodies is in its natural place, that all of them are where they are only in consequence of some force being employed, and that force must of necessity have had a beginning." (Ibid)

OMAR KHAYYAM (1048–1131)

Mathematician, astronomer, music theorist, and poet, Omar Khayyam was a Persian polymath. A philosopher of the nature of existence, epistemology, being and necessity, and the structures and ranks of life, Khayyam was an Aristotelian and commentator. He was heavily influenced by his countryman Ibn Sina and the Greek philosopher Plato (427–347) as well. He helped to create an astronomical observatory from which the Persian Jalali Calendar was derived. As a mathematician, he wrote *Problems of Arithmetic* and *Explanation of the Difficulties in the Postulates of Euclid.* He was interested in cubic equations and used geometry to solve some quadratic equations. Like other medieval mathematicians, Khayyam was influenced by Asian Indian mathematics: "In his *Algebra* Khayyam writes that methods for calculating square and cube roots come from India, and that he has extended them to the determination of roots of any order" ("Umar Khayyam").

Khayyam's poem, the *Rubaiyat*, reveals his astronomical interest:

Wake! For the Sun, who scatter'd into Flight
The Stars before him from the Field of Night,
Drives Night along with them from Heav'n, and strikes,
The Sultan's Turret with a Shaft of Light. (Trans. Fitzgerald)

NASIR AL-DIN AL-TUSI (1201–1274)

The Persian scholar Al-Tusi was a mathematician, astronomer, logician, and student of medicine. Present at the Mongol capture of Alamat in Persia, he became the scientific advisor to the Mongol conqueror Hulegu Khan and was present when the Mongols captured Baghdad in 1258. Hulegu Khan (r. 1256–1265) ruled out of

Maragheh in northwestern Persia; Al-Tusi continue as the court scientist and helped to construct an astronomical observatory. At the observatory, Al-Tusi constructed scientific instruments and made extensive observations, publishing his observations in the *Ilkhanic Tables*. He was a student of Ptolemy's astronomy, wrote commentaries on Ptolemy's work, and made some changes to Ptolemy's theories of lunar motion. He wrote commentaries on a variety of ancient mathematicians, such as Aristarchus (310–230), Euclid (326–265) and Archimedes (287–212). He was one of the originators of trigonometry as a mathematical discipline. He was interested in alchemy and medicine. Under his influence, the observatory at Maragheh became an important center of medieval Islamic science.

SOUTHERN ASIA/INDIAN SUBCONTINENT

Cultures of medieval Asia and Europe were in mutual contact during the years from 500 to 1500 in part because of cosmopolitan centers of trade of goods and ideas, such as Damascus and Constantinople, in which astronomical, mathematical, geographic, medical, and philosophical European, Islamic, Hindu, and Chinese ideas were exchanged. The expansion of Islam into North Africa, western Asia, and central Asia greatly accelerated the exchange of these ideas.

For example, the Persian Muslim scholar Al-Biruni provided an extensive assessment of the nature of culture and science throughout the known world. His *History of India* discussed medieval Hindu science. He described the Hindu practice of alchemy as "restricted to certain operations, drugs, and compound medicines, most of which are taken from plants. Its principles restore the health of those who were ill beyond hope, and give back youth to fading old age, so that people become again what they were in the age near puberty. . . . A famous representative of this art was Nagarjuna. . . . He excelled in it, and composed a book which contains the substance of the whole literature on this subject, and is very rare" (trans. Sachau).

Al-Biruni condemned Hindu astronomers for their reliance upon religion over science. He believed that they were willing to entertain nonsense as long as it pacified their religious sensibilities. Hence they believed the Himalayas, and the tallest mountain, Meru, were under the North Pole. Hindus and Buddhists throughout southern Asia built artificial mountains in imitation of Meru. Nevertheless, they recognized a circular earth, the center of the universe. The Northern Hemisphere, they thought, was dominated by land, while the Southern Hemisphere was dominated by water. According to the philosopher Brahmagupta (598–670), heaven is a globe as well. He wrote further, according to Al-Biruni, that "all people on earth stand upright, and all heavy things fall down to the earth by a law of nature, for it is the nature of the earth to attract and to keep things, as it is the nature of water to flow, that of fire to burn, and that of the wind to set in motion" (trans. Sachau).

Such Hindu astronomers believed, continued Al-Biruni, that "the wind makes the sphere of the fixed stars revolve; the two poles keep it in its place, and its motion appears to the inhabitants of Mount Meru as a motion from left to right." Brahmagupta wrote that "the sphere [of the earth] has been created as moving with the greatest rapidity possible about two poles without ever slackening." He said as well that "the wind makes all the fixed stars and the planets revolve towards the west in one and the same revolution; but the planets move also in a slow pace towards the east, like a dust-atom moving on a potter's-wheel in a direction opposite to that in which the wheel is revolving. That motion of this atom which is visible is identical with the motion which drives the wheel round, whilst its individual motion is not perceived" (trans. Sachau).

Al-Biruni was relatively familiar with the works of several prominent Hindu mathematicians and astronomers. Some of the leading medieval Hindu scholars included Aryabhata the Elder, Varahamirira, Bhaskara I, and Brahmagupta. Aryabhata the Elder (476–550) lived in northern India. He was the author of *Aryabhatiya,* a work on astronomy and mathematics, in which he explored algebra, trigonometry, pi, fractions, quadratic equations, place values, and zero. His astronomy was sophisticated: he reasoned that the motion of the planets and stars was in part due to the earth's rotation; he approximated, quite accurately, the circumference of the earth; he believed the planets orbited in ellipses; he understood the causes of eclipses; he made a relatively accurate approximation of the length of the day. Varahamihira (505–587) lived in western India, and authored *Pancasiddhantika,* a work on astronomy. Influenced by work of the Greeks, such as Hipparchus, he also knew of the Metonic cycle of nineteen years and the theory of epicycles. He wrote on mathematics and made contributions to trigonometry. Bhaskara I (c. 600–680) lived in western India. He authorized commentaries of his predecessor Aryabhata in mathematics and astronomy. Brahmagupta (598–660), born in western India, was the author of *Brahmasphutasiddhanta* and *Khandakhadyaka.* In astronomy, he worked on eclipses, conjunctions of planets, and diurnal rotation. In mathematics, he provided the solution of the quadratic equation, he understood the digital system, and he mastered the concept of zero, realizing it was a number used in arithmetic and algebraic calculations.

Other, later Indian medieval scientists included Mahavira (800–870), a contemporary of Al Khowarizmi, who worked with quadratic equations and commented on Brahmagupta's work. Vachaspati, who lived in the ninth or tenth centuries, developed coordinate geometry, understood that heat and light were made up of minute atomic particles, thought that sound occurred in aether and that sound occurred in longitudinal waves. Bhaskara II (1114–1185), from southern India, "invented the art of placing the numerator over the denominator in a fraction," according to Benoy Kumar Sarkar. He also worked with various types of right-angled triangles, the properties of circles, "the volume of a cone as one-third the volume of the cylinder," and "the volume of a pyramid as one-third the volume of the prism." Madhava (c. 1340–1425), founder of Kerala School of Astronomy and Mathematics in Kerala, was an astronomer and mathematician; he worked with infinite series, calculus, geometry, trigonometry, and algebra. The Kerala school was active from the fourteenth to the sixteenth centuries.

FURTHER READING

Alberuni's India: An Account of the Religion, Philosophy, Literature, Geography Chronology, Astronomy, Customs, Laws and Astrology of India about A. D. 1030, trans. Edward C. Sachau (London: Kegan Paul, Trench, Trubner, & Co., 1910).

Al-Biruni, *The Book of Instruction in the Elements of the Art of Astrology*, trans. R. Ramsay Wright (London: Luzac & Co, 1934; reprint Bel Air, MD: Astrology Classics, 2006).

Al-Biruni, *The Chronology of Ancient Nations*, trans. Edward C. Sachau (London: William H. Allen and Co., 1879).

Algebra, with Arithmetic and Mensuration, from the Sanscrit of Brahmegupta and Bhascara, trans. Henry Thomas Colebrooke (London: John Murray, 1817).

Brahmagupta, *Khandakhadyaka*, trans. Bina Chatterjee, vol. 1 (Delhi, India: Motilal Banarsidass, 1970).

Ibn Abi Usaibi'ah, *History of Physicians*, trans. L. Kopf (Bethesda, MD: National Library of Medicine, 1971).

Ibn Sina, *On Medicine*, https://sourcebooks.fordham.edu/source/1020Avicenna-Medicine.asp.

Omar Khayyam, *The Rubaiyat of Omar Khayyam*, trans. Edward Fitzgerald (San Francisco: Dodge Book and Stationery Co., 1896).

Henry Leicester, *The Historical Background of Chemistry* (New York: Dover Books, 1971).

John McGinnis, *Avicenna* (Oxford: Oxford University Press, 2010).

J. J. O'Connor and E. F. Robertson, *Aryabhata the Elder*, http://mathshistory.st-andrews.ac.uk/Biographies/Aryabhata_I.html.

J. J. O'Connor and E. F. Robertson, *Bhaskara I*, http://mathshistory.st-andrews.ac.uk/Biographies/Bhaskara_I.html.

J. J. O'Connor and E. F. Robertson, *Brahmagupta*, http://mathshistory.st-andrews.ac.uk/Biographies/Brahmagupta.html.

J. J. O'Connor and E. F. Robertson, "Nasir al-Din al-Tusi," http://mathshistory.st-andrews.ac.uk/Biographies/Al-Tusi_Nasir.html.

J. J. O'Connor and E. F. Robertson, *Varahamihira*, http://mathshistory.st-andrews.ac.uk/Biographies/Varahamihira.html.

Sajjad H. Rizvi, *Avicenna (Ibn Sina)*, Internet Encyclopedia of Philosophy, https://www.iep.utm.edu/avicenna/#H2.

Benoy Kumar Sarkar, *Hindu Achievements in Exact Science: A Study in the History of Scientific Development* (London: Longmans, Green and Co., 1918).

Emilie Savage-Smith, *Islamic Culture and the Medical Arts*, National Library of Medicine (1994), https://www.nlm.nih.gov/exhibition/islamic_medical/index.html#toc.

"Umar Khayyam," Stanford Encyclopedia of Philosophy, https://plato.stanford.edu/entries/umar-khayyam/#KhaMatSci.

John Wippel and Allan Wolter, eds., *Medieval Philosophy* (New York: Free Press, 1969).

30

Science in Medieval East and Southeast Asia (400–1644 CE)

Confucianism mixed with Taoism and new Buddhist elements dominated culture, including science, in China after the demise of the Han Dynasty and through subsequent Chinese dynasties, such as the Sui, T'ang, Song, Yuan, and Ming dynasties. Following the decline of the Han Dynasty, there was a period of disorder and conflict (222–581); during this time Buddhism made its way by means of the Silk Road from India through Tibet into China. Chinese Buddhist thinkers adapted their teachings to Taoism and Confucianism. For example, the *Sutra of Forty-Two Sayings* was a Buddhist text that had a huge impact on Buddhist influence in China. Buddhist adaptation to Taoism is revealed in this passage: "A monk asked the Buddha: 'What is good, and what is great?' The Buddha answered: 'Good is to practise the Way and to follow the truth. Great is the heart that is in accord with the Way.'" Buddhism adapted to traditional Confucian doctrines as in this passage: "The Buddha said: 'You should think of the four elements [earth, air, fire, water] of which the body is composed. Each of them has its own name, and there is no such thing there known as ego. As there is really no ego, it is like unto a mirage'" (trans. Suzuki). The four, or five, basic elements dominated Chinese thinking. They were part of the mysteries of existence that Chinese scientists tried to discern. The mathematician Zhu Shijie (1260–1320), for example, in the fourteenth century wrote an algebraic text, *Precious Mirror of the Four Elements* [*Unknowns*] in which he tried to uncover the nature of unknowns through mathematical problems.

During the T'ang Dynasty (618–906) and Song Dynasty (960–1279) the Confucian scholar bureaucracy, followed by a reinvigorated Confucianism, continued to dominate Chinese thought. During the neo-Confucianism of the Song, an intuitive element was added to Confucianism, which previously had been focused more on reason, ritual, and behavior. Neo-Confucianists still focused on personal behavior, on ritual, on reason, on morality, but they also emphasized a personal conflict going on between the ideal expression of self, the transcendent essence of self, *Li*, and the bodily expression of time, *Ch'i*. Li became the intuitive source of good, Ch'i the outside influences that could corrupt a person. The key is to cultivate Li by means of *Jen*, which one does by focusing on the Confucian classics, the teachings of

Confucius (551–479), seeking enlightenment. Such were the teachings of the Confucian scholar, Zhu Xi (Chu Hsi, 1130–1200). During the T'ang Dynasty, the Buddhist engineer and astronomer Yi Xing (c. 683–727) developed sophisticated astronomical instruments. The polymath Su Song (1020–1101), during the eleventh century, also worked on astronomical instruments such as the astronomical water clock. Outside of China, in Korea for example, Confucianism also dominated learning and the state. One of the great medieval Korean Confucian scholars was Kim Busik (1075–1151).

SHEN KUO (1031–1095)

A representative scientist of the Song Dynasty was Shen Kuo, a polymath who studied many branches of sciences but not from the standpoint of the modern scientist, that is, seeing science as based on particular disciplines that are secular and esoteric. Shen Kuo, like most scientists worldwide before the modern age, believed that science was the general pursuit of knowledge. He studied physics, history, biology, geology, meteorology, medicine, cartography, and astronomy. He is noteworthy for being the first person to describe the properties of the compass and its orientation toward true north. The director of the Astronomy Bureau under the emperor Shenzong (r. 1067–1085), Shen Kuo, per Chinese tradition, devised a new lunar calendar to conform to the new reign. He was interested in land reclamation projects. He wrote at length on medicine. Like many medieval scientists, Shen Kuo was willing to entertain the intuitive and mystical in science. He could believe in prodigies, superstition, divination, and magic as well as more logical and rational pursuits. Divination could provide a sense of the truth as well as scientific study. "Access to one's personal future," he thought, in the words of Nathan Sivin,

> whether by visionary foresight or by divination, is a perfectly natural phenomenon. It merges into the moral faculties, whose choices condition the future. It merges at the other end into the rational comprehension of the natural order as it is reflected in any authentic experience. Shen did not confuse introspection and observation, nor did he draw a clear line between them. Nor did he need to compare the importance of these two ways of knowing. What finally united the sciences, in other words, was the universal system of knowledge—uniting intellection, imagination, and intuition—of which they constituted only a part. (Sivin, p. 51)

The impact of the old Confucian and Taoist idea of yin-yang continued to influence medieval Chinese thinkers. Liu Chou in the sixth century CE described the wave theory of yin-yang: "When the Yang has reached its highest point the Yin begins to rise, and when the Yin has reached its highest point the Yang begins to rise. Just as when the sun has reached its greatest altitude it begins to decline, and when the moon has waxed to its full it begins to wane. This is the changeless Tao of Heaven. When forces have reached their climax, they begin to weaken . . . , and when natural things have become fully agglomerated they begin to disperse" (quoted in Needham, vol. 4, p. 7).

Medieval Chinese astronomers were beholden to the ideas of the impact of the heavens on human affairs. Even as late as the twelfth century, Chhen Chhang Fang explained the impact of the yin on the phases as well as the light of the moon. The influence of the tao on nature, on time, on self, had ongoing relevance for medieval Chinese thinkers. The seventeenth-century Jesuit Athanasius Kircher (1602–1680), in *China Illustrata*, observed that the Chinese were fascinated by time and chronologies and constantly sought to perfect their lunar calendars, which were extremely important to the lives of people, emperors as well as commoners. "The Chinese are the most superstitious people in the world," he wrote, "and so they use astronomy to decide what to do, and what not to do. They have no knowledge about the movement of the planets, and right up until our [Jesuit] fathers' arrival [in China] they believed that all the stars of whatever sphere are an equal distance from the earth" (trans. Van Tuyl). Kircher described the continuing impact of Confucianism, Buddhism, and Taoism on the Chinese even at the close of the medieval period.

Kircher was writing at the time of the Manchu Dynasty, after several centuries in which dramatic changes had occurred in China due to the Mongol invasions. With the Mongol invasions of the thirteenth century, control of China shifted, and the Yuan Dynasty was founded by the grandson of Genghis Khan, Kublai Khan, in 1279. After a century of Mongol rule, the Ming Dynasty (1368–1644) came to power in China, still heavily influenced by neo-Confucianism. Beijing became the capital of China in 1420. Increasingly in the thirteenth, fourteenth, and fifteenth into the sixteenth centuries, European traders were trying to establish greater trading links with the Chinese. European influence increased in China by means of traders and missionaries.

One of the most famous of these European travelers was Marco Polo, who in *Travels* provided a general overview of Chinese culture and society based on personal observation as well as many anecdotal stories. He noted the importance of astrology in the lives of people in medieval Asia. In Tangut, between Tibet and Cathay, he wrote:

> In respect to the dead, likewise, these idolaters have particular ceremonies. Upon the decease of a person of rank, whose body it is intended to burn, the relations call together the astrologers, and make them acquainted with the year, the day, and the hour in which he was born; whereupon these proceed to examine the horoscope, and having ascertained the constellation or sign, and the planet therein presiding, declare the day on which the funeral ceremony shall take place. If it should happen that the same planet be not then in the ascendant, they order the body to be kept a week or more, and sometimes even for the space of six months, before they allow the ceremony to be performed. In the hope of a propitious aspect, and dreading the effects of a contrary influence, the relations do not presume to burn the corpse until the astrologers have fixed the proper time. It being necessary on this account that, in many cases, the body should remain long in the house, in order to guard against the consequences of putrefaction, they prepare a coffin made of boards a palm in thickness, well fitted together and painted, in which they deposit the corpse, and along with it a quantity of sweet-scented gums, camphor, and other drugs; the joints or seams they smear with a mixture of pitch and lime, and the whole is then covered with silk. During this period the table is spread every day with bread, wine, and other provisions, which remain so long as is necessary for a convenient meal, as well as for the spirit of the deceased, which they suppose to be present on the occasion, to satisfy itself with the fumes of the victuals. Sometimes the astrologers

> signify to the relations that the body must not be conveyed from the house through the principal door, in consequence of their having discovered from the aspect of the heavens, or otherwise, that such a course would be unlucky, and it must therefore be taken out from a different side of the house. In some instances, indeed, they oblige them to break through the wall that happens to stand opposite to the propitious and beneficent planet, and to convey the corpse through that aperture; persuading them that if they should refuse to do so, the spirit of the defunct would be incensed against the family and cause them some injury. Accordingly, when any misfortune befalls a house, or any person belonging to it meets with an accident or loss, or with an untimely death, the astrologers do not fail to attribute the event to a funeral not having taken place during the ascendency of the planet under which the deceased relative was born, but, on the contrary, when it was exposed to a malign influence, or to its not having been conducted through the proper door. (Trans. Marsden)

Although the people of China knew solar calendars, they relied heavily upon lunar calendars, in particular because of their fascination with astrology. Marco Polo stated,

> It should be observed that the Tartars [Chinese] compute their time by a cycle of twelve years; to the first of which they give the name of the lion; to the second year, that of the ox; to the third, the dragon; to the fourth, the dog; and so of the rest, until the whole of the twelve have elapsed. When a person, therefore, is asked in what year he was born, he replies, In the course of the year of the lion, upon such a day, at such an hour and minute; all of which has been carefully noted by his parents in a book. Upon the completion of the twelve years of the cycle, they return to the first, and continually repeat the same series. (Trans. Marsden)

Parents of children were particularly concerned with knowing precisely not only the date but the hour and minute of a child's birth in order to cast the child's horoscope. Before any big decision, such as traveling or marriage, an astrologer must be consulted to discover if it was propitious to do so or not. Not surprisingly, Polo found astrologers everywhere in China. In the city of Khan-balik (Peking, Beijing), Marco Polo described the number of forecasters that Kublai Khan (r. 1260–1294) relied on:

> There are in the city of Kanbalu, amongst Christians, Saracens, and Cathaians, about five thousand astrologers and prognosticators, for whose food and clothing the grand khan provides in the same manner as he does for the poor families above mentioned, and who are in the constant exercise of their art. They have their astrolabes, upon which are described the planetary signs, the hours (at which they pass the meridian), and their several aspects for the whole year. The astrologers (or almanac-makers) of each distinct sect annually proceed to the examination of their respective tables, in order to ascertain from thence the course of the heavenly bodies, and their relative positions for every lunation. They discover therein what the state of the weather shall be, from the paths and configurations of the planets in the different signs, and thence foretell the peculiar phenomena of each month: that in such a month, for instance, there shall be thunder and storms; in such another, earthquakes; in another, strokes of lightning and violent rains; in another, diseases, mortality, wars, discords, conspiracies. As they find the matter in their astrolabes, so they declare it will come to pass; adding, however, that God, according to his good pleasure, may do more or less than they have set down. They write their predictions for the year upon certain small squares, which are called takuini, and these

> they sell, for a groat apiece, to all persons who are desirous of peeping into futurity. Those whose predictions are found to be the more generally correct are esteemed the most perfect masters of their art, and are consequently the most honoured. When any person forms the design of executing some great work, of performing a distant journey in the way of commerce, or of commencing any other undertaking, and is desirous of knowing what success may be likely to attend it, he has recourse to one of these astrologers, and, informing him that he is about to proceed on such an expedition, inquires in what disposition the heavens appear to be at the time. The latter thereupon tells him, that before he can answer, it is necessary he should be informed of the year, the month, and the hour in which he was born; and that, having learned these particulars, he will then proceed to ascertain in what respects the constellation that was in the ascendant at his nativity corresponds with the aspect of the celestial bodies at the time of making the inquiry. Upon this comparison he grounds his prediction of the favourable or unfavourable termination of the adventure. (Trans. Marsden)

Clearly over the space of several millennia the Chinese believed in the importance of yin-yang, the cosmic forces, in the lives of the Chinese and the events in their many imperial dynasties.

CHINESE MEDICINE

The Silk Road was probably the means by which Western and Eastern medical practices were exchanged. Early Tibetans learned of the Galenic approach to medicine and perhaps made use of it in treating the sick. For example, records from Dunhuang, a city on the Silk Road in western China near the Gobi Desert, reveal that early medical practice focused on healing injured soldiers. It was on the Silk Road that Buddhism made its slow relentless entrance into China. The impact of Buddhist medicine on China included "Indian-inspired healing deities, rituals, occult practices, and hagiography," which "all proved to be enormously popular and permanent contributions to Chinese culture" (Salguero, p. 2). The Four Tantras of Tibetan medicine were Buddhist in origin, reputedly from the Buddha himself, focusing on a diagnostic, therapeutic, naturalistic approach to medicine where physicians adhered to certain ethics and practices. Buddhist healing practices and knowledge of one's physical and spiritual self were appealing ideas in medieval China.

During the period of the Northern and Southern dynasties in the fifth and sixth centuries, the first government hospital, devoted to charity, was established. During the Sui Dynasty (581–618), when Buddhism was so prevalent in China, emperors established a directorate of medical administration that included professors of medicine, massage, and exorcism. The Sui Dynasty established hospitals to treat epidemics. During the succeeding T'ang Dynasty, the influence of Buddhism began to decline, though the government continued to exercise control over hospitals and medical practices. During the Song Dynasty, under Confucian influence, medicine was highly systematized; the Confucians established infirmaries and hospitals, especially for the poor—even a leprosaria for the treatment of lepers. The eleventh-century poet and pharmacologist Su Shi (1037–1101) was well known for helping to establish a government hospital. Under the influence of the Confucian bureaucracy, where education in subjects necessary to the state was emphasized, there were medical professorships funded by the government. Often scientists and physicians were not necessarily from the aristocracy but from the gentry, those who had been educated for the civil service. Medical knowledge was therefore part of education for the bureaucracy.

Because the Chinese saw a connection between the macrocosm and microcosm in life, between the cosmos and humans, influenced by the five elements, there was some experimentation and practice of alchemy. An example of a medieval alchemist who was an "iatrochemist," using chemicals in medical treatment, was Sun Simiao (581–682) in the seventh century. During the course of the Middle Ages, Chinese alchemists became iatrochemists, and actually created some medicinal therapies for some illnesses. Sun Simiao was an expert on materia medica and created a *materia dietetica* for healthy eating. He wrote volumes on medical practice from the perspective of a Taoist who accepted Buddhist and Confucian principles. His most famous book was *Prescriptions for Emergencies Worth a Thousand Gold.* His text on alchemy was *Essentials of the Elixir Manuals for Oral Transmission.*

Marco Polo made note in *Travels* of some of the sources of traditional Chinese medicine, including the gall taken from the intestines of a crocodile, "which is most highly valued in medicine. In case of the bite of a mad dog, a pennyweight of it, dissolved in wine, is administered. It is also useful in accelerating delivery, when the labour pains of women have come on. A small quantity of it being applied to carbuncles, pustules, and other eruptions on the body, they are presently dispersed; and it is efficacious in many other complaints" (trans. Marsden).

SOUTHEASTERN ASIAN SCIENCE

The ancient Funan culture of Southeast Asia transitioned to the medieval Chenla empire (550–802) and culture from the sixth century to the ninth century. This was a complex culture influenced by the Chinese and Asian Indians, with sophisticated political structures, trade, and rice agriculture. By the tenth century, the Khmer people of Cambodia built Hindu-like pyramids. The most notable was Angkor Wat, a ceremonial center with pyramids completed during the twelfth century. Scholars debate whether or not the idea of such pyramids was indigenous or borrowed. Hindu and Buddhist India and Buddhism from China influenced the Khmer culture.

Angkor Wat was an astronomical/religious center. The measurements of its structures were precise and, like other monoliths throughout the world, had astronomical significance. Astronomer-priests used the structures for lunar and solar observations. Like the Chinese, secular rulers relied heavily on the movements of the heavens, astrology, the calendar, the seasons, solstices, and eclipses. Also like the Chinese, the Khmer people used lunar and solar calendars, and, as other peoples have throughout the world, were concerned with dating and celebrating the spring equinox.

Megalithic structures are found on islands of Oceania, and they represent the movement of peoples bringing sophisticated cultural ideas, political and social organization, and engineering skills from other places. One of the Polynesian islands of Tonga had a sufficiently sophisticated society to construct a trilithon, a monolithic structure of two vertical slabs supporting a top horizontal slab. This structure, Ha'amonga 'a Maui, was constructed in the thirteenth century. The megalithic structures of Nan Madol were constructed on one of the islands of Micronesia in the twelfth century. The statues on Easter Island are a visible and beautiful reminder of the skill of medieval Polynesian peoples.

FURTHER READING

Subhuti Dharmananda, *Sun Simiao: Author of the Earliest Chinese Encyclopedia for Clinical Practice*, http://www.itmonline.org/arts/sunsimiao.htm.

John K. Fairbank, Edwin O. Reischauer, and Albert M. Craig, *East Asia: Tradition and Transformation* (Boston: Houghton Mifflin, 1978).

Elizabeth Hsu, *Innovation in Chinese Medicine* (Cambridge: Cambridge University Press, 2001).

Athanasius Kircher, *China Illustrata*, trans. Charles D. Van Tuyl (Muskogee, OK: Indian University Press, 1986).

Joseph Needham, *Science and Civilisation in China: Volume 4: Physics and Physical Technology: Part 3: Civil Engineering and Nautics* (Cambridge: Cambridge University Press, 1971).

Joseph Needham, *Science and Civilisation in China: Volume 6, Biology and Biological Technology, Part VI: Medicine* (Cambridge: Cambridge University Press, 2000).

Marco Polo, *The Travels*, trans. R. E. Latham (Harmondsworth, England: Penguin Books, 1958).

Marco Polo, *The Travels of Marco Polo, the Venetian, Revised from Marsden's Translation and Edited by Manuel Komroff* (New York: Boni & Liveright, 1926).

Jennifer C. Ross and Sharon R. Steadman, *Ancient Complex Societies* (New York: Routledge, 2017).

C. Pierce Salguero, *Translating Buddhist Medicine in Medieval China* (Philadelphia: University of Pennsylvania Press, 2014).

Robert M. Schoch, *Voyages of the Pyramid Builders: The True Origins of the Pyramids from Lost Egypt to Ancient America* (New York: Penguin Books, 2004).

Soyen Shaku, *Zen for Americans*, trans. Daisetz Teitaro Suzuki (New York: Dorset Press, 1987).

Nathan Sivin, "Why the Scientific Revolution Did Not Take Place in China—Or Didn't It?," *Chinese Science* 5 (1982): 45–66.

Sources of Tibetan Tradition, ed. Kurtis R. Schaeffer, Matthew T. Kapstein, and Gray Tuttle (New York: Columbia University Press, 2013).

Robert Stencel, Fred Gifford, and Eleanor Moron, "Astronomy and Cosmology at Angkor Wat," *Science* 193 (1976): 181–87, https://science.sciencemag.org/content/193/4250/281.

Arthur F. Wright, *Buddhism in Chinese History* (Stanford, CA: Stanford University Press, 1950).

31

Science and Philosophy during the European Middle Ages (500–1300 CE)

Logic and piety, reason and faith, describe thought and religion of the European Middle Ages. The various kingdoms that emerged in western Europe from the declining Roman Empire embraced Christianity as well as the Greco-Roman legacy of the liberal arts. Christian faith rarely gave way to secular reason, observation, and logic, but the European Middle Ages were hardly a time when only unthinking superstition reigned. There were significant thinkers during this period from about 400 to 1300 CE, such as Augustine, Boethius, Alcuin, Einhard, Anselm, Roger Bacon, William of Occam, Thomas Aquinas, John Scotus Eriugena, and Albertus Magnus.

The problem of when and how—and even if—the Roman Empire declined and fell is exacerbated by the varied dimensions of cultural change in the fourth and fifth centuries CE. In both the Western Roman Empire, and subsequent European kingdoms, and the Eastern Roman Empire, and the subsequent Byzantine Empire, polytheistic, superstitious, pantheistic pagans who watched constantly for divine signs to indicate the course of the future became monotheistic, similarly superstitious Christians who conceived of a variety of supernatural forces of both good and evil waging war over the Christian soul. There were more similarities than differences between paganism and Christianity, so it was common to find Christians who, like the philosopher Boethius, could not quite rid themselves of their pagan proclivities, and pagans who, like the emperor Constantine, were sufficiently attracted to Christianity to approach full conversion.

The Roman Empire began a precipitous decline in the third century, when there were twenty-six emperors during a fifty-year period. Disorder, chaos, civil war, agricultural decline, malnutrition, disease, and an interrupted trade and food supply were temporarily halted by the twenty-year reign of the emperor Diocletian (r. 284–205). Diocletian established a firm autocratic government. He divided the empire into two halves and established a clear line of succession. Henceforth, there was an Eastern Roman Empire with an emperor (Augustus) and a named successor

(Caesar). In the East, this was Diocletian and Galerius. In the Western Roman Empire, there was likewise an emperor (Augustus) and a named successor (Caesar): Maximian and Constantius. Diocletian abdicated in 305; he forced Maximian to abdicate as well. Maximian's son Maxentius thought he should succeed his father, but he was not chosen. Constantius I became Augustus; his son Constantine thought he should be Caesar, but he was not selected. In 306, Constantius I died, and his troops declared for his son Constantine (r. 306–337). This resulted in another civil war, which would not be halted until 325 and then only temporarily.

During this time, Christianity had grown in the Roman Empire. The religion was well organized; Christian cities had local officials called bishops; increasingly these bishops were starting to look to the bishop of Rome as the ultimate leader of the church. Pagans believed that Christianity was threatening the order of Rome, threatening the old ways, threatening the power of the emperor. As a result, Diocletian persecuted Christians; those who refused to renounce their faith were witnesses (martyrs).

But then, in a dramatic event in 312, Constantine experienced conversion to Christianity. In 313 he issued the Edict of Milan, which granted toleration for Christians. In 325, by this time sole emperor, Constantine founded the New Rome, Constantinople, the new capital of a Christian Roman Empire. In time Constantinople would become the capital of the Byzantine Empire (old Eastern Roman Empire). Also in 325, at the Council of Nicaea, Constantine solidified his power as head of church and state even as the bishop of Rome, the pope, was growing in power. Constantine's autocracy involved a growing bureaucracy, growing support of the church, increasing removal of the emperor from the people, and the growing power of secret police and eunuchs. During and after the reign of Constantine, the Roman Empire experienced a slow movement away from pagan and toward Christian influences in government, society, and culture. The West and East grew further apart; in the West, Germanic influences crept into government; the empire became more decentralized; Rome was no longer the center politically, but religiously it was the center of Christendom. In the East, increasing centralization politically and religiously around Christianity centered at Constantinople. The Western Roman Empire collapsed during the fifth century. In the several centuries after the invasions of the northern Germanic tribes and the sack of Rome by the Goths (in 410 CE), people pursuing happiness, peace, and order arranged themselves into various communal institutions. In Italy, France, and England, for example, primitive kingdoms emerged; warlords and great landowners provided protection for farmers, peasants who soon became entangled in the unbreakable cords of feudalism. These kings, such as the Frankish king Clovis, converted to Christianity, ensuring that Christianity would henceforth continue to dominate western Europe. Those with a religious bent retreated from the dangers of society to form isolated communities of ascetics. Some particularly zealous believers, such as St. Anthony, who lived in the third century, and St. Jerome (340–420) fled to the desert to live as hermits. Others, such as the anchorites, lived austere existences in the vain attempt to conquer the flesh so as to elevate the soul. Benedict of Nursia in the sixth century practiced such asceticism unsuccessfully before finding a balance between isolation and civilization in the Benedictine monastery.

These varied recluses were often the few thoughtful scholars who continued to think about the past and anticipate the future.

The European Dark Ages were dark from the perspective of the standards of civilization, in particular those of the cultivated and progressive cities of Renaissance Italy. Life in the Middle Ages was short and brutish; few could read the few books that survived war and conquest; great ideas vanished, as did schools; time, dates, age, and years were largely uncertain; ordered political structures were rare; the economy was agrarian and based on barter; towns were few, but cases of hunger and famine were not; death was frequent and familiar. Literacy all but vanished. Art and sculpture were primitive, anachronistic, and static. So, too, was thought—the philosophers and theologians of the Middle Ages tried to merge faith in the Scriptures with loyalty to ancient pagan sources such as Aristotle (384–322) and Virgil (70–19). They developed an intricate, esoteric approach to God and the universe that relied heavily on mind-numbing logic and ontological as well as nominalist approaches to knowledge.

The Christian Church at the end of the Roman Empire (c. 500 CE) was dominated by Roman Catholicism centered at Rome and Greek Orthodoxy centered at Constantinople. During subsequent centuries in western Europe, as small kingdoms replaced the crumbling Roman Empire, the Catholic Church grew to be the greatest institution throughout Europe. The church hierarchy, with the pope at the pinnacle and power disseminating down to cardinals, archbishops, bishops, priests, and deacons, controlled education, learning, literature, philosophy, and science. The great theologians of the Middle Ages, such as St. Augustine, St. Anselm, and St. Thomas Aquinas, emphasized the greatness of God, who is omniscient and omnipotent compared to humans, who are weak and helpless. Augustine claimed that reality was divided into the spiritual and corporeal realms, the City of God and the City of Man, the former perfect and eternal, the latter imperfect and doomed to failure. Christ saves the City of Man by means of his sacrifice. As the Middle Ages progressed, Catholic thought became more rational and inflexible. Scholastics such as Thomas Aquinas argued that human reason was sufficient to know God and his ways. The church grew confident that it was the body of Christ; church liturgy and ritual were the only ways. The Bible, translated into Latin by St. Jerome, was the infallible word of God understood best by the Catholic clergyman. Even though most medieval Christians could not understand Latin, the church continued to use Latin as the basis for communication, the Mass, and the Bible. The individual human could only approach God through the church.

Augustine (354–430) provides one of the first models for what we call medieval philosophy. In an essay written in 395, he sought to deduce the existence of God by means of his own understanding of *knowledge*. Augustine begins with three fundamental assumptions about himself that are proven through experience: that he exists; that he is alive, that he has understanding. Humans are separate from animals in possessing the capacity for understanding, that is, *reason*. Since humans display reason in their temporal lives, they can recognize the existence of that which transcends reason. Anticipating St. Anselm, Augustine argued that "if you find nothing above our reason save what is eternal and unchangeable, would you hesitate to call this God? Bodies, you know, are subject to change. . . . Reason itself proves mutable. . . . If reason, I say, catches sight of something eternal and immutable, ought it not at the same time recognize its own inferiority and proclaim this something to be its God?" Human reason is not eternal and immutable, but it does allow humans to know what is. Reason informs humans, for example, of the immutability of number. *One* is a fundamental reality, a singularity, not dependent upon our "bodily senses," standing "sure and unshakable, in common view of all who use their reason." Having established that because humans possess reason, they can conceive of that which transcends reason, because humans conceive of number, that the number *one* must represent the ultimate transcendence, Augustine went on to assume that since we seek wisdom and know humans who appear wise, something beyond our experience called *wisdom* necessarily exists. Humans have a basic awareness of happiness and wisdom, simply because humans lack them yet conceive or imagine them to exist. Like number, wisdom, the supreme good, is the same for all humans. Humans know that when they achieve the highest good, happiness will follow; but the highest good is disagreed upon. If wisdom and the good are perceived differently by different humans, then a number such as *one* is as well, which is clearly untenable, for one is a fundamental reality. There is a truth that all humans perceive differently. There are certain common assumptions that all humans share; such assumptions require wisdom to understand them; hence, all humans share in this wisdom. "Hence you would not deny that some unchanging truth exists which contains all of these things that are unchangeably true and that it cannot be called exclusively mine or yours or any man's." (trans. Wolter). The number 10 is a truth that is the same for all humans who conceive it, and we all share it, yet it is not part of us, nor is it God. Can we accept the unified consistency of number and not the unified consistency of life, love, and knowledge? If 5 + 5 = 10 is true, then an understanding of this truth reveals a part of the Truth itself. In short, we first know manifold truths dependent upon our own independent reason and reflection; realization of these temporal, limited truths, makes us realize that something similar yet transcendent, Truth, exists—this *Truth* is God. The experience of isolated truths, moments of wisdom, experiences perceived as good, lead us to a timeless truth outside of, beyond human experience, which is God.

Augustine's philosophy was useful in understanding how human experience relates to universal experience, that is, God. Providence is the relationship between the eternal and the temporal. If the eternal is knowledge, and the temporal consists of individual awareness or knowledge in time, then providence is based on the relationship between the two. Here the individual acts in time according to knowledge and awareness, and according to what is considered the absolute. How a person perceives this absolute is going to in part or fully dictate the person's action. It is in this sense that knowledge has a role in human behavior and action. Humans act according to knowledge, or they behave in willful ignorance. When they act according to knowledge, they act in conformance to knowledge. For Christians, this means that God wills all things, but humans nevertheless have the opportunity to act within God's will and in a way alter events consistent with God's will. Hence does Augustine's philosophy explain human behavior.

Augustine, like other medieval thinkers, was heavily influenced by Aristotelian syllogisms, such as

If I experience knowledge in time,
And knowledge in time relates to transcendent Knowledge,
Then my knowledge is related to transcendent Knowledge.

In *Confessions*, Augustine linked his personal life experiences to such logical syllogisms. Augustine proved the existence of Knowledge, or God, by his own being, the knowledge of his own experiences. How can he doubt knowledge of his own being? How can he doubt knowledge of universals that lie outside his own being? How can he doubt that knowing is shared by others, that his and their knowing is the same? How can he doubt this knowing, this knowledge, that is, knowledge in general? Knowledge is truth, truth is God, knowledge is God. By examining the experiences of himself and others, Augustine proved, at least for himself and many others, the existence of God.

BOETHIUS (480–524)

Another light in the darkness was Boethius, a transitional figure between ancient and medieval philosophy and science. He lived during the sixth century, serving under the Gothic king Theodoric (r. 493–526), who ruled the Italian remnant of the Western Roman Empire. Boethius was gifted in both Platonic and Aristotelian thought, referring in *The Consolation of Philosophy* to Aristotle as "my philosopher." Lady Philosophy, with whom he carried on an imaginary conversation in the *Consolation,* declared that Aristotle was her "disciple" and discussed many others as well, such as the Stoics Zeno (333–262) and Cicero (106–43), the Pythagoreans, the Platonists, the Epicureans, and the Eleatics. Boethius wrote Latin commentaries on the two great philosophers Plato and Aristotle, bringing to the Latin medieval West knowledge of classical metaphysics and Aristotelian logic. Boethius revealed his reliance upon ancient philosophy in one of the *Theological Tractates*, "How Substances May be Good." In this brief essay, Boethius assumed, following Plato (427–347), that there is an ultimate Good—goodness in essence and in fact. The Good, as the act to potential, the Creator of substances, imparts its goodness to these substances: "They are good in essence because one who was good willed them to be good" (trans. Wippel). Boethius aspired to translate the whole of the Platonic and Aristotelian corpus into Latin, though imprisonment and execution prevented him from doing so. But he did translate Aristotle's *Organon* into Latin, as well as *Isagoge*, by the Neoplatonist Porphyry (234–305). He was particularly interested in Aristotelian physics, ethics, and astronomy. He wrote on mathematics and music. Boethius was also a student, commentator, and translator (into Latin) of Ptolemy (100–170).

The *Consolation of Philosophy* is a book juxtaposing verse and prose, the prose being an imaginary conversation between Boethius and Lady Philosophy. Boethius, falsely imprisoned and condemned to die, wondering why such travail occurs to humans, is confronted by philosophy—that is, all of the lessons he has learned by reading the Greek and Roman philosophers—and reminded that philosophy is the means by which a person prepares for each moment, even death. The *Consolation* evokes the worldview of an ancient polymath, a thinker who has studied all of the elements of the liberal arts, the trivium and quadrivium, natural history, philosophy, mathematics, astronomy, logic, and the different schools of the thought of the ancient world: Ionic, Pythagorean, Platonic, Neoplatonic, Aristotelian, Peripatetic, Stoic, Epicurean, Skeptic, Christian. The expression of his scientific and philosophic worldview is revealed throughout, as in this verse:

The sun, refulgent source of day,
You trac'd o'er all his radiant way;
The moon that shines with borrowed light,
And cheers with radiance mild the night,
The silver moon's mysterious round
Was by your magic numbers bound;
The planets too that wand'ring go,
And seem no settled course to know,
Their periods, various and perplex'd,
Were, by your art victorious, fix'd;
Your tow'ring genius could resolve
What makes the heaven's vast frame revolve,
Whilst all the lights that gild the skies,
In order, daily set and rise;
You too could tell, where nature forms
Her mighty magazines of storms,
Which with impetuous fury roll,
And shake the earth from pole to pole;
Why spring awakes the genial hours,
And decks th' enamell'd field with flow'rs,
You knew;—and why kind Autumn's hand
Diffuses plenty o'er the land:
Thro' all her mazes your pursued
Coy Nature, and her secrets view'd. (Trans. Ridpath)

In the *Consolation of Philosophy*, Boethius dealt with some of the most perplexing questions confronting the ancient and medieval worlds, such as epistemology, human free will, God's foreknowledge, and the ultimate causes of all events throughout time. Like most ancient thinkers, Boethius believed that there was an ultimate supernatural cause for all things, which follow an inherent law; nothing is random. He therefore agreed with the Platonic and Aristotelian conception of an ultimate being or Logos. He was especially influenced by the thinking of St. Augustine in his understanding of divine foreknowledge and free will, arguing that God was the creator of time, beyond time, and therefore existed in the singular moment, able to see all events—past, present, and future—simultaneously. By seeing all events, God knows all events, even if in each singular moment of time creatures are acting freely according to their own wills. There is momentary cause and effect based on momentary instinct and reason within the scheme of God's knowledge of cause and effect, the product of his intelligence.

ISIDORE OF SEVILLE (560–636)

Other philosophic and scientific lights shining in the darkness of medieval western Europe included the work of Isidore of Seville, who became bishop of Seville in

600. Spain at this time was dominated by Orthodox Christianity. Roman institutions, laws, and some learning existed in a world turned upside down by the invasions of the Visigoths during previous centuries. Isidore was not an original thinker but, rather, a compiler of works from the past. He believed that the learning of the ancient Romans, and before them the Greeks, provided a path to knowledge alongside Christianity. Isidore sought to record as much learning as he could from the ancient past. He was a polymath, commentator, and encyclopediast. He wrote a variety of works, most of which were brought together in his massive *Etymologies*, in which he tried to collect the learning and wisdom of the ancients regarding a variety of objects of inquiry, ranging from astronomy and astrology to zoology and botany to geography and law. He wrote extensively on medicine, mathematics, and rhetoric. A compiler and commentator, Isidore helped to preserve the legacy of Cicero, Pliny, and Aristotle. His *Etymologies* owed much to earlier grammarians and compilers, such as Verrius Flaccus (c. 55 BCE–20 CE) and Sextus Pompeius Festus (fl. c. 200 CE). He was influenced by earlier encyclopediasts, notably Varro (116–27), Pliny the Elder (23–79), the biographer Suetonius (69–130), and the fourth-century grammarian Nonius Marcellus.

Etymology, "the derivation of words," was extremely important, Isidore believed, because "a knowledge of etymology is often necessary in interpretation, for, when you see whence a name has come, you grasp its force more quickly. For every consideration of a thing is clearer when its etymology is known." Isidore's *Etymologies* is one of the first full expositions of the trivium and quadrivium that so dominated European medieval learning. The first book, *Grammar*, is a detailed analysis of the parts of written and spoken language as well as a study of the Greek, Hebrew, and Latin alphabets. Chapter 2 of *Grammar*, "On the Seven Liberal Arts," explains: "The disciplines belonging to the liberal arts are seven. First, grammar, that is, practical knowledge of speech. Second, rhetoric, which is considered especially necessary in civil causes because of the brilliancy and copiousness of its eloquence. Third, dialectic, called also logic, which separates truth from falsehood by the subtlest distinctions. . . . Fourth, arithmetic, which includes the significance and the divisions of numbers. Fifth, music, which consists of poems and songs. . . . Sixth, geometry, which embraces measurements and dimensions. Seventh, astronomy, which contains the law of the stars" (trans. Brehaut).

Isidore's treatment of *Rhetoric* relied heavily on the works of earlier rhetoricians, and especially the writings of the sixth-century polymath Cassiodorus, who during his long life (485–585) produced works on chronology, the liberal arts, history, rhetoric, law, and institutions. Isidore's *Rhetoric* went into detail into the nature of the law and how cases were argued in court: analysis and causation. "There are three kinds of causes," he wrote, "deliberative, epideictic, judicial. The deliberative kind is that in which there is a discussion as to what ought or ought not to be done in regard to any of the practical affairs of life. The epideictic, in which a character is shown to be praiseworthy or reprehensible. . . The judicial, in which opinion as to reward or punishment with reference to an act of an individual is given" (trans. Brehaut).

Isidore's *Dialectic*, that is, a work on logic, was again heavily influenced by Cassiodorus and another late Roman compiler and liberal arts advocate, Martianus

Capella (c. 365–440), as well as the earlier Latin writer, Varro. Isidore analyzed the *Isagoge* (introduction) of Porphyry and Aristotle's *Categories* and *Interpretation*, as well as discussions of philosophy and syllogisms. "Dialectic," he wrote, "is the discipline elaborated with a view of ascertaining the causes of things." Of philosophy, he wrote: "Philosophy is the knowledge of things human and divine, united with a zeal for right living. It seems to consist of two things, knowledge and opinion." Of the latter, Isidore provided the example of speculations on the nature of the universe: "Whether the sun is [only] as large as it seems or greater than all the earth; likewise whether the moon is a sphere or concave; and whether the stars adhere to the heavens or pass in free course through the air; of what size the heaven itself is and of what material it is composed; whether it is quiet and motionless or revolves with incredible speed; how great is the thickness of the earth, or on what foundations it continues poised and supported." Philosophy as devised by the Greeks, Isidore argued, involves natural philosophy, ethics, and dialectics. Isidore believed the first Greek philosopher was Thales (626–545), whose work Plato took and "divided into four separate parts, namely, into arithmetic, geometry, music, and astronomy"—the quadrivium. Combining religious contemplation of the sixth-century Christian with the pagan philosophy of the Greeks, Isidore wrote that "natural philosophy is the name given when the nature of each and every thing is discussed, since nothing arises contrary to nature in life, but each thing is assigned to those uses for which it was purposed by the Creator, unless perchance by God's will it is shown that some miracle appears" (trans. Brehaut).

Isidore's *Arithmetic*, focusing on the art and science of numbers, relied on the work of Cassiodorus and, before him, Boethius, who translated *Introduction to Arithmetic*, by Nicomachus (60–120), into Latin. "Mathematics," Isidore wrote, "is called in Latin *doctrinalis scientia*. It considers abstract quantity. For that is abstract quantity which we treat by reason alone, separating it by the intellect from the material or from other non-essentials, as for example, equal, unequal, or the like. And there are four sorts of mathematics, namely, arithmetic, geometry, music and astronomy." Again, "Arithmetic is the science of numbers. For the Greeks call number [*arithmos*]. The writers of secular literature have decided that it is first among the mathematical sciences since it needs no other science for its own existence." Further, "they say that Pythagoras was the first among the Greeks to write of the science of number, and that it was later described more fully by Nicomachus, whose work Apuleius first, and then Boethius, translated into Latin." Isidore's comments in *Arithmetic* revealed his interest in etymology: how words were used to express numbers. He was also fascinated by the religious significance of numbers, as revealed in Scripture (trans. Brehaut).

Isidore's *Geometry* is a borrowed account from contemporaries such as Cassiodorus and Boethius. It has little originality but repeats what the thinkers of the Later Roman Empire recalled from the grandeur of the Greek geometers. Understanding the word to be based on measurements of the world, Isidore conceived of the origins of geometry in ancient thinkers attempting to measure parts of the earth, the whole of the earth, and the earth in relation to the universe. He wrote, "For, having their attention aroused, students began to search into the spaces of the heavens, after measuring the earth; how far the moon was from the earth, the sun itself from the moon, and how great a measure extended to the summit of the sky; and thus they laid off in numbers of stades with probable reason the very distances of the sky and the circuit of the earth" (trans. Brehaut).

Music, the seventh liberal art discussed by Isidore, has little to do with music as a technical art form using the human voice or instruments. Rather, Isidore, following upon the Roman understanding of music that was acquired from the Greek philosophers, conceived of music in a mathematical sense, as had Pythagoras (570–490). Isidore revealed this conception in his comment, "Just as this proportion exists in the universe, being constituted by the revolving circles, so also in the microcosm—not to speak of the voice—it has such great power than man does not exist without harmony" (trans. Brehaut).

The final liberal art, astronomy, is, similar to music, a compilation of information from the ancient Greek astronomers only vaguely understood by Isidore, his contemporaries, and subsequent medieval thinkers. The geocentric universe holds sway, and the ideas of Aristarchus of Samos (310–230) of a heliocentric universe are forgotten. Medieval thinkers knew of the great works of the astronomy of the past, particularly that of Ptolemy, but could add little to the science on their own. Christian conceptions of the universe had by this time mixed with the Ptolemaic conceptions. Astrology, considered a form of demonology, was viewed with suspicion by Isidore. Isidore's vague fascination with the subject is revealed in this comment: "Whoever was the discoverer, it was the movement of the heavens and his rational faculty that stirred him, and in the light of the succession of seasons, the observed and established courses of the stars, and the regularity of the intervals, he considered carefully certain dimensions and numbers, and getting a definite and distinct idea of them he wove them into order and discovered astrology." His simplistic conception of the universe is illustrated by this comment:

> The sphere of the heavens is rounded and its center is the earth, equally shut in on every side. This sphere, they say, has neither beginning nor end, for the reason that being rounded like a circle it is not easily perceived where it begins or where it ends. . . . The philosophers have brought in the theory of seven heavens of the universe, that is, globes with planets moving harmoniously, and they assert that by their circles all things are bound together, and they think that these, being connected, and, as it were, fitted to one another, move backward and are borne with definite motions in contrary directions. (Trans. Brehaut)

Isidore's *Astronomy* also included geodesy, the study of the shape of the earth and its relations to the heavens, as revealed in this comment: "There are five zones in the heavens, according to the differences of which certain parts of the earth are inhabitable, because of their moderate temperature, and certain parts are uninhabitable because of extremes of heat and cold. And these are called zones or circles for the reason that they exist on the circumference of the sphere." The medieval conception of the heavenly bodies, as seen through the work of Isidore, appears primitive: "The sun, being made of fire, heats to a whiter glow because of the excessive speed of its circular motion. And its fire, philosophers declare, is fed with water, and it receives the virtue of light and heat from an element opposed to it. Whence we see that it is often wet and dewy." Some astronomical information he reported correctly, such as the comparative distance of the moon (nearer) and sun (farther) to the earth; that a lunar eclipse is when the earth's shadow comes between the sun and moon; that the solar eclipse is when the moon intervenes between earth and sun; that the light of stars depends in part on their distance from the earth; that the wandering stars, so-called, are the planets. Other information he reported incorrectly, such as that the stars are lit only by the light of the sun, and that the stars are fixed in position and are motionless (trans. Brehaut).

In addition to his work on the liberal arts, Isidore compiled information on medical science from the Greeks and Romans. He particularly relied on the work of Caelius Aurelianus, a Late Roman physician and writer, who translated into Latin *On Acute and Chronic Diseases*, by the Greek physician Soranus of Ephesus, who lived in the second century CE. Isidore understood medicine to be studied and practiced by several schools of thought: the Methodist, Empiricist, and Logical. The latter, represented by Hippocrates (460–377), dominated the study of medicine because of the focus on the four humors, "blood, bile, black bile, and phlegm. Just as there are four elements so also there are four humors, and each humor imitates its element: blood, air; bile, fire; black bile, earth; phlegm, water." Later in the *Etymologies*, Isidore wrote that the human "body is made up of the four elements. For earth is in the flesh; air in the breath; moisture in the blood; fire in the vital heat." The Latin word for disease, *morbus*, he explained, is based on *mortis*, death; hence, "between health and disease the mean is cure, and unless it harmonizes with the disease it does not lead to health." Isidore included the absurd alongside the wise, as in his comment: "Every cure is wrought either by contraries or by likes. By contraries, as cold by warm and dry by moist, just as in man pride cannot be cured except by humility. . . . By likes, as a round bandage is put on a round wound, or an oblong one on an oblong wound." Bitters, bitter-tasting medicine, are prescribed for illness "because the bitterness of disease is dispelled by its bitterness." Isidore argued that an able physician must be well versed in all seven liberal arts (trans. Brehaut).

One of Isidore's greatest accomplishments was his book *On Laws*, which provides an overview of Roman law based on some of the same sources as the Justinian Code. Although the work of Tribonian (500–547), under Justinian, occurred a century before Isidore, the Spaniard was unfamiliar with the Justinian Code. However, he was familiar with the Theodosian Code, compiled in 438, as well as the commentators on Roman law Gaius, Paul, and Ulpian (170–228). "All laws," Isidore wrote, "are either divine or human. Divine laws depend on nature, human laws on customs; and so the latter differ, since different laws please different peoples." Human law, *jus*, "is either natural, or civil, or universal. . . . *Jus naturale* is what is common to all peoples, and what is observed everywhere by the instinct of nature rather than by any ordinance, as the marriage of man and woman" (trans. Brehaut).

Isidore was a historian of note, writer of the *Chronicon*, in which he examined, briefly, the history of humankind from the creation, as recorded in the book of Genesis, to his own time of the early seventh century. Based on the genealogies of the Old Testament, Isidore proclaimed that in 616 CE, the world was in its 5,815th year. None of this was original. He claimed that he was working in the light of Eusebius of Caesarea (264–340) and Julius Africanus (160–240). His work is similar to an extremely abbreviated chronology of the Roman world written by the fourth-century chronologist Eutropius. He discussed secular affairs of kings and leaders under the framework of religious interpretation. Isidore recognized the importance of chronology to understand human affairs. If a person can gauge the amount of time since a certain event, the person gains a historical perspective based on the relative distance in years. As many chronologists such as Eusebius, Jerome, and Dionysius Exiguus (470–540) understood, using numbers as a means to gauge existence—one's own and others'—was a means to understand personal history and existence.

In his *Etymologies*, Isidore included a chapter on "Time," in which he discussed the bases for calendars, the names of the days and what they signified, and various astronomical phenomena that aided humans in tracing days, months, and years. Once again, alongside some reasonable information he included the absurd: "Night is caused either because the sun is worn out with his long journey and is weary when he comes to the last stretch of heaven and blows out his weakened fires; or because he is driven under the lands with the same force with which he carried his light over them, and thus the shadow of the earth makes night" (trans. Brehaut).

Other parts of the *Etymologies* include chapters on "Theology," "Man and Monsters," "On Languages, Races, Empires," "Alphabetical List of Words," "Of Animals," "On the Universe and Its Parts," and "On the Earth and Its Parts." These are varied compilations with information neither original nor profound. The treatise "Of Animals" is a compilation of information Isidore had culled from various ancient writings, serious and otherwise. He described various domestic and wild animals, often telling fabulous stories to account for their names and habits. For example, "Castores (beavers) are so named from castrating. For their testicles are useful for medicine and therefore when they perceive a hunter, they castrate themselves and cut away their potency by a bite." Isidore had a special place in his heart for dogs, writing that "no creature is more sagacious than dogs, for they have more understanding than other animals. . . . For they alone recognize their names, love their masters, guard their masters' houses, risk their lives for their masters, of their own free will rush upon the prey with their master, do not abandon even their master's dead body. And finally their nature is such that they cannot exist without men. In dogs two things are to be regarded, courage and speed." Most of his account is not systematic, and often absurd: "It is said that when the asp begins to feel the influence of the wizard who summons her forth with certain forms of words suited thereto, in order that he may bring her out from her hole—when the asp is unwilling to come forth, she presses one ear against the earth, and the other she closes and covers up with her tail, and so refuses to hear those magical sounds, and does not come out at the incantation." Isidore retained the fable generated centuries before by Aristotle of spontaneous generation, as in worms. Yet the patient reader can sometimes cull from Isidore a fascinating fact, as when he wrote that "birds (*aves*) are so called because they have no definite roads (*viae*) but speed hither and thither through pathless (*avia*) ways." In "On the Universe and Its Parts," Isidore informed his readers that the *universe* means "motion" in Latin, "for no rest is permitted to its elements, and therefore it is always in motion." Isidore appeared to have accepted the Greek materialist philosophers' argument that the universe is comprised of atoms, as he provided an extensive discussion of atoms that make up material things, atoms that make up the years, and atoms that constitute numbers. Isidore provided an interesting summary of the early medieval European conception of the universe, where "ether is the place in which the stars are, and it signifies that fire which is separated on high from the whole universe." His discussion of meteorology was a combination of his fascination for the origins of words combined with ancient myth. He was fascinated by the mystical origins of things, such as the waters of the earth. "The abyss," he wrote, "is the deep water which cannot be penetrated; whether caverns of unknown waters from which springs and rivers flow; or the waters that pass secretly beneath, whence it is called abyss. For all waters or torrents return by secret channels to the abyss which is their source." In "On the Earth and Its Parts," Isidore wrote that "the earth is placed in the middle region of the universe, being situated like a center at an equal interval from all parts of heaven." He described Asia as including the Garden of Eden, and India, "so called from the river Indus, by which it is bounded on the west . . . stretches from the southern sea all the way to the sun-rise, and from the north all the way to Mount Caucasus, having many peoples and cities and the island of Taprobana, full of elephants, and Chryse and Argyra, rich in gold and silver, and Tyle, which never lacks leaves on its trees." Isidore maintained the fiction of the Antipodes: "A fourth part across the Ocean on the South, which is unknown to us on account of the heat of the sun, in whose boundaries, according to story, the Antipodes are said to dwell." He conceived of Thule: "Thyle is the furthest island in the ocean, between the region of North and that of West, beyond Britain, having its name from the sun, because there the sun makes its summer halt, and there is no day beyond it; whence the sea there is sluggish and frozen." And he believed in the Fortunate Isles, which "lie in the Ocean opposite the left of Mauretania, very near the West, and separated from one another by the sea" (trans. Brehaut). Isidore concluded his encyclopedic tour de force, *Etymologies,* with chapters "On Buildings and Fields," "On Stones and Metals," "On Agriculture," "On War and Amusements," "On Ships, Buildings, and Garments," and "On Provisions and Utensils of the Household and the Fields."

CAROLINGIAN RENAISSANCE

One of the brightest lights in the early Dark Ages occurred at Aix-la-Chapelle during the reign of Charles the Great, Charlemagne (766–814). The so-called Carolingian Renaissance was inspired by Charlemagne's interest in learning and in surrounding himself with able and intelligent counselors such as Alcuin (c. 732–804) and Einhard (770–840). It was a renaissance of sorts in the attempt to recover and preserve aspects of ancient learning. Alcuin tutored Charlemagne in the basics of philosophy, mathematics, and astronomy. Einhard, in *The Life of Charlemagne,* wrote that Charlemagne

> paid the greatest attention to the liberal arts; and he had great respect for men who taught them, bestowing high honours upon them. When he was learning the rules of

> grammar he received tuition from Peter the Deacon of Pisa, . . . but for all other subjects he was taught by Alcuin, . . . a man of the Saxon race who came from Britain and was the most learned man anywhere to be found. Under him the Emperor spent much time and effort in studying rhetoric, dialectic, and especially astrology. He applied himself to mathematics and traced the course of the stars with great attention and care. (Trans. Thorpe)

Charlemagne's mathematical study was primitive, focusing mostly on arithmetic. His astronomy was generally relegated to observing the stars out of wonder or, more practically, to understand the calendar. A surviving letter from one of Alcuin's students, Fridugis (d. 834), to Charlemagne is illustrative of the Carolingian Renaissance. In the letter, Fridugis argues that "nothing" exists, that it is, indeed, "something." He supported his argument with the logic of the syllogism as well as the testimony of Christian Scripture. This illustrates the reliance upon Aristotelian logic, as brought forward by Boethius and Isidore of Seville, to the thinkers of the Carolingian Renaissance, and afterward.

Carolingian intellects, following Isidore of Seville, focused particularly on the trivium, the arts of thought and communication. The *trivium* is Latin for, literally, "a place where three roads meet." The three roads are logic, grammar, and rhetoric. Logic refers to philosophy, to human thought, the fact that humans are thinkers, able to conceptualize ideas, to try to achieve a sense of what exists, or is, that is *being*, which is not subject to time. An example: Humans, like all animals, feel. When a human feels a sense of warmth toward another being, a sense of attachment, of not wanting to be separated, they conceptualize this feeling as *love*. Love is not material or physical; it is simply an idea based on conceptualized feelings. Next to logic is grammar, which refers to symbols that humans have invented to symbolize the concepts and ideas that they have conceptualized. These symbols can be learned and shared, which form the basis of communication. So for example, the concept of love can be designated by four symbols, L, O, V, E, joined together. Rhetoric involves all of the arts of communication that humans use to share their ideas and concepts by means of symbols. Humans can therefore share an idea or concept, such as love, by means of symbols that are expressed through writing or speaking.

Following upon the trivium in importance was the quadrivium, the arts of temporal and spatial reasoning. *Quadrivium* is Latin for "where four roads meet," as coined by the philosopher Boethius. The four roads are arithmetic, geometry, music, and astronomy. There is an arithmetic road because humans live in time and space, and they keep track of movement in time and space by counting and measurement. The counting function is arithmetic, and the spatial function is geometry. Arithmetic is the way to make sense of the multitude of things (quantities) and movement in our environment over time. Geometry is the way humans make sense of the multitude of things and movements in our environment that take up space. We observe various things at particular moments in time and can make sense of how they relate to us in terms of distance, volume, and dimension. Music is using arithmetic, counting things and movement in time, as it applies to sounds and harmony. Special notes are created to keep track of these sounds moving through time. Music symbolizes human creativity in different cultures. Astronomy is measuring space and its vast dimensions over time. We observe various things over time and can make sense of how they relate to us in terms of distance, volume, and dimension. Astronomy symbolizes the hard sciences, examining movement (physics), material substances (chemistry, geology), and organic substances (biology).

In addition to the trivium and quadrivium, the legacy of ancient science on Christian Europe and the Muslim Near East is largely the story of the growing number and sophistication of commentators on Aristotle. It is difficult to underestimate the impact this one scientist and philosopher had on the subsequent two millennia of thought. After the decline of Charlemagne's empire and the Carolingian Renaissance, amid the chaos of the ninth century, students of Aristotle continued to think and to speculate using the terms and techniques of ancient science. Before the twelfth century, most of Aristotle's works were unknown to the medieval West, and scientists often relied on compilations of ancient thought, in particular the works of the polymath Isidore of Seville.

JOHN SCOTUS ERIUGENA (C. 815–877)

The Carolingian Renaissance continued to inspire thinkers after the reign of Charlemagne. John Scotus Eriugena, a native of the British Isles, arrived at the court of Charles the Bald, the grandson of Charlemagne, about the mid-ninth century. He was a schoolmaster and philosopher heavily influenced by Aristotle, Plato, and Late Roman Christian thinkers, for example, Augustine. He believed that philosophy is the means by which absolute truths, as represented by Christian theology, could be discovered. Philosophy, the love of wisdom, involves not only rational thought and logic but observation as well. Faith, the basis of the pursuit of knowledge and wisdom, is fulfilled by rational understanding. Faith alone without rational understanding is insufficient to achieve salvation.

About 870, Eriugena wrote *On the Division of Nature,* in which he declared that Greek philosophy was of fundamental importance in knowing the actions of the Christian Word (Logos) in the generation of all things and the natural laws by which existence is ordered. Eriugena incorporated Aristotelian concepts such as the First Cause, dialectic, essence (*ousia*), nature (*physis*), and knowledge (*scientia*). The latter involves the search to discover the order of all things, how life can be categorized according to genera and species, the reflection of the First Cause in nature. The philosopher uses the tools of reason (*scientia*) to discover how nature (*physis*) relates to the original essence (*ousia*). Eriugena divided nature into five phenomena:

> a nature which creates but is not created; . . . a nature which creates and is created; . . . a nature which is created and does not create; . . . a nature which neither creates nor is created. . . . The fifth and last division is that of man into masculine and feminine. In him, namely in man, all visible and invisible creatures were constituted. Therefore he is called "that in which all things were fabricated," for all that was posterior to God is contained in man. For that reason he is also referred to as a mediator, for composed as he is of body and soul, he holds in himself widely divergent extremes, namely the spiritual and the corporeal, and gives them unity. (Trans. Wolter)

HONORIUS OF AUTUN (C. 1080–1154)

Eriugena, who set the stage for subsequent Christian Aristotelians to argue that *physics*, the study of nature, was the best complement to theology, had a profound impact on the twelfth-century Benedictine recluse Honorius of Autun, who wrote a compilation of undigested Greek geography, physics, and astronomy. His works included *The Key to Nature* and *The Picture of the World.* Honorius's conceptions owed a lot to the *Etymologies* of Isidore of Seville. Honorius believed that the use of words had an important scientific and religious value in conceptualizing the truth of God and nature. For example, the names of the four elements, fire, air, water, and earth, explained not only their origins in human thought but their significance in the physical world as created by God. Honorius wrote that the earth is a round orb; he used the work of Posidonius of Rhodes (135–50), Strabo (63 BCE–21 CE) and Ptolemy to estimate the circumference of the earth at 180,000 stadia (about 18,000 miles), an ancient Roman measurement the length of which not all scientists agreed upon. He followed earlier writers in describing the earth as divided in five zones: arctic, antarctic, and equatorial, all of which were uninhabited; and the regions of the summer solstice and winter solstice, the former clearly inhabited, the other unknown. The fiery realm of the heavenly bodies contained the moon, sun, planets, and stars. The planets, he wrote, were wanderers, going against the grain of the east-west motion of the stars, moon, and sun. His astronomical descriptions were a fascinating combination of fact and fantasy. He knew the moon was closest to the earth, and the smallest among the planets orbiting the earth, but he believed it was made of fire tempered with water. He believed that the moon is larger than the earth, and its fire would burn up the earth if said fire was not quenched by the water on the surface; the moon's facial characteristics were, in fact, clouds, he wrote. The appearance of the moon, like the sun, allowed scientists to forecast the weather on earth. Comets presage disastrous weather, or political calamities on earth. Following Pythagoras, Honorius argued for the harmony of the spheres. He compared the musical scale of notes to the planets, A representing the moon, B Mercury, and so on up the scale. The apparent distances between the planets accounted for the number of tones one could hear if the human ear were correctly attuned. He found that there were nine consonances, musical harmony, between the planets, which represented the nine Muses of the ancient Greeks. Meanwhile, the seven notes of the musical scale represent the seven harmonies of humankind, three for the soul, four for the body.

THEOLOGY AND SCIENCE

Thinkers such as Augustine, Boethius, Isidore, Eriugena, and Honorius relied on ancient philosophy to help them understand and develop Christian theology. Increasingly, the influence of Platonism, which fit so well with Christian thought, was supplemented by the ideas of Aristotle, which slowly increased in number and influence in Europe. The medieval worldview, as influenced by the philosophers Plato and Aristotle and the Christian theologians Paul (c. 5–67) and Augustine, developed a view of reality that was unified, orderly, structured, and divided between the spiritual and corporeal worlds. The idea of the chain of being was brought forward from the ancient world, the notion that all being is hierarchical, ranging from spiritual to physical beings. God is the ultimate expression of being, after which are other forms of spiritual being. At the opposite extreme of God is the lowest physical creature, an insect perhaps. Physical beings rise in hierarchy toward God to the ultimate expression of physical being, human, which is also the lowest expression of spiritual being. Humans are alone in having both body and soul, sin and glory, experiencing hell and heaven. God is the creator of all, a perfect being; everything conceived by God is created. All things exist, and in this perfect order, nothing goes away—there is no extinction.

During the Late Middle Ages, religious thinkers and architects combined to express the medieval worldview of the soaring of the corporeal to the spiritual, from the lowness of physical creatures to the height and glory of spiritual creatures, by means of the cathedral, the most famous examples being at Notre Dame in Paris, Canterbury in England, and Chartres in France. The building of cathedrals required many years, sometimes centuries, and the hard labor of masons and laborers. Stone was cut by hand, sometimes by mill, but mostly it was a laborious process of human labor. For many of the workers, it was a labor of love, an expression of faith, as the cathedral seemed to be the mediator between heaven and earth, the spirit and the body. The cathedral represented God above all humans, all believers unified in faith. The cathedral was a complex unit of competing and opposing forces, of gravity upon rising stone, thrusts and counterthrusts, pressures and violent contrasts—just like human existence. For many medieval thinkers and builders, the cathedral perfectly represented Augustine's argument of the City of God contrasted with the City of Man.

The building of the medieval cathedrals could not have occurred without the transformation that occurred in Europe after 1000 CE. Agricultural production and trade increased, which resulted in growing market centers that eventually became small and then large cities. Wealth was generated by a growing middle class of merchants, artisans, and shopkeepers. Such wealth enabled the building of vast architectural accomplishments.

PETER ABELARD (1079–1142)

Augustine was heavily influenced by Plato, as were Boethius and Eriugena; increasingly the ideas of Aristotle came into play. Peter Abelard was a Christian theologian who used Aristotle's logical methods to elaborate on Christian theology. Abelard was a theologian and logician who taught at various schools in twelfth-century Paris. Abelard was a critic of many schools of philosophy and wished to reduce thought to its basics, which is sometimes called *nominalism*. He argued that humans were limited in their ability to know reality, what might be called *universals*, because we use words to denote a particular phenomenon that we assume represents a universal phenomenon. But human knowledge is limited to perception of particulars, and humans cannot truly fathom the universal—only God knows the true universals.

ANSELM OF CANTERBURY (1033–1109)

Most of Aristotle's works were unknown to European intellectuals before their introduction by means of Latin translations of the work of Arabic scholars in the twelfth century. Until then, medieval thought tended to be dominated by Platonism or Neoplatonism. This was the case with Augustine, Boethius, and Anselm of Canterbury. Anselm argued that God is the being "something greater than which we can conceive of nothing." To understand that God *may* exist is to know that he *does* exist. Therefore, the fool who denies God, who claims not to understand God but who understands the concept of God, therefore understands God. If a person conceives of God not existing, it rests on the original conception of God—why would a person conceive of something that is nonexistent? Following from Anselm's argument, how could one conceive that one does not exist? For by saying that one does not exist, one has conceived of "I," hence recognized personal existence. Anselm said that to think and to understand are the same, and he argued a proof for the existence of God by his words and thoughts, which to him formed his understanding.

The type of deduction that Anselm relied on, which dominated medieval theology up to the twelfth century, was challenged by the introduction of the much more scientific thought of Aristotle during the course of the thirteenth century. Paris and Oxford were the centers of the controversy over Christian theology, supported by Platonic thinking, and Aristotelian science, which seemed to reflect a pagan approach to learning that was pre-Christian. Slowly, during the 1200s, Aristotelianism penetrated the university environment and brought many religious doubters to embrace what had been considered nontheological ideas. Bonaventure (John of Fidanza, 1217–1274), for example, was a Parisian university theologian and Franciscan monk who was wary of philosophy, yet he was devoted to the study of Aristotle, was a great student of Augustine, and imitated Anselm in his thinking. Another professor at the University of Paris, Albertus Magnus (c. 1200–1280) was a synthesizer who combined Aristotelian and Christian theological thought together. He believed that the encyclopedic knowledge of Aristotle and the ancients would benefit, rather than detract from, Christian theology. Albert, in the words of Ferdinand Van Steenberghen, recognized "the lawfulness of a natural knowledge distinct from supernatural wisdom" (p. 178).

THOMAS AQUINAS (1225–1274)

The greatest Aristotelian of the European Middles Ages, the Christian philosopher Thomas Aquinas, attempted to reconcile Greek philosophy and science with Christian theology. Originally from Italy, he taught for a time in Paris and studied under Albertus Magnus. The writings of Muslim commentators on Aristotle had become known through Latin translations by the early to mid-thirteenth century. Aquinas, therefore, had at his disposal a vast corpus of Aristotle's works. He made great use of them in his own writings, in particular the *Summa Theologica.* Aquinas relied heavily on Aristotelian methods to arrive at logical deductions about the existence and nature of God and God's works. Repeatedly Aquinas refers to

Aristotle as simply, "the Philosopher." Like the Philosopher, Aquinas used logical syllogisms of common everyday things, such as wood and fire, to arrive at correct answers to the questions he posed throughout the *Summa.* Aquinas's use of science was, of course, limited by his methodology—it was not empirical—and by his focus, on Christian theology.

In the *Summa Theologica,* Aquinas revealed his synthesis of the philosophy of Plato and Aristotle. Thomas argued that humans are a unified whole, body and soul. To be a knower, humans must know by means of a composite being, not just spirit. Humans know beings, not just essences. Humans know because of the use of their whole being, coordinated together. The senses of the body provide understanding of sensible existence. To know God, the First Cause, one must look at the creation, the works of God, or effects. An example of Thomas's thinking is as follows: Does God Exist? Yes. Why? Whatever is moved is moved by another. A mover and moved cannot be the same. All of this is "evident to the senses." Nothing can be both "potentiality" and "actuality." Fire is actually hot, while wood is potentially hot. Fire can transform wood, but then wood is no longer potential, but actual, fire. There must be an ultimate agent of change, which is God (trans. Pegis).

Thomas Aquinas's successors in the European Renaissance were quick to point out his shortcomings, as they attempted to use ancient scientific literature as the basis for a full study of all natural phenomena.

ROGER BACON (1219–1292)

Roger Bacon was an English Franciscan monk who lived and taught part of his life at Oxford. Of the many works he wrote, the most famous is the *Opus Majus, Greater Work.* Bacon was an Aristotelian commentator who advocated a more experiential, or empirical, approach to acquiring knowledge. The *Opus Majus* comprised a number of parts. Bacon dealt with "Human Ignorance," in which he examined the ways of acquiring knowledge in light of the teachings of ancient and medieval thinkers. He noted that many of the church fathers did not know most of the works of Aristotle, and, anyway, the early church was suspicious of pagan learning because of its tendencies toward superstition and magic. But in his day, he argued that pagan philosophy was actually the "mistress" of Christian theology. For one, the earliest philosophers were the Hebrew patriarchs, who in turn taught the Chaldeans and Egyptians, who in turn taught the Greeks. Even Plato learned from the Hebrew prophets. Moreover, philosophy, that is, science, by teaching humans of nature, teaches humans of the Creator. Bacon believed that the study of language—Greek, Latin, Hebrew—was an essential tool to understand both theology and philosophy. He also believed in the importance of mathematics, which provides the essence of philosophic truth. Mathematics is the handmaid to logic. Mathematics is the means by which humans understand astronomy, physics, optics, ophthalmology, geography, and astrology. But it is not metaphysics alone that informs us of the truths of the natural and spiritual worlds. It is experimental science as well. Armed with experimental science, the investigator studies the heavens and the earth. Bacon provided examples of such study, of the planets, of the human eye, of medicines for health and longevity, of alchemical experiments to increase the span of life, of astrology and the means by which the planets impact human existence, and of some of the forces on earth, such as magnetism. Experimental science can be extended to moral philosophy, he argued. A scientific study of virtue and vice, of laws and customs, of how humans can do what is good and right, is possible. Bacon was especially influenced by the writings of the ancient Roman Stoic philosopher Seneca (5–65). He concluded the book with a study of religion, particularly Christianity.

Roger Bacon has been lauded as one of the first influential advocates of experimental science, which during the latter Renaissance blossomed into the Scientific Revolution, led by his namesake Francis Bacon (1561–1626). "There are two modes of acquiring knowledge," Roger Bacon wrote, "namely by reasoning and experience. Reasoning draws a conclusion and makes us grant the conclusion, but does not make the conclusion certain, nor does it remove doubt so that the mind may rest on the intuition of truth, unless the mind discovers it by the path of experience" (trans. Burke). For example, one may have been taught to avoid fire, that fire burns, and may reason from various examples one has been told or shown of the danger of fire, but only when one puts one's hand forth into the flame, and feels its heat and accompanying pain, does one therefore know for certain, through experience, the danger of fire.

SUPERSTITION AND SCIENCE

By means of the Arabs, medieval Europeans learned of the possibilities of alchemical medicine. Roger Bacon had this interest, and the early sixteenth-century physician Paracelsus also practiced such medical alchemy. There continued to be a fair amount of superstition in Europe with the beginnings of the Renaissance in the early fourteenth century. Astrology was very popular, even among scientists and great thinkers such as Roger Bacon. Other examples include the fifteenth-century Italian Platonic philosopher Marsilio Ficino (1433–1499) and Dominican friar Girolamo Savonarola (1452–1498). White magic, an expression of science based on the magical manipulation of the environment predicated on the interaction of spiritual and physical forces in nature, was very popular in the Late Middles Ages/Renaissance in Europe. There was still the sense that the universe was spiritually alive, that Mind, Being, directed this universe. This belief, bordering on superstition and science, was due, in the words of historian Wayne Shumaker, to "the difficulty of understanding the physical universe in any other way at a time when hard scientific data were few" (p. 187). Thinkers in the Middle Ages and Renaissance accepted the writings of Hermes Trismegistus, a reputed contemporary of Moses and a pagan philosopher of great repute, to be legitimate, of the same authority as Plato and Aristotle. The Hermetic writings were magical and alchemical and came to Europe through Arab sources.

THE OXFORD CALCULATORS

At the same time during the fourteenth century at Oxford, there was a group of thinkers and scientists who eschewed the fantastic for logic, mathematics, astronomy, and physics. These were the so-called Oxford Calculators, which included Richard Swineshead (fl. 1350), who wrote the *Book of Calculations, Opus Aureum Calculationum*, which was the first book of arithmetic published in England. Swineshead advocated the use of mathematics in studying physical phenomena. Another mathematician and philosopher, Roger Swineshead (fl. 1350), wrote a treatise *On Natural Motions*, which, as the title suggests, examines the nature of motion and time. This work, heavily influenced by Aristotle, was a theoretical account of natural philosophy, not as rigorous as the work of Thomas Bradwardine (c. 1300–1349), *On the Proportions of Velocities in Motions*, which provided applied mathematics to the theory of the relation of force and resistance and resulting velocity. Richard Kilvington (c. 1302–1361), another Oxford professor and Aristotelian commentator, wrote a logic text for Oxford students, *Sophismata*, meaning difficult sentences that encourage logical analysis. He also lectured and wrote on *Questions on Aristotle's On Generation and Corruption, Questions on Aristotle's Physics, Questions on Aristotle's Ethics*, and *Questions on the Sentences*. In mathematics, he was interested in measuring the limits on things and extent of phenomena, the latitude of things, the different degrees and various qualities of phenomena, measuring speed in phenomena, and the nature of infinity. Not all of his objects of investigation were physical, and like other Oxford Calculators, much of his work involved theology.

WILLIAM OF OCCAM (1285–1347)

Likewise, William of Occam was an English Franciscan monk who studied at the University of Oxford and taught in Europe. He was an Aristotelian commentator, a philosopher who doubted universals, a believer in reducing all assumptions, all truth, to its basic form, and a believer in intuitive as well as rational thought. He wrote at length on logic, metaphysics, and theology within the context of the early Church Fathers and ancient thinkers, especially Aristotle.

William's approach to writing and thinking was logical in that he formed propositions and tried to reason from said propositions to find the truth. In this, his thinking had a scientific approach. He was especially concerned with the relationship between God and humans, between God's will and human will, between the timeless and time, between God's knowledge and human knowledge. He believed that the key to existence was to understand precisely what God's role was in the creation and in time: past, present, and future. Does God know all things, including the future? Does God, by knowing the future, therefore will the future? Is everything determined by God? Or are there contingencies in life subject to the will of creatures living in time? These were questions that had exercised the minds of ancient as well as medieval thinkers for millennia. William approached these questions from the perspective of a deep thinker seeking to find the simplest solution. He ultimately believed, in contrast to Aristotle, that human reason and logic were insufficient and that, ultimately, intuitive thought and feeling must be relied upon to understand the nature of reality, that is, the nature of God, who is the source of and essence of all things. William of Occam's work, therefore, cannot achieve the status of an Aristotle, who assumed the truth and that humans could discover it; therefore, William's work is ultimately unscientific, if science is based on the assumption that there is a truth and that humans can discover this truth. For William of Occam, ultimately, regarding time and existence, God alone *knows*.

RENAISSANCE EUROPE

The Aristotelian conception of nature still had such a hold over the medieval mind that it required a revolution to bring about change. The European Renaissance (1300–1600) was a time of political, cultural, and scientific rebirth of ancient learning. Whereas the medieval focus on ancient science was generally limited to an ongoing commentary on Aristotle's thought and writings, Renaissance thinkers developed a broader understanding of ancient thought because of the rediscovery of ancient texts, many of which had nothing to do with Aristotle. The great challenge, which would bring about the Scientific Revolution, was to question the universally recognized philosopher and scientist of ancient Greece: Aristotle.

FURTHER READING

Boethius, *Consolation of Philosophy*, trans. Philip Ridpath (London: C. Dilly, 1785).

Boethius, *The Consolation of Philosophy*, trans. V. E. Watts (Harmondsworth, England: Penguin Books, 1969).

Ernest Brehaut, trans., *An Encyclopedist of the Dark Ages: Isidore of Seville* (New York: Columbia University, 1912).

Einhard, *The Life of Charlemagne*, trans. Lewis Thorpe (Harmondsworth, England: Penguin Books, 1969).

Edward Grant and John E. Murdoch, eds., *Mathematics and Its Applications to Science and Natural Philosophy in the Middles Ages: Essays in Honor of Marshall Clagett* (Cambridge: Cambridge University Press, 1987).

Introduction to St. Thomas Aquinas, ed. and trans. Anton C. Pegis (New York: Modern Library, 1948).

Isidore of Seville, *Chronicon*, trans. Kenneth B. Wolf (2004), http://www.tertullian.org/fathers/isidore_chronicon_01_trans.htm.

Elzbieta Jung, "Richard Kilvington," *Stanford Encyclopedia of Philosophy*, https://plato.stanford.edu/entries/kilvington/.

Arthur O. Lovejoy, *The Great Chain of Being: A Study of the History of an Idea* (Cambridge, MA: Harvard University Press, 1950).

William Ockham, *Predestination, God's Foreknowledge, and Future Contingents*, trans. M. M. Adams and Norman Kretzmann (New York: Appleton-Century-Crofts, 1969).

The Opus Majus of Roger Bacon, trans. Robert Burke, vol. 1 (Philadelphia: University of Pennsylvania Press, 1928).

Wayne Shumaker, *The Occult Sciences in the Renaissance: A Study in Intellectual Patterns* (Berkeley: University of Los Angeles Press, 1972).

Richard E. Sullivan, *Aix-la-Chapelle in the Age of Charlemagne* (Norman: University of Oklahoma Press, 1963).

Edith D. Sylla, "Oxford Calculators," *Encyclopedia of Medieval Philosophy: Philosophy between 500 and 1500*, ed. Henrik Lagerlund (Dordrecht, Netherlands: Springer Reference, 2019), 903–8.

Fernand Van Steenberghen, *Aristotle in the West: The Origins of Latin Aristotelianism* (Louvain, Belgium: Nonwelaerts Publishing House, 1970).

John Wippel and Allan Wolter, eds., *Medieval Philosophy* (New York: Free Press, 1969).

PART II

Expansion of Science into New Worlds

32

Exploration and Geographic Knowledge in the Medieval World (500–1500 CE)

THE SILK ROAD

As in the ancient world, peoples in the medieval world, the thousand-year period from 500 to 1500 CE, traveled extensively, across oceans, along rivers, across broad plateaus, and over mountains. There were a variety of ports of call on land and sea where caravans would bring goods to trade. The most notable of these was the Silk Road, something of a metaphor for the many routes that caravan drivers and sea captains took that connected eastern Europe, western Asia, northern and eastern Africa, and central and southern Asia and that went across the mountains and steppes of eastern Asia and in the waters separating the Indian and Pacific Oceans. Chinese sources (*Sung-Shu, Wei-Shu*) written about 500 CE noted the difficulties in the land and sea routes connecting China, India, the Byzantine Empire, and Egypt, and the arrival of Chinese traders to the cities of Antioch and Babylon. A Chinese source written in the tenth century (the *Chiu-t'ang-shu*) provided a remarkably accurate description of the city of Constantinople. Another Chinese source written in the eleventh century (*Hsin-t'ang-shu*) provided a description of the people of the upper Nile River valley.

Chinese travelers using the Silk Road in the early Middle Ages included Faxian, Xuanzang, and Yijing. Often the travelers were Chinese Buddhist monks who sought more information about Buddhism and wanted to visit the places in which the Buddha had lived and the religion had begun in India. This was the case with Faxian (c. 337–422), a Buddhist who journeyed from Xinjiang China west across what is today Tajikistan and then south through what is today Afghanistan, crossing the Hindu Kush to the sources of the Indus River, and then continuing east across northern India to the sources of the Ganges River. Faxian wrote *A Record of the Buddhist Kingdoms*. After having descended the Ganges, he took ship through the Bay of Bengal along the east coast of Indian to Sri Lanka, where he stayed for several years before proceeding by ship east to Malaysia and Sumatra and then north through the South China Sea back to China. A few centuries later, the

Buddhist monk Xuanzang (c. 600–664) wrote *The Records of the Western Regions Visited during the Great Tang Dynasty*, based on his travels from China west through central Asia, what is today Kyrgyzstan and Uzbekistan and the ancient Silk Road trading city of Samarkand. He proceeded south through Afghanistan and east to the sources of the Indus River. He explored almost the entire peninsula of India, including the two great rivers, the Ganges and the Indus. He described the kingdom of Harsha (r. 606–647), a ruler of the Vrdhana Dynasty. Xuanzang's return journey was up the Indus River, through Afghanistan, east through northern Tibet, and back to the court of the Yang emperor Taizong (r. 626–649). A little over a generation later, the Buddhist traveler Yijing (635–713) journeyed from Yangzhou on the Yangtze River south through the China Sea to Sumatra, then west through the Strait of Malacca into the Indian Ocean, and then on to India and the Ganges River. He wrote *The Record of Buddhism as Practiced in India Sent Home from the Southern Seas* and *Memoirs of Eminent Monks Who Visited India and Neighboring Regions in Search on the Law during the Great Tang Dynasty.*

Various goods and ideas were exchanged via the different land and sea routes collectively known as the Silk Road. Europeans, West Asians, and Africans exported foodstuffs, horses, glassware, textiles, and slaves, while East Asians exported porcelain, tea, spices, medicines, gunpowder, and, of course, silk. Medieval Jewish merchants engaged in extensive trade throughout Eurasia, traveling throughout China, Indonesia, India, central Asia, western Asia, North Africa, the Mediterranean, and Europe. Ideas were also exchanged, as seen through the journeys of Faxian, Xuanzang, and Yijing.

COSMAS INDICOPLEUSTES (C. SIXTH CENTURY)

Meanwhile, the peoples of Europe and western Asia were actively seeking information on those places east across Asia. Cosimo Indicopleustes, or, as he referred to himself in his book *The Christian Topography, Cosmas, an Egyptian monk*, was such a traveler. Who precisely this Cosmas was and whether he was an Egyptian monk are difficult questions to answer. He was apparently a merchant and traveler during the reign of the Byzantine emperor Justinian. He traveled throughout eastern Africa and into the Indian Ocean, to India and Ceylon (Sri Lanka). He explored the lower and upper Nile River, what is today Egypt, Sudan, Ethiopia, and Somalia; explored the Sinai Desert; and sailed in the Red Sea, Gulf of Aden, and Arabian Sea. He also visited Jerusalem and retired to write his book in Alexandria, Egypt. He was a Nestorian Christian monk. Indeed, many Nestorian Christians had left the Byzantine Empire, where Greek Orthodoxy made their belief anathema to Byzantine Christians. The Nestorians became widespread travelers throughout Asia, going as far as the provinces of China.

Notwithstanding his travels, Cosmas had a strange conception of the earth, based not on observation but on Scripture. He considered the earth to be shaped like the Jewish tabernacle, a rectangle more wide, east to west, than long, north to south. The canopy of heaven hangs over the temporal part of the earth. The earth is at the bottom of the universe. The Ptolemaic conception of a spherical earth was absurd, Cosmas thought, as well as the notion of a geocentric universe. He believed the sun to be much smaller than ancient astronomers thought; it crossed before the face of the earth and then hid at night behind a mountain. Most of Cosmas's *Christian Topography* is an extensive attack on the Ptolemaic system and a defense of his own geographical system. In book 11, he digressed to a discussion of fauna he saw in Africa and Egypt, such as the rhinoceros, hippopotamus, giraffe, and of flora, such as the pepper tree. Cosmas called Ceylon, or Sri Lanka, Trapobane, after Megasthenes, the Greek explorer of India of the late fourth century BCE.

Notwithstanding Cosmas's attack, Claudius Ptolemy's *Geography* was the standard during the Later Roman Empire and throughout the European Middle Ages. Ptolemy (100–170), an astronomer as well as a geographer, developed standard directions (north, south, east, west) but made astonishing errors, overestimating the size of Asia, underestimating the circumference of Earth, and hypothesizing a land mass, the *terra incognita,* connecting Africa to Asia, making the Indian Ocean a landlocked sea.

Other European philosophers provided conceptions of the world that had a profound impact on medieval European understanding. In *Consolation of Philosophy,* by Boethius (480–524), the goddess Philosophy describes the nature of the world from the sixth-century perspective:

> You have learned from astronomy, that this globe of earth is but as a point, in respect to the vast extent of the heavens; that is, the immensity of the celestial sphere is such, that ours, when compared with it, is as nothing, and vanishes. You know likewise from the proofs that Ptolemy adduces, there is only one fourth part of this earth, which is of itself so small a portion of the universe, inhabited by creatures known to us. If from this fourth you deduct the space occupied by the seas and lakes, and the vast sandy regions which extreme heat and want of water render uninhabitable, there remains but a very small proportion of the terrestrial sphere for the habitation of men. Enclosed then and locked up as you are, in an unperceiveable point of a point, do you think of nothing, but of blazing far and wide your name and reputation? What can there be great or pompous in a glory circumscribed in so narrow a circuit? To this let me add, that even in this contracted circuit, there is a great variety of nations differing from one another in their languages, manners and customs, to whom, whether from the difficulty of travelling, or the diversity of tongues, or the want of commerce, the fame not only of particular persons, but even of great cities, cannot extend. In Cicero's time, as he tells us somewhere in his works, the renown of Rome herself, which she imagined was diffused every where, did not reach beyond Mount Caucasus, though the republick was then in her glory, and had rendered herself formidable to the Parthians, and to all the nations in their neighbourhood. (Trans. Ridpath)

VIKING EXPLORATIONS

The Scandinavian peoples of northern Europe, historically known by their violent exploring raids as *Vikings*, were a pagan, polytheistic culture north of the Roman Empire at the beginning of the first millennium CE. An agricultural, Iron Age people, they were traders in various goods, such as furs and slaves. There was some slight contact between the Scandinavian and Roman and Greek explorers and traders in the ancient world—it is probable that Pytheas's *Thule* was Scandinavia. As the Roman Empire declined in the West, the power of the Vikings increased. They developed their own writing, called runes, had a strong oral heroic tradition, worshipped violent gods, and engaged in violence themselves. A seagoing people, the Vikings explored the North Sea, Baltic Sea, Atlantic Ocean, and Mediterranean Sea. Up to about 800 CE, they explored seas and rivers in open boats propelled by oars. After about 800, their ships were more sophisticated, more seaworthy, and used sails and oars to propel them through the water.

The Vikings were superb explorers. From their lands of Norway, Denmark, and Sweden, they went east, south, and west in warships, intent on plunder and conquest. During the ninth century, they raided the British Isles, France, and Spain, and they settled Iceland. Swedes went east up the Vistula, Dnieper, and Volga, founding Novgorod and Kiev, reaching the Black Sea and trading with the Byzantines at Constantinople, and reaching the Caspian Sea and trading with the Arabs at Baghdad. Norse raiders attacked Muslim settlements in Spain in the mid-ninth century. The Vikings had merchant and war vessels, seagoing and coastal/river vessels. Out of sight of land, they navigated by the stars at night and the sun's position on the horizon during the day. Vikings also used sunstones (*solarstein*), crystals that produced polarized light, enabling the determination of the location of the sun even on cloudy days. Vikings also used the lodestone, magnetite, to magnetize an iron needle, which they could place in straw and float in a bowl filled with water, producing a compass. The Vikings were adept at overland trade as well, using wagons and sleds, and had a trade route from Bulghar to the Aral Sea.

The explorations of the Vikings from Iceland and Greenland about 1000 CE provide the first recorded evidence of Europeans crossing the Atlantic and exploring America. According to the *Vinland Sagas*, stories told for centuries of the Viking colonization of a land that they called Vinland (which could mean either land of wine, land of grapes, land of vine, or land of grain), one Erik the Red, a criminal from Iceland who established a colony in Greenland, had a son Leif (the Lucky), who crossed the Atlantic from Greenland to Baffin Island (Canada), becoming the first known European to visit America. The *Vinland Sagas* discuss subsequent voyages of the Norse to this new land. For many years, scholars assumed that the *Vinland Sagas* were based on legend and myth, until in the early 1960s, Helge Instad, a Norwegian archaeologist, discovered the remains of a Viking settlement at L'Anse de Meadows, on the northern tip of Newfoundland. Did the Vikings explore other parts of North America? Possibilities include the St. Lawrence River upstream to the Great Lakes, and the bays of the Maine coast. Viking explorers would have found the rivers of the St. Lawrence watershed as well as the many rocky inlets of the Maine coast similar to Scandinavian coastal and inland waters. Viking ships had a single mast with a square sail to propel it before the wind, and oars for navigation and extra propulsion against the wind or when going upstream. Ships had shallow drafts, excellent for exploring narrow rivers and intricate bays, such as along the Maine coast. The only concrete evidence of a Viking presence centuries ago in Maine is a silver coin dating from the eleventh century found at Naskeag Point, a peninsula jutting into the Atlantic at

Penobscot Bay. The environs of the bay, with its many islands and rocky points, would feel like home to the transplanted Scandinavians. As Naskeag Point was home to generations of Abenaki, the coin might have belonged to a native who acquired it by trade originating from the Norse settlements of Newfoundland. There are, moreover, throughout America museums and parks with runestones, runic being the writing script of the Vikings. A well-known example is the Kensington Stone in Minnesota.

The geographical and chronological extent of Viking settlements in North America is unknown; however, wherever they did establish camps and communities, they made contact, often violent, with the native peoples, whom the Vikings called, derisively, Skraelings—in Newfoundland these were either the Beothuk or Mi'kmaq tribe. One of the *Vinland Sagas*, the *Graenlendinga Saga*, describes a brief but violent conflict between the Vikings under one Thorvald and the natives, who attacked in "skin boats," canoes made of moose hide. Both sides suffered losses. Another saga tells the story of Thorfinn Karlsefni's voyage to Vinland, during which he established trade with the Indians, who offered skins in return for cow's milk. *Eirik's Saga* provides more details of Karlsefni's voyage, including an account of the approach of numerous Indians in skin boats: "The men in them were waving sticks which made a noise like flails, and the motion was sunwise [clockwise]." The Vikings took the waving of these rattlesticks as "a token of peace" and approached the natives with a white shield. The Indians "rowed towards them and stared at them in amazement as they came ashore. They were small and evil-looking, and their hair was coarse; they had large eyes and broad cheekbones. They stayed there for a while, marveling, and then rowed away south round the headland." Later they returned to trade, giving up skins in return for the red cloth of the Vikings. Still later they arrived in numerous skin boats, this time waving their sticks in a counter-clockwise fashion and screaming loudly, which the Vikings correctly interpreted as a war cry. The two sides came to blows. The natives attacked with small ballistic weapons: the Vikings "saw them hoist a large sphere on a pole; it was dark blue in colour. It came flying in over the heads of [the Vikings] and made an ugly din when it struck the ground." They also used arrows and other projectiles with sharp flint heads. At one point after the battle, the Vikings came upon "five Skraelings clad in skins, asleep; beside them were containers full of deer-marrow mixed with blood." The Vikings "reckoned that these five must be outlaws, and killed them." The marrow and blood were probably pemmican, meat mixed with marrow used by hunters on long hunts. The Vikings also captured five Indians: a bearded man, two women, and two boys; the man and women escaped, but the boys were adopted by the Vikings, baptized, taught the Icelandic language, and encouraged to tell about their culture. The boys told the Vikings that their people had "no houses" but "lived in caves or holes in the ground" (trans. Magnusson and Palsson).

The *Vinland Sagas* imply but are not explicitly detailed about the eventual abandonment of their American colonies by the Vikings. The Viking discovery of Vinland was well known in Iceland and Scandinavia but apparently not farther east and south in England, France, Spain, Portugal, and Italy.

ARAB EXPLORATIONS AND GEOGRAPHY

The eleventh-century Arab polymath Al-Biruni (971–1048) summarized the views of his time among Islamic scholars regarding the extent of the earth:

> The country south of the Line is not known, for the equatorial part of the earth is too much burned to be inhabitable. Parts of the inhabited world do not reach nearer the equator than to a distance of several days' journey. There the water of the sea is dense, because the sun so intensely vaporises the small particles of the water, that fishes and other animals keep away from it. Neither we nor any of those who care for those things have ever heard that any one has reached the Line or even passed the Line to the south. . . . As to the country beyond the Line, someone maintains that it is not inhabitable, because the sun, when reaching the perigee of his eccentric sphere, stands nearly in its utmost southern declination, and then burns all the countries over which he culminates, whilst all the countries of 65 degrees of southern latitude have the climate of the middle zone of the north. From that degree of latitude to the pole the world is again inhabitable. But the author of this theory must not represent this as necessary, because excessive heat and cold are not alone the causes which render a country uninhabitable, for they do not exist in the second quarter of the two northern quarters, and still that part of the world is not inhabited. So the matter is (and will be), because the apogee and perigee of the eccentric sphere, the sun's greater and less distance from the earth, are necessitated exclusively by the difference in the sun's rotation. (Trans. Sachau)

Al-Biruni, like other medieval Arab geographers, embraced Claudius Ptolemy's theory of the seven climatic zones, or seven climes, ranging across the north-south latitude of the earth. Some Arabs had reached the seventh clime, in northern Europe and Asia, in the tenth century. Ibn Fadlan (877–960), for example, a tenth-century explorer and geographer, reached the settlement of Bulghar near the confluence of the Kama and Volga Rivers in 922. Ibn Fadlan was a diplomat on a mission for the Abbasid caliph Al-Muqtadir (r. 908–932) to the Bulghars. An expedition traveled from Baghdad east, going near the southern shores of the Caspian Sea, reaching Bukhara in what is today Uzbekistan, and then proceeding north to the Aral Sea and farther north to the upper Volga and the settlement of Bulghar. Ibn Fadlan described the culture and customs of Arab, Turks, and the Rus that he encountered along the way. He described natural phenomena, such as the northern lights, which astonished the Arabs. While staying at Bulghar, Ibn Fadlan witnessed the customs of Viking traders, called the Rus, who had descended the Volga to trade. He described the elaborate ceremony of the funeral and immolation of a Russian noble in a Viking boat. Ibn Fadlan was also able to learn about the customs of the Khazars, Turkish people who lived along the Volga.

Abu Hamid (1080–1170) from Andalusia followed in the wake of Ibn Fadlan two centuries later. He was a scholar and traveler who journeyed across North Africa; explored the Nile River; lived in Cairo, Damascus, and Baghdad; and journeyed north to the Volga, where he lived in the city of Saqsin for about twenty years. He journeyed up the Volga as far north as Bulghar, traveled to the Dnieper River, journeyed down the Oka River to Kiev, and visited the kingdom of Bashghird (Hungary). He wrote *Exposition of Some of the Wonders of the West*, which was something of a medieval almanac. A devout Muslim, his book discussed the power of God reflected in the creation. He described the various peoples who lived at the northern edge of the Caspian Sea, such as the Khazars and Bulghars; the cities of Saqsin and Bulghar; natural wonders, such as the almost humanlike intelligence of beavers; and he proclaimed his view that knowledge derives from memorization, such as of the Qur'an. "Science is in the heart, it is not in book," he wrote (trans. Lunde and Stone).

One of the most well-traveled Arab explorers of the medieval world was Ibn Battuta (1304–1369), who wrote *A Gift to Those who Contemplate the Wonder of Cities and the Marvels of Traveling.* Over the course of many journeys, he traveled throughout North Africa and south into Saharan Africa, reaching the Niger River and Timbuktu as well as the African east coast down to Zanzibar; the Arabian Peninsula, Mesopotamia, and Persia; eastern Europe, Anatolia, the coasts of the Black Sea and Caspian Sea; central and southern Asia, including the islands of Sri Lanka and the Maldives; and Indonesia, the Philippines, and Southeast and East Asia. Like Marco Polo, Ibn Battuta narrated his *Travels* later in life; they were recorded by an Andalusian scholar, Ibn Juzayy (1321–1347).

MARCO POLO (1254–1324)

Marco Polo was the son of a Venetian merchant who traveled generally along the Silk Road to China in the late thirteenth century. He heralded himself as the most traveled man of his day in the generalized account of his *Travels* in Asia. He described the caravan route from Constantinople through Anatolia into Mesopotamia through Persia into Bactria through Tibet into China. He described many places he had personally seen but many more places that he learned of from hearsay. The *Travels* provides fascinating information on China under Mongol rule; specifically, Marco Polo met and served Kublai Khan (1215–1294) as an emissary to different places. As an explorer and observer, Marco Polo provided interesting geographical insights on peoples and places of central and eastern Asia. He noted the central trade entrepots and the goods that were produced, bought, and sold. He described the religious beliefs of peoples, separating them into Christians, Muslims, and "idolaters," that is, Buddhists and Hindus. The people of this time in central and eastern Asia, according to Marco Polo, were heavily reliant upon astrology to tell them about their fate and fortunes. Marco Polo claimed to have journeyed throughout China, to Khan-balik (Beijing) across the Yellow and Yangtze rivers, to Kinsai (Hangzhou), and south into southeast Asia. At Kinsai, he reported that there were "wise philosophers and natural physicians with a great knowledge of nature" (trans. Latham).

The Chinese had a long history of engaging in maritime trade. Marco Polo described the typical merchant ships as

> built of fir-timber. They have a single deck, and below this the space is divided into about sixty small cabins, fewer, or more, according to the size of the vessels, each of them affording accommodation for one merchant. They are provided with a good helm. They have four masts, with as many sails, and some of them have two masts, which can be set up and lowered again, as may be found necessary. Some ships of

> the larger class have as many as thirteen bulk-heads or divisions in the hold, formed of thick planks mortised into each other. . . . The ships are all double-planked; that is, they have a course of sheathing-boards laid over the planking in every part. These are caulked with oakum both inside and without, and are fastened with iron nails. They are not coated with pitch, as the country does not produce that substance, but the bottoms are smeared over with the following preparation. The people take quick-lime and hemp, which latter they cut small, and with these, when pounded together, they mix oil procured from a certain tree, making of the whole a kind of unguent, which retains its viscous proper ties more firmly, and is a better material than pitch. Ships of the largest size require a crew of three hundred men; others, two hundred; and some, one hundred and fifty only, according to their greater or less bulk. They carry from five to six thousand baskets, or mat bags, of pepper. (Trans. Marsden)

Marco Polo's *Travels* describes Japan, the China Sea, Indonesia, and Southeast Asia. Exactly where he might have traveled outside of northern China (Cathaia) and southern China (Manzi) is uncertain. He did describe a trip to Sumatra during which he was detained by weather for five months. Perhaps he traveled the water route from Indonesia to the Indian Ocean and into India; he described the Indian Peninsula at length. He claimed that the Asian Indians were just as devoted to soothsaying and astrology as the East Asians were. He provided a general description of Hindu religious leaders, such as the Brahmans. The Yogi, he claimed, lived for a very long time based on a sparing diet and an alchemical drink they made of quicksilver and sulfur. Malabar, he wrote, on the western coast of India, was a great trading center: "Ships come here from very many parts, notably from the great province of Manzi [southern China], and goods are exported to many parts. Those that go to Aden are carried thence to Alexandria" (trans. Latham). He described from hearsay the east coast of Africa, including the islands of Zanzibar and Madagascar.

RENAISSANCE EUROPEAN EXPLORATION

Largely in response to Marco Polo's *Travels*, European explorers of the fifteenth century were in quest of new trade routes to Asia (the East Indies). To discover a new route to Cathaia (which is what Marco Polo called China), the Portuguese (under King John II) sponsored several exploring voyages from Portugal south down the coast of Africa. King John's predecessor Henry the Navigator (1394–1460) had established something of a marine science center at the port city of Sagres, a peninsula jutting into the Atlantic at the southwestern corner of the country. Here the Portuguese experimented with ship design, developing the caravel, a small sailing vessel with triangular, lateen sails for maneuverability tacking into the wind. The Portuguese also explored the winds and currents off the coast of Africa, learning how to sail south along the African coast. During the reign of King John II (1481–1495), the Portuguese explored the west coast of Africa. In 1482, John sent the explorer Diego Cam along the west coast of Africa; Cam was able to cross the equator, proving the absurdity of the ancient notion of the ring of fire that separated the Northern and Southern Hemispheres; he also explored the

Congo River. Portuguese sailor Bartolomeu Dias (1450–1500) in 1488 reached the Cape of Good Hope, proving wrong the belief of ancient geographers, such as Claudius Ptolemy (100–170), that Africa was landlocked; Dias discovered that the continent was surrounded by oceans and, by extension, that there was more water on the earth's surface than hitherto recognized—another error made by ancient and medieval geographers. The Portuguese sailor Vasco da Gama (c. 1460–1524) in 1497 rounded the Cape of Good Hope, sailing from the Atlantic to the Indian Oceans, and crossed the Indian Ocean to India, founding a Portuguese colony. Pedro Álvares Cabral in 1500 sailed along the eastern coast of South America, claiming Brazil for Portugal. He circumnavigated Africa from west to east and crossed the Indian Ocean to the new Portuguese Indian port of Calicut.

Meanwhile, the Genoese explorer Christopher Columbus (1451–1506) relied on Claudius Ptolemy's geographical scheme to plan his "Enterprise of the Indies." Ptolemy's errors helped Columbus convince himself that a voyage from Spain west across the Atlantic to China was possible. He did not realize that America, unknown to the ancients, lay between. By the time Columbus developed his idea of sailing west across the Atlantic to China, it appeared to him and the Europeans to whom he conveyed it as a new idea. There is some evidence that Columbus heard of Vinland on voyages he made to England and Scandinavia, but if he did hear of the Viking voyages, such awareness was not recorded, and whether it influenced his developing ideas of exploration is not known.

Columbus, who was inspired by the scientific and exploring work of the Portuguese, was born in Genoa in 1451, went to sea as a boy, but had no formal education. He was a ship captain by age twenty. Self-taught, Columbus learned a variety of languages, including Latin, and techniques of cartography. During his career, Columbus captained ships that explored the Atlantic down to the Gulf of Guinea, and north as far as Thule (so called by the Europeans; perhaps Scandinavia, Iceland, or Greenland). Living in Portugal in the 1580s, he developed his idea of the Enterprise of the Indies. He tried to sell this plan to Italian princes and the kings of Portugal, Spain, France, and England. Columbus argued (erroneously) that geographic knowledge indicated that the quickest and most efficient way to get to Cathaia was to sail directly west from Spain with the trade winds, thereby reaching Asia in two months' time. After repeated attempts, Columbus finally convinced Queen Isabella of Spain (1451–1504) to support his voyage. Columbus set sail with three ships in August 1492. He made landfall in the Bahamas, October 12, 1492. In all, Columbus made four voyages (1492–1493, 1493–1496, 1498–1500, and 1502–1504) and was the first known European to discover the islands of the Caribbean, the coast of South America, and the coast of Central America. Columbus never sailed along the coast of North America. His journals of his voyages describe in detail what he saw and the impressions he had of the native peoples. As he wrote after first landing and seeing the native people,

> As I saw that they were very friendly to us, and perceived that they could be much more easily converted to our holy faith by gentle means than by force, I presented them with some red caps, and strings of beads to wear upon the neck, and many other trifles of small value, wherewith they were much delighted, and became wonderfully attached to us. Afterwards they came swimming to the boats, bringing

> parrots, balls of cotton thread, javelins, and many other things, which they exchanged for articles we gave them, such as glass beads and hawk's bells, which trade was carried on with the utmost good will. But they seemed on the whole to me to be a very poor people. They all go completely naked, even the women, though I saw but one girl. All whom I saw were young, not above thirty years of age, well made, with fine shapes and faces; their hair short and coarse like that of a horse's tail, combed toward the forehead, except a small portion which they suffer to hang down behind, and never cut. Some paint themselves with black, which makes them appear like those of the Canaries, neither black nor white; others with white, others with red, and others with such colours as they can find. Some paint the face, and some the whole body; others only the eyes, and others the nose. Weapons they have none, nor are they acquainted with them; for I showed them swords, which they grasped by the blades, and cut themselves through ignorance. They have no iron, their javelins being without it, and nothing more than sticks, though some have fish-bones or other things at the ends. They are all of a good size and stature, and handsomely formed. I saw some with scars of wounds upon their bodies, and demanded by signs the cause of them. They answered me in the same way, that there came people from the other islands in the neighbourhood who endeavoured to make prisoners of them, and they defended themselves. I thought then, and still believe, that these were from the continent. It appears to me that the people are ingenious, and would be very good servants; and I am of the opinion that they would readily become Christians, as they appear to have no religion. They very quickly learn such words as are spoken to them. If it please our Lord, I intend at my return to carry home six of them to your Highnesses, that they may learn our language. I saw no beasts in the island, nor any sort of animals except parrots. (Quoted by Adams)

In the wake of Columbus's voyages to America in the late 1400s and early 1500s, Europeans during the sixteenth and seventeenth centuries made contact with, traded and fought with, and eventually conquered many of the Indigenous tribes of South, Central, and North America.

FURTHER READING

Charles Kendall Adams, *Christopher Columbus: His Life and Work* (New York: Dodd, Mead and Company, 1892).

Al-Biruni, *The Chronology of Ancient Nations*, trans. C. Edward Sachau (London: William H. Allen and Co., 1879).

Boethius, *Consolation of Philosophy*, trans. Philip Ridpath (London: C. Dilly, 1785).

Daniel Boorstin, *The Discoverers* (New York: Random House, 1983).

Chinese Accounts of Rome, Byzantium and the Middle East, c. 91 B.C.E.–1643 C.E., East Asian History Sourcebook, https://sourcebooks.fordham.edu/eastasia/romchin1.asp.

The Christian Topography of Cosmas, an Egyptian Monk, trans. J. W. McCrindle (London: Hakluyt Society, 1897).

Ibn Fadlān, *Ibn Fadlān and the Land of Darkness: Arab Travellers in the Far North*, trans. Paul Lunde and Caroline Stone (London: Penguin Books, 2012).

Helge Instead, "Vinland Ruins Prove Vikings Found the New World," *National Geographic*, November 1964.

Howard La Fay, "The Vikings," *National Geographic*, April 1970.

Joshua J. Mark, "Silk Road," *Ancient History Encyclopedia*, https://www.ancient.eu/Silk_Road.

Marco Polo, *The Travels*, trans. R. E. Latham (Harmondsworth, England: Penguin Books, 1958).

Marco Polo, *The Travels of Marco Polo, The Venetian*, revised from Marsden's Translation and Edited by Manuel Komroff (New York: Boni & Liveright, 1926).

Tansen Sen, "The Travel Records of Chinese Pilgrims Faxian, Xuanzang, and Yijin: Sources for Cross-Cultural Encounters between Ancient China and Ancient India," *Education about Asia* 11, no. 3 (Winter 2006), http://afe.easia.columbia.edu/special/travel_records.pdf.

The Travels of Ibn Batuta, trans. Samuel Lee (London: Oriental Translation Committee, 1829).

The Vinland Sagas, trans. Magnus Magnusson and Hermann Palsson (New York: Penguin Books, 1987).

David M. Wilson, *The Vikings and their Origins: Scandinavia in the First Millennium* (London: Thames and Hudson, 1989).

33

Science in the Americas (500–1500 CE)

In the wake of Columbus's voyages to America and the European discovery of hitherto unknown people living in hitherto unknown continents between the Atlantic and Pacific Oceans, thinkers debated the origins of the peoples of America. Were the so-called Indians truly Indigenous peoples? If not, where did they come from? And when and how did peoples of the Old World (Europe, Asia, Africa) come into contact with them?

From the beginning, Europeans and Americans assumed that the Indigenous peoples of America were indigenous insofar as they came to America long before recorded memory. The Christian worldview dominated thought up through the nineteenth century, and Christianity taught, based on the book of Genesis in the Old Testament, that all humans were descended from one act of the special creation of God in the Garden of Eden. Since all humans derived ultimately from Adam and Eve, their descendants had spread from the Garden, in Mesopotamia, throughout Asia to Africa, to Europe, and clearly to America as well. As it is written in the New Testament book of Acts (17, 26–27, RSV), "And he made from one every nation of men to live on all the face of the earth, having determined allotted periods and the boundaries of their habitation, that they should seek God, in the hope that they might feel after him and find him." The American Indians had to have come to America from another place—otherwise Scripture would be incorrect. So from where, then, did the Indians derive? There were possibilities from all three of the Old World continents. Some thinkers believed that an ancient Welsh explorer, Madoc, came to American long ago, and the Indians were descended from the Welsh. Even up to the 1800s, explorers tried to find evidence of the Welsh Indians. Other thinkers favored the theory that one of the sons of Noah, or one of the sons of Jacob, might have led the Israelites to America. An advocate of this idea was the American Puritan minister Roger Williams (1603–1683), who published a study of the American Indian, *A Key into the Language of America*, in 1643:

> First, others (and my selfe) have conceived some of their words to hold affinities with the *Hebrew*. Secondly, they constantly *annoint* their *heads* as the *Jewes* did. Thirdly, they give *Dowries* for their wives, as the *Jewes* did. Fourthly (and which

> I have not so observed amongst other *Nations* as amongst the *Jewes*, and *these*:) they constantly separate their Women (during the time of their monthly sicknesse) in a little house along by themselves foure or five dayes, and hold it an *Irreligious thing* for either *Father* or *Husband* or any *Male* to come neere them. They have often asked me if it bee so with women of other *Nations*, and whether they are so *separated*: and for their practice they plead *Nature* and *Tradition*.

Then there were those such as the English historian William Robertson (1721–1793), who believed that the ancient Phoenicians sailed to America by way of their western Mediterranean colony Carthage in North Africa. Ancient Greek and Roman writers had recorded accounts of Phoenician sailors in the Atlantic Ocean and Carthaginian sailors discovering a giant island far west in the Atlantic. Some fanciful thinkers considered this island to be Atlantis, the ancient civilization that according to legend was swallowed up by the sea, and which was the namesake of the great ocean. Or perhaps Atlantis was, in fact, America.

The most reasonable argument was that the Indians were descended from the ancient Tartars, as they were called, that is, the ancient Chinese. Besides the similarities of physiognomy, hair, and stature, it seemed plausible that the people of ancient Siberia could have made the crossing of the Bering Sea and journeyed into America. The first advocate of this point of view was Jose de Acosta, who wrote *The Natural and Moral History of the Indies* in 1590. Others took up this argument, notably the greatest eighteenth-century American scientist, Thomas Jefferson, who in his *Notes on* Virginia, written originally in French in 1782, includes the following passage:

> Great question has arisen from whence came those aboriginal inhabitants of America? Discoveries, long ago made, were sufficient to shew that a passage from Europe to America was always practicable, even to the imperfect navigation of ancient times. In going from Norway to Iceland, from Iceland to Groenland, from Groenland to Labrador, the first traject is the widest: and this having been practiced from the earliest times of which we have any account of that part of the earth, it is not difficult to suppose that the subsequent trajects may have been sometimes passed. Again, the late discoveries of Captain Cook, coasting from Kamschatka to California, have proved that, if the two continents of Asia and America be separated at all, it is only by a narrow streight. So that from this side also, inhabitants may have passed into America: and the resemblance between the Indians of America and the Eastern inhabitants of Asia, would induce us to conjecture, that the former are the descendants of the latter, or the latter of the former: excepting indeed the Eskimaux, who, from the same circumstance of resemblance, and from identity of language, must be derived from the Groenlanders, and these probably from some of the northern parts of the old continent. A knowledge of their several languages would be the most certain evidence of their derivation which could be referred to. How many ages have elapsed since the English, the Dutch, the Germans, the Swiss, the Norwegians, Danes and Swedes have separated from their common stock? Yet how many more must elapse before the proofs of their common origin, which exist in their several languages, will disappear? It is to be lamented, then, very much to be lamented, that we have suffered so many of the Indian tribes already to extinguish, without our having previously collected and deposited in the records of literature, the general rudiments at least of the languages they spoke. Were vocabularies formed of all the languages spoken in North and South America, preserving their

> appellations of the most common objects in nature, of those which must be present to every national barbarous or civilized, with the inflections of their nouns and verbs, their principles of regimen and concord, and these deposited in all the public libraries, it would furnish opportunities to those skilled in the languages of the old world to compare them with these, now, or at any future time, and hence to construct the best evidence of the derivation of this part of the human race. (Jefferson, pp. 225–26)

Jefferson's supposition that the American Indians were not indigenous to America but, rather, were migrants from northeast Asia and, perhaps, mariners from Asia or elsewhere, finds support among modern archaeologists, who place the origin of the Indians in America at between twelve thousand and forty thousand years ago. According to this view, migrants by land followed herds of the mastodon from Siberia across a land bridge now covered by the Bering Strait to Alaska and then proceeded south, over the millennia spreading throughout North, Central, and South America, forming distinctive cultures.

The debate over the origins of the Indians in America and how their distinctive cultures have developed continues among scholars to the present day. Modern scholars differ in their explanations for the emergence of culture in America. Some believe it is indigenous, others that it is borrowed. *Isolationists* argue that American culture is indigenous, that cultures worldwide emerge in isolation through parallel development. For example, the reason why many world cultures built pyramids was because of the human attempt to reach the sky, to approach the divine: a universal idea that finds concrete realization in the monumental architecture of many cultures. Some isolationists embrace the theories of Carl Jung, that there is a shared universal human consciousness that would explain how humans living in isolation from one another can still have the same ideas. *Diffusionists*, however, believe that cultural characteristics diffuse from one culture to another. For example, the idea of the pyramid began in ancient Egypt (the mastaba, or primitive pyramid), spread to Mesopotamia (alternatively, that the idea began in Mesopotamia and spread to Egypt), and then spread to India (Harappan step pyramids in the Indus River valley) and finally to America, where step pyramids are found throughout the cultures of Central and South America. Norwegian archaeologist Thor Heyerdahl made several dramatic attempts to prove the diffusion of cultures. He built several boats along the lines of ancient models and made successful attempts at crossing the Pacific and the Atlantic. Heyerdahl argued that the indirect evidence of his voyages supported the theory that ancient Peruvians could have sailed across the Pacific from South America to South Pacific Islands, and that ancient Egyptians could have sailed west across the Atlantic to America.

Heyerdahl, in *Early Man and the Ocean*, argues that ancient peoples worldwide were much greater mariners than historians have otherwise suspected. The earth's oceans have prevailing winds and currents that will take any craft under sail, or even adrift, in a set direction—west across the southern Pacific if the mariner launches off the coast of Chile; northeast toward the Aleutian Islands if the mariner begins off the coast of northeastern Asia; west toward the Caribbean if the mariner launches off the coast of northwestern Africa. To support this theory, Heyerdahl in 1947 built a raft, christened the *Kon-Tiki* after an ancient Peruvian god and made of balsa wood from the forests of Peru, and set sail west from Peru into the southern Pacific, eventually reaching the islands of Polynesia. In 1969 and 1970 Heyerdahl constructed two ships from papyrus, the tall reed that grew along the Nile and other rivers of the Middle East. He discovered archaeological evidence that ancient peoples constructed seagoing vessels of papyrus, and he decided to try it out. In 1970 Heyerdahl and a small crew successfully sailed such a reed ship, christened the *Ra* after the Egyptian sun god, from northwestern Africa across the Atlantic to the Caribbean. Heyerdahl was careful in his claims about the visits of ancient Africans and Asians to ancient America. Other diffusionists have made more extravagant claims. Barry Fell, in *America B.C.* and *Bronze Age America*, argues that throughout America there are archaeological sites of the ancient Greeks, ancient Romans, ancient Phoenicians, and ancient Egyptians. The Irish legend of a Catholic monk, St. Brendan, setting sail for paradise across the ocean west of Ireland has spawned attempts to re-create his voyage to America. There is a tourist site in New Hampshire that allows the credulous visitor to wander the ruins of ancient Irish buildings that are 1,300 to 1,400 years old. The ancient Greek mariner Pytheas of Massilia set sail around 300 BCE west through the Strait of Gibraltar and north along the European coast to Britain to trade in tin. But he apparently explored even farther, to a distant land called Thule, which could have been Scandinavia, Iceland, Greenland, or even farther west: to America The ancient Carthaginians, following the great seafaring traditions of the Phoenicians (Carthage was a colony of Phoenicia, founded about 1000 BCE) explored the Atlantic; one story has it that Carthaginian mariners came to a large island with great rivers that unfortunately they refused to explore further. Perhaps ancient Egyptian mariners did sail to America 4,000 years ago, followed by the Carthaginians 2,600 years ago and the Greeks 300 years later. But if so, what impact did they have on the native peoples? Did they return to Africa, Asia, and Europe with knowledge gained from their experiences in America?

NORTH AMERICAN INDIANS

Native American communities extended throughout the North American continent. Along the east coast, south of the St. Lawrence River, the natives were divided into many tribes that spoke different dialects that were all part of the Algonquian language. Up the St. Lawrence and west to the Appalachian Mountains and the trans-Appalachian region to the Mississippi River, tribes spoke dialects of the Iroquoian and Siouan languages. Further south, from the mid-Atlantic to the Gulf of Mexico, tribes were part of the Muskogean language family. West of the Mississippi River, in the Great Plains, were Cadoan and Siouan language groups. In all of these regions, the Native Americans lived off the land through hunting, gathering, and agriculture. They typically lived in modest communities in wooden and skin homes. They had a strong community ethos, yet each tribe was fiercely independent.

The first contacts between European explorers and Native American tribes for which we have recorded evidence were sometimes peaceful, other times warlike, but always unequal due to the superior technology of the explorers from across the Atlantic. Whereas Native American culture and society were based on a comparatively primitive technology of tools and weapons fabricated from stone, wood, and animal bones, Europeans had long before discovered metallurgy and by the second millennium CE had produced sophisticated tools and weapons out of a variety of different elements, particularly iron, copper, and tin. Whereas Native Americans were at least part of the time nomadic, living in loosely structured tribes dominated typically by warriors, Europeans lived in settled communities with a patriarchal government, with a citizenry with some freedoms and a growing sense of equality.

North American Indian tribes did not independently invent writing; as a result, Native American historical experiences are often known to us today through the recording of their customs, habits, stories, and experiences by means of outside observers: explorers, adventurers, scientists, and missionaries who made contact with and traveled among or lived with the native peoples. Examples are numerous of these European and American sources of information about the American Indians. The sources range in quality from being excellently written and scientifically detached to being poorly written and biased against Indians.

European explorers and colonists in North America were impressed by the practical arts and natural knowledge exhibited by Indigenous peoples. Samuel de Champlain, John Smith, Cadwallader Colden, John Lawson, John Josselyn, and many others noted the skillfully made Indian birch bark canoes—their lightness, sturdiness, and buoyancy. Algonquian fishing weirs for catching salmon were intricate, practical, and useful. Snowshoes for traversing northern climes allowed for rapid movement through the winter. Indian culheags, natural traps made of wood, were effective ways to acquire furs and pelts. The Indian materia medica became the basis for the materia medica adopted by the European colonists in the sixteenth, seventeenth, eighteenth, and nineteenth centuries.

Thomas Nuttall (1786–1859), an English explorer, botanist, and natural historian, journeyed throughout North America on half a dozen occasions in the early nineteenth century. It disturbed Nuttall that many of the ancient tribes of North America no longer existed: "Their extinction will ever remain in the utmost mystery. The agency of this destruction is, however, fairly to be attributed to the Europeans, and the present hostile Indians who possess the country. It is from these exterminating and savage conquerors, that we in vain inquire of the unhappy destiny of this great and extinguished population, and who, like so many troops of assassins, have concealed their outrages by an unlimited annihilation of their victims." Because of such extinction, their language, customs, and culture had fairly vanished. Nuttall was afraid that many of the tribes he visited along the Missouri, Mississippi, and Arkansas rivers would experience the same fate, so he did what he could to record aspects of the language and culture, including their proclivity toward scientific understanding. Nuttall believed that many ceremonies and beliefs that the Indians had respecting the natural environment were similar to those of the ancient Europeans and Asians, such as those described by the Greek historian Herodotus (490–430 BCE), of millennia past. "Habitual observation," he wrote about the Osage Indians in his *Journal of Travels in the Arkansa Territory*, "had taught them [Osages] that the pole star remains stationary, and that all the others appear to revolve around it; they were acquainted with the Pleiades, for which they had a peculiar name, and remarked the three stars of Orion's belt. The planet Venus they recognized as the Lucifier or harbinger of day; and, as well as the Europeans, they called the Galaxy the heavenly path or celestial road. The filling and waning of the moon regulated their minor periods of time, and the number of moons, accompanied by the concomitant phenomena of the seasons, pointed out the natural duration of the year." Respecting their superstition, Nuttall wrote, "In no part of North America have we ever met with that kind of irrational adoration called idolatry. All the natives acknowledged the existence of a great, good, and indivisible Spirit, the author of all created being. Believing also in the immortality of the soul, and in the existence of invisible agencies, they were often subjected to superstitious fears, and the observance of omens and dreams, the workings of perturbed fancy. By these imaginary admonitions, they sometimes suffered themselves to be controlled in their most important undertakings, relinquishing every thing which was accidentally attended by any inauspicious presage of misfortune." Nuttall perceived the Cherokees as on the road to civilization, as among the "imperfectly civilized nations" of the world, as in Asia. Like many European naturalists, Nuttall discovered that the Indigenous people had an extensive materia medica. He met an Aricara medicine man who possessed down of cattail for burns and skin problems; wormwood for intestinal problems and malaria; black-eyed Susan and green-headed coneflower for burns, skin irritations, and snake bite.

Another nineteenth-century European naturalist who held the same beliefs as Nuttall about preserving knowledge of ancient Indian customs and traditions, including scientific knowledge, was Jean Louis Berlandier (c. 1800–1851), who traveled throughout Mexico and southern Texas in the 1820s, 1830s, and 1840s. Berlandier was a natural historian who recorded the scientific proclivities exhibited by American Indian tribes in the region of what is today Texas. Indians such as the Tonkawa, Comanche, Caddo, Tawakoni, Carrizo, and Lipan Apache possessed an extensive materia medica. For example, the Tonkawa used a species of holly, typically called yaupon, as a purgative and emmenagogue. The Texas mountain laurel, which Berlandier called the *frijolilla*, were used by Indians as a preventive against lice, as a general purgative for cleansing the system, and as a means to treat eye diseases. The Comanche, Berlandier discovered, used the root of a grass that they called *puip* as a cure-all. "No indigene exists who does not carry some pieces with him," he wrote. "When one of them is wounded, they pound the root, which is very long, and squeeze the sap (or saliva) into the wound. . . . I have often heard it said that serious wounds have often been radically cured by that means in a very short time." Many tribes used the western mugwort as a means to treat fever, wounds, sick stomach, headache, irritations on the skin, aches and pains, and asthma. Many tribes, such as the Comanche, had in their materia medica plant species to treat venereal diseases such as syphilis. Like Nuttall, Berlandier was impressed by the knowledge of the astronomical phenomena that Indians possessed. The Comanches frequently referred to the patterns of constellations in the night sky by which they kept track of time passing till dawn. Contrary to what most Euro-Americans believed, Berlandier believed that the Indigenous peoples of North America revealed the ability to engage in "careful observation of natural phenomena." This included abstract thinking, some elementary mathematics, approximations of the passage of time, and techniques of long-distance communication. Comanche used "the smoke of a fire, allowed to escape in puffs by removing the burning brands and then suddenly dropping them, its direction, and so on, [to] communicate to other Comanche groups the news of victory or defeat in pursuit of an enemy, the direction an expedition is taking, its progress, etc." (quotes from Lawson, pp. 92, 102).

Indian science was not about the distinction of object and subject, as European science was becoming, of analysis and experimentation on something outside of oneself; rather, it was a symbiotic connection of subject and object, of humans with the natural environment. Like the Comanche, the Plains Indian Pawnee tribe was captivated by the night sky. The Pawnee considered themselves people of the stars, born of the stars. Their villages were arranged to conform to pattern of the stars. They devised a star chart to mirror their nightly observations. As in many cultures, the morning star (Venus) was sacred, and Pawnee astro-religious rituals included the human sacrifice of female virgin. The Pueblo Indians of the North American Southwest likewise engaged in what scholars term *archaeoastronomy*. Peoples of the Uto-Aztecan language group, Pueblo forebears, were the Anasazi Indians, who developed a culture thousands of years ago based on agriculture. The Pueblo lived in mud-brick buildings constructed on the sides of hills in a desert environment of stark sunshine that dominated human existence. They observed the living sun going across the sky, day after day. They paid strict attention to solstices to begin ceremonies. Some rooms in Pueblo buildings were solstice markers, catching the rays of the sun on precise days of solstices. The Pueblo Indians explained solstices as the sun needing rest and Indian ceremonies providing it energy. This was observation and an attempted understanding of a living, spiritual, natural world. A more precise astronomical awareness is found at Fajada Butte at Chaco Canyon in New Mexico. There, among the petroglyphs, or rock carvings, are found drawings marking the movements of the sun and moon. At the summer solstice, a dagger from the light of the sun penetrates the center of a spiral petroglyph. At the winter solstice, sun daggers appear on each side of the spiral petroglyph. Likewise, a dominant motif among many North American tribes is the medicine wheel, which aligns celestial phenomena with the spokes on a wheel; it is *medicine* because the wheel has an astronomical as well as spiritual significance. The medicine wheel simultaneously indicates the basic elements of existence with the human life span. The directions of medicine wheel are: north—body, plants and animals, infancy of humans; east—mind, air, adolescence of humans; south—fire, heat, adulthood of humans; west—spirit, water, final stages of life.

INDIANS OF MEXICO

Agriculture, worldwide, is the basis for all settled, *civilized* societies. The domestication of corn in the Western Hemisphere occurred perhaps nine thousand to ten thousand years ago in Central America. The first domesticated plant used was a wild grass, teosinte. About three thousand years ago, sunflower and ragweed were domesticated in North America. Agriculture in America derived from testing of the cultivation of various seeds, irrigation, and a systematic practice based on observation. The Indians of Central America and Mexico, like those in North America (north of the Rio Grande and Gulf of Mexico) developed, based on their observation and consequent knowledge, an extensive materia medica. On his journeys into northern Mexico, for example, Jean Louis Berlandier discovered that the Carrizo Indians used *cenizilla* to combat fever, such as that associated with yellow fever. They also employed the leaves of the black willow to create a medicine that acted as a fever reducer. In his study, *Des Plantes Usuelles chez les Indiens du Mexique*, Berlandier wrote that Mexican Indians used the juice of the *Begonia alata* to relieve pain; as a purgative (after being injected into the anus); for dyspepsia; and to combat the effects of syphilis.

On his first visit to Mexico City in the 1820s, Berlandier observed the remnants of the ancient civilization of the Toltecs, a people who thrived in central Mexico from 900 to 1150 CE, basing their culture in part on the earlier Mayan and Olmec civilizations to the east. The Toltec centers of civilization were Culhuacan and Tollan, in central Mexico. Near Culhuacan, Berlandier visited the pyramid of Xochicalco; he observed that "the sides of the parallelogram situated at the summit are perfectly oriented towards the four cardinal points of the compass" (quoted in Lawson, p. 47). Little is known of the Toltec civilization. It appears as a step from the ancient Olmec and Mayan civilizations toward the Aztec civilization of the mid-second millennium CE.

The Olmec civilization developed toward the end of the second millennium BCE along the Gulf of Mexico west of the Yucatan Peninsula. Little is known of the Olmecs, which has led to much speculation about their origins and significance. Most scholars believe that the Olmecs were the initial American civilization, from which other civilizations (Mayan, Toltec, Aztec, Inca) derived. Olmec culture was centered at La Venta, a hot and humid jungle environment at the western terminus of the Atlantic Canary current, where the winds and currents come from Northwest Africa. Thor Heyerdahl, among others, noting that the Olmec civilization appearing in America about the same time that the Phoenicians were colonizing the northwest coast of Africa is perhaps not coincidence, argued that the Olmecs, who had writing and calendars and built pyramids, were influenced by Phoenician voyagers from Africa. Heyerdahl noted that the Phoenician colony of Lixus had, like the Olmecs, sun worship and gigantic megaliths. Astronomers, architects, scribes, and potters lived at Lixus, and perhaps at La Venta as well.

That the Maya civilization had writing, cities, calendars, sophisticated astronomy, and pyramids leads some scholars such as Heyerdahl to speculate that the Mayans descended from the Olmecs, who were influenced by cultures across the Atlantic. Whether or not such a trans-Atlantic crossing occurred, it is probable that the Olmecs did have a significant impact on subsequent civilizations in Mexico and, perhaps, South America.

People of the Mayan civilization of the first millennium CE, like other civilizations, developed extensive food production and trade based in urban areas. One of the most significant was Teotihuacan, which thrived in the centuries before and after the birth of Christ. The people of the city erected several significant public buildings, pyramids, and temples. Toward the end of the first millennium, the leading center, near where Veracruz is today, was El Tajín; another city on the Yucatan Peninsula was Chichen Itza. The people of these cities had a sophisticated view of astronomy. They created calendars based on their astronomical knowledge. They aligned public buildings and temples according to their understanding of the relationship of humans to the heavens. At Chichen Itza, for example, an astronomical tower called El Caracol has windows oriented toward solstices and the path of Venus through the heavens. During equinoxes, the shadows at the El Castillo pyramid at Chichen Itza descend steps to appear like a serpent. At another Mayan city, Uxmal, the Palace of the Governor was aligned toward phases of Venus; walls of the palace were decorated with symbols of the constellations.

Subsequent to the Mayas were the Aztecs of central Mexico. Aztec civilization was founded in the thirteenth century CE on older Toltec and even older Olmec civilizations. The capital at Tenochtitlán in the region or state of Tlaxcala was a large city of 250,000 people at its height, built on an island in Lake Texcoco (1325). The city was named for the Aztec chief who led his people there, Chief Tenoch. Aztecs were the Nahuatl people, their mode of communication was oral, no writing. They were polytheistic and violent. According to tradition the god Huitzilopochtli led the Aztecs to Lake Texcoco, and in exchange for their prosperity and power, they had to sacrifice humans (war captives) to him. Kings were literally "speakers" of the people, with perhaps a divine, incarnate aspect. Montezuma II (1466–1520), who was conquered by Cortes, had this quasi-divine aspect.

Hernan Cortes arrived in 1519 to Tenochtitlán and subsequently described the city to the Emperor Charles V:

> This great city of Tenochtitlán is built on the salt lake, and no matter by what road you travel there are two leagues from the main body of the city to the mainland. There are four artificial causeways leading to it, and each is as wide as two cavalry lances. The city itself is as big as Seville or Córdoba. The main streets are very wide and very straight; some of these are on the land, but the rest and all the smaller ones are half on land, half canals where they paddle their canoes. All the streets have openings in places so that the water may pass from one canal to another. Over all these openings, and some of them are very wide, there are bridges. . . . There are, in all districts of this great city, many temples or houses for their idols. They are all very beautiful buildings. . . . Amongst these temples there is one, the principal one, whose great size and magnificence no human tongue could describe, for it is so large that within the precincts, which are surrounded by very high wall, a town of some five hundred inhabitants could easily be built. All round inside this wall there are very elegant quarters with very large rooms and corridors where their priests live. There are as many as forty towers, all of which are so high that in the case of the largest there are fifty steps leading up to the main part of it and the most important of these towers is higher than that of the cathedral of Seville. (Trans. Pagden)

The riches of the civilization derived from not only conquest but also extensive agriculture and trade. The Spanish were impressed by the engineering and urban design achievements of the Aztecs, especially the architecture of temples and megaliths.

SOUTH AMERICAN INDIANS

Some of the world's most ancient pyramids were constructed in the fourth and third millennia BCE in what is today Peru. At Caral and Paraiso, archaeologists have discovered the remains of ancient societies (often called Norte Chico) with sufficient sophistication to raise monolithic structures for religious purposes. The civilization in the Supe Valley near the Pacific in Peru, called Caral, featured extensive agriculture and trade suggesting an urban environment as well as monumental architecture. These early civilizations, largely unknown to modern scientists today, were the first in a series of civilizations that emerged in the Andean highlands along the west coast of South America. The Chavin civilization, first millennium BCE, and the Nazca civilization of southern Peru, first millennium, CE, were forerunners of the Tiahuanaco (Tiwanaku) civilization that dominated the Andes around modern Peru and Bolivia in the first millennium CE. The Tiahuanaco culture was sufficiently sophisticated and wealthy, from agriculture and trade, to construct pyramidal temples and monoliths, such as the Gateway of the Sun.

The Inca civilization, built in part upon previous cultures such as the Tiahuanaco, dominated the Andes along a long and narrow region covering much of western South America over the course of a century—from the fifteenth to the sixteenth—when Spanish conquistadores conquered and took control. From the capital of Cusco, the Incas dominated millions of peoples and ruled an organized, efficient empire. The land afforded great wealth in terms of crops and precious metals. Lacking wheeled vehicles, nevertheless the Incas built a vast network of roads to connect their empire and support trade. The Incas were master builders who built walled communities and temples to honor their pantheon of deities. When the explorer Hiram Bingham reached Machu Pichu in the early 1900s, he was astonished by the workmanship of the ruins:

> The wall followed the natural curvature of the rock and was keyed to it by one of the finest examples of masonry I have ever seen. This beautiful wall, made of carefully matched ashlars of pure white granite, especially selected for its fine grain, was the work of a master artist. The interior surface of the wall was broken by niches and square stone-pegs. The exterior surface was perfectly simple and unadorned. The lower courses, of particularly large ashlars, gave it a look of solidity. The upper courses, diminishing in size toward the top, lent grace and delicacy to the structure. The flowing lines, the symmetrical arrangement of the ashlars, and the gradual gradation of the courses, combined to produce a wonderful effect, softer and more pleasing than that of the marble temples of the Old World. Owing to the absence of mortar, there are no ugly spaces between the rocks. They might have grown together. The elusive beauty of this chaste, undecorated surface seems to me to be due to the fact that the wall was built under the eye of a master mason who knew not the straight edge, the plumb rule, or the square. He had no instruments of precision, so he had to depend on his eye. He had a good eye, an artistic eye, an eye for symmetry and beauty of form. His product received none of the harshness of mechanical and mathematical accuracy. The apparently rectangular blocks are not really rectangular. The apparently straight lines of the courses are not actually straight in the exact sense of that term. To my astonishment I saw that this wall and its adjoining semicircular temple over the cave were as fine as the finest stonework in the far-famed Temple of the Sun in Cuzco. (Bingham, pp. 320–21)

Sacred places such as the Temple of the Sun, Coricancha, were also astronomical observation sites, as the Incas had a fascination for constellations, equinoxes, and lunar movements.

Thor Heyerdahl argued that the cultures of America stretching from Mexico to Chile were the product of a "narrow Central American bridge between La Venta and Tiahuanaco," a trade and cultural corridor that allowed for the movement of sophisticated ideas deriving initially from the Olmecs (Heyerdahl, p. 70). If so, then Inca civilization derived from Tiahuanaco, Chavin, Nazca, and Mochica cultures, which were likewise very sophisticated and had contact with central American and Mexican civilizations—Aztecs, Toltecs, Mayas, all of which derived from the shadowy Olmecs.

FURTHER READING

A. F. Aveni, *Archaeoastronomy in the New World: American Primitive Astronomy* (Cambridge: Cambridge University Press, 1982).

Hiram Bingham, *Inca Land: Exploration in the Highlands of Peru* (Boston: Houghton Mifflin Company, 1922).

David M. Braun, "Corn Domesticated from Mexican Wild Grass 8,700 Years Ago," National Geographic Society, https://blog.nationalgeographic.org/2009/03/23/corn-domesticated-from-mexican-wild-grass-8700-years-ago.

Mark Cartwright, "Chichen Itza," *Ancient History Encyclopedia*, https://www.ancient.eu/Chichen_Itza.

Mark Cartwright, "Inca Civilization," *Ancient History Encyclopedia*, https://www.ancient.eu/Inca_Civilization.

Mark Cartwright, "Teotihuacan," *Ancient History Encyclopedia*, https://www.ancient.eu/Teotihuacan.

Mark Cartwright, "Tiwanaku," *Ancient History Encyclopedia*, https://www.ancient.eu/Tiwanaku.

Hernan Cortes, *Letters from Mexico*, trans. Anthony Pagden (New Haven, CT: Yale University Press, 1986).

"First City in the World?" *Smithsonian Magazine*, August 2002, https://www.smithsonianmag.com/history/first-city-in-the-new-world-66643778.

Thomas Hariot, *A Brief and True Report of the New Found Land of Virginia, Reproduced in Facsimile from the First Edition of 1588* (New York: Dodd, Mead & Company, 1903).

Thor Heyerdahl, *Early Man and the Ocean: A Search for the Beginning of Navigation and Seaborne Civilizations* (New York: Vintage Books, 1980).

Thomas Jefferson, "Notes on Virginia," in *The Life and Selected Writings of Thomas Jefferson*, ed. Adrienne Koch and William Peden (New York: Modern Library, 1972).

Russell M. Lawson, *Frontier Naturalist: Jean Louis Berlandier and the Exploration of Northern Mexico and Texas* (Albuquerque: University of New Mexico Press, 2012).

Ramiro Matos Mendieta and José Barreiro, eds., *The Great Inka Road: Engineering an Empire* (Washington, DC: Smithsonian Institution, 2015).

S. C. McCluskey, "The Astronomy of the Hopi Indians," *Journal for the History of Astronomy* 8 (1977): 174–95.

Thomas Nuttall, *Journal of Travels in the Arkansa Territory* (Philadelphia: Palmer, 1821).

Christopher A. Pool, *Olmec Archaeology and Early Mesoamerica* (Cambridge: Cambridge University Press, 2007).

Robert M. Schoch, *Voyages of the Pyramid Builders: The True Origins of the Pyramids from Lost Egypt to Ancient America* (New York: Penguin Books, 2004)

Virgil Vogel, *American Indian Medicine* (Norman: University of Oklahoma Press, 1990).

Roger Williams, *A Key unto the Language of America: Or, an Help to the Language of the Natives in that Part of American, Called New England* (London: Gregory Dexter, 1643).

PART 12

The Impact of Ancient and Medieval Science upon Modern Science

34

Conclusion: The Scientific Revolution

THE SCIENTIFIC REVOLUTION

The European Renaissance (1300–1600) spawned a rediscovery of classical knowledge generating a new approach to learning, and new discoveries in the sciences, that resulted in what many historians have called the Scientific Revolution. The great thinkers at the dawn of modern science—Copernicus, Galileo, Kepler, da Vinci, Bacon, Harvey, and Vesalius—worked in the shadow of Ptolemy, Hipparchus, Aristotle, Pliny, Hippocrates, and Galen.

Some scholars debate the relevance of the term *Scientific Revolution* to describe the period of the sixteenth and seventeenth centuries in Western European science. Indeed, some of the developments in European science during this time were anticipated by earlier Islamic, Chinese, Greek, and Roman scientists. What made the approach to science, the methods and philosophy directed toward science by sixteenth- and seventeenth-century Europeans seem *revolutionary* was the impact that science had on the wider culture. The Scientific Revolution slowly changed the worldview of Europeans, from a focus on divine revelation and providence to a secular focus on human reason and will. *Man is the measure of all things* became the mantra of thinkers at the end of the Renaissance and of the larger culture in subsequent centuries. Previously, in ancient and medieval science, humans worked to understand a natural environment that had been clearly brought forth by something divine, even omniscient and omnipotent. Clearly, these beliefs were often extremely superstitious, and religion frequently got in the way of scientific thinking. For better or worse, the Scientific Revolution changed the way humans thought about themselves in relation to the universe. Humans came to see themselves (not God) as the ultimate expression of reason, the ultimate source of power. Barring any unfortunate human-made disasters, because of what was scientific, humans believed they had the ability to discover the true nature of all things and alter their lives and society accordingly.

It began with Renaissance philologists, who engaged in the painstaking work of studying the varied surviving handwritten copies of ancient works, attempting and succeeding in providing accurate texts close to the original. Few scholars could read classical Greek, but there was enough demand that some Italian printers issued editions of Aristotle (384–322), his Peripatetic followers, the geographer Strabo (64 BCE–21 CE), and similar works in the original Greek. Latin was the language of scholarship and learning during the Renaissance; printers issued Latin translations of Greek writers such as Euclid (325–265), Ptolemy (100–170), Galen (130–200), and Plato (327–347). All of these added to the growing corpus of knowledge about ancient Greek science.

The trivium and quadrivium continued to orient Renaissance academic studies. (As in the ancient world, specialization and professionalization of science were still long in the future.) From this comes the idea of the "Renaissance man" who engaged in the study of all phenomena, human as well as natural. The study of Aristotle and Plato continued to orient Renaissance pursuits of knowledge. Aristotelian logic was considered by some Renaissance scholars to be "scholastic," a derogatory term for a medieval thinker who relied on frozen formulas of thought. Indeed, there was more fluidity to Renaissance thinking, more openness and a broadening range of interests, not all of them religious. Christianity still dominated the Renaissance worldview, though there was more freedom to inquire into secular topics, perhaps because the world itself—its expanding trade, growing cities, and increasing wealth—was becoming more secular.

Most important, the Renaissance became a great time of questioning. Intellectuals of the fourteenth century, such as Francesco Petrarca (1304–1374), not only studied the ancient classics, but questioned the foundations of ancient thought—indeed, questioned ancient thinkers themselves. Petrarch gained a literary as well as vicarious personal familiarity with his hero Cicero (106–43), the Roman republican and Stoic philosopher of the first century BCE. Petrarch discovered many of Cicero's letters, which gave him a true sense of Cicero's personality and opened the doors to a critical understanding of Cicero and his work. In time, others imitated Petrarch in their willingness to question the ancients.

Modern psychological study continues to learn from the ancients thanks to the experiences of Renaissance intellectuals such as Francesco Petrarca and Michel de Montaigne (1533–1592). The philosophy and psychology of the ancient world inspired Petrarch to write his incredible works of self-examination, such as the *Secret*, and Montaigne's great essay on self, the *Essays*. Petrarch learned from his study of Augustine (354–430) that the path to contentment lies in self-examination in the context of a full understanding of time and the impact of the fleeting nature of time on the human psyche. Montaigne learned from Stoic thinkers such as Seneca (5–65) and Platonists such as Plutarch (46–120) a similar lesson in finding in the present sufficient contentment from the active recollection of the past and passive anticipation of the future. Montaigne gained assistance in combating his depression and anxiety by engaging in a dialogue with the past, conversing with past humans, by means of which Montaigne was able to see his life as hardly unique, as a common human experience of guilt, regret, suffering, and unhappiness. Accepting his *humanness* helped Montaigne to appropriate living, which he discovered was the means of contentment.

The initial realization of possible errors in ancient scientific thought occurred during the fifteenth century, when Renaissance explorers began to break from the bonds of ancient geographic thinking to arrive at a new and more accurate picture of the world. Portuguese and Italian explorers from the mid- to the late fifteenth century showed the willingness and courage to question the legends and myths of the world initiated during antiquity and accepted as truth during the Middle Ages. The ancient picture of the world was limited to three continents, Europe, Asia, and Africa, and two oceans, the Atlantic and the Indian. Ancient Greek philosophers had established the sphericity of the earth and its hemispheric nature. But there were many misconceptions in Greek geography. There was the notion of a fiery barrier that separated the Northern and Southern Hemispheres, through which no person or ship could pass. Claudius Ptolemy, whose works, translated into Latin, were more available by the fifteenth century, taught that Africa and Asia were joined by a terra incognita, an unknown land to the south, which made the Indian Ocean an inland sea. Ptolemy also overestimated the size of Asia and underestimated the circumference of the earth, making it appear that the Atlantic Ocean—that is, the distance from Europe to Asia—was much shorter than it is. The Portuguese proved in the late fifteenth century that the equatorial zone of fire was a myth, that one could sail from north to south and vice versa, and that Ptolemy was wrong in assuming that Africa could not be circumnavigated. The four voyages of Christopher Columbus (1451–1506), the Genoese sailor, showed that Ptolemy's geography of the earth was erroneous, that the distance from Europe to Asia was much greater than Ptolemy thought, and that there were peoples and continents—North and South America—unknown to the ancients. Michel de Montaigne wondered in one of his essays what Plato, who imagined the lost city of Atlantis, would have made of the inhabitants of the Americas and the rich civilizations of the Aztecs and Incas. The Portuguese sailor Ferdinand Magellan (c. 1480–1521) showed, in circumnavigating South America and sailing across the Pacific Ocean, just how rudimentary was ancient geographical knowledge.

Laying the basis for the new focus on science during the late European Renaissance were Italian cities such as Florence, Padua, Pisa, Genoa, Bologna, and Venice; these cities were the most flourishing commercial and cultural centers of the Renaissance. Padua, for example, was a center of Aristotelian studies, particularly as applied to medicine. In Florence at the end of the fifteenth century, Leonardo da Vinci (1452–1519) studied Greek philosophy and science, Aristotle and Galen, which encouraged his studies in physics and anatomy. Nicholas Copernicus (1473–1543), as a student in Bologna, came in contact with scientists and philosophers rediscovering the importance of Pythagoras (570–490) and Plato in the history of rational and mathematical thought. Indeed Copernicus, in the dedication to Pope Paul III that opened his landmark *On the Revolutions of the Heavenly Spheres*, commented on his debt to Claudius Ptolemy and the inspiration that he received at the hands of Plutarch, who had recorded the Pythagorean hypothesis of a moving and orbiting earth. Copernicus knew as well the heliocentric theory of Aristarchus of Samos (310–230). "Some think that the Earth is at rest," he wrote in the dedication to Pope Paul III, quoting Plutarch, "but Philolaus the Pythagorean says that it moves around the fire with an obliquely circular motion, like the sun and moon. Herakleides of Pontas and Ekphantus the Pythagorean do not give the Earth any movement of locomotion, but rather a limited movement of rising and setting around its centre, like a wheel" (trans. Wallis). Even so, Copernicus was largely beholden to Aristotle and his disciples, such as Claudius Ptolemy, in that Copernicus accepted the finite universe of sun, moon, earth, five planets, and outside starry vault of the fixed stars. But Copernicus believed that this conception of the universe worked better if one assumed it was heliocentric rather than geocentric. Copernicus's theory of a sun-centered universe caused such a stir that his book was condemned by the church but vigorously supported by courageous scientists such as Galileo Galilei and Johannes Kepler.

Johannes Kepler (1571–1630) was a student of Ptolemy in imitation of his mentor Tycho Brahe (1546–1601), the Danish astronomer who was a lifelong defender of Ptolemy's world system. Kepler, however, converted to Copernicus's world system partly because he was also convinced by Platonic and Pythagorean theories of harmony and mathematics. Like other Renaissance Neoplatonists, Kepler believed that geometric forms mirrored Plato's ideal forms and that the patterns of the universe reflected both. Kepler wrote *The Harmony of the Spheres,* revealing his belief that Pythagorean harmonies were reflected in the movement of the planets. Like the Neoplatonists of the ancient world, Kepler assumed that the sun, the source of light and power, must be the center of all things.

Galileo Galilei (1564–1642), who perhaps more than any other Renaissance scientist inaugurated the Scientific Revolution, used the works of Archimedes (287–212), Ptolemy, and Aristotle as the foundation for his own discoveries and repudiation of ancient theories about motion and the heavens. Galileo was Italian, a native of Pisa. A gifted mathematician, he mastered Euclid's *Elements* as a student. He reputedly was the first scientist to study the solar system and Milky Way with the telescope. Although many of his discoveries contradicted Aristotle, Galileo sympathized with the ancient scientist, believing that had Aristotle had the advantages of seventeenth-century thought, he, too, would have discovered the errors of his theories of motion and the universe.

Galileo, in *Dialogue on the Two Chief World Systems* (1632), provided empirical arguments, based on his observations using the telescope, to support the heliocentric universe. But more, Galileo showed that the entire ancient conception of the universe was incorrect. His studies of the sun revealed sunspots when it had long been assumed that the sun was a perfect, unchanging heavenly body. His observations of the moon revealed craters rather than the ideal, godlike moon of the ancient astronomers. Galileo discovered the moons of Jupiter, contradicting the ancient notion that the universe had a set number of heavenly bodies moving in set spherical paths. He turned his telescope to the stars and discovered that the Milky Way was a vast, seemingly infinite realm not limited to a starry vault.

With Galileo in the lead, other scientists took up the cause of empiricism. Francis Bacon (1561–1626) arrogantly tossed aside the ancients even as he relied on them for his initial assumptions. In *Novum Organum,* Bacon proposed a "new science" based on purely empirical methods. Andreas Vesalius (1514–1564) was an ardent student of Galen, using the Roman's works and theories in the process of making new discoveries to undermine them. William Harvey (1578–1657) likewise developed the theory of the circulation of the blood by first wondering whether Galen's theories were correct. Pierre Gassendi (1592–1655), a Skeptic who embraced the theories of the ancient philosophers Sextus Empiricus (160–210) and Pyrrho (365–279), declared emphatically that Aristotelian philosophy is not science. Gassendi and Robert Boyle (1627–1691) resurrected the ancient atomic theory—the philosophy of the universe of the ancient Epicureans and Stoics—arguing that the universe is comprised of invisible corpuscles in constant motion. René Descartes (1596–1650), like Francis Bacon, declared revolution from ancient science and philosophy as well, rather as a child rebels from the parental strictures of the past.

Next to the Greek Hellenistic Age, the most prolific period in the development of mathematics occurred after 1500 in Europe, beginning with the Scientific Revolution of the sixteenth century. European mathematicians invented a series of tools to help solve increasingly difficult problems. Scotsman John Napier (1550–1617), for example, invented logarithms, which allowed for advanced calculations using exponents. French philosopher René Descartes developed analytical geometry based on a coordinate system, which allowed for two- and three-dimensional geometric problems to be solved algebraically. Gottfied Leibniz (1646–1716) and Isaac Newton (1642–1727) in the late seventeenth century each independently developed a system, called calculus, to solve mathematical problems involving extremely large (infinitesimal) quantities and rates of change over time. Differential calculus is used to calculate change over time, while integral calculus is used to calculate problems involving quantities.

The greatest seventeenth-century thinker, Isaac Newton, provided an empirical, mathematical approach to explaining the universe in *Principia Mathematica* (1687). Newton proposed three laws of motion that explained the fundamental principle of universal gravitation. Newton's laws showed why the planets orbited the sun in elliptical motion, how the competing push and pull of gravitation kept the planets and other space phenomena in the position moving according to the same rate and direction, and why time and space were universal constants.

The same year that Isaac Newton published his *Principia Mathematica*, 1687, the curriculum at America's best college, Harvard, continued to be devoted to Aristotle in logic and physics and Ptolemy in astronomy. Copernicus had made little headway in the American colonies, although a few almanacs were beginning to include descriptions. Even into the eighteenth century, Newton was thought to be difficult reading for American scientists, and college curricula struggled to abandon the influence of classical physics and astronomy. Medicine continued its relationship with Galen. Aristotle's *Politics* was read alongside Thomas Hobbes. Stoic thinkers such as Cicero and Seneca continued to intrigue American philosophers. Plutarch was still the biographer of choice. That Thomas Jefferson, arguably the most brilliant and revolutionary eighteenth-century American thinker, was also the most learned student of the Greek and Roman classics, might seem ironic today but not for his time, when fluency in Greek and Latin was still the mark of the educated person.

Much of European science was metaphysical, and direct applications of theory into practice were rare. American scientists provided the practical complement to European metaphysics. Ben Franklin (1706–1790), for example, took European ideas on electricity and provided practical proofs and applications. Franklin built a kite made of silk to which he attached a small iron rod; he held the kite by a silk string, to which was also attached a brass key. Flying the kite in a thunderstorm, Franklin discovered that the brass key was electrified, which proved to him that lightning is in fact an electric charge. Later Franklin invented a lightning rod, which was designed to draw the electric charge from storm clouds to prevent buildings being directly hit by lightning.

The meeting between theory and practice in science has come to fruition in the past two centuries. Taking up where Franklin left off, Joseph Priestly (1733–1804) explained how electric force is comparable to a gravitation force. Charles Coulomb (1736–1806) explored the positive and negative charges of an electric charge. Hans Oersted (1777–1851) discovered the relationship between the electrical and magnetic forces. Andre-Maria Ampere (1775–1836) explained the nature of electric force and movement. Englishman Michael Faraday (1791–1867) in 1821 proved how electrical energy can become mechanical energy, which set the stage for the many inventions of the nineteenth and twentieth centuries, such as the invention of the electromagnet by Englishman William Sturgeon (1783–1850), the discovery of alternating current by Serbian American Nikola Tesla (1856–1943), and those of many other scientists and inventors, culminating in the work of Thomas Alva Edison (1847–1931), who invented the incandescent light bulb.

During the nineteenth and twentieth centuries, developments in mathematics included the discovery of Brownian motion (random variables); chaos theory (fractal geometry); group theory (algebraic geometry); game theory and cybernetics; and non-Euclidean geometry and subfields such as differential geometry, which go beyond Euclid's geometry of space to examine the geometry of curved surfaces and space as altered by gravity. Modern mathematics was increasingly tied to the development of sophisticated technology. The development of computers, for example, derived from the work of nineteenth- and twentieth-century mathematicians and statisticians such as the Englishman Charles Babbage (1791–1871) and the American Herman Hollerith (1860–1929). Babbage devised a "difference engine" to make calculations from mathematical tables using punch cards. Hollerith devised a similar machine (Hollerith Electric Tabulating System) for use at the U.S. Census Bureau, where he was employed; his machine was successfully used to tabulate 1890 census results. MIT mathematician Vannevar Bush (1890–1974) developed in 1931 an analog computer, the differential analyzer, which could rapidly solve complex problems in differential equations.

Another example of theoretical and applied science is atomic energy, the discovery and use of which began with French scientist Marie Curie (1867–1934) and her work with radium; New Zealand scientist Ernest Rutherford (1871–1937) and his work with uranium; Danish scientist Niels Bohr (1885–1962) and his work with the structure of the atom; and German scientist Albert Einstein (1879–1955) and his work with the relationship between energy and matter. The theoretical possibilities of the release of energy by the physical transformation, through fission or fusion, of the atom of an unstable element such as uranium, were expressed in the 1930s by Hungarian physicist Leo Szilard (1898–1964), who conceived of the possibility of the fission and consequent chain reaction of uranium atoms resulting in a massive release of energy. This set the stage for experimental evidence, supplied by the German Otto Hahn (1879–1968) with his discovery of fission in 1938 and the Italian Enrico Fermi (1901–1954) and his discovery of the chain reaction in 1942. The Manhattan Project, which brought together the best scientific minds in the United States from 1942 to 1945, culminated in the building of the atomic bomb.

Mathematician John von Neumann (1903–1957) in 1945 conceived of a flexible hard disk to perform complicated functions in computers, which set the stage for developments in computer applications. Von Neumann was also involved in the defense projects during World War II, such as studying the designs and trajectories of explosions; he contributed to the Manhattan Project. After the war Von Neumann developed game theory and organized the Computer Project at the Institute for Advanced Study. He published, in 1958, *The Computer and the Brain*. Von Neumann, Vannevar Bush, and Herman Hollerith typified the merging of mathematics and technology during the twentieth century.

Modern science in the twenty-first century is a global process of the search for, communication of, and application of knowledge. Whereas science was for centuries the province of the talented amateur and aristocrat, the rise of the modern university system has democratized science while at the same time making it a close-knit profession based on standards of training, communication, and intellectual honesty. The close association of research and application has also led to expanding links between research universities and governments. Many developments in science, such as in genetics, nuclear power, and computer science, have threatened the traditional moral standards of civilization, requiring a reassessment of the relationship between science and morality.

The modern world has in many ways never entirely broken from the influence of ancient scientists and philosophers. The liberal arts education still promoted in colleges and universities derives from Greek models of education developed 2,500 years ago and then resurrected during the Renaissance. Monumental architecture is still classical. Historical inquiry remains beholden to the likes of Thucydides and Tacitus. The philosophers and artists upon which we base our cultural expression and institutions—past masters such as Shakespeare, Locke, Montaigne, and Jefferson—were themselves heavily dependent upon Plutarch, Aristotle, Cicero, and Pliny. Historian of science I. Bernard Cohen noted in *The Birth of a New Physics* that our conceptions of the world are still Aristotelian, notwithstanding we live in a world where science is dominated by the Newtonian and Einsteinian paradigms. So subtle has been the influence of ancient science that, try as we might, we still cannot help but think that the sun rises and sets, the moon benevolently shines down upon us, that when we stand still, we are motionless, at rest on a still Earth, and that heavy objects fall faster than lighter ones. Modern science appears to contradict experience, what we daily observe and sense, which explains why it required a revolution in thought to break the spell that ancient science had cast upon the unconscious and conscious minds of humans. One wonders whether the works of ancient scientists and philosophers will ever cease to impose a hypnotic effect upon the modern mind.

FURTHER READING

Marie Boas, *The Scientific Renaissance, 1450–1630* (New York: Harper & Brothers, 1962).

I. Bernard Cohen, *Benjamin Franklin's Science* (Cambridge, MA: Harvard University Press, 1990).

I. Bernard Cohen, *The Birth of a New Physics* (New York: W. W. Norton, 1985).

Nicolaus Copernicus, *On the Revolution of the Heavenly Spheres*, trans. Charles G. Wallis (Chicago: Encyclopedia Britannica, 1952).

Michael J. Crowe, *Theories of the World from Antiquity to the Copernican Revolution* (New York: Dover Publications, 1990).

Galileo Galilei, *Discoveries and Opinions of Galileo*, trans. Stillman Drake (Garden City, NY: Anchor Books, 1957).

Thomas Goldstein, *Dawn of Modern Science* (Boston: Houghton Mifflin Company, 1988).

Thomas S. Kuhn, *The Copernican Revolution: Planetary Astronomy in the Development of Western Thought* (Cambridge: Harvard University Press, 1957).

Russell M., Lawson, ed., *Research and Discovery: Landmarks and Pioneers in American Science*, 3 vols. (Armonk, NY: M. E. Sharpe, 2008).

Donald Wilcox, *In Search of God & Self: Renaissance and Reformation Thought* (Prospect Heights, IL: Waveland Press, 1987).

Chronology

BEFORE THE COMMON ERA

2,600,000
Pleistocene Period begins

1,900,000
Homo erectus humans

500,000
Fire first used

200,000
Homo sapiens humans

200,000
Denisovan humans

200,000
Neanderthal humans

100,000
Spoken language developed

12,000
Holocene Period begins

9600
Jericho founded in Palestine

9000
Neolithic Era begins in Mesopotamia: domestication of plants and animals

8000
Copper Age in the Near East

8000–7000
Corn domesticated in the Western Hemisphere

7000
Çatal Hüyük founded in Turkey

7000
Neolithic Period begins in Europe

6000
Pottery developed in Mesopotamia

6000
Neolithic Period begins in East Asia

3500
Bronze Age in Mesopotamia: cities, irrigation, social structure, labor specialization

3300
Life and death of Ötzi, the Iceman

3200
Norte Chico civilization in Peru

3100
First Dynasty in Egypt

3100
Cuneiform developed in Mesopotamia

3000
Hieroglyphics in Egypt

3000
Trade between Mesopotamia and Egypt

3000–1000
Indo-European invasions

2800
First construction of Stonehenge in England

2700
Construction of first step pyramid, in Egypt under supervision of architect Imhotep

2654
Lunar eclipse determined by the Egyptian historian Manetho

2650
Construction of Great Pyramid of Cheops in Egypt

2500
Indus River civilization emerging

2350
Sargon of Akkad creates first world empire, in Mesopotamia

2112–2195
Reign of Sumerian king Ur-Nammu

2100–1600
Xia Dynasty in China

2000
The Epic of Gilgamesh written after having first been orally composed

2000
Bronze developed in Crete

2000
Achaean migration into Greece

1900
Mycenaean Age in Greek Peloponnesus

1800
Civilization developing in Chinese Yellow River region

1792–1750
Reign of Hammurabi of Babylon

1650
Linear A script developed in Crete

1600–1046
Shang Dynasty in China

1570
New Kingdom begins in Egypt

1500–500
Vedic Age in India

1450
Rise of Hittite Empire in Turkey and Syria

1450
Decline of Cretan power in Mediterranean

1450
Mycenaean control of Greece and Aegean Sea region

1400
Appearance of Linear B script in Greece

1379–1362
Reign of Amenhotep IV (Akhenaten), pharaoh of Egypt

1300
Development of Chinese solar calendar

1250
Appearance of Egyptian *Book of the Dead*

1200
Invasions of sea peoples throughout eastern Mediterranean; Dorian invasion of Greece

1200/1150–800
Greek Dark Ages

1200
Olmec civilization in Mexico

1046–256
Zhou Dynasty in China

1000
Homer's *Iliad* and *Odyssey* composed orally

1000–962
Reign of Hebrew king David

911–609
Assyrian Empire

900–270
Second kingdom of Kush

800–500
Greek Archaic Age

800
Carthage founded by Phoenicians; Greek poleis (city-states) dominating Mediterranean

776
Olympic games begin

753
Legendary founding of Rome

750–700
Appearance of written *Iliad* and *Odyssey*

750–330
Hellenic Age in Greece

733
Syracuse in Sicily founded by Corinth

722–481
Period of the *Spring and Autumn Annals*

700
Hesiod's poems *Theogony* and *Works and Days* appear

672
Height of Assyrian Empire in Mesopotamia

668–631
Reign of Assyrian king Ashurbanipal

650
Byzantion founded at the Bosporus

640–560
Athenian lawgiver Solon

626–545
Ionian philosopher and scientist Thales

610–540
Ionian scientist and philosopher Anaximander

604–562
Reign of Nebuchadnezzar, king of Babylon

600
Massilia founded in southern France

600
Mahajanapada kingdoms in India

600
Phoenician circumnavigation of Africa

600–530
Cyrus the Great, king of Persia

590
Solon reforms laws of Athens

586
Babylonians destroy Jerusalem

586–525
Milesian scientist and philosopher Anaximenes

585
Eclipse reputedly predicted by Thales of Miletus

570–490
Philosopher and mathematician Pythagoras

570–478
Philosopher and scientist Xenophanes

563–483
Indian philosopher Siddhartha Gautama

560
Architect and designer of Temple of Artemis at Ephesus, Chersiphron of Knossos, flourishes

559
Achaemenid Empire in Persia

551–479
Chinese philosopher K'ung fu-tzu (Confucius)

550–476
Geographer Hecataeus of Miletus

540–480
Philosopher and scientist Heraclitus of Ephesus

525–456
Playwright Aeschylus

510
Scylax of Caryanda explores the Indus River, Arabian Sea, and Red Sea

510–428
Philosopher and scientist Anaxagoras

509
Roman republic founded

500–331
Greek Classical Age

500
Explorer Euthymenes explores the west coast of Africa

500
Carthaginian Hanno explores the west coast of Europe and Africa

498–408
Milesian urban designer Hippodamus

496–406
Playwright Sophocles

495–435
Philosopher Empedocles

495–429
Pericles of Athens

491–406
Historian Hellanicus of Lesbos

490–430
Greek historian Herodotus

490–420
Astronomer Oenopides of Chios

490
Greek defeat of Persians at Marathon

485
Philosopher Parmenides flourishes

485–415
Philosopher Protagoras

484–407
Playwright Euripides

480–425
Sculptor Phidias

480
Greek defeat of Persians at Salamis

470–385
Philolaus of Croton

469–399
Athenian philosopher Socrates

460–370
Epicurean Democritus

460–377
Greek physician Hippocrates

460–400
Greek historian Thucydides

456
Completion of Temple of Zeus at Olympus by Libon of Ellis

447–438
Construction of Parthenon under supervision of Callicrates

445–365
Cynic philosopher Antisthenes

432
Euctemon the Athenian astronomer flourishes

431–404
Peloponnesian War

430–427
Plague in Athens

430–355
Philosopher and historian Xenophon

427–347
Athenian philosopher Plato

425
Atomist Leucippus flourishes

408–355
Student of Plato: Eudoxus of Cnidus

406–339
Athenian philosopher Speusippus

401–399
March of the Ten Thousand led by Xenophon

401–310
Chinese physician Bian Que

400–340
Chinese astronomer Gan De

398–323
Indian philosopher Calanus

396–314
Athenian philosopher Xenocrates

387–312
Astronomer Heraclides of Pontus

384–322
Athenian philosopher and scientist Aristotle

384–322
Athenian orator Demosthenes

380–320
Astronomer Manaechmus

380–320
Aristander of Telmessos, historian and soothsayer

380–320
Sophist Anaxarchus of Abdera

375–301
Soothsayer and philosopher Aristobulus of Cassandreia

372–289
Chinese philosopher Mencius

370–286
Student of Aristotle: Theophrastus

370–310
Astronomer Callippus of Cyzicus

369–286
Chinese philosopher Chuang Tzu

367–283
Ptolemy I Soter, Alexander's companion and pharaoh of Egypt

365–279
Skeptic Pyrrho of Elis

360–312
Explorer Nearchus of Crete

360–328
Philosopher Callisthenes of Olynth

360–290
Astronomer Autolycus of Pitane

356
Temple of Artemis at Ephesus destroyed

356–323
Alexander of Macedon

355–350
Construction of Mausoleum of Halicarnassus by Pythius of Halicarnassus

350
Completion of Mausoleum in Halicarnassus

350–280
Peripatetic philosopher Demetrius of Phalerum

345–250
Athenian historian Timaeus

343
Aristotle becomes tutor to Alexander of Macedon

341–271
Founder of Epicureanism: Epicurus

335–269
Head of the Lyceum: Strato

334
Temple of Artemis at Ephesus redesigned by Dinocrates of Rhodes

334–146
Hellenistic Macedonian Kingdom

333–262
Founder of Stoicism: Zeno

332
Alexander of Macedon crowned pharaoh of Egypt

331–30
Greek Hellenistic Age

331
Founding of Alexandria in Egypt

330–230
Stoic philosopher Cleanthes

325
Urban designer Dinocrates flourishes

327
Alexander led the Macedonian army in the invasion of the Indus River valley

325
Voyage of Nearchus along the southern coast of Asia from the Indus to the Persian Gulf

325–265
Greek geometer Euclid

325–255
Alexandrian anatomist Herophilus

325–260
Director of Library of Alexandria: Zenodotus

323
Death of Alexander in Babylon

322–185
Maurya Empire in India

320–230
Skeptic philosopher Timon

315–240
Astronomer Aratus of Soli

315–240
Arcesilaus, head of Middle Academy

312
Construction of Appian Way (Via Appia)

312–63
Seleucid kingdom

310–240
Callimachus, librarian at Alexandria

310–235
Chinese philosopher Hsün Tzu

310–230
Astronomer and mathematician Aristarchus of Samos

305–30
Ptolemaic Kingdom

302
Megasthenes's journey to upper India

300
Massilian explorer Pytheas explores North Atlantic

292–280
Construction of Colossus of Rhodes under direction of Chares of Lindos

287–212
Syracusan mathematician and engineer Archimedes

285–222
Ctesibius the Alexandrian pneumatic engineer

285–210
Stoic philosopher Sphaerus

280
Completion of the Colossus of Rhodes

280–220
Astronomer Conon of Samos

280–220
Greek scientist Philo of Byzantium

279
Completion of the Lighthouse of Alexandria

279–206
Stoic philosopher Chryssipus

276–195
Astronomer Eratosthenes

275–194
Physician Erasistratus

262–190
Mathematician Apollonius of Perga

247
Founding of Parthian Empire

234–149
Roman agriculturalist Marcus Porcius Cato

214–139
Carneades, head of the New Academy

208–126
Greek historian Polybius

202–220 CE
Han Dynasty in China

200
Chinese mathematical work *Book on Numbers and Compilations* written

190–120
Astronomer Hipparchus

190–120
Mathematician Hypsicles of Alexandria

187–110
Clitomachus, head of New Academy

185–110
Stoic philosopher Panaetius

180–120
Historian Apollodorus of Athens

164–113
Zhang Qian, Chinese traveler

160–90
Geometer and astronomer Theodosius of Tripolis

159–84
Philo of Larissa, Athenian Academic

156–74
Chinese thinker K'ung An-kuo

146
Roman destruction of Carthage

145–90
Chinese historian Sima Qian

140–70
Epicurean philosopher Phaedrus

135–50
Stoic scientist Posidonius of Rhodes

130–70
Chinese astronomer Luoxia Hong

125–68
Antiochus of Ascalon, Athenian Academic

120–40
Physician Asclepiades of Bithynia

116–27
Roman polymath Varro

106–43
Roman philosopher Cicero

105–35
Alexander Polyhistor, historian and geographer

100–44
Geographer and conqueror Julius Caesar

98–55
Epicurean Lucretius

90–30
Historian and polymath Diodorus Siculus

90–20
Roman architect Vitruvius

86–35
Roman historian Sallust

75–10
Peripatetic philosopher Boethus of Sidon

74–8
Stoic philosopher Athenodorus

70–19
Roman poet Virgil

64–17 CE
Gaius Julius Hyginus, agricultural writer

63–21 CE
Geographer Strabo

63–14 CE
Octavian (Augustus) Caesar

59–17 CE
Roman historian Livy

58–50
Caesar's conquest of Gaul

55–20 CE
Grammarian Verrius Flaccus

49–45
Civil war between Caesar and Pompey

44
Assassination of Julius Caesar

43–17 CE
Roman poet Ovid

31
Battle of Actium; Augustus begins 45-year reign (31 BCE–14 CE)

30–180 CE
Pax Romana

25 BCE–50 CE
Roman physician Celsus

25
Aelius Gallus explores the Arabian Peninsula

20 BCE–40 CE
Philosopher Philo of Alexandria

5 BCE–60 CE
Roman agriculturalist Columella

10 BCE–60 CE
Astronomer and mathematician Geminus

4(?)
Birth of Jesus of Nazareth

COMMON ERA

5–65
Roman Stoic Seneca

10–75
Inventor and mathematician Heron of Alexandria

14–37
Reign of Roman emperor Tiberius

23–79
Roman polymath Pliny the Elder

27–100
Chinese philosopher Wang Chong

30
Crucifixion and death of Jesus

30–101
Stoic philosopher Musonius Rufus

30–90
Greco-Roman physician Dioscorides

32–92
Chinese historian Ban Gu

35–100
Roman orator Quintilian

37–100
Jewish historian Josephus

37–41
Reign of Roman emperor Caligula

41–54
Reign of Roman emperor Claudius

43
Roman conquest of Britain

45
Geographer Pomponius Mela flourishes

46–120
Biographer and essayist Plutarch

50–135
Stoic philosopher Epictetus

54–68
Reign of Roman emperor Nero

56–117
Roman historian Tacitus

60–120
Mathematician Nicomachus of Gerasa

61–113
Pliny the Younger

62–152
Engineer Hero of Alexandria

64
Great Fire at Rome

69–130
Biographer Suetonius

69–96
Flavian Dynasty of Roman emperors (Vespasian, Titus, Domitian)

70–130
Geographer Marinus of Tyre

70–130
Mathematician Menelaus of Alexandria

71
Roman destruction of Jerusalem

74–130
Roman historian Florus

78–139
Chinese astronomer Zhang Heng

79
Eruption of Mount Vesuvius; death of Pliny the Elder

88–175
Biographer of Alexander, Arrian

95–165
Roman historian Appian

96–193
Antonine Dynasty of Roman emperors (Nerva, Trajan, Hadrian, Antoninus Pius, Marcus Aurelius, Commodus)

97
Journey of Gan Ying to the Roman Empire

100–165
Theologian Justin Martyr

100–167
Friend and tutor of Marcus Aurelius, Fronto

100–170
Astronomer, mathematician, geographer Claudius Ptolemy

110–170
Roman jurist Salvius Julianus

117–181
Student of dreams Aelius Aristides

120–175
Astrologer Vettius Valens

121–180
Stoic philosopher Marcus Aurelius

129–210
Chinese astronomer Liu Hong

130–200
Roman physician Galen

132–192
Chinese astronomer Cai Yong

134
Death of philosopher Aristides

140–212
Roman jurist Papinian

150–215
Theologian and commentator Clement of Alexandria

150–219
Chinese physician Zhang Zhongjing

160–227
Chinese astronomer Su Yue

160–210
Pyrrhonist philosopher Sextus Empiricus

160–240
Historian Julius Africanus

163–225
Roman historian Dio Cassius

170–228
Roman jurist Ulpian

170–250
Sophist Philostratus

180–235
Commentator Hippolytus

180–270
Chinese physician Wang Shuhe

180–565
Later Roman Empire

184–254
Theologian Origen

190–265
Chinese physician Hua Tho

193–235
Severan Dynasty of Roman emperors (Septimius Severus, Caracalla, Gela, Elagabalus, Severus Alexander)

200–900
Mayan civilization in Mexico centered at Teotihuacan

200–284
Alexandrian mathematician Diophantus

205–270
Founder of Neoplatonism: Plotinus

213–273
Philosopher Cassius Longinus

215–282
Chinese physician Huangfu Mi

216–276
Life of Mani, founder of Manicheism

220–280
Chinese astronomer Liu Hui

224
Founding of Sasanian Empire in Persia

234–305
Neoplatonist and disciple of Plotinus: Porphyry

235–284
Civil war in the Roman Empire: twenty-six emperors

242–325
Neoplatonist Sopater

250
Compiler Diogenes Laertius flourishes

250–325
Neoplatonist Iamblichus

251–356
Desert hermit St. Anthony

264–340
Church historian Eusebius of Caesarea

265
Beginning of Jin Dynasty in China

283–343
Chinese alchemist Ge Hong

284–305
Reign of Diocletian and administrative reform of the Roman Empire

290–350
Mathematician Pappus of Alexandria

290–364
Aristotelian commentator Gaius Marcus Victorinus

306–337
Reign of Roman emperor Constantine I

310–372
Pseudoscientist Maximus of Ephesus

312
Edict of Religious Toleration

312
Conversion of Constantine

314–393
Orator Libanius

317–388
Aristotelian commentator and orator Themistius

320
Gupta Empire in India

320–400
Physician Oribasius

325
Council of Nicaea

325–395
Historian Ammianus Marcellinus

329–390
Theologian Gregory Nazianzus

330–379
Theologian Basil of Caesarea

330
Dedication of Constantinople as capital of the Christian Roman Empire

335–395
Theologian Gregory of Nyssa

335–405
Mathematician Theon of Alexandria

337–361
Reign of Constantine's son Constantius II

337–422
Buddhist monk and traveler Faxian

340–402
Orator Symmachus

340–420
Theologian and Bible translator St. Jerome

345–414
Greek philosopher Eunapius

349–407
Theologian John Chrysostom

354–430
St. Augustine of Hippo

361–363
Reign of Roman emperor Julian

363–392
Reigns of Roman emperors Jovian, Valentinian I and II, Gratian, Valens

365–440
Compiler Marianus Capella

370–413
Theologian and philosopher Synesius of Cyrene

370–415
Alexandrian mathematician Hypatia

378
Battle of Adrianople

379–395
Reign of Roman emperor Theodosius the Great

393–457
Theologian Theodoret

395–408
Reign of Roman emperor Arcadius

395–423
Reign of Roman emperor Honorius

395
Permanent political division of Roman Empire (Western Roman Empire and Eastern Roman Empire)

395
Christianity made official religion of Roman Empire

397
Augustine becomes bishop of Hippo in North Africa

398
Augustine completes the writing of *Confessions*

398–445
Chinese historian Fan Ye

410
Sack of Rome by the Goths

412–485
Commentator Proclus

420–479
Liu Song Dynasty in China

426
Augustine completes writing the *City of God*

429–501
Chinese astronomer Zu Chongzhi

430
Augustine dies at Hippo while the city is under siege

438
Publication of Theodosian Code

440–520
Commentator Ammonius of Alexandria

442–537
Architect Isidorus of Miletus

470–528
Philosopher Procopius of Gaza

470–540
Chronologist Dionysius Exiguus

474–534
Architect Anthemius of Tralles

476
Abdication of last Western Roman emperor Romulus Augustulus

476–550
Aryabhata the Elder, Hindu scholar

479–502
Southern Qi Dynasty in China

480–524
Philosopher Boethius

485–585
Polymath Cassiodorus

490–570
Commentator John Philoponus

500–565
Historian Procopius of Caesarea

500
Agricultural writer Palladius flourishes

500
Philosopher Marinus of Neapolis flourishes

500–540
Commentator Simplicius

500–547
Roman jurist Tribonian

503–557
Liang Dynasty in China

505–587
Varahamihira, Hindu astronomer

527–565
Byzantine emperor Justinian

538–594
Historian Gregory of Tours

545
Traveler Cosmas Indicopleustes flourishes

550
Confucian scholar Liu Chou flourishes

560–636
Encyclopedist Isidore of Seville

570–632
Prophet Muhammad

580–640
Byzantine historian Theophylat Simocatta

581–682
Chinese alchemist Sun Simiao

581–618
Sui Dynasty in China

585
Dionysius Exiguus develops Easter table with December 25 date of Christ's birth

598–670
Indian philosopher Bramagupta

600
City of Cahokia in Mississippi River valley

600–664
Buddhist monk and traveler Xuanzang

600–680
Bhaskara I, Hindu scholar

619–906
T'ang Dynasty in China

622
Hegira of Muhammad

632–661
First four Muslim caliphs

635–713
Buddhist traveler Yijing

651
Muslim conquest of Sasanian Empire

661–721
Chinese historian Liu Chih-Chi

661–750
Umayyad Caliphate

673–735
English historian Bede

683–727
Chinese astronomer Yi Xing

688–741
Carolingian Dynasty founded by Charles Martel

895–785
Astrologer Theophilus

721–815
Alchemist Jabir ibn Hayyan (Geber)

732
Battle of Poitiers

732–804
English scholar Alcuin of York

742–814
Charlemagne, king of the Franks

750–1258
Abbasid Dynasty

770–840
Scholar and biographer Einhard

787–886
Abu Ma'shar (Albumasar), astrologer and mathematician

790–850
Persian mathematician Al-Khwarizmi

800–870
Mahavira, Hindu mathematician

801–873
Aristotelian scholar Al-Kindi

809–873
Scientific translator Hunayn ibn Ishaq

810–893
Byzantine scholar Photius

815–877
English philosopher John Scotus Eriugena

820–912
Translator Qusta ibn Luqa al-Ba'lbakki

820–911
Geographer Ibn Khurradadhbih

828
Death of Jabril ibn Bukhtishu

835–899
Polymath Al-Sarakhsi

836–901
Astronomer Thabit ibn Qurra

851–929
Mathematician Al-Battani

854–925
Persian polymath Al-Razi (Rhazes)

872–950
Mathematician Al-Farabi

877–960
Arab explorer Ibn Fadlan

895–979
North African physician Ibn al-Jazzar

896–946
Historian Al Mas'udi

900
Ibn 'Imran, Tunisian physician, flourishes

900–1150
Toltec civilization in central Mexico

913–959
Rule of scholar-emperor Constantine VII Porphyrogenitus

927
Cordoban physician Ibn al-Samina dies

930–994
Persian physician Al-Majusi (Haly Abbas)

943–994
Physician Ibn Juljul

950–1007
Cordoban mathematician Al-Mairiti

960–1279
Song Dynasty in China

965–1040
Egyptian polymath, Ibn al-Haytham (Alhazen)

971–1048
Persian polymath Al-Biruni

979–1035
Ibn al-Samh, Andalusian mathematician

980–1037
Philosopher and physician Ibn Sina (Avicenna)

980–1043
Scientific commentator Ibn al-Tayyib

987–1076
Historian Ibn Hayyan

988–1061
Egyptian physician and astrologer Ali ibn Ridwan

992
Viking Leif Erikson's exploration of America

996–1021
Al-Kirmani, mathematician and physician

997–1074
Andalusian pharmacologist Ibn al-Wafid (Abenguefit)

998–1062
Polymath Ali ibn Ridwan

1001–1038
Baghdad scientist Ibn Butlan

1018–1078
Byzantine polymath Michael Psellus

1019–1097
Egyptian mathematician Ibn Fatik

1020–1101
Chinese astronomer Su Song

1029–1070
Sa'id al-Andalusi, judge and scientist

1031–1095
Chinese polymath Shen Kuo

1033–1109
Theologian Anselm of Canterbury

1037–1101
Poet and pharmacologist Su Shi

1041
Invention of moveable type in China

1143–1180
Rule of Byzantine emperor and scientist Manuel I Comnenus

1044
Invention of gunpowder in China

1048–1131
Poet and astronomer Omar Khayyam

1062
Byzantine historian Nicephorus Bryennius

1071
Battle of Manzikert

1075–1151
Korean Confucian scholar Kim Busik

1079–1142
Theologian Peter Abelard

1080–1170
Arab explorer Abu Hamid

1080–1154
Polymath Honorius of Autun

1080–1152
Arabic translator Adelard of Bath

1083–1153
Byzantine empress and historian Anna Comnena

1085–1138
Andalusian physician Ibn Bajja (Avempace)

1089–1167
Jewish astrologer Abraham ibn Ezra

1094–1162
Ibn Zuhr (Avenzoar), Andalusian physician

1095–1155
Historian Geoffrey of Monmouth

1099–1165
Cartographer Muhammad al-Idrisi

1100–1180
Arabic translator John of Seville

1114–1187
Arabic translator Gerard of Cremona

1114–1185
Bhaskara II, Hindu mathematician

1126–1198
Aristotelian commentator Ibn Rushd (Averroes)

1130–1200
Zhu Xi, Confucian scholar

1138–1204
Jewish polymath Maimonides

1150–1218
Historian Geoffroy de Villehardouin

1153–1213
Mathematician Sharaf al-Din al-Tusi

1156–1242
Polymath Kamal-al-Din ibn Yunus

1162–1231
Baghdad scientist Abd al-Latif al-Baghdadi

1170–1230
Physician Al-Dakhwar

1180
Death of Ibn Harun, physician and teacher of Ibn Rushd

1197–1248
Physician and botanist Ibn al-Baitar

1198
Death of Jewish physician Ibn Jumay'

1200
Aztec civilization emerging in central Mexico

1200–1280
Theologian and philosopher Albertus Magnus

1201–1274
Polymath Nasir al-Din al-Tusi

1203–1270
Physician and medical historian Ibn Abi Usaibi'ah

1213–1288
Physician Ibn al-Nafis

1215–1294
Chinese emperor Kublai Khan

1217–1274
Theologian Bonaventure

1219–1292
English philosopher Roger Bacon

1225–1274
Philosopher and theologian Thomas Aquinas

1254–1324
Venetian traveler Marco Polo

1260–1320
Chinese mathematician Zhu Shijie

1271–1295
Travels of Marco Polo in China

1279
Yuan Dynasty founded by Kublai Khan

1285–1347
Philosopher William of Occam

1300–1600
European Renaissance

1300–1349
English mathematician Thomas Bradwardine

1302–1361
English logician Richard Kilvington

1304–1369
Scholar and explorer Ibn Battuta

1304–1374
Renaissance humanist Francesco Petrarca

1321–1347
Arab scholar Ibn Juzayy

1332–1406
Historian and social scientist Ibn Khaldun

1340–1425
Founder of Kerala School of Science and Mathematics, Madhava

1342–1400
Teller of tales Geoffrey Chaucer

1345
English mathematician Richard Swineshead flourishes

1350
English philosopher Roger Swineshead flourishes

1368–1644
Ming Dynasty in China

1394–1460
Portuguese king Henry the Navigator

1400
Inca civilization flourishing in Peru

1433–1499
Philosopher Marsilio Ficino

1450
Great Law of Peace and the Iroquois Confederacy

1451–1506
Italian explorer Christopher Columbus

1452–1519
Italian polymath Leonardo da Vinci

1452–1498
Theologian Girolamo Savonarola

1453
Fall of Constantinople to the Turks

1473–1543
Polish mathematician Nicholas Copernicus

1480–1521
Portuguese explorer Ferdinand Magellan

1482
Portuguese explorer Diego Cam's journey along West Africa

1488
Portuguese explorer Bartholomew Diaz's voyage to the Cape of Good Hope

1492
Christopher Columbus's first voyage to America

1497
Portuguese explorer Vasco Da Gama's circumnavigation of the Cape of Good Hope

1500
Portuguese explorer Pedro Álvares Cabral's voyage along the eastern coast of South America

1514–1564
Flemish anatomist Andreas Vesalius

1546–1601
Danish astronomer Tycho Brahe

1549
Publication of Gonzalo Oviedo's *General and Natural History of the Indies*

1550–1617
Scotch mathematician John Napier

1561–1625
Empiricist Francis Bacon of England

1564–1642
Italian astronomer and physicist Galileo Galilei

1571–1630
German astronomer Johannes Kepler

1578–1657
English anatomist William Harvey

1596–1650
French mathematician René Descartes

Bibliography

PRIMARY SOURCES

Ancient Near East

Budge, E. A. Wallis, trans. *The Book of the Dead: The Papyrus of Ani in the British Museum.* London: British Museum, 1895.

Budge, E. A. Wallis, trans. *Papyrus of Ani: Egyptian Book of the Dead.* New York: Dover Books, 1967.

Cory, I. P. trans. *Ancient Fragments.* 1832. https://www.sacred-texts.com/cla/af/af01.htm.

The Epic of Gilgamesh. Translated by N. K. Sandars. London: Penguin Books, 1972.

The Fragments of Manetho. Translated by W. G. Waddell. Cambridge, MA: Harvard University Press, 1940. http://penelope.uchicago.edu/Thayer/E/Roman/Texts/Manetho/home.html.

Hoffner, Harry A., Jr. *The Laws of the Hittites: A Critical Edition.* Leiden, Netherlands: Brill, 1997.

Josephus. *Against Apion.* Translated by William Whiston. http://www.gutenberg.org/files/2849/2849.txt.

Josephus. *The Jewish War.* Translated by G. A. Williamson. Harmondsworth, England: Penguin Books, 1969.

Lichtheim, M. *Ancient Egyptian Literature.* Vol. 2. Berkeley: University of California Press, 1976: http://web.archive.org/web/19990221040703/http://puffin.creighton.edu/theo/simkins/tx/Aten.html

Neugebauer, O. *Astronomical Cuneiform Texts.* 3 vols. London: Institute for Advanced Study, 1955.

Pritchard, James B., ed. *Ancient Near Eastern Texts Relating to the Old Testament.* Princeton, NJ: Princeton University Press, 1969.

Thompson, R. Campbell. *Assyrian and Babylonian Literature: Selected Transactions.* With a critical introduction by Robert Francis Harper. New York: D. Appleton and Company, 1901.

Greece

Aelian. *On the Characteristics of Animals.* Translated by A. F. Scholfield. 3 vols. Cambridge, MA: Harvard University Press, 1971.

Aeschylus. *Prometheus Bound.* Translated by Rex Warner. In *Ten Greek Plays.* Boston: Houghton Mifflin, 1957.

Apollonius of Rhodes. *The Voyage of Argo.* Translated by E. V. Rieu. Harmondsworth, England: Penguin Books, 1971.

The Apology of Aristides, Translated from the Syriac. Translated by D. M. Kay. In *Ante-Nicene Fathers.* Vol. 9. Buffalo, NY: Christian Literature Publishing, 1896. http://www.newadvent.org/fathers/1012.htm.

Aristarchus. "On the Sizes and Distances of the Sun and Moon." In *Aristarchus of Samos*, by Thomas Heath. New York: Dover Books, 1981.

Aristotle. *Meteorology.* Translated by E. W. Webster. In *The Works of Aristotle.* Vol. 1. Chicago: Encyclopedia Britannica, 1952.

Aristotle. *On the Athenian Constitution.* Translated by Frederic G. Kenyon. London: G. Bell and Sons, 1891.

Aristotle. *On Sleep and Sleeplessness; On Prophesying by Dreams; On Memory and Reminiscence.* Translated by J. I. Beare. In *The Parva Naturalia.* Oxford: Clarendon Press, 1908.

Aristotle. *The Works of Aristotle.* 2 vols. Chicago: Encyclopedia Britannica, 1952.

Arrian. *Anabasis* and *Indica.* Translated by P. A. Brunt and E. Iliff Robson. 2 vols. Cambridge, MA: Harvard University Press, 1933, 1976.

Arrian. *The Campaigns of Alexander.* Translated by Aubrey de Selincourt. Harmondsworth, England: Penguin Books, 1971.

Athenaeus. *The Deipnosophists.* Translated by C. B. Gulick. 7 vols. Cambridge, MA: Harvard University Press, 1963.

Aurelius, Marcus. *Meditations.* Translated by Maxwell Staniforth. Harmondsworth, England: Penguin Books, 1964.

Bambrough, Renford, ed. and trans. *The Philosophy of Aristotle.* New York: New American Library, 1963.

Barnes, Jonathan, trans. *Early Greek Philosophy.* London: Penguin Books, 1987.

Burnet, John, ed. and trans. *Early Greek Philosophy.* London: Adam and C. Black, 1930.

Clement of Alexandria. *The Stromata.* Translated by William Wilson. In *Ante-Nicene Fathers.* Vol. 2. Buffalo, NY: Christian Literature Publishing, 1885. http://www.newadvent.org/fathers/02101.htm.

Diodorus Siculus. *The Library of History of Diodorus Siculus.* Translated by C. H. Oldfather. Vol. 1. London: Loeb Classical Library, 1933. http://penelope.uchicago.edu/Thayer/E/Roman/Texts/Diodorus_Siculus/home.html.

Diogenes Laertius. *The Lives and Opinions of Eminent Philosophers.* Translated by C. D. Yonge. London: Henry Bohn, 1853.

Diogenes Laertius. *Lives of the Philosophers.* Translated by R. D. Hicks. 2 vols. Cambridge, MA: Harvard University Press, 1931, 1938.

Epictetus. *The Discourses.* Edited by Christopher Gill. Everyman's Library. Rutland, VT: Tuttle Publishing, 2001.

Euclid. *The Thirteen Books of the Elements.* Translated by Thomas Heath. New York: Dover Books, 1925.

Grant, Michael. *Readings in the Classical Historians.* New York: Scribner's, 1992.

Greek Mathematical Works. Vol. 2. Translated by Ivor Thomas. Cambridge, MA: Harvard University Press, 1941.

Hermias the Philosopher. *Derision of Gentile Philosophers.* Translated by J. A. Giles. In *The Writings of the Early Christians of the Second Century, namely Athenagoras, . . . Hermias. . . .* 1857. http://www.tertullian.org/fathers/hermias_1_satire.htm.

Hero of Alexandria. *The Pneumatics.* Translated by Bennet Woodcroft. London: C. Whittingham, 1851.

Herodotus. *The Histories.* Translated by Aubrey de Selincourt. Harmondsworth, England: Penguin Books, 1972.

Hesiod. *Homeric Hymns, Epic Cycle, Homerica.* Translated by Hugh G. Evelyn-White. Cambridge, MA: Harvard University Press, 1936.

Hesiod. *Theogony and Works and Days.* Translated by Richard Lattimore. Ann Arbor: University of Michigan Press, 1959.

Hippolytus. *Refutation of the Heresies.* Translated by J. H. MacMahon. In *Ante-Nicene Fathers.* Vol. 5. Edited by Alexander Roberts, James Donaldson, and A. Cleveland Coxe. Buffalo, NY: Christian Literature Publishing, 1886. Revised and edited for New Advent by Kevin Knight. http://www.newadvent.org/fathers/050101.htm.

Homer. *Iliad.* Translated by Robert Fitzgerald. New York: Doubleday, 1989.

Homer. *Odyssey.* Translated by Robert Fitzgerald. New York: Random House, 1990.

Jones, W. H. S., trans. *Hippocrates.* Vol. 1. Cambridge, MA: Harvard University Press, 1923.

Jowett, Benjamin, trans. *The Portable Plato.* Harmondsworth, England: Penguin Books, 1976.

Katz, Joseph, ed. *The Philosophy of Plotinus.* New York: Appleton Century Crofts, 1950.

Marinus of Samaria. *Life of Proclus, or Concerning Happiness.* Translated by Kenneth S. Guthrie. 1925. http://www.tertullian.org/fathers/marinus_01_life_of_proclus.htm.

Pausanias. *Description of Greece.* Translated by W. H. S. Jones. Vol. 1. New York: G. P. Putnam's Sons, 1918.

Perrin, Bernadotte, trans. *Plutarch's Lives.* Vol. 7. London: William Heinemann, 1919.

The Philocalia of Origen: A Compilation of Selected Passages from Origen's Works made by St. Gregory of Nazianzus and St. Basil of Caesarea. Translated by George Lewis. Edinburgh: T. & T. Clark, 1911.

Philostratus. *Life of Apollonius of Tyana.* Translated by F. C. Conybeare. Cambridge, MA: Harvard University Press, 1912.

Philostratus. *Lives of the Sophists.* Translated by W. C. Wright. Cambridge, MA: Harvard University Press, 1921.

Plato. *The Dialogues.* Translated by Benjamin Jowett. Chicago: Encyclopædia Britannica, 1952.

Plato. *The Last Days of Socrates.* Translated by Hugh Tredennick. Harmondsworth, England: Penguin Books, 1959.

Plato. *The Republic of Plato.* Translated by Francis Cornford. Oxford: Oxford University Press, 1945.

Plato. *Timaeus.* Translated by Benjamin Jowett. http://www.gutenberg.org/files/1572/1572-h/1572-h.htm.

Plutarch. *Essays.* Translated by Robin Waterfield. London: Penguin Books, 1992.

Plutarch. *Essays and Miscellanies: The Complete Works.* Corrected and revised by William W. Goodwin. Vol. 3. https://www.gutenberg.org/files/3052/3052-h/3052-h.htm.

Plutarch. *The Lives of the Noble Grecians and Romans.* Translated by John Dryden. Revised by Arthur Hugh Clough, 1864. Reprint New York: Random House, 1992.

Plutarch. *Moralia.* 15 vols. Translated by Harold Cherniss and William Helmbold. Loeb Classical Library. Cambridge, MA: Harvard University Press, 1968–1976.

Polybius. *The Histories.* Translated by W. R. Paton. 6 vols. Cambridge, MA: Harvard University Press, 1922–1927.

Porphyry. "Letter to Anebo." In *Iamblichus on the Mysteries of the Egyptians, Chaldeans, and Assyrians.* Translated by Thomas Taylor. Chiswick, England: C. Whittingham, 1821.

Porphyry. *Life of Pythagoras.* Translated by Kenneth S. Guthrie. 1920. http://www.tertullian.org/fathers/porphyry_life_of_pythagoras_02_text.htm.

Porphyry. *On Images.* Translated by Edwin H. Gifford. In Eusebius, *Preparation for the Gospel.* Oxford: Clarendon Press, 1903.

Proclus. *On Motion.* Translated by by Kenneth S. Guthrie. 1925. http://www.tertullian.org/fathers/proclus_on_motion.htm.

Ptolemy. *The Almagest.* Translated by R. Catesby Taliaferro. Chicago: Encyclopedia Britannica, 1952.

Strabo. *Geography.* Translated by H. L. Jones. 8 vols. Cambridge, MA: Harvard University Press, 1917–1932. http://penelope.uchicago.edu/Thayer/E/Roman/Texts/Strabo/home.html.

Theophrastus. *Enquiry into Plants and Minor Works on Odours and Weather Signs.* Translated by Arthur Hort. 2 vols. Cambridge, MA: Harvard University Press, 1916.

Thucydides. *The Peloponnesian War.* Translated by Rex Warner. Harmondsworth, England: Penguin Books, 1972.

Wheelwright, Philip, ed. and trans. *Aristotle.* New York: Odyssey Press, 1951.

The Works of Philo Judaeus. Translated by C. D. Yonge. Vols. 1, 4. London: George Bell and Sons, and Henry G. Bohn, 1854, 1855.

The Works of the Emperor Julian. Translated by W. C. Wright. Cambridge, MA: Harvard University Press, 1962, 1969.

Xenophon. *A History of My Times.* Translated by Rex Warner. Harmondsworth, England: Penguin Books, 1979.

Xenophon. *The Persian Expedition.* Translated by Rex Warner. Harmondsworth, England: Penguin Books, 1972.

Rome

Ammianus Marcellinus. Translated by John C. Rolfe. 2 vols. Cambridge, MA: Harvard University Press, 1950.

Caesar. *The Civil War.* Translated by Jane F. Gardner. Harmondsworth, England: Penguin Books, 1976.

Caesar. *The Conquest of Gaul.* Translated by S. A. Handford. Harmondsworth, England: Penguin Books, 1951.

Cassius Dio. *Roman History.* Translated by Earnest Cary. Cambridge, MA: Harvard University Press, 1914–1927.

Celsus, A. Cornelius. *On Medicine.* Translated by W. G. Spencer. Cambridge, MA: Harvard University Press, 1938.

Cicero. *Basic Works.* Edited by Moses Hadas. New York: Modern Library, 1951.

Clarke, John, trans. *Physical Science in the Time of Nero: Being a Translation of the Quaestiones Naturales of Seneca.* New York: Macmillan, 1910.

Columella. *On Agriculture.* Translated by Edward H. Heffner. Cambridge, MA: Harvard University Press, 1989.

Dieting for an Emperor: A Translation of Books 1 and 4 of Oribasius' Medical Compilations with an Introduction and Commentary. Translated by Mark David Grant. Leiden, Netherlands: Brill, 1997.

Eunapius. *The Lives of the Philosophers and Sophists.* Translated by W. C. Wright. Cambridge, MA: Harvard University Press, 1921. http://www.tertullian.org/fathers/eunapius_02_text.htm.

Eusebius. *History of the Church.* Translated by G. A. Williamson. Harmondsworth, England: Penguin Books, 1965.

Frontinus. *The Aqueducts of Rome.* Translated by Charles E. Bennett. Cambridge, MA: Harvard University Press, 1925.

Galen. *On the Natural Faculties.* Translated by A. J. Brock. Cambridge, MA: Harvard University Press, 1916.

Galen. "On Sustaining Causes." Translated by R. J. Hankinson. In *The Cambridge Companion to the Stoics*, edited by Brad Inwood. Cambridge: Cambridge University Press, 2003.

Grant, Mark, ed. and trans. *Galen on Food and Diet.* New York: Routledge, 2000.

Holy Bible, Containing the Old and New Testaments. New York: American Bible Society, 1865.

Justinian. *The Digest of Roman Law.* Translated by C. F. Kolbert. Harmondsworth, England: Penguin Books, 1979.

Livy. *The Early History of Rome.* Translated by Aubrey de Selincourt. Harmondsworth, England: Penguin Books, 1971.

Livy. *Rome and the Mediterranean.* Translated by Henry Bettenson. Harmondsworth, England: Penguin Books, 1976.

Lucretius. *The Nature of the Universe.* Translated by R. E. Latham. Harmondsworth, England: Penguin Books, 1951.

Martyr, Justin. "Horatory Address to the Greeks." Translated by Marcus Dods. In *Ante-Nicene Fathers.* Vol. 1. Buffalo, NY: Christian Literature Publishing, 1885. http://www.newadvent.org/fathers/0129.htm.

The Natural History of Pliny. Translated by John Bostock and H. T. Riley. Vol. 1. London: Henry G. Bohn, 1855.

The New Oxford Annotated Bible with the Apocrypha. Revised Standard Version. Oxford: Oxford University Press, 1977.

Nicander of Colophon, *Alexipharmac and Georgica*. Translated by A. S. F. Gow and A. F. Scholfield. Attalus, 1953. http://www.attalus.org/poetry/nicander.html.

Nichomachus of Gerasa, *Introduction to Arithmetic*. Translated by Martin Luther D'Ooge. New York: Macmillan, 1926.

Palladius. *On Husbandrie*. Translated by Barton Lodge. London: Early English Text Society, 1879.

Pharr, Clyde, trans. *The Theodosian Code and Novels*. Princeton, NJ: Princeton University Press, 1952.

Pliny the Elder. *Natural History*. Translated by John F. Healy. London: Penguin Books, 1991.

Pliny the Elder. *Natural History*. Translated by H. Rackham. 2 vols. Cambridge, MA: Harvard University Press, 1938, 1947.

Pliny the Younger. *The Letters of the Younger Pliny*. Translated by Betty Radice. Harmondsworth, England: Penguin Books, 1963.

Romer, F. E., trans. *Pomponius Mela's Description of the World*. Ann Arbor: University of Michigan Press, 1998.

Rufus of Ephesus. *On Melancholy*. Edited by Peter E. Pormann. Tuebingen, Germany: Mohr Siebeck Publishing Company, 2008.

Seneca. *Letters from a Stoic*. Translated by Robin Campbell. London: Penguin Books, 1969.

Sextus Empiricus. *Outlines of Scepticism*. Edited by Julia Annas and Jonathan Barnes. Cambridge: University of Cambridge Press, 2000.

The Sphere of Marcus Manilius Made an English Poem with Annotations and an Astronomical Appendix. Translated by Edward Sherburne. London: Nathanael Brooke, 1675.

St. Augustine. *City of God*. Translated by Henry Bettenson. London: Penguin Books, 1984.

St. Augustine. *Confessions*. Translated by R. S. Pine-Coffin. Harmondsworth, England: Penguin Books, 1961.

Suetonius, *On Grammarians*. Translated by J. C. Rolfe. 1914. http://penelope.uchicago.edu/Thayer/E/Roman/Texts/Suetonius/de_Grammaticis*.html.

Suetonius. *The Twelve Caesars*. Translated by Robert Graves. Harmondsworth, England: Penguin Books, 1957.

Tacitus. *The Agricola and the Germania*. Translated by H. Mattingly and S. A. Handford. Harmondsworth, England: Penguin Books, 1970.

Tacitus. *The Histories*. Translated by Kenneth Wellesley. Harmondsworth, England: Penguin Books, 1972.

Thatcher, Oliver J. *The Roman World*. Vol. 3. In *The Library of Original Sources*, 286–92. Milwaukee, WI: University Research Extension, 1907.

Todd, Robert, trans. *Themistius: On Aristotle's On the Soul.* London: Bloomsbury, 1996.

Valesius, Henri de, trans. *The History of the Church . . . Written in Greek by Eusebius Pamphilus, Socrates Scholasticus, Evagrius.* 1692. Reprint London: Henry Bohn, 1853.

Vegetius. *Epitome of Military Science.* Translated by N. P. Milner. Liverpool, England: Liverpool University Press, 1993.

Vettius Valens. *Anthologies.* Translated by Mark Riley. https://www.csus.edu/indiv/r/rileymt/Vettius%20Valens%20entire.pdf.

Vitruvius. *On Architecture.* Translated by Joseph Gwilt. London: Priestley and Weale, 1826.

Vitruvius, *The Ten Books on Architecture.* Translated by Morris H. Morgan. Cambridge, MA: Harvard University Press, 1914.

Central, Southern, East Asia

The Athenaeum, a Magazine of Literary and Miscellaneous Information. London: Longman, Hurst, Rees, and Orme, 1808.

The Bhagavad Gita, or the Message of the Master. Translated by Yogi Ramacharaka. Chicago: Yogi Publication Society, 1907.

Chinese Accounts of Rome, Byzantium and the Middle East, c. 91 B.C.E.–1643 C.E. East Asian History Sourcebook. https://sourcebooks.fordham.edu/eastasia/romchin1.asp.

De Bary, William Theodore, trans. *Sources of Chinese Tradition.* Vol. 1. New York: Columbia University Press, 1960.

The Gospel of Buddha: Compiled from Ancient Records. Translated by Paul Carus. Chicago: Open Court Publishing Co., 1915.

Hill, John E., trans. *The Peoples of the West, from the Weilue*, by Yu Huan. 2004. http://depts.washington.edu/silkroad/texts/weilue/weilue.html.

Kaviraj Kunja Lal Bhishagratna, ed. *An English Translation of the Sushruta Samhita.* Vol. 1: *Sutrasthanam.* Calcutta: By the author, 1907.

McCrindle, J. W., trans. *Ancient India as Described by Megasthenes and Arrian.* London: Trubner & Co., 1877.

The Sacred Books of China: The Texts of Confucianism. Part 1: *The Shû King.* Translated by James Legge. Oxford: Clarendon Press, 1879.

The Sacred Books of the East. Edited by F. Max Müller. Vol. 1: *The Upanishads.* New York: Christian Literature Co., 1897.

The Sacred Books of the East. Edited by F. Max Müller. Vol. 4: *The Zend-Avesta.* Oxford: Clarendon Press, 1895.

The Sacred Books of China: The Texts of Taoism. Translated by James Legge. Oxford: Clarendon Press, 1891.

Sima Qian. *Records of the Grand Historian.* Translated by Burton Watson. New York: Columbia University Press, 2011.

Watson, Burton, trans. *The Tso chuan: Selections from China's Oldest Narrative History.* New York: Columbia University Press, 1989.

Medieval

Al-Biruni. *The Book of Instruction in the Elements of the Art of Astrology.* Translated by R. Ramsay Wright. London: Luzac & Co, 1934; Reprint Bel Air, MD: Astrology Classics, 2006.

Al-Biruni. *The Chronology of Ancient Nations.* Translated by C. Edward Sachau. London: William H. Allen and Co., 1879.

Alberuni's India: An Account of the Religion, Philosophy, Literature, Geography, Chronology, Astronomy, Customs, Laws and Astrology of India about A. D. 1030. Translated by Edward C. Sachau. London: Kegan Paul, Trench, Trubner, & Co., 1910.

The Algebra of Mohammed Ben Musa. Translated by Frederic Rosen. London: Oriental Translation Fund, 1831.

Algebra, with Arithmetic and Mensuration, from the Sanscrit of Brahmegupta and Bhascara. Translated by Henry Thomas Colebrooke. London: John Murray, 1817.

Alhacen's Theory of Visual Perception: A Critical Edition, with English Translation and Commentary, of the First Three Books of Alhacen's De Aspectibus, the Medieval Latin Version of Ibn al-Haytham's Kitab al-Manazir. Translated by A. Mark Smith, Vol. 2. Philadelphia: American Philosophical Society, 2001.

Bede. *A History of the English Church and People.* Translated by Leo Sherley-Price. Harmondsworth, England: Penguin Books, 1968.

Boethius. *Consolation of Philosophy.* Translated by Philip Ridpath. London: C. Dilly, 1785.

Boethius. *The Consolation of Philosophy.* Translated by V. E. Watts. Harmondsworth, England: Penguin Books, 1969.

Brahmagupta. *Khandakhadyaka.* Translated by Bina Chatterjee. Vol. 1. Delhi, India: Motilal Banarsidass, 1970.

Chaucer, Geoffrey. *The Canterbury Tales of Geoffrey Chaucer.* Translated by Percy Mackaye. New York: Duffield & Company, 1914.

Chinese Accounts of Rome, Byzantium and the Middle East, c. 91 B.C.E.–1643 C.E. East Asian History Sourcebook. https://sourcebooks.fordham.edu/eastasia/romchin1.asp.

The Christian Topography of Cosmas, an Egyptian Monk. Translated by J. W. McCrindle. London: Hakluyt Society, 1897.

Comnena, Anna. *The Alexiad.* Translated by E. R. A. Sewter. Harmondsworth, England: Penguin Books, 1969.

Copernicus, Nicolaus. *On the Revolution of the Heavenly Spheres.* Translated by Charles G. Wallis. Chicago: Encyclopedia Britannica, 1952.

Einhard. *The Life of Charlemagne.* Translated by Lewis Thorpe. Harmondsworth, England: Penguin Books, 1969.

Evans, Sebastian, trans. *Geoffrey of Monmouth.* London: J. M. Dent, 1904.

Galilei, Galileo. *Discoveries and Opinions of Galileo.* Translated by Stillman Drake. Garden City, NY: Anchor Books, 1957.

Geoffrey of Monmouth. *History of the Kings of Britain.* Translated by Lewis Thorpe. Harmondsworth, England: Penguin Books, 1966.

Gregory of Nyssa. *The Life of Moses.* Translated by Abraham J. Malherbe and Everett Ferguson. New York: Paulist Press, 1978.

Gregory of Tours. *The History of the Franks.* Translated by Lewis Thorpe. Harmondsworth, England: Penguin Books, 1974.

Hunwick, John O. *Timbuktu and the Songhay Empire: Al-Sa'di's Ta'ridh al-sudan Down to 1613 and Other Contemporary Documents.* Leiden, Netherlands: Brill, 2003.

Ibn Abi Usaibi'ah. *History of Physicians.* Translated by L. Kopf. Bethesda, MD: National Library of Medicine, 1971.

Ibn Fadlān. *Ibn Fadlān and the Land of Darkness: Arab Travellers in the Far North.* Translated by Paul Lunde and Caroline Stone. London: Penguin Books, 2012.

Ibn Sina. *On Medicine.* https://sourcebooks.fordham.edu/source/1020Avicenna-Medicine.asp.

Introduction to St. Thomas Aquinas. Edited by Anton C. Pegis. New York: Modern Library, 1948.

Isidore of Seville, *Chronicon.* Translated by Kenneth B. Wolf. 2004. http://www.tertullian.org/fathers/isidore_chronicon_01_trans.htm.

Joinville and Villehardouin. *Chronicles of the Crusades.* Translated by M. R. B. Shaw. Harmondsworth, England: Penguin Books, 1963.

Abd Ar Rahman bin Muhammed ibn Khaldun. *The Muqaddimah.* Translated by Franz Rosenthal. https://asadullahali.files.wordpress.com/2012/10/ibn_khaldun-al_muqaddimah.pdf.

Omar Khayyam. *The Rubaiyat of Omar Khayyam.* Translated by Edward Fitzgerald. San Francisco: Dodge Book and Stationery Co., 1896.

Kircher, Athanasius. *China Illustrata.* Translated by Charles D. Van Tuyl. Muskogee, OK: Indian University Press, 1986.

The Koran. Translated by J. M. Rodwell. London: J. M. Dent & Sons, 1909.

The Opus Majus of Roger Bacon. Translated by Robert Burke, Vol. 1. Philadelphia: University of Pennsylvania Press, 1928.

Polo, Marco. *The Travels.* Translated by R. E. Latham. Harmondsworth, England: Penguin Books, 1958.

Polo, Marco. *The Travels of Marco Polo, the Venetian, revised from Marsden's Translation and Edited by Manuel Komroff.* New York: Boni & Liveright, 1926.

Procopius. *The Secret History.* Translated by G. A. Williamson. London: Penguin Books, 1981.

Psellus, Michael. *Fourteen Byzantine Rulers: The Chronographia of Michael Psellus.* Translated by E. R. A. Sewter. Harmondsworth, England: Penguin Books, 1966.

Robert of Chester's Latin Translation of the Algebra of Al-Khowarizmi. Translated and Edited by Louis C. Karpinski. New York: Macmillan, 1915.

Sa'id al-Andalusi. *Science in the Medieval World: "Book of the Categories of Nations."* Translated by Sema'an I. Salem and Alok Kumar. Austin: University of Texas Press, 1991.

Sources of Tibetan Tradition. Edited by Kurtis R. Schaeffer, Matthew T. Kapstein, and Gray Tuttle. New York: Columbia University Press, 2013.

The Travels of Ibn Batuta. Translated by Samuel Lee. London: Oriental Translation Committee, 1829.

The Vinland Sagas. Translated by Magnus Magnusson and Hermann Palsson. New York: Penguin Books, 1987.

William of Ockham. *Predestination, God's Foreknowledge, and Future Contingents*. Translated by M. M. Adams and Norman Kretzmann. New York: Appleton-Century-Crofts, 1969.

Wippel, John, and Allan Wolter, eds. *Medieval Philosophy*. New York: Free Press, 1969.

Yules, Henry. Translator and Editor. *Cathay and the Way Thither; Being a Collection of Medieval Notices of China*. Volume 1. London: Hakluyt Society, 1866.

America

Bingham, Hiram. *Inca Land: Exploration in the Highlands of Peru*. Boston: Houghton Mifflin Company, 1922.

Cortes, Hernan. *Letters from Mexico*. Translated by Anthony Pagden. New Haven, CT: Yale University Press, 1986.

Hariot, Thomas. *A Brief and True Report of the New Found Land of Virginia, Reproduced in Facsimile from the First Edition of 1588*. New York: Dodd, Mead & Company, 1903.

Nuttall, Thomas. *Journal of Travels in the Arkansa Territory*. Philadelphia: Palmer, 1821.

Williams, Roger. *A Key unto the Language of America: Or, an Help to the Language of the Natives in that Part of American, Called New England*. London: Gregory Dexter, 1643.

SECONDARY SOURCES

General

Abell, George O. *Exploration of the Universe*. 3rd ed. New York: Holt, Rinehart Winston, 1975.

Ahrens, C. Donald. *Meteorology Today: An Introduction to Weather, Climate, and the Environment*. St. Paul, MN: West Publishing, 1988.

Ancient History Encyclopedia. https://www.ancient.eu.

Ball, W. W. Rouse. *A Short History of Mathematics*. 4th ed. London: Macmillan and Co., 1908.

Boardman, John, Jasper Griffin, and Oswyn Murray. Editors. *The Oxford History of the Classical World*. Oxford: Oxford University Press, 1986.

Boorstin, Daniel. *The Discoverers*. New York: Random House, 1983.

Bowersock, G. W. *Fiction as History: Nero to Julian*. Berkeley: University of California Press, 1994.

Breisach, Ernst. *Historiography: Ancient, Medieval and Modern.* Chicago: University of Chicago Press, 1983.

Campbell, Joseph. *The Power of Myth.* New York: Anchor Books, 1991.

Cary, M. *The Geographical Background of Greek and Roman History.* Oxford: Clarendon Press, 1949.

Cary, M., and E. H. Warmington. *The Ancient Explorers.* Harmondsworth, England: Penguin Books, 1963.

Cassirer, Ernst. *An Essay on Man.* New Haven, CT: Yale University Press, 1944.

Casson, Lionel. *Libraries in the Ancient World.* New Haven, CT: Yale University Press, 2001.

Casson, Lionel. *Ships and Seamanship in the Ancient World.* Princeton, NJ: Princeton University Press, 1971.

Childe, Gordon. *What Happened in History.* Harmondsworth, England: Penguin Books, 1946.

Crowe, Michael J. *Theories of the World from Antiquity to the Copernican Revolution.* New York: Dover Books, 1990.

Davies, Paul. *God and the New Physics.* New York: Simon and Schuster, 1983.

De Camp, L. Sprague. *Ancient Engineers.* New York: Doubleday, 1963.

De Montaigne, Michel. *Essays.* Translated by Donald Frame. Stanford, CA: Stanford University Press, 1957.

Dunbavin, Paul. *Under Ancient Skies: Ancient Astronomy and Terrestrial Catastrophism* Nottingham, UK: Third Millennium Publishing, 2005.

Dvorsky, George. "The Final Days of Ötzi the Iceman Revealed through New Analysis of His Tools." Gizmodo, June 20, 2018. https://gizmodo.com/final-days-of-otzi-the-iceman-revealed-through-new-anal-1826982899.

Encyclopedia Britannica. 24 vols. Chicago: Encyclopedia Britannica, 1962.

Finley, M. I. *Aspects of Antiquity: Discoveries and Controversies.* Harmondsworth, England: Penguin Books, 1977.

Gross, M. Grant. *Oceanography: A View of the Earth.* Englewood Cliffs, NJ: Prentice-Hall, 1987.

Hannah, Robert. *Greek and Roman Calendars.* London: Bloomsbury, 2005. https://books.google.com/books?id=qTWPAQAAQBAJ.

Hawkes, Jacquetta. *Prehistory: History of Mankind: Cultural and Scientific Developments.* Vol. 1, part 1. New York: Mentor Books, 1965.

Heyerdahl, Thor. *Early Man and the Ocean: A Search for the Beginning of Navigation and Seaborne Civilizations.* New York: Vintage Books, 1980.

Heyerdahl, Thor. *Kon-Tiki: Across the Pacific by Raft.* Translated by F. H. Lyon. New York: Washington Square Press, 1984.

Hyde, Douglas. *The Story of Early Gaelic Literature.* London: Unwin, 1905.

Joly, N. "The Early History of Fire." *Popular Science Monthly* 10 (1876). https://en.wikisource.org/wiki/Popular_Science_Monthly/Volume_10/November_1876/The_Early_History_of_Fire.

Jung, Carl. *Man and His Symbols.* New York: Doubleday, 1964.

Kelsey, Morton. *Healing and Christianity.* New York: Harper and Row, 1973.

Kuhn, Thomas S. *The Copernican Revolution: Planetary Astronomy in the Development of Western Thought.* Cambridge, MA: Harvard University Press, 1957.

Kuhn, Thomas S. *The Structure of Scientific Revolutions.* Chicago: University of Chicago Press, 1970.

Kuriyama, Shigehisa. *The Expressiveness of the Body and the Divergence of Greek and Chinese Medicine.* New York: Zone Books, 1999.

Leicester, Henry M. *The Historical Background of Chemistry.* New York: Dover Books, 1971.

Nutton, Vivian. *Ancient Medicine.* 2nd ed. New York: Routledge, 2013.

O'Connor, J. J., and E. F. Robertson. *History of Mathematics.* School of Mathematics and Statistics, University of St. Andrew's Scotland. http://www-history.mcs.st-andrews.ac.uk/history/Indexes/HistoryTopics.html.

Ogden, Daniel. *Greek and Roman Necromancy.* Princeton, NJ: Princeton University Press, 2001.

Otto, Rudolf. *The Idea of the Holy.* New York: Oxford University Press, 1968.

Oxford Companion to the Bible. Oxford: Oxford University Press, 1993.

Oxford English Dictionary. 20 vols. Oxford: Oxford University Press, 1989.

Radice, Betty. *Who's Who in the Ancient World.* Harmondsworth, England: Penguin Books, 1973.

Raven, Charles E. *Natural Religion and Christian Theology.* Cambridge: Cambridge University Press, 1953.

Redfield, Robert. *The Primitive World and Its Transformations.* Ithaca, NY: Cornell University Press, 1957.

Robson, Vivian. *The Fixed Stars and Constellations in Astrology.* Whitefish, MT: Kessinger Publishing, 2003.

Ross, Jennifer C., and Sharon R. Steadman. *Ancient Complex Societies.* New York: Routledge, 2017.

Russell, Bertrand. *A History of Western Philosophy.* New York: Simon and Schuster, 1945.

Sasson, Jack, John Baines, Gary Beckman, and Karen S. Robinson, eds. *Civilizations of the Ancient Near East.* 4 vols. New York: Charles Scribner's Sons, 1995.

Schmitz, Leonhard. *A Manual of Ancient Geography.* Philadelphia: Blanchard and Lea, 1859.

Schoch, Robert M. *Voyages of the Pyramid Builders: The True Origins of the Pyramids from Lost Egypt to Ancient America.* New York: Penguin Books, 2004.

Solly, Meilan. "What Ötzi the Iceman's Tattoos Reveal about Copper Age Medical Practices." *Smithsonian,* September 10, 2018. https://www.smithsonianmag.com/smart-news/what-otzi-icemans-tattoos-reveal-about-copper-age-medical-practices-180970244.

The Stanford Encyclopedia of Philosophy. https://plato.stanford.edu.

Tillinghast, William H. "The Geographical Knowledge of the Ancients Considered in Relation to the Discovery of America." In *Narrative and Critical History of America.* Vol. 1, edited by Justin Winsor. Boston: Houghton, Mifflin, 1889.

Warnock, Christopher. *The Mansions of the Moon: A Lunar Zodiac for Astrology and Magic.* Morrisville, NC: Lulu Press, 2019.

Wilcox, Donald. *The Measure of Time's Past: Pre-Newtonian Chronologies and the Rhetoric of Relative Time*. Chicago: University of Chicago Press, 1987.

Wilford, John N. "Lessons in Iceman's Prehistoric Medicine Kit." *New York Times*, December 8, 1998. https://www.nytimes.com/1998/12/08/science/lessons-in-iceman-s-prehistoric-medicine-kit.html.

Woolley, Leonard. *The Beginnings of Civilization: History of Mankind: Cultural and Scientific Developments*. Vol. 1, part 2. New York: Mentor Books, 1965.

Zimmerman, J. E. *Dictionary of Classical Mythology*. New York: Harper and Row, 1971.

Near East and Asia

Adamson, Peter. "Al-Sarakhsi, Ahmad ibn al-Tayyib." https://www.academia.edu/36578109/Encyclopedia_of_Medieval_Philosophy_-_Sarakhsi.

Aterman, K. "From Horus the Child to Hephaestus Who Limps: A Romp through History." *American Journal of Medical Genetics* 5 (1999). http://www.ncbi.nlm.nih.gov/pubmed/10076885.

Beasley, W. G., and E. G. Pulleyblank, eds. *Historians of China and Japan*. London: Oxford University Press, 1961.

Budd, Paul. "Recasting the Bronze Age." *New Scientist* (October 1993). https://www.newscientist.com/article/mg14018964-100-recasting-the-bronze-age.

Chandler, David. *A History of Cambodia*. 4th ed. New York: Routledge, 2018.

Cimok, Fatih. *Istanbul*. Istanbul: A Turizm Yayinlari, 1989.

Daniel, Glyn. *The First Civilizations: The Archaeology of Their Origins*. New York: Thomas Y. Crowell, 1968.

Downey, Glanville. *Constantinople in the Age of Justinian*. Norman: University of Oklahoma Press, 1960.

Downey, Glanville. *Gaza in the Early Sixth Century*. Norman: University of Oklahoma Press, 1963.

Downey, Glanville. "*Philanthropia* in Religion and Statecraft in the Fourth Century after Christ." *Historia* 4 (1955): 199–208.

Erman, Adolf. *Life in Ancient Egypt*. Translated by H. M. Tirard. New York: Dover Books, 1894.

Fairbank, John K., Edwin O. Reischauer, and Albert M. Craig. *East Asia: Tradition and Transformation*. Boston: Houghton Mifflin, 1978.

Fairservis, Walter A. *Harrapan Civilization and Its Writing*. Leiden, Netherlands: Brill, 1992.

Gebel Barkal and the Sites of the Napatan Region, UNESCO. https://whc.unesco.org/en/list/1073.

Ghirshman, R. *Iran: From the Earliest Times to the Islamic Conquest*. Harmondsworth, England: Penguin Books, 1954.

Glassner, Jean-Jacques. *The Invention of Cuneiform: Writing in Sumer*. Translated by Zainab Bahrani and Marc Van De Mieroop. Baltimore, MD: Johns Hopkins University Press, 2003.

Grant, Michael. *Jesus: An Historian's Review of the Gospels*. New York: Charles Scribner's Sons, 1977.

Hallo, William, and William Simpson. *The Ancient Near East.* New York: Harcourt Brace Jovanovich, 1971.

Hoernle, F. Rudolf. *Studies in the Medicine of Ancient India. Part 1: Osteology or the Bones of the Human Body.* Oxford: Clarendon Press, 1907.

Hunger, Hermann, and David Pingree. *Astral Sciences in Mesopotamia Handbook of Oriental Studies: The Near and Middle East.* Leiden, Netherlands, and Boston: Brill, 1999.

Jacobsen, Thorkild. *The Treasures of Darkness: A History of Mesopotamian Religion.* New Haven, CT: Yale University Press, 1976.

Kalota, Narain Singh. *India as Described by Megasthenes.* Delhi, India: Concept Publishing Co., 1978.

Kenrick, John. *Phoenicia.* London: B. Fellowes, 1855.

Kramer, Samuel Noah. *History Begins at Sumer.* Philadelphia: University of Pennsylvania Press, 1980.

Kublin, Michael, and Hyman Kublin. *India.* New York: Houghton Mifflin Co., 1991.

Lin, Daren. "A Foundation of Western Ophthalmology in Medieval Islamic Medicine." *University of Western Ontario Medical Journal* 78 (2008): 41–45. https://www.medievalists.net/2008/11/a-foundation-of-western-ophthalmology-in-medieval-islamic-medicine/

Lindberg, David C. *Theories of Vision from Al-kindi to Kepler.* Chicago: University of Chicago Press, 1981.

Mirrazavi, Firouzeh. "Academy of Gundishapur." *Iran Review.* http://www.iranreview.org/content/Documents/_Academy_of_Gundishapur.htm.

Mote, Frederick W. *Intellectual Foundations of China.* New York: Knopf, 1971.

Neugebauer, O. *The Exact Sciences in Antiquity.* New York: Dover Books, 1969.

Nilakanta Sastri, K. A. *Age of the Nandas and Mauryas.* Delhi: Motilal Banarsidass, 1967.

Oates, Joan. *Babylon.* London: Thames and Hudson, 1986.

Orlinsky, Harry M. *Ancient Israel.* Ithaca, NY: Cornell University Press, 1960.

Pedley, John G. *Sardis in the Age of Croesus.* Norman: University of Oklahoma Press, 1968.

Price, John F. *Applied Geometry of the Sulba Sutras.* http://chaturpata-atharvan-ved.com/spiritual-books-section/spiritual-books/acharya-literature/scientist-acharya-of-ancient-india/SulbaSutras-Applied-Geometry-by-John-Price-EN.pdf.

Riefstahl, Elizabeth. *Thebes in the Time of Amunhotep III.* Norman: University of Oklahoma Press, 1964.

Ruben, R. J. "Otology at the Academy of Gondishapur, 200–600 CE." *Otology and Neurotology* 38 (2017): 1540–45.

Sandars, N. K. *The Sea Peoples.* London: Thames and Hudson, 1985.

Sarkar, Benoy Kumar. *Hindu Achievements in Exact Science: A Study in the History of Scientific Development.* London: Longmans, Green and Co., 1918.

Savage-Smith, Emilie. *Islamic Culture and the Medical Arts.* National Library of Medicine. 1994. https://www.nlm.nih.gov/exhibition/islamic_medical/index.html#toc.

Schmitt, Rüdiger. "Democedes." *Encyclopaedia Iranica*. http://www.iranicaonline.org/articles/democedes.

Scholz, Piotr O. *Ancient Egypt: An Illustrated Historical Overview.* Hauppauge, NY: Barron's Education Press, 1997.

Singh, Upinder. *A History of Ancient and Early Medieval India: From the Stone Age to the 12th Century.* Delhi, India: Pearson Education, 2008.

Steiner, Rudolph. *Christianity as a Mystical Fact.* New York: Anthroposophic Press, 1947.

Vasiliev, A. A. *History of the Byzantine Empire.* 2 vols. Madison: University of Wisconsin Press, 1958.

Violatti, Cristian. "Neolithic Period." *Ancient History Encyclopedia.* https://www.ancient.eu/Neolithic.

Watson, Burton. *Early Chinese Literature.* New York: Columbia University Press, 1962.

Watson, Burton. *Ssu-Ma Ch'ien: Grand Historian of China.* New York: Columbia University Press, 1958.

White, Jon Manchip. *Everyday Life in Ancient Egypt.* New York: Capricorn Press, 1967.

Greece

Barnes, Jonathan. *Aristotle*. Oxford: Oxford University Press, 1982.

Botsford, George, and Charles A. Robinson. *Hellenic History.* Revised by Donald Kagan. New York: Macmillan Publishing, 1969.

Brunschwig, Jacques, and Geoffrey Lloyd. *Greek Thought: A Guide to Classical Knowledge.* Cambridge, MA: Harvard University Press, 2000.

Burkert, Walter. *Structure and History in Greek Mythology and Ritual.* Berkeley: University of California Press, 1979.

Burn, A. R. *The Pelican History of Greece.* Harmondsworth, England: Penguin Books, 1974.

Burn, A. R. *The World of Hesiod.* Harmondsworth, England: Penguin Books, 1936.

Burnet, John. *Early Greek Philosophy.* London: Adam and Charles Black, 1892.

Collard, Christopher. *Aeschylus: Persians and Other Plays.* Oxford: Oxford University Press, 2008.

Doumas, Christos G. *Thera: Pompeii of the Ancient Aegean.* London: Thames and Hudson, 1983.

Durant, Will. *The Life of Greece.* New York: Simon and Schuster, 1939.

Ehrenberg, Victor. *Alexander and the Greeks.* Oxford: Oxford University Press, 1938.

Empereur, Jean-Yves. *Alexandria Rediscovered.* New York: George Braziller, 1998.

Faraone, Christopher A. *Magika Hiera: Ancient Greek Magic and Religion.* New York: Oxford University Press, 1991.

Finley, M. I. *The World of Odysseus.* Harmondsworth, England: Penguin Books, 1972.

Fowler, David. *The Mathematics of Plato's Academy: A New Reconstruction.* New York: Oxford University Press, 1999.

Fox, Robin Lane. *The Search for Alexander.* Boston: Little, Brown, 1979.
Freeman, Kathleen. *Greek City-States.* New York: W. W. Norton, 1950.
French, A. "The Economic Background to Solon's Reforms." *Classical Quarterly* 6 (1956): 11–25.
Gow, James. *A Short History of Greek Mathematics.* Cambridge: Cambridge University Press, 2010.
Grant, Michael. *From Alexander to Cleopatra: The Hellenistic World.* New York: History Book Club, 2000.
Graves, Robert. *The Greek Myths.* Vol. 1. Harmondsworth, England: Penguin Books, 1960.
Hadas, Moses. *Humanism: The Greek Ideal and Its Survival.* New York: Mentor Books, 1972.
Hamilton, J. R. *Alexander the Great.* Pittsburgh: University of Pittsburgh Press, 1974.
Hare, R. M. *Plato.* Oxford: Oxford University Press, 1982.
Heath, Thomas. *Aristarchus of Samos.* New York: Dover Books, 1981.
Hussey, E. *The Presocratics.* London: Duckworth, 1972.
Kerenyi, Karl. *Prometheus: Archetypal Image of Human Existence.* New York: Pantheon Books, 1963.
Lloyd, G. E. R. *Methods and Problems in Greek Science: Selected Papers.* Cambridge: Cambridge University Press, 1991.
Mayor, Adrienne. *Greek Fire, Poison Arrows, and Scorpion Bombs: Biological and Chemical Warfare in the Ancient World.* Woodstock, NY: Overlook Duckworth, 2003.
Netz, Reviel. *The Shaping of Deduction in Greek Mathematics: A Study in Cognitive History.* Cambridge: Cambridge University Press, 1999.
Phillips, E. D. *Greek Medicine.* London: Thames and Hudson, 1973.
Robinson, Charles A., Jr. *Athens in the Age of Pericles.* Norman: University of Oklahoma Press, 1959.
Schmitt, Charles B. *Aristotle and the Renaissance.* Cambridge, MA: Harvard University Press, 1983.
Starr, Chester. *The Awakening of the Greek Historical Spirit.* New York: Knopf, 1968.
Turner, William. "Aristotle." *Catholic Encyclopedia.* New York: Encyclopedia Press, 1913.
Vandvik, Eirik. *The Prometheus of Hesiod and Aeschylus.* Oslo: I Kommisjon hos J. Dybwad, 1943.
Warmington, E. H. *Greek Geography.* London: J. M. Dent, 1934.

Rome

Barrow, R. H. *Plutarch and His Times.* New York: AMS Press, 1979.
Barrow, R. H. *The Romans.* Harmondsworth, England: Penguin Books, 1949.
Birley, Anthony. *Marcus Aurelius.* Boston: Little, Brown, 1966.
Bowder, Diana. *The Age of Constantine and Julian.* London: Paul Elek, 1978.

Bowersock, G. W. *Julian the Apostate.* Cambridge, MA: Harvard University Press, 1978.

Brown, Peter. *Augustine of Hippo.* Berkeley: University of California Press, 1967.

Brown, Peter. *The Making of Late Antiquity.* Cambridge, MA: Harvard University Press, 1978.

Browning, Robert. *The Emperor Julian.* Berkeley: University of California Press, 1976.

Bujalkova, M. "Rufus of Ephesus and His Contribution to the Development of Anatomical Nomenclature." *Acta Medico-Historica Adriatica* 9, no. 1 (2011): 89–100.

Carrier, Richard. *Science Education in the Early Roman Empire.* Durham, NC: Pitchstone Publishing, 2016.

Chadwick, Henry. *Augustine.* Oxford: Oxford University Press, 1986.

Chadwick, Nora. *The Celts.* Harmondsworth, England: Penguin Books, 1970.

Church, Alfred J. *Carthage, or the Empire of Africa.* London: T. Fisher Unwin, 1886.

Dal Maso, Leonard B. *Rome of the Caesars.* Translated by Michael Hollingsworth. Florence, Italy: Bonechi-Edizioni, 1974.

Dodds, E. R. *Pagan and Christian in an Age of Anxiety.* New York: W. W. Norton, 1965.

Downey, Glanville. *Constantinople in the Age of Justinian.* Norman: University of Oklahoma Press, 1960.

Downey, Glanville. "Education in the Christian Roman Empire: Christian and Pagan Theories under Constantine and His Successors." *Speculum* 32 (1957): 48–61.

Downey, Glanville. *Gaza in the Early Sixth Century* Norman: University of Oklahoma Press, 1963.

Durant, Will. *Caesar and Christ.* New York: Simon and Schuster, 1944.

Gibbon, Edward. *The Decline and Fall of the Roman Empire.* New York: Modern Library, 1932.

Glauthier, Patrick. "Repurposing the Stars: Manilius, Astronomica 1, and the Aratean Tradition." *American Journal of Philology* 138, no. 2 (Summer 2017): 267–303. https://www.academia.edu/34309524/Repurposing_the_Stars_Manilius_Astronomica_1_and_the_Aratean_Tradition.

Grant, Michael. *The Climax of Rome.* Boston: Little, Brown, 1968.

Jones, A. H. M. *The Decline of the Ancient World.* London: Longman, 1966.

Jones, A. H. M. *The Greek City from Alexander to Justinian.* Oxford: Clarendon Press, 1940.

Jones, A. H. M. *The Later Roman Empire.* 2 vols. Norman: University of Oklahoma Press, 1964.

Jones, Tom B. *In the Twilight of Antiquity.* Minneapolis: University of Minnesota Press, 1978.

Karamanou, Marianna, Gregory Tsoucalas, George Creatas, and George Androutsos. "The Effect of Soranus of Ephesus (98–139) on the Work of Midwives." *Women and Birth* 26 (2013): 226–28.

Kleijn, Gerda de. *The Water Supply of Ancient Rome: City Area, Water, and Population.* Amsterdam: Gieben, 2001.
MacMullen, Ramsey. *Constantine.* New York: Harper and Row, 1971.
McCluskey, Stephen C. *Astronomies and Cultures in Early Medieval Europe.* Cambridge: Cambridge University Press, 1998.
Meagher, Robert. *Augustine: An Introduction.* New York: Harper and Row, 1978.
Muller, H. *Christian and Pagan in the Fourth Century A.D.* Pretoria, South Africa: Union Booksellers, 1946.
Ogilvie, R. M. *Roman Literature and Society.* Harmondsworth, England: Penguin Books, 1980.
Prioreschi, Plinio. *Roman Medicine.* Omaha, NE: Horatius Press, 1998.
Riddle, John M. *Dioscorides on Pharmacy and Medicine.* Austin: University of Texas Press, 1985.
Rowell, Henry Thompson. *Ammianus Marcellinus, Soldier Historian of the Late Roman Empire.* Cincinnati: University of Cincinnati Press, 1964.
Rowell, Henry Thompson. *Rome in the Augustan Age.* Norman: University of Oklahoma Press, 1962.
Sandbach, F. H. *The Stoics.* New York: W. W. Norton, 1975.
Sinnigen, William, and Arthur Boak. *A History of Rome.* 6th ed. New York: Macmillan Publishing, 1977.
Smith, John H. *Constantine the Great.* New York: Scribner, 1971.
Swift, Louis J., and James H. Oliver. "Constantius II on Flavius Philippus." *American Journal of Philology* 83 (1962): 247–64.
Taylor, Lily Ross. *Party Politics in the Age of Caesar.* Berkeley: University of California Press, 1949.
Thompson, E. A. *The Historical Work of Ammianus Marcellinus.* Cambridge: Cambridge University Press, 1947.
Vanags, Patricia. *The Glory That Was Pompeii.* New York: Mayflower Books, 1979.
Von Hagen, Victor W. *Roman Roads.* London: Werdenfeld and Nicholson, 1966.

Medieval

Adams, Charles Kendall. *Christopher Columbus: His Life and Work.* New York: Dodd, Mead and Company, 1892.
Boas, Marie. *The Scientific Renaissance, 1450–1630.* New York: Harper & Brothers, 1962.
Brehaut, Ernest. *An Encyclopedist of the Dark Ages: Isidore of Seville.* New York: Columbia University, 1912.
Cohen, I. Bernard. *The Birth of a New Physics.* New York: W. W. Norton, 1985.
Daiber, Hans, Anna Akasoy, and Emilie Savage-Smith, eds. *Islamic Philosophy, Theology and Science.* Vol. 83. Leiden, Netherlands: Koninklijke Brill, NV, 2012.
Deen, S. M. *Science under Islam: Rise, Decline and Revival.* Morrisville, NC: Lulu, 2010.

Dharmananda, Subhuti. *Sun Simiao: Author of the Earliest Chinese Encyclopedia for Clinical Practice*. http://www.itmonline.org/arts/sunsimiao.htm.

Enan, Mohammad Abdullah. *Ibn Khaldun: His Life and Works*. Kuala Lumpur, Malaysia: The Other Press, 2007.

Fancy, Nahyan. *Science and Religion in Mamluk Egypt: Ibn al-Nafis, Pulmonary Transit and Bodily Resurrection*. London: Routledge, 2013.

Goldstein, Thomas. *Dawn of Modern Science*. Boston: Houghton Mifflin Company, 1988.

Grant, Edward, and John E. Murdoch, eds. *Mathematics and Its Applications to Science and Natural Philosophy in the Middles Ages: Essays in Honor of Marshall Clagett*. Cambridge: Cambridge University Press, 1987.

Hsu, Elizabeth. *Innovation in Chinese Medicine*. Cambridge: Cambridge University Press, 2001.

Jung, Elzbieta. "Richard Kilvington." *Stanford Encyclopedia of Philosophy*. https://plato.stanford.edu/entries/kilvington.

Kraemer, Joel. L. *Maimonides: The Life and World of One of Civilization's Greatest Minds*. New York: Doubleday, 2008.

Lovejoy, Arthur O. *The Great Chain of Being: A Study of the History of an Idea*. Cambridge, MA: Harvard University Press, 1950.

Mark, Joshua J. "Silk Road." *Ancient History Encyclopedia*. https://www.ancient.eu/Silk_Road.

Needham, Joseph. *Science and Civilisation in China: Volume 4: Physics and Physical Technology: Part 3: Civil Engineering and Nautics*. Cambridge: Cambridge University Press, 1971.

Needham, Joseph. *Science and Civilisation in China: Volume 6, Biology and Biological Technology, Part 6: Medicine*. Cambridge: Cambridge University Press, 2000.

Nor Afifah Borhan, Nuramirah Mohd Nor, and Aminuddin Ruskam. "Ibn Al-Baitar: the Pioneer of [*sic*] Botanist and Pharmacist." http://eprints.utm.my/id/eprint/60916/1/AminuddinRuskam2014_IbnAlBaitarthePioneerofBotanist.pdf.

The Oxford Handbook of Byzantine Studies. Edited by Elizabeth Jeffreys, John Haldon, Robin Cormak. Oxford: University Press, 2008.

Rizvi, Sajjad H. *Avicenna (Ibn Sina)*. Internet Encyclopedia of Philosophy. https://www.iep.utm.edu/avicenna/#H2.

Rosner, Fred. "The Life of Moses Maimonides, a Prominent Medieval Physician." *Einstein Journal of Biology and Medicine* 19 (125–128). http://www.einstein.yu.edu/uploadedFiles/EJBM/19Rosner125.pdf.

Salguero, C. Pierce. *Translating Buddhist Medicine in Medieval China*. Philadelphia: University of Pennsylvania Press, 2014.

Savage-Smith, Emilie. *Islamic Culture and the Medical Arts*. National Library of Medicine. 1994. https://www.nlm.nih.gov/exhibition/islamic_medical/index.html#toc.

Sen, Tansen. "The Travel Records of Chinese Pilgrims Faxian, Xuanzang, and Yijin: Sources for Cross-Cultural Encounters between Ancient China and

Ancient India." *Education about Asia* 11, no. 3 (Winter 2006). http://afe.easia.columbia.edu/special/travel_records.pdf.

Shaku, Soyen. *Zen for Americans*. Translated by Daisetz Teitaro Suzuki. New York: Dorset Press, 1987.

Shumaker, Wayne. *The Occult Sciences in the Renaissance: A Study in Intellectual Patterns*. Berkeley: University of Los Angeles Press, 1972.

Sivin, Nathan. "Why the Scientific Revolution Did Not Take Place in China—Or Didn't It?" *Chinese Science* 5 (1982): 45–66.

Stencel, Robert, Fred Gifford, and Eleanor Moron. "Astronomy and Cosmology at Angkor Wat." *Science* 193 (1976): 181–87. https://science.sciencemag.org/content/193/4250/281.

Sullivan, Richard E. *Aix-la-Chapelle in the Age of Charlemagne*. Norman: University of Oklahoma Press, 1963.

Sylla, Edith D. "Oxford Calculators." *Encyclopedia of Medieval Philosophy: Philosophy between 500 and 1500*. Edited by Henrik Lagerlund, 903–8. Dordrecht, Netherlands: Springer Reference, 2019.

"Umar Khayyam." *Stanford Encyclopedia of Philosophy*. https://plato.stanford.edu/entries/umar-khayyam/#KhaMatSci.

Van Steenberghen, Fernand. *Aristotle in the West: The Origins of Latin Aristotelianism*. Louvain, Belgium: Nonwelaerts Publishing House, 1970.

White, Lynn, Jr. *Medieval Technology and Social Change* (Oxford: Clarendon Press, 1962).

Wilcox, Donald. *In Search of God & Self: Renaissance and Reformation Thought*. Prospect Heights, IL: Waveland Press, 1987.

Wilson, David M. *The Vikings and their Origins: Scandinavia in the First Millennium*. London: Thames and Hudson, 1989.

Wright, Arthur F. *Buddhism in Chinese History*. Stanford, CA: Stanford University Press, 1950.

America

Aveni, A. F. *Archaeoastronomy in the New World: American Primitive Astronomy*. Cambridge: Cambridge University Press, 1982.

Braun, David M. "Corn Domesticated from Mexican Wild Grass 8,700 Years Ago." National Geographic Society. https://blog.nationalgeographic.org/2009/03/23/corn-domesticated-from-mexican-wild-grass-8700-years-ago.

Cartwright, Mark. "Chichen Itza." *Ancient History Encyclopedia*. https://www.ancient.eu/Chichen_Itza.

Cartwright, Mark. "Inca Civilization." *Ancient History Encyclopedia*. https://www.ancient.eu/Inca_Civilization.

Cartwright, Mark. "Teotihuacan." *Ancient History Encyclopedia*. https://www.ancient.eu/Teotihuacan.

Cartwright, Mark. "Tiwanaku," *Ancient History Encyclopedia*: https://www.ancient.eu/Tiwanaku/

"First City in the World?" *Smithsonian Magazine*, August 2002. https://www.smithsonianmag.com/history/first-city-in-the-new-world-66643778.

Lawson, Russell M. *Frontier Naturalist: Jean Louis Berlandier and the Exploration of Northern Mexico and Texas.* Albuquerque: University of New Mexico Press, 2012.

Lawson, Russell M., ed. *Research and Discovery: Landmarks and Pioneers in American Science.* 3 vols. Armonk, NY: M. E. Sharpe, 2008.

McCluskey, S. C. "The Astronomy of the Hopi Indians." *Journal for the History of Astronomy* 8 (1977): 174–95.

Mendieta, Ramiro Matos, and José Barreiro, eds. *The Great Inka Road: Engineering an Empire.* Washington, DC: Smithsonian Institution, 2015.

Pool, Christopher A. *Olmec Archaeology and Early Mesoamerica.* Cambridge: Cambridge University Press, 2007.

Vogel, Virgil. *American Indian Medicine.* Norman: University of Oklahoma Press, 1990.

Index

Abbasid Caliphate, 420, 422–423, 445, 458, 461
Abelard, Peter (theologian), 496, 554
Abu Hamid (explorer), 510, 554
Abu Ma'shar (Albumasar), 430, 461–463, 552
Achaemenian Dynasty, 59–60
Achaemenid Empire, 60–63, 542
Acosta, Jose de, 517
Acropolis (Athens), 132–133, 234, 323
Aelian, Claudius (commentator), 70–71, 376–377
Aeschylus (playwright), 133, 177, 234, 542
  The Persians, 132
  Prometheus Bound, 35–36
Aghlabid Dynasty, 452
Akhenaten (Amenhotep IV), 26, 45, 97, 540
Alchemy
  in ancient Chinese medicine, 83–84
  in *Canterbury Tales* (Chaucer), 206
  in Hellenistic Age, 229–230, 322–323
  in medieval Chinese medicine, 479
  in medieval Europe, 498–499
  in medieval Islamic thought, 44, 442, 460, 464, 467, 471
  origins of gunpowder (China), 339
Alcuin of York, 354, 492–493, 552
Alexander of Aphrodisias, 258–260
Alexander of Macedon (Alexander the Great), 19, 68–69, 159, 166–168, 219–222, 234–235, 327, 544
  Alexandria and, 223–226, 323–325
  Al-Biruni on, 59
  explorations and expeditions under, 168–171, 174, 222–223
  geography under, 171–172
  influences on, 167
  invasions and conquests of, 60, 63, 69, 166–168, 175, 178, 180, 219, 235
  Plutarch on, 68, 167, 220, 224, 254–255
  student of Aristotle, 138, 141, 142, 145–146, 166, 219–221, 222, 224
Alexandria, 223–226, 323–325
Ali ibn Ridwan, 359, 428, 453–455, 553
Amenhotep IV (Akhenaten), 26, 45, 97, 540
American Indians
  Aztec civilization, 523–524, 533, 555
  Chavin civilization, 525
  Inca civilization, 523, 524, 525–526, 533, 555
  Mayan civilization, 214, 316, 523–524, 548
  of Mexico, 522–524
  Nazca civilization, 525
  of North America, 519–522
  Olmec civilization, 188, 316, 523–524, 541
  of South America, 524–526
  Tiahuanaco (Tiwanaku) civilization, 525, 526
  Toltec civilization, 523–524, 552
Ampere, Andre-Maria, 536
Anatolius (philosopher), 208, 382
Anaxagoras (philosopher and scientist), 95, 96, 97–98, 109, 114–115, 131–132, 135–137, 148, 196, 541
Anaximander of Miletus (scientist and philosopher), 91, 93–95, 97, 99, 103, 160, 163, 195, 233, 243, 291
Anaximenes of Miletus (scientist and philosopher), 93, 95–96, 97, 98, 148, 388, 542
Andronicus of Rhodes (commentator), 208, 258, 382, 426–427

Animal sacrifice, 38, 103, 307
Anselm of Canterbury, 497
Anthropomorphism
of ancient Celts, 38
break from, 109, 114, 118, 137, 255, 387
in Greek religion, 44
in Hindu religion, 64
smith gods and, 318
in works of Hesiod, 43
in works of Homer, 33–34, 43–44, 234, 255
Antiochus of Ascalon (Academic), 140, 252, 254, 546
Antonine Dynasty (of Roman emperors), 298, 350, 548
Antoninus Pius (emperor), 180, 298, 309, 386, 548
Apollodorus of Athens (historian), 210, 237, 348, 545
Apollonius of Perga (mathematician), 198, 226, 293, 294, 295, 406, 407, 545
Apollonius of Rhodes, 37, 163, 177, 228
Apollonius of Tyana, 376
Aquinas, Thomas (philosopher and theologian), 153, 233, 399, 467, 481, 483, 497–498, 555
Arcesilaus (Skeptic philosopher), 140, 153, 198, 544
Archelaus (teacher of Socrates), 97, 98, 135–136, 137, 388
Archimedes of Syracuse (mathematician and engineer), 230, 327–328, 545
Archimedes's screw, 230, 328
on Aristarchus, 198
death of, 328
discovery of *pi,* 230, 294
on hydraulics, 230, 327–328
influence of, 294, 295, 300, 327–328, 534
influences on, 152, 293
on mathematics, 230, 290, 293–294, 407, 471
on military science, 328
siege of Syracuse and, 328
Aristarchus of Samos (astronomer and mathematician), 197–199, 203, 209–210, 225, 227, 294, 489, 533, 545
Aristides, Aelius (student of dreams), 278, 359, 376, 548
Aristides of Athens (philosopher), 386–387, 548
Aristotle (philosopher and scientist), 140–148, 543
Alexander of Aphrodisias on, 258, 259–260
Arab translations of works of, 153, 413, 422
Aristotelian school at Rhodes, 226
on behavioral science, 144–145
Categories, 382
"defining principle" of, 145, 237, 241, 242
on Earth's sphericity, 163
first principles of, 142–143
on geocentric universe, 197, 198
on human psychology, 357
influence of, 152–153, 169, 230, 258, 334, 369–370, 381–383, 399, 422
influence on Alexander, 138, 141, 142, 145–146, 166, 219–221, 222, 224
influence on Euclid, 292–293
influence on Galen, 284, 287
John Philoponus on, 408
Justin Martyr on, 386
Logic, 422, 423, 426
Lyceum (Athens), 118, 141, 142, 148, 150, 151, 152, 221, 270, 292
on mathematics, 292–293
on metallurgy, 322
on metaphysics, 118, 141, 143, 232, 233
on meteorology, 197
Nichomachean Ethics, 118, 141, 144, 147
on oceans, 175
On Interpretation, 382
On Memory and Reminiscence, 357
On the Heavens and Meteorology, 147–148
Peripatetic philosophy of, 151, 235, 258, 270, 370, 382
Physics, 118, 141, 144–145, 382
Poetics, 118, 141, 143–144, 342, 343
Politics, 118, 141, 147, 341
Themistius on, 398–399
Theophrastus on, 148–150
on zoology, 145–146
Arrian (biographer of Alexander), 68, 168–169, 351–352, 411, 547
Anabasis, 168, 169, 222, 351, 352
as historian of science, 352
Indica, 168, 169–170, 351, 352
Asclepiades of Bithynia (physician), 229, 278–279, 281, 284, 287, 310, 546
Asclepiads, 266–267

Astrology, 16–18, 36–37, 80, 82, 201–207, 300, 398, 414, 423, 425, 462, 469–470, 476–477, 489, 511–512
Astronomy
in Achaemenid Empire, 61–62
Aristotle on, 147–148, 152–153
in Babylonia, 193–194
in China, 80–82
in Egypt, 22, 23, 194–195
in Greece, 195–199, 226
Herodotus on, 100, 101
in Mesopotamia, 16–17, 18–19
in North Africa and Asia, 214–215
in Plato's Academy, 138, 139, 140
in Rome, 200–201
in the Vedic Age, 67, 73
Atomic energy, 536
Atomic theory, 121–122, 249, 535
Atomic weapons, 536–537
Atomists, 67–68, 97, 119, 247–249, 270–271
Aristotle on, 123
Hellenistic alchemy and, 230
Roman pragmatic medicine and, 278–279
See also Democritus; Epicurus; Kanada; Leucippus; Lucretius
Augustine of Hippo, 483–485, 495–496, 550
City of God, 361, 392, 452, 551
Confessions, 245, 361, 392, 485, 551
conversion to Christianity, 356
influence of, 361, 392, 486
influenced by Aristotelian syllogisms, 484–485
influenced by Varro, 302
medieval philosophy of, 483–485, 495–496
on nature of time, 245
Neoplatonism of, 152, 373
Augustus (title), 297, 395, 398, 481, 482
Augustus (Octavian) Caesar, 180–181, 211, 297, 298, 300, 302–303, 331, 332, 426–427, 546
Aurelius Augustine. *See* Augustine of Hippo
Aztec civilization, 523–524, 533, 555

Babbage, Charles, 536
Bacon, Francis (empiricist), 534–535, 556
Bacon, Roger (philosopher), 462–463, 498–499, 555
Al-Baghdadi, Abd al-Latif (scientist), 428–429, 555
Al-Baitar (physician and botanist), 457, 555
Ban Gu (historian), 355, 547
Al-Battani (Albategnius), 425, 444, 448, 553
Bede (historian), 354, 552
Behavioral science, 355–363
Asian psycho-philosophical treatises, 361–363
Augustine on, 361
dreams, 357–359
Homeric psychology, 356–357
Plutarch on, 359–360
Porphyry on, 360
Berlandier, Jean Louis, 521–522, 523
Berossus, 17, 300
Al-Biruni (polymath), 468–470, 510–511, 553
on astrology, 203–206, 469–470
on astronomical observations, 192–193
The Book of Instruction in the Elements of the Art of Astrology, 206, 469–470
The Chronology of Ancient Nations, 59, 73, 203, 206, 208–209, 354, 406, 468, 469–470
condemnation of Hindu astronomers, 471
on February as leap-month, 208
on Hippocrates, 277
History of India, 471
influenced by Ptolemy, 194, 468, 469, 510
on Ramadan, 214–215
on science, 470
on theories of the deluge, 461–462
on timekeeping and calendars, 62, 194, 208–209, 214–215, 406
Boethius (philosopher), 485–486, 550
Consolation of Philosophy, 485–486, 507
influenced by Aristotle, 485, 486, 493
influenced by Augustine, 486
influenced by Plato, 152, 485, 486, 496, 497
paganism and, 481
"quadrivium" coined by, 493
translation of *Introduction to Arithmetic* (Nicomachus), 488
Bohr, Niels, 121, 536

Book of the Dead (Papyrus of Ani), 22, 37, 42, 540
Bow drill (fire drill), 4, 35
Boyle, Robert, 119, 122, 535
Bradwardine, Thomas (mathematician), 500, 555
Bronze
  Colossus of Rhodes (bronze statue), 325–326
  Greek and Roman metallurgy, 322–323, 325–326, 329
  history and development of, 76, 317
  production of, 16, 317–318, 319
  properties of, 319
  weapons and tools, 13, 317–318, 319, 321
Bronze Age, 317–320
  agriculture in, 318–319
  civilizations and cultures, 318, 319–320
  definition and dates of, 317–318
  Early Bronze Age, 12
  Harappan civilization (Indus River valley), 318, 319
  Hesiod on, 43
  Homer and, 34
  Middle Bronze Age, 14–15
  Minoan civilization (Crete), 318, 319–320, 540
  Mycenaean civilization (Greece), 318, 320
  Nile River valley, 318
  Sumeria (Mesopotamia), 12, 13, 14–15, 16, 318–319, 539
Budge, Wallis, 42
Bush, Vannevar, 536, 537

Cabral, Pedro Álvares (explorer), 513, 555
Caligula (emperor), 298, 350, 547
Callippus of Cyzicus (astronomer), 196–197, 209, 226, 469, 543
Cam, Diego (explorer), 512–513, 556
Campbell, Joseph, 31–32
Canals and canal systems
  in eastern Asia, 77
  in Egypt, 24, 317, 325
  in Greece, 324
  in Mesopotamia, 12, 315, 318–319
  during the Pax Romana, 307
  in the Persian Empire, 180
  in southern Asia, 64
Carolingian Renaissance, 337–338, 492–494
Carthaginians, 188–190
Çatal Hüyük (Turkey), 7, 11, 539
Cathedrals, 496
Cato, Marcus Porcius (agriculturalist), 281, 301, 382, 545
Celsus, Aurelius Cornelius (physician), 278, 279, 301, 310, 401, 548
Celts, 9–10, 38–39, 390
Chain of being, 146, 496
Chares of Lindos, 325, 545
Charlemagne (Charles the Great), 354, 492–494, 552
Charles Martel, 338, 552
Chaucer, Geoffrey, 206, 554
Chavin civilization, 525
Chhen Chhang Fang, 475–476
Christianity (in later Roman Empire), 383–392
  Aristides of Athens and, 386–387, 548
  Clement of Alexandria and, 388–390, 548
  Hermias the philosopher and, 385, 387–388, 407
  Justin Martyr and, 386, 548
  Origen and, 373, 385, 391, 406, 549
  paganism and, 385
  Paul of Tarsus and, 49, 384–385, 387, 388, 420, 495
  See also Augustine of Hippo; Hippolytus
Chronology and chronologists, 210–212
Chryssipus (philosopher), 236, 251, 545
Chuang Tzu (Zhuang Zhou), 362
Cicero, Marcus Tullius (philosopher), 252, 308, 546
  as "citizen of the world," 235
  On Divination, 308
  influence of, 532, 535, 537
  influences on, 140, 172, 251
  On the Nature of the Gods, 252, 308
  On the Republic, 175, 252
  "Scipio's Dream," 175, 252
Cincinnatus (writer), 301
Civilization
  agriculture in development of, 12, 20–21
  in Ancient Mesopotamia, 11–19
  definition of, 11–12
  first civilization (Sumer), 12
  mastery of fire in development of, 20
  in the Nile River Valley, 20–27
Claudius (emperor), 174, 253, 280, 298, 303, 308

Cleanthes (philosopher), 236, 251, 544
Clement of Alexandria (theologian and commentator), 388–390, 548
Colossus of Rhodes, 325–326, 545
Columbus, Christopher (explorer), 176, 513–514, 516, 533, 555
Columella (agriculturalist), 301, 302, 382, 547
Commentators, 375–381. *See also* Aelian, Claudius; Diogenes Laertius; Proclus of Lydia
Commodus (emperor), 284, 548
Confucius (K'ung-Fu-tzu), 78–79, 355, 474–475, 542
Constantine I (emperor), 212, 214, 330, 375, 395–399, 427, 482, 549
Constantine VII Porphyrogenitus (scholar-emperor), 413, 553
Constantius I (emperor), 482
Constantius II (emperor), 397–398, 401, 404, 550
Constellations, 212–214
Copernicus, Nicholas (mathematician), 233, 531, 533–534, 535, 556
Cortes, Hernan, 524
Cosmas Indicopleustes (traveler), 412–413, 506–507, 552
Coulomb, Charles, 536
Cultural emergence, theories of, 518–519
Curie, Marie, 536, 537

Demeter (Greek fertility goddess), 21, 25, 32, 43, 45, 243
Democracy, 132
Democritus (Epicurean and atomist), 122–125, 126, 129, 145, 174–175, 233, 249, 376, 543
Denisovan humans, 539
Descartes, René (mathematician), 119, 535, 556
Dias, Bartolomeu, 513
Diffusionist theory of cultural emergence, 518–519
Dinocrates (Deinocrates) of Rhodes (urban designer), 224, 323–325, 327, 544
Dio Cassius (historian), 303, 307–308, 350, 549
Diocletian (emperor), 373, 395, 396, 397, 481–482, 549
Diodorus Siculus (historian and polymath), 348–349
  on Alexander the Great, 324–325
  on astrology, 201
  on the Chaldeans, 18, 201
  dates of, 546
  on Egypt and Egyptians, 20–22, 70, 349
  on funeral pyre for Hephaestion, 324–325
  on the Hanging Gardens of Babylon, 326–327
  on the Hyperboreans, 9–10
  influence on Ammianus Marcellinus, 405
  influenced by Nearchus of Crete, 171
  on Oenopides of Chios, 208
  philosophy of, 348–349
  on the Phoenicians, 186–188
  travels of, 348–349
  on universal histories, 348
Diogenes Laertius (compiler), 22, 26, 58, 89–90, 91–92, 94–95, 110–111, 377–378, 549
Diogenes of Apollonia (materialist), 125–126
Dionysius Exiguus (chronologist), 212, 491, 551, 552
Dioscorides (physician), 280–281, 282, 310–311, 445–446, 547
Domitian (emperor), 298, 309, 548
Druids, 38–39, 390, 404–405

Easter, 212, 551
Edison, Thomas Alva, 536
Einhard (scholar and biographer), 354, 492–493, 552
Einstein, Albert, 233–234, 242, 536
Elder Pliny. *See* Pliny the Elder
Embalming, 25–26
Empedocles (philosopher), 107, 116, 120–121, 130, 233, 249, 270, 386, 542
Enuma Elish (Sumerian epic of creation), 41
Epic of Gilgamesh, 11, 12, 13, 32, 160, 176, 356, 461, 540
Epictetus (Stoic philosopher), 251, 253, 308–309, 351–352, 547
Epicureanism, 119, 122, 129–130, 140, 237, 241, 244, 247–251, 371
Epicurus (founder of Epicureanism), 126–129, 152, 153, 247–248, 298, 544
Eratosthenes of Cyrene (geographer), 169, 172, 173, 210, 225, 228, 251, 294, 405
Eriugena, John Scotus (philosopher), 494, 495, 496, 553

Euclid (geometer), 111, 152, 196, 225, 226, 290, 291, 293–295, 532, 534
Eudoxus of Cnidus (student of Plato), 151–152, 160, 173, 196–197, 198, 209, 226, 228, 290, 376, 469, 543
Eunapius (philosopher), 288, 371, 373, 374, 376, 397, 401, 402, 403, 550
European Renaissance. *See* Renaissance, European
Eusebius of Caesarea (church historian), 23, 25–26, 183, 211–212, 490–491, 549

Fan Ye (historian), 355, 551
Al-Farabi (mathematician), 425–427, 553
Faraday, Michael, 536
Fatimid Dynasty, 453
Fermi, Enrico, 536–537
Fertile Crescent (Jericho), 7, 11, 12
Ficino, Marsilio (philosopher), 152, 499, 556
Fire, first use of, 539
Fire drill (bow drill), 4, 35
First Dynasty of Egypt, 15, 23
Flavian Dynasty (of Roman emperors), 298, 350, 548
Fourth Dynasty of Egypt, 24
Franklin, Ben, 535–536

Galen of Pergamum (physician), 284–288
- Alexander of Aphrodisias on, 259, 260
- Ali ibn Ridwan on, 453, 454–455
- on Asclepius, 285–286
- *The Book of Principles,* 454–455
- on dreams, 358, 359
- on education of physicians, 299
- on good health and humors, 286–287
- Hippocratic approach of, 271, 274–275, 284, 286, 287, 311
- Ibn Abi Usaibi'ah on, 430–442
- Ibn al-Tayyib on, 427–428
- Ibn Rushd on, 450
- influence of, 272, 287, 311, 369, 370, 421–422, 430–431, 432, 446–447, 450, 453, 454–455, 456, 464–465, 478, 531, 534, 535
- influences on, 271, 272, 274–275, 282, 284–286
- John Philoponus on, 408
- list of works translated into Arabic, 432–442
- Maimonides on, 456
- *On the Natural Faculties,* 287, 311
- Oribasius of Pergamon and, 400–401
- physician to Marcus Aurelius, 278, 284, 358
- Rhazes as "Arab Galen," 463
- Rhazes on, 464
- Roman medicine after time of, 287–288
- Stoicism of, 271

Galileo Galilei (astronomer and physician), 119, 124, 153, 531, 534, 556
Gama, Vasco da (explorer), 513, 556
Gassendi, Pierre, 534–535
Ge Hong (alchemist), 84, 549
Geodesy, 163–164, 172, 251, 294, 305, 489
Geoffrey of Monmouth (historian), 9, 354–355, 554
Geoffroy de Villehardouin (historian), 354, 555
Geography
- Arab explorations and, 510–511
- continents, zones, and seas, 175
- in Hellenistic Age, 171–172, 228
- Homer and, 162–163
- in Pax Romana, 302–308
- in Rome, 173–175, 302–308
- *See also* Ptolemy, Claudius

Gibbon, Edward, 369, 410
Great Pyramid of Cheops, 24, 316, 325, 540
Greek Classical Age, 113–115
- Democritus and, 122–125
- Diogenes of Apollonia and, 125–126
- Empedocles and, 120–121
- Epicurean school and, 129–130
- Epicurus and, 126–129
- Heraclitus of Ephesus and, 115–117
- idealists and, 117–118
- Leucippus and, 121–122
- materialists and, 118–119
- Protagoras, 125

Gregory of Nyssa (theologian), 403–404, 406, 550
Gregory of Tours (historian), 212, 354, 551

Hadrian (emperor), 33, 179, 281, 298, 299, 308, 331–332, 386, 548
Hadrian's Wall, 179–180
Hahn, Otto, 536
Han Dynasty, 76, 79, 80–84, 180–181, 355, 362, 474, 545

Hanging Gardens of Babylon, 325, 326–327
Harvey, William (anatomist), 531, 534, 556
Hecataeus of Miletus (geographer), 95, 98–99, 100, 163, 174, 175, 244, 542
Hellanicus of Lesbos (historian), 99, 163, 244, 542
Hellenism and Hellenistic Age, 219–230, 234–235, 244, 544
  alchemy in, 229–230
  astronomy in, 197–199, 226–228
  cult of Asclepius and, 278
  definition of, 235
  division of Platonic and Aristotelian thought in, 370
  Euclid and, 293
  geography in, 171–172, 228
  influence of Aristotle on, 152
  inventiveness in, 230
  Julian and, 399
  mathematics and, 294
  mathematics in, 17–18, 226–228, 293–295
  medicine in, 229
  metallurgy in, 322
  respect for physicians in, 47
  See also Alexander of Macedon (Alexander the Great)
Hephaestus (smith god), 20, 23, 25, 34, 90, 211, 285, 318
Heraclitus of Ephesus (philosopher and scientist), 107, 115–117, 233, 241, 373, 542
Hermes Trismegistus, 499
Hermias the philosopher, 385, 387–388, 407
Herodotus (historian), 342, 520
  on Egypt and Egyptians, 20, 22, 24–25, 26, 195, 268
  on Great Pyramid of Cheops, 325
  Histories, 24, 60, 99–101, 113, 133, 186, 342
  influence of, 376, 405
  influences on, 244
  on Phoenicians, 185–186
Hesiod (poet), 35–36, 109, 162, 249, 342
  Theogony, 32, 43, 541
  Works and Days, 32, 114, 195, 541
Heyerdahl, Thor, 161, 188, 518–519, 523, 526
Hipparchus (astronomer), 172, 173, 175, 198–199, 209, 213, 227, 545
Hippocrates (physician), 51–52, 66, 267–277, 421–422, 425, 427–428, 543
Hippocratic Oath, 52, 271
Hippocratic Corpus, 267–269, 272, 432
Hippodamus of Miletus (urban designer), 133, 224, 324, 411, 542
Hippolytus (commentator), 390, 438
  on Anaximander, 94, 96, 243
  on Archelaus, 135–136
  on Democritus, 123–124
  on Ecphantus of Syracuse, 109
  on Empedocles, 120
  on Epicurus, 128–129
  on Heraclitus of Ephesus, 115, 116
  on Hippo of Croton, 108–109
  on Indian Brahmins, 72–73
  on influence of Greek Pythagoreans on the Druids, 39
  on Leucippus, 122
  on Parmenides of Elea, 111
  on Pythagoras, 107
  Refutation of All Heresies, 94, 128–129, 243, 390
  on Thales, 92, 93
  on Xenophanes, 109, 110
Historiography, medieval, 353–355
Hittite Empire, 15–16, 320, 321, 540
Hobbes, Thomas, 535
Hollerith, Herman, 536, 537
Homo erectus, 4, 539
Homo sapiens, 4–5, 539
Honorius of Autun (polymath), 495, 554
House of Wisdom, 420, 421, 422, 423, 448
Hunayn ibn Ishaq (scientific translator), 280–281, 421–422, 445, 553
  Anecdotes of Philosophers and Savants, 421
  The Book of Questions, 421
  The Ten Treatises on the Eye, 421
Hyginus, Gaius Julius (geographer), 302–303, 546
Hypatia of Alexandria (mathematician), 295, 406–409, 550
Hyperboreans, 9–10, 174
Hypsicles of Alexandria (mathematician), 203, 213, 226, 294, 545

Iamblichus (Neoplatonist), 108, 370, 374–375, 397, 399, 400, 402, 403, 549

Ibn Abi Usaibi'ah, 429–442, 555
on Aristotle, 143–144
on Asclepius, 430–431
on Al-Biruni, 468
on Dioscorides, 280–281
on Al-Farabi, 425–427
on Galen, 284–285, 287, 358–359
on Hippocrates, 272–273, 431
History of Physicians, 272–273, 280, 354, 407–408, 428, 429, 430–431, 432–442, 465–467
on Hunayn ibn Ishaq, 421, 422
on Ibn al- Samina, 444
on Ibn al-Tayyib, 428
on Ibn Bajja (Avempace), 449
on Ibn Fatik, 455
on Ibn Juljul, 445
on Ibn Jumay‘, 455–456
on Ibn Rushd, 449
on Ibn Sina, 465–467
on John Philoponus, 407–409
list of works by Alexander of Aphrodisias, 258–260
list of works by Galen, 432–442
list of works by John Philoponus, 408–409
list of works by Rhazes, 463–464
list of works by Rufus of Ephesus, 282–283
on Al-Majriti, 444
on origins of medicine, 59
on Thabit Ibn Qurra, 425
Ibn al-Haytham (Alhazen) (polymath), 453, 553
Ibn al-Jazzar (physician), 453, 553
Ibn al-Nafis (physician), 429, 457, 555
Ibn al-Samina (physician), 444, 448, 553
Ibn al-Tayyib (scientific commentator), 427–428, 553
Ibn al-Wafid (Abenguefit) (pharmacologist), 447–448, 554
Ibn Bajja (Avempace) (physician), 449, 554
Ibn Battuta (scholar and explorer), 354, 511, 555
Ibn Butlan (scientist), 428–429, 554
Ibn Fadlan (explorer), 510, 553
Ibn Fatik (mathematician), 273–277, 455, 554
Ibn Hayyan (historian), 354, 553
Ibn Juljul (physician), 273, 444–447, 553
Ibn Jumay‘ (physician), 455–456, 554
Ibn Khaldun (historian and social scientist), 354, 423, 457–460, 555
Ibn Khurradadhbih (geographer), 354, 553
Ibn Rushd (Averroes) (commentator), 449–451, 554, 555
Ibn Rusta (historian), 354
Ibn Sina (Avicenna) (philosopher and physician), 450, 465–467, 470, 553
Ibn Yunus, Kamal al-Din (polymath), 425, 555
Ibn Zuhr (Avenzoar) (physician), 449, 554
Idealists, 117–118
Al-Idrisi, Muhammad (cartographer), 355, 554
Iliad (Homer)
anachronism in, 243
anthropomorphism in, 33, 43–44, 255
Asclepios in, 265
first written records of, 541
geography in, 162
influence on Alexander the Great, 166, 220
influence on Strabo, 173
mythology in, 34–35
oral composition of, 33, 113, 541
Plutarch on, 90
psychology in, 356
role of fate in, 37
Imhotep, 24, 42, 265, 540
Immortality, search for, 26–27
Inca civilization, 523, 524, 525–526, 533, 556
Iron Age, 15–16, 320–322, 507–508
Isabella of Spain, 513
Ishaq Ibn `Imran (physician), 452, 553
Ishaq Ibn Sulayman (physician), 452–453
Isidore of Seville, 486–492, 552 (encyclopedist)
Arithmetic, 488
Astronomy, 489–490
Chronicon, 490
Dialectic, 487–488
Etymologies, 302, 487, 490–492, 495
Geometry, 488–489
Grammar, 487
influenced by Aristotle, 487, 493, 494
influenced by Varro, 302

On Laws, 490
on liberal arts, 487–490
on magi, 58–59
on medical science, 490
on music, 489
Rhetoric, 487
Isis (Egyptian fertility goddess), 20–22, 23, 25, 45, 195, 225, 349
Isolationist theory of cultural emergence, 518

Jabir ibn Hayyan (Geber) (alchemist), 442, 552
Jefferson, Thomas, 517–518, 535, 537
Jibril ibn Bukhtishu, 420–421
Jin Dynasty, 549
John II of Portugal, 512
Julian (emperor), 45, 139, 235, 288, 373–374, 375, 376, 397–398, 399–405, 442, 550
Julio-Claudian Dynasty, 297–298
Julius Africanus (historian and chronologist), 23, 211, 490, 548
Julius Caesar, 209–210, 546
assassination of, 297, 302, 546
calendar reform, 209
Gallica (Gallic War), 39, 174, 210
influenced by Epicurus, 247
on water clocks, 210
Jung, Carl, 50, 145, 357, 361, 372, 518
Justin Martyr (theologian), 386, 548
Justinian (emperor), 353–354, 396, 407, 409–413, 452, 490, 506, 551
Justinian Code, 490

Kanada (physicist), 67–68
Kepler, Johannes (astronomer), 531, 534, 556
Khayyam, Omar (poet and astronomer), 470, 553
Khmer culture, 479
Al-Khwarizmi (mathematician), 420, 423–425, 448, 472, 552
Kilvington, Richard (logician), 500, 555
Kim Busik (Confucian scholar), 475, 554
Al-Kindi, Ya'qub ibn Ishaq (Aristotelian scholar), 422–423, 448, 552
Kircher, Athanasius, 476

Lao Tzu (founder of Taoism), 52, 77, 362
Leibniz, Gottfied, 535
Leucippus (atomist), 121–122, 126, 129, 248, 291, 357, 388, 543
Liang Dynasty, 551
Libon of Ellis, 327, 542
Library of Alexandria, 224–226, 544
Lighthouse of Alexandria (Pharos), 223–224, 326, 545
Liu Song Dynasty, 551
Liu Chih-Chi (historian), 355, 552
Liu Chou (Confucian scholar), 475, 552
Livy (historian), 178, 210, 301, 303, 345, 348, 349–350, 546
Logic, 142–143
Logos, 50–51, 241–242
Lucretius (Epicurean), 247–251, 546
as atomist, 119, 244, 247–248
as Epicurean, 119, 125, 247–248, 250
influences on, 96, 122, 247
on mountains, 178–179
On the Nature of Things, 119, 244, 247–250, 308
on time, 244

Macedonian Kingdom, 219–223, 413, 544
Magellan, Ferdinand (explorer), 533, 556
Magi of Persia, 58–60
Magic, 36–38
Magna Graecia
definition of, 103
Parmenides and, 111–112
Pythagoras and, 103–109
Xenophanes and, 109–111
Mahavira (mathematician), 472, 552
Maimonides (polymath), 449, 455, 456–457, 554
Al-Majriti (mathematician), 444, 448
Al-Majusi (Haly Abbas) (physician), 464–465, 553
Al-Ma'mun (Abbasid caliph), 420, 423
Manchu Dynasty, 476
Manetho (chronologist), 22, 23–24, 25–26, 210–211, 540
Epitome of Physical Doctrines, 25–26
History of Egypt, 23
Manhattan Project, 536–537
Al-Mansur (Abbasid caliph), 420, 450
Manuel I Comnenus (emperor and scientist), 414, 553
Marcellinus, Ammianus (historian), 258, 288, 350, 353, 382, 404–406, 550

Marcus Aurelius (emperor and Stoic philosopher), 244–245, 548
Antonine Dynasty, 298, 548
as emperor, 138–139, 278, 284, 358
Fronto and, 548
Galen and, 278, 284, 358
on holy spirit, 242
influences on, 278, 308
Meditations, 244–245, 309–310
Stoicism of, 119, 138–139, 241, 244–245, 251, 308, 309–310, 385
Massilian explorers, 164–166
Materialists, 118–119
Mausoleum at Halicarnassus, 325, 326, 544
Maxentius (emperor), 330, 482
Maximian (emperor), 396, 482
Maximus of Ephesus (pseudoscientist), 370, 374, 402–403, 550
Mayan civilization, 214, 316, 523–524, 549
Megalithic structures, 188, 192–193, 479–480, 523, 524
Melissus of Samos, 15, 111, 112, 244
Menelaus of Alexandria (mathematician), 226, 294, 548
Metals and metallurgy
in Egypt, 317
in Greece and Rome, 322–323
Pliny the Elder on, 305–306, 322, 323
See also Bronze; Bronze Age; Iron Age
Metonic cycle, 67, 207–209, 382, 472
Military discipline, 335–336
Military science, 328–329, 334
Ming Dynasty, 474, 476, 554
Mithridates of Pontus, King, 279
Montaigne, Michel de, 253, 254, 255, 256–257, 309, 532, 533
Mountains and mountaineering, 176–179
Muhammad (prophet of Islam), 214, 419–420, 552
Museum at Alexandria, 224–225

Napier, John (mathematician), 535, 556
Nazca civilization, 525
Neanderthal humans, 4, 539
Nearchus of Crete (explorer), 168–171, 174, 175, 221, 228, 345, 351, 352
Nebuchadnezzar II of Babylon, 16, 47, 326–327, 357
Neolithic Age, 6–10
agriculture in, 6–7, 11, 539
anthropomorphism in, 118
Çatal Hüyük (Turkey), 7, 11, 539
change from Paleolithic Age to, 7
definition and dates of, 5, 6, 539
development of Magic in, 36
Fertile Crescent (Jericho), 7, 11, 12
history of, 6
magic in, 6–7, 36, 118
monolithic cultures, 9–10
as New Stone Age, 6
Ötzi, the Iceman, 8–9, 539
Stonehenge, 9–10, 193, 540
transition from Paleolithic Age to, 7–8
worldview in, 8–9
Neolithic Revolution, 315
Neoplatonism, 370–375. *See also* Iamblichus; Plotinus; Porphyry
Nero (emperor), 253, 280, 298, 308, 332, 350, 547
Nerva (emperor), 298, 331, 548
Netz, Reviel, 294
New Stone Age, 6. *See also* Neolithic Age
New Testament, 49–52
dreams in, 50
Jesus the healer in, 51–52
Jesus the Logos in, 50–51
medicine in, 51
Newton, Isaac, 141, 233, 242, 535

Octaëteris, 199, 207–209
Octavian (Augustus) Caesar, 180–181, 297–298, 300, 302–303, 330, 331, 332, 426, 546
Odyssey (Homer)
agriculture in, 114
anthropomorphism in, 33, 43–44
brimstone in, 339
first written records of, 541
geography in, 162, 183, 347
influence on Strabo, 173
mythology in, 35–36
Ogygia (island) in, 164
oral composition of, 33, 113, 541
Plutarch on, 90
psychology in, 356–357

Oersted, Hans, 536
Old Stone Age, 5. *See also* Paleolithic Age
Old Testament, 45–49
  diviners and physicians in, 46–47
  psychology in, 48
Olmec civilization, 188, 316, 523–524, 541
Oribasius of Pergamon (physician), 287–288, 359, 399, 400–402, 442, 550
Origen (theologian), 373, 385, 391, 406, 549
Osiris (Egyptian fertility god), 20–21, 22, 23, 25, 37, 42, 70, 195, 225, 349
Ötzi, the Iceman, 8–9, 539
Oxford Calculators, 499–500
  Bradwardine, Thomas, 500, 555
  Kilvington, Richard, 500, 555
  Swineshead, Richard, 500, 555
  Swineshead, Roger, 500, 556

Paleolithic Age, 5–6
  anthropomorphism in, 118
  art and sculpture in, 6
  definition and dates of, 5
  magic in, 36, 118
  as Old Stone Age, 5
  social organization in, 6
  technology in, 5
  tools in, 5
  transition to Neolithic Age from, 7–8
Palladius (agricultural writer), 258, 302, 382–383, 551
Panodôrus (monk), 23
Pappus of Alexandria (mathematician), 294, 295, 549
Parmenides of Elea (philosopher), 111–112, 542
  cofounder of Eleatic school of philosophy, 111–112
  on geodesy, 163
  influence on Leucippus, 121
  influence on Melissus, 115
  On Nature, 111–112
  as philosopher of being, 44
  philosophy and beliefs of, 233, 248, 388
  Xenophanes and, 109, 111
Parthenon, 133, 234, 323, 327, 542
Paul III, Pope, 533
Paul of Tarsus, 49, 384–385, 387, 388, 420, 495
Pax Romana
  agriculture in, 300–302
  Antonine Dynasty, 298, 350, 548
  architecture in, 300
  dates of, 235, 547
  ecumene in, 235
  education in, 298–299
  Flavian Dynasty, 298, 350, 547
  history and geography in, 302–308
  Julio-Claudian Dynasty, 297–298
  jurists in, 409–410
  Plutarch and, 90, 258, 359
  Roman medicine in, 271, 310–311
  Roman pragmatism in, 299–300
  Skeptics and, 310
  Stoicism in, 271
  Stoics and, 308–310
Peloponnesian War, 60, 132, 136, 341, 343, 344, 543
Pericles of Athens, 97, 115, 131–133, 153, 234, 323, 341, 542
Peripateticism, 151
  definition of, 151
  Hellenism and, 235
  influence of, 258
  on Logos, 115
  Platonism and, 254, 255
  See also Alexander of Aphrodisias; Aristarchus; Aristotle; Strato; Theophrastus; Xenarchus
Persian Wars, 97, 99, 342
Petrarca (Petrarch), Francesco (humanist), 179, 532, 555
Phidias (sculptor), 234, 323, 327, 542
Philip II of Macedon, 141, 254
Philo of Alexandria (philosopher), 48–49, 241–242, 406, 547
Philo of Byzantium (scientist), 325–327, 545
Philo of Larissa (academic), 140, 252, 254, 546
Pishdadian Dynasty, 62
Plato (philosopher), 137–140
  Academy of, 139–140
  on ideal forms, 48, 118, 136, 137, 138, 371, 534
  influence of, 152–153
  on political science, 138–139
Platonism, 254–258
Pleistocene Period, 539
  Late Pleistocene, 4–5
  Middle Pleistocene, 4

Pliny the Elder (polymath), 281, 304–306, 546
on astronomy, 200, 305
on Babylon, 15, 17
on eruption of Mount Vesuvius, 179, 304
on flora, 306
on Great Pyramid of Cheops, 325
influence of, 307–308, 404, 405, 487, 537
influences on, 301, 302
on King Mithridates, 279
on Lighthouse of Alexandria, 326
on magi and magic, 58
materia medica of, 281
on metals and metallurgy, 306, 322, 323
on mythical islands, 164
Natural History, 173–174, 200, 303, 304–306, 405
on Phoenicians, 183
on reed ships, 161
on Roman geography, 173–174
Tacitus and, 350, 351
on Temple of Artemis, 327
Pliny the Younger (consul, senator, and lawyer), 210, 304, 306–308, 350, 547
Plotinus (founder of Neoplatonism), 258, 370–373, 374, 382, 385, 400, 402, 549
Plutarch of Chaeronea (biographer and essayist)
on Alexander, 68, 167, 220, 224, 254–255
on Anaxagoras, 97, 98
on Anaximenes of Miletus, 95, 96
on Aristotle, 220
on calendars and dating systems, 207, 209, 210
on capacities of women, 298
dates of, 547
on Diogenes Laertius, 378
early life and education of, 254
on Greek reverence for Poseidon, 160
on happiness, 257
on Homer, 33–34, 90–91, 92, 104, 109–110, 114, 120, 195, 200, 250–251, 255–256, 265–266
influence of, 258, 405, 532, 533, 535, 537
on islands, 164
The Life and Poetry of Homer, 90, 104, 200, 253–254, 265–266
Life of Alexander, 68, 167, 220, 254–255, 258
Life of Camillus, 257
Life of Timoleon, 255
on mathematical sciences, 291
on meteorology, 200, 257–258
Moralia, 253, 254, 303, 359
Natural Questions, 98, 200, 253, 299
Parallel Lives, 254–255, 258, 303, 359
Platonic philosophy of, 250–251, 254–257, 532
as polymath, 303
on psychology, 256–257, 359–360
on Pythagoras, 104
on Roman diviners, 37
on the siege of Syracuse, 328
on Solon, 20, 91, 134, 207
on Stoicism, 253–254
on Theophrastus, 150
on time and history, 257
Polo, Marco (explorer), 476–477, 479, 511–512, 555
Polybius (historian), 346–348
on the Alps, 178
on the *castra,* 336, 337
on chronology, 210
on Eratosthenes of Cyrene, 172
on geography, 347
Histories, 346, 347
influences on, 228
on military discipline, 336
on Pytheas of Massilia, 165
on Roman government, 345–346, 347–348
Stoicism of, 251, 346
Strabo on, 173
on zone of fire, 175
Polytheism
in ancient Rome, 351
in ancient southern Asia, 64
in Aztec culture, 524
in Celtic culture, 38
definition of, 41
in Mesopotamia, 41, 419
monotheism compared with, 45, 387
in Viking culture, 507
Pomponius Mela (geographer), 174–175, 183, 303, 547
Porphyry (Neoplatonist), 360–361, 370, 372, 373–374, 400, 549
Isagoge, 382, 409, 427, 485, 488

Posidonius of Rhodes (Stoic scientist), 172, 173, 199, 226, 251–252, 258, 294, 322, 351, 495
Priestly, Joseph, 536
Proclus of Lydia (commentator), 291, 292, 293, 378–381, 551
Procopius of Caesarea (historian), 353–354, 406, 412, 550
Prometheus (Titan god of fire), 33, 35–36, 43, 44, 70, 177, 234, 250
Psellus, Michael (historian), 354, 413–414, 554
Ptolemaic Dynasty, 223–226, 270
Ptolemaeus (Ptolemy), Claudius (astronomer, mathematician, and geographer), 175–176, 199, 547
 Almagest, 226, 294, 295, 407, 414, 427, 444, 448
 Cosimo's critique of, 506–507
 Geography, 175–176, 424, 507, 533
 influence on Al-Biruni, 194, 468, 469, 510
 influence on Copernicus, 533–534
 influence on European exploration, 513, 533
 influence on Galileo, 534
 influence on Hypatia of Alexandria, 406, 407
 influence on Al-Tusi, 471
 influenced by Aristotle, 152–153
 influenced by Babylonian mathematics and astronomy, 17, 194
 influenced by Eratosthenes, 172
 influenced by Hipparchus, 172
 influenced by Marinus of Tyre, 175
 on mathematics, 295
 Tetrabiblos, 202
Ptolemy I Soter, 224, 225
Ptolemy II Philadelphus, 224–225
Ptolemy III, 225
Punic Wars, 188, 230, 328, 329
Pyramids, 24, 316–317
 in the Americas, 518, 523–525
 corbel vault building technique, 24
 cultural theories of, 518
 of Egypt, 24, 42, 92, 316, 325, 331, 540
 etymology of the word, 405
 Great Pyramid of Cheops, 24, 316, 325, 540
 of Inca civilization, 524–525
 of Mayan civilization, 523–524
 measuring height of, 92
 of Mesopotamia, 315–316
 of southeast Asia, 479
 step pyramid (ziggurat), 12, 15, 24, 176, 315–316, 326, 518, 540
 Tower of Babel, 316
 ziggurat of Ur, 12, 15
 See also Imhotep
Pyrrho of Elis (philosopher), 221, 222–223, 310, 390, 535, 543
Pythagoras (philosopher and mathematician), 103–109, 542
 Alcmaeon and, 107–108
 Aristotle on, 243
 on central fire, 197
 on geography, 175
 Hippo of Croton and, 108–109
 Iamblichus on, 374–375
 influences on, 104–105
 Philolaus of Croton and, 108
 Porphyry on, 184, 185
 on spheres, 196
 on transcendent *number,* 104, 105–106, 111, 233, 291, 372
 on transmigration of souls, 111, 360
Pythagorean school, 107–109, 115, 123, 291–292, 377
Pythagorean theorem, 17–18, 67, 82, 290, 295, 300
Pytheas of Massilia (explorer), 159–160, 164, 165–166, 173, 175, 347, 508, 519
Pythius of Halicarnassus, 326, 544

Quadrivium, 48, 153, 284, 298–300, 391, 411, 413, 414, 487, 488, 493–494, 532
Quintilian, 298–299, 547
Qur'an, 419, 420, 424–425, 448, 449, 460, 465, 511

Ramadan, 214–215
Rashidun Caliphate, 461
Al-Razi (Rhazes) (polymath), 283–284, 448, 456, 463–464, 465, 553
Renaissance, Carolingian, 337–338, 492–494
Renaissance, European
 Athenian philosophy and, 152–153
 exploration, 176, 512–514
 humanism, 235
 Scientific Revolution, 499, 501, 531–537

Rhazes (polymath), 283–284, 448, 456, 463–464, 465, 553
Robertson, William, 517
Rufus of Ephesus, 282–284
Rutherford, Ernest, 121

Sa'id al-Andalusi (Judge Sa'id) (judge and scientist), 447–449, 554
Al-Sarakhsi (polymath), 423, 552
Sarkar, Benoy Kumar, 68, 421, 472
Sasanian Empire, 63, 461, 548, 552
Savonarola, Girolamo (theologian), 499, 556
Scientific Revolution, 499, 501, 531–537
- mathematics, 535
- philologists, 531–532
- "Renaissance man," 532

Second Dynasty of Egypt, 23
Seleucid Kingdom, 69, 180, 219, 545
Seneca (Stoic), 253–254, 304, 308, 547
- *Epistles,* 308
- influence of, 405, 499, 532, 535
- *Medea,* 164
- *Moral Letters,* 253
- *Natural Questions,* 253
- *Problems in Natural Science,* 308
- stoicism of, 253–254, 308

Seven wonders of the ancient world, 224, 325–327
- Colossus of Rhodes, 325–326, 544
- Great Pyramid of Cheops, 24, 316, 325, 540
- Hanging Gardens of Babylon, 325, 326–327
- Lighthouse of Alexandria (Pharos), 223–224, 326, 545
- Mausoleum at Halicarnassus, 325, 326, 544
- Temple of Artemis, 325, 327, 541, 543, 544
- Temple of Zeus, 325, 327, 543

Severan Dynasty (of Roman emperors), 549
Sextus Empiricus, 115, 223, 310, 535, 548
Shang Dynasty, 76
Shen Kuo (polymath), 475, 554
Silk Road, 474, 478, 505–506, 511
- Faxian (traveler), 505, 550
- Xuanzang (traveler), 505–506, 552
- Yijing (traveler), 506, 552

Sima Qian (historian), 79–80, 82–83, 84, 355, 546
Sivin, Nathan, 475
Skeptics and Skepticism, 310. *See also* Arcesilaus; Gassendi, Pierre; Hermias; Pyrrho of Elis; Saturninus; Sextus Empiricus; Timon
Socrates (philosopher), 136–137, 543
- on application of mathematics, 292
- Aristotle and, 151
- on cyclical nature of time, 244
- Democritus and, 123
- on ideal forms, 117–118, 136
- influence of, 136
- influences on, 98, 112, 135–136
- Plato and, 117, 136–137, 138–140, 141
- on "the Good," 141
- Xenophon and, 344–345

Solon (lawgiver), 91–92, 132, 133–135, 147, 164, 207, 341, 378, 541
Song Dynasty, 474–475, 478, 551
Soranus of Ephesus, 281–282, 490
Sostratos of Cnidus, 224, 326
Southern Qi Dynasty, 551
Stoics and Stoicism, 251–254, 308–310
- Greek influence in the Roman Empire, 251–253
- origins and philosophy of, 235–241
- during the Pax Romana, 308–310
- *See also* Antipater of Tyre; Chryssipus; Cicero, Marcus Tullius; Cleanthes; Epictetus; Marcus Aurelius; Posidonius of Rhodes; Seneca; Zeno of Citium

Stonehenge (England), 9–10, 193, 540
Strabo (geographer), 172–173, 235, 546
- on Alexandria, 324
- on Athens and the Parthenon, 323
- on Chaldeans, 18–19
- on Eratosthenes of Cyrene, 172
- on eruption of Vesuvius, 303–304
- *Geography,* 172–173, 303–304
- influence of, 404–405, 495
- influences on, 172–173
- on Lighthouse of Alexandria, 326
- on Mesopotamians, 315–316
- on naphtha (type of asphalt), 14
- on Phoenician people of Sidon, 163
- on Polybius, 347
- on Pytheas, 165
- Stoicism of, 172–173

Sturgeon, William, 536
Su Song (astronomer), 475, 554
Sui Dynasty, 474, 478, 551
Sulba Sutras, 66–67
Sun worship, 20, 193, 403, 523
Sutra of Forty-Two Sayings, 474
Swineshead, Richard (mathematician), 500, 555
Swineshead, Roger (philosopher), 500, 556
Syracuse, siege of, 328
Szilard, Leo, 536

Tacitus (historian), 350–351, 547
  Agricola, 350, 351
  Annals, 350–351
  Germania, 39, 303, 350, 351
  Histories, 350, 351, 404
  influence on Ammianus Marcellinus, 353, 404
T'ang Dynasty, 474–475, 478, 551
Temple of Artemis, 325, 327, 541, 543, 544
Temple of Zeus, 325, 327, 543
Tesla, Nikola, 536
Thabit ibn Qurra (astronomer), 425, 553
Thales of Miletus (philosopher and scientist), 91–93, 160, 194, 226, 233–234, 291, 378, 389–390, 488, 542
Themistius (commentator and orator), 382, 397, 398–399, 402, 550
Theodorus Priscianus, 288
Theodosius of Tripolis (geometer and astronomer), 174, 294, 546
Theodosius the Great, 550
Theophrastus (student of Aristotle), 107, 118, 126, 148–151, 197, 198, 201, 258, 270, 306, 322, 411, 543
Third Dynasty of Egypt, 23, 24
Thucydides (historian), 132, 329, 341, 342, 343–344, 543
  influence of, 344, 345, 346, 537
  on invention of the trireme, 329
  The Peloponnesian War, 99, 133, 210, 234, 343, 344
Tiahuanaco (Tiwanaku) civilization, 525, 526
Tiberius (emperor), 298, 304, 350, 547
Timekeeping, 209–210
Timon (Skeptic philosopher), 223, 544
Titus (emperor), 298, 304–305, 332, 337, 548
Titus Livius. *See* Livy (Roman historian)
Toltec civilization, 523–524, 553
Trade and cultural exchange, 179–181. *See also* Silk Road
Trajan (emperor), 281, 298, 303, 306, 308, 330, 331, 386, 548
Trivium, 48, 153, 284, 298–300, 391, 411, 413, 487, 493–494, 532
Trojan War, 43, 105, 343, 348
Al-Tusi, Nasir al-Din (polymath), 425, 470–471, 555
Al-Tusi, Sharaf al-Din (mathematician), 425, 555

Umayyad Caliphate, 420, 444, 461, 552

Varro (polymath), 301–302, 487, 488, 546
Vedic Age, 64–73
Vesalius, Andreas (anatomist), 531, 534–535, 556
Vespasian (emperor), 298, 304, 337, 351, 548
Vesuvius, eruption of Mount, 179, 303–304, 305, 307–308, 548
Viking explorations, 507–509
Vinland Sagas, 508–509
Virgil (poet), 297, 483, 546
Vitruvius (architect), 300, 332–334, 546
  on Alexander the Great, 323–324
  on animal sacrifice, 38
  On Architecture, 38, 300, 332–334
  on Berossus, 17
  on construction of theaters, 333
  on Dinocrates of Rhodes, 323–324
  on education of engineers, 299
  influenced by Theophrastus, 148
  in invention of battering ram, 329
  on military science and discipline, 329, 334
  on Pythius of Halicarnassus, 326
  on Theodosius of Tripolis, 174
  on urban planning, 333–334
Von Neumann, John, 537

Western Jin Dynasty, 83
William of Occam (philosopher), 500–501, 555

Williams, Roger, 516–517
World War II, 537

Xenarchus (Stoic philosopher), 173
Xenophanes (philosopher and historian), 44, 98, 103, 109–111, 542
  influence of, 109–110
  on mind as basis of all things, 233
  on oceans, 160–161
  philosophy of, 109–111
Xenophon (philosopher and historian), 117, 133, 136, 344–345, 352, 543
Xia Dynasty, 76

Yin-yang duality, 77, 82–84, 361, 475, 478
Yuan Dynasty, 474, 476, 554

Zeno of Citium (founder of Stoicism), 153, 222, 235–236, 241, 251, 308, 309, 485, 545
Zeno of Cyprus (physician), 287–288, 401
Zeno of Elea (follower of Parmenides), 111
Zhou Dynasty, 76, 78, 81
Zhu Shijie (mathematician), 474, 555
Zhu Xi (Confucian scholar), 475, 554

About the Author

Russell M. Lawson holds a PhD from the University of New Hampshire. He is the author/editor of *Servants and Servitude in Colonial America* (Praeger, 2018), *Science in the Ancient World: An Encyclopedia* (ABC-CLIO, 2004), *Poverty in America*, along with coeditor Benjamin Lawson (Greenwood, 2008), *Encyclopedia of American Indian Issues Today* (Greenwood, 2013), and *Race and Ethnicity in America: From Pre-contact to the Present*, along with coeditor Benjamin Lawson (Greenwood, 2019). Dr. Lawson teaches and writes on scientists and explorers; the history of America, Europe, and the world; the history of ideas; and the history of social, cultural, and political issues. He has taught at schools in New England, Oklahoma, and Ontario.